Subcellular Biochemistry

Volume 99

The book series SUBCELLULAR BIOCHEMISTRY is a renowned and well recognized forum for disseminating advances of emerging topics in Cell Biology and related subjects. All volumes are edited by established scientists and the individual chapters are written by experts on the relevant topic. The individual chapters of each volume are fully citable and indexed in Medline/Pubmed to ensure maximum visibility of the work.

J. Robin Harris • Jon Marles-Wright
Editors

Macromolecular Protein Complexes IV

Structure and Function

Springer

Editors
J. Robin Harris
Institute of Molecular Physiology
University of Mainz
Mainz, Germany

Jon Marles-Wright
Biosciences Institute
Newcastle University
Newcastle upon Tyne, UK

ISSN 0306-0225 ISSN 2542-8810 (electronic)
Subcellular Biochemistry
ISBN 978-3-031-00795-8 ISBN 978-3-031-00793-4 (eBook)
https://doi.org/10.1007/978-3-031-00793-4

This Springer imprint is published by the registered company Springer Nature Switzerland AG
The registered company address is: Gewerbestrasse 11, 6330 Cham, Switzerland

Preface

The *Subcellular Biochemistry* series is moving toward Volume 100! Having started as a quarterly journal in 1973, under the editorship of the late Donald Roodyn, it evolved in 1978 to a book series. The diversity of content over the past ~50 years has been considerable, whilst retaining the underlying broad aim of presenting review chapters on the diverse subcellular components, from the organelle to the molecular level. Recent years have seen a strong emphasis upon the structure and function of macromolecular complexes, resulting in several thematically related volumes.

Volume 82 (2017) was devoted to Fibrous Proteins, followed by Volume 83, Macromolecular Protein Complexes. Because of the large amount of data available at this time, and the willingness of scientists to contribute to the series, Volume 87 and Volume 88 (2018) were published, on Membrane and Virus Proteins Complexes, respectively. Subsequently, a further book (Volume 93, 2019) entitled *Macromolecular Protein Complexes II* pulled together *all aspects of the broad field* and was followed by *Macromolecular Protein Complexes III* (2021). Several earlier related books have been included in the series, generating an almost encyclopedic coverage, with some macromolecular topics already being updated.

This volume has continued this approach and now contains defined sub-sections: Soluble Protein Complexes, Membrane Protein Complexes, Fibrous Protein Complexes, and Virus Protein Complexes. Owing to the exceptionally difficult COVID-19 times faced by scientists in academia and in research institutes, a number of initially agreed interesting chapters have been lost during production, but the remaining seventeen chapters presented here have enabled us to compile a strong and interesting book.

As in previous volumes, x-ray crystallography and cryo-electron microscopy of the structural biology of protein complexes predominates, with functional considerations also apparent in most chapters. Indeed, the impact of molecular genetics and the study of dynamic/flexible states and assembly/disassembly of macromolecular structures have become increasingly important, along with benefits for the understanding of disease states.

Rather than expanding here upon some individual chapters, we simply wish to emphasize the on-going strength of this group of chapters and refer the reader to the chapter list of the book. We have been impressed by the quality and outstanding interest of the material in these chapters and the enthusiasm shown by our authors. We hope that this book, as print version and available online as the e-book and individual e-chapters, will be of use and interest to the bioscience community.

Mainz, Germany J. Robin Harris
Newcastle upon Tyne, UK Jon Marles-Wright

Contents

Chapter 1
Fatty Acid Synthase: Structure, Function, and Regulation

Aybeg N. Günenc, Benjamin Graf, Holger Stark, and Ashwin Chari

Abstract Fatty acid (FA) biosynthesis plays a central role in the metabolism of living cells as building blocks of biological membranes, energy reserves of the cell, and precursors to second messenger molecules. In keeping with its central metabolic role, FA biosynthesis impacts several cellular functions and its misfunction is linked to disease, such as cancer, obesity, and non-alcoholic fatty liver disease. Cellular FA biosynthesis is conducted by fatty acid synthases (FAS). All FAS enzymes catalyze similar biosynthetic reactions, but the functional architectures adopted by these cellular catalysts can differ substantially. This variability in FAS structure amongst various organisms and the essential role played by FA biosynthetic pathways makes this metabolic route a valuable target for the development of antibiotics. Beyond cellular FA biosynthesis, the quest for renewable energy sources has piqued interest in FA biosynthetic pathway engineering to generate biofuels and fatty acid derived chemicals. For these applications, based on FA biosynthetic pathways, to succeed, detailed metabolic, functional and structural insights into FAS are required, along with an intimate knowledge into the regulation of FAS. In this review, we summarize our present knowledge about the functional, structural, and regulatory aspects of FAS.

Keywords Fatty acid biosynthesis · Fatty acid synthase · ACP · FAS structure · FAS regulation · Biofuels · Oleochemicals

A. N. Günenc · B. Graf · A. Chari (✉)
Research Group for Structural Biochemistry and Mechanisms, Max Planck Institute for Multidisciplinary Sciences, Göttingen, Germany
e-mail: agunenc@mpinat.mpg.de; benjamin.graf@mpinat.mpg.de; ashwin.chari@mpinat.mpg.de

H. Stark
Department of Structural Dynamics, Max Planck Institute for Multidisciplinary Sciences, Göttingen, Germany
e-mail: hstark1@mpinat.mpg.de

J. R. Harris, J. Marles-Wright (eds.), *Macromolecular Protein Complexes IV*, Subcellular Biochemistry 99, https://doi.org/10.1007/978-3-031-00793-4_1

The Importance of Fatty Acids and Fatty Acid Biosynthesis

Fatty acids (FAs) are essential molecules of all living organisms and their biosynthesis is essential for the survival of all organisms. In bacteria, yeasts, and plants, the de novo biosynthesis of FAs is both necessary and sufficient for survival (Schweizer and Hofmann 2004; White et al. 2005; Chan and Vogel 2010). Metazoans, on the other hand, additionally rely upon dietary uptake of some fatty acids and make proportionally less use of de novo FA biosynthesis pathways, for example, for the biosynthesis of some cofactors (Burr and Brown 1933; Kastaniotis et al. 2017). FAs primarily act as compact sources for energy storage, especially in the form of triglycerides, occupying less space than carbohydrates of similar energy content (Nelson and Cox 2013). In almost all organisms (except Archaea) (Jain et al. 2014), cellular membranes are made of phospholipids, composed of a phosphate containing hydrophilic head group containing two FAs (Nelson and Cox 2013). A variety of fatty acids can coexist in any given single organism, e.g., different chain lengths, differing functional groups such as alkenes, isopropyl, and methyl groups. These differences cause alterations in membranes when incorporated, or provide different values of energy when metabolized (Nelson and Cox 2013). Membrane fluidity is increased by incorporating phospholipids containing unsaturated FAs in comparison to when phospholipids containing saturated FAs are inserted (Vagelos 1971; White et al. 2005; Cronan and Thomas 2009). The degradation of unsaturated FAs to mobilize energy reserves is also energetically more favorable (Nelson and Cox 2013).

In addition, FAs serve as precursors to second messenger molecules and therefore control cellular signaling, intercellular communication, and hormonal regulation (Graber et al. 1994). This is accomplished by low-density lipoprotein (LDL) and high-density lipoprotein (HDL) formation that are also crucial to ensure steroid transport throughout the bodies of higher animals (Nelson and Cox 2013). The maintenance and equilibrium of these varieties need strict and self-responsive regulation.

Consequently, FA production is vital for all living organisms. It is therefore not surprising that subtle changes in FA biosynthesis impact numerous essential cellular functions, and are thus linked to ailments like cancer, obesity, and non-alcoholic fatty liver disease (Ronnett et al. 2006; López et al. 2007; Wang and DuBois 2012). The essential requirement of de novo FA biosynthesis, and the substantial architectural differences of the biosynthetic enzymes as discussed below, make this metabolic pathway a desirable target for antibiotic discovery (Campbell and Cronan 2001; Yao and Rock 2017).

Beyond academic interest in FA biosynthetic mechanisms, strained petroleum supplies have sparked interest in these cellular processes (Kallio et al. 2014; Zhang et al. 2021; Liu et al. 2021). An active quest is afoot to define alternative, renewable energy sources to replace liquid fuels by exploiting, and modifying FA biosynthetic mechanisms. This search is not limited to fuels, but also to replace the large foundry

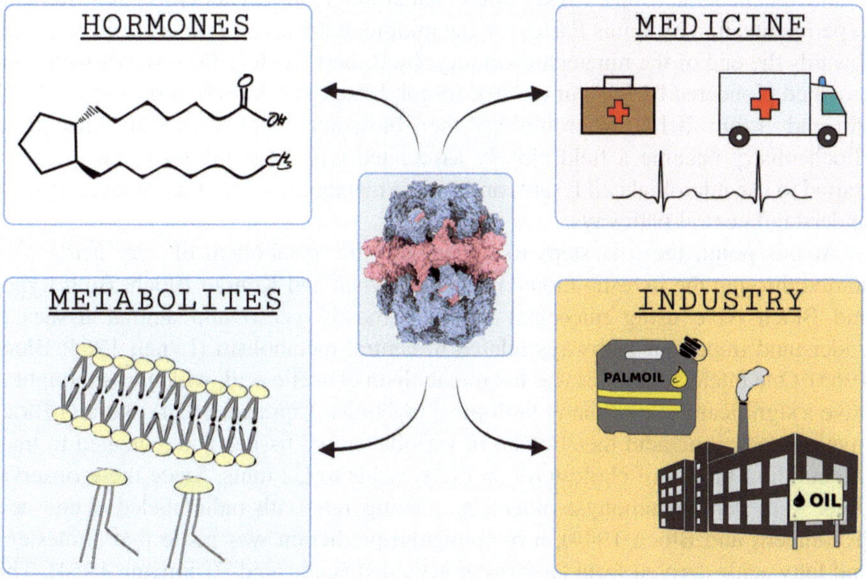

Fig. 1.1 The essential roles played by fatty acid biosynthetic mechanisms and fatty acid synthases in Biology, Medicine, and Biotechnology. Fatty acid synthases, represented by the yeast fatty acid synthase, are the cellular powerhouses which produce fatty acids (center). Fatty acids are essential building blocks of phospho- and glycolipids which form biological membranes (bottom left). They also represent precursors to second messenger molecules, which are paramount to cellular and intercellular signalling (top left). Cells need to maintain a fine homeostatic balance with regard to their fatty acid profiles, and imbalances can elicit disease (top right). Fatty acid biosynthetic mechanisms are exploited to generate fuels and precursors for chemical industry from renewable sources (bottom right)

of oleochemicals, FA-derived compounds. The essential functions of FAs in Biology, Medicine, and Biotechnology are summarized in Fig. 1.1.

Historical Background

Many different classes of lipids, defined as water-insoluble biological substances, were known and utilized since ancient times. FAs, an important component of lipids, are carboxylic acids with long aliphatic chains initially isolated from animal tissues, and first described by the French Biochemist Michel Eugène Chevreul (Chevreul 1823). At this time in the early half of the nineteenth century, Biochemistry and Organic Chemistry were considered only to be branches of analytical chemistry, since biological material was thought to be of “vital essence” that cannot be reproduced synthetically by humans. Thorough study of enzymes, investigation of plant and animal tissue extracts and secretions have now allowed Biochemistry to be a more mechanistic subject. After implementation of Germ Theory led by systematic experiments, first by Louis Pasteur in the middle of the nineteenth century, and later, towards the end of the nineteenth century, by Robert Koch (1882), sterile technique emerged pioneered by Pasteur (1879), Joseph Lister, and Joseph Snow (Snow 1849; Pitt and Aubin 2012). Microbiology then blossomed into a scientific discipline. Biochemistry became a field closely associated with Microbiology and scientists started to use microbial cell lysates and isolate metabolites from microbial cultures to understand central pathways.

At this point, the first steps to understand the metabolism of fatty acids were started through the investigations of Feodor Lynen and Konrad Bloch. Both Lynen and Bloch were using microbial cultures (mostly yeast) and animal tissues to understand important pathways related to central metabolism (Lynen 1964; Bloch 1965). One interesting point was the metabolism of acetic acid, which was thought to have a significant role in many biological reactions (Lipmann 1954). Konrad Bloch investigated acetic acid metabolism in various animal tissues and managed to trace the building blocks of cholesterol and fatty acids to C2 units. Since these observations were derived amongst others by feeding rats with radiolabeled acetic acid (Rittenberq and Bloch 1944), a monumental prediction was made that cholesterol and fatty acids derived from "a form of activated acetic acid" (Lipmann 1954). This activated form of acetic acid was found to be acetyl-coenzyme A (acetyl-CoA) by Feodor Lynen (Lynen et al. 1951). Lynen inferred, based on the seminal studies of Fritz Lipman on pantothenic acid (Klein and Lipmann 1953a, b), that not just acetate, but Acetyl-CoA, was an important component of FA degradation cycle and tried to elucidate the cycle further (Lynen 1964). From his investigations, he speculated that FA biosynthesis is just the reversal of the FA degradation cycle (Lynen 1964). It was proven wrong by Wakil, Vagelos and colleagues, who found out that synthesis takes place in the cytoplasm, while degradation takes place in mitochondria (Gibson et al. 1958; Smith et al. 2003). Lynen, along with others, later focused on finding and characterizing these biosynthetic enzymes. Both Bloch and Lynen shared the Nobel Prize for Medicine and Physiology in 1964 for their discoveries in "Fatty Acid and Cholesterol metabolism" (Lynen 1964; Bloch 1965).

Chemical characterization of fatty acid synthase (FAS) enzymes took substantially more time. The *Escherichia coli* (*E. coli*) acyl carrier protein (ACP), AcpP, was the first FAS protein to be discovered by Wakil, Vagelos and colleagues and the complete amino acid sequence to be documented (MAJERUS et al. 1965; Pugh and Wakil 1965). Later on, catalytic components of the *E. coli* system were also discovered and characterized by a combination of genetic and biochemical approaches (Volpe and Vagelos 1973). In yeast, the FAS complex was found and characterized in the early 1960s also led by Feodor Lynen (Sumper et al. 1969; Lynen et al. 1980). Metazoan FAS has a different story: biochemical and genetic evidence suggested all necessary catalytic components to be present in a single polypeptide. Wakil and colleagues fractionated the system to homogeneity and spear-headed the biochemical characterization (Mattick et al. 1983a, b; Wong et al. 1983; Tsukamoto et al. 1983).

Biochemical Steps in FA Biosynthesis

The cellular machinery that is responsible for FA biosynthesis in the various kingdoms of life can differ substantially, as discussed below. Yet notably, the biochemical reactions leading to saturated FA biosynthesis steps, at their core, are conserved (Fig. 1.2).

A central aspect in FA biosynthesis is that no substrates, intermediates, and products diffuse freely into the individual enzyme active sites, but only as covalent adducts to the acyl carrier protein (ACP) (Chan and Vogel 2010; Crosby and Crump 2012). To be able to function in FA biosynthesis, the ACP has to be modified on a defined serine side-chain by a post-translational modification (Beld et al. 2014). The post-translational modification is achieved with the help of Coenzyme A (CoA) by a group of enzymes called phosphopantetheinyl-transferases (PPT) (Fig. 1.3). The resulting 4′-phosphopantetheine (PPant) arm of the ACP had been termed as a "swinging arm" in the late 1960s by Lynen and colleagues (Sumper et al. 1969). Notably, bacteria possess enzymes, holo-ACP hydrolases (AcpH), that are able to catalyze the inverse reaction of removing the ACP PPant arm (Thomas and Cronan 2005).

FA biosynthesis can be sub-divided into 4 temporal cycles: activation, priming, elongation, and termination and requires the co-substrates acetyl-CoA, malonyl-

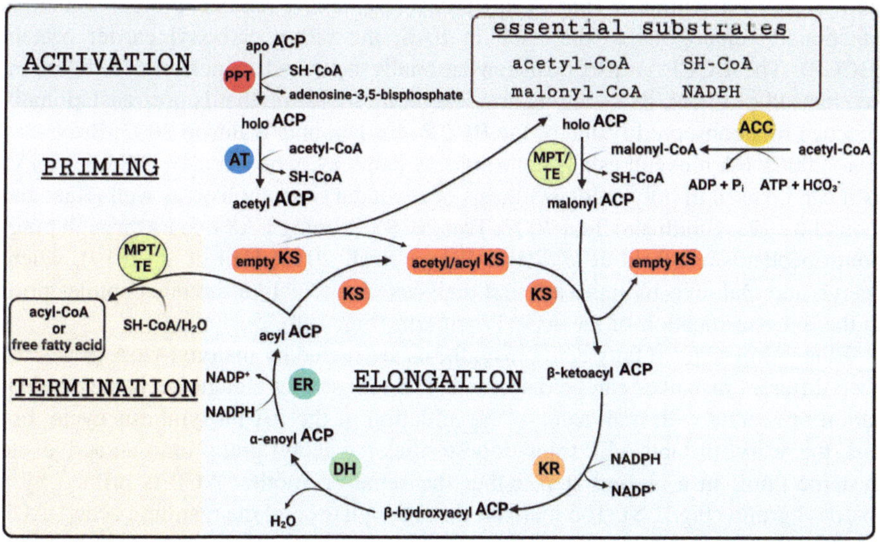

Fig. 1.2 Overview of the de novo FA biosynthesis cycle. FA biosynthesis requires CoA, acetyl-CoA, malonyl-CoA, and NADPH for functions. Malonyl-CoA is generated by the condensation of acetyl-CoA with CO_2 by the enzyme ACC. Four phases catalyzed by FAS, consisting of activation, priming, elongation, and termination, can be discerned in the FA biosynthesis cycle. A key feature is that all chemical reactions occur, when substrates and intermediates are affixed as chemical adducts to the ACP. This general FA biosynthesis cycle is used by virtually all organisms

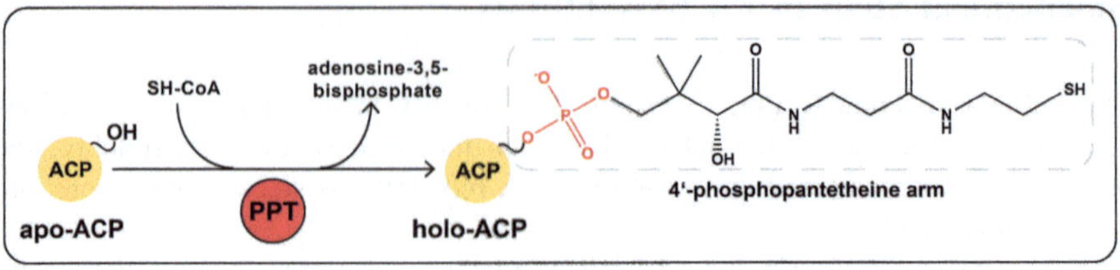

Fig. 1.3 The activation step of the FA biosynthesis cycle. The apo- acyl carrier protein (ACP) is modified on a defined serine side chain with the help of coenzyme A (CoA) by phosphopantetheinyl-transferases (PPT). This endows the ACP with a 4′-phosphopantetheine arm and a unique thiol on which substrates, intermediates, and products are attached as covalent (thioester-) adducts

CoA, and the reduced form of Nicotinamide Adenine Dinucleotide Phosphate (NADPH) (Fig. 1.2). The starting point of a fatty acid is the "activated form of acetic-acid" discovered by Lynen, acetyl-CoA, which is produced as a result of glycolysis (Lynen 1964; Perham 2000). Acetate (or acetyl-CoA) can also be considered a fatty acid (or fatty-acyl-CoA) with a two-carbon chain length. Malonyl-CoA is not bound to metabolic flux and is produced by an acetyl-CoA carboxylase enzyme (ACC) (Cronan and Waldrop 2002; Brownsey et al. 2006). The ACC reaction uses acetyl-CoA, ATP, and carbonate and represents the first committed step in FA biosynthesis (Fig. 1.4a). ACC contains a carrier protein which is functionally analogous to the ACP in FAS, the biotin carboxyl carrier protein (BCCP). The BCCP is also post-translationally activated to perform its function, yet, instead of PPant, its swinging arm is a biotin molecule that is postranslationally attached to a conserved lysine of the BCCP (Cronan and Waldrop 2002; Brownsey et al. 2006). FA biosynthesis is not a solitary pathway, where malonyl-CoA is used, as it can be used in polyketide synthesis of secondary metabolism as well (Katz and Donadio 1993; Smith and Tsai 2007; Tsai 2018). Priming FAS occurs through ping-pong mechanisms (Beld et al. 2015; Finzel et al. 2015; Heil et al. 2019), where acetyl- and malonyl- moieties are first transferred onto active site nucleophiles prior to the thioesterification of the ACP PPant arm (Fig. 1.4b, c).

Elongation steps in fatty acid biosynthesis are fueled by malonyl-CoA (Fig. 1.5). Two different instances can be discerned: The first step in elongation is the condensation of acetate with malonate, or the initiation of the FA biosynthesis cycle. For this, the acetyl-primed ACP trans-thioesterifies the acetyl group onto an active site cysteine thiol. In a second step, either the same or another ACP is primed by a malonyl group (Fig. 1.5). The malonyl decarboxylates and the resulting acetyl-ACP anion condenses with the thioacetyl, the result is a β-ketoacyl-ACP product and carbon dioxide as a side-product (Wakil et al. 1983; Smith et al. 2003; Beld et al. 2015; Finzel et al. 2015). In the second instance, an intermediate fatty acyl-ACP trans-thioesterifies onto an active site cysteine thiol (Fig. 1.5). The malonyl-primed ACP then again decarboxylates, and the resulting acetyl-ACP anion condenses with the thioacyl group. In any case, this reaction adds stoichiometrically two carbons to

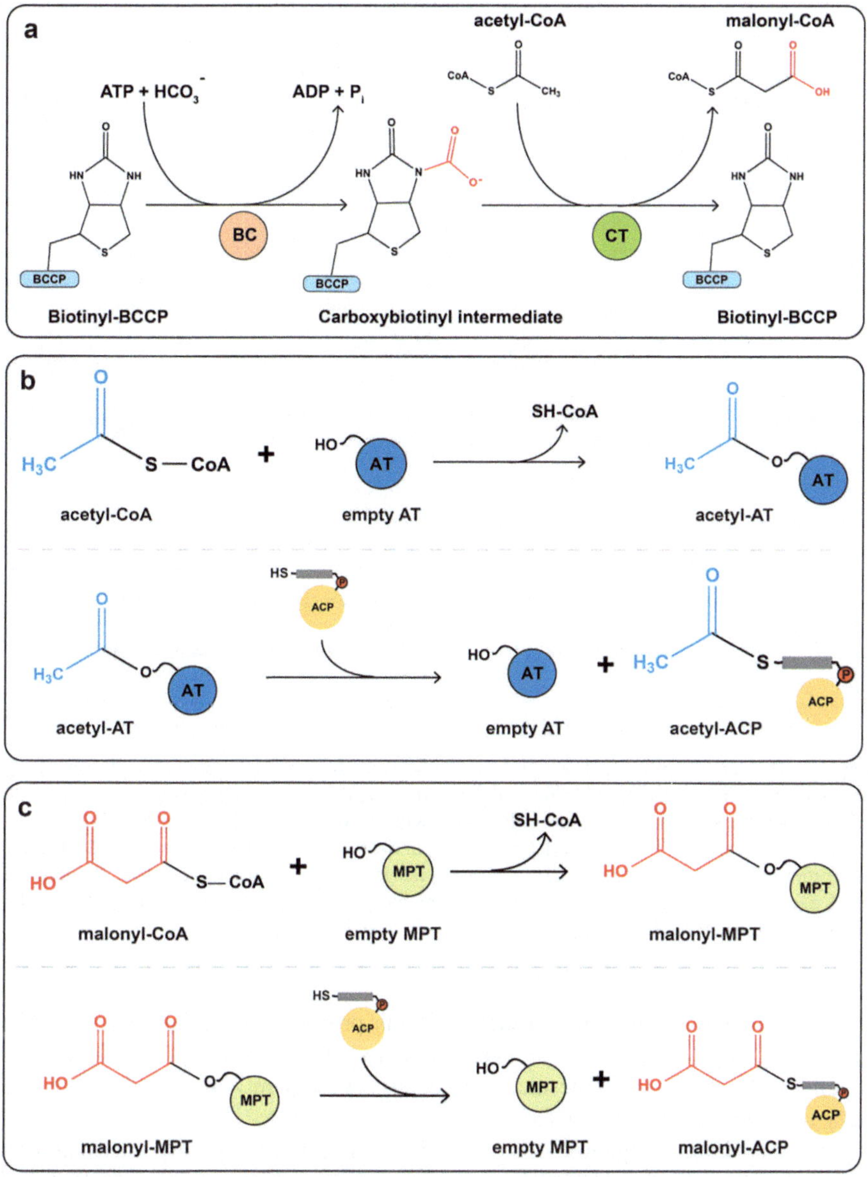

Fig. 1.4 Priming reactions in FA biosynthesis as employed in yeast FAS. (**a**) Generation of malonyl-CoA by acetyl-CoA carboxylases (ACC). The BCCP component of ACC is modified by a biotin swinging arm. Initially, the biotin carboxylase (BC) part of ACC phosphorylates carbonate to carbonyl phosphate, which is then used by BC to carboxylate biotin. The carboxyltransferase (CT) of ACC then transfers the carboxyl group to acetyl-CoA to form the malonyl-CoA co-substrate of FA biosynthesis. (**b**) Acetyl priming of FAS. The acetyl transferase (AT) of FAS uses an acetyl-CoA molecule to esterify a serine active site nucleophile by acetyl (acetyl-AT). This acetyl is then trans-esterified to the ACP PPant thiol. (**c**) Malonyl priming of FAS. The malonyl–/palmitoyl-transferase (MPT) of FAS uses a malonyl-CoA molecule to esterify a serine active site nucleophile by malonyl (malonyl-MPT). This malonyl group is then trans-esterified to the ACP PPant thiol

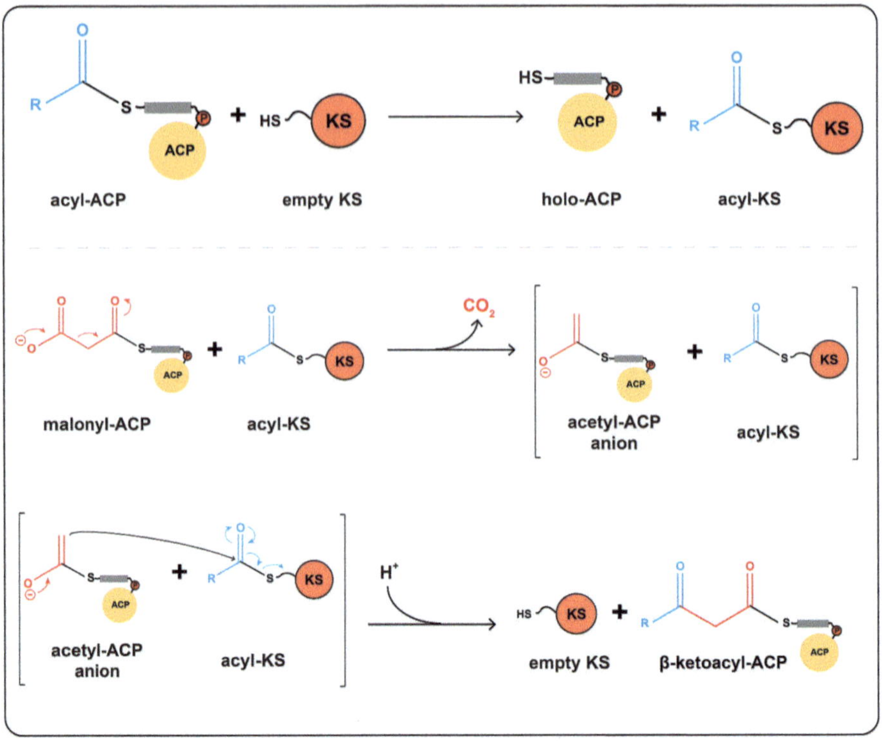

Fig. 1.5 The condensation reaction in FA biosynthesis. First, either the acetyl- or acyl-primed ACP trans-thioesterifies its acyl cargo (blue) onto an active site thiol of the empty ketosynthase (KS; top panel). Then the same or a different ACP primed with malonyl enters the acyl-KS active site. First the malonyl-primed ACP decarboxylates resulting in the formation of CO_2 as a side product and the generation of a transient acetyl-ACP anion (middle panel). The transient acetyl-ACP anion then condenses with the acetyl- or acyl-KS to regenerate the empty KS thiol and generates a two-carbon elongate β-ketoacyl-ACP

the growing fatty acid chain. Because of this process, biologically relevant fatty acids are always composed of an even-numbered carbon skeleton. The number can be odd, if the de novo synthesized fatty acid is subsequently modified with certain enzymes (Nelson and Cox 2013).

In the ensuing elongation steps (Fig. 1.6), the β-ketoacyl-ACP intermediate is then reduced in the first NADPH-consuming step to a β-hydroxyacyl-ACP. This is then followed by a dehydration reaction, where the removal of water generates an α-enoyl-ACP. Finally, in the second NADPH-consuming step, which reduces the α-enoyl-ACP to the fatty acyl-ACP in saturated form that is now elongated by a two-carbon unit. The enoyl reduction can occur by the direct transfer of reducing equivalents from NADPH to the α-enoyl-ACP, or can employ, in certain instances, flavin mononucleotide (FMN) as intermediary. In the latter cases, the reduced form

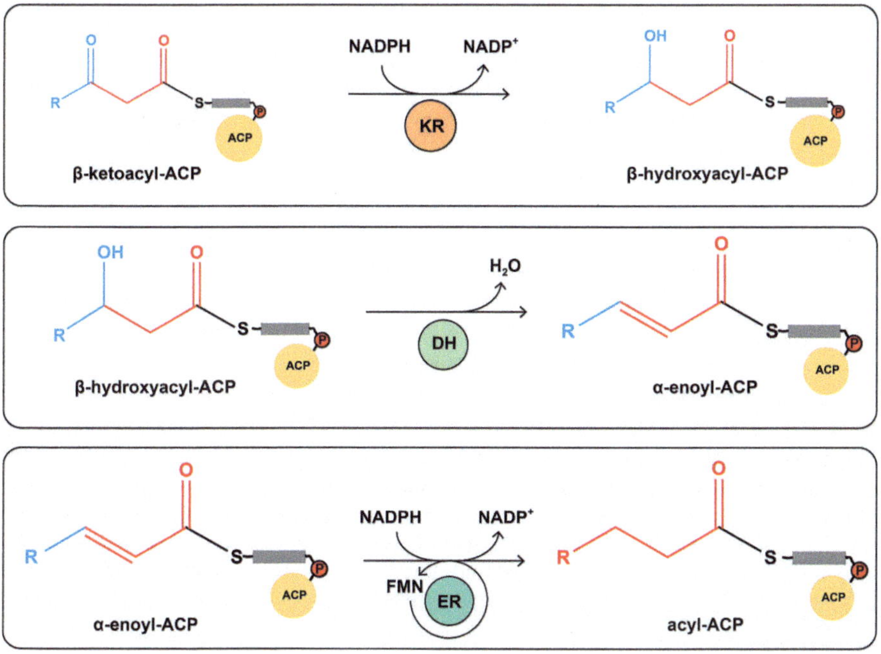

Fig. 1.6 Elongation steps in FA biosynthesis as employed in yeast FAS. The β-ketoacyl-ACP is initially reduced by NADPH in the ketoreductase (KR) to β-hydroxyacyl-ACP (top panel). Subsequent water removal from the β-hydroxyacyl-ACP by the dehydratase (DH) generates α-enoyl-ACP. The α-enoyl-ACP is then reduced by the enoyl reductase (ER) to the elongated acyl-ACP. Two versions of this reactions have evolved in nature: the one depicted here from yeast utilizes FMN as a co-factor to reduce the α-enoyl-ACP and NADPH to regenerate FMN. In other instances NADPH directly acts as a reductant for α-enoyl-ACP

of FMN ($FMNH_2$) reduce the α-enoyl-ACP and the reducing equivalents of NADPH are utilized to re-generate FMN to $FMNH_2$ (Fig. 1.6).

Malonyl-priming, condensation, and elongation cycles continue elongating the fatty acyl-ACP by sequential two-carbon units until the desired fatty acid length is obtained. The termination step also uses two distinct mechanisms and therefore generates two different products as a result of FA biosynthesis (Fig. 1.7). Some organisms transfer the fatty acyl-ACP to a free CoA molecule, resulting in acyl-CoAs as products (such as yeasts) (Tehlivets et al. 2007), others hydrolyze fatty acyl-ACP with water to generate free fatty acids (animalia) (Smith et al. 2003; Smith and Tsai 2007; Maier et al. 2010; Herbst et al. 2018). Bacterial, mitochondrial, and chloroplastic systems do not appear to release fatty acyl chains from the ACP, but rather fatty acyl-ACPs directly act as substrates in downstream reactions (Ohlrogge and Jaworski 1997; Cronan and Thomas 2009). Considering the importance of the ACP in all FAS biochemical reactions, it is justified to consider the acylated PPant arm together with ACP as the true substrate for FAS. As this consideration extends to the ACC BCCP and the biotin swinging arm, and the lipoyl domains of pyruvate

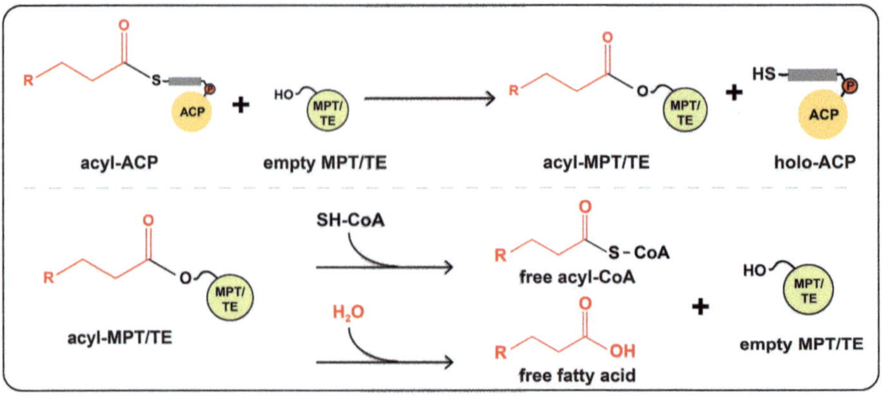

Fig. 1.7 Termination reactions in FA biosynthesis. When the fatty acyl-ACP has reached the appropriate chain length, it is transferred onto the termination enzyme, either malonyl–/palmitoyl-transferase (MPT in yeast) or thioesterase (TE in mammalia). In the respective termination enzyme, the fatty acyl-ACP thioester is hydrolyzed by an active site serine side chain generating acyl-MPT or acyl-TE and holo-ACP (top panel). At this point, the mechanisms and resulting products diverge, the acyl-MPT reacts with the thiol of a free CoA molecule to give rise to an empty MPT and a free acyl-CoA. The acyl-TE is hydrolyzed by a water molecule to generate an empty TE and a free fatty acid product (bottom panel)

dehydrogenase complexes, Perham, in the beginning of the twenty-first century, justifiably extended the swinging arm hypothesis to a swinging domain hypothesis (Perham 2000).

Type I and Type II FAS Architectures: Enzyme- and Enzymatic Domain Functions

As follows from the discussion of biochemical steps in FA biosynthesis, most reactions appear to be highly conserved. While differences are apparent, these seem to be variations on a theme rather than substantially divergent pathways. One would therefore be inclined to assume that the FA biosynthetic machinery likewise bears a striking resemblance in the various kingdoms of life. However, even from the very initial attempts to fractionate FA biosynthetic enzymes from various organisms, they appeared to differ. Some, primarily bacterial, mitochondrial, and chloroplastic systems, were strongly dependent on the addition of exogenous holo-ACP for the reconstitution of biosynthetic activity. On the other hand, some other systems appeared to be self-sufficient, only reliant on the addition of an ample supply of co-substrates for efficient function (Bloch 1969, 2006; Lynen et al. 1980; Wakil et al. 1983; Schweizer and Hofmann 2004; White et al. 2005; Smith and Tsai 2007; Cronan and Thomas 2009; Maier et al. 2010). These varying functional properties hinted at architectural and structural differences (Fig. 1.8).

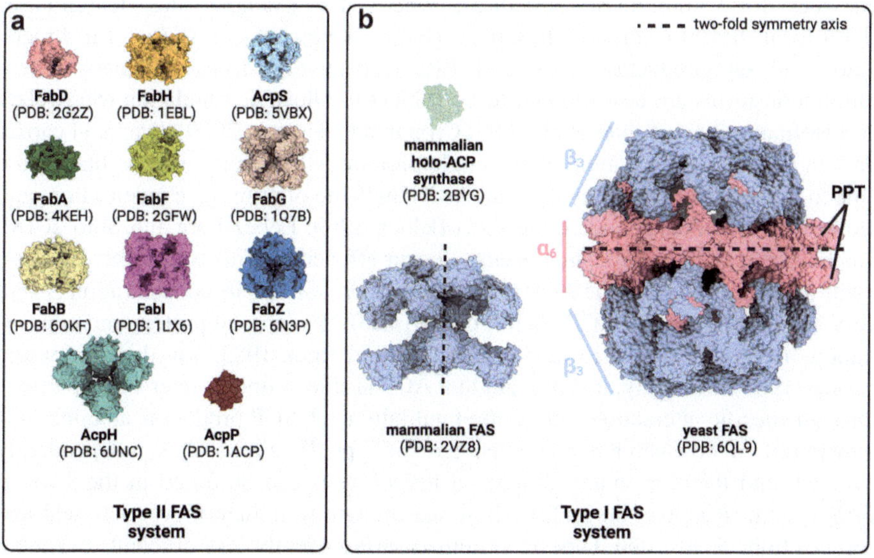

Fig. 1.8 Architectures of Type I and Type II FAS systems. (**a**) Type II systems predominate in bacteria, mitochondria, and chloroplasts. Representatively, in this figure, we show structures of the *E. coli* enzymes. The holo-ACP synthase (AcpS, PDB-ID: 5VBX) is the PPT responsible for the generation of the PPant arm in holo-ACP, the ACP (AcpP, PDB-ID: 1ACP), and the holo-ACP hydrolase (AcpH, PDB-ID: 6UNC) affect ACP function. The malonyl-CoA ACP transacylase (FabD, PDB-ID: 2G2Z), the β-ketoacyl-ACP synthase (KAS) III (FabH, PDB-ID: 1EBL), which are together responsible for the initiation of FA biosynthesis, the first priming ACP and the second performing the first condensation step. The KAS I enzyme (FabB, PDB-ID: 6OKF) is responsible for condensation steps leading to C6-C14 fatty acyl-ACPs, whereas KAS II (FabF, PDB-ID: 2GFW) generates C6-C16 fatty acyl-ACPs. The β-ketoacyl-ACP reductase (FabG, PDB-ID: 1Q7B), the two β-hydroxyacyl-ACP dehydratases (FabZ, PDB-ID: 6N3P and FabA, PDB-ID 4KEH), and the α-enoylacyl-ACP reductase (FabI, PDB-ID: 1LX6) comprise the elongation module of Type II systems. (**b**) Two different Type I FAS systems have evolved in nature. Left panel: In the mammalian system, a α_2 homodimeric X-shaped FAS (PDB-ID: 2VZ8) is found which is 540 kDa in size. ACP and TE domains are not resolved in the structure, but are present as C-terminal fusions on each of the α polypeptide chains. The mammalian PPT (holo-ACP synthase, PDB-ID: 2BYG) is separate and the only exogenous protein required for FA biosynthesis in the mammalian system. Right panel: The most integrated FA biosynthetic machinery is represented by the Type I yeast FAS. It exists as a 2-2.5 MDa homo-hexameric barrel structure, which is composed of either six copies of a single polypeptide chain $(\alpha)_6$ or two polypeptide chains $(\alpha\beta)_6$. In the *S. cerevisiae* FAS shown here (PDB-ID: 6QL9), where the building block is a αβ heterodimer, the 6 α-chains (colored in purple) form a central wheel of the structure, whereas the 6 β-chains (colored in blue) form the domes of the barrel. All enzymatic activities are present in the yeast FAS barrel, with ACP tethered as domains within the barrel, all enzyme active sites are pointing into the barrel cavity, and the PPT is located on the outside of the barrel at the central wheel

Structurally, fatty acid synthases can be divided into two different groups, called Type I and Type II systems (Fig. 1.8). In Type II systems, all enzymes needed for fatty acid synthesis are expressed as individual proteins (Schweizer and Hofmann 2004; White et al. 2005; Cronan and Thomas 2009). What is observed is that these enzymes form multimeric structures; where, for example, the dehydratases (DH) form dimers (FabA) or hexamers (FabZ), ketosynthases (KS) form dimers, keto- and enoylreductase (KR and ER, respectively) form tetramers. These multimeric forms are also thought to be subject to allosteric regulation (Schweizer and Hofmann 2004; White et al. 2005; Cronan and Thomas 2009). The acyl carrier protein (ACP), is also a monomeric single protein. All enzymes and facilitators are able to diffuse freely in the cytoplasm. As the half-life of priming, condensation, and elongation intermediates are quite short (Bloch 1969; Lopez-Lara and Soto 2018), this hints that ACP modification needs to occur efficiently. This is facilitated *via* two main-mechanisms: First, the ACP is an abundant protein and its concentration is kept very high. In *E. coli* the ACP, AcpP, comprises 0.25% of total protein and concentrations reaching up to 0.15 mM (Jackowski and Rock 1983; van den Boom and Cronan 1989). Secondly, it is thought that ACP is able to find its interaction partners through specific interactions. Once the final fatty acyl-ACP product is attained, it is transferred to downstream reactions. Every Type II enzyme has independently evolved, and therefore many functional homologues can be found in the Type II FAS system (Ruppe et al. 2020). These are thought to influence product yield and composition. Hence, the Type II system is an apparently less efficient, but more versatile pathway that readily lends itself to produce fatty acids of greater diversity (van den Boom and Cronan 1989; White et al. 2005; Cronan and Thomas 2009).

Since each enzyme is separate in Type II systems, it makes it simpler to explain enzyme functions in these systems. In the following, we will use the *E. coli* FAS for this purpose (Fig. 1.9a). As mentioned earlier, during the entire FA biosynthetic process, all substrates, intermediates, and products are always covalent adducts to ACP, the AcpP protein in *E. coli*. The PPant arm in AcpP is affixed to Ser37, the PPT is AcpS, and *E. coli*, like some other bacteria, contains an enzyme, AcpH, that is able to catalyze the hydrolysis of the PPant arm to generate apo-AcpP (Fig. 1.8a) (Vagelos and Larrabes 1967; Thomas and Cronan 2005). The malonyl-CoA generating ACC in *E. coli* is composed of four subunits: AccA, AccB, AccC, and AccD (Cronan and Waldrop 2002). In *E. coli* FA biosynthesis, ACP is initially primed by the malonyl-CoA-ACP transacylase FabD, whereas acetyl-CoA is not used to prime ACP. The ketoacyl-ACP synthase (KAS) III protein FabH performs the first condensation reaction between malonyl-ACP and acetyl-CoA to generate acetoacetyl-CoA. Later condensation steps are performed by the KAS I enzyme FabB to generate C_6-C_{14}-β-ketoacyl-ACP, whereas the KAS II enzyme FabF yields C_6-C_{18}-β-ketoacyl-ACP (Cronan and Thomas 2009). The β-ketoacyl-ACP reductase FabG utilizes the reducing equivalents of NADPH to generate β-hydroxyacyl-ACP. The two β-hydroxyacyl-ACP dehydratases, FabA and FabZ, remove a water molecule to yield α-enoyl-ACP, where the double-bond is in trans-configuration. A unique feature of FabA is that it can isomerize the α-enoyl-decanoyl-ACP to a β-enoyl-ACP, with a cis-double bond. This β-enoyl-ACP can directly re-enter a condensation reaction with malonyl-ACP and either FabB or FabF to generate a mono-unsaturated FA (Cronan and Thomas 2009). Elongation continues with the NADPH-dependent reduction of α-enoyl-ACP by the α-enoyl-ACP reductase FabI to the two-carbon elongated which is deposited onto the active site thiol of either

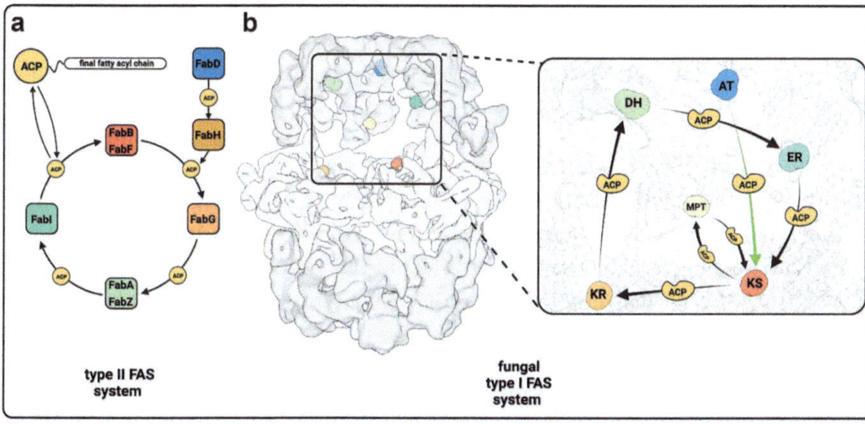

Fig. 1.9 Fatty acid biosynthesis cycles of Type I and Type II systems. (**a**) Type II systems strongly depend on the exogenous addition of holo-ACP (reflected by solid lines). An additional feature is that multiple enzymes can have seemingly redundant activities in the Type II system. This is illustrated by FabB and FabF KS enzymes, as well as FabA and FabZ DH enzymes. (**b**) In the left panel, the yeast Type I FAS barrel is shown, with active site clefts of enzymatic domains colored in the same scheme as in (**a**). The right panel indicates a zoom-in of the window depicted in the left panel. FA biosynthesis cycles in Type I FAS systems are independent and self-sufficient, requiring only addition of co-substrates for activity. Based on the spatial proximity enzyme active site clefts, elongation cycles have been proposed to run in clock-wise direction as indicated by the arrows

FabB or FabF for yet another condensation with malonyl-ACP. FabI is considered to be the last irreversible reaction of the fatty acid synthesis cycle. Thus, it is vital for the cycle to continue, as other intermediates tend to dissociate quickly. That is why FabI is considered to be a promising drug target for novel antimicrobials (Lu and Tonge 2008). Once the fatty-acyl-ACP chain has reached a suitable length (usually C_{12}-C_{16}-acyl-ACP), the mature fatty acid chain can be used for downstream pathways (Fig. 1.9a). In bacterial systems, the fatty acid stays bound to the ACP and when needed interacts with downstream partners such as those required to produce phospholipids or triglycerides (Bartholow et al. 2021). In *E. coli*, AcpP is known to have more partner proteins other than FAS enzymes, which is also an active area of research. As of today, *E. coli* AcpP is known to have ~30 interaction partners in the *E. coli* proteome (Gully et al. 2003; Gully and Bouveret 2006).

Type I systems, in contrast, are composed of large, multi-domain polypeptides. They form multimeric assemblies that contain all necessary enzymatic activities in the same macromolecular complex. The ACP is incorporated as a fusion into the polypeptide chains of these assemblies and therefore becomes an inherent domain of these FAS complexes which shunts the growing FA chain to the various enzymatic domains of the complex. Type I systems themselves come in two different flavors (Fig. 1.8b): Mammalian FAS is a homodimer, where each monomer contains all necessary domains. Its dimeric structure is found to be X-shaped and works in an open environment, as suggested by its 3.2 Å crystal structure (Maier et al. 2008).

Unfortunately, neither ACP nor the terminating thioesterase (TE) domains are resolved in the structure. However, it is well understood that the final FA product is released from the complex as free fatty acid (Smith et al. 2003; Maier et al. 2010; Herbst et al. 2018). Electron microscopic studies have shed light on the conformational flexibility of the whole enzyme and revealed that substantial conformational motion is required for FA biosynthesis to occur (Brink et al. 2004; Asturias et al. 2005; Brignole et al. 2009). Mammalian Type I FAS systems will not be the focus of this review. However, the interested reader is referred to several excellent reviews, where functional aspects of mammalian Type I FAS systems and the structural basis of their function are discussed in great detail (Smith and Tsai 2007; Maier et al. 2010; Herbst et al. 2018; Heil et al. 2019).

The yeast-like Type I FAS, on the other hand, is a hexameric barrel-like structure (Figs. 1.8b and 1.9b). In the yeast *Saccharomyces cerevisiae*, for example, all activities are present on two polypeptides and the hexameric barrel is of a $(\alpha\beta)_6$-Type (Jenni et al. 2007; Leibundgut et al. 2007; Lomakin et al. 2007; Johansson et al. 2008; Gipson et al. 2010; Joppe et al. 2020; Singh et al. 2020). In *S. cerevisiae* FAS, the six α-subunit copies form a central wheel which divides the barrel equatorially, three β-subunits on each side of this equatorial divide form the two domes of the hexameric barrel (Fig. 1.8b). Thus, the αβ-heterodimer can be considered the building block of the yeast FAS hexamer. This has profound implications for the assembly of yeast FAS, which occurs co-translationally (Shiber et al. 2018). The ACP domain resides within the barrel, where all enzymatic sites are also facing, and is tethered by two anchors to the α-subunit. The PPT responsible for activating the ACP with the PPant arm (on Ser 180 of the α-subunit) is encoded on the C-terminal end of the α-subunit and resides on the outside of the central wheel in monomeric form in the assembled particle. However, for activity, the PPT must oligomerize (Johansson et al. 2009), therefore holo-ACP formation is thought to occur during assembly in yeast (Fischer et al. 2020). All enzymatic domains in yeast FAS are arranged in a manner that the active sites' cavities are opened to the inner cavities of the dome (Fig. 1.9b).

FA biosynthesis in yeast FAS starts with the priming of ACP to acetyl-ACP at the acetyltransferase (AT) domain of the β-subunit which is located at the top of the domes (Fig. 1.9b). The acetyl-ACP then transfers the acetyl group onto the active site cysteine of the α-subunit ketosynthase (KS) domain. The ACP is then primed to malonyl-ACP by the malonyl−/palmitoyl-transferase (MPT) domain, which is formed by both α- and β-subunits and might be another explanation for the co-translational assembly of yeast FAS (Shiber et al. 2018; Fischer et al. 2020). The malonyl-ACP condenses with the acetyl-KS to form β-ketoacyl-ACP, which then proceeds to the α-subunit encoded ketoreductase (KR) domain. The KR domain reduces the β-ketoacyl-ACP to β-hydroxyacyl-ACP using the reducing equivalents of NADPH. The β-subunit encoded dehydratase (DH) domain then removes water to generate α-enoylacyl-ACP. The α-enoylacyl-ACP is subsequently reduced by the $FMNH_2$ cofactor of the β-subunit encoded enoylreductase (ER) domain to the two-carbon elongated fatty acyl-ACP, which is then transferred onto the KS active site thiol. The FMN is then regenerated to the $FMNH_2$ by NADPH and the acyl-KS can react yet again with a malonyl-primed-ACP. The cycle of condensation, ketoreduction, dehydration, and enoylreduction is repeated 6–7 times to generate the preferred C_{16}/C_{18}-fatty-acyl products of the yeast FAS. At this stage, the acyl-CoA is transferred to the MPT, where the fatty acyl product reacts with CoA and is released as a fatty acyl-CoA molecule.

This closed environment is specific to the fungal FAS and it allows for an estimated local ACP concentration of about 1 mM (Lynen et al. 1980; Maier et al. 2010), enables a rate of 18 FA biosynthetic cycles a minute (Fichtlscherer et al. 2000), and allows for strict regulation (Lynen et al. 1980; Maier et al. 2010; Gajewski et al. 2017a, b; Herbst et al. 2018). Some other yeasts, fungi, and most notably the Corynebacteria/Mycobacteria/Nocardia (CMN) bacteria have FAS organized as α_6-Type homohexamers where the $\alpha\beta$-heterodimer is fused to a single α-chain (Schweizer and Hofmann 2004; Ciccarelli et al. 2013; Elad et al. 2018). Type I systems are known to be more efficient and exhibit narrower product ranges than Type II Systems.

Phylogeny of FAS Architectures

As discussed in the earlier sections, FA biosynthesis is central in the eubacterial, protistic, plant, fungal, and animal kingdoms of life (López-Lara and Geiger 2010). Archaebacteria, on the other hand, are thought to be lacking a FA biosynthesis mechanism, or at least FAs appear not to be essential for them (Lombard et al. 2012). Viewed phylogenetically (Fig. 1.10), one can conclude that Type I systems most likely evolved from the concatenation of dissociated Type II FAS enzymes to multidomain multienzyme polypeptides. Type II FAS enzymes can be observed in eubacteria, as well as organelles of prokaryotic origin (mitochondria, chloroplasts, apicoplasts). In archaea, FA biosynthesis is not a central part of their metabolic cycle since they possess membranes composed of isoprenoids, instead of phospholipids. As a consequence, for a long while, archaebacteria were thought to be devoid of FA biosynthetic enzymes and possess no FAS. Nevertheless, recent observations indicate that they do indeed utilize FAs. These are attached to either their proteins as post-translational modifications, or are found in membranes. Moreover, archaea appear to encode for FA biosynthetic enzymes that are of Type II origin, suggesting the existence of FAS in several archaeal species (Lombard et al. 2012). However, the significance of these findings is yet to be determined.

When considering the eukaryota branch, one significant event is the endosymbiosis of an unknown alpha-proteobacterium which evolved into today's mitochondria. It has been observed that mitochondria still possess an FA biosynthesis system reminiscent of the bacterial Type II FAS. This system is proposed to be involved in mitochondrial membrane remodeling but also production of certain metabolic intermediates, such as lipoic acid (Kastaniotis et al. 2017). The second endosymbiosis event that led to the incorporation of plastid-like organelles encompasses the photosynthetic eukaryotes, and other protists of the TSAR superphylum. Similar to the

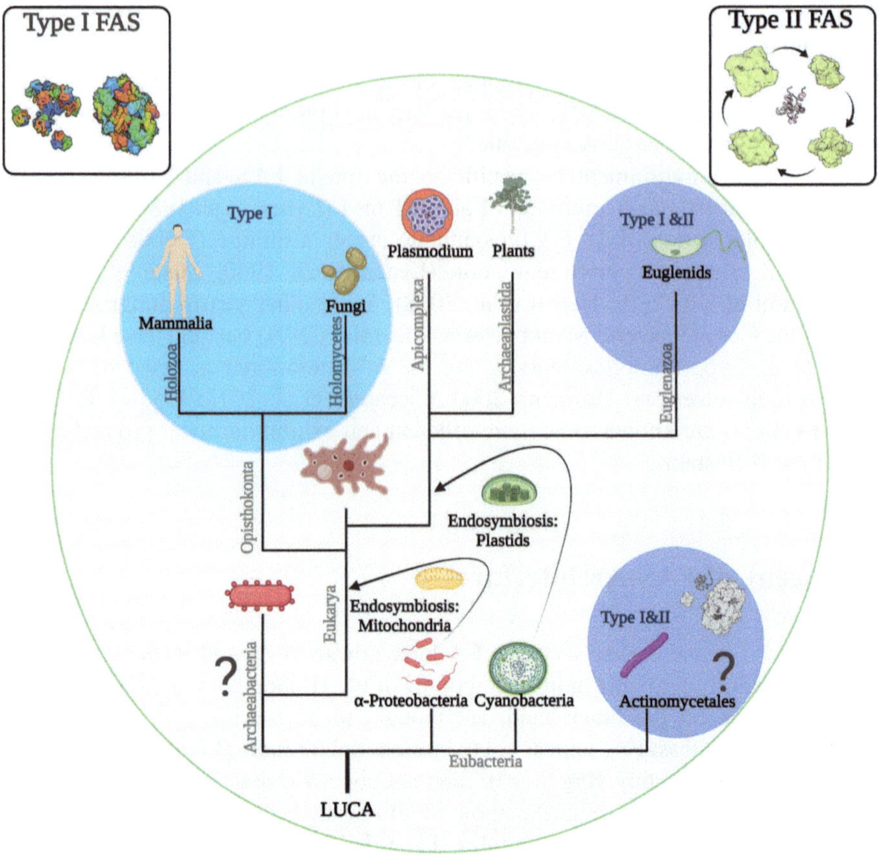

Fig. 1.10 Phylogeny of FAS architectures. Dissociated Type II enzymes appear to be the early forms of FA biosynthetic enzymes, whereas the multimeric Type I enzymes emerge later in the phylogenetic tree. Type II enzymes are widely found in all genera of life. In Archaea, FAS enzymes are not essential components of metabolism, since their membranes are mostly composed of isoprenoid variants. Yet, archaea possess Type II FAS enzymes since it has been observed to be important for certain metabolic processes. As a consequence of their endosymbiotic origin, mitochondria and plastids also possess Type II enzymes, like their prokaryotic ancestors. For plants, FA biosynthesis mainly occurs in chloroplasts. For various species within the SAR superphylum, different plastid Types exist for different metabolic processes. For example, Apicomplexan species such as Plasmodium falciparum and Toxoplasma gondii possess plastid organelles that are called apicoplasts, which are important for certain stages in their life cycle. Type I enzymes have been observed to occur within opisthokonts such as holozoa and holomycota. In certain situations, different branches have separately evolved both systems within their metabolism, because of their atypical fatty acid derivatives. The CMN bacteria possess both Type I and Type II systems to produce their very long and branched mycolic acids, aided by a polyketide synthase pks13 in mycobacterium tuberculosis. Poly-unsaturated fatty acid producing organisms can have Type I, II, and PKS-like enzymes to continue the process

situation in mitochondria, the endosymbiont also brought their own bacterial-Type Type II FAS systems which have remained present in these plastids. For plants, both carbon fixation and electron coupling *via* photolysis happen in the chloroplast, which also makes it the primary organelle where energy and carbon source surplus can occur (Ohlrogge and Jaworski 1997). This is in accordance with the fact that FA synthesis takes place in the chloroplasts of such organisms (Ohlrogge and Jaworski 1997).

Under the TSAR superphylum, a group called apicomplexans, which also includes toxoplasma and plasmodium species, are known to possess plastid organelles that are called as apicoplastids. These organelles do not perform photosynthesis, but they are responsible for important metabolic functions such as fatty acid synthesis. For plasmodium species, which is the causative agent of malaria, it is known that this fatty acid synthesis pathway is crucial for the sporozoite to merozoite transformation in the liver (Shears et al. 2015). This switch in the life cycle of plasmodium is well known to be a key event leading to the acute stages of malarial disease (Phillips et al. 2017), also being important for immune recognition (Mazier et al. 1990). For Plasmodium falciparum, it was later observed that the Type II FAS is rather more important for the sporozoite development in the midgut of mosquitos (van Schaijk et al. 2014). Consequently, the plasmodium FAS is a validated and valuable target for anti-malarial drugs.

Type I systems appear independently in several phyla. Under the phylum Unikonta, under which animal and fungal species are found, we observe the appearance of Type I systems. As mentioned earlier, the Type I architecture in animals is in an X-shaped homodimeric form, whereas in fungi, the Type I architecture is homohexameric where the building block is either a αβ-heterodimer or a single α chain. The single α chain homohexamer predominant in microbial FAS of some bacteria also is present in CMN bacteria, alongside with a Type II system (Marrakchi et al. 2014). CMN-bacteria are known to produce extremely long, dual-branched fatty acids that are termed mycolic acids, which are significant for their cell wall structures. In these organisms, the Type I system is described to exhibit a bi-modal product distribution of $C_{16/18}$-CoA and $C_{22\text{-}26}$-CoA (Marrakchi et al. 2014). The $C_{20\text{-}24}$-CoA comprise the FA branch of mycolic acids, the other mero-branch stems from the Type II system elongating C_{16}-CoA to $C_{42\text{-}62}$-CoA (Marrakchi et al. 2014). The FA branch is activated by a specialized ACC and fused with the mero branch in a specialized polyketide synthase pks13 (Marrakchi et al. 2014). This makes both Types of FAS in Mycobacterium tuberculosis targets of vital importance in antimycobacterial strategies (Bosch et al. 2021).

Regulation of FA Biosynthesis

Fatty acid synthesis is a crucial but cumbersome process. It needs a high energy-containing carbon source and ATP to synthesize its three co-substrates acetyl-CoA, malonyl-CoA, and NADPH. For the synthesis of a C_{16} FA, for example, cellular metabolism has to produce the energy that is contained in 122 ATP molecules (Nelson and Cox 2013). As alluded to earlier, FAs in the form of triglycerides are compact sources for energy storage that occupy less physical space than the energetic equivalent in the form of carbohydrates. Therefore, when carbon sources are abundant, microorganisms tend to favor FA biosynthesis to build new membranes and store excess carbon. Metazoans more strongly rely on their diet to acquire FAs rather than de novo biosynthesis to produce their own (Burr and Brown 1933). Nevertheless, FA biosynthesis, which does not occur in all cells and tissues of a metazoan, does take place in some specialized cell types. These include adipocytes and hepatocytes that regulate blood sugar levels by taking up excess monosaccharides, and the lactating mammary gland (Lane and Moss 1971; Suburu et al. 2014).

The three co-substrates of FAS (acetyl-CoA, malonyl-CoA, and NADPH) are provided from different sources. Whenever there is a surplus of carbon sources, acetyl-CoA is obtained through glycolysis in the citric acid cycle by the pyruvate dehydrogenase complex which oxidatively decarboxylates pyruvate to produce acetyl-CoA (Perham 2000). The citric acid cycle also elevates the eponymous citrate concentrations which can be used by ATP-citrate lyase to produce acetyl-CoA (Verschueren et al. 2019; Wei et al. 2019; Granchi 2019). In low carbon flux, acetyl-CoA synthetase, alcohol dehydrogenase, and the degradation of branched-chain ketogenic amino acids such as valine, leucine, and isoleucine generate acetyl-CoA (López-Lara and Geiger 2010). Also at low carbon abundance, the degradation of FA energy stores induces the reversal of FA biosynthesis, the FA β-oxidation pathway also yields acetyl-CoA (Lopez-Lara and Soto 2018). Likewise, NADPH is linked to respiration and is primarily generated in the pentose-phosphate pathway, in the first step glucose-6-phosphate dehydrogenase. Other sources of NADPH include the Ferrodoxin-NADP+ reductase, which is a major source in photosynthetic organisms, and malic enzyme, isocitrate dehydrogenase, and glutamate dehydrogenase (Ohlrogge and Jaworski 1997; Lopez-Lara and Soto 2018). Both these processes therefore put FA biosynthesis in connection with metabolic flux. Malonyl-CoA, on the other hand, is not directly bound to metabolic flux. As shown in Figs. 1.2 and 1.4a, it is produced from acetyl-CoA by ACC, which also provides malonyl-CoA for polyketide synthesis. For this reason, the production of malonyl-CoA is considered the first and only committed step in FA biosynthesis (Wakil et al. 1983; Cronan and Waldrop 2002; Tong 2013).

Considering the intimate linkage of FA biosynthesis, and therefore FAS activity, to the metabolic state of the cell, two immediate consequences arise. First, FA biosynthesis is subject to feedback regulation from metabolic pathways and the ensuing flux. Second, were FAS activity to continue unhindered and unregulated when carbon sources are limiting, FA biosynthesis has the potential to jeopardize the survival of cells. It would deplete ATP, acetyl-CoA, and NADPH stores under conditions where already diminishing energy and carbon stores make these co-substrates sparse and withdraw them from processes vital for the subsistence of cells. This downward cycle would cause cells to starve and ultimately drive them to death. To circumvent this impending doom, FA biosynthesis is subject to feedback metabolic regulation cycles, including transcriptional feedback cycles and multiple layers of biochemical regulation (Lust and Lynen 1968; Wakil et al. 1983; Schweizer and Hofmann 2004).

Genetic regulatory pathways of FA biosynthesis pathways have been elucidated over the years *via* microbiological, biochemical and genetic methods (Parsons and Rock 2013). In *E. coli*, for instance, the first global regulatory system uncovered was related to the abundance of the global long-chain acyl-ACP and long-chain acyl-CoA pool, which is low under normal homeostatic conditions. However, when, for example, phospholipid biosynthesis is compromised by either poor activity of the glycerolphosphate acyltransferase PlsB system or the stringent response following starvation, this pool is elevated. The effector of the stringent response appears to be the alarmone ppGpp, which inhibits PlsB. There also exists a complex transcriptional regulation of FA biosynthesis through the transcription factors FadR and FabR (Lopez-Lara and Soto 2018; Gago et al. 2019). FadR is tightly bound to DNA under conditions of low long-chain acyl-CoA concentrations and represses transcription from the genes of FA β-oxidation pathway (Cronan John Jr and Subrahmanyam 1998). On the other hand, under low long-chain acyl-CoA concentrations, it enhances transcription from several FA biosynthetic genes, including fabA, fabB, the fabHDG operon, fabF, fabI, acpP, and the accA-D genes. Under high long-chain acyl-CoA concentrations, it dissociates from DNA, eliciting the opposite transcriptional response. FabR acts through a different mechanism, unsaturated long-chain acyl-CoAs or unsaturated long-chain acyl-ACPs promote binding to DNA, shutting of the transcription of fabA/fabB genes (Zhu et al. 2009). Binding of either saturated long-chain acyl-CoAs or saturated long-chain acyl-ACPs to FabR elicit dissociation from DNA and switch on the transcription of fabA and fabB genes. In the yeast *S. cerevisiae*, it has been shown that the expression of FAS and ACC is regulated to FA and inositol supplementation (Schüller et al. 1992; Chirala 1992; Chirala et al. 1994). It also requires the INO2 and INO4 proteins which stimulate transcription, and the effects of FA and inositol are cumulative (Schüller et al. 1992; Chirala et al. 1994). In mammalian cells, sterol regulatory element-binding proteins (SREBPs) play a role in the regulation of FAS, ACC, and stearoyl-CoA desaturase gene expression (Bennett et al. 1995; Lopez et al. 1996; Kim and Spiegelman 1996; Tabor et al. 1999). ACC in mammals is also inhibited by glucagon and epinephrine, while insulin activates the enzyme through transcription regulatory pathways (Witters et al. 1988; Mabrouk et al. 1990; Andreolas et al. 2002).

In addition to genetic regulation, biochemical regulatory processes are in play with regard to FA biosynthesis. Acetyl-CoA synthesis, generated by the oxidative decarboxylation of glycolytic pyruvate, drives fatty acid synthesis (Nelson and Cox 2013). The same acetyl-CoA also promotes the flux through the citric acid cycle. In eukaryotes, the enzymes of the citric acid cycle are located in the mitochondrion and, when citrate levels are high, they inhibit the pyruvate dehydrogenase and the succinate dehydrogenase complexes (Iacobazzi and Infantino 2014). Superfluous citrate is also exported back into the cytosol, where it negatively impacts several glycolytic enzymes, but positively regulates fructose-1,6-bisphosphatase, and ACC in mammals (Nelson and Cox 2013; Iacobazzi and Infantino 2014). The positive effector function of citrate promotes polymerization of mammalian ACC, and therefore the production of malonyl-CoA (Hunkeler et al. 2018). It, however, appears not to impact yeast ACC (Wei et al. 2016) and in bacterial ACC, where cellular compartmentalization does not exist (Nelson and Cox 2013). However, as mentioned earlier, ATP-citrate lyase is cytoplasmic and can break down citrate to oxaloacetate and acetyl-CoA (Verschueren et al. 2019; Wei et al. 2019; Granchi 2019). Thus, either by the elevated production of malonyl-CoA and/or acetyl-CoA, citrate exerts a positive effect on FA biosynthesis. When carbon sources are low, AMP is generated and activates AMP kinase, which, in turn, universally phosphorylates ACC inhibiting malonyl-CoA production (Wakil et al. 1983; Wei et al. 2016). Another, universal biochemical regulatory process lies in the products of FA biosynthesis, long-chain fatty acids, long-chain fatty-acyl-CoAs, and long-chain acyl-ACPs, exerting feedback inhibition on ACC and FAS (Wakil et al. 1983; Tong 2013; Hunkeler et al. 2018).

Direct Regulation of ACP

The regulatory pathways described in the previous section were indirectly affecting the FA biosynthesis pathway, but no regulatory cues impacting FAS directly were mentioned. Another layer of biochemical regulation relates to the fact that one has to consider the ACP together with its acylated PPant arm as the true substrate for FAS. It is worth mentioning that bacterial and metazoan ACPs are composed of a four-helix bundle (termed the canonical domain), which is extended in fungal FAS by a structural four helix domain (Fig. 1.11). Considering the central pivotal role of the ACP, one could conceive it also to be a central point of regulation. In a sense, it has to fulfill the role of a universal recognition platform for all FAS active site (Chan and Vogel 2010; Beld et al. 2015; Finzel et al. 2015). On the other hand, it also has to protect its growing hydrophobic cargo from the aqueous environment of the cell

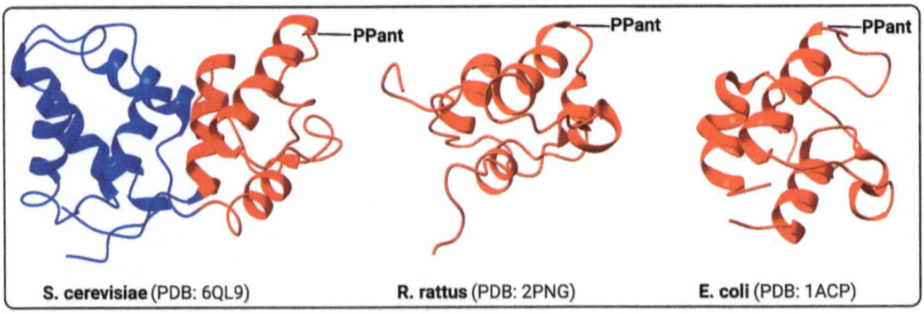

Fig. 1.11 ACP structural conservation. Shown are ACP structures from the Type I yeast FAS (left, PDB-ID: 6QL9), the Type I mammalian FAS (middle, PDB-ID: 2PNG), and the Type II bacterial FAS (right, PDB-ID: 1ACP). Colored in red is the canonical domain of ACPs, which is a four-helix bundle and conserved in all kingdoms of life. The fungal Type I FAS structure contains a four-helix extension called the structural domain (colored in blue)

(Chan and Vogel 2010; Beld et al. 2015; Finzel et al. 2015). To achieve this, a "switchblade" mechanism involving its PPant arm comes into play (Chan and Vogel 2010; Beld et al. 2015; Finzel et al. 2015).

Considering the essential nature of ACP in substrate shuttling and recognition, it is surprising that very little structural knowledge has been gained about its recognition by FAS. Most ACP–FAS interactions seem to be centered around the PPant arm and involve electrostatic interactions (Leibundgut et al. 2007; Lomakin et al. 2007; Gipson et al. 2010; Nguyen et al. 2014; Lou et al. 2019; Lou and Mazhab-Jafari 2020; Mindrebo et al. 2020; Singh et al. 2020). However, if this universally applies to all ACP–FAS enzyme interactions remains to be seen. At least in the yeast Type I FAS, the additional structural domain of the ACP appears to be involved in interactions to ER, KS, DH, and AT domains (Leibundgut et al. 2007; Lomakin et al. 2007; Gipson et al. 2010; Lou et al. 2019; Lou and Mazhab-Jafari 2020; Singh et al. 2020). These interactions involve the structural domain of the ACP and architectural domains of the Type I yeast FAS. To better understand the regulation of FAS activity by ACP regulation, clearly more structures of ACP bound to each FAS enzyme-active site cleft are needed. Ideally, these structures also are of sufficiently high-resolution that the molecular basis of ACP and PPant arm recognition can be deciphered. This raises the issue of biochemically trapping FAS such that homogenous preparations of FA biosynthetic intermediate-ACP adducts can be obtained in preparative amounts. Activity-based probes, based on CoA derivatives as proposed by Burkart and co-workers and employed for Type II FAS systems, appear to be pre-destined for this purpose (Nguyen et al. 2014; Beld et al. 2015; Finzel et al. 2015; Mindrebo et al. 2020).

The importance of ACP regulation is exemplified by the fact that overexpression of apo-ACP is toxic to bacteria (Keating et al. 1995). Thus, PPT activity has to occur faithfully and spatially and makes this group of enzymes crucial for FA biosynthesis and, conversely, valuable drug targets. This is highlighted by recent findings that a mycobacterial PPT, PptT protein, has recently been validated as a drug target and an inhibitory compound has been discovered (Ballinger et al. 2019). Aside from PPT, as shown in Fig. 1.7, holo-ACP hydrolases exist that are able to generate apo-ACP by removing the PPant arm. The *E. coli* enzyme is termed AcpH, and notably the mycobacterial counterpart confers resistance to PptT inhibitors (Ballinger et al. 2019). In mycobacteria, where Type I and Type II FAS systems co-exist, there is a special Type II found called as AcpM that contains an additional C-terminal domain that is unfolded (Wong et al. 2002). What the function of this domain is remains enigmatic until today. Two hypotheses exist: one suggests that this domain helps to carry the longer fatty acid chains that are synthesized by mycobacteria, the second is that this domain mediates the interaction of AcpM with cognate enzymes (Wong et al. 2002). Considering the importance of ACP in FAS, and its modifying enzymes emerging as drug targets, more detailed mechanistic studies into ACP function are warranted.

Discovery of a Regulatory γ-Subunit of the Yeast FAS

The regulatory pathways described above affect FA biosynthesis indirectly. They depend either on metabolic transcriptional feedback circuits, biochemical feedback regulation on ACC and FAS through products of the FA biosynthetic pathway which are relayed through direct binding or post-translational modifications, and/or factors affecting the modification of the ACP PPant arm. However, until very recently, no regulatory proteins have been described which directly bind to FAS enzymes or enzymatic domains and regulate FAS activity. Should such proteins exist, they would be expected to bind transiently, and in a low-affinity, labile manner. As a consequence, these proteins would be expected to spontaneously dissociate from FAS when either the concentration of FAS drops below a critical threshold value in the course of purification, and/or stringent conditions such as high-ionic strength are applied during purification. To circumvent these issues in the purification of large macromolecular complexes, we recently developed and patented chromatography-free purification procedures that rely upon fractionated polyethylene glycol precipitation, and orthogonal size fractionation on density-gradients (Schrader et al. 2016; Haselbach et al. 2017; Henneberg and Chari 2021). The primary advantage of these procedures lies in the ability to maintain constant pH and ionic strength throughout purification, and maintaining high concentrations of the proteins of interest in intermediate steps of purification. Another potential advantage lies in avoiding dissociating shear forces that might arise through the use of chromatographic solid supports (Henneberg and Chari 2021).

These chromatography-free purification procedures were also applied to the purification of the Type I FAS from *S. cerevisiae* (Singh et al. 2020). In these conditions, we found a protein with an apparent molecular weight of 20 kDa on SDS-PAGE to reproducibly co-purify with the $(\alpha\beta)_6$ *S. cerevisiae* Type I FAS (Fig. 1.12a). This protein was identified by tandem mass-spectrometry as the product of the YDL110C gene and was re-named the γ-subunit. The γ-subunit no longer co-purified with FAS when the YDL110C open reading frame was deleted from the *S. cerevisiae* genome (Fig. 1.12a). Association of the γ-subunit is readily abolished in ionic strength greater than 150 mM KCl and exhibits an apparent affinity for FAS of only low μM values, partially explaining why its discovery had eluded earlier studies. In the absence of the γ-subunit, FAS immediately establishes its steady-state rate, whereas in its presence, a slower initial rate is readily identified before FAS establishes steady-state (Fig. 1.12b). This kinetic hysteresis is more pronounced in low NADPH concentrations, and in high NADPH concentration, it is hardly observed (Fig. 1.12b). In addition, in high NADPH concentrations, FAS activity is enhanced by about 25% in comparison to FAS activity in the absence of the γ-subunit (Singh et al. 2020). The γ-subunit also suppresses the FMN-mediated formation of reactive oxygen species. Therefore, the *S. cerevisiae* γ-subunit, to the best of our knowledge, is the first protein to directly bind FAS and regulate its activity. The γ-subunit tailors FAS activity in response to metabolic changes, decelerating at low NADPH concentrations, while accelerating at high NADPH

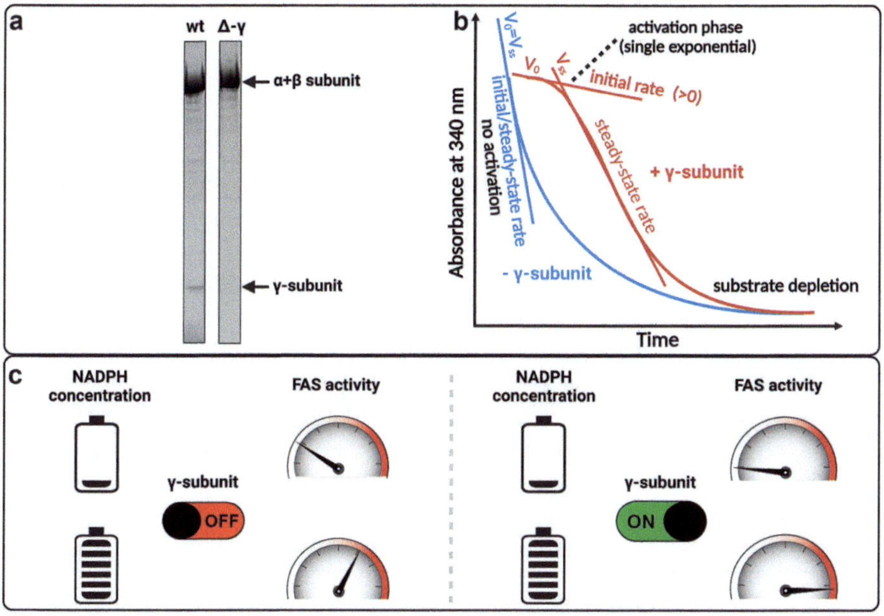

Fig. 1.12 Discovery of a γ-subunit of the Type I *S. cerevisiae* FAS. (**a**) SDS-PAGE of purified FAS from a wild-Type strain (left) and a YDL110C-deletion strain (right). Note, the lack of co-purified regulatory γ-subunit from the YDL110C-deletion strain. (**b**) Kinetic hysteresis in FAS activity elicited by the γ-subunit. In its absence, FAS immediately establishes steady-state (blue), whereas in the absence of the γ-subunit, a slower activation phase is encountered before FAS establishes its steady-state rate (red). This activation phase is more pronounced in low NADPH concentrations, while in high NADPH, it is almost non-existent. (**c**) The γ-subunit regulates the activity of FAS in response to the cellular abundance of the co-substrate NADPH. In the presence of the γ-subunit (top left), FAS activity in low NADPH concentration (depicted by the empty battery) is lower than in its absence (top right). On the other hand, at high NADPH concentrations (shown by the full battery) FAS activity is about 25% higher with (bottom right) than without (bottom left) the γ-subunit

concentrations (Fig. 1.12c). As it also affects the binding of the other FAS co-substrates acetyl-CoA, and malonyl-CoA, it may well represent a safeguard that prevents FAS from depleting cellular pools of its co-substrates vital for the survival of cells. Six γ-subunits can bind to a $\alpha_6\beta_6$ *S. cerevisiae* Type I FAS; however, it remains to be seen if this is the case in all metabolic states, which may fine-tune its function.

To elucidate the structural basis of γ-subunit function, structures of the *S. cerevisiae* Type I FAS were solved by both X-ray crystallography and cryo-EM, with and without the bound γ-subunit (Singh et al. 2020). In the absence of the γ-subunit, FAS was found to predominantly adopt a non-rotated state with ACP bound to the central wheel adjacent to KS domains (Fig. 1.13a), as previously reported (Leibundgut et al. 2007; Lomakin et al. 2007; Johansson et al. 2008). However, in the presence of the γ-subunit, a rotated conformation characterized by

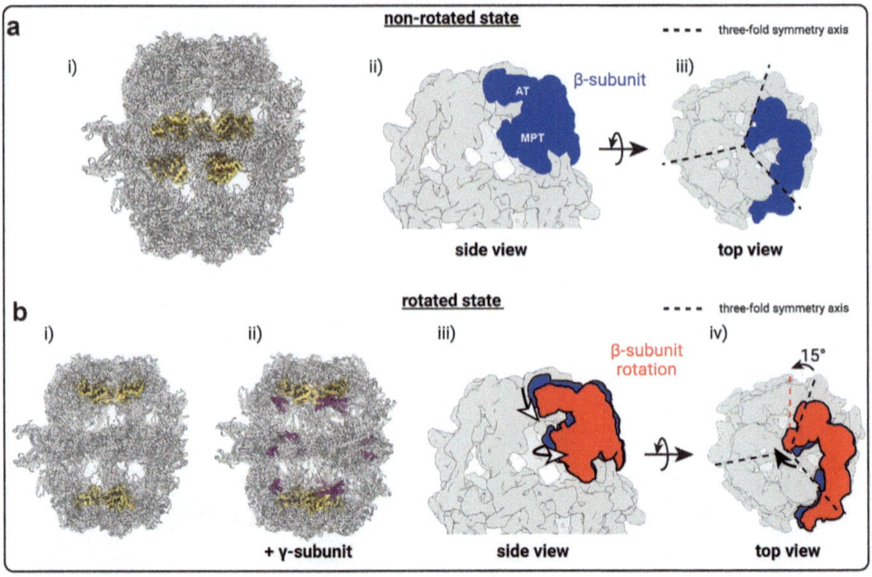

Fig. 1.13 Conformational changes in the *S. cerevisiae* Type I FAS, stabilized by the γ-subunit. (**a**) A non-rotated β-subunit state (blue) is predominantly found when the γ-subunit is absent, this state is correlated with ACP domains (yellow) occupying positions juxtaposed to KS domains. (i) A cutout is shown to view the ACP domains, (ii) side view, and (iii) top view of the Type I yeast FAS structure surface. (**b**) (i) and (ii) cutout to allow a view into the FAS domes shows that ACP domains (yellow) are adjacent to AT domains in the rotated state. (ii) The ACP domains at this position are additionally stabilized by the γ-subunit (violet). (iii) Side view, and (iv) top view of the Type I yeast FAS structure surface to indicate the rotation of the β-subunit (blue to red). This entails a rotation of 15° with respect to the three-fold symmetry axis (indicated by a dashed line) and an inward reorientation of the AT domain, along with an outward motion of the MPT domain

a 15° swivel around the three-fold axis at the top of the dome. This results in a corresponding inward shift of the AT domain, and an outward shift of the MPT domain (Fig. 1.13b). This rotated conformation had been previously visualized by cryo-EM, albeit not in a homogenous conformational state (Gipson et al. 2010). In the rotated conformation, ACP domains rearrange from the central wheel to sites adjacent to the AT domain. Thus, the rotated state is representative of the ACP priming reaction with acetyl. Further in the rotated state, densities were found in proximity of the rearranged ACP domains which could not be explained by any segments of either α- or β-subunits. Cross-linking mass spectrometry-guided interpretation of the densities established them as representing the γ-subunit. The γ-subunit appears to stabilize ACP at positions adjacent to the AT domains and the rotated conformation. By energy landscape analysis (Haselbach et al. 2017; Singh et al. 2020), this was found to be established by the γ-subunit introducing an energy barrier in transition from rotated to non-rotated conformations, which would be sufficient to explain the kinetic hysteresis elicited by the γ-subunit.

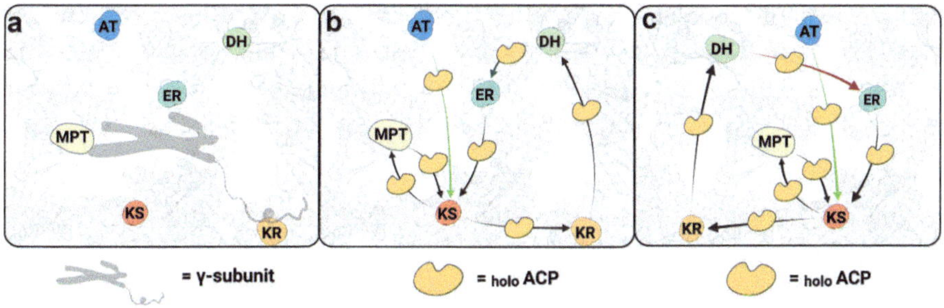

Fig. 1.14 The interaction of the γ-subunit defines a functional compartment in *S. cerevisiae* Type I FAS. (**a**) The γ-subunit (grey) binds adjacent to and occludes the ER domain active with its absolute N-terminus. The following γ-subunit coiled-coil domain points an acidic loop towards the MPT active site. The γ-subunit coiled-coil domain is followed by poorly resolved linker segment that leads up to the absolute C-terminus of the γ-subunit which occupies the KR active site cleft. (**b**) Based on this binding mode and the fact that the γ-subunit contacts many elongation domains and the MPT domain, it can be thought of defining a functional compartment within the yeast Type I FAS dome. This notably shortens the path from DH to ER domain in the elongation cycle by approximately 25 Å, which might explain the higher FAS activity evidenced in the presence of the γ-subunit at high NADPH concentrations. (**c**) Shown is the formerly proposed functional compartment which is based on the spatial proximity of active site clefts. Note, the DH-ER path which is longer than in the γ-subunit defined functional compartment (indicated by a red arrow)

Closer examination of the γ-subunit interaction mode with FAS indicates that the N-terminus forms a helix which occludes the ER active site (Fig. 1.14a). This is followed by a coiled-coil domain which appears to position negatively charged residues adjacent to the MPT active site. The C-terminus of the coiled-coil domain transitions into a poorly resolved loop segment until the absolute C-terminus of the γ-subunit occupies the KR active site cleft. Especially, ER and KR binding immediately rationalize the profound impact on NADPH binding and processing, the kinetic hysteresis, and the suppression of ER futile cycles by the γ-subunit. Moreover, by simultaneously contacting MPT, ER, and KR sites, as well as stabilizing the ACP at the AT domain, one could infer another function for the γ-subunit. It shuts off all other FAS active sites promoting acetyl priming of ACP and synchronizing the function of FAS. Thus, acetyl priming of ACP would represent a rate-limiting, regulatory step in the yeast FA biosynthesis cycle, and low acetyl-CoA concentrations would be sensed by arresting FAS in the rotated state. In addition, the γ-subunit binding mode could be interpreted as defining a functional compartment within FAS (Fig. 1.14b). In this γ-subunit-defined functional compartment, the condensation and elongation cycle ACP paths would progress in a counter-clockwise motion. In contrast, the FAS functional compartment that was proposed earlier (Figs. 1.9 and 1.14c), by exclusively considering the spatial proximity of active site-clefts, hypothesized that condensation and elongation cycles would run in clockwise direction (Jenni et al. 2007; Leibundgut et al. 2007; Lomakin et al. 2007; Maier et al. 2010). Comparing the two functional compartments, one realizes that in the γ-subunit-defined functional compartment the DH-ER path is shortened by about 25 Å. This would explain the γ-subunit induced 25% higher FAS activity observed in higher NADPH concentrations. The discovery of the γ-subunit and the establishment of its structure–function relationships may open new venues for the engineering FAS function in biofuel or oleochemical production.

Conclusions and Outlook

The study of FA biosynthesis has a long history, and much information has been gained in the last 75 years. This includes studies on metabolic flux, biochemical mechanisms, an inventory of the enzymes and co-factors involved, feedback regulatory process, and structural information of increasing resolution and complexity. Nevertheless, considering the importance of FA biosynthetic processes in all organisms and the role played in disease, the study of FA biosynthesis is not predicted to abate any time soon. Type I and Type II FAS enzymes are increasingly recognized as important drug targets to fight infectious agents, cancer, and obesity. These days of strained petroleum resources have further increased the importance of the study of FA biosynthetic processes and FAS. While the basic mechanisms have been established, the investigation of regulatory FA biosynthetic and FAS regulatory mechanisms lags behind. The advent of genomic approaches in recent years should greatly assist in tweaking transcriptional regulatory networks; the same holds true for the recently made strides in pathway and genome engineering by various tools. However, detailed enzymatic and biochemical approaches are going to remain essential. As exemplified here by the discovery of the yeast FAS γ-subunit, new biochemical purification procedures might well become necessary to uncover regulatory proteins, along with procedures to arrest cells and organisms at defined metabolic states. These will have to be complemented by biochemical tools of ever-increasing complexity to arrest and characterize the many and transient interactions, especially of the ACP and putative regulatory proteins, in FAS functional cycles. Precisely these advances in biochemistry will aid the next-generation structural studies of transient intermediates, allosteric regulation, and ultimately enable time-resolved structural biology, where entire catalytic transformations can be visualized. The entirety of these approaches will be essential for the biotechnological exploitation of FAS and FA biosynthetic pathways, as well as pharmacological intervention in these processes.

Acknowledgements We thank the Max-Planck Society for continued financial support. We further apologize to our many colleagues that we did not cite due to length constraints. Figures were created with Biorender.com and rendered using UCSF Chimera X (Goddard et al. 2018) or Pymol (Schrödinger 2015).

References

Andreolas C, Da Silva XG, Diraison F et al (2002) Stimulation of acetyl-CoA carboxylase gene expression by glucose requires insulin release and sterol regulatory element binding protein 1c in pancreatic MIN6 β-cells. Diabetes 51:2536–2545. https://doi.org/10.2337/diabetes.51.8.2536

Asturias FJ, Chadick JZ, Cheung IK et al (2005) Structure and molecular organization of mammalian fatty acid synthase. Nat Struct Mol Biol 12:225–232. https://doi.org/10.1038/nsmb899

Ballinger E, Mosior J, Hartman T et al (2019) Opposing reactions in coenzyme A metabolism sensitize Mycobacterium tuberculosis to enzyme inhibition. Science (80-) 363:eaau8959. https://doi.org/10.1126/science.aau8959

Bartholow TG, Sztain T, Young MA et al (2021) Protein-protein interaction based substrate control in the: *E. coli* octanoic acid transferase, LipB. RSC Chem Biol 2:1466–1473. https://doi.org/10.1039/d1cb00125f

Beld J, Sonnenschein EC, Vickery CR et al (2014) The phosphopantetheinyl transferases: catalysis of a post-translational modification crucial for life. Nat Prod Rep 31:61–108. https://doi.org/10.1039/C3NP70054B

Beld J, Lee DJ, Burkart MD (2015) Fatty acid biosynthesis revisited: structure elucidation and metabolic engineering. Mol BioSyst 11:38–59. https://doi.org/10.1039/c4mb00443d

Bennett MK, Lopez JM, Sanchez HB, Osborne TF (1995) Sterol regulation of fatty acid synthase promoter: coordinate feedback regulation of two major lipid pathways. J Biol Chem 270:25578–25583. https://doi.org/10.1074/jbc.270.43.25578

Bloch K (1965) The biological synthesis of cholesterol

Bloch K (1969) Enzymic synthesis of monounsaturated fatty acids. Acc Chem Res 2:193–202. https://doi.org/10.1021/ar50019a001

Bloch K (2006) Control mechanisms for fatty acid synthesis in mycobacterium smegmatis. Adv Enzymol Relat Areas Mol Biol 45:1–84. https://doi.org/10.1002/9780470122907.ch1

Bosch B, DeJesus MA, Poulton NC et al (2021) Genome-wide gene expression tuning reveals diverse vulnerabilities of M. tuberculosis. Cell 184:4579–4592.e24. https://doi.org/10.1016/j.cell.2021.06.033

Brignole EJ, Smith S, Asturias FJ (2009) Conformational flexibility of metazoan fatty acid synthase enables catalysis. Nat Struct Mol Biol 16:190–197. https://doi.org/10.1038/nsmb.1532

Brink J, Ludtke SJ, Kong Y et al (2004) Experimental verification of conformational variation of human fatty acid synthase as predicted by normal mode analysis. Structure 12:185–191. https://doi.org/10.1016/j.str.2004.01.015

Brownsey RW, Boone AN, Elliott JE et al (2006) Regulation of acetyl-CoA carboxylase. Biochem Soc Trans 34:223–227. https://doi.org/10.1042/BST0340223

Burr GO, Brown WR (1933) On the fatty acids essential in nutrition. Proc Soc Exp Biol Med 30: 1349–1352. https://doi.org/10.3181/00379727-30-6927

Campbell JW, Cronan JE (2001) Bacterial fatty acid biosynthesis: targets for antibacterial drug discovery. Annu Rev Microbiol 55:305–332. https://doi.org/10.1146/annurev.micro.55.1.305

Chan DI, Vogel HJ (2010) Current understanding of fatty acid biosynthesis and the acyl carrier protein. Biochem J 430:1–19. https://doi.org/10.1042/BJ20100462

Chevreul M-E (1823) Recherches chimique sur les corps gras d'origine animale. Chez F.G. Levrault, Paris

Chirala SS (1992) Coordinated regulation and inositol-mediated and fatty acid-mediated repression of fatty acid synthase genes in Saccharomyces cerevisiae. Proc Natl Acad Sci U S A 89:10232–10236. https://doi.org/10.1073/pnas.89.21.10232

Chirala SS, Zhong Q, Huang W, Al-Feel W (1994) Analysis of FAS3/ACC regulatory region of Saccharomyces cerevisiae: identification of a functional UAS INO and sequences responsible for fatty acid mediated repression. Nucleic Acids Res 22:412–418. https://doi.org/10.1093/nar/22.3.412

Ciccarelli L, Connell SR, Enderle M et al (2013) Structure and conformational variability of the mycobacterium tuberculosis fatty acid synthase multienzyme complex. Structure 21:1251–1257

Cronan John E Jr, Subrahmanyam S (1998) FadR, transcriptional co-ordination of metabolic expediency. Mol Microbiol 29:937–943. https://doi.org/10.1046/j.1365-2958.1998.00917.x
Cronan JE, Thomas J (2009) Bacterial fatty acid synthesis and its relationships with polyketide synthetic pathways. Methods Enzymol 459:395–433. https://doi.org/10.1016/S0076-6879(09)04617-5
Cronan JE, Waldrop GL (2002) Multi-subunit acetyl-CoA carboxylases. Prog Lipid Res 41:407–435. https://doi.org/10.1016/S0163-7827(02)00007-3
Crosby J, Crump MP (2012) The structural role of the carrier protein – active controller or passive carrier. Nat Prod Rep 29:1111–1137. https://doi.org/10.1039/C2NP20062G
Elad N, Baron S, Peleg Y et al (2018) Structure of Type-I Mycobacterium tuberculosis fatty acid synthase at 3.3 Å resolution. Nat Commun 9:3886. https://doi.org/10.1038/s41467-018-06440-6
Fichtlscherer F, Wellein C, Mittag M, Schweizer E (2000) A novel function of yeast fatty acid synthase. Eur J Biochem 267:2666–2671. https://doi.org/10.1046/j.1432-1327.2000.01282.x
Finzel K, Lee DJ, Burkart MD (2015) Using modern tools to probe the structure-function relationship of fatty acid synthases. Chembiochem 16:528–547. https://doi.org/10.1002/cbic.201402578
Fischer M, Joppe M, Mulinacci B et al (2020) Analysis of the co-translational assembly of the fungal fatty acid synthase (FAS). Sci Rep 10:895. https://doi.org/10.1038/s41598-020-57418-8
Gago G, Arabolaza A, Diacovich L, Gramajo H (2019) Components and key regulatory steps of lipid biosynthesis in actinomycetes. In: Biogenesis of fatty acids, lipids and membranes, pp 409–433. doi:https://doi.org/10.1007/978-3-319-50430-8_65
Gajewski J, Buelens F, Serdjukow S et al (2017a) Engineering fatty acid synthases for directed polyketide production. Nat Chem Biol 13:1–28. https://doi.org/10.1038/nchembio.2314
Gajewski J, Pavlovic R, Fischer M et al (2017b) Engineering fungal de novo fatty acid synthesis for short chain fatty acid production. Nat Commun 8:1–8. https://doi.org/10.1038/ncomms14650
Gibson DM, Titchener EB, Wakil SJ (1958) Studies on the mechanism of fatty acid synthesis V. Bicarbonate requirement for the synthesis of long-chain fatty acids. Biochim Biophys Acta 30:376–383. https://doi.org/10.1016/0006-3002(58)90063-5
Gipson P, Mills DJ, Wouts R et al (2010) Direct structural insight into the substrate-shuttling mechanism of yeast fatty acid synthase by electron cryomicroscopy. Proc Natl Acad Sci U S A 107:9164–9169. https://doi.org/10.1073/pnas.0913547107
Goddard TD, Huang CC, Meng EC et al (2018) UCSF ChimeraX: meeting modern challenges in visualization and analysis. Protein Sci 27:14–25. https://doi.org/10.1002/pro.3235
Graber R, Sumida C, Nunez EA (1994) Fatty acids and cell signal transduction. J Lipid Mediat Cell Signal 9:91–116
Granchi C (2019) Discovery of allosteric inhibition of human ATP-citrate lyase. Trends Pharmacol Sci 40:364–366. https://doi.org/10.1016/j.tips.2019.04.008
Gully D, Bouveret E (2006) A protein network for phospholipid synthesis uncovered by a variant of the tandem affinity purification method in *Escherichia coli*. Proteomics 6:282–293. https://doi.org/10.1002/pmic.200500115
Gully D, Moinier D, Loiseau L, Bouveret E (2003) New partners of acyl carrier protein detected in *Escherichia coli* by tandem affinity purification. FEBS Lett 548:90–96. https://doi.org/10.1016/S0014-5793(03)00746-4
Haselbach D, Schrader J, Lambrecht F et al (2017) Long-range allosteric regulation of the human 26S proteasome by 20S proteasome-targeting cancer drugs. Nat Commun 8:15578. https://doi.org/10.1038/ncomms15578
Heil CS, Wehrheim SS, Paithankar KS, Grininger M (2019) Fatty acid biosynthesis: chain-length regulation and control. Chembiochem 20:2298–2321. https://doi.org/10.1002/cbic.201800809
Henneberg F, Chari A (2021) Chromatography-free purification strategies for large biological macromolecular complexes involving fractionated PEG precipitation and density gradients. Life 11

Herbst DA, Townsend CA, Maier T (2018) The architectures of iterative Type I PKS and FAS. Nat Prod Rep 35:1046–1069. https://doi.org/10.1039/C8NP00039E
Hunkeler M, Hagmann A, Stuttfeld E et al (2018) Structural basis for regulation of human acetyl-CoA carboxylase. Nature 558:470–474. https://doi.org/10.1038/s41586-018-0201-4
Iacobazzi V, Infantino V (2014) Citrate-new functions for an old metabolite. Biol Chem 395:387–399. https://doi.org/10.1515/hsz-2013-0271
Jackowski S, Rock CO (1983) Ratio of active to inactive forms of acyl carrier protein in *Escherichia coli*. J Biol Chem 258:15186–15191
Jain S, Caforio A, Driessen AJM (2014) Biosynthesis of archaeal membrane ether lipids. Front Microbiol 5:641
Jenni S, Leibundgut M, Boehringer D et al (2007) Structure of fungal fatty acid synthase and implications for iterative substrate shuttling. Science (80-) 316:254–261. https://doi.org/10.1126/science.1138248
Johansson P, Wiltschi B, Kumari P et al (2008) Inhibition of the fungal fatty acid synthase Type I multienzyme complex. Proc Natl Acad Sci U S A 105:12803–12808. https://doi.org/10.1073/pnas.0805827105
Johansson P, Mulinacci B, Koestler C et al (2009) Multimeric options for the auto-activation of the Saccharomyces cerevisiae FAS Type I megasynthase. Structure 17:1063–1074. https://doi.org/10.1016/j.str.2009.06.014
Joppe M, D'Imprima E, Salustros N et al (2020) The resolution revolution in cryoEM requires high-quality sample preparation: a rapid pipeline to a high-resolution map of yeast fatty acid synthase. IUCrJ 7:220–227. https://doi.org/10.1107/S2052252519017366
Kallio P, Pásztor A, Akhtar MK, Jones PR (2014) Renewable jet fuel. Curr Opin Biotechnol 26:50–55. https://doi.org/10.1016/j.copbio.2013.09.006
Kastaniotis AJ, Autio KJ, Kerätär JM et al (2017) Mitochondrial fatty acid synthesis, fatty acids and mitochondrial physiology. Biochim Biophys Acta Mol Cell Biol Lipids 1862:39–48. https://doi.org/10.1016/j.bbalip.2016.08.011
Katz L, Donadio S (1993) Polyketide synthesis: prospects for hybrid antibiotics. Annu Rev Microbiol 47:875–912
Keating DH, Carey MR, Cronan JE Jr (1995) The unmodified (Apo) form of *Escherichia coli* acyl carrier protein is a potent inhibitor of cell growth. J Biol Chem 270:22229–22235. https://doi.org/10.1074/jbc.270.38.22229
Kim JB, Spiegelman BM (1996) ADD1/SREBP1 promotes adipocyte differentiation and gene expression linked to fatty acid metabolism. Genes Dev 10:1096–1107. https://doi.org/10.1101/gad.10.9.1096
Klein HP, Lipmann F (1953a) The relationship of coenzyme A to lipide synthesis. II. Experiments with rat liver. J Biol Chem 203:101–108
Klein HP, Lipmann F (1953b) The relationship of coenzyme A to lipide synthesis. I. Experiments with yeast. J Biol Chem 203:95–99
Koch R (1882) Die Ätiologie der Tuberkulose
Lane MD, Moss J (1971) Regulation of fatty acid synthesis in animal tissues, Chapter 2. In: Vogel HJBT-MR, 3rd edn. Academic, pp 23–54
Leibundgut M, Jenni S, Frick C, Ban N (2007) Structural basis for substrate delivery by acyl carrier protein in the yeast fatty acid synthase. Science (80-) 316:288–290. https://doi.org/10.1126/science.1138249
Lipmann F (1954) Development of the acetylation problem, a personal account. Science (80-) 120:855–865
Liu Y, Cruz-Morales P, Zargar A et al (2021) Biofuels for a sustainable future. Cell 184:1636–1647. https://doi.org/10.1016/j.cell.2021.01.052
Lomakin IB, Xiong Y, Steitz TA (2007) The crystal structure of yeast fatty acid synthase, a cellular machine with eight active sites working together. Cell 129:319–332. https://doi.org/10.1016/j.cell.2007.03.013

Lombard J, López-García P, Moreira D (2012) phylogenomic investigation of phospholipid synthesis in Archaea. Archaea 2012:630910. https://doi.org/10.1155/2012/630910

Lopez JM, Bennett MK, Sanchez HB et al (1996) Sterol regulation of acetyl coenzyme A carboxylase: a mechanism for coordinate control of cellular lipid. Proc Natl Acad Sci U S A 93:1049–1053. https://doi.org/10.1073/pnas.93.3.1049

López M, Lelliott CJ, Vidal-Puig A (2007) Hypothalamic fatty acid metabolism: a housekeeping pathway that regulates food intake. Bio Essays 29:248–261. https://doi.org/10.1002/bies.20539

López-Lara IM, Geiger O (2010) Formation of fatty acids. Handbook of hydrocarbon and lipid microbiology, pp 385–393. doi:https://doi.org/10.1007/978-3-540-77587-4_26

Lopez-Lara IM, Soto MJ (2018) Fatty acid synthesis and oxidation. Biogenesis of fatty acids, lipids and membranes, pp 1–16

Lou JW, Mazhab-Jafari MT (2020) Steric occlusion regulates proximal interactions of acyl carrier protein domain in fungal fatty acid synthase. Commun Biol 3:274. https://doi.org/10.1038/s42003-020-0997-y

Lou JW, Iyer KR, Hasan SMN et al (2019) Electron cryomicroscopy observation of acyl carrier protein translocation in Type I fungal fatty acid synthase. Sci Rep 9:12987. https://doi.org/10.1038/s41598-019-49261-3

Lu H, Tonge PJ (2008) Inhibitors of FabI, an enzyme drug target in the bacterial fatty acid biosynthesis pathway. Acc Chem Res 41:11–20. https://doi.org/10.1021/ar700156e

Lust G, Lynen F (1968) The inhibition of the fatty acid synthetase multienzyme complex of yeast by long-chain acyl coenzyme A compounds. Eur J Biochem 7:68–72. https://doi.org/10.1111/j.1432-1033.1968.tb19575.x

Lynen F (1964) The pathway from "activated acetic acid" to the terpenes and fatty acids. Proc R Caroline Inst:103–138

Lynen F, Reichert E, Rueff L (1951) Zum biologischen Abbau der Essigsäure VI. "Aktivierte Essigsäure", ihre Isolierung aus Hefe und ihre chemische Natur. Justus Liebigs Ann Chem 574: 1–32. https://doi.org/10.1002/jlac.19515740102

Lynen F, Engeser H, Foerster E-C et al (1980) On the structure of fatty acid synthetase of yeast. Eur J Biochem 112:431–442. https://doi.org/10.1111/j.1432-1033.1980.tb06105.x

Mabrouk GM, Helmy IM, Thampy KG, Wakil SJ (1990) Acute hormonal control of acetyl-CoA carboxylase. The roles of insulin, glucagon, and epinephrine. J Biol Chem 265:6330–6338. https://doi.org/10.1016/s0021-9258(19)39330-5

Maier T, Leibundgut M, Ban N (2008) The crystal structure of a mammalian fatty acid synthase. Science (80-) 321:1315–1322. https://doi.org/10.1126/science.1161269

Maier T, Leibundgut M, Boehringer D, Ban N (2010) Structure and function of eukaryotic fatty acid synthases. Q Rev Biophys 43:373–422. https://doi.org/10.1017/S0033583510000156

Majerus PW, Alberts AW, Vagelos PR (1965) Acyl carrier protein. IV. The identification of 4′-phosphopantetheine as the prosthetic group of the acyl carrier protein. Proc Natl Acad Sci U S A 53:410–417. https://doi.org/10.1073/pnas.53.2.410

Marrakchi H, Lanéelle MA, Daffé M (2014) Mycolic acids: structures, biosynthesis, and beyond. Chem Biol 21:67–85. https://doi.org/10.1016/j.chembiol.2013.11.011

Mattick JS, Nickless J, Mizugaki M et al (1983a) The architecture of the animal fatty acid synthetase. II. Separation of the core and thioesterase functions and determination of the NC orientation of the subunit. J Biol Chem 258:15300–15304

Mattick JS, Tsukamoto Y, Nickless J, Wakil SJ (1983b) The architecture of the animal fatty acid synthetase. I. Proteolytic dissection and peptide mapping. J Biol Chem 258:15291–15299

Mazier D, Goma J, Pied S et al (1990) Hepatic phase of malaria: a crucial role as "go-between" with other stages. Bull World Health Organ 68(Suppl):126–131

Mindrebo JT, Patel A, Kim WE et al (2020) Gating mechanism of elongating β-ketoacyl-ACP synthases. Nat Commun 11:1727. https://doi.org/10.1038/s41467-020-15455-x

Nelson DL, Cox MM (2013) Lehninger - principles of biochemistry

Nguyen C, Haushalter RW, Lee DJ et al (2014) Trapping the dynamic acyl carrier protein in fatty acid biosynthesis. Nature 505:427–431. https://doi.org/10.1038/nature12810

Ohlrogge JB, Jaworski JG (1997) Regulation of fatty acid synthesis. Annu Rev Plant Physiol Plant Mol Biol 48:109–136. https://doi.org/10.1146/annurev.arplant.48.1.109

Parsons JB, Rock CO (2013) Bacterial lipids: metabolism and membrane homeostasis. Prog Lipid Res 52:249–276. https://doi.org/10.1016/j.plipres.2013.02.002

Pasteur L (1879) Théorie physiologique de la Fermentation

Perham RN (2000) Swinging arms and swinging domains in multifunctional enzymes: catalytic machines for multistep reactions. Annu Rev Biochem 69:961–1004. https://doi.org/10.1146/annurev.biochem.69.1.961

Phillips MA, Burrows JN, Manyando C et al (2017) Malaria. Nat Rev Dis Prim 3:17050. https://doi.org/10.1038/nrdp.2017.50

Pitt D, Aubin J-M (2012) Joseph Lister: father of modern surgery. Can J Surg 55:E8–E9. https://doi.org/10.1503/cjs.007112

Pugh EL, Wakil SJ (1965) Studies on the mechanism of fatty acid synthesis: XIV. The prosthetic group of acyl carrier protein and the mode of its attachment to the protein. J Biol Chem 240: 4727–4733. https://doi.org/10.1016/S0021-9258(18)97016-X

Rittenberq D, Bloch K (1944) The utilization of acetic acid for fatty acid synthesis. J Biol Chem 154:311–312. https://doi.org/10.1016/S0021-9258(18)71971-6

Ronnett GV, Kleman AM, Kim E-K et al (2006) Fatty acid metabolism, the central nervous system, and feeding. Obesity 14:201S–207S. https://doi.org/10.1038/oby.2006.309

Ruppe A, Mains K, Fox JM (2020) A kinetic rationale for functional redundancy in fatty acid biosynthesis. Proc Natl Acad Sci U S A 117:23557–23564. https://doi.org/10.1073/pnas.2013924117

Schrader J, Henneberg F, Mata RA et al (2016) The inhibition mechanism of human 20S proteasomes enables next-generation inhibitor design. Science (80-) 353:594–598. https://doi.org/10.1126/science.aaf8993

Schrödinger L (2015) The PyMOL molecular graphics system, Version 1.8

Schüller HJ, Hahn A, Tröster F et al (1992) Coordinate genetic control of yeast fatty acid synthase genes FAS1 and FAS2 by an upstream activation site common to genes involved in membrane lipid biosynthesis. EMBO J 11:107–114. https://doi.org/10.1002/j.1460-2075.1992.tb05033.x

Schweizer E, Hofmann J (2004) Microbial type I fatty acid synthases (FAS): major players in a network of cellular FAS systems. Microbiol Mol Biol Rev 68:501–517. https://doi.org/10.1128/MMBR.68.3.501-517.2004

Shears MJ, Botté CY, McFadden GI (2015) Fatty acid metabolism in the Plasmodium apicoplast: drugs, doubts and knockouts. Mol Biochem Parasitol 199:34–50. https://doi.org/10.1016/j.molbiopara.2015.03.004

Shiber A, Döring K, Friedrich U et al (2018) Cotranslational assembly of protein complexes in eukaryotes revealed by ribosome profiling. Nature 561:268–272. https://doi.org/10.1038/s41586-018-0462-y

Singh K, Graf B, Linden A et al (2020) Discovery of a regulatory subunit of the yeast fatty acid synthase. Cell 180:1130–1143.e20. https://doi.org/10.1016/j.cell.2020.02.034

Smith S, Tsai S-C (2007) The Type I fatty acid and polyketide synthases: a tale of two megasynthases. Nat Prod Rep 24:1041–1072. https://doi.org/10.1039/B603600G

Smith S, Witkowski A, Joshi AK (2003) Structural and functional organization of the animal fatty acid synthase. Prog Lipid Res 42:289–317. https://doi.org/10.1016/S0163-7827(02)00067-X

Snow J (1849) On the mode of communication of Cholera

Suburu J, Shi L, Wu J et al (2014) Fatty acid synthase is required for mammary gland development and milk production during lactation. Am J Physiol Endocrinol Metab 306:E1132–E1143. https://doi.org/10.1152/ajpendo.00514.2013

Sumper M, Riepertinger C, Lynen F, Oesterhelt D (1969) Die Synthese verschiedener Carbonsäuren durch den Multienzymkomplex der Fettsäuresynthese aus Hefe und die Erklärung ihrer Bildung. Eur J Biochem 10:377–387. https://doi.org/10.1111/j.1432-1033.1969.tb00701.x

Tabor DE, Kim JB, Spiegelman BM, Edwards PA (1999) Identification of conserved cis-elements and transcription factors required for sterol-regulated transcription of stearoyl-CoA desaturase 1 and 2 *. J Biol Chem 274:20603–20610. https://doi.org/10.1074/jbc.274.29.20603

Tehlivets O, Scheuringer K, Kohlwein SD (2007) Fatty acid synthesis and elongation in yeast. Biochim Biophys Acta Mol Cell Biol Lipids 1771:255–270. https://doi.org/10.1016/j.bbalip.2006.07.004

Thomas J, Cronan JE (2005) The enigmatic acyl carrier protein phosphodiesterase of *Escherichia coli*: genetic and enzymological characterization. J Biol Chem 280:34675–34683. https://doi.org/10.1074/jbc.M505736200

Tong L (2013) Structure and function of biotin-dependent carboxylases. Cell Mol Life Sci 70:863–891. https://doi.org/10.1007/s00018-012-1096-0

Tsai SC (2018) The structural enzymology of iterative aromatic polyketide synthases: a critical comparison with fatty acid synthases. Annu Rev Biochem 87:503–531. https://doi.org/10.1146/annurev-biochem-063011-164509

Tsukamoto Y, Wong H, Mattick JS, Wakil SJ (1983) The architecture of the animal fatty acid synthetase complex. IV. Mapping of active centers and model for the mechanism of action. J Biol Chem 258:15312–15322

Vagelos PR (1971) Regulation of fatty acid biosynthesis. Curr Top Cell Regul 4:119–166. https://doi.org/10.1016/B978-0-12-152804-1.50010-6

Vagelos PR, Larrabes AR (1967) Acyl carrier protein. IX. Acyl carrier protein hydrolase. J Biol Chem 242:1776–1781

van den Boom T, Cronan JE (1989) Genetics and regulation of bacterial lipid metabolism. Annu Rev Microbiol 43:317–343. https://doi.org/10.1146/annurev.mi.43.100189.001533

van Schaijk BCL, Kumar TRS, Vos MW et al (2014) Type II fatty acid biosynthesis is essential for Plasmodium falciparum sporozoite development in the midgut of Anopheles mosquitoes. Eukaryot Cell 13:550–559. https://doi.org/10.1128/EC.00264-13

Verschueren KHG, Blanchet C, Felix J et al (2019) Structure of ATP citrate lyase and the origin of citrate synthase in the Krebs cycle. Nature 568:571–575. https://doi.org/10.1038/s41586-019-1095-5

Volpe JJ, Vagelos PR (1973) Saturated fatty acid biosynthesis and its regulation. Annu Rev Biochem 42:21–60. https://doi.org/10.1146/annurev.bi.42.070173.000321

Wakil SJ, Stoops JK, Joshi VC (1983) Fatty acid synthesis and its regulation. Annu Rev Biochem 52:537–579. https://doi.org/10.1146/annurev.bi.52.070183.002541

Wang D, DuBois RN (2012) Associations between obesity and cancer: the role of fatty acid synthase. J Natl Cancer Inst 104:343–345. https://doi.org/10.1093/jnci/djs010

Wei J, Zhang Y, Yu T-Y et al (2016) A unified molecular mechanism for the regulation of acetyl-CoA carboxylase by phosphorylation. Cell Discov 2:16044. https://doi.org/10.1038/CELLDISC.2016.44

Wei J, Leit S, Kuai J et al (2019) An allosteric mechanism for potent inhibition of human ATP-citrate lyase. Nature 568:566–570. https://doi.org/10.1038/s41586-019-1094-6

White SW, Zheng J, Zhang Y-M, Rock CO (2005) The structural biology of type II fatty acid biosynthesis. Annu Rev Biochem 74:791–831. https://doi.org/10.1146/annurev.biochem.74.082803.133524

Witters LA, Watts TD, Daniels DL, Evans JL (1988) Insulin stimulates the dephosphorylation and activation of acetyl-CoA carboxylase. Proc Natl Acad Sci U S A 85:5473–5477. https://doi.org/10.1073/pnas.85.15.5473

Wong H, Mattick JS, Wakil SJ (1983) The architecture of the animal fatty acid synthetase. III. Isolation and characterization of beta-ketoacyl reductase. J Biol Chem 258:15305–15311. https://doi.org/10.1016/S0021-9258(17)43807-5

Wong HC, Liu G, Zhang Y-M et al (2002) The solution structure of acyl carrier protein from *Mycobacterium tuberculosis*. J Biol Chem 277:15874–15880. https://doi.org/10.1074/jbc.M112300200

Yao J, Rock CO (2017) Bacterial fatty acid metabolism in modern antibiotic discovery. Biochim Biophys Acta Mol Cell Biol Lipids 1862:1300–1309. https://doi.org/10.1016/j.bbalip.2016.09.014

Zhang J, Chen Y, Fu L et al (2021) Accelerating strain engineering in biofuel research via build and test automation of synthetic biology. Curr Opin Biotechnol 67:88–98. https://doi.org/10.1016/j.copbio.2021.01.010

Zhu K, Zhang Y-M, Rock CO (2009) Transcriptional regulation of membrane lipid homeostasis in *Escherichia coli*. J Biol Chem 284:34880–34888. https://doi.org/10.1074/jbc.M109.068239

Chapter 2
ATP-Binding Cassette Transporters: Snap-on Complexes?

Iqra Younus, Sofia Kochkina, Cheri C. Choi, Wenjuan Sun, and Robert C. Ford

Abstract ATP-binding cassette (ABC) transporters are one of the largest families of membrane proteins in prokaryotic organisms. Much is now understood about the structure of these transporters and many reviews have been written on that subject. In contrast, less has been written on the assembly of ABC transporter complexes and this will be a major focus of this book chapter. The complexes are formed from two cytoplasmic subunits that are highly conserved (in terms of their primary and three-dimensional structures) across the whole family. These ATP-binding subunits give rise to the name of the family. They must assemble with two transmembrane subunits that will typically form the permease component of the transporter. The transmembrane subunits have been found to be surprisingly diverse in structure when the whole family is examined, with seven distinct folds identified so far. Hence nucleotide-binding subunits appear to have been bolted on to a variety of transmembrane platforms during evolution, leading to a greater variety in function. Furthermore, many importers within the family utilise a further external substrate-binding component to trap scarce substrates and deliver them to the correct permease components. In this chapter, we will discuss whether assembly of the various ABC transporter subunits occurs with high fidelity within the crowded cellular environment and whether promiscuity in assembly of transmembrane and cytoplasmic components can occur. We also discuss the new AlphaFold protein structure prediction tool which predicts a new type of transmembrane domain fold within the ABC transporters that is associated with cation exporters of bacteria and plants.

Keywords ABC transporters · Quaternary structure · Protein complex · Protein complex assembly · Co-translational assembly

I. Younus · S. Kochkina · C. C. Choi · W. Sun · R. C. Ford (✉)
Faculty of Biology, Medicine and Health, School of Biological Sciences, The University of Manchester, Manchester, UK
e-mail: iqra.younus@manchester.ac.uk; sofia.kochkina@student.manchester.ac.uk; wenjuan.sun@postgrad.manchester.ac.uk; robert.ford@manchester.ac.uk

J. R. Harris, J. Marles-Wright (eds.), *Macromolecular Protein Complexes IV*, Subcellular Biochemistry 99, https://doi.org/10.1007/978-3-031-00793-4_2

Introduction

ATP-binding cassette (ABC) transporters represent one of the largest families of membrane proteins in prokaryotes. A broad range of complexes can be formed from the basic building blocks: two transmembrane subunits (or domains, if fused with other components) and two nucleotide-binding subunits (or domains, if fused), (see Fig. 2.1) (Higgins 1992; Ford and Beis 2019). In this chapter, we will use the terms 'subunit' and 'domain' to distinguish between isolated versus fused proteins where relevant, and likewise for abbreviations. However since the domain is the same whether fused or not, we will also use the term 'domain' when considering the

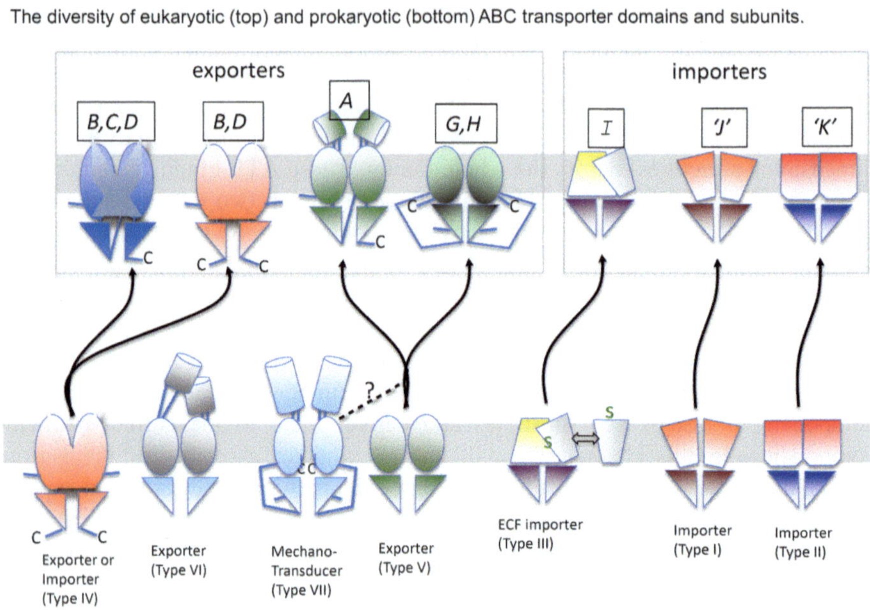

Fig. 2.1 Highly conserved cytoplasmic NBDs (triangles) associate with diverse TMDs (ovals, bucket shapes) embedded in a lipid bilayer (grey band). Fusion of genes and addition of linking regions (blue lines) has led to completely or partially fused ABC transporters (C-terminus is indicated for these cases). Type IV transporter TMDs also show a domain swap such that each NBD interacts with cytoplasmic loops coming from each TMD, not just the one it is fused to. Structural comparisons of the varied TMDs have suggested origins for the eukaryotic ABC transporter families *A* to *K* that are indicated by the curved arrows. The evolutionary relationship between Type VII and Type V ABC transporters is uncertain (dashed line). In most cases, gene fusion gives an N-terminal TMD followed by an NBD, but the sequence can be reversed (e.g. Type VII (prokaryotes) and in the G family (eukaryotes). Lastly, fusions between two NBS genes or between two TMS genes are found in prokaryotes, but this does not appear to have been passed on into eukaryotes, except for in the ABCE and ABCF NBD-only families (not shown). Type I and II importers appear to be present in some photosynthetic eukaryotes, with the proposal to classify these members as ABCJ and ABCK families. Members of the ECF importer (Type III) have exchangeable 'S' subunits (lime green bucket shape) that bind different substrates (S)

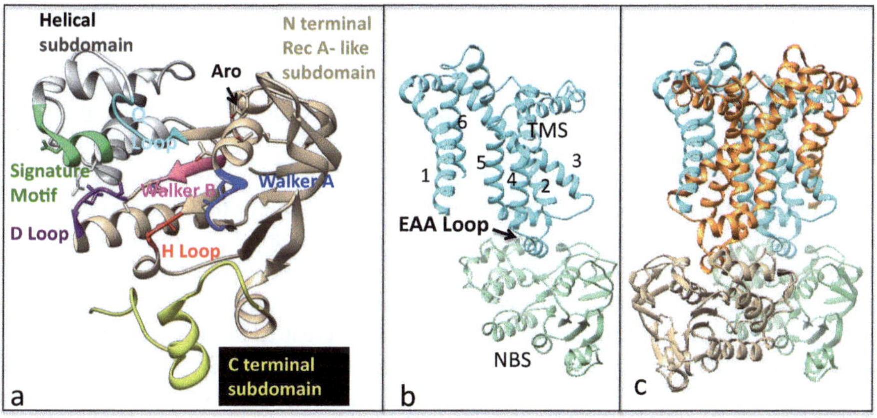

Fig. 2.2 (**a**) General Nucleotide Binding Domain (NBD) structure with conserved sequences and various subdomains highlighted in different colours. Nucleotide (ADP, ATP) is bound in a pocket formed by the Walker A, Walker B and H loop residues and the purine ring pi-pi stacks with the aromatic residue in the first β-strand (Aro), which in this case is a tyrosine residue. When the two NBDs assemble into a dimer in the presence of nucleotide, the Signature motif and D loop are close to the nucleotide binding site of the opposing NBD. This forms a so-called 'sandwich dimer'. The Q-loop region interacts with a cytoplasmic loop extending down from the transmembrane domain (TMD) which would be positioned above the NBD in this depiction. (**b**) Illustration of the interaction of the transmembrane subunit TMS (cyan) with the NBS (green) in Type I ABC transporters where the EAA loop is involved. The order of the six transmembrane helices is indicated. The NBS is oriented as in panel a, with the nucleotide-binding site towards the observer. (**c**) The arrangement of all four subunits is shown with the additional TMS (orange) and NBS (taupe) closer to the observer. The NBS shown in panel a is EcfA' from the *Thermatoga maritima* folate importer (PDBID 4HLU) (Karpowich and Wang 2013)

general properties of a subunit/domain, as exemplified in the next few paragraphs. Hence the terms 'nucleotide binding subunit' (NBS) and 'transmembrane subunit' (TMS) will only be used where there is a need to indicate that these are separate polypeptides. As much of this chapter considers the assembly of ABC transporter complexes, it will sometimes be important to distinguish between subunits and domains.

Whilst amino acid sequences within nucleotide binding domains (NBDs) have remained fairly conserved across evolutionary timescales (Fig. 2.2a), the transmembrane domains (TMDs) display a range of sequences and three-dimensional folds (Dean and Allikmets 2001; Dean et al. 2001). Hence, whilst it is the NBD sequences that define the ABC transporter family, it is the structure of the TMD that is increasingly being employed to classify the family into different sub-types (Thomas et al. 2020; Ford and Beis 2019). Here, we will adopt the classification system proposed by Thomas and co-authors, which has seven types of ABC transporters (Thomas et al. 2020). Of the different types of prokaryotic ABC transporters (Fig. 2.1), only exporters were originally thought to have made the leap into eukaryotes, where later gene fusion events led to half-transporter complexes consisting of two copies of fused TMD–NBD polypeptides, or full transporters consisting of a single polypeptide chain where all four subunits are fused (Dean and Annilo 2005). However, recent reports have suggested that ABC importers have also made their way into eukaryotes (Choi and Ford 2021; Ford et al. 2020) and, at least five of the seven bacterial sub-types appear to have homologues in eukaryotes (Fig. 2.1).

In this chapter, we will consider the seven types of complexes formed by ABC transporters, with a particular focus on their structures and (when known) how they assemble from their component subunits. Co-translational assembly of the various types of ABC transporter complexes is considered versus simple diffusion and molecular recognition-based models for their mechanism of formation at the various membranes within the cell. A second feature of interest to us was whether (and to what extent) ABC transporters (and their eukaryotic descendants) are able to use different subunits interchangeably. The idea of snap-on complexes has already been established for transporters of the type III (Energy Coupling Factor) ABC transporter family where some members have a common 'T' transmembrane subunit but different 'S' transmembrane subunits that select for alternative imported substances (Setyawati et al. 2020; Rempel et al. 2019; Majsnerowska et al. 2015). Similarly, it is known that different ABC importers of types I and II can employ different external substrate-binding proteins (SBPs) to bring alternative metabolites and co-factors to a common TMD permease component (Higgins and Ames 1981; Leonard et al. 1996; Park et al. 1998; Thomas 2010). If promiscuity of subunit composition were to widely apply to NBSs and TMSs, then it would mean that the family is a much more complicated dynasty than previously thought.

In summary, inspection of both primary and tertiary structures indicates highly conserved NBD folds (Fig. 2.2) that have associated with a range of different TMD folds to energise transport (Fig. 2.1). This evolutionary teamwork is not unique to the transporters within the ABC family—there are also examples of NBDs that have associated with other soluble proteins to energise a specific function, e.g. Ribosome-associated complexes (Chen et al. 2014; Qu et al. 2018). However, the transporters in the family share a common mechanism for coupling energy encapsulated in ATP to transport of substances, as elaborated in the alternating access model of Jardetzky (1966).

Assembly of Protein Complexes

The textbook teaches us that the assembly of protein complexes follows a well-defined pathway where pre-folded subunits are sequentially incorporated into the growing complex. Much of the early knowledge on assembly pathways was gleaned from studies in the 1970s and 1980s on the small (30S) and large (50S) subunits of the prokaryotic ribosome (Nomura 1973; Herold and Nierhaus 1987). After isolation of these two abundant cellular complexes, all their components (polypeptides and ribosomal RNA species) could be dissociated, separately purified and then reassembled (Dohme and Nierhaus 1976; Stoffler et al. 1968). The good correspondence between the relationships discovered in these landmark biochemical experiments and the later prize-winning determination of the three-dimensional organisation of the ribosome (Nissen et al. 2000; Ban et al. 1998, 2000; Yonath et al. 1998; Ramakrishnan and White 1998) gave further credence to the idea that well-defined assembly pathways existed. Although some of the experimental steps in the in vitro ribosome assembly pathway were somewhat non-physiological (Herold and Nierhaus 1987), there appeared to be no reason to doubt that these steps were representative of the in vivo mechanisms.

Later studies showed that chaperones were essential in aiding the assembly of some complexes such as the purinosome (French et al. 2013) and SNARE complexes (Joglekar and Hay 2005) and proteomics analysis suggested that chaperones, such as Hsp90, were involved in regulating the assembly of many cellular protein complexes (Makhnevych and Houry 2012). Recent studies have been interpreted as showing a major role for co-translational assembly of protein complexes (Shieh et al. 2015). Here, it is thought that the close proximity of nascent chains emerging from ribosomes that are translating the same mRNA chain (encoding multiple polypeptide subunits of the same complex) allows the efficient assembly of said subunits. Moreover, unfolded or partially folded nascent chains may be important in facilitating this process, apparently in contradiction to the textbook view of protein complex assembly.

Molecular Recognition in ABC Transporter Complex Formation

Prokaryotic ABC transporters are usually heteromeric, typically consisting of separately expressed transmembrane and water-soluble nucleotide-binding subunits (TMS and NBS). In eukaryotes, the opposite is the case: transmembrane and nucleotide-binding components are usually fused and hence are considered as domains rather than subunits and are termed transmembrane domain (TMD) and nucleotide binding domain (NBD). In prokaryotes, ABC transporter subunits will almost always be encoded in tandem within an operon in the bacterial genome that also contains genes related to the specific function of the transporter (such as substrate-binding proteins, enzymes and other regulatory proteins) (Higgins et al. 1982; Davis et al. 2020; Higgins and Ames 1982). Some prokaryotic transporters have fused transmembrane (TMD) and cytosolic (NBD) domains, partly solving the problem of diffusion and molecular recognition for assembly—at least, for these two components. However, even in this simplified case, two TMD–NBD fusion peptides (whether homodimeric or heterodimeric) must still find their correct partner and assemble within the crowded cellular milieu. The current dogma is that once the subunits have folded in the cytoplasm (or inserted and folded in the membrane *via* the Sec machinery), then complex formation can take place *via* the processes of diffusion and precise docking through molecular recognition at large subunit/subunit interfaces. For membrane proteins such as the ABC transporters, diffusion within the membrane will be slow due to microviscosity of the lipid bilayer. However, limiting diffusion to two dimensions within the membrane will likely favour TMS–TMS encounters. In contrast, NBS–NBS associations and NBS–TMS associations will occur within the complex and crowded environment of the cytoplasm, a much more challenging proposition.

For type I and II ABC transporters (see Figs. 2.1 and 2.2), the process of molecular recognition is thought to involve residues in a cytoplasmic loop extending down from the TMD. This is sometimes termed the 'EAA loop' (Saurin et al. 1994; Kerppola and Ames 1992) based on its (somewhat) conserved Glutamate and Alanine residues. The EAA loop contacts the 'Q-loop' in the NBD, named for its conserved Glutamine residue, which is found between the fifth β-strand and second α-helix in the core NBD fold (Fig. 2.2a, b). Indeed, the Q-loop region could be considered to be a universal NBD 'socket' into which various types of TMD 'balls' can be fitted. For example, type I, II, V, VI and VII ABC transporters' TMDs (Fig. 2.1) all have a single loop extending down into the NBD interface to contact the Q-loop region. Other contacts are also provided by N-terminal and C-terminal regions of the TMDs (Fig. 2.2b, c). In contrast, type III transporters have contacts primarily made by the 'T' subunit (TMS) *via* a very long loop that forms two sequential helices between the fourth and fifth transmembrane spans. These two helices are arranged as a hairpin lying along the cytoplasmic surface of the lipid bilayer. The end of the first helix slots into the Q-loop region of the opposing NBS, whilst the second helix turns back and contacts the Q-loop region of the NBS underlying the 'T' subunit. Lastly, type IV ABC transporter TMDs each have *two* loops interacting with the NBDs; one loop crosses over from one TMD to contact the Q-loop region of the opposite NBD, whilst the other loop contacts the helix following the Walker A motif in the underlying NBD. Hence the TMD-contacting Q-loop region structure has remained fairly constant across the entire ABC transporter family, whilst the NBD-contacting regions of the TMDs display a variety of structures. Given this, it seems somewhat surprising that no mistakes are made in molecular recognition between TMS and NBS (or NBS and NBS) in the complex mixture of newly synthesised NBSs in the cytoplasm. For example, Fig. 2.3 illustrates the sequence homology for some subunits of *E. coli* type I ABC transporters which have distinct substrates. Their NBSs associate with different TMSs, but have very similar sequences. Another conundrum illustrated by Fig. 2.3 is that many ABC transporters in prokaryotes are encoded by three core subunit genes rather than two or four. The three-core subunit transporters have an intrinsic symmetry mismatch: for example, in the arginine importer in *E. coli* K12, there are two separate TMSs (ArtM, ArtQ) that must assemble with two copies of the same NBS (ArtP). This poses an interesting question: if ArtP can readily bind to both ArtM and ArtQ, which have non-identical amino acid sequences (30% sequence identity), then what prevents ArtP from also binding to, for example, TcyL and HisM, which have higher (33% and 32%) sequence identities with ArtM than ArtQ does? Is it the case that the EAA loops of ArtM and ArtQ share common features that are subtly different to close homologues such as TcyL and HisM? From examination of their primary

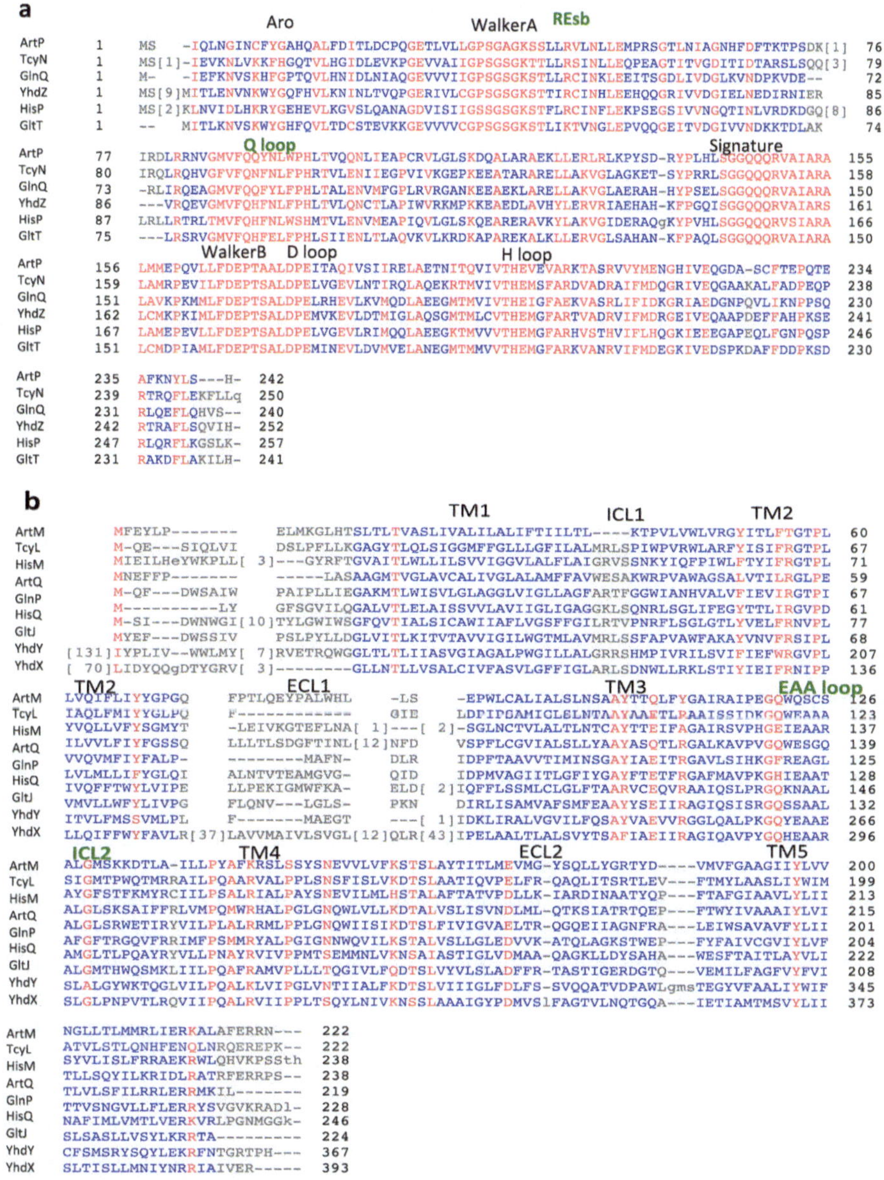

Fig. 2.3 (**a**) Sequence alignment of some Type I ABC transporters in *E. coli* K12. The NBS sequences of importers (of different amino acids) had their sequences aligned with ArtP, with sequence identity ranging from 47% (TcyN) to 43% (GltT). Identical or structurally related residues are highlighted in red (Cobalt alignment, conservation setting: 3 bit). Conserved motifs involved in ATP binding or hydrolysis are indicated (black text), conserved motifs involved in TMD recognition are in green. A conserved arginine residue that forms a salt-bridge with the glutamate in the TMD EAA loop in these transporters is indicated as 'REsb'. (**b**) The corresponding TMD sequences were aligned versus ArtM, with percentage identity ranked from 33% (TcyL/YecS) to 24% (YhdX). Transmembrane (TM) alpha helices are numbered along with extracellular loops (ECL) and intracellular loops (ICL). The EAA motif that is part of ICL2 is associated with binding to the NBD Q-loop region

structures, there does not appear to be any clear explanation (Fig. 2.3). It follows that there are many aspects of molecular recognition and subunit assembly in ABC transporters that we do not yet understand.

Molecular Recognition and Subunit Assembly in ABC Transporters

In this section, we will cover the different types of ABC transporters that have been characterised based on the fold of their TMD; and (when applicable) experiments that have addressed molecular recognition between subunits or domains and the assembly of the transporter subunits into functional complexes will be discussed. We have taken the liberty of making the section on type III (ECF) ABC transporters somewhat longer as we considered that these transporters have been relatively overlooked in the past and because the subsequent sections in the chapter that focus on *E. coli* K12 ABC transporters are inevitably lacking any mention of type III ABC transporters.

Type I ABC Transporters

Type I importers are thought to mostly transport abundant metabolites such as amino acids and sugars, but there are exceptions such as the molybdate importer ModBC (Hollenstein et al. 2007). There are a few papers that have tested models of type I ABC transporter complex formation in vitro by purification of individual subunits and also by mutagenesis. In the study by Eckey and coworkers (2010), ArtM (TMS) ArtP (NBS) and ArtJ (SBP) from *Geobaccilus stearothermophilus* were expressed separately and then individually purified using hexa-Histidine affinity tags. ArtM was purified in detergent to solubilise it. After purification, its affinity tag was cleaved off. When the untagged ArtM and the tagged ArtP were added together, they formed a tight, functional $ArtM_2P_2$ complex in vitro, as shown by the retention of ArtM on a cobalt affinity column and that the eluted complex showed strong stimulation of ATPase activity by the addition of ArtJ that had been loaded with its substrate, the amino acid Arginine. Mutations made in the Q-loop of ArtP had no effect on complex formation, but mutation did reduce the amount of stimulation of ATP hydrolysis caused by the addition of ArtJ/Arginine. These experiments were in line with the idea promulgated above, that the Q-loop is a universal socket and will not be affected greatly by mutation. However, the data implied that the Q-loop has an allosteric role in coupling ATPase activity to the presence of its transport client. Mutations were then made in the EAA loop of ArtM that reduced complex formation greatly, and (inevitably) the ArtJ-stimulated ATPase activity was lost. Experiments where both EAA loop and Q-loop mutated versions of ArtM and ArtP were added together yielded no complexes (and zero stimulation of the ATPase activity). These experiments therefore demonstrated the importance of the TMDs and the EAA loop in determining complex formation in vitro, in agreement with the structural and sequence considerations rehearsed in the previous section. They also showed that fully folded subunits were competent to assemble together in vitro, in accordance with the textbook conceptualisation of protein complex formation. However, the experiments on the ArtJMP complex were done with purified components, hence the fidelity of molecular recognition of ArtM by ArtP was untested for conditions similar to those existing in the crowded environment of the cytoplasm.

In earlier in vitro experiments with the $MalFGK_2$ importer, the challenge of cellular complexity was cleverly included (Sharma et al. 2005). In this work, MalF,G & K proteins were expressed simultaneously and membranes containing the in vivo assembled transporter were isolated. MalK (NBS) was removed from the membranes by treatment with 6M urea and then the MalFG heterodimeric complexes were solubilised in detergent and purified using an affinity tag present on MalF. Sharma and coworkers demonstrated that the affinity-immobilised MalFG was able to specifically fish out MalK from a complex cytosolic extract obtained from a MalK-overexpressing *E. coli* strain. This demonstrated that the MalF and G polypeptides together encode a recognition surface that is sufficient to drive the assembly of MalK. Moreover, after elution of the in vitro assembled complexes, they were reconstituted into liposomes so that transport of maltose could be assayed. MalE, loaded with maltose, stimulated ATPase activity and the rate of maltose transport for the in vitro assembled complex was found to be similar to that of the in vivo complex. These latter experiments suggest that the MalFG TMSs can specifically assemble with two correct NBSs to form a fully functional ABC transporter complex and provide support for the idea that ABC transporters' subunits can find each other accurately, even with two different TMSs.

Type II ABC Transporters

Type II ABC importers are thought to transport rare co-factors and trace elements such as vitamin B12 and transition metal ions needed for enzyme catalysis. Exceptions to this general rule are likely to exist, however (see later). Similar in vitro studies to those described in the previous section were set up to investigate the assembly of the vitamin B12 importer complex BtuCD (Di Bartolo et al. 2011). In these experiments, BtuCD was first expressed and affinity-purified according to the same procedure employed in its structure determination (Locher et al. 2002). As for the experiments with MalFGK, the NBS (BtuD) was removed from the immobilised BtuC by using a 5M urea concentration which was sufficient to disrupt BtuC–BtuD interactions, but not so high as to denature BtuC. After elution of 'naked' BtuC, reassembly of the complex was possible by adding ATP and BtuD that had been separately expressed and purified. Surprisingly, Di Bartolo and co-workers found that both assembly of the complex and functional activity (ATPase activity was measured) were enhanced when one, or both, subunits were partially *unfolded*. The enhancement of BtuCD reassembly by partial unfolding in vitro was unexpected, and was explained in the paper in terms of the potential involvement of chaperones for the assembly of protein complexes in vivo (where the role of the chaperones would presumably be to stabilise partially unfolded states of an assembly-competent subunit). Within the context of this book chapter, however, these experiments with BtuCD could also be taken as supporting the idea of a co-translational mechanism, where partially folded nascent polypeptide chains of different subunits initiate assembly of the complex. Moreover, the in vitro assembly experiments described for type I ABC transporters also involved possible partial denaturation of the TMSs with either urea (MalF,G) or detergent (ArtM). Hence, it seems possible to argue that unfolding or partial unfolding of subunits may be linked to assembly of ABC transporter complexes in general.

Type III ABC Transporters or Energy Coupling Factor Transporters (ECF)

Energy coupling factor proteins (ECF), a family of ABC transporters of type III in the new nomenclature, import vitamins and a variety of micronutrients that are required in trace amounts in prokaryotes, such as Co^{2+}, Ni^{2+} and others (Rodionov et al. 2006, 2009; Hebbeln et al. 2007; Erkens et al. 2012; Berntsson et al. 2012; Karpowich and Wang 2013; Xu et al. 2013; Wang et al. 2013; Zhang et al. 2010). In this respect, they share some functional overlap with the type II importers such as BtuCD. Each ECF transporter contains four components: the S-component (EcfS), which is a multi-pass transmembrane domain that also doubles as a substrate-binding protein, two ATP-binding subunits often termed A components (EcfA and EcfA′), and a core transmembrane domain, i.e. T component (EcfT) (ter Beek et al. 2011; Wang et al. 2013; Zhang et al. 2010) (Fig. 2.4). Two EcfA, EcfA′ and one EcfT altogether, are known as the energising module or ECF module. ECF transporters are importers, but differ from other ABC importers in that one of the membrane-embedded components (S-component) is responsible for substrate binding; it grabs the substrate and then binds with the energising module to form a functional transporter (Chai et al. 2013; Henderson et al. 1979). The size of these S-components is 20–25 kDa, and they are assumed to have 4–6 hydrophobic membrane-spanning domains, based on known structures, but their sequences are quite dissimilar, and their evolutionary relationship is unknown (Erkens et al. 2011). The lack of sequence conservation among S-components is surprising, especially where they share the same energising module.

ECF transporters are categorised into two groups: group I ECF transporters have a dedicated S-component for a specific ECF module. The best-categorised group I ECF transporters are the biotin transporter (BioMNY), cobalt transporter (CbiMNQO) and nickel transporter (NikMNQO) (Hebbeln et al. 2007; Okamoto

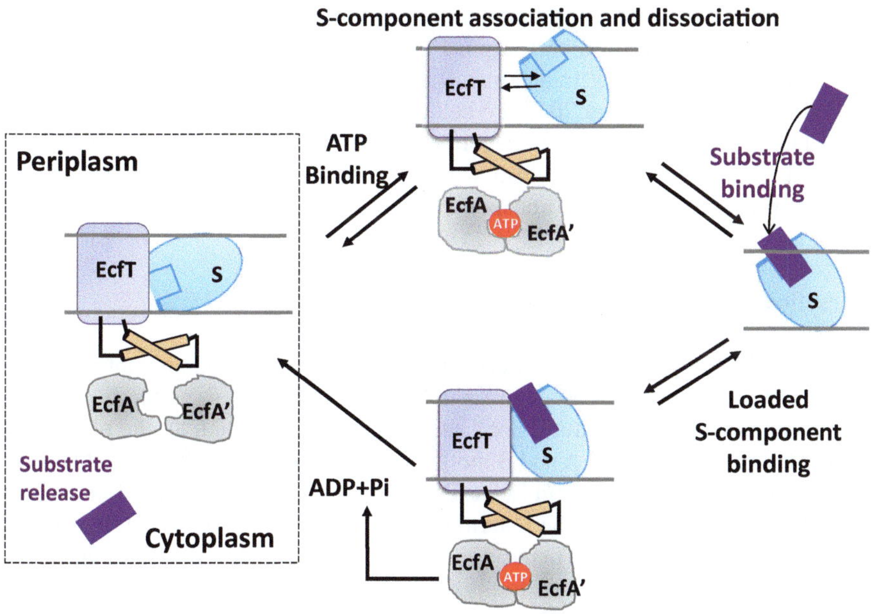

Fig. 2.4 A model for transport in ECF (type III) transporters: ATP binding to the A–A′ NBD heterodimer has been proposed to be associated with release and re-association of the S-component. ATP hydrolysis may be associated with (or triggered by) toppling of the S-component and this leads to substrate translocation across the membrane. The dashed box indicates the specific ECF complex and conformational state that has so far been resolved at high resolution in structural studies. The grey lines indicate the approximate boundaries of the lipid bilayer

and Eltis 2011; Lebrette et al. 2014). In contrast, group II ECF transporters use the same ECF module in combination with several different S-components with different substrate specificities, allowing chemically diverse substrates to be transported by the same complex (Fisher et al. 2012). For example, in *Lactococcus lactis,* eight S-components engage with the same energising module (ter Beek et al. 2011; Zhang et al. 2014), while in *Lactobacillus brevis*, it is seven S-components. So far, three full-length structure of group II ECF transporters from *L. brevis* have been determined, all sharing the same ECF module, but a different S-component: ECF–FolT, ECF–PanT and ECF–HmpT (Xu et al. 2013; Wang et al. 2013; Zhang et al. 2014). Only one high-resolution structure of a group I S-component, i.e. NikM, has been determined to date (Yu et al. 2014). On the other hand, several group II EcfS structures have been determined, such as ThiT (thiamine specific), FolT (folate specific), RibU (riboflavin specific), HmpT (hydroxyMethyl pyrimidine specific), BioY (biotin specific), PanT (pantothenate specific). Furthermore, the stoichiometry of ECF subunits is debatable (Finkenwirth et al. 2010; Neubauer et al. 2011). Structural atomic models with one T, S, and A, A′ (1:1:1:1) have been reported (Rodionov et al. 2009; ter Beek et al. 2011; Zhang et al. 2010; Erkens et al. 2012), and these are taken as the most likely *status quo*. However, other findings imply that the functional transporter can also contain several T and S subunits, and a stoichiometry of 1A:1A′:2T:2S has been argued based on experimental observations in a *Thermotoga maritima* ECF transporter (Finkenwirth et al. 2010; Neubauer et al. 2011; Karpowich and Wang 2013). Whether different substrate-specific S subunits can be integrated into the same transporter complex at the same time has also been discussed (Siche et al. 2010), which would raise the possibility of taking up two substrate molecules in a single transport cycle.

Interactions Between EcfA–A′ and EcfT

Like all ABC transporter NBDs, EcfA and EcfA′ have the predominantly β-sheet RecA domain containing the Walker A and B motifs and the predominantly α-helical domain containing the Signature motif and Q-loop. As expected, all the previously identified motifs, i.e. Walker A and B motifs, the Signature motif, the D and Q-loops, and the His switch are retained in EcfA–EcfA′ (Xu et al. 2013). A third domain also exists in the EcfA/A′ NBDs, the C-terminal domain (CTD) (Karpowich and Wang 2013). EcfA and EcfA′ form a heterodimer, with the CTD helping to form the dimer interface. The C-terminal domain forms the primary subunit interface in the absence of ATP, and deletion of the CTD completely abolished the association of the EcfA–EcfA′ heterodimer (Karpowich and Wang 2013).

As described earlier, EcfT proteins have two long and well-conserved coupling helices termed CH2 and CH3 that are arranged as a hairpin (Eitinger et al. 2011; Neubauer et al. 2009). The CH2 and CH3 helices contain a motif known as the XRX motif, which binds to a conserved acidic groove generated between the EcfA–A′ heterodimer and may serve as anchor to the complex. The CH2 and CH3 helices must not only mediate complex formation but also allow transfer of the ATP-driven EcfA–EcfA′ conformational changes to EcfS *via* EcfT. The conserved EcfA–A′ groove is made up of the conserved Q-loop helix that interacts with the T subunits, but is not directly engaged in ATP binding and hydrolysis. Interestingly, a stable complex of EcfT–EcfA proteins of group I can be formed, whereas such a sub-complex is difficult to obtain for group II ECF transporters (Neubauer et al. 2011).

Interactions Between EcfS and EcfT Modules

The S-components have different C-terminal domains which bind the various substrates as well as a structurally conserved N-terminal domain that interacts with the ECF module. The degree of structural conservation of the N-terminal domain of Group II S-components should be compatible with the shared use of the energising module, although clearly, some induced fit must also be taken into account. So far, molecular recognition subtleties have not yet been systematically explored; however, the conserved alanine motif (AXXXA, X: hydrophobic amino acids) in transmembrane helix 1 (SM1) has been proposed to play a role in interaction with

EcfT (Berntsson et al. 2012). The hydrophobic groove of EcfS and the hydrophobic surface of SM1 establish hydrophobic interactions with the coupling helices CH2 and CH3 of EcfT. The extensive, but weak, van der Waals interactions, rather than hydrogen bonds, may facilitate the requirement for EcfS movement relative to EcfT during the proposed transport mechanism. This unusual structural characteristic may be necessary for EcfS transport activity. In group II ECF transporters, EcfS proteins from the same species contain similar transmission elements and sliding surfaces between the EcfS and EcfT proteins, and hence this may explain how they can share a single ECF module (Finkenwirth et al. 2010; Fisher et al. 2012).

Proposed Transport Mechanism of ECF Transporters

A 'toppling' or 'elevator' transport method has been proposed based on comparison of full ECF transporter structures with isolated S-components. In this mechanism (Fig. 2.4), (1) Membrane-embedded EcfS recognises and captures the substrate. (2) The bound substrate and S-component migrate in the lipid bilayer to the correct ECF module. (3) Association with the ECF component induces a reorientation of the S-component in the membrane that brings the substrate partway across the membrane. (4) The substrate dissociates from the S-component and becomes available to the inside of the cell. (5) ATP binding and/or hydrolysis cause further conformational changes in EcfA–EcfA′ that are transferred to EcfS *via* the coupling helices of EcfT, resulting in a significant conformational shift in EcfS allowing substrate release (Fig. 2.4) (Swier et al. 2016). It is clear from the crystal structures of group II ECF transporters that the initial reorientation of the S-component appears to be associated with release of substrate. Although structures are lacking, it has been argued that the various conformational phases do not result in the complete separation of the S-component from the ECF module in group I ECF importers (Finkenwirth et al. 2015). All EcfS proteins have a positively charged intracellular surface and a more hydrophobic/uncharged extracellular surface (Slotboom 2014). It is tempting to speculate that this structural feature could be responsible for the rapid reorientation of S-components in the membrane.

Genetic Organisation of the ECF Operons

It could be argued that group I ECF importers, which have a dedicated ECF module for a specific S-component will have genes encoding the S-component and the ECF module in the same operon. On the other hand, one might expect group II genes encoding ECF modules and S-components to be found in different regions of the genome. A search of the PATRIC (Pathosystems Resource Integration Center) tool showed that group I ECF components tend to be translated by the same operon, whereas for group II transporters, the energising module is translated by a single promoter, but S-protein genes can be found all over the genome.

Type IV ABC Transporters

Type IV ABC transporters perform various functions, e.g. lipid flippases, importers and exporters, as well as ion channel regulators and (in one case) an ion channel (Table 2.1). They probably represent the most diverse group in terms of their function, although the basic fold of their TMDs is very similar. The mammalian ABC transporters' subfamilies ABCB–ABCD are also in type IV in this proposed classification system (Fig. 2.1).

Type IV ABC transporters are characterised by their relatively long α-helical elements in the TMDs (6 per TMD) (Dawson and Locher 2006). The membrane-spanning helices extend out beyond the lipid bilayer on the cytoplasmic side of the transporters, hence the NBDs are relatively further away from the cytoplasmic surface of the lipid bilayer. The topology of the type IV TMDs means that there

Table 2.1 Examples of Type IV ABC transporters for which high resolution structures have been determined, and their range of functions

Type IV ABC transporter	Functions
Prokaryotic members	
Sav1866	Multidrug export (Dawson and Locher 2007)
MsbA	Lipid A/ LPS flopping (Zhou et al. 1998)
NaAtm1	Export of Glutathione complexes (Lee et al. 2014)
TM287/288	Daunorubicin export (Hohl et al. 2012)
McjD	Antimicrobial peptide export (Choudhury et al. 2014)
PCAT1	Polypeptide export (Lin et al. 2015)
PglK	Export of lipid-linked oligosaccharides (Perez et al. 2015)
TmrAB	Peptide export (Nöll et al. 2017)
PrtD	Polypeptide type 1 secretion system (Morgan et al. 2017)
YbtPQ	Metal-siderophore import (Rodriguez and Smith 2006)
Rv1819c	Import of cobalamin and bleomycin (Gopinath et al. 2013)
IrtAB	Iron-siderophore import (Rodriguez and Smith 2006)
Eukaryotic members	
ABCB1	Multidrug export (Sharom 2006)
ABCB2/3	Peptide export (Phillips et al. 2003)
ABCB4	Lipid export (Elferink and Paulusma 2007)
ABCB6	Porphyrin transporter (Paterson et al. 2007)
ABCB8	Potassium channel regulator (Zutz et al. 2009)
ABCB10	Biliverdin export (Shum et al. 2021)
ABCB11	Bile salt export (Stieger 2009)
ABCC1	Multidrug, leukotriene, sphingolipid export (Cole et al. 1992; Gradhand and Kim 2008)
ABCC7	Chloride channel (Yee et al. 2000; Riordan et al. 1989)
ABCC8	Regulatory module of K_{ATP} channel (Lehman et al. 2008)
ABCD4	Cobalamin, Fatty acid transport (Kashiwayama et al. 2009)

are two (long) intracellular loops and three (usually short) extracellular loops, with both N- and C-termini on the cytoplasmic side of the membrane. As stated earlier, interactions with the NBDs occur *via* the two cytoplasmic loops and the second loop (connecting between transmembrane α-helices 4 and 5 in the sequence) crosses over to interact with the Q-loop region in the NBD on the opposite side of the transporter (Dawson and Locher 2006). The inward-facing conformation of type IV ABC transporters houses a large internal cavity that penetrates into the hydrophobic region of the lipid bilayer (Ward et al. 2007; Aller et al. 2009). This cavity is accessible from the cytoplasm but may also be accessed by hydrophobic molecules in the inner leaflet of the lipid bilayer *via* gaps formed between transmembrane α-helices 4 and 6 in the sequence. The internal cavity has been found to be where transported substrates bind, and its large internal volume and its alternative access points may explain why type IV transporters have such a wide range of substrates and functions (Aller et al. 2009).

So far, all characterised type IV transporters have the NBDs and TMDs fused as a single polypeptide, which may be relevant in terms of assembly of the transporter. Since the NBDs are relatively distant from the stabilising influence of the membrane compared to other types of ABC transporters (Fig. 2.1), it may be that fusion of the NBD to the TMD is an absolute necessity in order to avoid the NBDs diffusing off into the cytoplasm. Whether NBD–TMD fusion is *de rigueur* in type IV transporters is a moot point. In *E. coli* K12, all type IV transporters are formed from fused NBD–TMD subunits. However, SbmA and the related BacA are homodimeric exporters of anti-microbial peptides with transmembrane folds that very closely resemble the type IV ABC transporters (Ghilarov et al. 2021; Corbalan et al. 2013). SbmA transport of peptides is energised by the proton gradient rather than by ATP, and there does not appear to be any cognate NBD for it. However, the mere existence of SbmA and BacA implies that type IV ABC transporter TMDs *may* exist in some organisms as separately expressed subunits along with a cognate NBS for energising the transport function.

Type V ABC Transporters

Type V transporters have 2 TMDs with 6 (short) TM α-helices each and a coupling loop between TM2 and TM3 that contacts the Q-loop region in the NBD (Lee et al. 2016; Taylor et al. 2017; Qian et al. 2017). In some cases, the first extracellular loop can be very large, for example, in the ABCA family of eukaryotic transporters (Qian et al. 2017; Xie et al. 2021). Other distinctive features of type V transporters are two extracellular re-entrant helices that dip into the outer lipid leaflet of the membrane between TM5 and TM6 and an amphipathic "elbow helix" at the N-terminus on the cytoplasmic side of the transporter (Caffalette et al. 2019; Lee et al. 2016; Bi et al. 2018). The latter feature is also found in type IV and type VII transporters. Almost all characterised type V ABC transporters are exporters. In prokaryotes, structures are available for two that act as transporters of the outer membrane O-antigen

[Wzm-Wzt (Bi et al. 2018)] and teichoic acid [TarGH (Chen et al. 2020)]. In eukaryotes, Type V transporters are exemplified by the ABCA and ABCG sub-families and have NBD and TMD fused together and sometimes all four domains fused together (Ford et al. 2020) predominantly in the ABCA family and a few in yeast ABCG transporters. They are associated with the export of a wide range of substrates, including phospholipids and cholesterol [ABCA1 (Qian et al. 2017)], sterol [ABCG5/8 (Lee et al. 2016; Taylor et al. 2017)] as well as conferring multidrug resistance *via* export of a range of drugs and xenobiotics [ABCG2 (Taylor et al. 2017)]). Type V transporters may also include a few members that are importers, e.g. the retina-specific flippase ABCA4 (Xie et al. 2021; Quazi et al. 2012; Jumper et al. 2021).

The assignment of the Mla lipid exporter (MlaEF) into the new classification system is somewhat equivocal as it resembles a type V ABC transporter in terms of its general topology (Zhou et al. 2021), but it lacks the last TM span and also the re-entrant helices of the extracellular portion. Interestingly, in *E. coli* K12, YadGH and YbrFRS are predicted to have type V transporter folds by the AlphaFold server (Jumper et al. 2021) and they show some ability to bind each other's subunits (see later sections).

Type VI ABC Transporters

Type VI ABC transporters display a TMD fold that is somewhat similar to type V transporters, and the NBD-contacting loop of the TMD also occurs between the second and third membrane-spans. For these transporters, however, there is a large extracellular loop in the TMD between the third and fourth transmembrane spans which forms a folded extracellular domain (Luo et al. 2017). In the well-characterised $LptB_2FG$ lipopolysaccharide (LPS) transporter (Freinkman et al. 2012), these, in turn, interact with an adaptor protein (LptC) that couples the transporter to periplasmic (LptA) and outer membrane components (LptD, LptE) involved in translocating the LPS to the outer leaflet of the outer membrane (Li et al. 2019; Tang et al. 2019; Luo et al. 2017). Closure of the NBD dimer and concomitant change in the arrangement of the TMDs appears to push the single transmembrane α-helix of LptC away from the TMDs of LptF&G, possibly causing release of the LPS. Density for the LPS substrate can be observed in the nucleotide-free state, but not when nucleotide is present (Li et al. 2019; Tang et al. 2019).

Unusually, the complex between the $LptB_2FG$ ABC transporter and the periplasmic and outer membrane components (seven proteins in total) can be studied after solubilisation of the bacterial membranes (Chng et al. 2010), but so far, a structure for the whole complex is lacking. The periplasm-spanning bridge between the outer and inner membrane components has been modelled as a stack of 40–50 antiparallel β-sheet strands formed by the extracellular 'jellyroll' domains of LptF and G, followed by the β-sheet domain of LptC, the β-sheet domain of LptA and the N-terminal β-sheet domain of LptD (Freinkman et al. 2012). Regulation of assembly of the complex (by LptE) is thought to be involved in overall regulation of LPS transport in this system (Freinkman et al. 2012).

As for type I and type II ABC transporters, mutational analysis of the Q-loop region of LptB was carried out and loss-of-function rescue mutations in the TMD were examined (Simpson et al. 2016). Changes in the NBD-contacting loop in LptF and LptG (the equivalent of the 'EAA' loop) were shown to rescue loss of a positive charge in LptB (R91S). These involved complementary changes in a negatively charged residue (LptF, E84A; LptG, E88A). These glutamate residues appear to form stabilising TMD–NBD salt-bridges with R91 in LptB, when the structure of the transporter is examined (Li et al. 2019; Luo et al. 2017; Tang et al. 2019). However, substitution of E84 or E88 with aspartate was also detrimental, suggesting that it is not only charge, but also packing at the interface that mediates the stability of the NBD–TMD interactions (Simpson et al. 2016). These experiments again describe the importance of salt-bridges in the assembly of ABC transporter subunits into a functional complex, but add extra subtlety to the earlier descriptions.

Type VII ABC Transporters (or ABC Mechanotransducers)

Type VII ABC transporters show a fold similar to the type V transporters, with a large extracellular domain between the first and second membrane spans and an NBD-interacting loop located between the second and third membrane spans. However, type VII transporter TMDs are somewhat truncated and display only four transmembrane α-helices per TMD. Experiments with the $LolCD_2E$ lipoprotein exporter system (Kanamaru et al. 2007) showed that a functional complex could be reconstituted in vitro using just lipid and purified components. Interestingly LolE (TMD) and LolD (NBD) were also able to form a functional unit. This artificial complex therefore lacked LolC (the other TMD), providing a hint that subunit promiscuity may exist in cells and be compatible with function. Mutation of the 'Q-loop' residues in the LolD recognition region for the TMDs resulted in loss of function that could be rescued by suppressor mutations in the TMDs (Ito et al. 2006). The recent availability of the LolCDE structure (Tang et al. 2021) has allowed these mutations' locations to be visualised. Many, but not all, suppressor mutations in the LolE TMS localised to residues close to the TMS–NBS interface, as expected. However, suppressor mutations in the LolC TMS were found throughout the protein and few were at the TMS–NBS interface. These studies imply that mutation of the Q-loop region can be rescued by residues throughout the TMDs, and emphasise the importance of the Q-loop region in mediating allosteric changes in the transporter. It is also noteworthy from a molecular recognition perspective that LolD is very close to the structure of BtuD, but the interacting loops from the TMDs are entirely different, again pointing to the apparent ability of the Q-loop region to adapt to multiple types of TMD interactions.

Type VII ABC transporters are thought to be mechanotransducers that convert the chemical potential of ATP hydrolysis in the cytoplasm into mechanical force in the periplasm. For MacB, probably the best studied type VII ABC transporter, ample structural data has been obtained (Okada et al. 2017; Fitzpatrick et al. 2017), including cryo-EM data for the full tripartite complex with MacA, MacB and TolC (Fitzpatrick et al. 2017). For the latter study, formation and stabilisation of the complex was favoured by inserting short linkers between MacA and MacB subunits using genetic engineering. For this, the C-terminus of MacB and the N-terminus of the (predominantly) periplasmic adaptor protein MacA were fused together and then the MacAB fusion was coexpressed with TolC. In another approach, disulphide linkages between MacA and MacB were engineered by inserting cysteine residues at expected interfacial positions. These two separate approaches yielded *the same* complex consisting of the expected stoichiometry from the preceding X-ray crystallographic studies of the individual components, i.e. there was a trimeric TolC oligomer bolted on to a hexameric MacA adaptor ring, which, in turn, was associated with a MacB homodimer. Surprising insights into the assembly of the functional complex were revealed by these studies: in the former (fusion) approach, there would have been four MacB monomers associated with the complex that were surplus to requirements for the 3:6:2 stoichiometry of the tripartite export system. None of these 'surplus' MacB monomers could be identified within the 3D density map, implying that they were disordered and that the assembly of the functional MacB homodimer was strongly dependent on the presence of the MacA hexamer. Similarly, in the cross-linked complex, only two of the MacA subunits are positioned to form disulphide bonds with MacB, and again no density for any additional cross-linked MacB subunits could be detected (Fitzpatrick et al. 2017). These experiments suggest that correct folding of MacB monomers is in some way linked with the assembly of the MacB homodimer complex. Again, this could be taken as a hint that ABC transporter complex assembly may involve disordered or partially disordered subunits.

Thermodynamics of ABC Transporter Complex Formation

When structures of complexes have been determined by crystallography to reasonably high resolution, they can be analysed using bioinformatics tools such as PISA (Proteins, Interfaces, Structures and Assemblies) (Krissinel and Henrick 2007). Structures of complexes determined by cryo-EM cannot be fully analysed yet using the PISA tool; however, it can be used to determine interface surfaces for these latter-day structures. The TMDs of ABC transporters are generally predicted by PISA to be 'held together' *via* large interfaces and with dissociation of these TMDs being strongly disfavoured (large +ve ΔG values) (Kos and Ford 2009) (see also Table 2.2). These thermodynamic energy predictions do not vary much whether the protein is in the inward-facing or outward-facing state (although examples of ABC transporter structures in the outward-facing state are rare and presumably reflect the instability of this state). In contrast, the NBDs within ABC transporters are predicted to associate weakly together, with ΔG values for dissociation close to

Table 2.2 ABC transporter complex stability: thermodynamic parameters were calculated from structures of ABC transporters determined by X-ray crystallography using the PISA server. A free energy for dissociation (ΔG_{diss}) that is large and positive indicates a stable complex

		ΔGdiss (kcal/mol)				
Bacterial ABC Transporter	PDB ID (best Res.)	TMD–TMD	NBD–NBD	TMD–NBD	All (+SBP,	-SBP)
$MalFGK_2$	3FH6	62.1	3	0.8	12.6	
$MalFGK_2$–MalE	4JBW	89.1	4.7	2.8	2.1	17.2
ModBC–ModA	2ONK	80	11.6	Unstable	22.1	
HmuUV	4G1U	30.5	1.8	3.9	19	
BtuCD	4R9U	20.1	16.7	3.3	17.9	
BtuCD–BtuF	4FI3	32.5	14.5	2.6	16.4	16.6
MetNI	3DHW	23.8	3.4	Unstable	9.3	
MetNI–MetQ	6CVL	18.1	4.4	Unstable	2.7	6.3
MolBC (H10796)	2NQ2	26.7	17.8	10.7	32.7	
Sav1866	2ONJ	n/a	n/a	93		
MsbA	6O30	n/a	n/a	94.4		
Wzm–Wzt	6OIH	5.6	Unstable	Unstable	10.6	
$Art(QN)_2$	4YMV	21.6	Unstable	Unstable	23.3	
AlgM1M2SS-Q2	4TQU	82	2.5	3.4	12	12.2
BhuUV–BhuT	5B58	24.6	1.5	1.8	14.7	15
EcfTAA′–FolT	5D3M	53.8	7.6	Unstable	54.7	2.1
EcfTAA′–PdxU2	4HZU	59.4	7.4	5.8	14.5	
LbECF–PanT	4RFS	51.2	7.5	2.4	10.7	
CbiMQO	5X41	61	Unstable	7.7	9.5	
ECF–CbrT	6FNP	41.4	5.3	1.6	7.4	
NaAtm1	6PAO	n/a	n/a	62.9		
TM287/288	3QF4	n/a	n/a	100.1		
TmrAB	5MKK	n/a	n/a	96.4		
MacB	5WS4	n/a	n/a	17		
McjD	5OFR	n/a	n/a	108.1		
PCAT1	4RY2	n/a	n/a	95.7		
PglK	5C76	n/a	n/a	95.2		
LptB2FG–LptC	6MIT	18.8	Unstable	(F)8.2 (G) 10.8	15.7	Unstable
LptB2FG	5 L75	21.5	Unstable	(F)5.7 (G) 1.7	9.3	
PrtDEF	5 L22	n/a	n/a	97.6		
IrtAB	6ETJ	Unstable (−1.9)				
MacB–ATPyS	5LJ6	n/a	n/a	32.9		
BmrA–E504A	6R72	n/a	n/a	91.4		

zero (Kos and Ford 2009). This probably reflects the rarity of the NBD-dimerised state in vivo (Kos and Ford 2009). These free energy values do not increase by much upon ATP and Mg^{2+} binding (Table 2.2), and even in the few NBD-only dimeric structures, the NBD–NBD interactions are still relatively weak. Similarly, the TMD–NBD interactions are weak, with ΔG values for dissociation of around zero (Table 2.2).

Strong TMD–TMD and weak NBD–NBD interactions within an ABC transporter seem appropriate given the need for reversible NBD dimerisation during the transport cycle (Ford and Hellmich 2020; Callaghan et al. 2006). One would hypothesise that, provided the NBDs remain anchored to the TMDs, they would be part of a stable complex; and that this would be the case throughout the transport cycle. However, the prediction of weak TMD–NBD interactions is surprising, as it suggests that NBDs could readily dissociate from their cognate TMDs, leading to deficiencies in transporter function. Weak TMD–NBD interactions would also pose a problem for correct molecular recognition: for example, could a given TMD pair (that had correctly assembled together) associate just as easily with the wrong NBDs as the correct ones? If so, would such a chimeric transporter be dysfunctional? These questions at first sight appear sufficiently radical as to merit immediate disqualification. However, thermodynamic considerations of purified transporters may not be sufficient to gain an understanding of stability in vivo: even if dissociation of TMD and NBD is slightly favoured, kinetic considerations also need to be considered; and if there is a high activation barrier for dissociation (e.g. a major unfolding event is needed) then the process might not occur within a physiologically relevant lifetime. Furthermore, the presence of a charged lipid monolayer close to the TMD-interacting surface of the NBD may provide an additional stabilising force that keeps the NBD in close contact with the membrane and cognate TMD. For example, pivotal studies of the histidine transporter in the 1990s showed that the NBS (HisP) associated with the cell membrane even in the absence of HisM and HisQ (the TMSs) (Kerppola et al. 1991). However, there is important evidence that multi-task NBDs (that can associate with more than one TMD) do exist in some bacteria (Leisico et al. 2020; Quentin et al. 1999), as discussed later, implying that reversible NBD dissociation may be a real phenomenon, at least for some systems.

The Role of ATP in NBD Dimer Formation and Substrate Transport

ATP binding is associated with NBD dimerisation and the driving of the inward-facing state to the outward-facing state (for exporters and mechanotransducers), or vice versa for importers. ATP is often included in the purification of ABC transporters and their subunits as the ATP-bound form of the NBD is usually found to be more stable than the apo form. However, structural studies concur with predictions from thermodynamic calculations described above and show that the assembly of an NBD dimer is difficult. Even the NBD dimer structures that have been observed by crystallography are usually only observed after mutation has been carried out (Ford and Hellmich 2020). Moreover, NBDs that have bound Mg^{2+} and ATP (MgATP) can exist as monomers and full MgATP-bound ABC transporters can also exist in the inward-facing state with the NBDs separated. A proposal discussed elsewhere was that NBD dimerisation and the formation of the outward-facing state is favoured, but not imposed by MgATP (Ford and Hellmich 2020). In this conceptualisation of the mechanism, the equilibrium between the inward- and outward-facing states is shifted towards the outward-facing state in the presence of MgATP, but the inward-facing state still predominates, even in the presence of nucleotide.

Fused Subunits of Prokaryotic ABC Transporters and Assembly

Fusion of genes/subunits is encountered occasionally in prokaryotic ABC transporters, and presumably where this occurs, assembly of the full complex is likely to be facilitated. In the simplest case, fusion of TMD and NBD subunit genes can allow a single gene to encode a homodimeric ABC transporter complex such as MsbA and MacB. However, even in this simplest case, the two partnering polypeptides must team up to form the functional complex. In prokaryotes, fused NBD–NBD and TMD–TMD polypeptides are also encountered, but these do not seem to have persisted in eukaryotes. Fused NBD subunits have been structurally characterised for the ABCE family of NBD-only proteins (that are not transporters), surprisingly, there are no structures for fused NBDs from ABC transporters, neither in the presence, nor absence of their TMDs. Based on sequence and structural alignments, the linkers fusing the NBD subunits appear to be longer (~25–100 residues) than in the ABCE family (~25–40 residues). In the latter family, the linker is found to be structurally ordered and wraps around the surface of the opposing NBD, apparently holding the two ABCE domains in close proximity (Ford and Hellmich 2020). It seems possible that longer linkers in fused transporter NBDs may be structurally disordered, making crystallisation more problematic. Further information may arise from future cryo-EM studies of the purified full transporters such as AlsAC, RbsAC, YbhFRS and YejBEF. Local regions of disorder are much less likely to affect structure determination with cryo-EM (Thonghin et al. 2018).

A similar deficit of structural data also applies to the fused TMD–TMD ABC transporters. Although there are many structures for fully fused ABC transporters of type IV in eukaryotes, the long linker regions in these transporters are always between NBD1 and TMD2 and are absent from 3D density maps, implying they are mostly disordered (Ford et al. 2020). The short linker between transmembrane helix 6 in the TMD and the start of the NBD in these transporters (and in prokaryotic half transporters such as MsbA) are usually ordered, and wrap around the surface of the connecting NBD. Hence there are currently no clues about the nature of prokaryotic TMD–TMD linkers such as those found in AfuB, FhuB, PhnE and

YnjC (Table 2.3). Sequence alignment for TMD regions is much more difficult because of the low homology, but for FhuB, such analysis implies that it probably arose from a fusion of two type II TMDs like FecC and FecD, and that the linker between the two domains will be roughly 30 residues long.

Although experimental structures are not available for fused NBD–NBD and fused TMD–TMD proteins, this book chapter has been written at the start of an era of accurate protein structure prediction. For example, the AlphaFold structure prediction server has been used to generate models for all of the ABC proteins from *E. coli* K12 (Jumper et al. 2021). The model generated for YnjC (a TMD–TMD fusion) is highly homologous to a type I ABC transporter TMD fold with 2×6 transmembrane α-helices (Choi and Ford 2021; Thomas et al. 2020). The model predicts a short cytoplasmic linker of about 10 residues in length. Similarly, the FhuB protein (TMD–TMD) is predicted to be a type II ABC transporter, and here the cytoplasmic linker is roughly 20 residues in length. YbbP, the TMS of an uncharacterised ABC transporter is predicted by AlphaFold to be a fused type VII (mechanostransducer) with a TMD–TMD linker that re-inserts into the membrane as a hairpin of two additional membrane-spanning α-helices. Not surprisingly, the prediction confidence for all these linker regions is considerably lower than for the regions that are homologous to available structures of non-fused TMD subunits.

The AlphaFold predictions for NBD–NBD fusion proteins are also worth considering here: examination of the predictions for AlsA, GsiA, RbsA, YbhF and YejF (Fig. 2.5) shows that for AlsA and RbsA, which have short linkers of about 25 residues, the inclusion of secondary structural elements into the linker (by AlphaFold) produces a non-canonical NBD dimer with the two domains twisted relative to each other. As modelled, these dimers would not dock onto the cognate TMDs without some unwinding of the linker. Contrastingly, in the YejF model, where the 25-residue long linker has little secondary structure, it is long enough to be compatible with formation of a canonical NBD dimer. For GsiA, the 60 residue-long linker is predicted to have some α-helical and antiparallel β-strand elements, but it is still long enough to permit a canonical NBD dimer to be modelled. YbhF has a very long linker of ~100 residues and this is modelled as a small folded domain on the cytoplasmic side of the second NBD. Whether such a domain exists should be readily testable by cryo-EM.

Analysis of coevolution of residues in orthologues of the *E. coli* K12 fused NBD–NBD polypeptides using the *BIS2Analyzer* tool (Oteri et al. 2017), shows that residues in the linker regions do appear to be relatively conserved and are coevolving with residues in the core of the NBDs (see Fig. 2.5 with reference to Fig. 2.2). For example, D216 in RbsA, which is close to the Walker A region in NBD1, shows coevolution with V283, which is in the linker region. Similarly, in YejF, L105 in the Q-loop region appears to be coevolving with T253 in the linker and similarly, P175, in a turn preceding the Walker B-containing β-strand, shows coevolution with P261 in the linker. Surprisingly, these latter two proline residues are significantly separated in the (AlphaFold modelled) 3D structure (Fig. 2.5). In YbhF, G262 in the predicted structured linker is associated with A44 in the NBD1 Walker A region and C376 in NBD2 which immediately follows the Walker A region.

Table 2.3 Alphabetical list of ABC transporters in *E. coli* strain K12, with NBDs and TMDs listed separately (columns 1 and 2). Alternative names are indicated in a few cases. ABC proteins that can interact with the transporter subunits are recorded in column 3 (source: BioGrid). Where the interaction is with cognate transporter proteins, the interacting pair is indicated using capital letters separated by a slash. Importer periplasmic binding proteins are listed in column 4. NBD components are in *italics*, fused TMD–NBD proteins are in **bold**, periplasmic substrate binding proteins are in brackets, periplasmic adaptor proteins are in bold and superscripted. NBD–NBD fusions and TMD–TMD fusions are followed by '(f)'. The interactors listed are where a direct binary interaction has been demonstrated, mostly *via* affinity-trapped bait proteins (Babu et al. 2018) or detected by a yeast two-hybrid assay (Rajagopala et al. 2014). ABC proteins that are not transporters (e.g. DNA and ribosome-associated, NBD-only proteins) such as EttA, RbbA, UvrA, Uup and YbiT are not listed in column 1

NBS	TMS	Physical interactors (ABC proteins)	SBP	Biological function	References
AfuC	AfuB (f)[a]		(AfuA)	Predicted ferric transporter	Chin et al. (1996)
AlsA(f)	AlsC		(AlsB)	D-allose transporter	Moussatova et al. (2008)
AraG	AraH	G/G, G/F, *ZnuA*	(AraF)	Arabinose ABC transporter	Moussatova et al. (2008)
ArtP	ArtM, ArtQ	M/Q, *HisP*	(ArtI), (ArtJ)[b]	Arginine transporter	Eitinger et al. (2011)
BtuD	BtuC		(BtuF)	Vitamin B12 transporter	Borths et al. (2005)
CcmA	CcmB, CcmC			CcmABC ABC transporter	Schulz et al. (1999)
CydC, CydD	**CydC, CydD**	C/D, (FepB), **MsbA**, (UgpB), *UgpC, LivG,* [**MacA**], *YecC,*		Fused cysteine transporter	Kana et al. (2001)
CysA	CysT, CysW		(CysP)	Sulfate/thiosulfate transporter	Sirko et al. (1990)
DdpD, DdpF	DdpB, DdpC	*DppF*, DppD	(DdpA)	D-ala-D-ala transporter	(Aiba et al. 1996)
DppF, DppD	DppB, DppC	C/D, B/D, C/F, D/F, FtsX, *FtsE,* **MsbA,** OppB	(DppA)	Dipeptide ABC transporter	Moussatova et al. (2008), Aiba et al. (1996)
FecE	FecC, FecD		(FecB)	Iron dicitrate ABC transporter	Staudenmaier et al. (1989)
FepC	FepD, FepG	C/C, D/G	(FepB)	Ferric enterobactin ABC transporter	Chenault and Earhart (1992)
FetA/ YbbL	FetB / YbbM	A/A, *UgpC*		Predicted transporter	Nicolaou et al. (2013)
FhuC	FhuB (f)	B/C, B/D, C/D, *LolD,* (RbsB)	(FhuD)	Iron (III) hydroxamate ABC transporter	Köster and Böhm (1992)
FtsE	FtsX	E/E, E/X, DppB, DppC, (OppA), (PstS),*UgpC,* **MsbA,** (ArtI), *GlnQ,*		FtsE/FtsX ABC mechano-transducer	de Leeuw et al. (1999)

(continued)

Table 2.3 (continued)

NBS	TMS	Physical interactors (ABC proteins)	SBP	Biological function	References
GlnQ	GlnP	P/Q, *RbsA*, (RbsB), (PstS), FtsX, *YehX*	(GlnH)	Glutamine ABC transporter	Moussatova et al. (2008)
GltL	GltK, GltJ	L/J, L/I, *YecC*	(GltI)	Glutamate ABC transporter	Moussatova et al. (2008)
GsiA(f)	GsiC, GsiD	A/C, A/D, C/B, C/D, *ZnuC*, YejB, YejE	(GsiB)	GsiABCD glutathione ABC transporter	Wang et al. (2017)
HisP	HisQ, HisM	ArtQ,	(HisJ)	Histidine/lysine/arginine/ornithine transporter	Liu and Ames (1998)
LivG, LivF	Liv H, LivM	F/G, F/K, G/K, **CydD**	(LivK, LivJ)[b]	Leucine/isoleucine/val-ine transporter	Adams et al. (1990)
LolD	LolC, LolE	C/D, C/E, D/E, DppB, *FhuC, FtsE,* PotC, (PotD), (PstS), (RbsB), *UgpC*, (ZnuA)		Lipoprotein mechanotransducer	Narita et al. (2002)
LptB	LptG, LptF	F/G, **MsbA**		LptA/LptB ABC transporter	Thélot et al. (2020)
LsrA	LsrC, LsrD		(LsrB)	Fused AI2 transporter	Lennen (2007)
MacB	**MacB**			Macrolide mechano-transducer	Crow et al. (2017)
MalK	MalF, MalG	K/G, F/G, (TcyJ),UgpA, UgpE, (UgpB), PstC	(MalE)	Fused maltose transport subunit	Hsu et al. (2018), Fabre et al. (2017)
MdlA, MdlB	**MdlA, MdlB**	A/B, (PstS),		Fused predicted multidrug transporter	Tsou et al. (1990)
MetN	MetI	N/I, *ArtP, YbhF(f)*	(MetQ)	DL-methionine transporter	Merlin et al. (2002)
MglA	MglC	*YecC*	(MglB)	Fused methyl-galactoside transporter	Stamm et al. (1996)
ModC	ModB	(OppA)	(ModA)	Molybdate transporter	Neubauer et al. (1999)
MsbA	**MsbA**	**CydC,** FtsX, DppB, PotC,		Fused lipid transporter	Cruz-Ramos et al. (2004)
NikD, NikE	NikB, NikC	D/A, E/E, (GsiB)	(NikA)	Nickel transporter subunit	Bruntner et al. (1999)
OppD, OppF	OppB, OppC	B/D, C/D, B/F, C/F, B/C, *YejF*, (UgpB), **MacA**, DppC	(OppA)	Oligopeptide transporter	Jones et al. (2014)
PhnC	PhnE (f)[a]	(UgpB)	(PhnD)		Bardin et al. (1996)

(continued)

Table 2.3 (continued)

NBS	TMS	Physical interactors (ABC proteins)	SBP	Biological function	References
				Phosphonate/ organophosphate ester transporter	
PotA	PotB, PotC	A/B, A/C, A/D, B/C, **CydC, MsbA**	(PotD)	Polyamine transporter	Matthysse et al. (1996)
PotG	PotH, PotI	G/H	(PotF)	Putrescine transporter	Matthysse et al. (1996)
ProV	ProW	V/V, (RbsB)	(ProX)	Glycine betaine transporter	Haardt and Bremer (1996)
PstB	PstA, PstC	(TcyJ), MalG,	(PstS)	Phosphate transporter	Braibant et al. (1996)
RbsA(f)	RbsC	A/A, *GlnQ*, TcyL,	(RbsB)	Fused D-ribose transporter	Bell et al. (1986)
SapF, SapD	SapB, SapC	B/C, B/F, C/F, D/F, *YbiT,*		Predicted antimicrobial peptide transporter	Parra-Lopez et al. (1993)
SsuB	SsuC		(SsuA)	Alkanesulfonate transporter	Koch et al. (2005)
TauB	TauC		(TauA)	Taurine transporter	van der Ploeg et al. (1996)
ThiQ	ThiP		(ThiB)	Thiamin transporter	Keseler et al. (2013)
UgpC	UgpA, UgpE	A/B, A/C, A/E, C/E, (TcyJ), *FtsE*, FtsX, (LivK), (MalE), *MalK,* (PstS) (RbsB), *UvrA*, *ZnuA*	(UgpB)	Glycerol-3-phosphate transporter	Chandravanshi et al. (2016)
XylG	XylH		(XylF)	Fused D-xylose transporter	Keseler et al. (2013)
YadG	YadH	*SufC*, *YbhF,* YbhR		YadG/YadH ABC transporter	(Daley et al. 2005)
YbbA	YbbP	*YrbF*		YbbA/YbbP ABC transporter	Daley et al. (2005)
YbhF (f)	YbhR, YbhS	F/R, F/S, R/S, MetI, YadH,		Putative ABC-type transporter	Moussatova et al. (2008)
YcjV[c]	YcjO, YcjP		(YcjN)	Predicted sugar transporter subunit	Mukherjee et al. (2019)
YdcT	YdcU, YdcV	T/U, T/V	(YdcS)	Predicted spermidine/putrescine transporter subunit	Saier Jr. et al. (2014)
YddA	**YddA**			Fused predicted multidrug transporter	

(continued)

Table 2.3 (continued)

NBS	TMS	Physical interactors (ABC proteins)	SBP	Biological function	References
YecC/ TcyN	YecS/ TcyL	**CydC, CydD,** *MglA, RbsA*	(TcyJ/ FliY)	Predicted transporter	Saier Jr. et al. (2016)
YehX	YehY, YehW	Y/Y, *GlnQ, MetN, YbiT,* DppB,	(YehZ)	Predicted transporter	Kertesz (2001)
YejF(f)	YejB, YejE	B/F, B/E, E/F, *OppD,* OppC, *GsiA*	(YejA)	Fused predicted oligopeptide transporter	Checroun and Gutierrez (2004)
YhdZ	YhdX, YhdY		(YhdW)	Predicted amino-acid transporter	Linton and Higgins (1998)
YhhJ	**YhhJ**			Predicted transporter	Liu et al. (2018)
YnjD	YnjC (f)		(YnjB?)	Predicted transporter	Moussatova et al. (2008), Saier Jr. et al. (2016)
YojI	**YojI**			Fused predicted multidrug transport	Feng et al. (2018)
YphE	YphD		(YphF)	Fused predicted sugar transporter	Feng et al. (2018)
YrbF/ MlaF	YrbE/ MlaE	F/D[d], (DdpA), DppB, (TcyJ), **MsbA**		Predicted toluene transporter	Malinverni and Silhavy (2009)
YtfR	YtfT, YjfF		(YtfQ)	Predicted sugar transporter	Horler et al. (2009)
ZnuC	ZnuB	B/C, *GsiA*	(ZnuA)	Zinc transporter	Patzer and Hantke (1998)

[a]AfuB and PhnE are severely truncated in *E. coli* strain K12 and likely to be nonfunctional. Hence *AfuC and PhnC* have become orphan NBDs in this strain. In other non-lab *E. coli* strains, AfuB and PhnE both appear to be Type I ABC transporter permeases with their two TMDs fused
[b]LivFGHM and ArtMPQ have two substrate-binding proteins in their operons
[c]*YcjV* is unusual in its operon by being separated by five genes from the nearest TMD, YcjP
[d]YrbD is the periplasmic adaptor protein

The short linker of AlsA does not show any significant coevolution with other residues, but Q152, which is close to the signature sequence in NBD1, and R438, which is close to the Walker B region in the opposing NBD, points to possible allosteric effects in the fused NBDs. Similarly, K13, which is in the first beta strand and sits close to the aromatic residue (Aro, Fig. 2.2) that pi-pi stacks with the adenine ring of ATP, is linked with K445, which lies just after the Walker B region in the opposing NBD.

Quaternary structure of protein complexes is still a challenge for modern prediction algorithms, even AlphaFold. Hence the reliability of models for functional ABC transporters still depends on structural characterisation by X-ray crystallography and cryo-EM. TMD–TMD fusions in *E. coli* K12 all interact with a *single* NBD

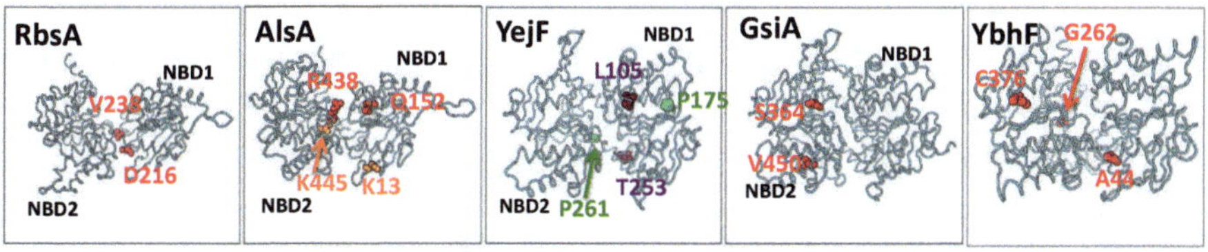

Fig. 2.5 Coevolution in fused NBD–NBD subunits. Coevolution of residues within different proteins were analysed using the BIS2Analyzer server: the polypeptide chains are shown in grey ribbon representation and top-scoring residues are displayed in spacefill representation, with different colours for different clusters of coevolving residues. For RbsA, V283 in the linker is associated with D216 in the penultimate beta strand of NBD1 (close to the Walker A region). For AlsA, no linker residues were identified as significant; however, Q152, close to the signature sequence in NBD1, and R438, close to the Walker B region in NBD2, are significantly correlated. Similarly, K13 in the first β-strand of NBD1, and K445, just after the Walker B region, are identified. In YejF, L105 is in the Q-loop region and T253 is in the linker. P175 is in a turn preceding the Walker B region and P261 is also in the linker. For GsiA, S364 is in the Walker A region and V450 is in the helical domain of NBD2. YbhF shows coevolution of G262 in the predicted structured linker with A44 in the NBD1 Walker A region and C376 which immediately follows the Walker A region in NBD2

component, meaning that they will be akin to the 3-core subunit transporters (Table 2.3) where asymmetry exists: here the single NBD subunit will associate slightly differently with the two fused and non-identical halves of the TMD. For NBD–NBD fusions, two of them (GsiA and YbhF) interact with two separate TMD subunits (Table 2.3), so it follows that these are akin to the 4-core subunit transporters: here one would expect each half of the NBD fusion to interact specifically with only one of the TMD subunits.

AlphaFold may be able to predict existing folds of ABC transporter TMDs with some confidence, and the predictions are shown in Fig. 2.6 for *E. coli* K12, with categorisation into the various types proposed in the new classification system. This laboratory-coddled strain of *E. coli* does not appear to have any type III ABC transporters as well as losing two TMDs due to truncation mutations. AlphaFold predicts that some type II ABC transporters have a function in sugar import, for example, YtfFRT (galactofuranose), XylGH (xylose) and RbsAC (ribose). To some extent, this goes against expectations for type II ABC importers that are thought to be predominantly involved in import of rare co-factors and ions. Conversely the nickel and molybdenum importers are Type I.

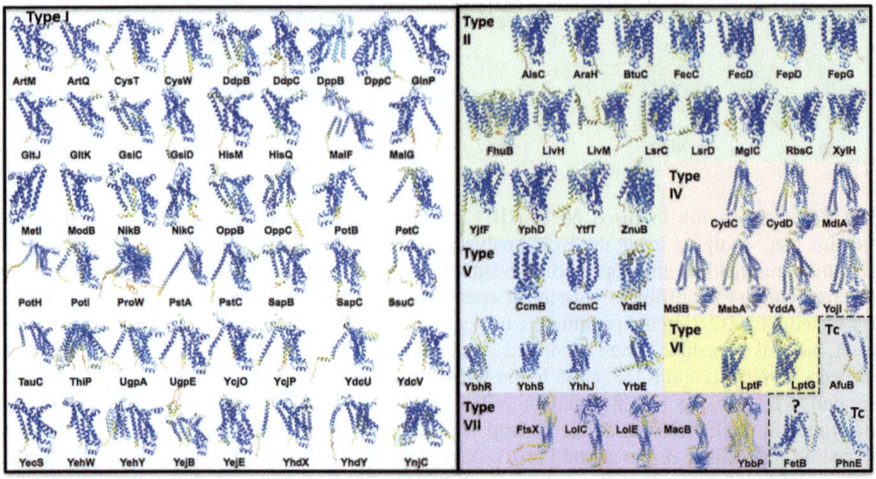

Fig. 2.6 Classification of *E. coli* K12 ABC transporter TMDs. AlphaFold predictions for all TMDs are shown, including fused TMD–TMD subunits. Classification into the seven types proposed by Thomas et al. (2020) is indicated. AfuB and PhnE are truncated (Tc) in *E. coli* K12 and are presumably inoperative (bottom right). One TMS (FetB) has an AlphaFold prediction that does not fit within any of the seven experimentally determined types of folds (bottom right). TMDs are coloured according to confidence level with blue, very high; cyan, high; yellow, low and red, very low confidence

AlphaFold Predictions of All *E. coli* K12 ABC Transporter Structures

The AlphaFold artificial intelligence (AI) server hosted by the European Bioinformatics Institute (EBI) was accessed to evaluate predictions for all *E. coli* ABC transporters and to look for novel TMD folds. The reliability of the AlphaFold predictions for ABC transporters can be assessed by comparison of the root mean squared de*via*tion (RMSD) between existing experimentally determined structures of ABC transporters and the AlphaFold predictions. For many of the experimentally determined structures, there is not yet a directly comparable AlphaFold model (i.e. with sequence identity of 100%), but fortunately, there are sufficient examples with 100% identity to assess the software's success. Table 2.4 illustrates this structural comparison between predicted and experimentally determined structures for individual core subunits as well as fused TMD–NBD proteins. It is clear that AlphaFold models are not always exactly the same as the experimentally determined structure, even where there is a direct comparison possible and where 100% sequence identity exists. This lack of exact structural identity is probably a reflection of the 'fuzzy' modelling algorithms, but also could reflect natural flexibility within the proteins and any structural constraints placed by the crystallisation conditions and the presence of other subunits within the complex. For example, the prediction of the MalG structure appears good (RMSD of 2.8Å), whilst MalF, less so (RMSD

Table 2.4 The Cα root-mean-square de*via*tion (RMSD) value highlighted in green between experimentally determined structures of ABC transporters (left column) and the closest predicted structure from AlphaFold. AlphaFold predicts each monomer by fetching the individual similar protein models from UniProt, listed under the column UniProt Name. The percentage identity (% Id) in the amino acid sequences are also shown. The smaller the RMSD value, the greater the similarity of the two models, and this is roughly correlated with a higher % Id. Where two RMSD values are given for the same protein, each corresponds to different molecules in the crystal asymmetric unit. For type III (ECF) transporters, the TMS comparisons were done with the T component

ABC Transporter	UniProt Name NBS	RMSD (Å)	% Id	UniProt Name TMS	RMSD (Å)	% Id
MalFGK2(-E)	MALK_ECOLI	2.16	100	MALF_ECOLI	7.83, 4.66	100
				MALG_ECOLI	2.75	100
ModB2C2(-A)	WTPC_METJA	4.16	53	MODB_ECOLI	5.92	33
MetNI(-Q)	METN_ECOLI	4.37	100	METI_ECOLI	2.18	100
Art(QN)2	TCYN_ECOLI	1.64	58	Q2FX86_STAA8	2.76	49
AlgM1M2SS-Q2	YCJV_ECOLI	2.48	54	(M1) MALG_ECOLI	7.07	29.3
				(M2) MALG_ECOLI	5.40	32
BtuC2D2(-F)	BTUD_ECOLI	0.40	99	BTUC_ECOLI	1.59	98
H10796	DDPF_ECOLI	4.92	26	BTUC_ECOLI	1.91	97.9
HmuUV	AB2F_ARATH	9.24	24	FECD_ECOLI	2.83	38
BhuUV(-T)	O06138_MYCTU	6.60	37	FEPD_ECOLI	2.45	35
EcfTAA'-FolT	ECFA1_STAA8	2.56	50	Q2FW36_STAA8	2.71	38
	ECFA2_STAA8	1.64	41			
LbECF-PanT	ECFA2_STAA8	1.94	43	Q2FW36_STAA8	4.99	43
	Q5ZD09_ORYSJ	5.58	40			
EcfTAA'-PdxU2	ECFA2_STAA8	1.79	43	Q2FW36_STAA8	3.98	41
	Q5ZD09_ORYSJ	4.82	41			
CbiMQO	ECFA_METJA	2.22	47	CBIM_METJA	4.07	55
	ECFA_METJA	4.00	45			
ECF-CbrT	ECFA1_STAA8	3.19	50	Q2FW36_STAA8	6.08	41
	ECFA2_STAA8	3.02	41			
LptB2FG-LptC	LPTB_ECOLI	1.51	95	LPTF_ECOLI	5.48	88
				LPTG_ECOLI	1.38	96
Wzm-WztN	Y873_METJA	6.52, 8.68	29	YADH_ECOLI	6.15	24
TarGH	TAGH_STAA8	3.01	53	YADH_ECOLI	7.37	25
		TMD-NBD 1			TMD-NBD 2	
Sav1866	Y2003_STAA8	2.88	100			
MsbA	MSBA_ECOLI	6.14	96			
NaAtm1	ABCB7_MOUSE	5.59	45			
TM287/288	ABCBB_RAT	5.71	25	ABCB2_DICDI	5.31	29
McjD	K7KAE3_SOYBN	7.31	23			
PCAT1	MDR3_HUMAN	11.30	27	MDR3_HUMAN	10.35	28
PglK	K7MEJ8_SOYBN	10.90	23			
TmrAB	AB14B_ARATH	5.04	28	MITOS_HUMAN	4.29	32
PrtD	I1N600_SOYBN	6.89	23			
YbtPQ	DA1D6J0E8_MAIZE	6.07	26	Q2FVJ2_STAA8	7.17	27
Rv1819c	BACA_MYCTU	1.60	100			
IrtAB	IRTA_MYCTU	2.63	74	IRTB_MYCTU	5.57	69
MacB	MACB_ECOLI	9.20	54			

7.8 and 4.7Å), Table 2.4. Examination of the MalF discrepancies shows that the largest variation is for the N-terminal MalF transmembrane region and periplasmic domain for which there is no equivalent in MalG. The overall structure and topology of this N-terminal region is correctly predicted by AlphFold, but the whole region is rotated as a solid body versus the core of the MalF protein when compared to the experimentally determined structure.

The conclusion from this exercise is that AlphaFold appears to be accurate at predicting protein topology/fold in ABC transporters, although the relationship between separate folded domains (where they exist) may differ from experimentally determined structures. Clearly, AlphaFold will, to some extent, rely on existing structural information; however, it should be pointed out that the structure for SbmA became available after the release of the AlphFold *E. coli* K12 predictions (Ghilarov et al. 2021). In this case, SbmA's topology was correctly predicted by AlphaFold before its publication, and the overall RMSD between predicted and experimental structures is 6.8Å.

Novel TMD Folds

Does AlphaFold predict any novel folds in the *E. coli* K12 set of ABC transporters? The stamp album of AlphaFold-predicted TMDs of *E. coli* K12 transporters in Fig. 2.6 shows that nearly every TMD falls into one of the seven categories of fold although type III ABC transporters are significant by their absence in this strain. There appears to be just one predicted new fold in *E. coli* K12, which is the model for FetB. The FetB TMS model has a distinct 'EAA' connecting loop (EKL in FetB) between TM5 and 6 that is presumably contacting the Q-loop region in the cognate NBS (FetA). This cytoplasmic loop and its connecting TM spans in FetB are strongly reminiscent of the type I ABC transporter fold, and over this region, and there is a good structural alignment with, for example, the ModB protein (rmsd of 4.3Å across 65 Cα atoms). However, the topology of TM helices 1–4 and 7 in the FetB model differs radically from that of the type I ABC transporters (overall rmsd versus ModB is 18.8Å across all 216 Cα comparisons, much higher than any value shown in Table 2.4). Functionally, the FetAB transporter is predicted to be an exporter (of iron) (Nicolaou et al. 2013) that allows resistance to oxidative stress in the presence of iron. Its role in *E. coli* K12 remains to be elaborated. The topology of the FetB fold is illustrated in more detail in Fig. 2.7. FetB has no homologues in the sequences of proteins deposited in the protein data bank (PDB), in support of the novelty of its fold. However, a psi-BLAST search (Altschul and Koonin 1998) of the Uniprot-annotated sequences reveals prokaryotic homologues in *Bacillus subtilis* (Altschul and Koonin 1998), *Methanocaldococcus janaschii* (a thermophile) and *Synechocystis* sp. 6803 (a cyanobacterium). Perhaps more revealingly, homologues in plants are also identified (*Arabidopsis thaliana* and rice) that have been associated with conferring tolerance to Aluminium (Al^{3+}) in acidic soils (Huang et al. 2009). The FetB homologue in plants is termed STAR2/ABCI16, and it associates with an

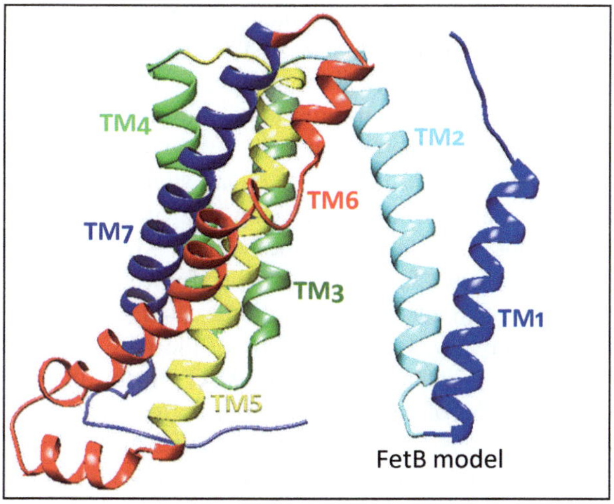

Fig. 2.7 AlphaFold-modelled novel topology for FetB is indicated with the text and colours. The closest 3D fit is type I, with transmembrane spans (TM) 3&4 and the EAA loop of, e.g. ModB aligning with TM5&6 of FetB. However, with this alignment, the topology of FetB is quite different, with FetB TM7 in the same position as TM1 of ModB, whilst ModB TM5 has no equivalently placed span in FetB (see also Fig. 2.2b)

NBS termed STAR1/ALS1/ABCI12, although their categorisation within the ABCI sub-family may be incorrect as the STAR2 sequence and AlphaFold prediction do not bear similarity to known ECF T nor ECF S subunits (see Fig. 2.1 and section "Type III ABC Transporters or Energy Coupling Factor Transporters (ECF)"). STAR1 appears to have a long N-terminal extension that may represent a targeting sequence, although the TargetP-2.0 server does not give a high score for either mitochondrial or chloroplast targeting (Emanuelsson et al. 2007). Similarly, STAR2 does not score highly for targeting to the chloroplast, nor the mitochondrion. In conclusion, it seems possible that FetB/STAR2 represents a novel TMD fold and that future efforts to resolve the FetAB structure should be of considerable interest.

Promiscuity in ABC Transporter Subunit Interactions and Assembly

As discussed in previous sections of this chapter, considerations of NBD conservation versus TMD plasticity give rise to questions about the degree of promiscuity in NBS–TMS interactions in the prokaryotic cell. For example, how often do NBSs interact with the incorrect TMS platform or the incorrect partner NBS? From prior sections on thermodynamic considerations, promiscuous interactions are predicted to occur. We asked ourselves whether a review of the interactome databases would pick up such promiscuity, and again the *E. coli* K12 strain provided an experimental playground for this: Table 2.3 shows a list of *E. coli* ABC transporters, with the core components of the NBS(s) and the TMS(s) listed in columns 1 and 2. Interactome data from the BioGrid database for *E. coli* strain K 12 is shown in column 3. This lists any (expected) pairwise interactions between components of the given ABC transporter complex (capital letters separated by a slash). It also lists other (unexpected) *E. coli* ABC proteins that have been found to interact with any of the proteins in columns 1 and 2. The data in the table are restricted to physical interactions only, and are based on either yeast in vivo two-hybrid probes, or, more extensively, on in vitro studies using affinity-tagged proteins as bait and an *E. coli* extract as the test sample.

For about half the ABC proteins studied, the correct, cognate NBS or TMS subunits were identified as being one of the interacting proteins. However, for many transporters, this was not the case. Although somewhat surprising, it is likely that experimental factors explain this: for example, the position of the affinity tag on the bait protein may account for some cases where there was no detection of a binary complex in vitro. However, many of the ABC transporter NBS and TMS subunits were found to interact with other ABC proteins that would not normally be considered to be part of the functional transporter complex. Sometimes, these interactions are with closely related transporters (in terms of function): for example, Arginine (ArtMPQ) and Histidine (HisMPQ) transporter subunits can interact with their cousin under the experimental constraints of the proteomics experiments. Other interactions that were discovered are harder to rationalise, however: for example, subunits from the mechanotransducing ABC tranporters LolCDE and FtsEX appear to interact with large cohorts of functionally unrelated ABC transporter subunits, including some SBPs, whilst MacB, another mechanotransducer, displayed no unexpected physical interactions with other ABC transporter subunits. Not shown in Table 2.3, column 3 are non-ABC proteins that are found to interact with the ABC transporter subunits *via* these two methods and are documented in BioGrid. Reassuringly, many of these other interactors appear to be physiological and include components of the translation and folding machinery as well as (for the TMDs) components of the membrane insertion (Sec) machinery.

Questions arise from examination of the ABC transporter interactome data summarised in Table 2.3: are the interacting ABC proteins listed in column 3 examples of incorrect molecular recognition? Or do the data indicate natural and useful promiscuity in ABC transporter complex formation within the cell? If the *latter* is correct, then it would imply that a huge mixture of ABC components can assemble to form a range of alternative complexes. This could allow an expanded range of functions and/or substrates to be transported. If the f*ormer* is correct, (mistaken recognition within the cell), then one must ask how prevalent this defective behaviour is under typical cellular conditions and whether it can be corrected: for example, is there a group of proteins that recognise incorrectly assembled ABC transporter complexes and target them for disassembly and/or degradation? A third question must also be posed, *procaciter*: are interactomics data, in general, (or at least for ABC transporters) likely to point to some interactions that only exist in vitro and hence cannot be taken as reflective of the situation in the cell?

Orphan/Multitask NBSs and Promiscuity of Complex Formation

ABC transporter components are usually encoded by genes in operons or clustered together in bacterial genomes, and their expression is frequently regulated in a coordinated manner (see later section). However, in some ABC transporters, genes encoding the substrate-binding proteins (SBPs) and permease subunits (TMSs) occasionally lack genes encoding the nucleotide-binding subunits (NBSs) in their proximity. Similarly, genes encoding ABC transporter NBSs sometimes lack any obvious TMS gene in their vicinity on the given bacterial genome. For example, the *Bacillus subtilis* genome encodes five ATP-binding proteins without any gene coding for TMS protein in their neighbourhood (Quentin et al. 1999). Only one of the five orphan NBSs, MsmX (PDB id: 6YIR), has been characterised structurally so far (Leisico et al. 2020).

MsmX is thought to energise six incomplete carbohydrates importers in *B. subtilis*: arabino-oligossacharide (AraNPQ), maltodextrin (MdxEFG), galacto-oligossacharids (GanSPQ), galacturonic acid oligomer and/or rhamnose-galacturonic acid disaccharide (YesOPQ and YtcPQ-YteP) transporters (Ferreira and Sá-Nogueira 2010; Ferreira et al. 2017; Morabbi Heravi et al. 2019). Because of its ability to engage with and energise several TMSs from different ABC transporters, MsmX was given the term ‘multitask ATPase’ (Ferreira and Sá-Nogueira 2010). Although multifunctional ABC ATPases have been discovered in other bacteria such as *Streptomyces lividans* and *Streptococcus suis*, the multipurpose ABC NBS phenomenon is rare (Hurtubise et al. 1995; Tan et al. 2015). To date, multipurpose NBSs have been found only in bacterial ABC sugar importers belonging to the carbohydrate uptake transporter 1 (CUT1) subfamily, which transport a variety of di-, tri- and higher oligosaccharides as well as polyols (Saurin et al. 1999; Saier Jr 2000).

The genome of the pathogen *Streptococcus pneumoniae* TIGR4 encodes six incomplete carbohydrate ABC importers that lack the ATPase required to energise the transport. It was discovered that the orphan ATPase MsmK, which is encoded by a gene at a different location in the genome, energises four of the six importers (Marion et al. 2011; Linke et al. 2013). MsmK is functionally exchangeable with MsmX from *Bacillus subtilis* that was discussed above (Ferreira and Sá-Nogueira 2010). Three orphan genes for NBDs have been discovered in *Listeria monocytogenes*. Lmo0278 is the most likely equivalent of MsmX, as it shares 72% identity and 83% similarity with it (Gopal et al. 2010).

Co-Translational Assembly of ABC Transporter Complexes

The idea that protein complex assembly can occur co-translationally rather than *via* random encounters of already folded subunits has begun to gain ground (Kramer et al. 2019). Formation of simple complexes of proteins during translation on ribosomes was demonstrated in 2002 using a system of chaperone-assisted folding of Caspase complexes (Sakahira and Nagata 2002). The idea that co-translational folding of multi-domain eukaryotic proteins was important for the correct assembly of the separate domains had already been discussed (Netzer and Hartl 1998). In subsequent years, larger and more intricate complexes were studied: for example, for the histone methylase complex, the individual protein subunits (and mRNA component) were shown to assemble together co-translationally (Halbach et al. 2009). A wider study of co-translational assembly of protein complexes was carried out in yeast, using co-purification of mRNA species that encoded proteins that were partners in various complexes (Duncan and Mata 2011). A surprisingly large number of the proteins examined (38%) were found to show this behaviour. A ribosome profiling study in yeast suggested an even higher proportion of complexes that assembled co-translationally (75%), whilst the exceptions (25%) needed chaperones to assist assembly (Shiber et al. 2018). That co-translational assembly of complexes could also apply to membrane proteins was demonstrated in various studies, such as in a study of heteromeric human ether à-go-go (hERG) ion channels (Liu et al. 2016).

The operon structure in prokaryotes, where proteins within a given complex tend to have their genes clustered near to each other on the chromosome, may be taken as support for the role of co-translational assembly of complexes in prokaryotes. This has been explored recently in *E. coli* for bacterial luciferase genes that were tagged with yellow fluorescent protein (YFP) or cyan fluorescent protein (CFP) to allow fluorescence resonance energy transfer (FRET) to monitor their physical interactions in vivo. The results suggested that co-localisation of the genes and (by implication) co-translational assembly of the LuxA and LuxB proteins significantly increased (2×) the efficiency of LuxA/B heterodimer assembly compared to a strain where LuxA and LuxB could assemble only *via* diffusion through the cytoplasm. Separate experiments also suggested that when their genes were co-localised, LuxA was able to assemble with partner LuxB after only 120 out of the total 325 total residues of the LuxB chain were synthesised (Shieh et al. 2015). Co-co- assembly of homodimers (or larger homomers) has been described recently (Bertolini et al. 2021). Here, direct interactions occur between nascent polypeptide chains being synthesised on two or more nearby ribosomes.

Operon Structure for ABC Transporter Complexes

As detailed above, co-translational assembly of ABC transporter complexes is dependent on co-localisation of nascent polypeptide chains as they emerge from linked ribosomes (polysomes). Hence ABC transporter gene-containing operons' organisation may be particularly revealing in this respect. A systematic review of the PATRIC database for the operons of 61 ABC transporters of *E. coli* K-12 revealed common features in the organisation of the ABC transporter genes that can be considered within the context of the core subunits of the complexes. A summary is given in Table 2.5. NBSs, TMSs and SBP (where present) forming a particular ABC transporter are mostly positioned on the same operon and on the same strand, in agreement with the idea of co-translational assembly. The exceptions are, for instance, MalFGK, ZnuBC, SsuBC, YcjCD, FtsEX and BtuCD. Two have operons where TMS, NBS and SBP genes are clustered, but one component is on a different DNA strand (ZnuBC and $MalFGK_2$). The rest have the SBP remotely located relative to the TMS and NBS subunits. As SBP components are likely to reversibly associate with the extracytoplasmic surface of their cognate TMDs, operons with remote SBP genes do not pose a problem for the co-translational assembly hypothesis. Interestingly, the MalK operon structure resembles that of other three core-component transporters, but as if an inversion happened at some point (Fig. 2.8).

In most of the cases, NBSs and TMSs are adjacent to each other, but there are exceptions in which TMS and NBS are separated by the SBP (GsiD, ArtP, FhuC, PhnC) or other genes (BtuCD, FetAB). In general, it can be summarised that for two-component core domains (consisting of just one NBS gene and one TMS gene), type II importers mostly have NBS first, TMS second, while type I importers can have either order. Three-component systems (with one NBS gene and two TMS genes) could have both NBS first, then the TMSs or vice versa, regardless of the transporter type, but the two TMS genes are always in tandem with just one exception, YehWXY. This transporter's operon has the NBS in between the two TMSs. Interestingly, four-component operons with two NBS genes and two TMS genes are always organised with the SBP gene followed by the two TMS genes and then the two NBS genes. Frustratingly, this provides no obvious clues about which NBS might associate with which TMS in these four-component systems. A few ABC transporter operons are, themselves, positioned in clusters in the *E. coli* K12 genome, for instance LivFGHM and UgpACE are in the same vicinity.

In summary, as for many essential cellular processes in bacteria, the operon organisation will be important for regulation and coordination of these processes. For ABC transporters, it would be important to allow coordination of responses to deficits or surpluses of metabolites and xenobiotics. However, recent work would also support the argument that the operon structure for ABC transporters points to a possible role for the operon organisation in co-translational assembly of the correct subunits in the ABC transporter complex. It seems reasonable to propose that confining the process of complex formation to a small region within the cell will

Table 2.5 ABC transporter operon structures. Type I importers are in **bold**, type II importers are in *italics*, type IV exporters are highlighted, type V transporters have a wave underline, the type VI transporter is in brackets, type VII transporters are underlined

Operon structure (subunits or domains in the same order they appear in the operon)	**ABC transporters of different types** (subunit letters in the same order they appear in the operon)
Single core gene transporters	
TMD+NBD	MsbA, YddA, YhhJ, YojI,
NBD+TMD	MacB
Two core gene transporters	
TMS, NBS	**ThiBPQ, AfuBC, ModABC, YnjBCD, YecSC, SsuACB**, GlnHPQ
NBS, TMS	**MetNIQ,** YbbAP, *AraFGH, AlsBAC, XylFGH, YphFED, RbsACB*, **ProVWX,** YadGH, CcmAB, **TauABC,** FetAB, *MglBAC,* FtsEX, MlaFE=YrbFE, *ZnuCB*
NBS, SBP, TMS	*FhuCDB,* **PhnCDE**
SBP, big gap, TMS, small gap, NBS	*BtuCDF*
TMD+NBD, TMD+NBD	MdlAB, CydCD
Three core gene transporters	
SBP, TMS1, TMS2, NBS	**PstSCAB, HisJQMP, CysPTWA,** *FecBCDE,* **GltIJKL, UgpBAEC, YejABEF, YhdWXYZ**
SBP, small gap, TMS1, TMS2, NBS	*FepBCDG*
NBS, TMS1, TMS2	*LsrACD,* **PotFGHI, YdcSTUV,** YbhGFSR, **PotABCD,** *YtfQRT-YjfF*
NBS, big gap, TMS1, TMS2	(LptBFG)
NBS, SBP1, TMS1, TMS2	**ArtPIQMJ, MalFGK*, GsiABCD**
TMS1, NBS, TMS2	**YehWXYZ,** LolCDE
TMS1, TMS2, small gap, NBS	**YcjNOP**
Four core gene transporters	
SBP, TMS1, TMS2, NBS1, NBS2	**DdpABCDF, DppABCDF,** *LivJ-LivKHMGF,* **SapABCDF, OppABCDF, NikABCDE**

[a]MalFGK operon has the gene for the NBD (MalK) on the opposite strand

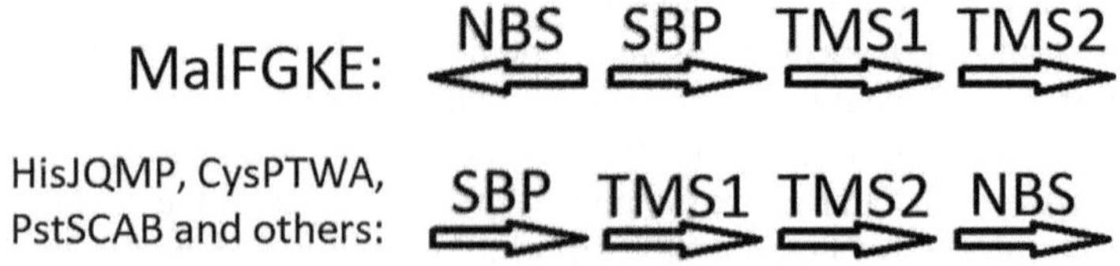

Fig. 2.8 Comparison of the organisation of MalK and other three-component operons

minimise incorrect interactions between TMSs and NBSs of other ABC transporters as well as greatly enhancing the efficiency of assembly of the correct partners.

Conclusions

The variety of ABC transporter complexes in prokaryotic and eukaryotic organisms indicates that the bolting on of energy-release modules (NBDs) to permease modules (TMDs) has been a frequent event during evolution. The abundance of ABC transporter subunits in prokaryotes may pose problems for efficient molecular recognition and assembly of the functional complexes. Indeed proteomics readily identifies promiscuous interactions between ABC transporter subunits in vitro and in vivo. Textbook models for complex assembly *via* diffusion of pre-folded subunits may not be sufficient to explain these data. Co-translational assembly of ABC transporter subunits can be considered to be more compatible with the apparent high fidelity of ABC transporter assembly under normal physiological conditions. Whilst a few ABC transporter subunits may be naturally promiscuous in terms of their interactions, the co-location of ABC transporter genes, coupled with co-translational assembly processes, seems to ensure the rigour of assembly in most cases in prokaryotic organisms. There are now seven types of ABC transporters that have been identified based on the experimentally determined structures and conserved topological folds of the TMDs. However, even in a well-studied laboratory strain of *E. coli*, it seems possible that more variety exists, hinting at new horizons for ABC transporter research in the vast prokaryotic and eukaryotic landscapes.

Acknowledgements We are grateful to Dr. Kostas Beis (ICSTM, UK) for helpful discussion and for kindly allowing insights into the SbmA structure ahead of publication. We would also like to thank Profs. Lutz Schmitt (Heinrich Heine University, Germany) and Ian Collinson (Bristol University, UK) for answering basic questions about ABC transporters and insertion machinery with equanimity. Lastly, we thank Dr. Steve Prince for reading this chapter prior to submission and Robin Harris for his editing of the book. Iqra Younus is a recipient of a PEEF (Punjab Educational Endowment Fund) scholarship.

Competing Interests The authors declare no competing interests with this work.

References

Adams MD, Wagner LM, Graddis TJ, Landick R, Antonucci TK, Gibson AL, Oxender DL (1990) Nucleotide sequence and genetic characterization reveal six essential genes for the LIV-I and LS transport systems of Escherichia coli. J Biol Chem 265(20):11436–11443

Aiba H, Baba T, Hayashi K, Inada T, Isono K, Itoh T, Kasai H, Kashimoto K, Kimura S, Kitakawa M, Kitagawa M, Makino K, Miki T, Mizobuchi K, Mori H, Mori T, Motomura K, Nakade S, Nakamura Y, Nashimoto H, Nishio Y, Oshima T, Saito N, Sampei G, Horiuchi T et al

(1996) A 570-kb DNA sequence of the Escherichia coli K-12 genome corresponding to the 28.0-40.1 min region on the linkage map. DNA Res 3(6):363–377. https://doi.org/10.1093/dnares/3.6.363

Aller SG, Yu J, Ward A, Weng Y, Chittaboina S, Zhuo R, Harrell PM, Trinh YT, Zhang Q, Urbatsch IL, Chang G (2009) Structure of P-glycoprotein reveals a molecular basis for poly-specific drug binding. Science 323(5922):1718–1722

Altschul SF, Koonin EV (1998) Iterated profile searches with PSI-BLAST--a tool for discovery in protein databases. Trends Biochem Sci 23(11):444–447. https://doi.org/10.1016/s0968-0004(98)01298-5

Babu M, Bundalovic-Torma C, Calmettes C, Phanse S, Zhang Q, Jiang Y, Minic Z, Kim S, Mehla J, Gagarinova A, Rodionova I, Kumar A, Guo H, Kagan O, Pogoutse O, Aoki H, Deineko V, Caufield JH, Holtzapple E, Zhang Z, Vastermark A, Pandya Y, Lai CC, El Bakkouri M, Hooda Y, Shah M, Burnside D, Hooshyar M, Vlasblom J, Rajagopala SV, Golshani A, Wuchty S, Greenblatt JF, Saier M, Uetz P, Moraes TF, Parkinson J, Emili A (2018) Global landscape of cell envelope protein complexes in Escherichia coli. Nat Biotechnol 36(1): 103–112. https://doi.org/10.1038/nbt.4024

Ban N, Freeborn B, Nissen P, Penczek P, Grassucci RA, Sweet R, Frank J, Moore PB, Steitz TA (1998) A 9 A resolution X-ray crystallographic map of the large ribosomal subunit. Cell 93(7): 1105–1115. https://doi.org/10.1016/s0092-8674(00)81455-5

Ban N, Nissen P, Hansen J, Moore PB, Steitz TA (2000) The complete atomic structure of the large ribosomal subunit at 2.4 A resolution. Science 289(5481):905–920. https://doi.org/10.1126/science.289.5481.905

Bardin S, Dan S, Osteras M, Finan TM (1996) A phosphate transport system is required for symbiotic nitrogen fixation by Rhizobium meliloti. J Bacteriol 178(15):4540–4547. https://doi.org/10.1128/jb.178.15.4540-4547.1996

Bell AW, Buckel SD, Groarke JM, Hope JN, Kingsley DH, Hermodson MA (1986) The nucleotide sequences of the rbsD, rbsA, and rbsC genes of Escherichia coli K12. J Biol Chem 261(17): 7652–7658

Berntsson RP-A, ter Beek J, Majsnerowska M, Duurkens RH, Puri P, Poolman B, Slotboom D-J (2012) Structural divergence of paralogous S components from ECF-type ABC transporters. Proc Natl Acad Sci U S A 109(35):13990–13995

Bertolini M, Fenzl K, Kats I, Wruck F, Tippmann F, Schmitt J, Auburger JJ, Tans S, Bukau B, Kramer G (2021) Interactions between nascent proteins translated by adjacent ribosomes drive homomer assembly. Science 371(6524):57–64. https://doi.org/10.1126/science.abc7151

Bi Y, Mann E, Whitfield C, Zimmer J (2018) Architecture of a channel-forming O-antigen polysaccharide ABC transporter. Nature 553(7688):361–365. https://doi.org/10.1038/nature25190

Borths EL, Poolman B, Hvorup RN, Locher KP, Rees DC (2005) In vitro functional characterization of BtuCD-F, the Escherichia coli ABC transporter for vitamin B12 uptake. Biochemistry 44(49):16301–16309. https://doi.org/10.1021/bi0513103

Braibant M, Lefèvre P, de Wit L, Peirs P, Ooms J, Huygen K, Andersen AB, Content J (1996) A Mycobacterium tuberculosis gene cluster encoding proteins of a phosphate transporter homologous to the Escherichia coli Pst system. Gene 176(1–2):171–176. https://doi.org/10.1016/0378-1119(96)00242-9

Bruntner C, Lauer B, Schwarz W, Möhrle V, Bormann C (1999) Molecular characterization of co-transcribed genes from Streptomyces tendae Tü901 involved in the biosynthesis of the peptidyl moiety of the peptidyl nucleoside antibiotic nikkomycin. Mol Gen Genet 262(1): 102–114. https://doi.org/10.1007/pl00008637

Caffalette CA, Corey RA, Sansom MSP, Stansfeld PJ, Zimmer J (2019) A lipid gating mechanism for the channel-forming O antigen ABC transporter. Nat Commun 10(1):824. https://doi.org/10.1038/s41467-019-08646-8

Callaghan R, Ford RC, Kerr ID (2006) The translocation mechanism of P-glycoprotein. FEBS Lett 580(4):1056–1063. https://doi.org/10.1016/j.febslet.2005.11.083

Chai C, Yu Y, Zhuo W, Zhao H, Li X, Wang N, Chai J, Yang M (2013) Structural basis for a homodimeric ATPase subunit of an ECF transporter. Protein Cell 4(10):793–801
Chandravanshi M, Gogoi P, Kanaujia SP (2016) Computational characterization of TTHA0379: A potential glycerophosphocholine binding protein of Ugp ATP-binding cassette transporter. Gene 592(2):260–268. https://doi.org/10.1016/j.gene.2016.07.017
Checroun C, Gutierrez C (2004) Sigma(s)-dependent regulation of yehZYXW, which encodes a putative osmoprotectant ABC transporter of Escherichia coli. FEMS Microbiol Lett 236(2): 221–226. https://doi.org/10.1016/j.femsle.2004.05.046
Chen B, Boel G, Hashem Y, Ning W, Fei J, Wang C, Gonzalez RL Jr, Hunt JF, Frank J (2014) EttA regulates translation by binding the ribosomal E site and restricting ribosome-tRNA dynamics. Nat Struct Mol Biol 21(2):152–159. https://doi.org/10.1038/nsmb.2741
Chen L, Hou WT, Fan T, Liu B, Pan T, Li YH, Jiang YL, Wen W, Chen ZP, Sun L, Zhou CZ, Chen Y (2020) Cryo-electron microscopy structure and transport mechanism of a wall teichoic acid ABC transporter. mBio 11(2):e02749-19. https://doi.org/10.1128/mBio.02749-19
Chenault SS, Earhart CF (1992) Identification of hydrophobic proteins FepD and FepG of the Escherichia coli ferrienterobactin permease. Microbiology 138(10):2167–2171. https://doi.org/10.1099/00221287-138-10-2167
Chin N, Frey J, Chang CF, Chang YF (1996) Identification of a locus involved in the utilization of iron by Actinobacillus pleuropneumoniae. FEMS Microbiol Lett 143(1):1–6. https://doi.org/10.1111/j.1574-6968.1996.tb08452.x
Chng SS, Gronenberg LS, Kahne D (2010) Proteins required for lipopolysaccharide assembly in Escherichia coli form a transenvelope complex. Biochemistry 49(22):4565–4567. https://doi.org/10.1021/bi100493e
Choi CC, Ford RC (2021) ATP binding cassette importers in eukaryotic organisms. Biol Rev Camb Philos Soc 96:1318–1330. https://doi.org/10.1111/brv.12702
Choudhury HG, Tong Z, Mathavan I, Li Y, Iwata S, Zirah S, Rebuffat S, Van Veen HW, Beis K (2014) Structure of an antibacterial peptide ATP-binding cassette transporter in a novel outward occluded state. Proc Natl Acad Sci U S A 111(25):9145–9150
Cole SP, Bhardwaj G, Gerlach JH, Mackie JE, Grant CE, Almquist KC, Stewart AJ, Kurz EU, Duncan AM, Deeley RG (1992) Overexpression of a transporter gene in a multidrug-resistant human lung cancer cell line. Science 258(5088):1650–1654
Corbalan N, Runti G, Adler C, Covaceuszach S, Ford RC, Lamba D, Beis K, Scocchi M, Vincent PA (2013) Functional and structural study of the dimeric inner membrane protein SbmA. J Bacteriol 195(23):5352–5361. https://doi.org/10.1128/JB.00824-13
Crow A, Greene NP, Kaplan E, Koronakis V (2017) Structure and mechanotransmission mechanism of the MacB ABC transporter superfamily. Proc Natl Acad Sci U S A 114(47): 12572–12577. https://doi.org/10.1073/pnas.1712153114
Cruz-Ramos H, Cook G, Wu G, Cleeter M, Poole R (2004) Membrane topology and mutational analysis of Escherichia coli CydDC, an ABC-type cysteine exporter required for cytochrome assembly. Microbiology 150:3415–3427. https://doi.org/10.1099/mic.0.27191-0
Daley DO, Rapp M, Granseth E, Melén K, Drew D, Von Heijne G (2005) Global topology analysis of the Escherichia coli inner membrane proteome. Science 308(5726):1321–1323
Davis JJ, Wattam AR, Aziz RK, Brettin T, Butler R, Butler RM, Chlenski P, Conrad N, Dickerman A, Dietrich EM, Gabbard JL, Gerdes S, Guard A, Kenyon RW, Machi D, Mao C, Murphy-Olson D, Nguyen M, Nordberg EK, Olsen GJ, Olson RD, Overbeek JC, Overbeek R, Parrello B, Pusch GD, Shukla M, Thomas C, VanOeffelen M, Vonstein V, Warren AS, Xia F, Xie D, Yoo H, Stevens R (2020) The PATRIC Bioinformatics Resource Center: expanding data and analysis capabilities. Nucleic Acids Res 48(D1):D606–D612. https://doi.org/10.1093/nar/gkz943
Dawson RJ, Locher KP (2006) Structure of a bacterial multidrug ABC transporter. Nature 443(7108):180–185
Dawson RJ, Locher KP (2007) Structure of the multidrug ABC transporter Sav1866 from Staphylococcus aureus in complex with AMP-PNP. FEBS Lett 581(5):935–938

de Leeuw E, Graham B, Phillips GJ, ten Hagen-Jongman CM, Oudega B, Luirink J (1999) Molecular characterization of Escherichia coli FtsE and FtsX. Mol Microbiol 31(3):983–993. https://doi.org/10.1046/j.1365-2958.1999.01245.x

Dean M, Allikmets R (2001) Complete characterization of the human ABC gene family. J Bioenerg Biomembr 33(6):475–479

Dean M, Annilo T (2005) Evolution of the ATP-binding cassette (ABC) transporter superfamily in vertebrates. Annu Rev Genomics Hum Genet 6:123–142. https://doi.org/10.1146/annurev.genom.6.080604.162122

Dean M, Rzhetsky A, Allikmets R (2001) The human ATP-binding cassette (ABC) transporter superfamily. Genome Res 11(7):1156–1166. https://doi.org/10.1101/gr.184901

Di Bartolo ND, Hvorup RN, Locher KP, Booth PJ (2011) In vitro folding and assembly of the Escherichia coli ATP-binding cassette transporter, BtuCD. J Biol Chem 286(21):18807–18815. https://doi.org/10.1074/jbc.M110.176891

Dohme F, Nierhaus KH (1976) Total reconstitution and assembly of 50 S subunits from Escherichia coli Ribosomes in vitro. J Mol Biol 107(4):585–599. https://doi.org/10.1016/s0022-2836(76)80085-x

Duncan CD, Mata J (2011) Widespread cotranslational formation of protein complexes. PLoS Genet 7(12):e1002398. https://doi.org/10.1371/journal.pgen.1002398

Eckey V, Landmesser H, Schneider E (2010) Studying subunit-subunit interactions in a bacterial ABC transporterby in vitro assembly. Biochim Biophys Acta 1798(6):1250–1253. https://doi.org/10.1016/j.bbamem.2010.03.001

Eitinger T, Rodionov DA, Grote M, Schneider E (2011) Canonical and ECF-type ATP-binding cassette importers in prokaryotes: diversity in modular organization and cellular functions. FEMS Microbiol Rev 35(1):3–67

Elferink RPO, Paulusma CC (2007) Function and pathophysiological importance of ABCB4 (MDR3 P-glycoprotein). Pflügers Arch Eur J Physiol 453(5):601–610

Emanuelsson O, Brunak S, von Heijne G, Nielsen H (2007) Locating proteins in the cell using TargetP, SignalP and related tools. Nat Protoc 2(4):953–971. https://doi.org/10.1038/nprot.2007.131

Erkens GB, Berntsson RP, Fulyani F, Majsnerowska M, Vujičić-Žagar A, Ter Beek J, Poolman B, Slotboom DJ (2011) The structural basis of modularity in ECF-type ABC transporters. Nat Struct Mol Biol 18(7):755–760

Erkens GB, Majsnerowska M, ter Beek J, Slotboom DJ (2012) Energy coupling factor-type ABC transporters for vitamin uptake in prokaryotes. Biochemistry 51(22):4390–4396

Fabre L, Bao H, Innes J, Duong F, Rouiller I (2017) Negative stain single-particle EM of the maltose transporter in nanodiscs reveals asymmetric closure of MalK(2) and catalytic roles of ATP, MalE, and Maltose. J Biol Chem 292(13):5457–5464. https://doi.org/10.1074/jbc.M116.757898

Feng Y, Chen CS, Ho J, Pearce D, Hu S, Wang B, Desai P, Kim KS, Zhu H (2018) High-throughput chip assay for investigating Escherichia coli interaction with the blood-brain barrier using microbial and human proteome microarrays (dual-microarray technology). Anal Chem 90(18):10958–10966. https://doi.org/10.1021/acs.analchem.8b02513

Ferreira MJ, Sá-Nogueira I (2010) A multitask ATPase serving different ABC-type sugar importers in Bacillus subtilis. J Bacteriol 192(20):5312–5318

Ferreira MJ, Mendes AL, de Sa-Nogueira I (2017) The MsmX ATPase plays a crucial role in pectin mobilization by Bacillus subtilis. PLoS One 12(12):e0189483

Finkenwirth F, Neubauer O, Gunzenhäuser J, Schoknecht J, Scolari S, Stöckl M, Korte T, Herrmann A, Eitinger T (2010) Subunit composition of an energy-coupling-factor-type biotin transporter analysed in living bacteria. Biochem J 431(3):373–381

Finkenwirth F, Sippach M, Landmesser H, Kirsch F, Ogienko A, Grunzel M, Kiesler C, Steinhoff H-J, Schneider E, Eitinger T (2015) ATP-dependent conformational changes trigger substrate capture and release by an ECF-type biotin transporter. J Biol Chem 290(27):16929–16942

Fisher DJ, Fernández RE, Adams NE, Maurelli AT (2012) Uptake of biotin by Chlamydia spp. through the use of a bacterial transporter (BioY) and a host-cell transporter (SMVT)
Fitzpatrick AWP, Llabres S, Neuberger A, Blaza JN, Bai XC, Okada U, Murakami S, van Veen HW, Zachariae U, Scheres SHW, Luisi BF, Du D (2017) Structure of the MacAB-TolC ABC-type tripartite multidrug efflux pump. Nat Microbiol 2:17070. https://doi.org/10.1038/nmicrobiol.2017.70
Ford RC, Beis K (2019) Learning the ABCs one at a time: structure and mechanism of ABC transporters. Biochem Soc Trans 47(1):23–36. https://doi.org/10.1042/BST20180147
Ford RC, Hellmich UA (2020) What monomeric nucleotide binding domains can teach us about dimeric ABC proteins. FEBS Lett 594(23):3857–3875. https://doi.org/10.1002/1873-3468.13921
Ford RC, Marshall-Sabey D, Schuetz J (2020) Linker domains: why ABC transporters 'live in fragments no longer'. Trends Biochem Sci 45(2):137–148. https://doi.org/10.1016/j.tibs.2019.11.004
Freinkman E, Okuda S, Ruiz N, Kahne D (2012) Regulated assembly of the transenvelope protein complex required for lipopolysaccharide export. Biochemistry 51(24):4800–4806. https://doi.org/10.1021/bi300592c
French JB, Zhao H, An S, Niessen S, Deng Y, Cravatt BF, Benkovic SJ (2013) Hsp70/Hsp90 chaperone machinery is involved in the assembly of the purinosome. Proc Natl Acad Sci U S A 110(7):2528–2533. https://doi.org/10.1073/pnas.1300173110
Ghilarov D, Inaba-Inoue S, Stepien P, Qu F, Michalczyk E, Pakosz Z, Nomura N, Ogasawara S, Walker GC, Rebuffat S, Iwata S, Heddle JG, Beis K (2021) Molecular mechanism of SbmA, a promiscuous transporter exploited by antimicrobial peptides. Sci Adv 7(37):eabj5363. https://doi.org/10.1126/sciadv.abj5363
Gopal S, Berg D, Hagen N, Schriefer E-M, Stoll R, Goebel W, Kreft J (2010) Maltose and maltodextrin utilization by Listeria monocytogenes depend on an inducible ABC transporter which is repressed by glucose. PLoS One 5(4):e10349
Gopinath K, Venclovas Č, Ioerger TR, Sacchettini JC, McKinney JD, Mizrahi V, Warner DF (2013) A vitamin B12 transporter in Mycobacterium tuberculosis. Open Biol 3(2):120175
Gradhand U, Kim RB (2008) Pharmacogenomics of MRP Transporters (ABCC1-5) and BCRP (ABCG2). Drug Metab Rev 40(2):317–354
Haardt M, Bremer E (1996) Use of phoA and lacZ fusions to study the membrane topology of ProW, a component of the osmoregulated ProU transport system of Escherichia coli. J Bacteriol 178:5370–5381. https://doi.org/10.1128/jb.178.18.5370-5381.1996
Halbach A, Zhang H, Wengi A, Jablonska Z, Gruber IML, Halbeisen RE, Dehé P-M, Kemmeren P, Holstege F, Géli V, Gerber AP, Dichtl B (2009) Cotranslational assembly of the yeast SET1C histone methyltransferase complex. EMBO J 28(19):2959–2970
Hebbeln P, Rodionov DA, Alfandega A, Eitinger T (2007) Biotin uptake in prokaryotes by solute transporters with an optional ATP-binding cassette-containing module. Proc Natl Acad Sci U S A 104(8):2909–2914
Henderson GB, Zevely EM, Huennekens F (1979) Mechanism of folate transport in Lactobacillus casei: evidence for a component shared with the thiamine and biotin transport systems. J Bacteriol 137(3):1308–1314
Herold M, Nierhaus KH (1987) Incorporation of six additional proteins to complete the assembly map of the 50 S subunit from Escherichia coli ribosomes. J Biol Chem 262(18):8826–8833
Higgins CF (1992) ABC transporters: from microorganisms to man. Annu Rev Cell Biol 8:67–113
Higgins CF, Ames GF (1981) Two periplasmic transport proteins which interact with a common membrane receptor show extensive homology: complete nucleotide sequences. Proc Natl Acad Sci U S A 78(10):6038–6042. https://doi.org/10.1073/pnas.78.10.6038
Higgins CF, Ames GF (1982) Regulatory regions of two transport operons under nitrogen control: nucleotide sequences. Proc Natl Acad Sci U S A 79(4):1083–1087. https://doi.org/10.1073/pnas.79.4.1083

Higgins CF, Haag PD, Nikaido K, Ardeshir F, Garcia G, Ames GF (1982) Complete nucleotide sequence and identification of membrane components of the histidine transport operon of S. typhimurium. Nature 298(5876):723–727

Hohl M, Briand C, Grütter MG, Seeger MA (2012) Crystal structure of a heterodimeric ABC transporter in its inward-facing conformation. Nat Struct Mol Biol 19(4):395–402

Hollenstein K, Frei DC, Locher KP (2007) Structure of an ABC transporter in complex with its binding protein. Nature 446(7132):213–216

Horler RS, Müller A, Williamson DC, Potts JR, Wilson KS, Thomas GH (2009) Furanose-specific sugar transport: characterization of a bacterial galactofuranose-binding protein. J Biol Chem 284(45):31156–31163. https://doi.org/10.1074/jbc.M109.054296

Hsu WL, Furuta T, Sakurai M (2018) The mechanism of nucleotide-binding domain dimerization in the intact maltose transporter as studied by all-atom molecular dynamics simulations. Proteins 86(2):237–247. https://doi.org/10.1002/prot.25433

Huang CF, Yamaji N, Mitani N, Yano M, Nagamura Y, Ma JF (2009) A bacterial-type ABC transporter is involved in aluminum tolerance in rice. Plant Cell 21(2):655–667. https://doi.org/10.1105/tpc.108.064543

Hurtubise Y, Shareck F, Kluepfel D, Morosoli R (1995) A cellulase/xylanase-negative mutant of Streptomyces lividans 1326 defective in cellobiose and xylobiose uptake is mutated in a gene encoding a protein homologous to ATP-binding proteins. Mol Microbiol 17(2):367–377

Ito Y, Matsuzawa H, Matsuyama S, Narita S, Tokuda H (2006) Genetic analysis of the mode of interplay between an ATPase subunit and membrane subunits of the lipoprotein-releasing ATP-binding cassette transporter LolCDE. J Bacteriol 188(8):2856–2864. https://doi.org/10.1128/JB.188.8.2856-2864.2006

Jardetzky O (1966) Simple allosteric model for membrane pumps. Nature 211(5052):969–970. https://doi.org/10.1038/211969a0

Joglekar AP, Hay JC (2005) Evidence for regulation of ER/Golgi SNARE complex formation by hsc70 chaperones. Eur J Cell Biol 84(5):529–542. https://doi.org/10.1016/j.ejcb.2004.12.028

Jones MM, Johnson A, Koszelak-Rosenblum M, Kirkham C, Brauer AL, Malkowski MG, Murphy TF (2014) Role of the oligopeptide permease ABC Transporter of Moraxella catarrhalis in nutrient acquisition and persistence in the respiratory tract. Infect Immun 82(11):4758–4766. https://doi.org/10.1128/iai.02185-14

Jumper J, Evans R, Pritzel A, Green T, Figurnov M, Ronneberger O, Tunyasuvunakool K, Bates R, Zidek A, Potapenko A, Bridgland A, Meyer C, Kohl SAA, Ballard AJ, Cowie A, Romera-Paredes B, Nikolov S, Jain R, Adler J, Back T, Petersen S, Reiman D, Clancy E, Zielinski M, Steinegger M, Pacholska M, Berghammer T, Bodenstein S, Silver D, Vinyals O, Senior AW, Kavukcuoglu K, Kohli P, Hassabis D (2021) Highly accurate protein structure prediction with AlphaFold. Nature 596(7873):583–589. https://doi.org/10.1038/s41586-021-03819-2

Kana BD, Weinstein EA, Avarbock D, Dawes SS, Rubin H, Mizrahi V (2001) Characterization of the cydAB-encoded cytochrome bd oxidase from Mycobacterium smegmatis. J Bacteriol 183(24):7076–7086. https://doi.org/10.1128/JB.183.24.7076-7086.2001

Kanamaru K, Taniguchi N, Miyamoto S, Narita S, Tokuda H (2007) Complete reconstitution of an ATP-binding cassette transporter LolCDE complex from separately isolated subunits. FEBS J 274(12):3034–3043. https://doi.org/10.1111/j.1742-4658.2007.05832.x

Karpowich NK, Wang D-N (2013) Assembly and mechanism of a group II ECF transporter. Proc Natl Acad Sci 110(7):2534–2539

Kashiwayama Y, Seki M, Yasui A, Murasaki Y, Morita M, Yamashita Y, Sakaguchi M, Tanaka Y, Imanaka T (2009) 70-kDa peroxisomal membrane protein related protein (P70R/ABCD4) localizes to endoplasmic reticulum not peroxisomes, and NH2-terminal hydrophobic property determines the subcellular localization of ABC subfamily D proteins. Exp Cell Res 315(2):190–205

Kerppola RE, Ames GF (1992) Topology of the hydrophobic membrane-bound components of the histidine periplasmic permease. Comparison with other members of the family. J Biol Chem 267(4):2329–2336

Kerppola RE, Shyamala VK, Klebba P, Ames GF (1991) The membrane-bound proteins of periplasmic permeases form a complex. Identification of the histidine permease HisQMP complex. J Biol Chem 266(15):9857–9865

Kertesz MA (2001) Bacterial transporters for sulfate and organosulfur compounds. Res Microbiol 152(3–4):279–290. https://doi.org/10.1016/s0923-2508(01)01199-8

Keseler IM, Mackie A, Peralta-Gil M, Santos-Zavaleta A, Gama-Castro S, Bonavides-Martínez C, Fulcher C, Huerta AM, Kothari A, Krummenacker M, Latendresse M, Muñiz-Rascado L, Ong Q, Paley S, Schröder I, Shearer AG, Subhraveti P, Travers M, Weerasinghe D, Weiss V, Collado-Vides J, Gunsalus RP, Paulsen I, Karp PD (2013) EcoCyc: fusing model organism databases with systems biology. Nucleic Acids Res 41(Database issue):D605–D612. https://doi.org/10.1093/nar/gks1027

Koch DJ, Rückert C, Rey DA, Mix A, Pühler A, Kalinowski J (2005) Role of the ssu and seu genes of Corynebacterium glutamicum ATCC 13032 in utilization of sulfonates and sulfonate esters as sulfur sources. Appl Environ Microbiol 71(10):6104–6114. https://doi.org/10.1128/aem.71.10.6104-6114.2005

Kos V, Ford RC (2009) The ATP-binding cassette family: a structural perspective. Cell Mol Life Sci 66(19):3111–3126. https://doi.org/10.1007/s00018-009-0064-9

Köster W, Böhm B (1992) Point mutations in two conserved glycine residues within the integral membrane protein FhuB affect iron(II) hydroxamate transport. Mol Gen Genet MGG 232(3): 399–407. https://doi.org/10.1007/BF00266243

Kramer G, Shiber A, Bukau B (2019) Mechanisms of cotranslational maturation of newly synthesized proteins. Annu Rev Biochem 88:337–364. https://doi.org/10.1146/annurev-biochem-013118-111717

Krissinel E, Henrick K (2007) Inference of macromolecular assemblies from crystalline state. J Mol Biol 372(3):774–797. https://doi.org/10.1016/j.jmb.2007.05.022

Lebrette H, Brochier-Armanet C, Zambelli B, de Reuse H, Borezée-Durant E, Ciurli S, Cavazza C (2014) Promiscuous nickel import in human pathogens: structure, thermodynamics, and evolution of extracytoplasmic nickel-binding proteins. Structure 22(10):1421–1432

Lee JY, Yang JG, Zhitnitsky D, Lewinson O, Rees DC (2014) Structural characterization of heavy metal detoxification by a bacterial Atm1-family ABC transporter

Lee JY, Kinch LN, Borek DM, Wang J, Wang J, Urbatsch IL, Xie XS, Grishin NV, Cohen JC, Otwinowski Z, Hobbs HH, Rosenbaum DM (2016) Crystal structure of the human sterol transporter ABCG5/ABCG8. Nature 533(7604):561–564. https://doi.org/10.1038/nature17666

Lehman T, Mallireddy V, Seddon M, Modali R, Ratnasinghe L (2008) Association between an ABCC8/SUR1 polymorphism and breast cancer risk. AACR

Leisico F, Godinho LM, Goncalves IC, Silva SP, Carneiro B, Romao MJ, Santos-Silva T, de Sa-Nogueira I (2020) Multitask ATPases (NBDs) of bacterial ABC importers type I and their interspecies exchangeability. Sci Rep 10(1):19564. https://doi.org/10.1038/s41598-020-76444-0

Lennen R (2007) Inhibitors of autoinducer-2 quorum sensing and their effect on bacterial biofilm formation

Leonard BA, Podbielski A, Hedberg PJ, Dunny GM (1996) Enterococcus faecalis pheromone binding protein, PrgZ, recruits a chromosomal oligopeptide permease system to import sex pheromone cCF10 for induction of conjugation. Proc Natl Acad Sci U S A 93(1):260–264. https://doi.org/10.1073/pnas.93.1.260

Li Y, Orlando BJ, Liao M (2019) Structural basis of lipopolysaccharide extraction by the LptB2FGC complex. Nature 567(7749):486–490. https://doi.org/10.1038/s41586-019-1025-6

Lin DY, Huang S, Chen J (2015) Crystal structures of a polypeptide processing and secretion transporter. Nature 523(7561):425–430. https://doi.org/10.1038/nature14623

Linke CM, Woodiga SA, Meyers DJ, Buckwalter CM, Salhi HE, King SJ (2013) The ABC transporter encoded at the pneumococcal fructooligosaccharide utilization locus determines the ability to utilize long-and short-chain fructooligosaccharides. J Bacteriol 195(5):1031–1041

Linton KJ, Higgins CF (1998) The Escherichia coli ATP-binding cassette (ABC) proteins. Mol Microbiol 28(1):5–13

Liu PQ, Ames GF (1998) In vitro disassembly and reassembly of an ABC transporter, the histidine permease. Proc Natl Acad Sci U S A 95(7):3495–3500. https://doi.org/10.1073/pnas.95.7.3495

Liu F, Jones DK, de Lange WJ, Robertson GA (2016) Cotranslational association of mRNA encoding subunits of heteromeric ion channels. Proc Natl Acad Sci U S A 113(17): 4859–4864. https://doi.org/10.1073/pnas.1521577113

Liu J, Hou LL, Zhao CY (2018) Effect of YHHJ on the expression of the hepatocellular bile acid transporters multidrug resistance-associated protein 2 and bile salt export pump in ethinylestradiol-induced cholestasis. Exp Ther Med 15(4):3699–3704. https://doi.org/10.3892/etm.2018.5891

Locher KP, Lee AT, Rees DC (2002) The E. coli BtuCD structure: a framework for ABC transporter architecture and mechanism. Science 296(5570):1091–1098

Luo Q, Yang X, Yu S, Shi H, Wang K, Xiao L, Zhu G, Sun C, Li T, Li D, Zhang X, Zhou M, Huang Y (2017) Structural basis for lipopolysaccharide extraction by ABC transporter LptB2FG. Nat Struct Mol Biol 24(5):469–474. https://doi.org/10.1038/nsmb.3399

Majsnerowska M, Ter Beek J, Stanek WK, Duurkens RH, Slotboom DJ (2015) Competition between different S-components for the shared energy coupling factor module in energy coupling factor transporters. Biochemistry 54(31):4763–4766. https://doi.org/10.1021/acs.biochem.5b00609

Makhnevych T, Houry WA (2012) The role of Hsp90 in protein complex assembly. Biochim Biophys Acta 1823(3):674–682. https://doi.org/10.1016/j.bbamcr.2011.09.001

Malinverni J, Silhavy T (2009) An ABC transport system that maintains lipid asymmetry in the Gram-negative outer membrane. Proc Natl Acad Sci U S A 106:8009–8014. https://doi.org/10.1073/pnas.0903229106

Marion C, Aten AE, Woodiga SA, King SJ (2011) Identification of an ATPase, MsmK, which energizes multiple carbohydrate ABC transporters in Streptococcus pneumoniae. Infect Immun 79(10):4193–4200

Matthysse AG, Yarnall HA, Young N (1996) Requirement for genes with homology to ABC transport systems for attachment and virulence of Agrobacterium tumefaciens. J Bacteriol 178(17):5302–5308. https://doi.org/10.1128/jb.178.17.5302-5308.1996

Merlin C, Gardiner G, Durand S, Masters M (2002) The Escherichia coli metD locus encodes an ABC transporter which includes Abc (MetN), YaeE (MetI), and YaeC (MetQ). J Bacteriol 184(19):5513–5517. https://doi.org/10.1128/JB.184.19.5513-5517.2002

Morabbi Heravi K, Watzlawick H, Altenbuchner J (2019) The melREDCA operon encodes a utilization system for the raffinose family of oligosaccharides in Bacillus subtilis. J Bacteriol 201(15):e00109–e00119

Morgan JLW, Acheson JF, Zimmer J (2017) Structure of a type-1 secretion system ABC transporter. Structure 25(3):522–529. https://doi.org/10.1016/j.str.2017.01.010

Moussatova A, Kandt C, O'Mara ML, Tieleman DP (2008) ATP-binding cassette transporters in Escherichia coli. Biochim Biophys Acta Biomembr 1778(9):1757–1771. https://doi.org/10.1016/j.bbamem.2008.06.009

Mukherjee K, Huddleston JP, Narindoshvili T, Nemmara VV, Raushel FM (2019) Functional characterization of the ycjQRS gene cluster from Escherichia coli: a novel pathway for the transformation of d-gulosides to d-glucosides. Biochemistry 58(10):1388–1399. https://doi.org/10.1021/acs.biochem.8b01278

Narita S, Tanaka K, Matsuyama S, Tokuda H (2002) Disruption of lolCDE, encoding an ATP-binding cassette transporter, is lethal for Escherichia coli and prevents release of lipoproteins from the inner membrane. J Bacteriol 184(5):1417–1422. https://doi.org/10.1128/jb.184.5.1417-1422.2002

Netzer WJ, Hartl FU (1998) Protein folding in the cytosol: chaperonin-dependent and -independent mechanisms. Trends Biochem Sci 23(2):68–73. https://doi.org/10.1016/s0968-0004(97)01171-7

Neubauer H, Pantel I, Lindgren PE, Götz F (1999) Characterization of the molybdate transport system ModABC of Staphylococcus carnosus. Arch Microbiol 172(2):109–115. https://doi.org/10.1007/s002030050747

Neubauer O, Alfandega A, Schoknecht J, Sternberg U, Pohlmann A, Eitinger T (2009) Two essential arginine residues in the T components of energy-coupling factor transporters. J Bacteriol 191(21):6482–6488

Neubauer O, Reiffler C, Behrendt L, Eitinger T (2011) Interactions among the A and T units of an ECF-type biotin transporter analyzed by site-specific crosslinking. PLoS One 6(12):e29087

Nicolaou SA, Fast AG, Nakamaru-Ogiso E, Papoutsakis ET (2013) Overexpression of fetA (ybbL) and fetB (ybbM), encoding an iron exporter, enhances resistance to oxidative stress in Escherichia coli. Appl Environ Microbiol 79(23):7210–7219. https://doi.org/10.1128/AEM.02322-13

Nissen P, Hansen J, Ban N, Moore PB, Steitz TA (2000) The structural basis of ribosome activity in peptide bond synthesis. Science 289(5481):920–930. https://doi.org/10.1126/science.289.5481.920

Nöll A, Thomas C, Herbring V, Zollmann T, Barth K, Mehdipour AR, Tomasiak TM, Brüchert S, Joseph B, Abele R (2017) Crystal structure and mechanistic basis of a functional homolog of the antigen transporter TAP. Proc Natl Acad Sci U S A 114(4):E438–E447

Nomura M (1973) Assembly of bacterial ribosomes. Science 179(4076):864–873. https://doi.org/10.1126/science.179.4076.864

Okada U, Yamashita E, Neuberger A, Morimoto M, van Veen HW, Murakami S (2017) Crystal structure of tripartite-type ABC transporter MacB from Acinetobacter baumannii. Nat Commun 8(1):1336. https://doi.org/10.1038/s41467-017-01399-2

Okamoto S, Eltis LD (2011) The biological occurrence and trafficking of cobalt. Metallomics 3(10):963–970

Oteri F, Nadalin F, Champeimont R, Carbone A (2017) BIS2Analyzer: a server for co-evolution analysis of conserved protein families. Nucleic Acids Res 45(W1):W307–W314. https://doi.org/10.1093/nar/gkx336

Park JT, Raychaudhuri D, Li H, Normark S, Mengin-Lecreulx D (1998) MppA, a periplasmic binding protein essential for import of the bacterial cell wall peptide L-alanyl-gamma-D-glutamyl-meso-diaminopimelate. J Bacteriol 180(5):1215–1223. https://doi.org/10.1128/JB.180.5.1215-1223.1998

Parra-Lopez C, Baer MT, Groisman EA (1993) Molecular genetic analysis of a locus required for resistance to antimicrobial peptides in Salmonella typhimurium. EMBO J 12(11):4053–4062

Paterson JK, Shukla S, Black CM, Tachiwada T, Garfield S, Wincovitch S, Ernst DN, Agadir A, Li X, Ambudkar SV (2007) Human ABCB6 localizes to both the outer mitochondrial membrane and the plasma membrane. Biochemistry 46(33):9443–9452

Patzer SI, Hantke K (1998) The ZnuABC high-affinity zinc uptake system and its regulator Zur in Escherichia coli. Mol Microbiol 28(6):1199–1210. https://doi.org/10.1046/j.1365-2958.1998.00883.x

Perez C, Gerber S, Boilevin J, Bucher M, Darbre T, Aebi M, Reymond J-L, Locher KP (2015) Structure and mechanism of an active lipid-linked oligosaccharide flippase. Nature 524(7566):433–438

Phillips RB, Zimmerman A, Noakes MA, Palti Y, Morasch MR, Eiben L, Ristow SS, Thorgaard GH, Hansen JD (2003) Physical and genetic mapping of the rainbow trout major histocompatibility regions: evidence for duplication of the class I region. Immunogenetics 55(8):561–569

Qian H, Zhao X, Cao P, Lei J, Yan N, Gong X (2017) Structure of the human lipid exporter ABCA1. Cell 169(7):1228–1239 e1210. https://doi.org/10.1016/j.cell.2017.05.020

Qu L, Jiang Y, Cheng C, Wu D, Meng B, Chen Z, Zhu Y, Shaw N, Ouyang S, Liu ZJ (2018) Crystal structure of ATP-bound human ABCF1 demonstrates a unique conformation of ABC proteins. Structure 26(9):1259–1265 e1253. https://doi.org/10.1016/j.str.2018.05.019

Quazi F, Lenevich S, Molday RS (2012) ABCA4 is an N-retinylidene-phosphatidylethanolamine and phosphatidylethanolamine importer. Nat Commun 3:925. https://doi.org/10.1038/ncomms1927

Quentin Y, Fichant G, Denizot F (1999) Inventory, assembly and analysis of Bacillus subtilis ABC transport systems. J Mol Biol 287(3):467–484

Rajagopala SV, Sikorski P, Kumar A, Mosca R, Vlasblom J, Arnold R, Franca-Koh J, Pakala SB, Phanse S, Ceol A, Hauser R, Siszler G, Wuchty S, Emili A, Babu M, Aloy P, Pieper R, Uetz P (2014) The binary protein-protein interaction landscape of Escherichia coli. Nat Biotechnol 32(3):285–290. https://doi.org/10.1038/nbt.2831

Ramakrishnan V, White SW (1998) Ribosomal protein structures: insights into the architecture, machinery and evolution of the ribosome. Trends Biochem Sci 23(6):208–212. https://doi.org/10.1016/s0968-0004(98)01214-6

Rempel S, Stanek WK, Slotboom DJ (2019) ECF-type ATP-binding cassette transporters. Annu Rev Biochem 88:551–576. https://doi.org/10.1146/annurev-biochem-013118-111705

Riordan JR, Rommens JM, Kerem B, Alon N, Rozmahel R, Grzelczak Z, Zielenski J, Lok S, Plavsic N, Chou JL et al (1989) Identification of the cystic fibrosis gene: cloning and characterization of complementary DNA. Science 245(4922):1066–1073

Rodionov DA, Hebbeln P, Gelfand MS, Eitinger T (2006) Comparative and functional genomic analysis of prokaryotic nickel and cobalt uptake transporters: evidence for a novel group of ATP-binding cassette transporters. J Bacteriol 188(1):317–327

Rodionov DA, Hebbeln P, Eudes A, Ter Beek J, Rodionova IA, Erkens GB, Slotboom DJ, Gelfand MS, Osterman AL, Hanson AD (2009) A novel class of modular transporters for vitamins in prokaryotes. J Bacteriol 191(1):42–51

Rodriguez GM, Smith I (2006) Identification of an ABC transporter required for iron acquisition and virulence in Mycobacterium tuberculosis. J Bacteriol 188(2):424–430

Saier MH Jr (2000) Families of transmembrane sugar transport proteins: microreview. Mol Microbiol 35(4):699–710

Saier MH Jr, Reddy VS, Tamang DG, Västermark A (2014) The transporter classification database. Nucleic Acids Res 42(Database issue):D251–D258. https://doi.org/10.1093/nar/gkt1097

Saier MH Jr, Reddy VS, Tsu BV, Ahmed MS, Li C, Moreno-Hagelsieb G (2016) The Transporter Classification Database (TCDB): recent advances. Nucleic Acids Res 44(D1):D372–D379. https://doi.org/10.1093/nar/gkv1103

Sakahira H, Nagata S (2002) Co-translational folding of caspase-activated DNase with Hsp70, Hsp40, and inhibitor of caspase-activated DNase. J Biol Chem 277(5):3364–3370. https://doi.org/10.1074/jbc.M110071200

Saurin W, Koster W, Dassa E (1994) Bacterial binding protein-dependent permeases: characterization of distinctive signatures for functionally related integral cytoplasmic membrane proteins. Mol Microbiol 12(6):993–1004

Saurin W, Hofnung M, Dassa E (1999) Getting in or out: early segregation between importers and exporters in the evolution of ATP-binding cassette (ABC) transporters. J Mol Evol 48(1):22–41

Schulz H, Fabianek RA, Pellicioli EC, Hennecke H, Thöny-Meyer L (1999) Heme transfer to the heme chaperone CcmE during cytochrome c maturation requires the CcmC protein, which may function independently of the ABC-transporter CcmAB. Proc Natl Acad Sci U S A 96(11):6462–6467. https://doi.org/10.1073/pnas.96.11.6462

Setyawati I, Stanek WK, Majsnerowska M, Swier L, Pardon E, Steyaert J, Guskov A, Slotboom DJ (2020) In vitro reconstitution of dynamically interacting integral membrane subunits of energy-coupling factor transporters. elife 9:e64389. https://doi.org/10.7554/eLife.64389

Sharma S, Davis JA, Ayvaz T, Traxler B, Davidson AL (2005) Functional reassembly of the Escherichia coli maltose transporter following purification of a MalF-MalG subassembly. J Bacteriol 187(8):2908–2911. https://doi.org/10.1128/JB.187.8.2908-2911.2005

Sharom FJ (2006) Shedding light on drug transport: structure and function of the P-glycoprotein multidrug transporter (ABCB1). Biochem Cell Biol 84(6):979–992

Shiber A, Doring K, Friedrich U, Klann K, Merker D, Zedan M, Tippmann F, Kramer G, Bukau B (2018) Cotranslational assembly of protein complexes in eukaryotes revealed by ribosome profiling. Nature 561(7722):268–272. https://doi.org/10.1038/s41586-018-0462-y

Shieh YW, Minguez P, Bork P, Auburger JJ, Guilbride DL, Kramer G, Bukau B (2015) Operon structure and cotranslational subunit association direct protein assembly in bacteria. Science 350(6261):678–680. https://doi.org/10.1126/science.aac8171

Shum M, Shintre CA, Althoff T, Gutierrez V, Segawa M, Saxberg AD, Martinez M, Adamson R, Young MR, Faust B, Gharakhanian R, Su S, Chella Krishnan K, Mahdaviani K, Veliova M, Wolf DM, Ngo J, Nocito L, Stiles L, Abramson J, Lusis AJ, Hevener AL, Zoghbi ME, Carpenter EP, Liesa M (2021) ABCB10 exports mitochondrial biliverdin, driving metabolic maladaptation in obesity. Sci Transl Med 13(594). https://doi.org/10.1126/scitranslmed.abd1869

Siche S, Neubauer O, Hebbeln P, Eitinger T (2010) A bipartite S unit of an ECF-type cobalt transporter. Res Microbiol 161(10):824–829

Simpson BW, Owens TW, Orabella MJ, Davis RM, May JM, Trauger SA, Kahne D, Ruiz N (2016) Identification of residues in the lipopolysaccharide ABC transporter that coordinate ATPase activity with extractor function. mBio 7(5):e01729-16. https://doi.org/10.1128/mBio.01729-16

Sirko A, Hryniewicz M, Hulanicka D, Böck A (1990) Sulfate and thiosulfate transport in Escherichia coli K-12: nucleotide sequence and expression of the cysTWAM gene cluster. J Bacteriol 172(6):3351–3357. https://doi.org/10.1128/jb.172.6.3351-3357.1990

Slotboom DJ (2014) Structural and mechanistic insights into prokaryotic energy-coupling factor transporters. Nat Rev Microbiol 12(2):79–87

Stamm LV, Young NR, Frye JG (1996) Sequences of the Salmonella typhimurium mglA and mglC genes. Gene 171(1):131–132. https://doi.org/10.1016/0378-1119(96)00072-8

Staudenmaier H, Van Hove B, Yaraghi Z, Braun V (1989) Nucleotide sequences of the fecBCDE genes and locations of the proteins suggest a periplasmic-binding-protein-dependent transport mechanism for iron(III) dicitrate in Escherichia coli. J Bacteriol 171(5):2626–2633. https://doi.org/10.1128/jb.171.5.2626-2633.1989

Stieger B (2009) Recent insights into the function and regulation of the bile salt export pump (ABCB11). Curr Opin Lipidol 20(3):176–181

Stoffler G, Rudloff V, Wittman HG (1968) Ribosomal proteins. 3. Preparative separation of Escherichia coli and yeast ribosomal proteins by means of gel filtration. Mol Gen Genet 101(1):70–81. https://doi.org/10.1007/BF00434813

Swier LJ, Guskov A, Slotboom DJ (2016) Structural insight in the toppling mechanism of an energy-coupling factor transporter. Nat Commun 7(1):1–11

Tan M-F, Gao T, Liu W-Q, Zhang C-Y, Yang X, Zhu J-W, Teng M-Y, Li L, Zhou R (2015) MsmK, an ATPase, contributes to utilization of multiple carbohydrates and host colonization of Streptococcus suis. PLoS One 10(7):e0130792

Tang X, Chang S, Luo Q, Zhang Z, Qiao W, Xu C, Zhang C, Niu Y, Yang W, Wang T, Zhang Z, Zhu X, Wei X, Dong C, Zhang X, Dong H (2019) Cryo-EM structures of lipopolysaccharide transporter LptB2FGC in lipopolysaccharide or AMP-PNP-bound states reveal its transport mechanism. Nat Commun 10(1):4175. https://doi.org/10.1038/s41467-019-11977-1

Tang X, Chang S, Zhang K, Luo Q, Zhang Z, Wang T, Qiao W, Wang C, Shen C, Zhang Z, Zhu X, Wei X, Dong C, Zhang X, Dong H (2021) Structural basis for bacterial lipoprotein relocation by the transporter LolCDE. Nat Struct Mol Biol 28(4):347–355. https://doi.org/10.1038/s41594-021-00573-x

Taylor NMI, Manolaridis I, Jackson SM, Kowal J, Stahlberg H, Locher KP (2017) Structure of the human multidrug transporter ABCG2. Nature 546(7659):504–509. https://doi.org/10.1038/nature22345

ter Beek J, Duurkens RH, Erkens GB, Slotboom DJ (2011) Quaternary structure and functional unit of energy coupling factor (ECF)-type transporters. J Biol Chem 286(7):5471–5475

Thélot F, Orlando BJ, Li Y, Liao M (2020) High-resolution views of lipopolysaccharide translocation driven by ABC transporters MsbA and LptB(2)FGC. Curr Opin Struct Biol 63:26–33. https://doi.org/10.1016/j.sbi.2020.03.005

Thomas GH (2010) Homes for the orphans: utilization of multiple substrate-binding proteins by ABC transporters. Mol Microbiol 75(1):6–9. https://doi.org/10.1111/j.1365-2958.2009.06961.x

Thomas C, Aller SG, Beis K, Carpenter EP, Chang G, Chen L, Dassa E, Dean M, Duong Van Hoa F, Ekiert D, Ford R, Gaudet R, Gong X, Holland IB, Huang Y, Kahne DK, Kato H, Koronakis V, Koth CM, Lee Y, Lewinson O, Lill R, Martinoia E, Murakami S, Pinkett HW, Poolman B, Rosenbaum D, Sarkadi B, Schmitt L, Schneider E, Shi Y, Shyng SL, Slotboom DJ, Tajkhorshid E, Tieleman DP, Ueda K, Varadi A, Wen PC, Yan N, Zhang P, Zheng H, Zimmer J, Tampe R (2020) Structural and functional diversity calls for a new classification of ABC transporters. FEBS Lett 594(23):3767–3775. https://doi.org/10.1002/1873-3468.13935

Thonghin N, Kargas V, Clews J, Ford RC (2018) Cryo-electron microscopy of membrane proteins. Methods 147:176–186. https://doi.org/10.1016/j.ymeth.2018.04.018

Tsou AY, Ransom SC, Gerlt JA, Buechter DD, Babbitt PC, Kenyon GL (1990) Mandelate pathway of Pseudomonas putida: sequence relationships involving mandelate racemase, (S)-mandelate dehydrogenase, and benzoylformate decarboxylase and expression of benzoylformate decarboxylase in Escherichia coli. Biochemistry 29(42):9856–9862. https://doi.org/10.1021/bi00494a015

van der Ploeg JR, Weiss MA, Saller E, Nashimoto H, Saito N, Kertesz MA, Leisinger T (1996) Identification of sulfate starvation-regulated genes in Escherichia coli: a gene cluster involved in the utilization of taurine as a sulfur source. J Bacteriol 178(18):5438–5446. https://doi.org/10.1128/jb.178.18.5438-5446.1996

Wang T, Fu G, Pan X, Wu J, Gong X, Wang J, Shi Y (2013) Structure of a bacterial energy-coupling factor transporter. Nature 497(7448):272–276

Wang Z, Zhang M, Shi X, Xiang Q (2017) Purification and characterization of an ATPase GsiA from Salmonella enterica. Biomed Res Int 2017:3076091. https://doi.org/10.1155/2017/3076091

Ward A, Reyes CL, Yu J, Roth CB, Chang G (2007) Flexibility in the ABC transporter MsbA: alternating access with a twist. Proc Natl Acad Sci U S A 104(48):19005–19010

Xie T, Zhang Z, Fang Q, Du B, Gong X (2021) Structural basis of substrate recognition and translocation by human ABCA4. Nat Commun 12(1):3853. https://doi.org/10.1038/s41467-021-24194-6

Xu K, Zhang M, Zhao Q, Yu F, Guo H, Wang C, He F, Ding J, Zhang P (2013) Crystal structure of a folate energy-coupling factor transporter from Lactobacillus brevis. Nature 497(7448):268–271

Yee K, Robinson C, Hurlock G, Moss RB, Wine JJ (2000) Novel cystic fibrosis mutation L1093P: Functional analysis and possible Native American origin. Hum Mutat 15(2):208–208

Yonath A, Harms J, Hansen HA, Bashan A, Schlunzen F, Levin I, Koelln I, Tocilj A, Agmon I, Peretz M, Bartels H, Bennett WS, Krumbholz S, Janell D, Weinstein S, Auerbach T, Avila H, Piolleti M, Morlang S, Franceschi F (1998) Crystallographic studies on the ribosome, a large macromolecular assembly exhibiting severe nonisomorphism, extreme beam sensitivity and no internal symmetry. Acta Crystallogr A 54(Pt 6 Pt 1):945–955. https://doi.org/10.1107/s010876739800991x

Yu Y, Zhou M, Kirsch F, Xu C, Zhang L, Wang Y, Jiang Z, Wang N, Li J, Eitinger T (2014) Planar substrate-binding site dictates the specificity of ECF-type nickel/cobalt transporters. Cell Res 24(3):267–277

Zhang P, Wang J, Shi Y (2010) Structure and mechanism of the S component of a bacterial ECF transporter. Nature 468(7324):717–720

Zhang M, Bao Z, Zhao Q, Guo H, Xu K, Wang C, Zhang P (2014) Structure of a pantothenate transporter and implications for ECF module sharing and energy coupling of group II ECF transporters. Proc Natl Acad Sci U S A 111(52):18560–18565

Zhou Z, White KA, Polissi A, Georgopoulos C, Raetz CR (1998) Function of Escherichia coli MsbA, an essential ABC family transporter, in lipid A and phospholipid biosynthesis. J Biol Chem 273(20):12466–12475

Zhou C, Shi H, Zhang M, Zhou L, Xiao L, Feng S, Im W, Zhou M, Zhang X, Huang Y (2021) Structural insight into phospholipid transport by the MlaFEBD complex from P. aeruginosa. J Mol Biol 433(13):166986. https://doi.org/10.1016/j.jmb.2021.166986

Zutz A, Gompf S, Schägger H, Tampé R (2009) Mitochondrial ABC proteins in health and disease. Biochim Biophys Acta Bioenerg 1787(6):681–690

Chapter 3
The Fork Protection Complex: A Regulatory Hub at the Head of the Replisome

Daniel B. Grabarczyk

Abstract As well as accurately duplicating DNA, the eukaryotic replisome performs a variety of other crucial tasks to maintain genomic stability. For example, organizational elements, like cohesin, must be transferred from the front of the fork to the new strands, and when there is replication stress, forks need to be protected and checkpoint signalling activated. The Tof1–Csm3 (or Timeless–Tipin in humans) Fork Protection Complex (FPC) ensures efficient replisome progression and is required for a range of replication-associated activities. Recent studies have begun to reveal the structure of this complex, and how it functions within the replisome to perform its diverse roles. The core of the FPC acts as a DNA grip on the front of the replisome to regulate fork progression. Other flexibly linked domains and motifs mediate interactions with proteins and specific DNA structures, enabling the FPC to act as a hub at the head of the replication fork.

Keywords Fork protection complex (FPC) · DNA replication · Replisome · Cohesin · Helicase · Polymerase · Timeless · Tipin

The Eukaryotic Replisome

The faithful duplication of genetic material is a fundamental determinant of life. In eukaryotes, the replication of DNA is performed by a large and dynamic multi-protein machine known as the 'replisome' (Fig. 3.1a) (O'Donnell and Li 2016; Bell and Labib 2016; Yuan and Li 2020), which has been best characterized in budding yeast (*Saccharomyces cerevisiae*). The main process catalysed by the replisome is the coordinated unwinding of the parental duplex DNA and synthesis of two new daughter DNA strands. The unwinding reaction is catalysed by the CMG (Cdc45-

D. B. Grabarczyk (✉)
Rudolf Virchow Center for Integrative and Translational Bioimaging, Institute for Structural Biology, University of Würzburg, Würzburg, Germany

Research Institute of Molecular Pathology, Vienna BioCenter, Vienna, Austria
e-mail: daniel.grabarczyk@imp.ac.at

J. R. Harris, J. Marles-Wright (eds.), *Macromolecular Protein Complexes IV*, Subcellular Biochemistry 99, https://doi.org/10.1007/978-3-031-00793-4_3

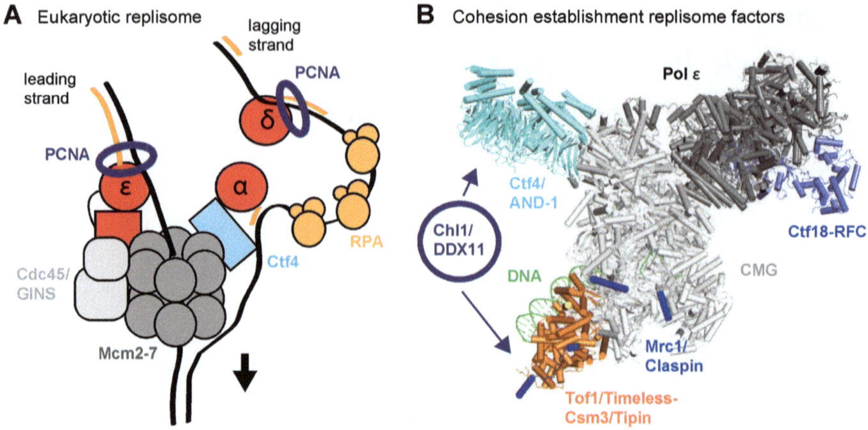

Fig. 3.1 The eukaryotic replisome. (**a**) A schematic model of the main unwinding and DNA synthesis activities in the eukaryotic replisome. (**b**) A structural model of where cohesion establishment factors bind to the replisome. Structures of full-length Pol e (PDB 6WJV) (Yuan et al. 2020), and the Ctf18-1-8/Pol ε complex (PDB 6S2E) (Stokes et al. 2020) are aligned onto the structure of the FPC-bound human replisome (PDB 7PFO) (Jones et al. 2021). For each factor, the budding yeast name is followed by the name in humans, when these are different

MCM-GINS) replicative helicase, where an MCM (Mini Chromosome Maintenance) AAA+ ATPase core, composed of the six Mcm2–7 subunits, is stably associated with Cdc45 and the GINS (go-ichi-ni-san) complex (Gambus et al. 2006; Yuan et al. 2016). Three polymerase activities then work on the two new single-stranded DNA ends which emerge from this helicase. As synthesis only occurs in the 5′ to 3′ direction, the two strands must be duplicated in opposite directions. The simpler leading strand can be replicated continuously on the 3′ to 5′ single-stranded DNA (ssDNA) that comes straight out of the helicase. This is performed by DNA polymerase (Pol) ε, which is tethered *via* its non-catalytic C-terminal domain and accessory subunit Dpb2 to the GINS complex and Mcm5 (Goswami et al. 2018; Langston et al. 2014; Sengupta et al. 2013). The lagging strand, in contrast, is synthesized in smaller stretches, named Okazaki fragments and then stitched back together (Okazaki et al. 1968). The bulk of lagging strand synthesis is performed by Pol δ which is generally thought not to be stably bound to the core replisome, but instead dynamically disassociates and rebinds to each new fragment (Bell and Labib 2016). However, some recent evidence suggests that Pol δ may be retained for multiple fragments (Lewis et al. 2020). Both bulk polymerases δ and ε require the processivity factor PCNA, a sliding clamp (Schauer and O'Donnell 2017; Lancey et al. 2020; Yeeles et al. 2017). This closed clamp is loaded and unloaded in an active process by AAA+ ATPase RFC clamp loader complexes (Bowman et al. 2004). DNA synthesis cannot occur de novo, and so each reaction requires an RNA primer synthesized by the primase module of Pol α (Perera et al. 2013; Klinge et al. 2009). This is then transferred to the polymerase module to initiate DNA synthesis before transfer to DNA pol δ, and for the leading strand from DNA pol δ to DNA pol ε (Garbacz et al. 2018). DNA pol α is flexibly tethered to the core helicase *via* the Ctf4 trimeric adaptor (Simon et al. 2014).

Replication-Associated Processes

Beyond the activities required directly for DNA synthesis, a large array of other factors are required to regulate this process, respond to problems and maintain the organization of the genome. DNA must only be replicated once per cell cycle, and thus the assembly and disassembly is tightly regulated. Inactive MCM double hexamers are loaded onto origins by the Origin Recognition Complex, Cdc6 and Cdt1 (Li et al. 2018; Douglas et al. 2018; Yuan et al. 2017). In S phase, the CDK and DDK kinases initiate the conversion of the MCM double hexamer into two CMG helicases operating in opposite directions (Yeeles et al. 2015). Termination occurs when two replisomes meet (Dewar et al. 2015), and requires helicase activity to allow fork convergence (Deegan et al. 2019) followed by ubiquitination of Mcm7 by SCF^{Dia2} for extraction of the MCM helicase by Cdc48 (Maric et al. 2014; Mukherjee and Labib 2019; Deegan et al. 2020). DNA replication can be prematurely blocked or stalled by depletion of nucleotides needed for synthesis, or by physical blocks—such as DNA lesions and proteins bound to the DNA (Zeman and Cimprich 2014). Response factors need to be recruited, such as components of a suitable DNA repair pathway for a lesion. There are many ways to recover from or bypass a break, such as translesion synthesis, template switching, fork reversal, repriming and firing dormant origins (Zeman and Cimprich 2014; Guilliam 2021). Throughout stalling and slowing, the replisome must be protected (Errico and Costanzo 2012). In particular, the unwinding and synthesis of two daughter strands must be tightly coupled to prevent the generation of exposed stretches of mutagenic single-stranded DNA. When there is sufficient global replication stress, the S phase checkpoint is activated to block the firing of new origins, stop the cell cycle, protect forks and upregulate nucleotide synthesis (Pardo et al. 2017). Stalling is sensed by the build-up of RPA (Replication Protein A)-coated ssDNA by the Mec1 kinase *via* Ddc2 (Deshpande et al. 2017). Various pathways then mediate the Mec1-dependent phosphorylation of the Rad53 checkpoint kinase, which blocks cell cycle progression and late origin firing (Moriel-Carretero et al. 2019; Pardo et al. 2017).

One aspect of genome organization the replisome performs is the transfer of histones and epigenetic marks from the front of the fork to the newly synthesized DNA. There is a complex system of mechanisms of either reloading (Sauer et al. 2018) or transfer for both the leading (He et al. 2017) and lagging strands (Gan et al. 2018). Another problem is the generation of supercoiling ahead of the fork during replication. Topoisomerases must be recruited to relax this (Pommier et al. 2016). One of the most enigmatic replication-associated processes at a mechanistic level is cohesion establishment (Uhlmann and Nasmyth 1998). Cohesion is the process which links the two daughter chromatids with cohesin molecules (Michaelis et al.

1997). This organizational element ensures that one copy of the genetic material is transferred to each daughter cell during the following mitosis. Cohesin is loaded during the G1 phase of the cell cycle (Ciosk et al. 2000), but is in a loose dynamic form which can be disassociated from DNA (Bernard et al. 2008). Only once replication forks pass loaded cohesin do they transform it into a tight form which can no longer dissociate through the cohesin acetylation function of Eco1 (Zhang et al. 2008; Rowland et al. 2009; Unal et al. 2008). The coordination of replication and cohesion establishment ensures that the two new daughter strands are locked together and only separated in mitosis.

Replication-Associated Cohesion Establishment Factors

A number of non-essential factors were discovered which play a role in cohesion establishment and associate with DNA replication forks (Lengronne et al. 2006). These include Ctf4, the alternative clamp loader Ctf18-RFC, the helicase Chl1, the adaptor Mrc1 and the fork protection complex (FPC) comprising Tof1 and Csm3 (Mayer et al. 2004; Hanna et al. 2001; Petronczki et al. 2004). Interestingly, the same proteins were also shown to have roles in many of the other replication-associated processes: the S phase checkpoint (Crabbé et al. 2010; Foss 2001), histone transfer (Foltman et al. 2013; Gan et al. 2018), fork protection (Noguchi et al. 2003; Calì et al. 2016), replication regulation (Hodgson et al. 2007) and countering supercoiling (Schalbetter et al. 2015). This suggests that these factors have fundamental roles in the replisome where they can work in all of these processes. This idea was confirmed by a wealth of recent structural data (Fig. 3.1b). Ctf4 was shown to be an organizational hub at the rear of the fork which links the replicative helicase and primase (Simon et al. 2014), but can also transiently recruit other factors when required (Samora et al. 2016). The functions of Ctf18-RFC require it to be tethered to the leading strand DNA Pol ε *via* its Ctf18-1-8 module (Grabarczyk et al. 2018; Stokes et al. 2020), and it is known to be the main regulator of PCNA levels on the leading strand (Liu et al. 2020). Curiously, despite the sequence and phenotypic conservation, Chl1 is thought to be recruited to forks by Ctf4 in yeast (Samora et al. 2016), but by the Tof1 orthologue Timeless in humans (Calì et al. 2016; Cortone et al. 2018). It is believed to act as a scaffold in cohesion establishment (Samora et al. 2016), while its function in replication stress involves its helicase activity (Stokes et al. 2020), probably for unwinding unusual DNA secondary structures which may impede the fork (Lerner et al. 2020; Guo et al. 2015; Calì et al. 2016; Bharti et al. 2013). Mrc1 is a long unstructured protein which links together many of the proteins in the replisome (Lou et al. 2008; Baretić et al. 2020; Komata et al. 2009), and thus might be a way to coordinate and regulate it, particularly the speed of replication (Hodgson et al. 2007; Yeeles et al. 2017; Lewis et al. 2017). The most prominent function of Mrc1 is as a mediator of the replication checkpoint where it recruits Mec1 to the global effector kinase Rad53 (Chen and Zhou 2009; Osborn and Elledge 2003; Alcasabas et al. 2001). Tof1 and Csm3 form a constitutive complex (Bando et al. 2009; Mayer et al. 2004; Gotter 2003) which binds to the front of the replicative helicase (Eickhoff et al. 2019; Baretić et al. 2020), where it performs a myriad of replication-associated functions, as discussed in detail in the rest of this chapter.

Despite progress in understanding where these factors are located on the replisome and, for some, what their molecular role is during normal replication, it is largely unclear at a mechanistic level how they perform any of the replication-associated functions for which they were originally discovered. Uncovering how these proteins function is not only important because of the fundamental biological importance of DNA replication but because of their relevance to human disease. Mutations in genes important for cohesion can lead to inherited diseases called cohesinopathies (Cucco and Musio 2016). These are severe recessive developmental diseases characterized by precocious chromosome separation during mitosis, leading to daughter cells missing chromosomes, or having additional chromosomes or fragments. One example is Warsaw breakage syndrome caused by mutations in the Chl1 orthologue DDX11 (van der Lelij et al. 2010). However, the major medical motivation to study these proteins is for their role in cancer formation and potential as targets for therapeutic intervention. When cells divide more rapidly, they must also replicate their DNA more quickly and with less controls and are more likely to encounter blocks such as DNA damage, which overall lead to replication stress. If the replication stress response is not sufficient, further DNA lesions, and double strand breaks, resulting in gross chromosome rearrangements. Therefore, replication stress is one of the hallmarks of cancer (Gaillard et al. 2015; Macheret and Halazonetis 2015). Due to the intense replication stress in cancer cells, these replication-associated factors become essential for cell survival and thus are excellent cancer-specific therapeutic targets (Yamaguchi et al. 2014; Bianco et al. 2019; Bhattacharya et al. 2012). This conditional essentiality is clear from their synthetic lethality with each other and further genes involved in maintaining genome stability (McLellan et al. 2009).

Functions of the Tof1-Csm3 Fork Protection Complex

In budding yeast, Csm3 was discovered due to its cohesion establishment defect phenotype (Rabitsch et al. 2001), while Tof1 was found as a topoisomerase interacting factor (Park and Sternglanz 1999). Later studies showed that Csm3 and Tof1 work in the same cohesion establishment pathway as each other (Mayer et al. 2004), plus Ctf4 and Chl1, and these were synthetic lethal with a parallel pathway involving Ctf18-RFC and Mrc1 (Xu et al. 2007). A recent study suggested that the pathway involving the Tof1-Csm3 complex works to transfer cohesin from the front of the fork to the newly synthesized DNA, while the parallel pathway loads cohesin de novo (Srinivasan et al. 2020).

Besides cohesion establishment, one of the most prominent roles of the Tof1-Csm3 complex is in the DNA replication checkpoint (Foss 2001; Noguchi et al. 2003; Katou et al. 2003; Sommariva et al. 2005). This is a response to stalled forks, which blocks late firing origins and acts in parallel to the DNA damage checkpoint operating *via* Rad9 and Rad24 to activate Rad53 *via* Mec1 (Crabbé et al. 2010; Alcasabas et al. 2001). Tof1 and Mrc1 are epistatic in this pathway, and show synthetic lethality with damage checkpoint genes (Foss 2001). Tof1 has a smaller role in the checkpoint than Mrc1, and it is thought that its main function is to form part of the interacting surface which recruits Mrc1 to the replisome to perform its checkpoint function (Baretić et al. 2020; Shimmoto et al. 2009).

Mrc1 is sometimes included in the definition of the Fork Protection Complex. However, it is clear that their functions are separable (Bando et al. 2009; Razidlo and Lahue 2008; Mohanty et al. 2006; Lewis et al. 2017; Tourrière et al. 2005; Hodgson et al. 2007). One key checkpoint-independent role for these proteins is fork progression. Replication forks move slower in the absence of these proteins both in vivo (Hodgson et al. 2007; Tourrière et al. 2005) and in reconstituted in vitro replication forks (Lewis et al. 2017; Yeeles et al. 2017). Their other main function is to protect paused forks, allowing them to restart, hence the name of 'Fork Protection Complex'. One aspect of this function is to prevent the uncoupling of the helicase and polymerase activities of the fork (Nedelcheva et al. 2005). If the helicase continued while nucleotide depletion prevented polymerase function, then a very large amount of toxic ssDNA would be generated. The role of the Tof1–Csm3 in this was first characterized in *Schizosaccharomyces pombe*, where the Tof1–Csm3 orthologues Swi1–Swi3 were shown to be important for pausing forks during nucleotide depletion-induced or natural barriers stalling to prevent uncoupling (Noguchi et al. 2003; Sommariva et al. 2005; Vengrova and Dalgaard 2004; Dalgaard and Klar 2000). Protein barriers also block the fork, and it has been shown that Tof1 is required for pausing at these protein blocks, which leads to the recruitment of the Rrm3 helicase, enabling the fork to overcome the barrier (Calzada et al. 2005). This activity has also been demonstrated using an in vitro reconstituted DNA replication system (Hizume et al. 2018). Additionally, Tof1–Csm3-dependent fork pausing is used to control Rrm3-regulated replication termination at the ends of rDNA (Mohanty et al. 2006; Bastia et al. 2016). The fork protection activity is also important at DNA-inverted repeat sequences, where hairpins would form and block fork progression (Voineagu et al. 2008; Voineagu et al. 2009). Indeed, the absence of Tof1 results in genome instability when hairpin-forming disease-relevant CAG sequence repeats are introduced (Gellon et al. 2019; Razidlo and Lahue 2008).

Another function of the complex is to regulate the rate of replication in response to oxidative stress *via* peroxiredoxin proteins (Huang and Kolodner 2005; Somyajit et al. 2017). Tof1 also recruits topoisomerases to the fork, and this has an important role in suppressing excessive fork rotation (Westhorpe et al. 2020; Schalbetter et al. 2015), and may also help the fork to overcome blocks (Shyian et al. 2020).

Human Timeless was originally discovered as the orthologue of the Drosophila Timeless protein involved in the circadian rhythm (Sehgal et al. 1994; Koike et al. 1998). Although there have been some reports linking human Timeless to circadian rhythm regulation (Unsal-Kaçmaz et al. 2005; Engelen et al. 2013), it is clear that the primary function of metazoan Timeless is as the orthologue of Tof1, and this is equivalent to drosophila Timeout (Gotter et al. 2000). The orthologue of Csm3,

Tipin, was discovered as a Timeless-interacting protein. Like in yeast, the metazoan Timeless-Tipin complex has roles in the replication checkpoint, cohesion establishment, fork progression, protecting stalled replisomes and redox regulation of replication (Gotter et al. 2007; Chan et al. 2003; Yoshizawa-Sugata and Masai 2007; Urtishak et al. 2009; Errico et al. 2007; Unsal-Kaçmaz et al. 2007; Chou and Elledge 2006). However, some additional metazoan-specific functions have emerged. Timeless–Tipin is involved in replication through difficult secondary structures such as G-quadruplexes by recruiting the DDX11 helicase (Lerner et al. 2020). It is recruited by PARP1 to DNA damage to promote homologous recombination (Xie et al. 2015; Young et al. 2015). It also interacts with the ssDNA-binding protein RPA (Gotter et al. 2007; Unsal-Kaçmaz et al. 2007), which may be important for its checkpoint function.

Functional Domains of the Fork Protection Complex

The central core of the Tof1–Csm3 complex is a capped alpha-helical repeat, where a tetrahelical bundle from the middle of Csm3 caps the armadillo repeat fold of the N-terminus of Tof1. Outside of this core, the proteins are mostly unstructured, with a few domains and motifs which vary between the yeast and human complexes (Fig. 3.2). Tof1/Timeless contains a PAB (PARP1-binding) domain which is known to interact with PARP1 in humans. Timeless additionally has two DNA-binding domains in between this PAB domain and the solenoid core. The very C-terminus of Tof1 is known to interact with Topoisomerase I, but the structural details of this are unknown. For Csm3, nothing is known about its C-terminal disordered region, whereas for human and plant Tipin, there is an RPA-interaction motif.

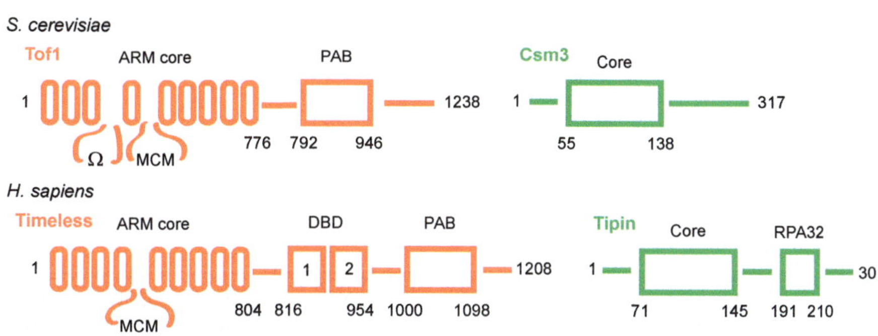

Fig. 3.2 Domain architecture of the FPC. Schematics of the domain architecture of Tof1 and Csm3 from *S. cerevisiae*, and human Timeless and Tipin. RPA32 indicates the RPA32-C interaction motif

The Solenoid Core

The first molecular detail of the solenoid core was a crystal structure of residues 1–463 of human Timeless which revealed an Armadillo (ARM) repeat alpha solenoid forming a horseshoe shape with a conserved pocket in its interior (Holzer et al. 2017). The authors also mapped an extended fragment from 1–814, which was required to pull-down the core folded region (57–160) of Tipin. Following on from this, two studies reported the structure of the entire core Tof1–Csm3 complex. One study used crystallography to determine the structure of a minimal complex from the thermophilic fungus *Chaetomium thermophilum*, using the previously mapped fragments with large predicted loops deleted (Fig. 3.3a) (Grabarczyk 2020). The other study used Cryo-electron microscopy to solve the structure of the entire Tof1–Csm3-bound yeast replisome (Baretić et al. 2020). This showed the same core Tof1–Csm3 structure, but also revealed the structure of many important loops extending from the solenoid, which became ordered upon binding to the replisome.

The solenoid core consists of nine ARM repeats. The final repeat forms the highly hydrophobic interface with Csm3 (Fig. 3.3b). Csm3 forms a tetrahelical bundle which is most closely related to DNA-binding viral proteins such as the SF6 small terminase (Greive et al. 2016). However, this bundle is flattened such that it does not have a hydrophobic core and, instead, resembles an inverted ARM repeat which caps the Tof1 solenoid. In particular, the high conserved α27 helix is inserted into the

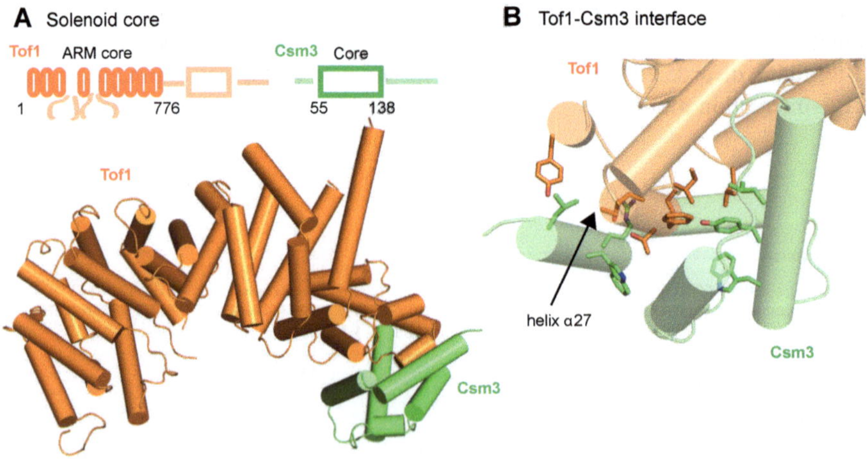

Fig. 3.3 The Tof1–Csm3 solenoid core. (**a**) Crystal structure of the Tof1–Csm3 solenoid core from the thermophilic fungus *Chaetomium thermophilum* (PDB 6XWX) (Grabarczyk 2020). The parts of the FPC which are resolved are indicated in the domain architecture schematic. Throughout, Tof1/Timeless is coloured in orange or yellow, and Csm3/Tipin is in green. (**b**) Details of the Tof1–Csm3 interface. Important residues which make up the predominantly hydrophobic interface are highlighted in stick representation

concave Csm3 bundle (Fig. 3.3b). Interestingly, mutations of this helix have previously been characterized in the *S. pombe* Tof1 orthologue Swi1 and shown to have strong phenotypes (Tanaka et al. 2010; Dalgaard and Klar 2000), similar to a mutation which truncates Tof1 in the middle of this helix (Westhorpe et al. 2020). The other major de*via*tions from the ARM solenoid structure are two large insertions—the Ω-loop between ARM repeats 3 and 4 and the MCM plugin between ARM repeats 4 and 5. These form DNA and CMG interaction sites and will be discussed in further detail below. Importantly, the drosophila Timeless protein, which is involved in circadian rhythm regulation, has neither these key insertions nor conservation of the α27 helix. This confirms that this protein is a structural homologue, but has a different function (Gotter et al. 2000).

Anchoring onto the CMG Helicase

An initial negative stain EM study investigated where the FPC binds to the replisome using 2D classification and showed clear density forming at the front of the helicase where incoming dsDNA is fed in (Eickhoff et al. 2019). The cryo-EM structure of the Tof1–Csm3-bound replisome fully revealed the details of this recruitment (Fig. 3.4a) (Baretić et al. 2020). The interface only involves Tof1, which directly interacts with multiple MCM subunits. Firstly, the exposed N-terminal face of the solenoid packs onto the N-terminal extension of Mcm2, while ARM3 makes some contacts with the Zinc Finger domain of Mcm7. The remainder of the interactions are formed by the two loops which come out of the solenoid. The Tof1 Ω-loop interacts with the zinc finger domains of Mcm6 and Mcm4 (Fig. 3.4b), while the MCM-plugin loop forms an extended structure which wraps around one side of the helicase interacting with Mcm4, Mcm6 and Mcm7 (Fig. 3.4c). The first section of the MCM-plugin is a long helix which bridges Mcm6 and Mcm4. Following this is a folded region which plugs into a pocket in Mcm4 between the OB domain, ZnF domain and helical domain, while also forming a composite antiparallel beta sheet with a strand from the N-terminal region of Mcm4. Interestingly, the Ω-loop has a similar composite beta sheet interaction, but with the N-terminus of Mcm6 (Fig. 3.4b). The final stretch of the MCM-plugin travels back under the Tof1 solenoid, and forms further interactions with N-terminal part of Mcm4.

A dsDNA Grip Enables Fork Pausing and Replisome Progression

2D classification of negative stain data of the replisome has shown that without Tof1–Csm3 the incoming dsDNA is rather flexible and occupies different angles with respect to the helicase (Eickhoff et al. 2019). However, when Tof1–Csm3 is

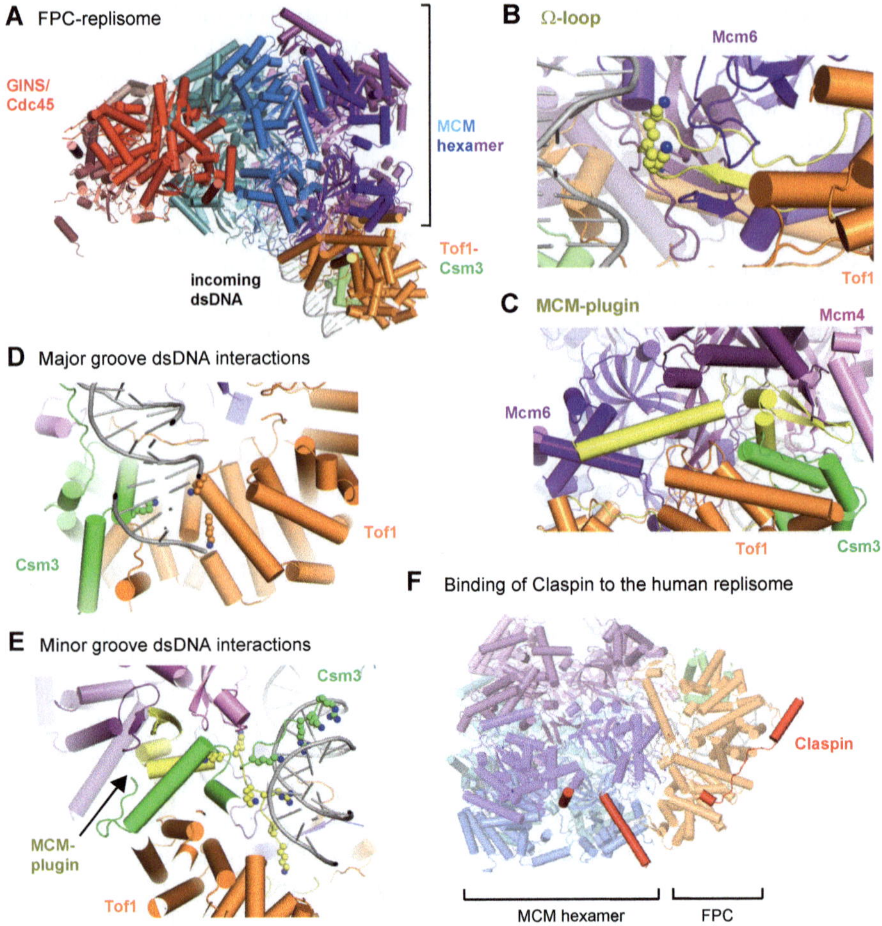

Fig. 3.4 Structure of the FPC within the replisome. (**a**) The Tof1–Csm3 complex binds to front of the yeast replisome (PDB 6SKL) (Baretić et al. 2020). Ctf4 is omitted for clarity. Cdc45 and the GINS complex are coloured in shades of red, while the MCM helicase is coloured in a gradient per subunit from cyan to violet. (**b**) Molecular interactions of the Ω-loop (in yellow) with Mcm6 (in purple) and dsDNA (in grey). Key DNA-interacting lysines are shown in sphere representation. (**c**) Molecular details of the MCM-plugin (yellow) and its interactions with Mcm6 (purple) and Mcm4 (violet). (**d**) dsDNA major groove interactions formed by the concave Tof1–Csm3 interface. Backbone-interacting basic residues are highlighted in sphere representation. (**e**) Minor groove interactions formed by the N-terminal portion of Csm3 (in green) and the MCM-plugin (in yellow). (**f**) Interactions of human Claspin (Mrc1 in yeast) with the replisome (PDB 7PFO) (Jones et al. 2021). Regions of Claspin which could be modelled are shown in red on the MCM helicase and Timeless–Tipin complex. GINS, Cdc45, AND-1 and Pol ε are omitted for clarity

added, the incoming DNA is bound by these proteins and now has a fixed position. This suggests a role for the fork protection complex to guide DNA into the helicase, explaining its role in fork progression. The high resolution features of this grip were revealed by the cryo-EM structure of the yeast replisome (Baretić et al. 2020). The central DNA binding region is a concave surface at the interface between Tof1 and Csm3. Here, the helix in Csm3 equivalent to a dsDNA binding helix in a standard helix-turn-helix fold, as well as Tof1 ARM7, interact with the DNA major groove (Fig. 3.4d). This interaction is further stabilized by a flexible but highly positively charged N-terminal section of Csm3 which binds to the minor groove, and from two loops from Tof1 (Fig. 3.4e). Interestingly, these two Tof1 loops bridge dsDNA with the MCM helicase, suggesting mechanistic importance in helicase activity. The Tof1 Ω-loop is wedged between the zinc finger domains of Mcm6 and Mcm4, which positions the highly basic tip of the loop into the DNA major groove (Fig. 3.4b). The Tof1 MCM-plugin loop is held in place by multiple MCM subunits to contact the minor groove with an arginine and lysine (Fig. 3.4e). Overall, the major- and minor-groove interactions result in three-quarters of a DNA turn being gripped by the Tof1–Csm3 complex.

The interaction mode observed in the cryo-EM structure was also detected biochemically in the paper reporting the crystal structure of the FPC (Grabarczyk 2020). Here, mutagenesis of three basic surface patches was performed to map an observed dsDNA-binding activity. The mutations which disrupted DNA binding were highly consistent with the cryo-EM structure. Baretić and colleagues went further to determine the importance of the dsDNA grip for replisome function both in vitro and in yeast (Baretić et al. 2020). A mutation, with five basic residues mutated to alanine in the DNA-binding regions of both Tof1 and Csm3, resulted in a decrease in the rate of DNA replication in an in vitro reconstituted assay. However, this was potentially due to a decrease in association of both Tof1–Csm3 and Mrc1 to the replisome. They also investigated fork pausing at a protein barrier, and showed this mutation resulted in a complete loss of pausing. The strength of this phenotype, compared to the weaker replication defect, suggested a more direct role for these residues in the pausing activity and not just weakened recruitment of Tof1–Csm3 and Mrc1 to the replisome. To test the importance of these residues in yeast, they used camptothecin to generate covalent topoisomerase I-DNA blocks, and saw a sensitivity approaching a *tof1* deletion strain. As there was no sensitivity to the general fork stalling agent hydroxyurea, this suggested a fork pausing defect of this mutation rather than a checkpoint defect caused by weakened Mrc1 replisome recruitment.

Interactions with the Checkpoint Mediator Mrc1

As discussed above, one key role of the FPC is to recruit Mrc1 to the replisome. The replisome imaged by Baretić and colleagues also included Mrc1, and gives some insight into this recruitment mechanism (Baretić et al. 2020). They saw several extended sections of density across the replisome, which they were unable to assign. To obtain more insight into the identity of the Mrc1 regions occupying these different densities, they performed cross-linking mass spectrometry. They observed that residues in the N-terminal part of Mrc1 (amino acids 300 and 322) interacted across the exterior of the Tof1 solenoid (amino acids 203, 499 and 616). Residues from the middle of Mrc1 interacted with the MCM helicase, and the very C-terminal part of Mrc1 interacted with Cdc45 and Ctf4. Mrc1 is an entirely unfolded protein, which would support this very extended Mrc1 structure. Indeed, an extended structure which bridges together regions of the replisome separated in space could be a way that Mrc1 regulates and coordinates the replisome—especially considering it also contacts the catalytic domain of the leading strand polymerase (Lou et al. 2008). In the *C. thermophilum structure*, the crystal packing resulted in the N-terminal purification tag occupying a conserved cleft in the interior of the solenoid at ARM1–3 (Grabarczyk 2020). A quadruple mutation of residues coordinating this purification tag disrupted the interaction between Tof1–Csm3 and a section of Mrc1 from amino acid 400–800, suggesting that the peptide may be mimicking a section of Mrc1. A truncation study of Tof1 showed that its first 627 amino acids are sufficient for its checkpoint function (Westhorpe et al. 2020), consistent with the structural data showing that the sections of Tof1 interaction with Mrc1 are N-terminal to residue 627.

Mechanistic Conservation in Humans

The Yeeles lab recently published a structure of the human replisome, including the Timeless–Tipin complex and Claspin, the human Mrc1 orthologue (Jones et al. 2021). This showed a highly similar overall structure, meaning that the mechanistic function of the fork protection complex is highly conserved across eukaryotes. Some small differences were observed, such as the lack of the Ω-loop in humans, but it is not clear how this affects function. Most importantly, the higher quality of the data allowed them to assign the sequence of Claspin to the additional observed density (Fig. 3.4f), and these assignments were consistent with their previous XL-MS data from the yeast replisome. They observed five additional stretches of density; however, only four of them were lost in a structure without Claspin. The largest Timeless–Claspin interaction involves a helix of Claspin from residue 284 to 299, followed by an extended region up to 319 (Fig. 3.4f). Unusually for an ARM repeat protein, this section binds on the other surface of the solenoid contacting ARM repeats 4, 5, 6 and 7. The majority of the other Claspin contacts are with the MCM helicase (Fig. 3.4f). Intriguingly, the density equivalent to the purification tag in the *C. thermophilum* Tof1–Csm3 crystal structure (Grabarczyk 2020) is not dependent on the presence of Claspin, and instead appears to be occupied by the N-terminus of Mcm2. It will be interesting to see in the yeast replisome whether Mrc1 or Mcm2 occupies this cleft and what the role of this interaction might be.

The PARP1-Binding PAB Domain

200 amino acids C-terminal to the solenoid core of human Timeless lies a PARP1 binding (PAB) domain (Fig. 3.5a) (Xie et al. 2015). This forms a bundle of five helices fronted by three strands and has some homology to PUB (Peptide:N-glycanase/UBA or UBX-containing proteins) protein-interaction domains. In isolation, this domain forms a homodimer, which could be seen in a crystal structure, with the interface comprising the loops between the strands and helix α3 (Fig. 3.5a). Co-immunoprecipitation and mass spectrometry showed that PARP1 was the main binder of this domain, and more specifically the catalytic domain of PARP1. These two domains interact tightly with a nanomolar affinity, and the complex was

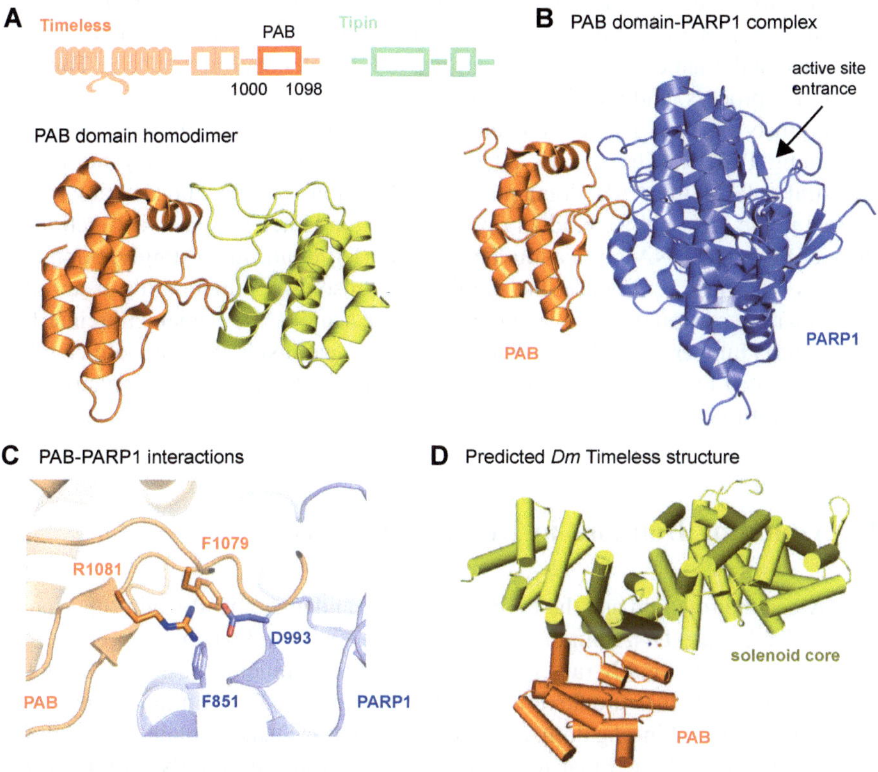

Fig. 3.5 The PARP1-binding timeless PAB domain. (**a**) Structure of the isolated PAB domain shows a homodimer (PDB 4XHT) (Xie et al. 2015). One monomer is coloured in orange, and the second is in yellow. The PAB domain is highlighted within the domain architecture schematic for context. (**b**) Crystal structure of the complex between the PAB domain (orange) and PARP1 (blue) (PDB 4XHU). (**c**) Important molecular interactions which stabilize the PAB domain-PARP1 complex. (**d**) Structure of Drosophila Timeless predicted by Alphafold (Jumper et al. 2021) showing an intramolecular interaction between the PAB domain (orange) and the solenoid core (yellow)

structurally characterized (Fig. 3.5b). Interestingly, the same interface for homodimerisation of the PAB is used for interacting with PARP1. The complex is stabilized by two large electrostatic interfaces as well as a stacking interaction between two phenylalanines (Fig. 3.5c). The interacting residues on PARP1 are not conserved in other PARP family members, and indeed the authors showed that PARP2 and PARP3 do not interact with Timeless. PARP1 is highly regulated by intramolecular interactions. However, the binding of the PAB domain did not affect the conformation of the PARP1 catalytic domain and did not block the active site. In agreement with these observations, the interaction does not affect PARP1 activity, and thus likely exists as a recruitment mechanism. Indeed, when PARP inhibitors were used to trap PARP1 onto DNA breaks, there was an increase in Timeless recruitment. This recruitment was dependent on PARP1 and could be abolished by an R1081G mutation in Timeless, which abrogated binding in vitro. Interestingly, this mutation resulted in defects in homologous recombination. A PARP1-mediated role of Timeless in homologous recombination was confirmed in another study published at the same time (Young et al. 2015). It is still not clear what role Timeless has in this process, and whether it relates to DNA replication.

Curiously, yeast Tof1 also contains a PAB domain, despite this organism not containing PARP1 (Grabarczyk 2020). This suggests that there may be a more ancestral role of the PAB domain in Tof1 function. In fact, the domain is also present in the drosophila circadian rhythm regulator Timeless, which is functionally distinct from the DNA replication Timeless/Tof1 protein. Interestingly, Alphafold predicts with high confidence that the PAB in drosophila Timeless forms an intramolecular interaction with the solenoid using the same PARP1-interacting interface (Fig. 3.5d) (Jumper et al. 2021). This observation, alongside the homodimerization of the isolated human PAB, could indicate that this domain has a regulatory role which is yet to be discovered.

DNA-binding Domains in Human Timeless

In between the solenoid and the PAB domain are tandem homeodomain superfamily three-helix bundle DNA-binding domains (DBDs) (Fig. 3.6a). Lerner et al. used NMR spectroscopy and X-ray crystallography to determine the structure of the two domains and show they are flexibly linked (Lerner et al. 2020). Unlike normal three-helix bundle DNA-binding domains, both the Timeless DBDs contain a fourth helix, which is positioned such that it would sterically clash with the dsDNA backbone when the helix-turn-helix motif is inserted in the major groove (Fig. 3.6b). Consistent with this, the domains had no preference for dsDNA over ssDNA, and bound both with a weak micromolar affinity. A previous study had shown that Timeless activates the helicase DDX11 to unwind G-quadruplex structures, and in fact recruits DDX11 to this particular DNA secondary structure (Calì et al. 2016). Following on from this, Lerner et al. showed that Timeless is required for processive replication past G-quadruplex structures in cells. Thus, they investigated whether the DBDs

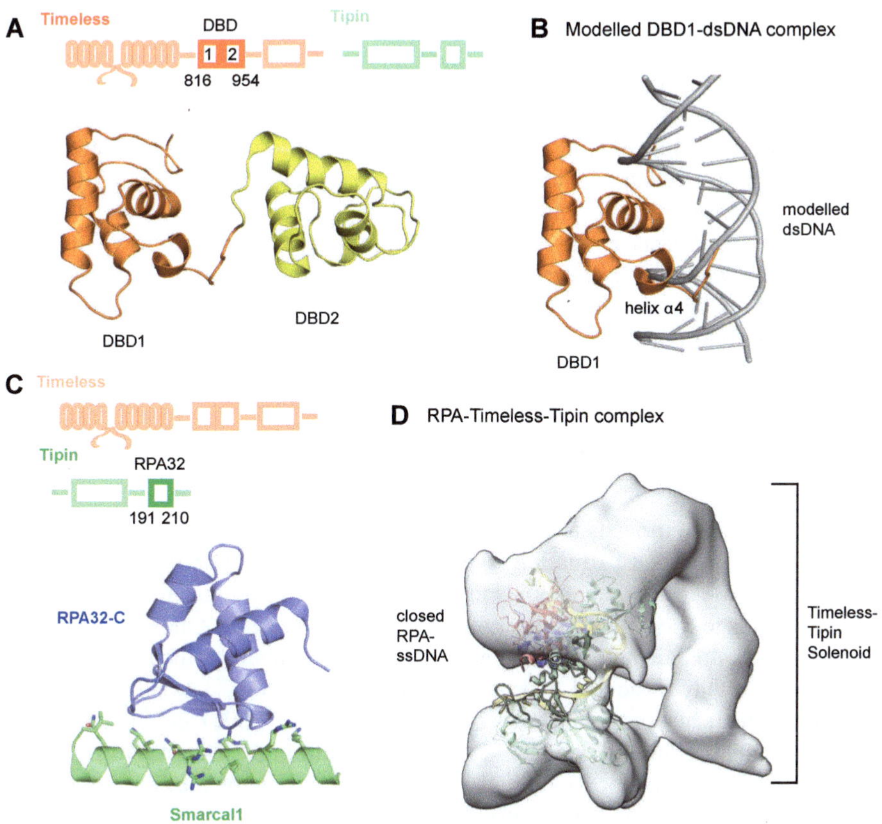

Fig. 3.6 Further interactions of the FPC. (**a**) NMR structure of the tandem DBDs found in Timeless (PDB 6TAZ) (Lerner et al. 2020). (**b**) Model of the Timeless DBD1 binding to dsDNA. DBD1 is overlaid on the structure of DNA-bound hTRF1 (PDB 1W0T) (Court et al. 2005). (**c**) The RPA32-C interacting motif. Above, the position of the RPA32 interacting motif on human Tipin within the context of its overall domain architecture is shown. Below this is the crystal structure of the complex between the C-terminal winged-helix domain of RPA32 (blue) and a peptide derived from Smarcal1, coloured in green (PDB 4MQV) (Xie et al. 2014). Residues which are either absolutely conserved or R:K exchanged between the Smarcal1 motif and the RPA32-interacting motif on Tipin are highlighted in stick representation. (**d**) Low resolution cryo-EM map of the Timeless–Tipin–RPA complex (EMDB 2789) (Witosch et al. 2014). The closed conformation of RPA, which is normally induced by binding to long stretches of ssDNA (PDB 4GNX) (Fan and Pavletich 2012), can be fit into the map, leaving unmodelled density reminiscent of the Timeless–Tipin alpha–solenoid

might recognize G-quadruplexes. They observed approximately ten-fold stronger binding to these structures compared to dsDNA and ssDNA. Additionally, they saw that the full Timeless–Tipin complex also bound tighter to DNA containing G-quadruplexes. Truncating Timeless after the solenoid caused a strong defect in processive replication though G-quadruplexes in cells. Curiously, both the DBD and

PAB could rescue this effect to some degree, meaning they both function in this process.

The interaction between Timeless and DDX11 has been mapped to an EYE sequence motif at residue 201 in a disordered loop of DDX11, and it has been shown that this binding is important for cohesion (Cortone et al. 2018) establishment as well as dealing with replication stress (Calì et al. 2016). The DDX11 EYE motif is also required for efficient replication through G-quadruplexes (Lerner et al. 2020). As the catalytic helicase activity of DDX11 is needed as well, it has been proposed that Timeless identifies fork-blocking G-quadruplexes with its DBDs, and recruits DDX11 to unwind these, allowing the replication fork to proceed.

Tipin-RPA Interactions

As ssDNA is toxic, it needs to be covered. In eukaryotes, this role is played by the RPA complex. As well as binding ssDNA with high affinity, RPA also contains various protein–protein interaction modules which are important for coordinating the DNA damage checkpoint, DNA repair and DNA replication. One such module is a winged helix domain at the C-terminus of the RPA32 subunit which interacts with an extended highly basic and hydrophobic helical motif (Fig. 3.6c). This has been structurally characterized in the cases of UNG2 (Mer et al. 2000) and Smarcal1 (Xie et al. 2014). This conserved helical motif is also found in the C-terminus of human Tipin (Fig. 3.6c), but not in fungal Csm3, and accordingly the Timeless–Tipin complex binds to RPA in vivo (Unsal-Kaçmaz et al. 2007; Gotter et al. 2007) and this is important in the checkpoint response (Kemp et al. 2010). NMR spectroscopy and ITC have been used to show that the helical peptide of Tipin binds to the RPA32 winged helix domain like UNG2 and Smarcal1 and with a low micromolar affinity (Xie et al. 2014).

To investigate the FPC–RPA interaction in more detail, Witosch and colleagues reconstituted the Timeless–Tipin–RPA complex (Witosch et al. 2014). Surprisingly, the RPA32 winged helix domain was not sufficient for the interaction, but rather it seemed to involve all the RPA subunits and multiple interfaces. This was further confirmed by a low resolution cryo-EM structure where the structure of RPA can be docked in its compact conformation (Fig. 3.6d). The remaining Timeless–Tipin density wraps around RPA, contacting it at several locations. The compact conformation of RPA is normally induced by it binding to long stretches of ssDNA so that all the DNA binding modules on RPA interact with the same DNA molecule (Fan and Pavletich 2012; Bochkareva et al. 2002). When RPA binds to shorter DNA fragments, it has a more extended conformation. Interestingly, addition of short ssDNA caused dissociation of the Timeless–Tipin–RPA complex, suggesting that DNA binds more tightly to RPA than Timeless–Tipin, and that the compact conformation of RPA enforced by Timeless–Tipin is incompatible with binding to short ssDNA (Witosch et al. 2014). These switches could regulate Timeless–Tipin function at the replication fork where there are varying amounts of RPA and ssDNA, depending on levels of replication stress.

Unresolved Interactions

Beyond the many structurally characterized interactions discussed so far are a number of binding events where the molecular details remain unclear. It is known that Topoisomerase I interacts with the C-terminus of Tof1 (Shyian et al. 2020; Westhorpe et al. 2020), but this has not been mapped more specifically. As previously mentioned, an EYE motif in a loop in DDX11 is known to bind Timeless and this is important for replication stress (Calì et al. 2016) and cohesion establishment (Cortone et al. 2018), but where it interacts on the side of Timeless has not been determined. Other interactors include the splicing factor SDE2 (Rageul et al. 2020), the redox regulator PRDX2 (Somyajit et al. 2017) and the Dbf4-dependent replication initiation DDK kinase (Murakami and Keeney 2014).

Outlook

Although there has been much progress recently in determining the structure of the Fork Protection Complex and the molecular mechanism of how it acts during DNA replication, much still remains to be resolved. Current structures of the FPC-bound replisome represent a minimalistic replisome undergoing DNA unwinding as would occur during normal replication. However, many of the functions of the FPC relate to specific replication-associated events at the fork, for example, cohesion establishment, S-phase checkpoint signalling and fork pausing. It will now be important to perform in vitro reconstituted DNA replication reactions under conditions where these events can be trapped and visualized by cryo-EM to reveal exactly how the FPC is involved. However, it is likely that many of these events are highly dynamic, and thus an integrative structural biology approach will be required.

Now that it is clear that the FPC is bound to the front of the fork, it is necessary to consider its previously characterized functions with this in mind – as an interaction hub at the head of DNA replication, which can also directly regulate the replicative helicase. Such a position is highly consistent with recruitment of topoisomerases to upstream dsDNA. Similarly, Tof1–Csm3 is well placed to transfer cohesin from the front of the fork to the daughter strands, together with Ctf4 and Chl1 which are located behind the helicase. However, the interaction of Tof1–Csm3 with RPA is harder to reconcile, considering that RPA would coat the ssDNA behind the helicase. One possibility is that Tof1–Csm3 might not always be bound to the front of the replisome. Indeed, PRDX2-mediated regulation of fork speed is proposed to involve displacing Timeless from the replisome. Together with structural biology and biochemistry, single-molecule methodology will be a valuable tool for resolving these types of questions.

References

Alcasabas AA, Osborn AJ, Bachant J, Hu F, Werler PJ, Bousset K, Furuya K, Diffley JF, Carr AM, Elledge SJ (2001) Mrc1 transduces signals of DNA replication stress to activate Rad53. Nat Cell Biol 3(11):958–965. https://doi.org/10.1038/ncb1101-958

Bando M, Katou Y, Komata M, Tanaka H, Itoh T, Sutani T, Shirahige K (2009) Csm3, Tof1, and Mrc1 form a heterotrimeric mediator complex that associates with DNA replication forks. J Biol Chem 284(49):34355–34365. https://doi.org/10.1074/jbc.M109.065730

Baretić D, Jenkyn-Bedford M, Aria V, Cannone G, Skehel M, Yeeles JTP (2020) Cryo-EM structure of the fork protection complex bound to CMG at a replication fork. Mol Cell 78(5): 926–940.e913. https://doi.org/10.1016/j.molcel.2020.04.012

Bastia D, Srivastava P, Zaman S, Choudhury M, Mohanty BK, Bacal J, Langston LD, Pasero P, O'Donnell ME (2016) Phosphorylation of CMG helicase and Tof1 is required for programmed fork arrest. Proc Natl Acad Sci U S A 113(26):E3639–E3648. https://doi.org/10.1073/pnas.1607552113

Bell SP, Labib K (2016) Chromosome duplication in Saccharomyces cerevisiae. Genetics 203(3): 1027–1067. https://doi.org/10.1534/genetics.115.186452

Bernard P, Schmidt CK, Vaur S, Dheur S, Drogat J, Genier S, Ekwall K, Uhlmann F, Javerzat JP (2008) Cell-cycle regulation of cohesin stability along fission yeast chromosomes. EMBO J 27(1):111–121. https://doi.org/10.1038/sj.emboj.7601955

Bharti SK, Sommers JA, George F, Kuper J, Hamon F, Shin-ya K, Teulade-Fichou M-P, Kisker C, Brosh RM (2013) Specialization among iron-sulfur cluster helicases to resolve G-quadruplex DNA structures that threaten genomic stability. J Biol Chem 288(39):28217–28229

Bhattacharya C, Wang X, Becker D (2012) The DEAD/DEAH box helicase, DDX11, is essential for the survival of advanced melanomas. Mol Cancer 11(1):82

Bianco JN, Bergoglio V, Lin YL, Pillaire MJ, Schmitz AL, Gilhodes J, Lusque A, Mazières J, Lacroix-Triki M, Roumeliotis TI, Choudhary J, Moreaux J, Hoffmann JS, Tourrière H, Pasero P (2019) Overexpression of Claspin and Timeless protects cancer cells from replication stress in a checkpoint-independent manner. Nat Commun 10(1):910. https://doi.org/10.1038/s41467-019-08886-8

Bochkareva E, Korolev S, Lees-Miller SP, Bochkarev A (2002) Structure of the RPA trimerization core and its role in the multistep DNA-binding mechanism of RPA. EMBO J 21(7):1855–1863. https://doi.org/10.1093/emboj/21.7.1855

Bowman GD, O'Donnell M, Kuriyan J (2004) Structural analysis of a eukaryotic sliding DNA clamp-clamp loader complex. Nature 429(6993):724–730. https://doi.org/10.1038/nature02585

Calì F, Bharti SK, Di Perna R, Brosh RM, Pisani FM (2016) Tim/Timeless, a member of the replication fork protection complex, operates with the Warsaw breakage syndrome DNA helicase DDX11 in the same fork recovery pathway. Nucleic Acids Res 44(2):705–717. https://doi.org/10.1093/nar/gkv1112

Calzada A, Hodgson B, Kanemaki M, Bueno A, Labib K (2005) Molecular anatomy and regulation of a stable replisome at a paused eukaryotic DNA replication fork. Genes Dev 19(16): 1905–1919. https://doi.org/10.1101/gad.337205

Chan RC, Chan A, Jeon M, Wu TF, Pasqualone D, Rougvie AE, Meyer BJ (2003) Chromosome cohesion is regulated by a clock gene paralogue TIM-1. Nature 423(6943):1002–1009. https://doi.org/10.1038/nature01697

Chen SH, Zhou H (2009) Reconstitution of Rad53 activation by Mec1 through adaptor protein Mrc1. J Biol Chem 284(28):18593–18604. https://doi.org/10.1074/jbc.M109.018242

Chou DM, Elledge SJ (2006) Tipin and Timeless form a mutually protective complex required for genotoxic stress resistance and checkpoint function. Proc Natl Acad Sci U S A 103(48): 18143–18147. https://doi.org/10.1073/pnas.0609251103

Ciosk R, Shirayama M, Shevchenko A, Tanaka T, Toth A, Nasmyth K (2000) Cohesin's binding to chromosomes depends on a separate complex consisting of Scc2 and Scc4 proteins. Mol Cell 5(2):243–254. https://doi.org/10.1016/s1097-2765(00)80420-7

Cortone G, Zheng G, Pensieri P, Chiappetta V, Tatè R, Malacaria E, Pichierri P, Yu H, Pisani FM (2018) Interaction of the Warsaw breakage syndrome DNA helicase DDX11 with the replication fork-protection factor Timeless promotes sister chromatid cohesion. PLoS Genet 14(10): e1007622. https://doi.org/10.1371/journal.pgen.1007622

Court R, Chapman L, Fairall L, Rhodes D (2005) How the human telomeric proteins TRF1 and TRF2 recognize telomeric DNA: a view from high-resolution crystal structures. EMBO Rep 6(1):39–45. https://doi.org/10.1038/sj.embor.7400314

Crabbé L, Thomas A, Pantesco V, De Vos J, Pasero P, Lengronne A (2010) Analysis of replication profiles reveals key role of RFC-Ctf18 in yeast replication stress response. Nat Struct Mol Biol 17(11):1391–1397. https://doi.org/10.1038/nsmb.1932

Cucco F, Musio A (2016) Genome stability: What we have learned from cohesinopathies. Am J Med Genet C Semin Med Genet 172(2):171–178. https://doi.org/10.1002/ajmg.c.31492

Dalgaard JZ, Klar AJ (2000) swi1 and swi3 perform imprinting, pausing, and termination of DNA replication in S. pombe. Cell 102(6):745–751. https://doi.org/10.1016/s0092-8674(00)00063-5

Deegan TD, Baxter J, Ortiz Bazán M, Yeeles JTP, Labib KPM (2019) Pif1-family helicases support fork convergence during DNA replication termination in eukaryotes. Mol Cell 74(2):231–244. e239. https://doi.org/10.1016/j.molcel.2019.01.040

Deegan TD, Mukherjee PP, Fujisawa R, Polo Rivera C, Labib K (2020) CMG helicase disassembly is controlled by replication fork DNA, replisome components and a ubiquitin threshold. Elife 9. https://doi.org/10.7554/eLife.60371

Deshpande I, Seeber A, Shimada K, Keusch JJ, Gut H, Gasser SM (2017) Structural basis of Mec1-Ddc2-RPA Assembly and activation on single-stranded DNA at sites of damage. Mol Cell 68(2):431–445.e435. https://doi.org/10.1016/j.molcel.2017.09.019

Dewar JM, Budzowska M, Walter JC (2015) The mechanism of DNA replication termination in vertebrates. Nature 525(7569):345–350. https://doi.org/10.1038/nature14887

Douglas ME, Ali FA, Costa A, Diffley JFX (2018) The mechanism of eukaryotic CMG helicase activation. Nature 555(7695):265–268. https://doi.org/10.1038/nature25787

Eickhoff P, Kose HB, Martino F, Petojevic T, Abid Ali F, Locke J, Tamberg N, Nans A, Berger JM, Botchan MR, Yardimci H, Costa A (2019) Molecular basis for ATP-hydrolysis-driven DNA translocation by the CMG helicase of the eukaryotic replisome. Cell Rep 28(10):2673–2688. e2678. https://doi.org/10.1016/j.celrep.2019.07.104

Engelen E, Janssens RC, Yagita K, Smits VA, van der Horst GT, Tamanini F (2013) Mammalian TIMELESS is involved in period determination and DNA damage-dependent phase advancing of the circadian clock. PLoS One 8(2):e56623. https://doi.org/10.1371/journal.pone.0056623

Errico A, Costanzo V (2012) Mechanisms of replication fork protection: a safeguard for genome stability. Crit Rev Biochem Mol Biol 47(3):222–235. https://doi.org/10.3109/10409238.2012.655374

Errico A, Costanzo V, Hunt T (2007) Tipin is required for stalled replication forks to resume DNA replication after removal of aphidicolin in Xenopus egg extracts. Proc Natl Acad Sci U S A 104(38):14929–14934. https://doi.org/10.1073/pnas.0706347104

Fan J, Pavletich NP (2012) Structure and conformational change of a replication protein A heterotrimer bound to ssDNA. Genes Dev 26(20):2337–2347. https://doi.org/10.1101/gad.194787.112

Foltman M, Evrin C, De Piccoli G, Jones RC, Edmondson RD, Katou Y, Nakato R, Shirahige K, Labib K (2013) Eukaryotic replisome components cooperate to process histones during chromosome replication. Cell Rep 3(3):892–904. https://doi.org/10.1016/j.celrep.2013.02.028

Foss EJ (2001) Tof1p regulates DNA damage responses during S phase in Saccharomyces cerevisiae. Genetics 157(2):567–577

Gaillard H, García-Muse T, Aguilera A (2015) Replication stress and cancer. Nat Rev Cancer 15(5): 276–289. https://doi.org/10.1038/nrc3916

Gambus A, Jones RC, Sanchez-Diaz A, Kanemaki M, van Deursen F, Edmondson RD, Labib K (2006) GINS maintains association of Cdc45 with MCM in replisome progression complexes at eukaryotic DNA replication forks. Nat Cell Biol 8(4):358–366. https://doi.org/10.1038/ncb1382

Gan H, Serra-Cardona A, Hua X, Zhou H, Labib K, Yu C, Zhang Z (2018) The Mcm2-Ctf4-Polα axis facilitates parental histone H3-H4 transfer to lagging strands. Mol Cell 72(1):140–151. e143. https://doi.org/10.1016/j.molcel.2018.09.001

Garbacz MA, Lujan SA, Burkholder AB, Cox PB, Wu Q, Zhou ZX, Haber JE, Kunkel TA (2018) Evidence that DNA polymerase δ contributes to initiating leading strand DNA replication in Saccharomyces cerevisiae. Nat Commun 9(1):858. https://doi.org/10.1038/s41467-018-03270-4

Gellon L, Kaushal S, Cebrián J, Lahiri M, Mirkin SM, Freudenreich CH (2019) Mrc1 and Tof1 prevent fragility and instability at long CAG repeats by their fork stabilizing function. Nucleic Acids Res 47(2):794–805. https://doi.org/10.1093/nar/gky1195

Goswami P, Abid Ali F, Douglas ME, Locke J, Purkiss A, Janska A, Eickhoff P, Early A, Nans A, Cheung AMC, Diffley JFX, Costa A (2018) Structure of DNA-CMG-Pol epsilon elucidates the roles of the non-catalytic polymerase modules in the eukaryotic replisome. Nat Commun 9(1): 5061. https://doi.org/10.1038/s41467-018-07417-1

Gotter AL (2003) Tipin, a novel timeless-interacting protein, is developmentally co-expressed with timeless and disrupts its self-association. J Mol Biol 331(1):167–176

Gotter AL, Manganaro T, Weaver DR, Kolakowski LF, Possidente B, Sriram S, MacLaughlin DT, Reppert SM (2000) A time-less function for mouse timeless. Nat Neurosci 3(8):755–756. https://doi.org/10.1038/77653

Gotter AL, Suppa C, Emanuel BS (2007) Mammalian TIMELESS and Tipin are evolutionarily conserved replication fork-associated factors. J Mol Biol 366(1):36–52. https://doi.org/10.1016/j.jmb.2006.10.097

Grabarczyk DB (2020) Crystal structure and interactions of the Tof1-Csm3 (Timeless-Tipin) fork protection complex. Nucleic Acids Res 48(12):6996–7004. https://doi.org/10.1093/nar/gkaa456

Grabarczyk DB, Silkenat S, Kisker C (2018) Structural basis for the recruitment of Ctf18-RFC to the replisome. Structure 26(1):137–144.e133. https://doi.org/10.1016/j.str.2017.11.004

Greive SJ, Fung HK, Chechik M, Jenkins HT, Weitzel SE, Aguiar PM, Brentnall AS, Glousieau M, Gladyshev GV, Potts JR, Antson AA (2016) DNA recognition for virus assembly through multiple sequence-independent interactions with a helix-turn-helix motif. Nucleic Acids Res 44(2):776–789. https://doi.org/10.1093/nar/gkv1467

Guilliam TA (2021) Mechanisms for maintaining eukaryotic replisome progression in the presence of DNA damage. Front Mol Biosci 8:712971. https://doi.org/10.3389/fmolb.2021.712971

Guo M, Hundseth K, Ding H, Vidhyasagar V, Inoue A, Nguyen CH, Zain R, Lee JS, Wu Y (2015) A distinct triplex DNA unwinding activity of ChlR1 helicase. J Biol Chem 290(8):5174–5189. https://doi.org/10.1074/jbc.M114.634923

Hanna JS, Kroll ES, Lundblad V, Spencer FA (2001) Saccharomyces cerevisiae CTF18 and CTF4 are required for sister chromatid cohesion. Mol Cell Biol 21(9):3144–3158. https://doi.org/10.1128/MCB.21.9.3144-3158.2001

He H, Li Y, Dong Q, Chang AY, Gao F, Chi Z, Su M, Zhang F, Ban H, Martienssen R, Chen YH, Li F (2017) Coordinated regulation of heterochromatin inheritance by Dpb3-Dpb4 complex. Proc Natl Acad Sci U S A 114(47):12524–12529. https://doi.org/10.1073/pnas.1712961114

Hizume K, Endo S, Muramatsu S, Kobayashi T, Araki H (2018) DNA polymerase ε-dependent modulation of the pausing property of the CMG helicase at the barrier. Genes Dev 32(19–20): 1315–1320. https://doi.org/10.1101/gad.317073.118

Hodgson B, Calzada A, Labib K (2007) Mrc1 and Tof1 regulate DNA replication forks in different ways during normal S phase. Mol Biol Cell 18(10):3894–3902. https://doi.org/10.1091/mbc.E07-05-0500

Holzer S, Degliesposti G, Kilkenny ML, Maslen SL, Matak-Vinkovíc D, Skehel M, Pellegrini L (2017) Crystal structure of the N-terminal domain of human Timeless and its interaction with Tipin. Nucleic Acids Res 45(9):5555–5563. https://doi.org/10.1093/nar/gkx139

Huang ME, Kolodner RD (2005) A biological network in Saccharomyces cerevisiae prevents the deleterious effects of endogenous oxidative DNA damage. Mol Cell 17(5):709–720. https://doi.org/10.1016/j.molcel.2005.02.008

Jones ML, Baris Y, Taylor MRG, Yeeles JTP (2021) Structure of a human replisome shows the organisation and interactions of a DNA replication machine. EMBO J 40:e108819. https://doi.org/10.15252/embj.2021108819

Jumper J, Evans R, Pritzel A, Green T, Figurnov M, Ronneberger O, Tunyasuvunakool K, Bates R, Žídek A, Potapenko A, Bridgland A, Meyer C, Kohl SAA, Ballard AJ, Cowie A, Romera-Paredes B, Nikolov S, Jain R, Adler J, Back T, Petersen S, Reiman D, Clancy E, Zielinski M, Steinegger M, Pacholska M, Berghammer T, Bodenstein S, Silver D, Vinyals O, Senior AW, Kavukcuoglu K, Kohli P, Hassabis D (2021) Highly accurate protein structure prediction with AlphaFold. Nature 596(7873):583–589. https://doi.org/10.1038/s41586-021-03819-2

Katou Y, Kanoh Y, Bando M, Noguchi H, Tanaka H, Ashikari T, Sugimoto K, Shirahige K (2003) S-phase checkpoint proteins Tof1 and Mrc1 form a stable replication-pausing complex. Nature 424(6952):1078–1083. https://doi.org/10.1038/nature01900

Kemp MG, Akan Z, Yilmaz S, Grillo M, Smith-Roe SL, Kang TH, Cordeiro-Stone M, Kaufmann WK, Abraham RT, Sancar A, Unsal-Kaçmaz K (2010) Tipin-replication protein A interaction mediates Chk1 phosphorylation by ATR in response to genotoxic stress. J Biol Chem 285(22):16562–16571. https://doi.org/10.1074/jbc.M110.110304

Klinge S, Núñez-Ramírez R, Llorca O, Pellegrini L (2009) 3D architecture of DNA Pol alpha reveals the functional core of multi-subunit replicative polymerases. EMBO J 28(13):1978–1987. https://doi.org/10.1038/emboj.2009.150

Koike N, Hida A, Numano R, Hirose M, Sakaki Y, Tei H (1998) Identification of the mammalian homologues of the Drosophila timeless gene, Timeless1. FEBS Lett 441(3):427–431. https://doi.org/10.1016/s0014-5793(98)01597-x

Komata M, Bando M, Araki H, Shirahige K (2009) The direct binding of Mrc1, a checkpoint mediator, to Mcm6, a replication helicase, is essential for the replication checkpoint against methyl methanesulfonate-induced stress. Mol Cell Biol 29(18):5008–5019. https://doi.org/10.1128/MCB.01934-08

Lancey C, Tehseen M, Raducanu VS, Rashid F, Merino N, Ragan TJ, Savva CG, Zaher MS, Shirbini A, Blanco FJ, Hamdan SM, De Biasio A (2020) Structure of the processive human Pol δ holoenzyme. Nat Commun 11(1):1109. https://doi.org/10.1038/s41467-020-14898-6

Langston LD, Zhang D, Yurieva O, Georgescu RE, Finkelstein J, Yao NY, Indiani C, O'Donnell ME (2014) CMG helicase and DNA polymerase ε form a functional 15-subunit holoenzyme for eukaryotic leading-strand DNA replication. Proc Natl Acad Sci U S A 111(43):15390–15395. https://doi.org/10.1073/pnas.1418334111

Lengronne A, McIntyre J, Katou Y, Kanoh Y, Hopfner KP, Shirahige K, Uhlmann F (2006) Establishment of sister chromatid cohesion at the S. cerevisiae replication fork. Mol Cell 23(6):787–799. https://doi.org/10.1016/j.molcel.2006.08.018

Lerner LK, Holzer S, Kilkenny ML, Šviković S, Murat P, Schiavone D, Eldridge CB, Bittleston A, Maman JD, Branzei D, Stott K, Pellegrini L, Sale JE (2020) Timeless couples G-quadruplex detection with processing by DDX11 helicase during DNA replication. EMBO J 39(18):e104185. https://doi.org/10.15252/embj.2019104185

Lewis JS, Spenkelink LM, Schauer GD, Hill FR, Georgescu RE, O'Donnell ME, van Oijen AM (2017) Single-molecule visualization of Saccharomyces cerevisiae leading-strand synthesis reveals dynamic interaction between MTC and the replisome. Proc Natl Acad Sci U S A 114(40):10630–10635. https://doi.org/10.1073/pnas.1711291114

Lewis JS, Spenkelink LM, Schauer GD, Yurieva O, Mueller SH, Natarajan V, Kaur G, Maher C, Kay C, O'Donnell ME, van Oijen AM (2020) Tunability of DNA polymerase stability during eukaryotic DNA replication. Mol Cell 77(1):17–25.e15. https://doi.org/10.1016/j.molcel.2019.10.005

Li N, Lam WH, Zhai Y, Cheng J, Cheng E, Zhao Y, Gao N, Tye BK (2018) Structure of the origin recognition complex bound to DNA replication origin. Nature 559(7713):217–222. https://doi.org/10.1038/s41586-018-0293-x

Liu HW, Bouchoux C, Panarotto M, Kakui Y, Patel H, Uhlmann F (2020) Division of labor between PCNA loaders in DNA replication and sister chromatid cohesion establishment. Mol Cell 78(4):725–738.e724. https://doi.org/10.1016/j.molcel.2020.03.017

Lou H, Komata M, Katou Y, Guan Z, Reis CC, Budd M, Shirahige K, Campbell JL (2008) Mrc1 and DNA polymerase epsilon function together in linking DNA replication and the S phase checkpoint. Mol Cell 32(1):106–117. https://doi.org/10.1016/j.molcel.2008.08.020

Macheret M, Halazonetis TD (2015) DNA replication stress as a hallmark of cancer. Annu Rev Pathol 10:425–448. https://doi.org/10.1146/annurev-pathol-012414-040424

Maric M, Maculins T, De Piccoli G, Labib K (2014) Cdc48 and a ubiquitin ligase drive disassembly of the CMG helicase at the end of DNA replication. Science 346(6208):1253596. https://doi.org/10.1126/science.1253596

Mayer ML, Pot I, Chang M, Xu H, Aneliunas V, Kwok T, Newitt R, Aebersold R, Boone C, Brown GW, Hieter P (2004) Identification of protein complexes required for efficient sister chromatid cohesion. Mol Biol Cell 15(4):1736–1745. https://doi.org/10.1091/mbc.e03-08-0619

McLellan J, O'Neil N, Tarailo S, Stoepel J, Bryan J, Rose A, Hieter P (2009) Synthetic lethal genetic interactions that decrease somatic cell proliferation in Caenorhabditis elegans identify the alternative RFC CTF18 as a candidate cancer drug target. Mol Biol Cell 20(24):5306–5313. https://doi.org/10.1091/mbc.E09-08-0699

Mer G, Bochkarev A, Gupta R, Bochkareva E, Frappier L, Ingles CJ, Edwards AM, Chazin WJ (2000) Structural basis for the recognition of DNA repair proteins UNG2, XPA, and RAD52 by replication factor RPA. Cell 103(3):449–456. https://doi.org/10.1016/s0092-8674(00)00136-7

Michaelis C, Ciosk R, Nasmyth K (1997) Cohesins: chromosomal proteins that prevent premature separation of sister chromatids. Cell 91(1):35–45. https://doi.org/10.1016/s0092-8674(01)80007-6

Mohanty BK, Bairwa NK, Bastia D (2006) The Tof1p-Csm3p protein complex counteracts the Rrm3p helicase to control replication termination of Saccharomyces cerevisiae. Proc Natl Acad Sci U S A 103(4):897–902. https://doi.org/10.1073/pnas.0506540103

Moriel-Carretero M, Pasero P, Pardo B (2019) DDR Inc., one business, two associates. Curr Genet 65(2):445–451. https://doi.org/10.1007/s00294-018-0908-7

Mukherjee PP, Labib KPM (2019) *In vitro* reconstitution defines the minimal requirements for Cdc48-dependent disassembly of the CMG helicase in budding yeast. Cell Rep 28(11):2777–2783.e2774. https://doi.org/10.1016/j.celrep.2019.08.026

Murakami H, Keeney S (2014) Temporospatial coordination of meiotic DNA replication and recombination *via* DDK recruitment to replisomes. Cell 158(4):861–873. https://doi.org/10.1016/j.cell.2014.06.028

Nedelcheva MN, Roguev A, Dolapchiev LB, Shevchenko A, Taskov HB, Stewart AF, Stoynov SS (2005) Uncoupling of unwinding from DNA synthesis implies regulation of MCM helicase by Tof1/Mrc1/Csm3 checkpoint complex. J Mol Biol 347(3):509–521. https://doi.org/10.1016/j.jmb.2005.01.041

Noguchi E, Noguchi C, Du LL, Russell P (2003) Swi1 prevents replication fork collapse and controls checkpoint kinase Cds1. Mol Cell Biol 23(21):7861–7874

O'Donnell M, Li H (2016) The eukaryotic replisome goes under the microscope. Curr Biol 26(6):R247–R256. https://doi.org/10.1016/j.cub.2016.02.034

Okazaki R, Okazaki T, Sakabe K, Sugimoto K, Sugino A (1968) Mechanism of DNA chain growth. I. Possible discontinuity and unusual secondary structure of newly synthesized chains. Proc Natl Acad Sci U S A 59(2):598–605. https://doi.org/10.1073/pnas.59.2.598

Osborn AJ, Elledge SJ (2003) Mrc1 is a replication fork component whose phosphorylation in response to DNA replication stress activates Rad53. Genes Dev 17(14):1755–1767. https://doi.org/10.1101/gad.1098303

Pardo B, Crabbé L, Pasero P (2017) Signaling pathways of replication stress in yeast. FEMS Yeast Res 17(2). https://doi.org/10.1093/femsyr/fow101

Park H, Sternglanz R (1999) Identification and characterization of the genes for two topoisomerase I-interacting proteins from Saccharomyces cerevisiae. Yeast 15(1):35–41. https://doi.org/10.1002/(SICI)1097-0061(19990115)15:1<35::AID-YEA340>3.0.CO;2-R

Perera RL, Torella R, Klinge S, Kilkenny ML, Maman JD, Pellegrini L (2013) Mechanism for priming DNA synthesis by yeast DNA polymerase α. Elife 2:e00482. https://doi.org/10.7554/eLife.00482

Petronczki M, Chwalla B, Siomos MF, Yokobayashi S, Helmhart W, Deutschbauer AM, Davis RW, Watanabe Y, Nasmyth K (2004) Sister-chromatid cohesion mediated by the alternative RF-CCtf18/Dcc1/Ctf8, the helicase Chl1 and the polymerase-alpha-associated protein Ctf4 is essential for chromatid disjunction during meiosis II. J Cell Sci 117(Pt 16):3547–3559. https://doi.org/10.1242/jcs.01231

Pommier Y, Sun Y, Huang SN, Nitiss JL (2016) Roles of eukaryotic topoisomerases in transcription, replication and genomic stability. Nat Rev Mol Cell Biol 17(11):703–721. https://doi.org/10.1038/nrm.2016.111

Rabitsch KP, Tóth A, Gálová M, Schleiffer A, Schaffner G, Aigner E, Rupp C, Penkner AM, Moreno-Borchart AC, Primig M, Esposito RE, Klein F, Knop M, Nasmyth K (2001) A screen for genes required for meiosis and spore formation based on whole-genome expression. Curr Biol 11(13):1001–1009. https://doi.org/10.1016/s0960-9822(01)00274-3

Rageul J, Park JJ, Zeng PP, Lee EA, Yang J, Hwang S, Lo N, Weinheimer AS, Schärer OD, Yeo JE, Kim H (2020) SDE2 integrates into the TIMELESS-TIPIN complex to protect stalled replication forks. Nat Commun 11(1):5495. https://doi.org/10.1038/s41467-020-19162-5

Razidlo DF, Lahue RS (2008) Mrc1, Tof1 and Csm3 inhibit CAG.CTG repeat instability by at least two mechanisms. DNA Repair (Amst) 7(4):633–640. https://doi.org/10.1016/j.dnarep.2008.01.009

Rowland BD, Roig MB, Nishino T, Kurze A, Uluocak P, Mishra A, Beckouët F, Underwood P, Metson J, Imre R, Mechtler K, Katis VL, Nasmyth K (2009) Building sister chromatid cohesion: smc3 acetylation counteracts an antiestablishment activity. Mol Cell 33(6):763–774. https://doi.org/10.1016/j.molcel.2009.02.028

Samora CP, Saksouk J, Goswami P, Wade BO, Singleton MR, Bates PA, Lengronne A, Costa A, Uhlmann F (2016) Ctf4 links DNA replication with sister chromatid cohesion establishment by recruiting the Chl1 helicase to the replisome. Mol Cell 63(3):371–384. https://doi.org/10.1016/j.molcel.2016.05.036

Sauer PV, Gu Y, Liu WH, Mattiroli F, Panne D, Luger K, Churchill ME (2018) Mechanistic insights into histone deposition and nucleosome assembly by the chromatin assembly factor-1. Nucleic Acids Res 46(19):9907–9917. https://doi.org/10.1093/nar/gky823

Schalbetter SA, Mansoubi S, Chambers AL, Downs JA, Baxter J (2015) Fork rotation and DNA precatenation are restricted during DNA replication to prevent chromosomal instability. Proc Natl Acad Sci U S A 112(33):E4565–E4570. https://doi.org/10.1073/pnas.1505356112

Schauer GD, O'Donnell ME (2017) Quality control mechanisms exclude incorrect polymerases from the eukaryotic replication fork. Proc Natl Acad Sci U S A 114(4):675–680. https://doi.org/10.1073/pnas.1619748114

Sehgal A, Price JL, Man B, Young MW (1994) Loss of circadian behavioral rhythms and per RNA oscillations in the Drosophila mutant timeless. Science 263(5153):1603–1606. https://doi.org/10.1126/science.8128246

Sengupta S, van Deursen F, de Piccoli G, Labib K (2013) Dpb2 integrates the leading-strand DNA polymerase into the eukaryotic replisome. Curr Biol 23(7):543–552. https://doi.org/10.1016/j.cub.2013.02.011

Shimmoto M, Matsumoto S, Odagiri Y, Noguchi E, Russell P, Masai H (2009) Interactions between Swi1-Swi3, Mrc1 and S phase kinase, Hsk1 may regulate cellular responses to stalled replication forks in fission yeast. Genes Cells 14(6):669–682. https://doi.org/10.1111/j.1365-2443.2009.01300.x

Shyian M, Albert B, Zupan AM, Ivanitsa V, Charbonnet G, Dilg D, Shore D (2020) Fork pausing complex engages topoisomerases at the replisome. Genes Dev 34(1–2):87–98. https://doi.org/10.1101/gad.331868.119

Simon AC, Zhou JC, Perera RL, van Deursen F, Evrin C, Ivanova ME, Kilkenny ML, Renault L, Kjaer S, Matak-Vinković D, Labib K, Costa A, Pellegrini L (2014) A Ctf4 trimer couples the CMG helicase to DNA polymerase α in the eukaryotic replisome. Nature 510(7504):293–297. https://doi.org/10.1038/nature13234

Sommariva E, Pellny TK, Karahan N, Kumar S, Huberman JA, Dalgaard JZ (2005) Schizosaccharomyces pombe Swi1, Swi3, and Hsk1 are components of a novel S-phase response pathway to alkylation damage. Mol Cell Biol 25(7):2770–2784. https://doi.org/10.1128/MCB.25.7.2770-2784.2005

Somyajit K, Gupta R, Sedlackova H, Neelsen KJ, Ochs F, Rask MB, Choudhary C, Lukas J (2017) Redox-sensitive alteration of replisome architecture safeguards genome integrity. Science 358(6364):797–802. https://doi.org/10.1126/science.aao3172

Srinivasan M, Fumasoni M, Petela NJ, Murray A, Nasmyth KA (2020) Cohesion is established during DNA replication utilising chromosome associated cohesin rings as well as those loaded de novo onto nascent DNAs. Elife 9:e56611. https://doi.org/10.7554/eLife.56611

Stokes K, Winczura A, Song B, Piccoli G, Grabarczyk DB (2020) Ctf18-RFC and DNA Pol ϵ form a stable leading strand polymerase/clamp loader complex required for normal and perturbed DNA replication. Nucleic Acids Res 48(14):8128–8145. https://doi.org/10.1093/nar/gkaa541

Tanaka T, Yokoyama M, Matsumoto S, Fukatsu R, You Z, Masai H (2010) Fission yeast Swi1-Swi3 complex facilitates DNA binding of Mrc1. J Biol Chem 285(51):39609–39622. https://doi.org/10.1074/jbc.M110.173344

Tourrière H, Versini G, Cordón-Preciado V, Alabert C, Pasero P (2005) Mrc1 and Tof1 promote replication fork progression and recovery independently of Rad53. Mol Cell 19(5):699–706. https://doi.org/10.1016/j.molcel.2005.07.028

Uhlmann F, Nasmyth K (1998) Cohesion between sister chromatids must be established during DNA replication. Curr Biol 8(20):1095–1101

Unal E, Heidinger-Pauli JM, Kim W, Guacci V, Onn I, Gygi SP, Koshland DE (2008) A molecular determinant for the establishment of sister chromatid cohesion. Science 321(5888):566–569. https://doi.org/10.1126/science.1157880

Unsal-Kaçmaz K, Mullen TE, Kaufmann WK, Sancar A (2005) Coupling of human circadian and cell cycles by the timeless protein. Mol Cell Biol 25(8):3109–3116. https://doi.org/10.1128/MCB.25.8.3109-3116.2005

Unsal-Kaçmaz K, Chastain PD, Qu PP, Minoo P, Cordeiro-Stone M, Sancar A, Kaufmann WK (2007) The human Tim/Tipin complex coordinates an Intra-S checkpoint response to UV that slows replication fork displacement. Mol Cell Biol 27(8):3131–3142. https://doi.org/10.1128/MCB.02190-06

Urtishak KA, Smith KD, Chanoux RA, Greenberg RA, Johnson FB, Brown EJ (2009) Timeless maintains genomic stability and suppresses sister chromatid exchange during unperturbed DNA replication. J Biol Chem 284(13):8777–8785. https://doi.org/10.1074/jbc.M806103200

van der Lelij P, Chrzanowska KH, Godthelp BC, Rooimans MA, Oostra AB, Stumm M, Zdzienicka MZ, Joenje H, de Winter JP (2010) Warsaw Breakage Syndrome, a Cohesinopathy Associated with Mutations in the XPD Helicase Family Member DDX11/ChlR1. Am J Hum Genet 86(2):262–266

Vengrova S, Dalgaard JZ (2004) RNase-sensitive DNA modification(s) initiates S. pombe mating-type switching. Genes Dev 18(7):794–804. https://doi.org/10.1101/gad.289404

Voineagu I, Narayanan V, Lobachev KS, Mirkin SM (2008) Replication stalling at unstable inverted repeats: interplay between DNA hairpins and fork stabilizing proteins. Proc Natl Acad Sci U S A 105(29):9936–9941. https://doi.org/10.1073/pnas.0804510105

Voineagu I, Surka CF, Shishkin AA, Krasilnikova MM, Mirkin SM (2009) Replisome stalling and stabilization at CGG repeats, which are responsible for chromosomal fragility. Nat Struct Mol Biol 16(2):226–228

Westhorpe R, Keszthelyi A, Minchell NE, Jones D, Baxter J (2020) Separable functions of Tof1/Timeless in intra-S-checkpoint signalling, replisome stability and DNA topological stress. Nucleic Acids Res 48(21):12169–12187. https://doi.org/10.1093/nar/gkaa963

Witosch J, Wolf E, Mizuno N (2014) Architecture and ssDNA interaction of the Timeless-Tipin-RPA complex. Nucleic Acids Res 42(20):12912–12927. https://doi.org/10.1093/nar/gku960

Xie S, Lu Y, Jakoncic J, Sun H, Xia J, Qian C (2014) Structure of RPA32 bound to the N-terminus of SMARCAL1 redefines the binding interface between RPA32 and its interacting proteins. FEBS J 281(15):3382–3396. https://doi.org/10.1111/febs.12867

Xie S, Mortusewicz O, Ma HT, Herr P, Poon RY, Poon RR, Helleday T, Qian C (2015) Timeless interacts with PARP-1 to promote homologous recombination repair. Mol Cell 60(1):163–176. https://doi.org/10.1016/j.molcel.2015.07.031

Xu H, Boone C, Brown GW (2007) Genetic dissection of parallel sister-chromatid cohesion pathways. Genetics 176(3):1417–1429. https://doi.org/10.1534/genetics.107.072876

Yamaguchi K, Yamaguchi R, Takahashi N, Ikenoue T, Fujii T, Shinozaki M, Tsurita G, Hata K, Niida A, Imoto S, Miyano S, Nakamura Y, Furukawa Y (2014) Overexpression of cohesion establishment factor DSCC1 through E2F in colorectal cancer. PLoS One 9(1):e85750. https://doi.org/10.1371/journal.pone.0085750

Yeeles JT, Deegan TD, Janska A, Early A, Diffley JF (2015) Regulated eukaryotic DNA replication origin firing with purified proteins. Nature 519(7544):431–435. https://doi.org/10.1038/nature14285

Yeeles JT, Janska A, Early A, Diffley JF (2017) How the eukaryotic replisome achieves rapid and efficient DNA replication. Mol Cell 65(1):105–116. https://doi.org/10.1016/j.molcel.2016.11.017

Yoshizawa-Sugata N, Masai H (2007) Human Tim/Timeless-interacting protein, Tipin, is required for efficient progression of S phase and DNA replication checkpoint. J Biol Chem 282(4):2729–2740. https://doi.org/10.1074/jbc.M605596200

Young LM, Marzio A, Perez-Duran P, Reid DA, Meredith DN, Roberti D, Star A, Rothenberg E, Ueberheide B, Pagano M (2015) TIMELESS forms a complex with PARP1 distinct from its complex with TIPIN and plays a role in the DNA damage response. Cell Rep 13(3):451–459. https://doi.org/10.1016/j.celrep.2015.09.017

Yuan Z, Li H (2020) Molecular mechanisms of eukaryotic origin initiation, replication fork progression, and chromatin maintenance. Biochem J 477(18):3499–3525. https://doi.org/10.1042/BCJ20200065

Yuan Z, Bai L, Sun J, Georgescu R, Liu J, O'Donnell ME, Li H (2016) Structure of the eukaryotic replicative CMG helicase suggests a pumpjack motion for translocation. Nat Struct Mol Biol 23(3):217–224. https://doi.org/10.1038/nsmb.3170

Yuan Z, Riera A, Bai L, Sun J, Nandi S, Spanos C, Chen ZA, Barbon M, Rappsilber J, Stillman B, Speck C, Li H (2017) Structural basis of Mcm2-7 replicative helicase loading by ORC-Cdc6 and Cdt1. Nat Struct Mol Biol 24(3):316–324. https://doi.org/10.1038/nsmb.3372

Yuan Z, Georgescu R, Schauer GD, O'Donnell ME, Li H (2020) Structure of the polymerase ε holoenzyme and atomic model of the leading strand replisome. Nat Commun 11(1):3156. https://doi.org/10.1038/s41467-020-16910-5

Zeman MK, Cimprich KA (2014) Causes and consequences of replication stress. Nat Cell Biol 16(1):2–9. https://doi.org/10.1038/ncb2897

Zhang J, Shi X, Li Y, Kim BJ, Jia J, Huang Z, Yang T, Fu X, Jung SY, Wang Y, Zhang P, Kim ST, Pan X, Qin J (2008) Acetylation of Smc3 by Eco1 is required for S phase sister chromatid cohesion in both human and yeast. Mol Cell 31(1):143–151. https://doi.org/10.1016/j.molcel.2008.06.006

Chapter 4
Ferritin-Like Proteins: A Conserved Core for a Myriad of Enzyme Complexes

Rahul Banerjee, Vivek Srinivas, and Hugo Lebrette

Abstract Ferritin-like proteins share a common fold, a four α-helix bundle core, often coordinating a pair of metal ions. Although conserved, the ferritin fold permits a diverse set of reactions, and is central in a multitude of macromolecular enzyme complexes. Here, we emphasize this diversity through three members of the ferritin-like superfamily: the soluble methane monooxygenase, the class I ribonucleotide reductase and the aldehyde deformylating oxygenase. They all rely on dinuclear metal cofactors to catalyze different challenging oxygen-dependent reactions through the formation of multi-protein complexes. Recent studies using cryo-electron microscopy, serial femtosecond crystallography at an X-ray free electron laser source, or single-crystal X-ray diffraction, have reported the structures of the active protein complexes, and revealed unprecedented insights into the molecular mechanisms of these three enzymes.

Keywords Ferritin-like superfamily · Methane monooxygenase · Ribonucleotide reductase · Aldehyde deformylating oxygenase · X-ray crystallography · Serial femtosecond crystallography · X-ray free electron laser · Cryo-electron microscopy

R. Banerjee
Department of Biochemistry, Molecular Biology and Biophysics, University of Minnesota, Minneapolis, MN, USA
e-mail: baner025@umn.edu

V. Srinivas
Department of Biochemistry and Biophysics, Stockholm University, Stockholm, Sweden
e-mail: vivek.srinivas@dbb.su.se

H. Lebrette (✉)
Department of Biochemistry and Biophysics, Stockholm University, Stockholm, Sweden

Laboratoire de Microbiologie et Génétique Moléculaires (LMGM), Centre de Biologie Intégrative (CBI), CNRS, UPS, Université de Toulouse, Toulouse, France
e-mail: hugo.lebrette@univ-tlse3.fr

J. R. Harris, J. Marles-Wright (eds.), *Macromolecular Protein Complexes IV*, Subcellular Biochemistry 99, https://doi.org/10.1007/978-3-031-00793-4_4

Introduction: The Ferritin-Like Superfamily

The ferritin-like superfamily currently groups 26 different protein families, including more than 84,000 domains in the Pfam database (Clan CL0044) (Mistry et al. 2021). Proteins of this superfamily all share a conserved structural domain consisting of a bundle of four α-helices known as the ferritin fold (Fig. 4.1). This protein architecture is characterized by a distinctive arrangement of α-helices in an up-down-down-up topology, with a crossover segment connecting the second and third helices (Lundin et al. 2012). The first description of the ferritin fold as a conserved and specific feature of the ferritin-like superfamily dates back to 1992 (Murzin and Chothia 1992).

The ferritin-like superfamily is named after the protein ferritin, which is an ancient and ubiquitous enzyme (Andrews 2010). Ferritins are large spherical multimeric assemblies made of 12 or 24 subunits of the characteristic four-helical bundle, arranged to form a hollow nanocage (Theil 2011). Ferritin is a key player in iron homeostasis (Plays et al. 2021). It can store large amounts of iron and therefore protect the cell against excess or deficit of iron. A ferroxidase center is usually located in the middle of the four-helical bundle of the ferritin subunit (Crichton and

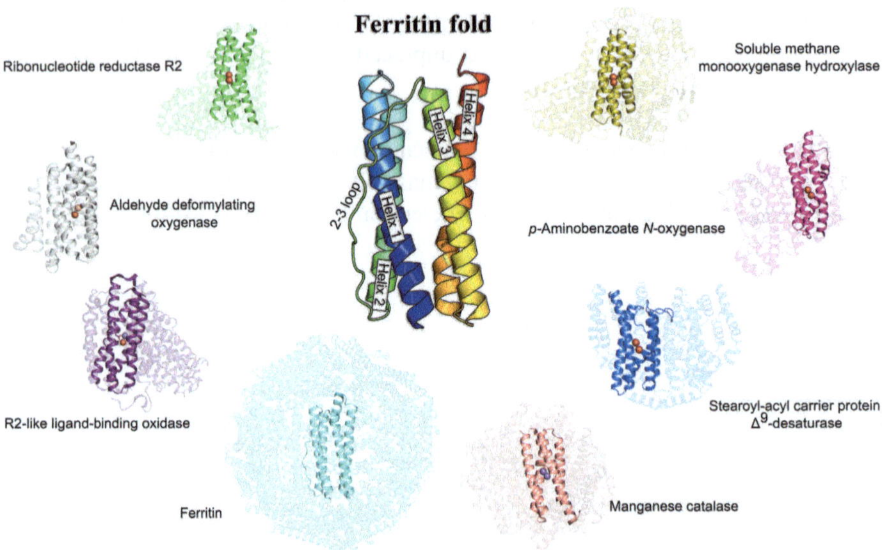

Fig. 4.1 The ferritin fold and several representative members of the ferritin-like superfamily. The four-helix ferritin fold is highlighted in crystal structures of ribonucleotide reductase R2 subunit (PDB ID: 1mxr) (Högbom et al. 2003), aldehyde deformylating oxygenase (PDB ID: 4rc5) (Jia et al. 2015), R2-like ligand-binding oxidase (PDB ID: 3ee4) (Andersson and Högbom 2009), ferritin (PDB ID: 1eum) (Stillman et al. 2001), manganese catalase (PDB ID: 1jku) (Barynin et al. 2001), stearoyl–acyl carrier protein Δ^9-desaturase (PDB ID: 1afr) (Lindqvist et al. 1996), *p*-aminobenzoate *N*-oxygenase (PDB ID: 3chh) (Choi et al. 2008), and soluble methane monooxygenase hydroxylase (PDB ID: 1mhy) (Elango et al. 1997)

Declercq 2010), but variations exist, as in the case of ferritin from *Mycoplasma*, in which the ferroxidase center has been found on the inner surface of the protein (Wang et al. 2021). This active site oxidizes Fe(II) to Fe(III) using molecular oxygen and/or hydrogen peroxide (Bou-Abdallah 2010). Then, Fe(III) is transported into the cavity of the spherical structure, to be stored in the form of iron oxide, and could later be released by reduction when iron is needed (Theil 2012). The iron pathway through the ferritin protein has been studied by anomalous X-ray diffraction using apo-ferritin crystals exposed to Fe(II) for different periods of time (Pozzi et al. 2015).

Despite structural similarity, members of the ferritin-like superfamily display little sequence conservation, and perform a variety of functions which can be mainly divided into the following four categories: iron storage, reactive oxygen species protection, substrate oxidation, and radical generation (Lundin et al. 2012). The ferritin fold has undergone an evolutionary diversification, and the number of functional categories is inclined to grow as more ferritin-like proteins are being characterized. Most ferritin-like proteins use this scaffold to house a non-heme dinuclear metal center coordinated by histidines and carboxylate residues in the middle of the four-helix bundle (Nordlund and Eklund 1995; Fox 2021). A notable exception is the protein Dps (DNA-binding protein from starved cells) which, instead exhibits metal-binding sites at the interface between subunits (Crichton and Declercq 2010). Some metal cofactors of ferritin-like enzymes are capable of catalyzing very challenging reactions upon dioxygen activation (Trehoux et al. 2016; Jasniewski and Que 2018) (Fig. 4.2). Interestingly, the versatility of the ferritin-like scaffold is further illustrated by the fact that it serves as a template in the design and engineering of artificial metalloenzymes (Yu et al. 2014; Chino et al. 2015; Nastri et al. 2016).

In this chapter, we will particularly focus on the soluble methane monooxygenase (sMMO), class I ribonucleotide reductase (RNR), and aldehyde deformylating oxygenase (ADO). All three are soluble enzyme systems harboring a non-heme dinuclear metal center—except the recently discovered subclass Ie RNR (Srinivas

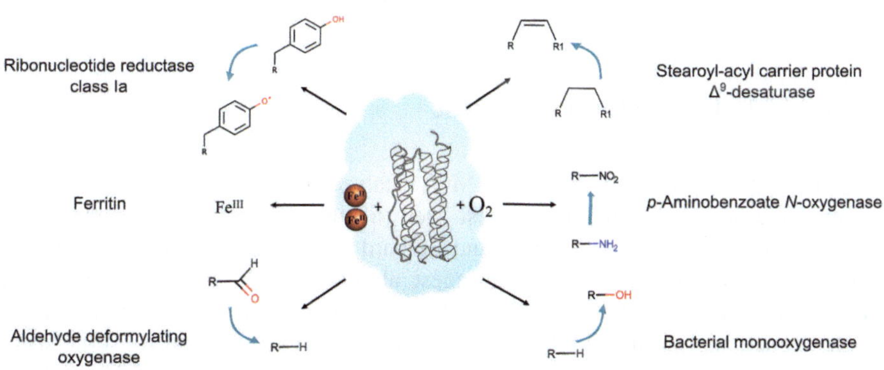

Fig. 4.2 Representative examples of reactions catalyzed by enzymes of the ferritin-like superfamily

et al. 2018; Blaesi et al. 2018)—which react with oxygen (Trehoux et al. 2016; Jasniewski and Que 2018). They can be grouped into a structural family within the ferritin-like superfamily which could be designated as ferritin-like diiron oxidases and oxygenases (FDOs) (Rajakovich et al. 2020)—although not all subclasses of class I RNR use a diiron cofactor (see section *Class I Ribonucleotide Reductase*). In FDOs, all four helices of the ferritin fold harbor one or more residues coordinating the dinuclear metal site. The metal-binding sites display conserved ligands, including two histidines (His) and four carboxylates (aspartate, Asp, and glutamate, Glu) forming two copies of the characteristic motif Glu/Asp-X_n-Glu/Asp-X-X-His (Andrews 2010; Rajakovich et al. 2020). RNR, sMMO, and ADO are structurally more closely related to each other than they are to ferritin, and they also exhibit a larger common structural core which extends on both the N- and C-termini of the four-helix bundle of the ferritin fold (Lundin et al. 2012). A recent work illustrated the close structural similarity between ADO and RNR R2 subunit, as well as the functional modularity of the ferritin-like scaffold. Using an approach, including machine learning and computational protein design, the authors were able to introduce ADO-like activity into the R2 subunit of the subclass Ia RNR from *Escherichia coli* by mutating only ten residues (Mak et al. 2020).

Using cutting-edge approaches in structural biology, recent breakthroughs have been achieved that have remarkably improved our understanding of sMMO, RNR, and ADO, at the molecular level. Firstly, high resolution crystal structures of the sMMO hydroxylase in complex with its regulatory protein MMOB were obtained at room temperature and in two oxidation states using serial femtosecond crystallography at an X-ray free electron laser (XFEL) source (Srinivas et al. 2020). Secondly, the long-awaited structure of the active holocomplex of class I RNR was solved by cryo-electron microscopy (cryo-EM) and provided an unprecedented snapshot of the radical transfer pathway (Kang et al. 2020). Thirdly, an elegant use of X-ray crystallography revealed the structure of ADO in complex with acyl-acyl carrier protein reductase (AAR) and enabled a visualization of their remarkable collaboration within the two-step alkane biosynthesis pathway (Gao et al. 2020). These milestones will be presented in the following sections.

Soluble Methane Monooxygenase

Soluble methane monooxygenase (sMMO) catalyzes the O_2-dependent conversion of methane to methanol in the first step of the C1 metabolic pathway of methanotrophic organisms (Tinberg and Lippard 2011; Banerjee et al. 2019). The ability to oxidatively cleave the strongest aliphatic C–H bond (methane bond dissociation energy = 105 kcal/mol), while avoiding the over-oxidation of the easily oxidized methanol product highlights the exceptional reactivity and substrate specificity of this enzyme. sMMO, along with a membrane-bound form of the methane monooxygenase enzyme (particulate MMO, pMMO), are the only catalysts capable of aerobically oxidizing methane at ambient temperature and pressure (Wang et al. 2017b).

sMMO is a multi-protein complex, comprising a hydroxylase protein containing a dinuclear iron cluster in the active site (sMMOH, 245 kDa), a cofactor-less regulatory protein (MMOB, 15 kDa), and a flavin and [2Fe-2S] containing reductase protein (MMOR, 38 kDa) (Green and Dalton 1986; Banerjee et al. 2019). The formation of discrete, transient protein complexes enables the reduction of the sMMOH diiron cluster by MMOR at the start of catalysis and subsequently, regulation of the catalytic cycle within sMMOH by MMOB. While another cofactor-less protein MMOD has been shown to bind sMMOH, it does not play a role in sMMO catalysis, and methane oxidation at physiological turnover rates can take place in the absence of MMOD (Merkx and Lippard 2002; Kim et al. 2019). sMMO mechanistic studies have been directed at the enzyme isolated from *Methylosinus trichosporium* OB3b (*Mt*) and *Methylococcus capsulatus* Bath (*Mc*). This research seeks to understand the molecular mechanisms by which Nature generates and controls hyper-reactive oxidants. Lessons learned from this investigation inform the generation of synthetic catalysts for strong C–H bond functionalization and natural gas to liquid fuel conversions (Conrado and Gonzalez 2014; Wang et al. 2017b). More immediate applications of the sMMO enzyme system are found in the fields of bioremediation and making value-added products from methane through synthetic biology (Singh and Singh 2017; Clomburg et al. 2017).

Chemical Mechanism of Methane Oxidation in sMMO

The dinuclear iron cluster in sMMOH is supported by two histidine and four glutamate residues from the four-helix bundle of the ferritin fold (Rosenzweig et al. 1993; Elango et al. 1997). In the as-isolated resting state of sMMOH, both iron atoms are in the Fe(III) oxidation state (Fig. 4.3) A couple of carboxylate side chains and hydroxo moieties act as bridging ligands and facilitate exchange-coupling between the iron atoms (Hendrich et al. 1990). Reduction by electrons derived from NADH through the MMOR protein generates an Fe_2(II) state and primes sMMOH for the catalytic cycle (Tinberg and Lippard 2011; Banerjee et al. 2019). Upon the subsequent formation of a protein complex between sMMOH and MMOB, dioxygen diffuses into the active site and binds both iron atoms in a bridging fashion. This O_2 molecule is activated by concerted electron transfers from the iron atoms to generate an Fe_2(III)-peroxo intermediate **P**, followed by homolytic O-O bond scission to yield an Fe_2(IV)-oxo species, with both processes being proton-dependent (Lee and Lipscomb 1999). This intermediate, termed compound **Q**, is observed to react with methane in a bimolecular, methane concentration-dependent reaction to generate methanol along with an oxo-bridged Fe_2(III) cluster in the product complex **T** (Lee et al. 1993a, b). The last step of catalysis is the rate-limiting release of methanol from the active site.

The unprecedented ability of **Q** to break the C-H and C-D bond of methane (bond dissociation energy of CD_4 = 107 kcal/mol) at atmospheric pressure and

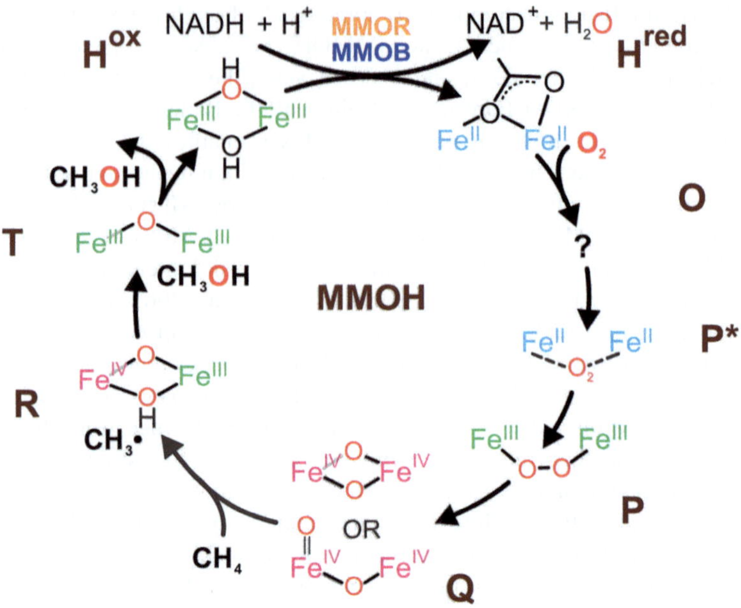

Fig. 4.3 sMMO catalytic cycle. Catalysis ensues upon the reduction of the dinuclear iron(III) cluster (**H^ox^**) by MMOR to the **H^red^** state. O_2 binding and activation leads to the generation of intermediates **O**, **P***, **P**, **Q**, **R**, and **T**. The current understanding of the structure and electronics of the intermediates is depicted as models of the iron cluster

temperatures as low as 4 °C makes compound **Q** Nature's most powerful oxidant (Nesheim and Lipscomb 1996). In order to understand this reactivity, the structure and electronics of **Q** have been probed with a breadth of spectroscopic techniques, including Mössbauer, X-ray Emission/Absorption spectroscopy (XES/XAS), Extended X-ray Absorption Fine Structure (EXAFS), resonance Raman, and Nuclear Resonance Vibrational spectroscopy (NRVS) (Lee et al. 1993a; Liu et al. 1995a; Shu et al. 1997; Banerjee et al. 2015; Castillo et al. 2017; Cutsail et al. 2018; Jacobs et al. 2021). The Mössbauer spectrum of **Q** clearly depicts both iron atoms to be present in an antiferromagnetically coupled, high-spin Fe(IV) configuration with the irons in similar electronic environments. The understanding of the geometric arrangement of the iron-bound, O_2-derived oxygen atoms and first-sphere ligands is, however, less complete. Resonance Raman and NRVS vibrational studies support a structural 'closed-core' model, where the oxygen atoms bridge the irons in the form of two vibrationally coupled μ-oxo moieties. On the other hand, the High Energy Resolution Fluorescence–EXAFS (HERFD–EXAFS) spectrum of **Q** reveals an Fe–Fe distance of ~3.4 Å, which fits an 'open-core' model where the oxygen atoms are arranged as a mono-μ-oxo bridge and a terminal oxo moiety to one of the iron atoms (Fig. 4.3). Recent density functional theory (DFT) calculations of the hydrogen atom transfer (HAT) reaction between **Q** and methane with various closed- and open-core structural models of **Q** have predicted that the activation energy barrier is lower with closed-core models (Jacobs et al. 2021). This stabilization arises from the provision of a greater thermodynamic driving force for HAT by concerted bond-length changes of two coupled bis-μ-oxo bridges, as opposed to the bond-length change of a single Fe(IV)=O bond. Post-HAT, the hydroxyl moiety of the diiron cluster rebounds onto the methyl radical to generate methanol (Banerjee et al. 2019). In order to provide further structural information and resolve the controversy between the closed- and open-core models of **Q**, the status of an ongoing study with a cutting-edge structural technique will be described later in this chapter.

The sMMO containing ferritin-like superfamily of enzymes catalyzes a diverse range of chemical conversions with largely similar di-nuclear metal clusters. Several factors can rationalize this divergence in reactivity, including: differences in the identity of the metal coordinated by first-sphere ligands; alterations in second-sphere ligands; variations in protein structure outside the ferritin-fold core; and differences in the type of complexes formed with accessory proteins (Lundin et al. 2012). These facets, especially the last, will be illustrated with sMMO in the three following sections. A picture will emerge where protein–protein interactions are key to catalysis, for example, by engendering substrate specificity for the smallest hydrocarbon with the strongest C-H bond and ensuring that **Q** is not quenched by aberrant electron transfer from MMOR.

sMMOH Protein Structure and Function

sMMOH is a homodimeric protein with an $(\alpha\beta\gamma)_2$ architecture and a predominantly α-helical secondary structure (Rosenzweig et al. 1993; Elango et al. 1997). It is the largest member of the ferritin-like superfamily of enzymes at the primary and tertiary structure levels. While both α- and β-subunits of sMMOH contain a ferritin fold stemming from a gene-duplication event, sequence divergence has led to the loss of the metal coordinating residues in the β-subunit (Lundin et al. 2012). The dinuclear iron cluster is buried within the active site in the α-subunit, approximately 12 Å away from the protein surface. An interior void is positioned alongside the face of the diiron cluster that is furthest from the histidine ligands. This active site cavity is predominantly lined by hydrophobic residues outside of the polar glutamate ligands to the iron atoms (Rosenzweig et al. 1993; Elango et al. 1997). The diiron cluster is present in an overall charge-neutral state in the diferric state ($2Fe^{3+}$, 4 Glu-COO^-, 2 OH^-, H_2O) and in the diferrous state ($2Fe^{2+}$, 4 Glu-COO^-, $2H_2O$), thereby alleviating the thermodynamic penalty of placing a charged metal cluster inside a hydrophobic environment (Fig. 4.4) (Rosenzweig et al. 1995). During iron reduction, the bridging hydroxide ligands are converted to water molecules and one of them is displaced by Glu243, which switches to a bidentate, bridging coordination mode.

The conservation of protein structural features in the ferritin family is reflected in the presence of rare π-helices, where two consecutive residues (i_n) possess intra-main chain hydrogen bonds to the $i + 5$ residue. π-helical sections are thermodynamically

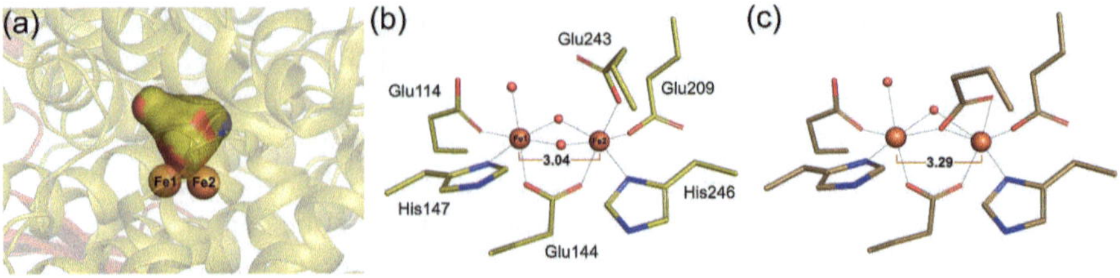

Fig. 4.4 Structure and environment of the dinuclear iron cluster in sMMOH. (**a**) Active site cavity along one face of the iron atoms is shown as a surface in the oxidized sMMOH:MMOB structure (PDB ID: 6yd0). (**b**) Structure of the oxidized diiron cluster with first-sphere ligands in the sMMOH:MMOB complex (PDB ID: 6yd0). (**c**) Structure of the reduced diiron cluster with first-sphere ligands in the sMMOH:MMOB complex (PDB ID: 6ydi)

higher in energy than α-helices, but are evolutionarily conserved on account of a strong association with protein function (Cooley et al. 2010). The α-subunit of sMMOH contains six π-helical sections, all of which contain strictly conserved residues (Jones et al. 2020). While their functional roles have not yet been probed by mutagenesis studies, four π-helices are directly associated with iron ligation, interactions with the regulatory MMOB protein, and gating substrate access tunnels (*vide infra*) (Srinivas et al. 2020).

Another structural feature of note in sMMOH involves a hydrogen bond network to the histidine iron ligands. A strictly conserved acidic residue (Asp/Glu) is observed to hydrogen bond to one or both histidine ligands in most ferritin-like enzymes. Additional hydrogen bond donors to this carboxylate residue in the form of strictly conserved, basic arginine/lysine and polar tyrosine residues are found to be uniquely present in sMMOH and the closely related bacterial multicomponent monooxygenase (BMM) enzymes (McCormick and Lippard 2011; Srinivas et al. 2020). The short donor acceptor distances of ~2.7 Å between the Asp-His pairs in sMMOH suggests that the histidine ligands could possess a charged imidazolate character. Compared to their commonplace conception as neutral imidazole moieties, imidazolate ligands can significantly alter the electronics at the diiron cluster. Such hydrogen bonding pairs to axial cysteine/histidine ligands of heme-iron centers in heme oxygenases and peroxidases have been shown to perturb the HAT reactivity of the Compound I oxidant in heme-P450 oxygenases (Poulos 2014; Onderko et al. 2017). The binding of MMOB to sMMOH has recently been shown to alter this hydrogen bond network (Srinivas et al. 2020).

sMMOH and MMOB Protein Complex

sMMOH is capable of catalyzing methane oxidation in isolation by either an O_2-mediated reaction after reduction by MMOR or by performing a peroxide-shunt reaction with H_2O_2 (Froland et al. 1992). However, physiological rates of methane oxidation are only obtained upon formation of a sMMOH:MMOB protein complex (Green and Dalton 1985; Liu et al. 1995b). Among the many regulatory roles of MMOB, the most important are: an increase in substrate specificity for methane; a doubling of product yield; an enhancement in the O_2 reactivity of sMMOH by 1000-fold; and an increase in the turnover number by 150-fold (Fox et al. 1989; Liu et al. 1995b; Wallar and Lipscomb 2001). A structural investigation of the protein complex of sMMOH and MMOB is, therefore, most germane to understanding sMMO catalysis.

A crystal structure of the *Mc* sMMOH:MMOB protein complex in the diferric state was one of the big advances made towards this goal (Lee et al. 2013). Albeit at limited resolution (2.9 Å), this structure described many of the conformational changes occurring in sMMOH and MMOB upon complexation and validated predictions made about the MMOB binding interface by prior NMR studies (Chang et al. 2001). However, the observation of Glu243 in the bidentate, bridging coordination mode to the iron atoms, as typically observed in structures of diferrous sMMOH (Rosenzweig et al. 1995), suggested that the iron atoms had been partially photo-reduced during the long-duration diffraction data collection process at the synchrotron.

Recent advances in X-ray generation technology have significantly alleviated the issue of radiation-mediated reduction of metal centers in enzymes. X-ray Free Electron Laser (XFEL) sources produce a pulsed, coherent X-ray beam that is approximately a billion times brighter than the X-ray beams generated at synchrotrons (McNeil and Thompson 2010). This short duration X-ray pulse (10–35 fs) elicits the generation of diffraction data faster than the reduction of metals by solvated electrons. This 'diffraction before damage' methodology also obviates the need to cryogenically cool crystals during data collection for minimizing radiation damage. An advantage of this ambient temperature data collection process is that crystallographic snapshots of an enzyme reaction are possible if the reaction can be initiated in crystallo (Chapman 2019). This serial femtosecond crystallography (SFX) technique promises, therefore, to not only provide crystal structures of metalloenzymes in authentic oxidation states but also structures of reaction intermediates, such as the radiation-sensitive, high-valent iron species in compound **Q**. In operation, an SFX experiment delivers a stream of microcrystals to the XFEL beam at a programmable rate. For collecting crystal structures of catalytic intermediates in O_2-activating metalloenzymes, reduced anaerobic microcrystals are exposed to O_2 after varied incubation times just before injection into the XFEL beam (Fuller et al. 2017). XES spectra of the metal centers can be collected concurrently with diffraction data and serves as a valuable, independent readout of the metal oxidation state (Fuller et al. 2017; Bergmann et al. 2021).

The SFX methodology has recently provided high-resolution (1.96 Å) crystal structures of the *Mt* sMMOH:MMOB protein complex in the diferric and diferrous oxidation states (Srinivas et al. 2020). The observation of a 'carboxylate shift' of Glu243 solely after chemical reduction of the protein crystals strongly supports the ability of the SFX technique to avoid aberrant metal reduction during diffraction data collection. A measurement 10 s after the O_2-exposure showed that the iron cluster had undergone a single turnover reaction and achieved its resting diferric state.

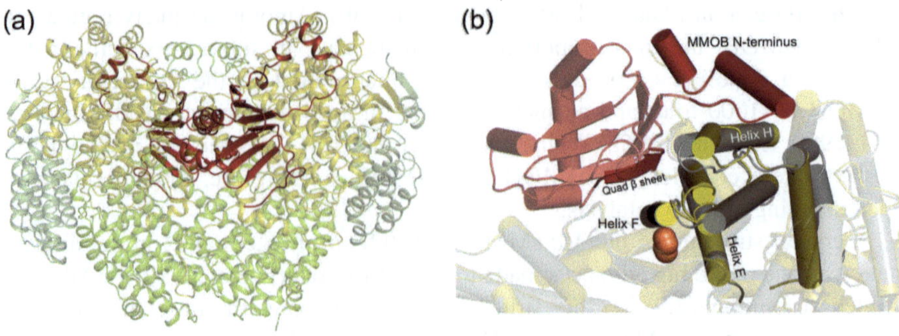

Fig. 4.5 Structure of the sMMOH:MMOB protein complex. (**a**) Quaternary structure shows two molecules of MMOB (magenta cartoon) bound the α-subunits of the heterodimeric sMMOH molecule (PDB ID: 6yd0). (**b**) An overlay of sMMOH (PDB ID: 1mhy, in grey) with the sMMOH:MMOB complex shows the reorganization of two helices of the four-helix bundle (Helices E and F) by interaction with the core region of MMOB. The organization of the N-terminus of MMOB on the surface of sMMOH instigates a reorganization of helices H and 4 in sMMOH. The sMMOH:MMOB interaction-mediated newly formed β-sheet in MMOB is labelled as 'Quad'

The high-resolution crystal structures of the sMMOH:MMOB complex have enabled a complete delineation of the structural changes in sMMOH and MMOB upon complexation and their catalytic relevance in the context of prior mutational, kinetic, and spectroscopic information (Jones et al. 2020; Srinivas et al. 2020). MMOB binds sMMOH with high affinity (K_D = 67 nM) in a process driven primarily by electrostatic interactions (Zhang et al. 2006). The stoichiometry of complex formation is 1:2 (sMMOH:MMOB), with each αβγ protomer of dimeric sMMOH binding a MMOB molecule (Fox et al. 1991). Large structural changes are observed in MMOB upon binding sMMOH, with the most dramatic being the organization of a disordered 35 residue-long N-terminal tail into a ring-like structure on the sMMOH surface (Fig. 4.5) (Lee et al. 2013; Srinivas et al. 2020). The structural changes in sMMOH upon binding MMOB are restricted to the α-subunit. Two of the four helices of the ferritin fold reorganize and reflect the most direct effect of binding of the well-folded core of MMOB upon the active site. Two adjacent sMMOH helices also undergo changes due to interactions with the organized N-terminal tail of MMOB. It is interesting to note that these conformational changes in sMMOH involve four of the six existing π-helices along with the generation of a new π-helical section. The connection between altered sMMOH protein structure and function and the regulatory roles of MMOB will be discussed in the following sections.

Enhancement of O_2 Reactivity by MMOB Binding to sMMOH

The most dramatic regulatory effect of MMOB upon sMMOH catalysis is the 1000-fold enhancement in the rate constant of O_2 binding to the reduced diiron cluster (Liu et al. 1995b). As a result, the rate-limiting step of the catalytic cycle is shifted to methanol release, thereby effecting a 150-fold increase in the sMMO turnover number (Fox et al. 1989). A faster reaction with dioxygen also results in the buildup of catalytic intermediates in a single turnover reaction. This phenomenon appears to arise from an increase in the rate-constant of intramolecular diffusion of O_2 into sMMOH and also from an increase in the rate-constant of binding to the iron atoms.

The MMOB-mediated structural reorganization of sMMOH, as observed in crystal structures of the sMMOH:MMOB complex, increases the hydrophobicity of the active site cavity (Srinivas et al. 2020). The rotamer change of a polar residue (Thr213) and the loss of discrete water molecules from the active site cavity are responsible for this change. This prepares sMMOH for reaction with O_2 and CH_4 by providing not only space but also an environment conducive to binding non-polar molecules. A 50 Å-long channel leading from bulk solvent to the active site cavity is also observed to be organized upon protein complex formation (Fig. 4.6) (Jones et al. 2020). This channel traverses the interface between sMMOH and MMOB and enters sMMOH past a gating residue (Trp308, W308) that gives this channel its name. A

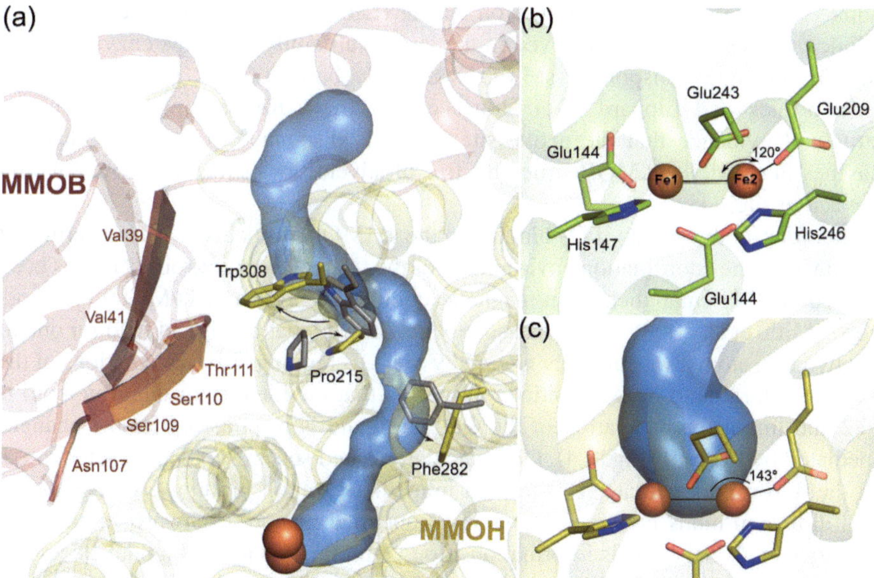

Fig. 4.6 MMOB-binding structural changes that enhance O_2 reactivity with sMMOH. (**a**) A small-molecule W308-tunnel (blue surface) is formed as a result of the movement of gating residues (stick representation) after iron reduction and sMMOH:MMOB complexation (PDB ID: 6ydi). The opaque orange β-sheets contain MMOB residues where mutations perturb the intramolecular diffusion of O_2 (Val39, Val41) and hydrocarbon substrates (Asn107, Ser109, Ser110, Thr111). (**b**) Glu209 sterically hinders the binding of oxygen to Fe2 and this block is reflected by the Fe1–Fe2–Glu209 angle of 120° (PDB ID: 1fyz). (**c**) The steric block on Fe2 is relieved by the formation of the sMMOH:MMOB complex and is reflected by the Fe1–Fe2–Glu209 angle of 143° (PDB ID: 6ydi)

comparison of diferric sMMOH, diferrous sMMOH, and diferrous sMMOH:MMOB structures shows that conformational changes of bulky, planar aromatic residue side chains open up contiguous interior space, leading to the construction of this transient access route (Jones et al. 2020; Banerjee and Lipscomb 2021). The movement of these gating residues is coincident with iron reduction and formation of the protein complex. This channel is predominantly narrow along its entire length, with constriction points where the diameter is close to the quantum mechanical diameter for O_2 (Banerjee and Lipscomb 2021). This physical characteristic has led to its designation as a tunnel that can only allow one-dimensional movement of molecules along its length. A tunnel can potentially enhance the binding rate of substrates compared to a wider channel because it precludes a three-dimensional search for a binding spot occurring in channels with large cavities. The W308-tunnel is lined by strictly conserved residues that are uniformly non-polar in character. Mutations at two of these residues in MMOB with bulkier, polar side chain residues (Val39Arg, Val41Arg) retard the rate-constant of O_2 binding to the diiron cluster 25,000-fold (Jones et al. 2020). This has provided experimental support for the W308-tunnel being used as a route for O_2 transport over the competing model of three large, interconnected cavities leading to the iron cluster (Lee et al. 2013; Wang et al. 2015). While these cavities are conserved in the larger subfamily of BMM enzymes, their role in substrate transport in sMMO is unlikely based upon a few observations (McCormick and Lippard 2011; Jones et al. 2020). Key amongst them is the structural evidence that the active site cavity is disconnected from the chain of cavities in sMMOH:MMOB structures free of adventitious substrates bound within the active site.

In order to investigate if O_2 binding to the sMMOH iron cluster is also perturbed by MMOB, the coordination environment of the O_2-reactive diferrous state has been probed with circular dichroism/magnetic circular dichroism (CD/MCD) spectroscopy (Mitić et al. 2008). These measurements showed that the ligand field for one of the two iron atoms was uniquely altered by MMOB binding. DFT calculations of the experimental ligand field transitions suggested that the identity of this iron atom is Fe2 and that a reorganization of the Glu209 ligand is responsible for the changes mediated by MMOB (Fig. 4.4). DFT calculations and spectroscopic characterization of the diiron intermediate (**P***) formed immediately upon binding O_2 have shown that dioxygen binds both iron atoms in a concerted manner (Schwartz et al. 2008; Banerjee et al. 2013). This information, along with the ligand field model calculated on the basis of the CD/MCD study, helped devise a model for the binding of O_2 with the diiron cluster. This model predicts that while O_2 binding to Fe1 in diferrous sMMOH is facile, it is sterically hindered at Fe2 by the Glu209 ligand. The MMOB-induced conformational change of Glu209 was predicted to relieve this steric block and enhance O_2 binding. The recent crystal structure of the diferrous sMMOH: MMOB complex supports this theoretical model with structural data (Srinivas et al. 2020). The Glu209 residue is indeed located on one of the two helices (Helix E) of the four-helix bundle that undergoes large structural changes upon direct interaction with MMOB (Fig. 4.6). In summary, it appears that MMOB engenders greater O_2-reactivity in sMMOH by organizing a non-polar tunnel for fast intramolecular diffusion into the active site and an altered metal cluster coordination geometry with open coordination positions on both iron atoms.

Engendering Substrate Specificity for Methane

One of the remarkable characteristics of sMMO is its substrate specificity for methane (Banerjee and Lipscomb 2021). This selectivity is observed despite the fact that sMMO is a promiscuous enzyme. The fact that the hydrocarbon substrate selectively enters the sMMOH active site after **Q** has been generated and that it only needs to be positioned adjacent to the iron atoms for oxidation makes sMMO capable of binding many hydrocarbon molecules. Combined with the oxidizing power of **Q**, this promiscuity means that any chemical species with a C–H or C=C bond that can access the active site is a potential substrate, and this fact is exemplified by the over 40 demonstrated alternative sMMO substrates (Green and Dalton 1989; Pilkington and Dalton 1990). Nevertheless, methane must be preferentially oxidized despite the significantly stronger C–H bond relative to alternative substrates because it is the only viable substrate for the methanotroph. One way to demonstrate this specificity for methane is the observation that the rate constants for C-H bond-breaking by **Q** are similar for methane and ethane (Brazeau and Lipscomb 2000). If, however, the C–H bond-breaking reactions were truly rate-limited by the substrate C–H bond strength, ethane oxidation would occur 4000-fold faster than methane (Banerjee and Lipscomb 2021). A clue to the origin of this partiality towards methane oxidation was obtained by measurements of the substrate deuterium kinetic isotope effect (D-KIE) on the reaction of **Q** with methane and alternative substrates. For methane, a D-KIE value of 50 is observed, demonstrating that the C-H bond-breaking reaction is rate-limiting (Nesheim and Lipscomb 1996; Ambundo et al. 2002). In contrast, a D-KIE of 1 is measured for other alternative substrates, suggesting that the rate-limiting step is not C-H bond cleavage (Brazeau and Lipscomb 2000). These observations were rationalized by a 'molecular sieve' model, which proposed that the intramolecular diffusion of substrates larger than methane into the sMMOH active site is slow and this rate-limits the oxidation reaction by **Q**. Experimental support for this model was obtained by the study of MMOB mutants ('Quad'—Asn107Gly/Ser109Ala/Ser110Ala/Thr111Ala, and 'DBL2'—Ser109Ala/Thr111Ala) that displayed a faster rate constant for oxidation of substrates larger than methane and a D-KIE value of 2 with ethane (Brazeau et al. 2001; Wallar and Lipscomb 2001; Brazeau and Lipscomb 2003; Zheng and Lipscomb 2006). This suggested that the tunnel for hydrocarbon substrate access into sMMOH was enlarged by MMOB mutations that are located at the interface with sMMOH.

When the position for these residues (Ser109, Thr111) is mapped onto the sMMOH:MMOB crystal structure, they appear in close vicinity to the W308-tunnel. Indeed, these residues and those where the mutations (Val39, Val41) perturb the O_2 intramolecular diffusion rate constant are present on adjacent β-sheets (Fig. 4.6). It has therefore been hypothesized that the W308-tunnel is also used to transport hydrocarbon substrates into the active site (Jones et al. 2020). The bimolecular kinetics of substrate oxidation by **Q** better fits the model of diffusion through a one-dimensional tunnel as opposed to the hyperbolic kinetics expected from diffusion through the large, chain of cavities that can pre-bind multiple substrate molecules. The constricted nature of the W308-tunnel is particularly well suited to playing the role of the molecular sieve. Indeed, the bottlenecks within the tunnel are so finely tuned that the intramolecular diffusion rate constant of methane is almost two orders of magnitude slower than that of the only slightly smaller oxygen molecule. This displays the impressive ability of the tunnels in sMMO to discriminate between small molecules such as oxygen, methane, and ethane that differ by less than 1 Å in diameter (Banerjee and Lipscomb 2021).

Once methane preferentially accesses the active site, the C-H bond-breaking reaction by **Q** is further enhanced by hydrogen atom quantum tunneling (Zheng and Lipscomb 2006). This is reflected in the large methane D-KIE value of ~50. When C-H bond-breaking become rate-limiting for ethane in complexes with MMOB variants, a D-KIE value of 2 is observed, thereby suggesting a reduced role for quantum tunneling in the HAT reaction with alternative substrates (Brazeau et al. 2001; Zheng and Lipscomb 2006). A quantum tunneling of the hydrogen atom during HAT typically requires optimal positioning of the donor and acceptor molecules, which is dependent upon the orientation of the substrate molecule by the active site (Klinman and Kohen 2013). It is therefore noteworthy that MMOB binding to sMMOH reduces the volume of the activity-site cavity by 25% (Srinivas et al. 2020). The W308-tunnel also enters the active site cavity close to equatorial plane of the diiron cluster, where the reactive acceptor μ-oxo moiety of **Q** is expected to be positioned (Fig. 4.6) (Banerjee et al. 2019). The restricted size of the cavity and the unfavorable contacts of non-polar methane with the polar carboxylate ligands to the irons are expected to localize methane in a precise position for reaction with **Q** (Banerjee and Lipscomb 2021). Experimental support for this model comes from observations that mutations in MMOB that alter the constrictions in the W308-tunnel reduce the quantum tunneling contribution to the methane oxidation reaction (Brazeau et al. 2001). In summary, MMOB binding to sMMOH engenders substrate specificity for methane by using the diminutive size of methane to enable it faster access to the active site and quantum tunneling during HAT to provide a kinetic advantage over alternative substrates.

Dynamic Equilibrium of sMMOH, MMOB, and MMOR Protein Complexes

A recurrent theme in metalloenzymes containing the ferritin fold is the initiation of O_2-mediated catalysis from the reduced state of the metal cluster (Wallar and Lipscomb 1996; Jasniewski and Que 2018). This necessitates reduction of the oxidized metal atoms between each turnover cycle. MMOR fulfills this electron transfer role in sMMO with flavin and plant-type [2Fe–2S] iron-sulfur cofactors (Banerjee et al. 2019). NMR structures of the flavin and ferredoxin domains show that MMOR is similar to members of the ferredoxin reductase and fer2 superfamily of proteins (Muller et al. 2002; Chatwood et al. 2004). Measurements of the reduction potential of the two cofactors support a model of electron flow from the two-electron reduced FAD to the [2Fe–2S] cluster, which subsequently passes them sequentially to the two iron atoms in sMMOH (Kopp et al. 2001). A strict requirement for this electron transfer process is that it should be confined to the resting, diferric state of sMMOH so that high-valent, electrophilic reaction intermediates such as **Q** are not quenched. This regulation is highlighted by observations that product yields from a single turnover reaction of sMMOH is lowered from 40% (relative to active site concentration) to 12% in the presence of reduced MMOR. However, when MMOB is added to this reaction mix, the product yields increase to 100% (Liu et al. 1997). This demonstrates that electron transfer regulation arises from an interplay of protein complex formation between sMMOH, MMOB, and MMOR.

The understanding of this regulation has been held back by the lack of clarity regarding the MMOR binding surface on sMMOH. Early chemical cross-linking studies of the protein components indicated that MMOR and MMOB had distinct binding surfaces on sMMOH (Fox et al. 1991). In conjunction with a small-angle X-ray scattering study of sMMO protein structure, this data pointed towards the formation of a ternary sMMOH:MMOB:MMOR complex (Gallagher et al. 1999). In this non-competitive model of MMOB and MMOR binding to sMMOH, a conformational change either at the sMMOH:MMOR interface or within the sMMOH protein prevents electron transfer from reduced MMOR, while catalysis ensues within the active site.

More recently, a hydrogen–deuterium exchange coupled to mass spectrometry (HDX-MS) investigation has provided insight into the *Mc* sMMOH surface that interacts with a truncated form of MMOR comprising the ferredoxin domain (Fd) (Wang et al. 2014b). Certain regions of sMMOH were identified where deuterium exchange from the solvent was reduced upon the formation of a protein complex with Fd. Indeed, this surface of sMMOH matched closely with the region that binds MMOB in crystal structures of the sMMOH:MMOB complex. This model for overlapping binding surfaces for MMOB and Fd was further supported by protein–protein binding experiments with fluorophore-labelled MMOB, which showed that Fd titration into an sMMOH:MMOB complex displaced MMOB (Wang et al. 2014b). Because Fd was more capable of displacing a truncated form of MMOB missing the N-terminal tail, a model was proposed where MMOB remains anchored to sMMOH during reduction and catalysis *via* the N-terminus. Discrete, diametrically opposing changes in the binding affinity of the MMOB core and MMOR for sMMOH in the oxidized and reduced state would facilitate an ordered sequence of binding events (Wang et al. 2014b, 2015). This ensures that reduced MMOR competitively binds only the resting diferric state of sMMOH. A more-tightly bound MMOB protein prevents the quenching of intermediates mid-catalysis by occluding the binding surface for reduced MMOR.

While this competitive model is an elegant proposal for temporally regulating electron transfer, it relies on dramatic changes in the affinity of the protein components between different redox states. The experimentally measured affinity constants for protein complexation are, however, inconsistent and vary significantly between the *Mc* and *Mt* sMMO enzymes (Fox et al. 1991; Gassner and Lippard 1999; Zhang et al. 2006; Wang and Lippard 2014; Jones et al. 2021). Considering the similarity of these homologous proteins, it is likely that the differences have arisen from the methods used to measure binding affinities. Binding constants have been obtained primarily by tracking fluorescence changes that occur upon protein complexation. A caveat of such studies is that large fluorophore molecules have been attached to environment sensitive positions on MMOB, potentially disrupting protein–protein interactions and the native affinity constants. The challenge of measuring binding affinities in the nanomolar range with minimal to no perturbation was recently overcome for *Mt* sMMO with a protein-observed ^{19}F-NMR (PrOF) technique (Jones et al. 2021).

The PrOF methodology combines the sensitivity of ^{19}F-NMR signals and its hyper-responsiveness to chemical environments with the bioisosteric replacement of an amino acid C-H bond with a C-^{19}F bond (Divakaran et al. 2019). In sMMO, the tryptophan residues of MMOB and MMOR were labelled with ^{19}F to prepare minimally perturbed and fully functional protein probes for binding assays. The formation of sMMOH:MMOB and sMMOH:MMOR complexes were measured by following the loss of intensity of unbound protein resonances and the appearance of resonances from large 275 kDa protein complexes. MMOB was observed to bind sMMOH tightly in both oxidized and reduced states ($K_D < 100$ nM). Titrating in MMOR into a pre-formed sMMOH:MMOB complex led to the release of MMOB from the protein complex as visualized through the appearance of the free MMOB resonance. The binding affinity of MMOR for sMMOH was measured from this competitive titration and was found to be similar to the affinity of MMOB for oxidized and reduced sMMOH. The ^{19}F labels positioned on both the N-terminal tail and core region of MMOB reported an equally effective displacement from the sMMOH surface when titrating MMOR. This PrOF investigation supported the competitive model of sMMO protein–protein interactions and delineated it further by showing that MMOB is completely displaced from sMMOH by MMOR binding. Furthermore, the dramatic redox-linked changes in protein affinities required for sequential protein binding events are absent.

A new model has been proposed to explain the regulation of electron transfer based upon this PrOF study (Jones et al. 2021). The competitive binding of MMOB and MMOR to sMMOH with similar affinities suggests an equilibrium between sMMOH:MMOB and sMMOH:MMOR complexes, which is biased towards MMOB due to its higher intracellular concentration (Fox et al. 1989). Starting at the oxidized state of sMMOH, this equilibrium is dynamically shifted towards the reduced form of sMMOH through the irreversible and thermodynamically downhill transfer of electrons from reduced MMOR (Fig. 4.7). This reduced sMMOH equilibrium state in MMOB- and MMOR-bound complexes is then further shifted towards MMOB-induced O_2-mediated catalysis by the irreversible steps of O_2

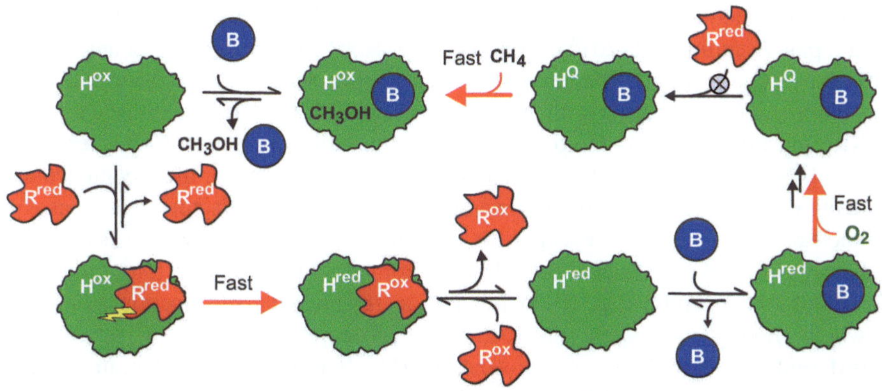

Fig. 4.7 Dynamic equilibrium model for the competitive binding of MMOB and MMOR to sMMOH. The equilibrium of sMMOH:MMOB and sMMOH:MMOR complexes in the oxidized and reduced states of sMMOH are moved forward through the catalytic cycle by the irreversible steps of iron reduction, O_2 activation, and CH_4 oxidation (red arrows)

activation and C-H bond cleavage, leading to methanol generation and the recycling of sMMOH back to the oxidized state. The irreversibility of the reduction and oxidation reactions is key to dynamically drive the equilibrium of sMMOH states through catalysis. The potential danger of high-valent reaction intermediates being quenched by MMOR is prevented by the completion of the catalytic cycle during the dwell-time of MMOB on sMMOH. This proposal has been borne out of observations that the k_{off} for MMOB release from sMMOH is slower than the rate constants in the sMMOH catalytic cycle up until methanol formation (Zhang et al. 2006). Additionally, single turnover kinetics of sMMOH turnover in the presence of a combination of MMOB mutants that dramatically alter the rate constants of different chemical steps have shown that the bound MMOB molecules do not exchange with each other during catalysis (Wallar and Lipscomb 2001). As a result, sMMO may engender the divergent chemistry of reduction and oxidation at the metal cluster by coupling it to the formation of mutually exclusive protein complexes and ensuring that the reaction half-cycles are completed during the lifetime of the complex. This model highlights the catalytic role played by the cofactor-less regulatory proteins in the BMM enzymes, which are unique amongst the proteins interacting with the ferritin-fold containing enzymes because they are not explicitly involved in delivering either substrate or electrons required for catalysis.

Class I Ribonucleotide Reductase

Ribonucleotide reductase (RNR) is an enzyme conserved in all domains of life. RNR represents the only pathway for de novo synthesis of the DNA building blocks by catalyzing the reduction of ribonucleotides to the corresponding deoxyribonucleotides. It plays a key role in the cell as it is critical for DNA synthesis and repair (Lundin et al. 2015). Consequently, RNR inhibitors serve in anti-cancer therapies (Aye et al. 2015) and constitute promising options for antimicrobial drug development (Miret-Casals et al. 2018; Greene et al. 2020). There are three classes of RNR, and all utilize a common radical-based reaction mechanism (Tomter et al. 2013; Högbom et al. 2020). This radical, necessary to catalyze the ribonucleotide reduction, is almost exclusively generated using a metal cofactor: class I uses a dinuclear metal center (except for subclass Ie, see below), class II an adenosylcobalamin cofactor, and class III an iron-sulfur cluster (Nordlund and Reichard 2006; Lundin et al. 2015). In addition, the three classes function differently towards molecular oxygen: class I is oxygen-dependent, class II is oxygen-indifferent, and class III is oxygen-sensitive (Högbom et al. 2020). Organisms can possess more than one RNR class in order to survive under different environmental conditions (Torrents 2014).

Class I RNR is a multi-component enzyme constituted of two homodimeric proteins, the catalytic subunit R1 (also called α) which performs the ribonucleotide reduction, and the radical-generating subunit R2 (or β) which displays a ferritin-like fold housing generally a metal cofactor. Upon oxygen activation, R2 produces a radical which is shuttled to R1 *via* a long-range reversible proton-coupled electron transfer (PCET) process. Consequently, a cysteine in R1 active site is transiently oxidized into a cysteinyl radical that initiates the reaction *via* abstraction of a hydrogen atom at the 3′ position of the ribonucleotide substrate. This is followed by the release of the 2’ OH-group as water and the two-electron reduction of the substrate concomitant with formation of a disulfide bridge between two additional cysteines in the active site. The reaction is completed by the return of the 3′ hydrogen atom back to the reduced substrate and the regeneration of the initial cysteinyl radical. Finally, the radical is transferred back to the R2 subunit active site and stored until the next turnover.

In the following sections, the ferritin-like R2 subunit will be presented in more detail. First, the different R2 subclasses and their respective strategy for radical generation will be summarized. In a second part, we will discuss the quaternary structure of two different complexes involving R2 proteins: the complex of a manganese-dependent R2 with the NrdI flavoprotein; the active form of the R1-R2 holocomplex.

Ferritin-Like R2 Subunits in Class I RNR

The R2 subunit of the multi-component class I RNR enzyme belongs to the ferritin-like superfamily and forms a heart-shaped homodimer (Fig. 4.8) (Nordlund et al. 1990; Lundin et al. 2012). Here, the core four-helix bundle characteristic of the ferritin fold harbors an active site capable of producing a stable radical used to catalyze ribonucleotide reduction. In class I RNR, five subclasses have been discovered to date (Ia to Ie), where their respective R2 subunits (R2a to R2e) employ

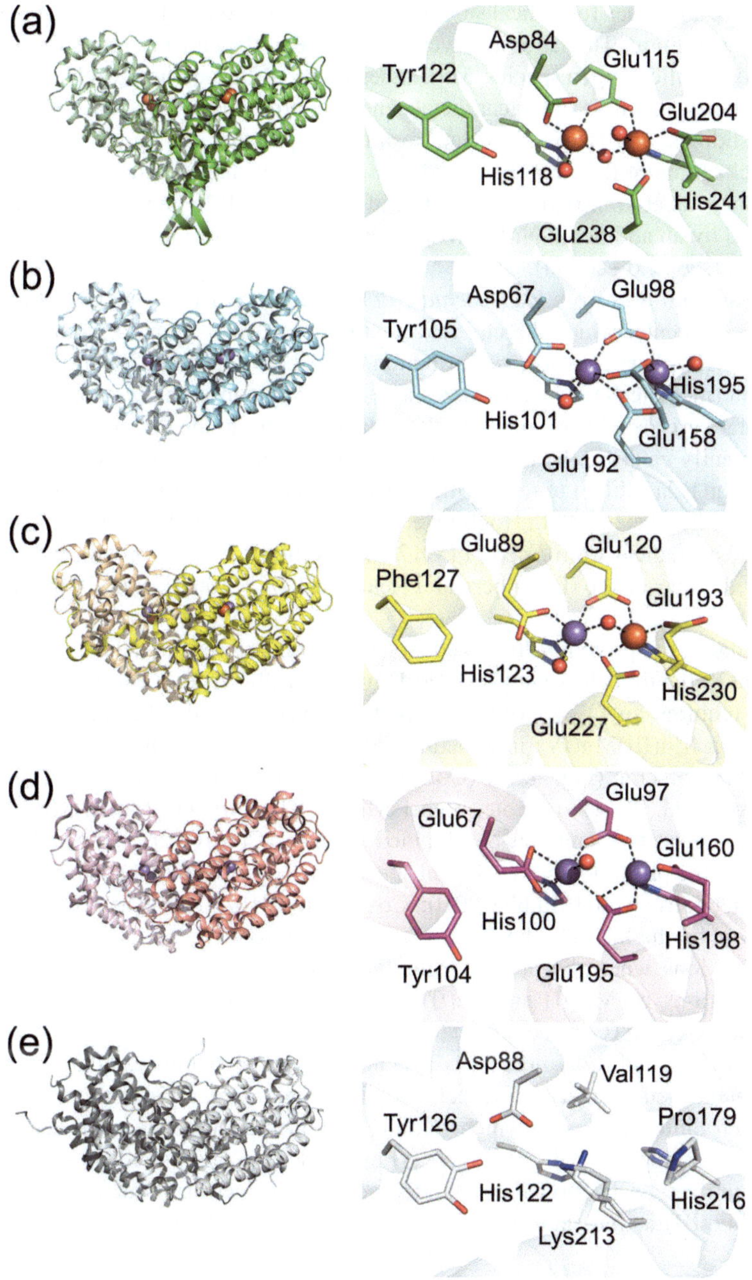

Fig. 4.8 Overall fold and active sites of the ferritin-like R2 subunits in class I RNR. (**a**) R2a homodimer from *E. coli* in the Fe(III)Fe(III) state (PDB ID: 1mxr) (Högbom et al. 2003). (**b**) R2b homodimer from *E. coli* in the Mn(II)Mn(II) state (PDB ID: 3n37) (Boal et al. 2010). (**c**) R2c homodimer from *Chlamydia trachomatis* in the Mn(II)Fe(II) state (PDB ID: 4m1i) (Dassama et al. 2013). (**d**) R2d homodimer from *Flavobacterium johnsoniae* in the Mn(II)Mn(II) state (PDB ID:

distinctive strategies for radical storage (metal-based or tyrosyl radical, Tyr•) and for radical generation. The latter involves various metal-dependent or metal-free cofactors and specific oxygen species. Subclasses R2a to R2d specifically rely on different iron and/or manganese dinuclear metal centers coordinated by six conserved ligands, two histidines, and four carboxylate residues (aspartates and glutamates), whereas R2e lacks three of the four conserved carboxylate residues and does not bind metal ions (Srinivas et al. 2018). Importantly, in the case of R2b and R2e, the oxidant is provided by an additional NrdI flavoprotein through the formation of a tight protein complex (see next section).

The subunit R2a from *Escherichia coli* has been extensively studied. It utilizes a Fe(II)Fe(II) center which directly reacts with molecular oxygen to produce a stable Fe(III)Fe(III)–Tyr• cofactor, *via* formation of the high-valent Fe(III)Fe(IV) (referred to as intermediate **X**), which oxidizes a neighboring conserved tyrosine (Fig. 4.8a). The cofactor is capable of initiating the radical transfer to the R1 subunit, which will subsequently catalyze the ribonucleotide reduction and shuttle the catalytic radical back to the R2a subunit to reform the active Fe(III)Fe(III)–Tyr• cofactor (Atkin et al. 1973; Nordlund et al. 1990; Högbom et al. 2003; Doan et al. 2015). Remarkably, the R2a tyrosyl radical from *E. coli* (Tyr122•) was the first stable protein radical to be observed 50 years ago (Ehrenberg and Reichard 1972; Sjöberg et al. 1978; Gräslund et al. 1985).

Similarly, the protein R2b (also denoted as NrdF) stores the radical on a tyrosine adjacent to the dinuclear metal center. However, R2b carries a Mn(II)Mn(II) center which requires an O_2-derived superoxide delivered by the flavoprotein NrdI to generate a Mn(III)Mn(IV) intermediate and finally produce the active Mn(III)Mn (III)–Tyr• cofactor (Fig. 4.8b) (Cotruvo and Stubbe 2010; Cotruvo et al. 2013; Berggren et al. 2014). To note, both Fe(III)Fe(III)–Tyr• and Mn(III)Mn(III)–Tyr• cofactors are active and can be obtained in vitro (Cotruvo Jr. and Stubbe 2012). However, the physiologically relevant species in vivo is the dimanganese form of the cofactor (Cotruvo and Stubbe 2010; Cox et al. 2010). This was recently further supported by the fact that the protein scaffold preferentially selects manganese ions over iron ions when the protein from *Bacillus anthracis* is incubated with equal concentrations of Mn(II) and Fe(II), irrespective of the presence or absence of molecular oxygen (Grāve et al. 2020).

Unlike R2a and R2b, there is no Tyr• in R2c or R2d, but instead the formation of high-valent metal centers which act as radical equivalents. R2c harbors a Mn(II)Fe (II) heterodinuclear center which initially provides all four electrons required for complete reduction of dioxygen (Fig. 4.8c). Consequently, a Mn(IV)Fe (IV) intermediate is formed and undergoes a one-electron reduction to yield the Mn(IV)Fe(III) active state of the protein (Högbom et al. 2004; Jiang et al. 2007;

Fig. 4.8 (continued) 6cwp) (Rose et al. 2018). (**e**) R2e homodimer from *Mesoplasma florum* in the DOPA state (PDB ID: 6gp2) (Srinivas et al. 2018). Iron and manganese ions are depicted as orange and purple spheres, respectively

Dassama et al. 2013; Fuller et al. 2017). In contrast, R2d utilizes a high-valent manganese cofactor Mn(III)Mn(IV) (Fig. 4.8d). This metal cofactor cannot be formed spontaneously from Mn(II) and molecular oxygen, but requires reduced oxygen species scavenged from solution, e.g., superoxide or hydrogen peroxide in vitro (Rozman Grinberg et al. 2018; Rose et al. 2018; Rozman Grinberg et al. 2019).

Finally, R2e is phylogenetically closely related to R2b; however, three of the conserved manganese-binding glutamates are substituted by valine, lysine, and proline residues (Fig. 4.8e). Instead of a di-manganese tyrosyl cofactor, the conserved tyrosine is post-translationally *meta*-hydroxylated to be converted into a 3,4-dihydroxyphenylalanine (DOPA) radical. This process is not yet well-characterized, but it necessitates the flavoprotein NrdI and aerobic conditions (Srinivas et al. 2018). Once the DOPA radical is formed, this RNR variant is able to catalyze the same ribonucleotide reduction reaction with specific activity comparable to typical RNRs (Blaesi et al. 2018; Srinivas et al. 2018). The occurrence of R2e could represent an adaptation to overcome the requirement of metal ions, and could provide an advantage in metal-restricted habitats (Srinivas et al. 2018).

Importantly, more than one type of class I RNR can be found in a single organism. Because RNRs have evolved unique specificities, they can be utilized interchangeably in response to environmental changes. For example, the genome of *E. coli* encodes RNRs from subclasses Ia, expressed under standard aerobic conditions, and Ib which is produced under oxidative stress or iron-limited conditions (Monje-Casas et al. 2001; McHugh et al. 2003). To note, *E. coli* possesses also a class III RNR which allows the growth in anaerobiosis (Torrents 2014; Högbom et al. 2020).

The R2-NrdI Complex in Subclass Ib RNR

Subclasses Ib and Ie are peculiar among class I RNRs because they require an additional flavoprotein, NrdI, to be able to generate a radical in presence of molecular oxygen. The *nrdI* gene coding for this protein is often included in the same operon as the *nrdE* and *nrdF* genes coding for subunits R1 and R2, respectively, in subclasses Ib and Ie (Roca et al. 2008; Cotruvo and Stubbe 2010; Srinivas et al. 2018). NrdI is a flavodoxin-like maturase (~15 kDa). It is able to use flavin mononucleotide (FMN) to perform a one-electron reduction of dioxygen to generate a superoxide radical which is necessary to produce the mature manganese-tyrosyl radical cofactor of R2b (Berggren et al. 2014). Crystal structures of NrdI have been obtained for different orthologs in subclass Ib (Røhr et al. 2010; Boal et al. 2010; Johansson et al. 2010; Makhlynets et al. 2014) and for one ortholog in subclass Ie (Blaesi et al. 2018). These structures revealed that the FMN cofactor is bound in a pocket at the surface of the protein, exposing the benzene ring of the isoalloxazine moiety to the solvent.

Several crystal structures of R2b were obtained in interaction with a dinuclear manganese center, including the proteins from *E. coli* (Boal et al. 2010), *Corynebacterium ammoniagenes* (Cox et al. 2010), *Bacillus subtilis* (Boal et al. 2012), *Bacillus cereus* (Hammerstad et al. 2014), *Streptococcus sanguinis* (Makhlynets et al. 2014), and *Bacillus anthracis* (Grāve et al. 2019). In R2b from *E. coli*, the radical-harboring residue is Tyr105, and the metal-binding ligands are constituted by two histidines (His101 and His195), one aspartate (Asp67), and three glutamates (Glu98, Glu158, and Glu192) (Fig. 4.8b). Interestingly, Glu158 exhibits an unusual conformation in the dimanganese form of the protein compared to the diiron form, bridging the two Mn(II) ions in a μ-1,3 coordination (Boal et al. 2010). Glu158 is part of a π-helix (in helix E), which is conserved in ferritin-like enzymes and could play a crucial role for function (Cooley et al. 2010). In addition, redox-induced structural rearrangements have been observed in helix E in R2b (Grāve et al. 2019) and might facilitate access to the active site (Eriksson et al. 1998). Owing to the configuration of Glu158 in *E. coli* R2b, the active site is exposed to a solvent channel connected to the protein surface. This channel is not present in R2a, but seems structurally conserved in other R2b proteins (Hammerstad et al. 2014; Makhlynets et al. 2014). It is hypothesized to be the route for transport of the oxidant species to the dimanganese center (Boal et al. 2010). However, the unusual conformation of the glutamate (Glu158 in *E. coli*) has not been observed in all dimanganese R2b proteins, and therefore it is unclear if different channels or channel-opening mechanisms exist in different organisms (Boal et al. 2012).

Crystal structures of R2b in complex with NrdI have been solved for *E. coli* and *B. cereus*, and the quaternary structure is conserved between the two organisms (Fig. 4.9a) (Boal et al. 2010; Hammerstad et al. 2014). The interaction of NrdI does not seem to trigger major structural changes in R2b as the overall fold and active site retain the same arrangement. The protein–protein interface covers an area of approximately 800 Å^2 in the *E. coli* complex, including hydrogen bonds and hydrophobic interactions between the subunits (Boal et al. 2010). The FMN cofactor is located at this interface. Positively charged residues from R2b, e.g., Lys18 and Arg25, are in the vicinity of the isoalloxazine group and could play a role in the reactivity of the cofactor. Importantly, the putative oxidant channel in R2b is prolongated toward the FMN, and is shielded from the solvent by NrdI residues such as Phe86. This extension of the tunnel is lined by conserved residues, including Asn83 and Asn85 from NrdI and Ser159, Tyr163, Lys260, and Asn264 from R2b (Boal et al. 2010; Hammerstad et al. 2014). Interestingly, Tyr163 is located on helix E below Glu158, and has been proposed to close the superoxide channel upon oxidation of the metal center (Grāve et al. 2019). In the crystal structure of the complex from *E. coli*, the channel accommodates a network of water molecules, connecting the FMN cofactor to the dimanganese center (Fig. 4.9b) (Boal et al. 2010).

No crystal structure of the complex between R2e and NrdI is yet available. However, small-angle X-ray scattering (SAXS) measurements made on the complex from *Mesoplasma florum* suggests that the quaternary structure is similar to the subclass Ib in solution (Srinivas et al. 2018). Furthermore, the crystal structure of the subclass Ie NrdI from *Aerococcus urinae* shows that the flavoprotein

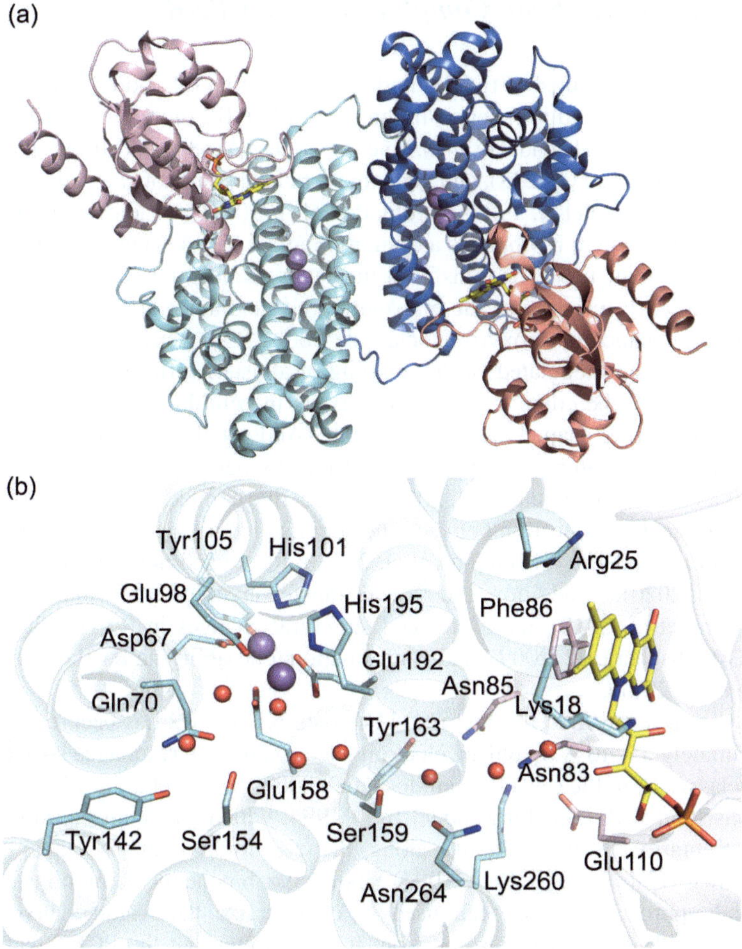

Fig. 4.9 Protein complex of R2b and NrdI. (**a**) Quaternary structure of the *E. coli* complex (PDB ID: 3n3a) (Boal et al. 2010). Protomers of the R2b dimer are colored in cyan and blue, NrdI proteins in pink and salmon, the FMN cofactors are depicted in yellow stick and manganese ions are shown as purple spheres. (**b**) The R2b–NrdI interface showing the network of water molecules (red spheres) and a selection of important residues

conserves the main functional characteristics. In addition, a docking model of the complex, from the same study, is compatible with the formation of an interface between the proteins comparable to subclass Ib. Indeed, residues putatively involved in the interactions are conserved and a solvent channel similar to the proposed oxidant route in R2b could be hypothesized, linking the FMN cofactor in NrdI and the active site in R2e (Blaesi et al. 2018).

The R1–R2 Active-State Complex in Class I RNR

Class I RNR is constituted of two homodimeric subunits: R1 and R2. In subclass Ia RNR from *E. coli*, the diiron center in the ferritin-like R2 protein is coordinated by two histidines (His118 and His241), one aspartate (Asp84) and three glutamates (Glu115, Glu204, and Glu238), and generates a radical stored on a tyrosine (Tyr122) (Fig. 4.8a) (Gräslund et al. 1985; Nordlund et al. 1990). This radical is translocated to R1 active site where it produces a cysteinyl radical (Cys439) used to catalyze the ribonucleotide reduction. The transfer of the catalytic radical is reversible and takes place during each turnover of the enzyme (Ravichandran et al. 2015). In order to maintain the balance of deoxyribonucleotide pools, RNR activity is modulated by oligomerization and regulated *via* the binding of allosteric effectors (Eriksson et al. 1997). In *E. coli*, the complex in solution exists in equilibrium between an active $\alpha_2\beta_2$ state and a ring-shaped inactive $\alpha_4\beta_4$ state (Ando et al. 2011). The formation of the active R1–R2 complex is transient and the dissociation rate constant (k_{off}) is higher than 60 s^{-1} (Minnihan et al. 2013a). The active complex assembles a pathway allowing radical translocation between the subunits *via* a long-range PCET process (Uhlin and Eklund 1994; Minnihan et al. 2013b; Reece and Seyedsayamdost 2017; Kang et al. 2020). In addition, the radical transfer involves conformational gating (Yokoyama et al. 2010), redox-induced structural changes in R2 (Högbom et al. 2003; Offenbacher et al. 2009), and conformational flexibility in R1 (Ge et al. 2003; Greene et al. 2017).

The distance travelled by the radical between R1 and R2 has been estimated to approximately 35–40 Å by pulsed electron–electron double resonance spectroscopy for subclasses Ia and Ic (Seyedsayamdost et al. 2007; Livada et al. 2015). This result was consistent with a docking model from 1994, which was built based on shape complementarity of the subunits along the two-fold symmetry axes of each homodimer (Uhlin and Eklund 1994). Extensive biochemical, biophysical, and structural approaches have been used to characterize the residues involved or associated with the radical translocation. It was shown that the PCET pathway is constituted of conserved redox-active amino acids, including Tyr730 and Tyr731 in R1, and Tyr356 (and possibly Trp48) in R2a from *E. coli* (Climent et al. 1992; Seyedsayamdost and Stubbe 2006; Minnihan et al. 2013b). To note, Tyr356 is believed to be crucial at the interface between R1 and R2 (Nick et al. 2017), but it has not been observable in X-ray crystal structures of R2 alone as it is located in the disordered C-terminus of the protein. Besides, additional residues were shown to be critical in conformational gating of the radical transfer, such as Glu52 (Lin et al. 2017) and Glu350 (Climent et al. 1992; Ravichandran et al. 2017). Importantly, a crystal structure of a holocomplex between R1 and R2 of subclass Ib from *Salmonella typhimurium* in an $\alpha_2\beta_2$ state was obtained at 4.0 Å resolution in 2006 (Uppsten et al. 2006). However, the interaction between R1 and R2 appeared loose and the full radical translocation pathway was not resolved as residues at the protein–protein interface (e.g., Tyr356 equivalent) were not visible. This structure was interpreted as

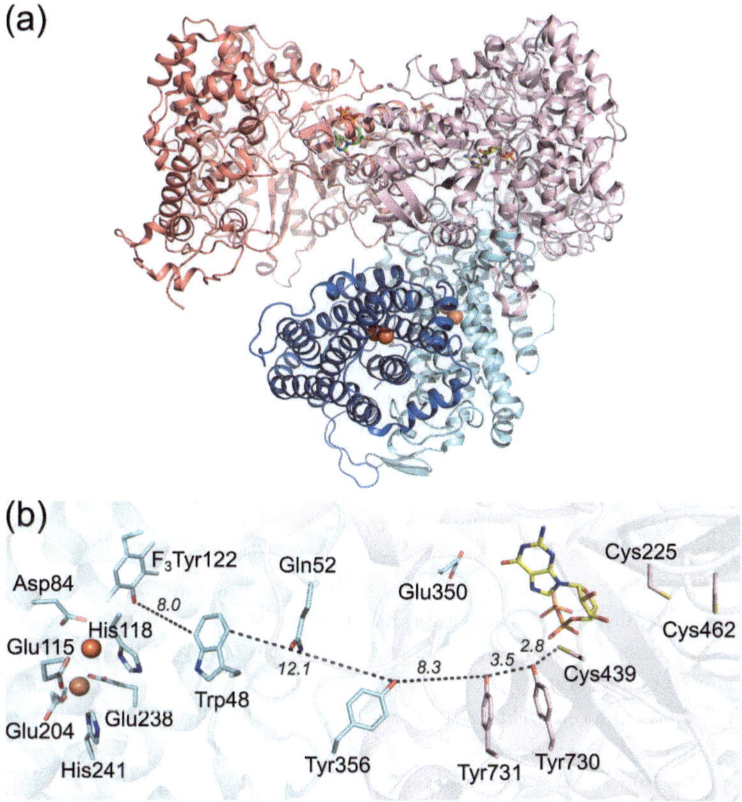

Fig. 4.10 Active complex of class I RNR. (**a**) The structure of the subclass Ia RNR holocomplex from *E. coli* captured by cryo-EM (PDB ID: 6w4x) (Kang et al. 2020). R1 subunits are colored in pink and salmon, and R2 subunits are in cyan and blue. The ribonucleotide substrate (guanosine diphosphate, GDP) is depicted in yellows stick, allosteric effectors (thymidine triphosphate, TTP) are in green, and iron ions are shown as orange spheres. (**b**) Proposed PCET pathway for the radical translocation between R2 and R1 subunits. Distances between residues putatively involved in the PCET pathway are shown in Å

an intermediate conformation trapped in a transitional stage in the reaction cycle (Uppsten et al. 2006; Logan 2011).

In 2020, the structure of the active R1-R2 holocomplex was solved by cryo-electron microscopy (cryo-EM) at an overall resolution of 3.6 Å (Kang et al. 2020) (Fig. 4.10). This milestone greatly improved our understanding of class I RNR, especially because it provided the first clear picture of the long-range radical transfer pathway. This achievement was made possible thanks to the recent transformative developments in single-particle cryo-EM, which has opened up new opportunities in structural biology and allowed the characterization of numerous macromolecular complexes at high resolution without the need of crystals (Cheng 2018). However, many macromolecular assemblies are short-lived or unstable. In the case of class I

RNR, the authors were able to trap the transient active complex by using a wild type R1 subunit in complex with a doubly substituted R2 protein, where Glu52 was mutated to glutamine and Tyr122 was replaced by a 2,3,5-trifluorotyrosine (F_3Tyr122) (Kang et al. 2020). Indeed, this double mutant was shown to slow down the dissociation of the protein complex into minutes, a timescale compatible with the single-particle cryo-EM approach (Lin et al. 2017).

Remarkably, the active $\alpha_2\beta_2$ RNR complex is asymmetric and the interface involves mainly one R1 subunit (Fig. 4.10a). This asymmetry contradicts the predicted docking model used until then as reference (Uhlin and Eklund 1994), but is consistent with the previous crystal structure obtained for subclass Ib (Uppsten et al. 2006) and recent SAXS data (Thomas et al. 2019). This configuration does not allow the coexistence of two PCET pathways simultaneously and consequently, only one radical can be translocated at a time between the two homodimeric subunits. Here, the complex was likely trapped after a single turnover event and the R1 subunit interacting with R2 seems to be in a pre-turnover state. Indeed, this R1 monomer is the only one binding a ribonucleotide substrate and exhibiting no disulfide bridge between Cys225 and Cys462, the two cysteines oxidized during catalysis (Kang et al. 2020).

For the first time, the radical transfer pathway connecting Tyr122 in R2 to the active site Cys439 in R1, measuring 32.4 Å in the structure, is completely visible, revealing the arrangement of putative residues involved in the PCET and potential proton acceptors and donors (Fig. 4.10b). Remarkably, the C-terminus tail of R2 is well-ordered and goes into R1 active site to position Tyr356 at the R1–R2 interface, at 8 Å from R1 Tyr731. In R2, Trp48 is located exactly between F_3Tyr122 and Tyr356, at 8 Å and 12 Å, respectively, suggesting its implication in the PCET even if no direct experimental evidence is available (Minnihan et al. 2013b). Distances between these residues are consistent with the hypothesis of a radical transfer taking place *via* orthogonal PCET in R2, i.e., coupling short-range proton transfers orthogonally to the long-range electron transfer (Reece et al. 2006; Reece and Seyedsayamdost 2017). In contrast, positions of Cys439, Tyr730, and Tyr731, approximately 3 Å apart, are compatible with the proposed mechanism of colinear PCET within R1 (Argirević et al. 2012; Nick et al. 2015). In addition, the location of Gln52 (Glu52 in wild type R2) at the interface between R1 and R2 along the PCET pathway could suggest a role for Glu52 in proton transfer and would explain its essential function in radical translocation (Lin et al. 2017; Kang et al. 2020). Besides, Glu350 seems to stabilize R2 C-terminus by interacting with R1, but variations might exist between subclasses because this residue is not completely conserved among R2 proteins (Ruskoski and Boal 2021). Finally, the fact that the R2 C-terminus tail extends into R1 active site prevents the release of the reaction product. Therefore, the R1–R2 complex has to dissociate and consequently the PCET pathway disassembles after each substrate conversion (Kang et al. 2020).

In closing, the cryo-EM structure of the class I RNR in the active state provides unprecedented structural insights. However, the mechanism triggering the radical translocation remains puzzling, in particular because of the use of an R2 variant where the radical-harboring tyrosine has been substituted by a 2,3,5-trifluorotyrosine

(Ruskoski and Boal 2021). We could imagine that a structural snapshot at atomic resolution of a wild type R2 in the radical state could help in understanding further the enzyme catalytic process.

Aldehyde Deformylating Oxygenase

In 2010, an alkane biosynthesis pathway was uncovered in cyanobacteria constituted of two enzymes: an acyl–acyl carrier protein (ACP) reductase (AAR) and an aldehyde deformylating oxygenase (ADO), formerly called aldehyde decarbonylase, which belongs to the ferritin-like superfamily (Schirmer et al. 2010). AAR produces C_n fatty aldehydes which are then converted to the corresponding C_{n-1} alkanes or alkenes by ADO. Genes coding for these two enzymes are often found in tandem throughout the cyanobacterial phylum but appear to have separate, specific promoters and therefore might be transcribed independently and even regulated differently (Klähn et al. 2014). Accordingly, ADO and AAR can catalyze their respective reaction independently from each other (Schirmer et al. 2010; Khara et al. 2013). Although the capacity of producing alkanes exists in organisms from all three domains of life, AAR and ADO seem to be present only in cyanobacteria (Klähn et al. 2014). This alkane biosynthesis pathway was originally uncovered by combining comparative biochemistry and genomic analyses, showing for example that heterologous co-expression of both genes caused *E. coli*, usually not producing hydrocarbons, to synthesize alkanes (Schirmer et al. 2010). Although the role of these enzymes in vivo remains unclear, they have attracted massive interest. Indeed, alkanes are major constituents of transportation fuels. Thus, this alkane biosynthesis pathway in cyanobacteria represents an attractive strategy for photosynthetic production of renewable hydrocarbon biofuels (Machado and Atsumi 2012; Zhou et al. 2018).

In this two-step alkane biosynthesis pathway, AAR and ADO act in series to catalyze the first and second step, respectively. Firstly, AAR converts the substrate fatty acyl-acyl carrier protein (acyl-ACP) or fatty acyl-coenzyme A (acyl-CoA) into the corresponding fatty aldehyde using NADPH. The reaction is catalyzed *via* the formation of an acyl-enzyme thioester intermediate where the fatty acyl group of the substrate is covalently bound to a cysteine residue in AAR, while the ACP or CoA moieties are released (Lin et al. 2013; Warui et al. 2015). Secondly, ADO catalyzes the conversion of this fatty aldehyde into formate and alkane (or alkene) that is one carbon atom smaller (Schirmer et al. 2010).

Below, we will present further the ferritin-like enzyme ADO. Firstly, we will look at structures obtained for this protein revealing a remarkable hydrophobic pocket to bind the substrate. Secondly, we will introduce the structure of the complex solved between AAR and ADO which suggests the existence of a potential aldehyde-transferring route between the enzymes.

The Ferritin-Like ADO

The ADO enzyme was originally named aldehyde decarbonylase as there was initially some confusion on the nature of the reaction catalyzed and whether oxygen was required for the reaction. However, later evidence established molecular oxygen as its co-substrate (Li et al. 2011). Although redox-neutral, the reaction catalyzed by ADO requires, in addition to dioxygen, an auxiliary reducing system to provide four electrons (Warui et al. 2011; Li et al. 2011, 2012). While the identity of the physiological reductant remains elusive, ADO activity was shown to be supported by different ferredoxin–ferredoxin reductase systems, including a potential cognate system from cyanobacteria (Zhang et al. 2013; Crépin et al. 2018). Moreover, the enzyme seems to have a wide specificity of substrates, including C_4 to C_{18} aldehydes (Khara et al. 2013).

The ferritin-like fold of ADO was confirmed by several crystal structures solved for different organisms, including *Prochlorococcus marinus* (*Pm*ADO) (Khara et al. 2013; Buer et al. 2014), *Synechococcus elongatus* (*Se*ADO) (Jia et al. 2015), *Synechocystis sp.* (*Sy*ADO) (Wang et al. 2017a), *Limnothrix* sp. KNUA012 (*Li*ADO) and *Oscillatoria* sp. KNUA011 (*Os*ADO) (Park et al. 2016). The enzyme architecture comprises a total of eight helices. ADO is monomeric in solution and crystal structures support the absence of oligomerization (Fig. 4.11a) (Buer et al. 2014; Chang et al. 2020). However, the existence of a homodimeric species has been reported for ADO from *Nostoc punctiforme* (*Np*ADO) (Warui et al. 2015; Hayashi et al. 2015).

ADO exhibits a dinuclear iron site which is coordinated by two histidine (His76 and His160, *Pm*ADO numbering) and four glutamate residues (Glu45, Glu73,

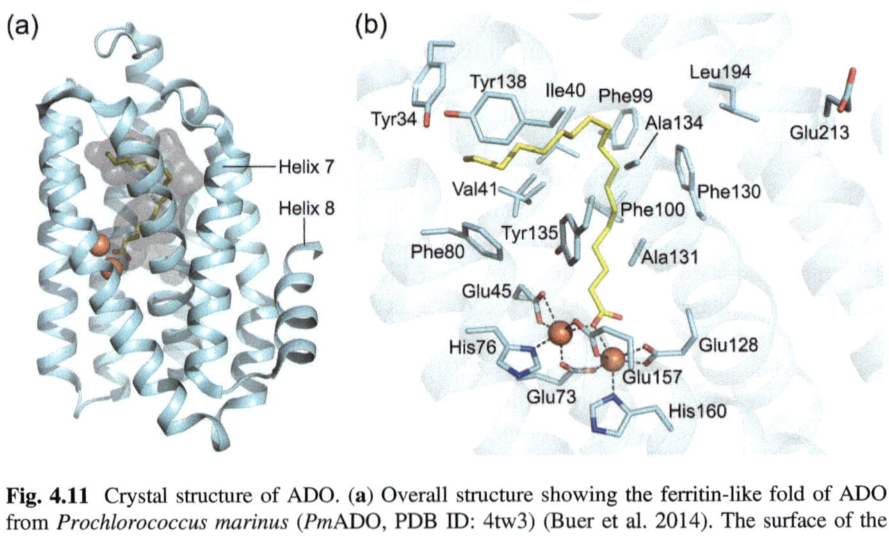

Fig. 4.11 Crystal structure of ADO. (**a**) Overall structure showing the ferritin-like fold of ADO from *Prochlorococcus marinus* (*Pm*ADO, PDB ID: 4tw3) (Buer et al. 2014). The surface of the hydrophobic pocket is shown in grey. (**b**) The stearic acid ligand interacts with the diiron center of ADO. Iron ions are depicted as orange spheres and the ligand as yellow stick

Glu128 and Glu157, *Pm*ADO numbering) (Fig. 4.11b). To note, Glu144 in *Se*ADO (equivalent to Glu157 in *Pm*ADO) is proposed to swing away from the diiron site during the catalytic process as it was observed in an outward conformation after treatment of the enzyme with hydrogen peroxide, possibly representing an oxidized state of the enzyme (Jia et al. 2015). Spectroscopic studies suggest that the two iron ions are in nearly identical environments, both in diferric and diferrous states (Aukema et al. 2013). ADO catalysis could proceed through a radical mechanism where the initial Fe(II)Fe(II) center turns into a Fe(III)Fe(III) cofactor after exposure to molecular oxygen (Das et al. 2014; Rajakovich et al. 2015; Wang et al. 2016). It has been suggested by spectroscopy that a diferric-peroxide intermediate is formed when ADO in complex with substrate reacts with dioxygen (Pandelia et al. 2013). To note, no experimental data was obtained so far to support the formation of a high-valent Fe(III)Fe(IV) intermediate, similar to intermediate **X** in R2a RNR (Rajakovich et al. 2020).

Surprisingly, a low iron content is reported for isolated ADO in solution (Buer et al. 2014; Jia et al. 2015). Accordingly, the metal-binding site is often partially occupied by iron ions or completely metal-free in several crystal structures (Buer et al. 2014; Jia et al. 2015; Park et al. 2016). The apparent labile nature of the diiron center is associated with a large conformational change in the middle of helix 5 shown to unfold into a solvent-exposed loop for *Pm*ADO, *Se*ADO, *Sy*ADO, and *Li*ADO in absence of iron (Buer et al. 2014; Jia et al. 2015; Park et al. 2016; Wang et al. 2017a; Basri et al. 2020). This segment comprises the two iron ligands Glu157 and His160 (*Pm*ADO numbering) from the second motif Glu/Asp-X_n-Glu/Asp-X-X-His characteristic of ferritin-like enzymes. Consequently, these two residues move away from the binding site, completely disassembling the second iron site while the first one remains intact. The flexibility of helix 5 is proposed as a mechanism allowing access to the active site for the iron ions (Buer et al. 2014) and for the co-substrate dioxygen (Jia et al. 2015). This conformational change is rather unusual in ferritin-like enzymes. For example, both sMMO and R2 in class I RNR show some variations for a few side chains in metal-free structures but metal-binding sites remain mainly intact (Aberg et al. 1993; Sazinsky et al. 2004). However, an analogous feature can be observed in another ferritin-like enzyme, namely, R2-like ligand-binding oxidase (R2lox). In the metal-free form of R2lox, the first metal-site is disordered and an α-helix is disrupted whereas the second site is preformed (Griese et al. 2013).

The coordination of the iron ions in ADO crystal structures is often completed by an additional exogenous ligand different from the actual substrate. ADOs are recombinantly produced in *E. coli* and co-purify with a mixture of ligands identified by mass spectrometry as mainly long-chain fatty acids, such as palmitic acid, stearic acid, and oleic acid (Khara et al. 2013; Jia et al. 2015). It is probable that these ligands mimic the binding of aldehyde substrates and are trapped from the host cells. The enzyme accommodates the substrate analogue in a long, tunnel-like hydrophobic pocket within the protein scaffold. The carboxylate functional group interacts with the metal site and different binding modes have been observed in crystal structures, e.g., the fatty acid carboxylate can bridge both iron ions in a μ-1,3 or a μ-1,1 fashion (Fig. 4.11b), or provide a monodentate coordination to a single iron (Fox 2021). Besides, the substrate pocket is lined by conserved hydrophobic residues, including Tyr34, Ile40, Val41, Phe80, Phe99, Phe100, Phe130, Ala131, Ala134, Tyr135, Tyr138 (*Pm*ADO numbering) (Basri et al. 2020). The entry mechanism of the substrate remains elusive, but a crystal structure obtained for ADO trapped with a second ligand provides an interesting piece of information. Here, Leu194 side chain and helix 8 at the C-terminus of the protein are displaced to allow the ligand seated inside the cavity to extend into bulk solvent, thereby unveiling the putative entrance to the channel (Buer et al. 2014).

The use of ADO for biotechnological purposes is attractive, but could be limited by the low apparent activity of the enzyme in vitro [$k_{cat} < 1\ min^{-1}$ (Wang et al. 2017a)] and the requirement of an auxiliary-reducing partner. Bioengineering is currently being performed on ADO in order to rationally design solutions. For instance, self-sufficient ADOs converting aldehydes into alkanes were generated by fusing the enzyme with different combinations of ferredoxin–ferredoxin reductase systems (Wang et al. 2014a). Besides, large mutational studies have demonstrated that ADO mutants with significantly improved enzymatic activity can be produced (Wang et al. 2017a; Kudo et al. 2019a). Furthermore, structure-based engineering was also able to identify residues critical for substrate specificity and to yield ADO variants with modified substrate preference towards different chain-length substrates (Khara et al. 2013; Bao et al. 2016). For example, the residue Ala134 lining the channel next to the C9 position of the ligand (Fig. 4.11b) was mutated into a bulkier phenylalanine residue to prevent the binding of long-chain aldehydes. Consequently, the ADO variant displays both enhanced selectivity and activity for short to medium chain aldehydes compared to wild type *Pm*ADO (Khara et al. 2013).

ADO in Complex with AAR

In the cyanobacterial alkane biosynthesis pathway, AAR performs the first, NADPH-dependent reaction, to reduce the fatty acyl group bound to ACP or CoA into a free fatty aldehyde. During the reaction, a thioester intermediate is generated involving the acyl moiety and residue Cys294 (*Se*AAR numbering) of the enzyme. Then, AAR uses NADPH to transfer a hydride to the thioester intermediate to release the aldehyde product (Lin et al. 2013; Warui et al. 2015). Similar to ADO, the biotechnological use of AAR for biofuel production could be limited by intrinsic properties of the enzyme that cause it to aggregate when recombinantly expressed in *E. coli* (Sharma et al. 2021) and exhibit slow turnover ($k_{cat} < 1\ min^{-1}$) (Lin et al. 2013). Protein engineering has been thus also applied to AAR in order to improve both its activity and solubility (Kudo et al. 2016, 2019b).

Long-chain aldehyde molecules are poorly soluble in water and assemble as micelles in aqueous environments (Warui et al. 2015). Therefore, the transfer of the AAR product as a substrate for ADO represents a crucial step for efficient alkane biosynthesis and may proceed through formation of a complex between the two enzymes. In solution, AAR and ADO form a complex in a 1:1 ratio with a K_D of approximately 2 μM and 3 μM for proteins from *S. elongatus* and *N. punctiforme*, respectively (Warui et al. 2015; Gao et al. 2020). Analytical size-exclusion chromatography and mutational studies showed that the interface between AAR and ADO appears to be stabilized through electrostatic interactions and involves, in particular, the critical ADO residue Glu200 in helix 7 (*Se*ADO numbering, corresponding to Glu213 in *Pm*ADO) (Fig. 4.12) (Gao et al. 2020; Chang et al. 2020). Interestingly, the use of a chimeric ADO–AAR fusion protein can increase alkane yield in bacteria (Rahmana et al. 2014).

The existence of a putative aldehyde-transferring pathway from AAR to ADO was recently supported by crystal structures of the protein complex from *S. elongatus* in interaction with NADPH and/or substrates (Gao et al. 2020). Three structures were obtained for the complex, between 2.1 and 3.0 Å resolution, each one captured at a putatively different intermediate state along the catalytic reaction. Using single-crystal X-ray diffraction, this was achieved by ingeniously combining different components during crystal preparation. First, the structure of the acyl-enzyme intermediate was obtained by incubating AAR with stearoyl-CoA, and then adding ADO but no NADPH prior to crystallization. This structure shows that a long hydrophobic tunnel in AAR accommodates a stearic chain bound to Cys294 *via* a thioester bond (Fig. 4.12). In order to solve the complex structure in interaction with NADPH, the same crystals were subsequently soaked in an NADPH-containing solution. Moreover, a structure of the complex in interaction with an intact substrate was obtained by incubating stearoyl-CoA with an AAR

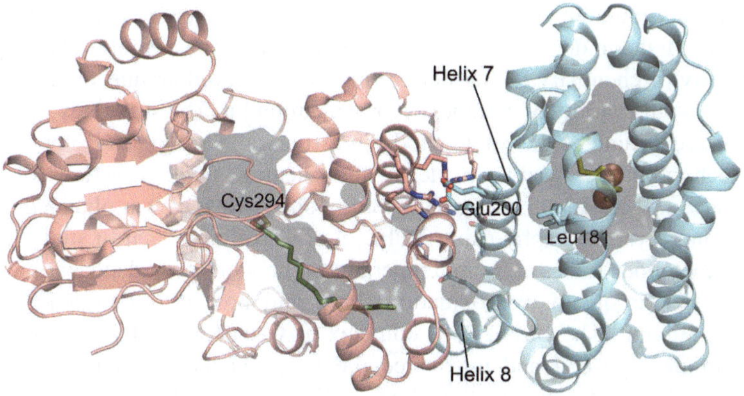

Fig. 4.12 Crystal structure of ADO in complex with AAR from *S. elongatus* (PDB ID: 6jzu) (Gao et al. 2020). The interface between ADO and AAR (in cyan and salmon, respectively) is stabilized by electrostatic interactions involving many basic and acidic residues shown in stick. The surface of the hydrophobic pockets in AAR and ADO are shown in grey. The stearic chain bound to Cys294 *via* a thioester linkage in AAR is shown in green stick. The ligand and iron ions in ADO are depicted as yellow stick and orange spheres. Leu181 and Glu200 in *Se*ADO correspond to Leu194 and Glu213 in *Pm*ADO, respectively

mutant where Cys294 is substituted by a serine residue, thereby preventing the formation of the thioester intermediate (Gao et al. 2020).

In the structures of the ADO–AAR complex, numerous salt bridges and hydrogen bonds can be observed between basic side chains from AAR and acidic residues from helix 7 of ADO, including Glu200, consistent with the data obtained in solution (Fig. 4.12) (Gao et al. 2020; Chang et al. 2020). Strikingly, the channel in AAR extends towards the intermolecular interface, next to helix 8 of ADO believed to undergo conformational changes to open the putative entrance of ADO hydrophobic pocket (Buer et al. 2014). Therefore, the arrangement of the proteins within the complex suggests the manner by which the aldehyde molecule could be delivered to ADO. However, ADO shows a closed conformation in all three complex structures (Gao et al. 2020). Consequently, the crystallographic data do not allow the visualization of a continuous channel between the enzymes where a hydrophobic substrate could be directly transferred without being exposed to bulk solvent.

Altogether, the results obtained by Gao et al. (2020) shows that the formation of the AAR–ADO complex is probably key in the alkane biosynthesis pathway and suggest the presence of a putative aldehyde-transferring channel between the subunits (Gao et al. 2020). These structures set the stage for future studies aiming to decipher the complete catalytic process of these enzymes.

Conclusion and Perspective

In this chapter, we have presented the current knowledge on structures of complexes involving three ferritin-like dinuclear metalloenzymes: sMMO, class I RNR and ADO. Recent groundbreaking structural studies on these enzymes provided new key information in the understanding of their macromolecular assembly. They highlight the diversity of chemistry taking place within a conserved architecture, the ferritin fold, and also the variety of complexes formed around it. The structures were solved using different state-of-the-art methods in structural biology and emphasize the recent technological advances made in this scientific field. Structures of sMMO hydroxylase sMMOH with its regulatory component MMOB was obtained by serial femtosecond crystallography using XFEL, a powerful approach to study radiation sensitive metalloenzymes at room temperature. The work provided clear pictures of the metal active site in two oxidation states and how it is reorganized by the regulatory subunit. The class I RNR complex between R1 and R2 was obtained by single-particle cryo-EM and uncovered the long-awaited structure of the active complex. It allowed for the first time the visualization of the PCET pathway between the subunits and gave structural grounds to better understand the catalytical mechanism of this enzyme. Finally, the structure of ADO interacting with AAR unraveled the intermolecular interactions and architecture of the complex through a more standard but efficient X-ray diffraction scheme. The results suggest a remarkable strategy used by the alkane biosynthesis pathway to transfer the substrate through a potential hydrophobic channel between the enzymes.

Table 4.1 Selection of representative protein complex structures in the ferritin-like superfamily

Enzyme	Protein complex	Function	PDB ID	References
sMMO	sMMOH–MMOB	Catalytic regulation/substrate transfer	6yd0	Srinivas et al. (2020)
RNR R2b	R2–NrdI	Superoxide transfer	3n3a	Boal et al. (2010)
RNR R2a	R1–R2	Radical transfer	6w4x	Kang et al. (2020)
ADO	ADO–AAR	Substrate transfer	6jzu	Gao et al. (2020)
Toluene monooxygenase	T4moH–T4moD	Catalytic regulation	3dhh	Bailey et al. (2008)
Toluene monooxygenase	T4moH–T4moC	Electron transfer	4p1b	Acheson et al. (2014)
Phenol hydroxylase	PHH–PHM	Catalytic regulation	2inp	Sazinsky et al. (2006)
Δ^9-desaturase	Desaturase–ACP	Substrate transfer	2xz1	Guy et al. (2011)

This book chapter focuses on three ferritin-like enzymes, but more protein complex structures have been obtained for this superfamily. A selection of representative structures is listed in Table 4.1. For example, we can cite members of the bacterial multicomponent monooxygenases (BMMs) family which include sMMO but also phenol hydroxylase (PH) and toluene 4-monoxygenase (T4MO). Crystal structures of PH and T4MO involved in different complexes have also been solved. A structure of the PH hydroxylase (PHH) was obtained in complex with its regulatory protein (PHM) (Sazinsky et al. 2006). Besides, the hydroxylase component of T4MO (T4moH) was solved in complex with its effector protein (T4moD) (Bailey et al. 2008) and in complex with its electron transfer ferredoxin (T4moC) (Acheson et al. 2014). Moreover, structures were obtained for the complex between the stearoyl–acyl carrier protein Δ^9-desaturase and acyl carrier protein (ACP) (Guy et al. 2011).

The rapid development of structural biology and increasing use of cryo-EM and XFEL approaches open unprecedented opportunities to decipher biological processes. Therefore, in the coming years, more complexes of ferritin-like enzymes will be solved and thus could provide, for example, detailed pictures of high-valent diiron intermediates for sMMO or quaternary assemblies of manganese-dependent and metal-free class I RNRs.

References

Aberg A, Nordlund P, Eklund H (1993) Unusual clustering of carboxyl side chains in the core of iron-free ribonucleotide reductase. Nature 361(6409):276–278. https://doi.org/10.1038/361276a0

Acheson JF, Bailey LJ, Elsen NL, Fox BG (2014) Structural basis for biomolecular recognition in overlapping binding sites in a diiron enzyme system. Nat Commun 5:5009. https://doi.org/10.1038/ncomms6009

Ambundo EA, Friesner RA, Lippard SJ (2002) Reactions of methane monooxygenase intermediate Q with derivitized methanes. J Am Chem Soc 124(30):8770–8771

Andersson CS, Högbom M (2009) A mycobacterium tuberculosis ligand-binding Mn/Fe protein reveals a new cofactor in a remodeled R2-protein scaffold. Proc Natl Acad Sci 106(14):5633–5638. https://doi.org/10.1073/pnas.0812971106

Ando N, Brignole EJ, Zimanyi CM, Funk MA, Yokoyama K, Asturias FJ, Stubbe J, Drennan CL (2011) Structural interconversions modulate activity of Escherichia coli ribonucleotide reductase. Proc Natl Acad Sci U S A 108(52):21046–21051. https://doi.org/10.1073/pnas.1112715108

Andrews SC (2010) The Ferritin-like superfamily: evolution of the biological iron storeman from a rubrerythrin-like ancestor. Biochim Biophys Acta 1800(8):691–705. https://doi.org/10.1016/j.bbagen.2010.05.010

Argirević T, Riplinger C, Stubbe J, Neese F, Bennati M (2012) ENDOR spectroscopy and DFT calculations: evidence for the hydrogen-bond network within α2 in the PCET of E. coli ribonucleotide reductase. J Am Chem Soc 134(42):17661–17670. https://doi.org/10.1021/ja3071682

Atkin CL, Thelander L, Reichard P, Lang G (1973) Iron and free radical in ribonucleotide reductase. Exchange of iron and Mössbauer spectroscopy of the protein B2 subunit of the Escherichia coli enzyme. J Biol Chem 248(21):7464–7472

Aukema KG, Makris TM, Stoian SA, Richman JE, Münck E, Lipscomb JD, Wackett LP (2013) Cyanobacterial aldehyde deformylase oxygenation of aldehydes yields n-1 aldehydes and alcohols in addition to alkanes. ACS Catal 3(10):2228–2238. https://doi.org/10.1021/cs400484m

Aye Y, Li M, Long MJC, Weiss RS (2015) Ribonucleotide reductase and cancer: biological mechanisms and targeted therapies. Oncogene 34(16):2011–2021. https://doi.org/10.1038/onc.2014.155

Bailey LJ, McCoy JG, Phillips GN, Fox BG (2008) Structural consequences of effector protein complex formation in a diiron hydroxylase. Proc Natl Acad Sci U S A 105(49):19194–19198. https://doi.org/10.1073/pnas.0807948105

Banerjee R, Lipscomb JD (2021) Small-molecule tunnels in metalloenzymes viewed as extensions of the active site. Acc Chem Res 54(9):2185–2195. https://doi.org/10.1021/acs.accounts.1c00058

Banerjee R, Meier KK, Münck E, Lipscomb JD (2013) Intermediate P* from soluble methane monooxygenase contains a diferrous cluster. Biochemistry 52(25):4331–4342

Banerjee R, Proshlyakov Y, Lipscomb JD, Proshlyakov DA (2015) Structure of the key species in the enzymatic oxidation of methane to methanol. Nature 518(7539):431–434. https://doi.org/10.1038/nature14160

Banerjee R, Jones JC, Lipscomb JD (2019) Soluble methane monooxygenase. Ann Rev. Biochemist 88:409–431. https://doi.org/10.1146/annurev-biochem-013118-111529

Bao L, Li J-J, Jia C, Li M, Lu X (2016) Structure-oriented substrate specificity engineering of aldehyde-deformylating oxygenase towards aldehydes carbon chain length. Biotechnol Biofuels 9(1):185. https://doi.org/10.1186/s13068-016-0596-9

Barynin VV, Whittaker MM, Antonyuk SV, Lamzin VS, Harrison PM, Artymiuk PJ, Whittaker JW (2001) Crystal structure of manganese catalase from Lactobacillus plantarum. Structure 9(8):725–738. https://doi.org/10.1016/s0969-2126(01)00628-1

Basri RS, Rahman RNZRA, Kamarudin NHA, Ali MSM (2020) Cyanobacterial aldehyde deformylating oxygenase: structure, function, and potential in biofuels production. Int J Biol Macromol 164:3155–3162. https://doi.org/10.1016/j.ijbiomac.2020.08.162

Berggren G, Duraffourg N, Sahlin M, Sjöberg B-M (2014) Semiquinone-induced maturation of Bacillus anthracis ribonucleotide reductase by a superoxide intermediate. J Biol Chem 289(46):31940–31949. https://doi.org/10.1074/jbc.M114.592535

Bergmann U, Kern J, Schoenlein RW, Wernet P, Yachandra VK, Yano J (2021) Using X-ray free-electron lasers for spectroscopy of molecular catalysts and metalloenzymes. Nat Rev Phys 3(4): 264–282. https://doi.org/10.1038/s42254-021-00289-3

Blaesi EJ, Palowitch GM, Hu K, Kim AJ, Rose HR, Alapati R, Lougee MG, Kim HJ, Taguchi AT, Tan KO, Laremore TN, Griffin RG, Krebs C, Matthews ML, Silakov A, Bollinger JM, Allen BD, Boal AK (2018) Metal-free class Ie ribonucleotide reductase from pathogens initiates catalysis with a tyrosine-derived dihydroxyphenylalanine radical. Proc Natl Acad Sci U S A 115(40):10022–10027. https://doi.org/10.1073/pnas.1811993115

Boal AK, Cotruvo JA, Stubbe J, Rosenzweig AC (2010) Structural basis for activation of class Ib ribonucleotide reductase. Science 329(5998):1526–1530. https://doi.org/10.1126/science.1190187

Boal AK, Cotruvo JA, Stubbe J, Rosenzweig AC (2012) The dimanganese(II) site of Bacillus subtilis class Ib ribonucleotide reductase. Biochemistry 51(18):3861–3871. https://doi.org/10.1021/bi201925t

Bou-Abdallah F (2010) The iron redox and hydrolysis chemistry of the ferritins. Biochim Biophys Acta 1800(8):719–731. https://doi.org/10.1016/j.bbagen.2010.03.021

Brazeau BJ, Lipscomb JD (2000) Kinetics and activation thermodynamics of methane monooxygenase compound Q formation and reaction with substrates. Biochemistry 39(44): 13503–13515

Brazeau BJ, Lipscomb JD (2003) Key amino acid residues in the regulation of soluble methane monooxygenase catalysis by component B. Biochemistry 42(19):5618–5631

Brazeau BJ, Wallar BJ, Lipscomb JD (2001) Unmasking of deuterium kinetic isotope effects on the methane monooxygenase compound Q reaction by site-directed mutagenesis of component B. J Am Chem Soc 123(42):10421–10422

Buer BC, Paul B, Das D, Stuckey JA, Marsh ENG (2014) Insights into substrate and metal binding from the crystal structure of cyanobacterial aldehyde deformylating oxygenase with substrate bound. ACS Chem Biol 9(11):2584–2593. https://doi.org/10.1021/cb500343j

Castillo RG, Banerjee R, Allpress CJ, Rohde GT, Bill E, Que L Jr, Lipscomb JD, DeBeer S (2017) High-energy-resolution fluorescence-detected X-ray absorption of the Q intermediate of soluble methane monooxygenase. J Am Chem Soc 139:18024–18033. https://doi.org/10.1021/jacs.7b09560

Chang SL, Wallar BJ, Lipscomb JD, Mayo KH (2001) Residues in Methylosinus trichosporium OB3b methane monooxygenase component B involved in molecular interactions with reduced- and oxidized-hydroxylase component: a role for the N-terminus. Biochemistry 40(32): 9539–9551

Chang M, Shimba K, Hayashi Y, Arai M (2020) Electrostatic interactions at the interface of two enzymes are essential for two-step alkane biosynthesis in cyanobacteria. Biosci Biotechnol Biochem 84(2):228–237. https://doi.org/10.1080/09168451.2019.1677142

Chapman HN (2019) X-Ray free-electron lasers for the structure and dynamics of macromolecules. Annu Rev Biochem 88(1):35–58. https://doi.org/10.1146/annurev-biochem-013118-110744

Chatwood LL, Mueller J, Gross JD, Wagner G, Lippard SJ (2004) NMR structure of the flavin domain from soluble methane monooxygenase reductase from Methylococcus capsulatus (Bath). Biochemistry 43(38):11983–11991

Cheng Y (2018) Single-particle cryo-EM-How did it get here and where will it go. Science 361(6405):876–880. https://doi.org/10.1126/science.aat4346

Chino M, Maglio O, Nastri F, Pavone V, DeGrado WF, Lombardi A (2015) Artificial Diiron enzymes with a de novo designed four-helix bundle structure: artificial diiron enzymes with a four-helix bundle structure. Eur J Inorg Chem 2015(21):3371–3390. https://doi.org/10.1002/ejic.201500470

Choi YS, Zhang H, Brunzelle JS, Nair SK, Zhao H (2008) In vitro reconstitution and crystal structure of p-aminobenzoate N-oxygenase (Aur F) involved in aureothin biosynthesis. Proc Natl Acad Sci U S A 105(19):6858–6863. https://doi.org/10.1073/pnas.0712073105

Climent I, Sjöberg BM, Huang CY (1992) Site-directed mutagenesis and deletion of the carboxyl terminus of Escherichia coli ribonucleotide reductase protein R2. Effects on catalytic activity and subunit interaction. Biochemistry 31(20):4801–4807. https://doi.org/10.1021/bi00135a009

Clomburg JM, Crumbley AM, Gonzalez R (2017) Industrial biomanufacturing: The future of chemical production. Science 355(6320):aag0804. https://doi.org/10.1126/science.aag0804

Conrado RJ, Gonzalez R (2014) Envisioning the bioconversion of methane to liquid fuels. Science 343(6171):621–623. https://doi.org/10.1126/science.1246929

Cooley RB, Arp DJ, Karplus PA (2010) Evolutionary origin of a secondary structure: π-helices as cryptic but widespread insertional variations of α-helices that enhance protein functionality. J Mol Biol 404(2):232–246. https://doi.org/10.1016/j.jmb.2010.09.034

Cotruvo JA Jr, Stubbe J (2012) Metallation and mismetallation of iron and manganese proteins in vitro and in vivo: the class I ribonucleotide reductases as a case study. Metallomics 4(10): 1020. https://doi.org/10.1039/c2mt20142a

Cotruvo JA, Stubbe J (2010) An active dimanganese(III)-tyrosyl radical cofactor in Escherichia coli class Ib ribonucleotide reductase. Biochemistry 49(6):1297–1309. https://doi.org/10.1021/bi902106n

Cotruvo JA, Stich TA, Britt RD, Stubbe J (2013) Mechanism of assembly of the dimanganese-tyrosyl radical cofactor of class Ib ribonucleotide reductase: enzymatic generation of superoxide is required for tyrosine oxidation *via* a Mn(III)Mn(IV) intermediate. J Am Chem Soc 135(10): 4027–4039. https://doi.org/10.1021/ja312457t

Cox N, Ogata H, Stolle P, Reijerse E, Auling G, Lubitz W (2010) A tyrosyl-dimanganese coupled spin system is the native metalloradical cofactor of the R2F subunit of the ribonucleotide reductase of Corynebacterium ammoniagenes. J Am Chem Soc 132(32):11197–11213. https://doi.org/10.1021/ja1036995

Crépin L, Barthe M, Leray F, Guillouet SE (2018) Alka(e)ne synthesis in *Cupriavidus necator* boosted by the expression of endogenous and heterologous ferredoxin–ferredoxin reductase systems. Biotechnol Bioeng 115(10):2576–2584. https://doi.org/10.1002/bit.26805

Crichton RR, Declercq J-P (2010) X-ray structures of ferritins and related proteins. Biochim Biophys Acta 1800(8):706–718. https://doi.org/10.1016/j.bbagen.2010.03.019

Cutsail GE, Banerjee R, Zhou A, Que L, Lipscomb JD, DeBeer S (2018) High-resolution extended X-ray absorption fine structure analysis provides evidence for a longer Fe···Fe distance in the Q intermediate of methane monooxygenase. J Am Chem Soc 140(48):16807–16820. https://doi.org/10.1021/jacs.8b10313

Das D, Ellington B, Paul B, Marsh ENG (2014) Mechanistic insights from reaction of α-oxiranyl-aldehydes with cyanobacterial aldehyde deformylating oxygenase. ACS Chem Biol 9(2): 570–577. https://doi.org/10.1021/cb400772q

Dassama LMK, Krebs C, Bollinger JM, Rosenzweig AC, Boal AK (2013) Structural basis for assembly of the Mn(IV)/Fe(III) cofactor in the class Ic ribonucleotide reductase from Chlamydia trachomatis. Biochemistry 52(37):6424–6436. https://doi.org/10.1021/bi400819x

Divakaran A, Kirberger SE, Pomerantz WCK (2019) SAR by (protein-observed) 19F NMR. Acc Chem Res 52(12):3407–3418. https://doi.org/10.1021/acs.accounts.9b00377

Doan PE, Shanmugam M, Stubbe J, Hoffman BM (2015) Composition and Structure of the Inorganic Core of Relaxed Intermediate **X** (Y122F) of *Escherichia coli* Ribonucleotide Reductase. J Am Chem Soc 137(49):15558–15566. https://doi.org/10.1021/jacs.5b10763

Ehrenberg A, Reichard P (1972) Electron spin resonance of the iron-containing protein B2 from ribonucleotide reductase. J Biol Chem 247(11):3485–3488

Elango NA, Radhakrishnan R, Froland WA, Wallar BJ, Earhart CA, Lipscomb JD, Ohlendorf DH (1997) Crystal structure of the hydroxylase component of methane monooxygenase from *Methylosinus trichosporium* OB3b: structure methane monooxygenase hydroxylase. Protein Sci 6(3):556–568. https://doi.org/10.1002/pro.5560060305

Eriksson M, Uhlin U, Ramaswamy S, Ekberg M, Regnström K, Sjöberg BM, Eklund H (1997) Binding of allosteric effectors to ribonucleotide reductase protein R1: reduction of active-site cysteines promotes substrate binding. Structure 5(8):1077–1092. https://doi.org/10.1016/s0969-2126(97)00259-1

Eriksson M, Jordan A, Eklund H (1998) Structure of Salmonella typhimurium nrdF ribonucleotide reductase in its oxidized and reduced forms. Biochemistry 37(38):13359–13369. https://doi.org/10.1021/bi981380s

Fox BG (2021) Diiron enzyme structure and catalysis. In: Comprehensive coordination chemistry III. Elsevier, pp 455–499

Fox BG, Froland WA, Dege JE, Lipscomb JD (1989) Methane monooxygenase from Methylosinus trichosporium OB3b. Purification and properties of a three-component system with high specific activity from a type II methanotroph. J Biol Chem 264(17):10023–10033

Fox BG, Liu Y, Dege JE, Lipscomb JD (1991) Complex formation between the protein components of methane monooxygenase from Methylosinus trichosporium OB3b. Identification of sites of component interaction. J Biol Chem 266(1):540–550

Froland WA, Andersson KK, Lee S-K, Liu Y, Lipscomb JD (1992) Methane monooxygenase component B and reductase alter the regioselectivity of the hydroxylase component-catalyzed reactions. A novel role for protein-protein interactions in an oxygenase mechanism. J Biol Chem 267(25):17588–17597

Fuller FD, Gul S, Chatterjee R, Burgie ES, Young ID, Lebrette H, Srinivas V, Brewster AS, Michels-Clark T, Clinger JA, Andi B, Ibrahim M, Pastor E, de Lichtenberg C, Hussein R, Pollock CJ, Zhang M, Stan CA, Kroll T, Fransson T, Weninger C, Kubin M, Aller P, Lassalle L, Bräuer P, Miller MD, Amin M, Koroidov S, Roessler CG, Allaire M, Sierra RG, Docker PT, Glownia JM, Nelson S, Koglin JE, Zhu D, Chollet M, Song S, Lemke H, Liang M, Sokaras D, Alonso-Mori R, Zouni A, Messinger J, Bergmann U, Boal AK, Bollinger JM, Krebs C, Högbom M, Phillips GN, Vierstra RD, Sauter NK, Orville AM, Kern J, Yachandra VK, Yano J (2017) Drop-on-demand sample delivery for studying biocatalysts in action at X-ray free-electron lasers. Nat Methods 14(4):443–449. https://doi.org/10.1038/nmeth.4195

Gallagher SC, Callaghan AJ, Zhao J, Dalton H, Trewhella J (1999) Global conformational changes control the reactivity of methane monooxygenase. Biochemistry 38(21):6752–6760

Gao Y, Zhang H, Fan M, Jia C, Shi L, Pan X, Cao P, Zhao X, Chang W, Li M (2020) Structural insights into catalytic mechanism and product delivery of cyanobacterial acyl-acyl carrier protein reductase. Nat Commun 11(1):1525. https://doi.org/10.1038/s41467-020-15268-y

Gassner GT, Lippard SJ (1999) Component interactions in the soluble methane monooxygenase system from Methylococcus capsulatus (Bath). Biochemistry 38(39):12768–12785

Ge J, Yu G, Ator MA, Stubbe J (2003) Pre-steady-state and steady-state kinetic analysis of E. coli class I ribonucleotide reductase. Biochemistry 42(34):10071–10083. https://doi.org/10.1021/bi034374r

Gräslund A, Sahlin M, Sjöberg BM (1985) The tyrosyl free radical in ribonucleotide reductase. Environ Health Perspect 64:139–149. https://doi.org/10.1289/ehp.64-1568609

Grāve K, Lambert W, Berggren G, Griese JJ, Bennett MD, Logan DT, Högbom M (2019) Redox-induced structural changes in the di-iron and di-manganese forms of Bacillus anthracis ribonucleotide reductase subunit NrdF suggest a mechanism for gating of radical access. J Biol Inorg Chem 24(6):849–861. https://doi.org/10.1007/s00775-019-01703-z

Grāve K, Griese JJ, Berggren G, Bennett MD, Högbom M (2020) The Bacillus anthracis class Ib ribonucleotide reductase subunit NrdF intrinsically selects manganese over iron. J Biol Inorg Chem 25(4):571–582. https://doi.org/10.1007/s00775-020-01782-3

Green J, Dalton H (1985) Protein B of soluble methane monooxygenase from Methylococcus capsulatus (Bath). A novel regulatory protein of enzyme activity. J Biol Chem 260(29): 15795–15801

Green J, Dalton H (1986) Steady-state kinetic-analysis of soluble methane mono-oxygenase from Methylococcus-capsulatus (Bath). Biochem J 236(1):155–162

Green J, Dalton H (1989) Substrate specificity of soluble methane monooxygenase. Mechanistic implications. J Biol Chem 264(30):17698–17703

Greene BL, Taguchi AT, Stubbe J, Nocera DG (2017) Conformationally dynamic radical transfer within ribonucleotide reductase. J Am Chem Soc 139(46):16657–16665. https://doi.org/10.1021/jacs.7b08192

Greene BL, Kang G, Cui C, Bennati M, Nocera DG, Drennan CL, Stubbe J (2020) Ribonucleotide reductases: structure, chemistry, and metabolism suggest new therapeutic targets. Annu Rev Biochem 89(1):45–75. https://doi.org/10.1146/annurev-biochem-013118-111843

Griese JJ, Roos K, Cox N, Shafaat HS, Branca RMM, Lehtio J, Graslund A, Lubitz W, Siegbahn PEM, Högbom M (2013) Direct observation of structurally encoded metal discrimination and ether bond formation in a heterodinuclear metalloprotein. Proc Natl Acad Sci U S A 110 (43):17189–17194. https://doi.org/10.1073/pnas.1304368110

Guy JE, Whittle E, Moche M, Lengqvist J, Lindqvist Y, Shanklin J (2011) Remote control of regioselectivity in acyl-acyl carrier protein-desaturases. Proc Natl Acad Sci U S A 108(40): 16594–16599. https://doi.org/10.1073/pnas.1110221108

Hammerstad M, Hersleth H-P, Tomter AB, Røhr AK, Andersson KK (2014) Crystal structure of Bacillus cereus class Ib ribonucleotide reductase di-iron NrdF in complex with NrdI. ACS Chem Biol 9(2):526–537. https://doi.org/10.1021/cb400757h

Hayashi Y, Yasugi F, Arai M (2015) Role of cysteine residues in the structure, stability, and alkane producing activity of cyanobacterial aldehyde deformylating oxygenase. PLoS One 10(4): e0122217. https://doi.org/10.1371/journal.pone.0122217

Hendrich MP, Münck E, Fox BG, Lipscomb JD (1990) Integer-spin EPR studies of the fully reduced methane monooxygenase hydroxylase component. J Am Chem Soc 112:5861–5865

Högbom M, Galander M, Andersson M, Kolberg M, Hofbauer W, Lassmann G, Nordlund P, Lendzian F (2003) Displacement of the tyrosyl radical cofactor in ribonucleotide reductase obtained by single-crystal high-field EPR and 1.4-Å x-ray data. Proc Natl Acad Sci U S A 100(6):3209–3214. https://doi.org/10.1073/pnas.0536684100

Högbom M, Stenmark P, Voevodskaya N, McClarty G, Gräslund A, Nordlund P (2004) The radical site in chlamydial ribonucleotide reductase defines a new R2 subclass. Science 305(5681): 245–248. https://doi.org/10.1126/science.1098419

Högbom M, Sjöberg B-M, Berggren G (2020) Radical enzymes. In: eLS. Wiley, pp 375–393

Jacobs AB, Banerjee R, Deweese DE, Braun A, Babicz JT Jr, Gee LB, Sutherlin KD, Bottger LH, Yoda Y, Saito M, Kitao S, Kobayashi Y, Seto M, Tamasaku K, Lipscomb JD, Park K, Solomon EI (2021) Nuclear resonance vibrational spectroscopic definition of the Fe(IV)2 intermediate Q in methane monooxygenase and its reactivity. J Am Chem Soc 143(39):16007–16029. https://doi.org/10.1021/jacs.1c05436

Jasniewski AJ, Que L (2018) Dioxygen activation by nonheme diiron enzymes: diverse dioxygen adducts, high-valent intermediates, and related model complexes. Chem Rev 118(5): 2554–2592. https://doi.org/10.1021/acs.chemrev.7b00457

Jia C, Li M, Li J, Zhang J, Zhang H, Cao P, Pan X, Lu X, Chang W (2015) Structural insights into the catalytic mechanism of aldehyde-deformylating oxygenases. Protein Cell 6(1):55–67. https://doi.org/10.1007/s13238-014-0108-2

Jiang W, Yun D, Saleh L, Barr EW, Xing G, Hoffart LM, Maslak M-A, Krebs C, Bollinger JM (2007) A manganese(IV)/iron(III) cofactor in Chlamydia trachomatis ribonucleotide reductase. Science 316(5828):1188–1191. https://doi.org/10.1126/science.1141179

Johansson R, Torrents E, Lundin D, Sprenger J, Sahlin M, Sjöberg B-M, Logan DT (2010) High-resolution crystal structures of the flavoprotein NrdI in oxidized and reduced states--an unusual flavodoxin. Structural biology. FEBS J 277(20):4265–4277. https://doi.org/10.1111/j.1742-4658.2010.07815.x

Jones JC, Banerjee R, Shi K, Aihara H, Lipscomb JD (2020) Structural studies of the Methylosinus trichosporium OB3b soluble methane monooxygenase hydroxylase and regulatory component complex reveal a transient substrate tunnel. Biochemistry 59(32):2946–2961. https://doi.org/10.1021/acs.biochem.0c00459

Jones JC, Banerjee R, Shi K, Semonis MM, Aihara H, Pomerantz WCK, Lipscomb JD (2021) Soluble methane monooxygenase component interactions monitored by 19F NMR. Biochemistry 60(25):1995–2010. https://doi.org/10.1021/acs.biochem.1c00293

Kang G, Taguchi AT, Stubbe J, Drennan CL (2020) Structure of a trapped radical transfer pathway within a ribonucleotide reductase holocomplex. Science 368(6489):424–427. https://doi.org/10.1126/science.aba6794

Khara B, Menon N, Levy C, Mansell D, Das D, Marsh ENG, Leys D, Scrutton NS (2013) Production of propane and other short-chain alkanes by structure-based engineering of ligand specificity in aldehyde-deformylating oxygenase. Chembiochem 14(10):1204–1208. https://doi.org/10.1002/cbic.201300307

Kim H, An S, Park YR, Jang H, Yoo H, Park SH, Lee SJ, Cho US (2019) MMOD-induced structural changes of hydroxylase in soluble methane monooxygenase. Sci Adv 5(10): eaax0059. https://doi.org/10.1126/sciadv.aax0059

Klähn S, Baumgartner D, Pfreundt U, Voigt K, Schön V, Steglich C, Hess WR (2014) Alkane biosynthesis genes in cyanobacteria and their transcriptional organization. Front Bioeng Biotechnol 2:24. https://doi.org/10.3389/fbioe.2014.00024

Klinman JP, Kohen A (2013) Hydrogen tunneling links protein dynamics to enzyme catalysis. Annu Rev Biochem 82:471–496. https://doi.org/10.1146/annurev-biochem-051710-133623

Kopp DA, Gassner GT, Blazyk JL, Lippard SJ (2001) Electron-transfer reactions of the reductase component of soluble methane monooxygenase from Methylococcus capsulatus (Bath). Biochemistry 40(49):14932–14941

Kudo H, Nawa R, Hayashi Y, Arai M (2016) Comparison of aldehyde-producing activities of cyanobacterial acyl-(acyl carrier protein) reductases. Biotechnol Biofuels 9:234. https://doi.org/10.1186/s13068-016-0644-5

Kudo H, Hayashi Y, Arai M (2019a) Identification of non-conserved residues essential for improving the hydrocarbon-producing activity of cyanobacterial aldehyde-deformylating oxygenase. Biotechnol Biofuels 12:89. https://doi.org/10.1186/s13068-019-1409-8

Kudo H, Hayashi Y, Arai M (2019b) Improving hydrocarbon production by engineering cyanobacterial acyl-(acyl carrier protein) reductase. Biotechnol Biofuels 12:291. https://doi.org/10.1186/s13068-019-1623-4

Lee SK, Lipscomb JD (1999) Oxygen activation catalyzed by methane monooxygenase hydroxylase component: proton delivery during the O-O bond cleavage steps. Biochemistry 38(14): 4423–4432. https://doi.org/10.1021/bi982712w

Lee SK, Fox BG, Froland WA, Lipscomb JD, Münck E (1993a) A transient intermediate of the methane monooxygenase catalytic cycle containing a FeIVFeIV cluster. J Am Chem Soc 115: 6450–6451

Lee SK, Nesheim JC, Lipscomb JD (1993b) Transient intermediates of the methane monooxygenase catalytic cycle. J Biol Chem 268(29):21569–21577

Lee SJ, McCormick MS, Lippard SJ, Cho U-S (2013) Control of substrate access to the active site in methane monooxygenase. Nature 494(7437):380–384. https://doi.org/10.1038/nature11880

Li N, Nørgaard H, Warui DM, Booker SJ, Krebs C, Bollinger JM (2011) Conversion of fatty aldehydes to alka(e)nes and formate by a cyanobacterial aldehyde decarbonylase: cryptic redox by an unusual dimetal oxygenase. J Am Chem Soc 133(16):6158–6161. https://doi.org/10.1021/ja2013517

Li N, Chang W-C, Warui DM, Booker SJ, Krebs C, Bollinger JM (2012) Evidence for only oxygenative cleavage of aldehydes to alk(a/e)nes and formate by cyanobacterial aldehyde decarbonylases. Biochemistry 51(40):7908–7916. https://doi.org/10.1021/bi300912n

Lin F, Das D, Lin XN, Marsh ENG (2013) Aldehyde-forming fatty acyl-CoA reductase from cyanobacteria: expression, purification and characterization of the recombinant enzyme. FEBS J 280(19):4773–4781. https://doi.org/10.1111/febs.12443

Lin Q, Parker MJ, Taguchi AT, Ravichandran K, Kim A, Kang G, Shao J, Drennan CL, Stubbe J (2017) Glutamate 52-β at the α/β subunit interface of Escherichia coli class Ia ribonucleotide reductase is essential for conformational gating of radical transfer. J Biol Chem 292(22): 9229–9239. https://doi.org/10.1074/jbc.M117.783092

Lindqvist Y, Huang W, Schneider G, Shanklin J (1996) Crystal structure of delta9 stearoyl-acyl carrier protein desaturase from castor seed and its relationship to other di-iron proteins. EMBO J 15(16):4081–4092

Liu KE, Valentine AM, Wang DL, Huynh BH, Edmondson DE, Salifoglou A, Lippard SJ (1995a) Kinetic and spectroscopic characterization of intermediates and component interactions in reactions of methane monooxygenase from Methylococcus capsulatus (Bath). J Am Chem Soc 117(41):10174–10185

Liu Y, Nesheim JC, Lee S-K, Lipscomb JD (1995b) Gating effects of component B on oxygen activation by the methane monooxygenase hydroxylase component. J Biol Chem 270(42): 24662–246625

Liu Y, Nesheim JC, Paulsen KE, Stankovich MT, Lipscomb JD (1997) Roles of the methane monooxygenase reductase component in the regulation of catalysis. Biochemistry 36(17): 5223–5233

Livada J, Martinie RJ, Dassama LMK, Krebs C, Bollinger JM, Silakov A (2015) direct measurement of the radical translocation distance in the class I ribonucleotide reductase from Chlamydia trachomatis. J Phys Chem B 119(43):13777–13784. https://doi.org/10.1021/acs.jpcb.5b04067

Logan DT (2011) Closing the circle on ribonucleotide reductases. Nat Struct Mol Biol 18(3): 251–253. https://doi.org/10.1038/nsmb0311-251

Lundin D, Poole AM, Sjöberg B-M, Högbom M (2012) Use of structural phylogenetic networks for classification of the ferritin-like superfamily. J Biol Chem 287(24):20565–20575. https://doi.org/10.1074/jbc.M112.367458

Lundin D, Berggren G, Logan D, Sjöberg B-M (2015) The origin and evolution of ribonucleotide reduction. Life 5(1):604–636. https://doi.org/10.3390/life5010604

Machado IMP, Atsumi S (2012) Cyanobacterial biofuel production. J Biotechnol 162(1):50–56. https://doi.org/10.1016/j.jbiotec.2012.03.005

Mak WS, Wang X, Arenas R, Cui Y, Bertolani S, Deng WQ, Tagkopoulos I, Wilson DK, Siegel JB (2020) Discovery, design, and structural characterization of alkane-producing enzymes across the ferritin-like superfamily. Biochemistry 59(40):3834–3843. https://doi.org/10.1021/acs.biochem.0c00665

Makhlynets O, Boal AK, Rhodes DV, Kitten T, Rosenzweig AC, Stubbe J (2014) Streptococcus sanguinis class Ib ribonucleotide reductase: high activity with both iron and manganese cofactors and structural insights. J Biol Chem 289(9):6259–6272. https://doi.org/10.1074/jbc.M113.533554

McCormick MS, Lippard SJ (2011) Analysis of substrate access to active sites in bacterial multicomponent monooxygenase hydroxylases: X-ray crystal structure of xenon-pressurized phenol hydroxylase from Pseudomonas sp. OX1. Biochemistry 50(51):11058–11069. https://doi.org/10.1021/bi201248b

McHugh JP, Rodríguez-Quinoñes F, Abdul-Tehrani H, Svistunenko DA, Poole RK, Cooper CE, Andrews SC (2003) Global iron-dependent gene regulation in Escherichia coli. A new mechanism for iron homeostasis. J Biol Chem 278(32):29478–29486. https://doi.org/10.1074/jbc.M303381200

McNeil BWJ, Thompson NR (2010) X-ray free-electron lasers. Nat Photonics 4:814–821. https://doi.org/10.1038/nphoton.2010.239

Merkx M, Lippard SJ (2002) Why orfY? Characterization of MMOD, a long overlooked component of the soluble methane monooxygenase from Methylococcus capsulatus (Bath). J Biol Chem 277(8):5858–5865

Minnihan EC, Ando N, Brignole EJ, Olshansky L, Chittuluru J, Asturias FJ, Drennan CL, Nocera DG, Stubbe J (2013a) Generation of a stable, aminotyrosyl radical-induced α2β2 complex of Escherichia coli class Ia ribonucleotide reductase. Proc Natl Acad Sci U S A 110(10): 3835–3840. https://doi.org/10.1073/pnas.1220691110

Minnihan EC, Nocera DG, Stubbe J (2013b) Reversible, long-range radical transfer in E. coli class Ia ribonucleotide reductase. Acc Chem Res 46(11):2524–2535. https://doi.org/10.1021/ar4000407

Miret-Casals L, Baelo A, Julián E, Astola J, Lobo-Ruiz A, Albericio F, Torrents E (2018) Hydroxylamine derivatives as a new paradigm in the search of antibacterial agents. ACS Omega 3(12):17057–17069. https://doi.org/10.1021/acsomega.8b01384

Mistry J, Chuguransky S, Williams L, Qureshi M, Salazar GA, Sonnhammer ELL, Tosatto SCE, Paladin L, Raj S, Richardson LJ, Finn RD, Bateman A (2021) Pfam: the protein families database in 2021. Nucleic Acids Res 49(D1):D412–D419. https://doi.org/10.1093/nar/gkaa913

Mitić N, Schwartz JK, Brazeau BJ, Lipscomb JD, Solomon EI (2008) CD and MCD studies of the effects of component B variant binding on the biferrous active site of methane monooxygenase. Biochemistry 47(32):8386–8397. https://doi.org/10.1021/bi800818w

Monje-Casas F, Jurado J, Prieto-Alamo MJ, Holmgren A, Pueyo C (2001) Expression analysis of the nrdHIEF operon from Escherichia coli. Conditions that trigger the transcript level in vivo. J Biol Chem 276(21):18031–18037. https://doi.org/10.1074/jbc.M011728200

Muller J, Lugovskoy AA, Wagner G, Lippard SJ (2002) NMR structure of the [2Fe-2S] ferredoxin domain from soluble methane monooxygenase reductase and interaction with its hydroxylase. Biochemistry 41(1):42–51

Murzin AG, Chothia C (1992) Protein architecture: new superfamilies. Curr Opin Struct Biol 2(6): 895–903. https://doi.org/10.1016/0959-440X(92)90116-O

Nastri F, Chino M, Maglio O, Bhagi-Damodaran A, Lu Y, Lombardi A (2016) Design and engineering of artificial oxygen-activating metalloenzymes. Chem Soc Rev 45(18): 5020–5054. https://doi.org/10.1039/C5CS00923E

Nesheim JC, Lipscomb JD (1996) Large kinetic isotope effects in methane oxidation catalyzed by methane monooxygenase: evidence for C-H bond cleavage in a reaction cycle intermediate. Biochemistry 35(31):10240–10247. https://doi.org/10.1021/bi960596w

Nick TU, Lee W, Kossmann S, Neese F, Stubbe J, Bennati M (2015) Hydrogen bond network between amino acid radical intermediates on the proton-coupled electron transfer pathway of E. coli α2 ribonucleotide reductase. J Am Chem Soc 137(1):289–298. https://doi.org/10.1021/ja510513z

Nick TU, Ravichandran KR, Stubbe J, Kasanmascheff M, Bennati M (2017) Spectroscopic evidence for a H bond network at Y356 located at the subunit interface of active E. coli ribonucleotide reductase. Biochemistry 56(28):3647–3656. https://doi.org/10.1021/acs.biochem.7b00462

Nordlund P, Eklund H (1995) Di-iron—carboxylate proteins. Curr Opin Struct Biol 5(6):758–766. https://doi.org/10.1016/0959-440X(95)80008-5

Nordlund P, Reichard P (2006) Ribonucleotide reductases. Annu Rev Biochem 75(1):681–706. https://doi.org/10.1146/annurev.biochem.75.103004.142443

Nordlund P, Sjöberg BM, Eklund H (1990) Three-dimensional structure of the free radical protein of ribonucleotide reductase. Nature 345(6276):593–598. https://doi.org/10.1038/345593a0

Offenbacher AR, Vassiliev IR, Seyedsayamdost MR, Stubbe J, Barry BA (2009) Redox-linked structural changes in ribonucleotide reductase. J Am Chem Soc 131(22):7496–7497. https://doi.org/10.1021/ja901908j

Onderko EL, Silakov A, Yosca TH, Green MT (2017) Characterization of a selenocysteine-ligated P450 compound I reveals direct link between electron donation and reactivity. Nat Chem 9(7): 623–628. https://doi.org/10.1038/nchem.2781

Pandelia ME, Li N, Nørgaard H, Warui DM, Rajakovich LJ, Chang W-C, Booker SJ, Krebs C, Bollinger JM (2013) Substrate-triggered addition of dioxygen to the diferrous cofactor of aldehyde-deformylating oxygenase to form a diferric-peroxide intermediate. J Am Chem Soc 135(42):15801–15812. https://doi.org/10.1021/ja405047b

Park AK, Kim I-S, Jeon BW, Roh SJ, Ryu M-Y, Baek H-R, Jo S-W, Kim Y-S, Park H, Lee JH, Yoon H-S, Kim H-W (2016) Crystal structures of aldehyde deformylating oxygenase from Limnothrix sp. KNUA012 and Oscillatoria sp. KNUA011. Biochem Biophys Res Commun 477(3):395–400. https://doi.org/10.1016/j.bbrc.2016.06.090

Pilkington SJ, Dalton H (1990) Soluble methane monooxygenase from Methylococcus capsulatus (Bath). Methods Enzym 188:181–190

Plays M, Müller S, Rodriguez R (2021) Chemistry and biology of ferritin. Metallomics 13(5): mfab021. https://doi.org/10.1093/mtomcs/mfab021

Poulos TL (2014) Heme enzyme structure and function. Chem Rev 114:3919–3962

Pozzi C, Di Pisa F, Lalli D, Rosa C, Theil E, Turano P, Mangani S (2015) Time-lapse anomalous X-ray diffraction shows how Fe $^{2+}$ substrate ions move through ferritin protein nanocages to oxidoreductase sites. Acta Crystallogr D Biol Crystallogr 71(4):941–953. https://doi.org/10.1107/S1399004715002333

Rahmana Z, Sung BH, Yi J-Y, Bui LM, Lee JH, Kim SC (2014) Enhanced production of n-alkanes in Escherichia coli by spatial organization of biosynthetic pathway enzymes. J Biotechnol 192 (Pt A):187–191. https://doi.org/10.1016/j.jbiotec.2014.10.014

Rajakovich LJ, Nørgaard H, Warui DM, Chang W, Li N, Booker SJ, Krebs C, Bollinger JM, Pandelia M-E (2015) Rapid reduction of the diferric-peroxyhemiacetal intermediate in aldehyde-deformylating oxygenase by a cyanobacterial ferredoxin: evidence for a free-radical mechanism. J Am Chem Soc 137(36):11695–11709. https://doi.org/10.1021/jacs.5b06345

Rajakovich LJ, Zhang B, McBride MJ, Boal AK, Krebs C, Martin Bollinger J (2020) Emerging structural and functional diversity in proteins with dioxygen-reactive dinuclear transition metal cofactors. In: Comprehensive natural products III. Elsevier, pp 215–250

Ravichandran KR, Minnihan EC, Wei Y, Nocera DG, Stubbe J (2015) Reverse electron transfer completes the catalytic cycle in a 2,3,5-trifluorotyrosine-substituted ribonucleotide reductase. J Am Chem Soc 137(45):14387–14395. https://doi.org/10.1021/jacs.5b09189

Ravichandran K, Minnihan EC, Lin Q, Yokoyama K, Taguchi AT, Shao J, Nocera DG, Stubbe J (2017) Glutamate 350 plays an essential role in conformational gating of long-range radical transport in Escherichia coli Class Ia ribonucleotide reductase. Biochemistry 56(6):856–868. https://doi.org/10.1021/acs.biochem.6b01145

Reece SY, Seyedsayamdost MR (2017) Long-range proton-coupled electron transfer in the Escherichia coli class Ia ribonucleotide reductase. Essays Biochem 61(2):281–292. https://doi.org/10.1042/EBC20160072

Reece SY, Hodgkiss JM, Stubbe J, Nocera DG (2006) Proton-coupled electron transfer: the mechanistic underpinning for radical transport and catalysis in biology. Philos Trans R Soc Lond Ser B Biol Sci 361(1472):1351–1364. https://doi.org/10.1098/rstb.2006.1874

Roca I, Torrents E, Sahlin M, Gibert I, Sjöberg B-M (2008) NrdI essentiality for class Ib ribonucleotide reduction in Streptococcus pyogenes. J Bacteriol 190(14):4849–4858. https://doi.org/10.1128/JB.00185-08

Røhr AK, Hersleth H-P, Andersson KK (2010) Tracking flavin conformations in protein crystal structures with Raman spectroscopy and QM/MM calculations. Angew Chem Int Ed Engl 49(13):2324–2327. https://doi.org/10.1002/anie.200907143

Rose HR, Ghosh MK, Maggiolo AO, Pollock CJ, Blaesi EJ, Hajj V, Wei Y, Rajakovich LJ, Chang W-C, Han Y, Hajj M, Krebs C, Silakov A, Pandelia M-E, Bollinger JM, Boal AK (2018) Structural basis for superoxide activation of Flavobacterium johnsoniae Class I ribonucleotide reductase and for radical initiation by its dimanganese cofactor. Biochemistry 57(18): 2679–2693. https://doi.org/10.1021/acs.biochem.8b00247

Rosenzweig AC, Frederick CA, Lippard SJ, Nordlund P (1993) Crystal structure of a bacterial non-haem iron hydroxylase that catalyses the biological oxidation of methane. Nature 366(6455):537–543

Rosenzweig AC, Nordlund P, Takahara PM, Frederick CA, Lippard SJ (1995) Geometry of the soluble methane monooxygenase catalytic diiron center in two oxidation states. Chem Biol 2(6): 409–418

Rozman Grinberg I, Lundin D, Hasan M, Crona M, Jonna VR, Loderer C, Sahlin M, Markova N, Borovok I, Berggren G, Hofer A, Logan DT, Sjöberg B-M (2018) Novel ATP-cone-driven allosteric regulation of ribonucleotide reductase *via* the radical-generating subunit. eLife 7: e31529. https://doi.org/10.7554/eLife.31529

Rozman Grinberg I, Berglund S, Hasan M, Lundin D, Ho FM, Magnuson A, Logan DT, Sjöberg B-M, Berggren G (2019) Class Id ribonucleotide reductase utilizes a Mn2(IV,III) cofactor and undergoes large conformational changes on metal loading. J Biol Inorg Chem 24(6):863–877. https://doi.org/10.1007/s00775-019-01697-8

Ruskoski TB, Boal AK (2021) The periodic table of ribonucleotide reductases. J Biol Chem 297(4): 101137. https://doi.org/10.1016/j.jbc.2021.101137

Sazinsky MH, Merkx M, Cadieux E, Tang S, Lippard SJ (2004) Preparation and X-ray structures of metal-free, dicobalt and dimanganese forms of soluble methane monooxygenase hydroxylase from Methylococcus capsulatus (Bath). Biochemistry 43(51):16263–16276. https://doi.org/10.1021/bi048140z

Sazinsky MH, Dunten PW, McCormick MS, DiDonato A, Lippard SJ (2006) X-ray structure of a hydroxylase-regulatory protein complex from a hydrocarbon-oxidizing multicomponent monooxygenase, Pseudomonas sp. OX1 phenol hydroxylase. Biochemistry 45(51): 15392–15404. https://doi.org/10.1021/bi0618969

Schirmer A, Rude MA, Li X, Popova E, del Cardayre SB (2010) Microbial biosynthesis of alkanes. Science 329(5991):559–562. https://doi.org/10.1126/science.1187936

Schwartz JK, Wei P, Mitchell KH, Fox BG, Solomon EI (2008) Geometric and electronic structure studies of the binuclear nonheme ferrous active site of toluene-4-monooxygenase: parallels with methane monooxygenase and insight into the role of the effector proteins in O2 activation. J Am Chem Soc 130(22):7098–7109

Seyedsayamdost MR, Stubbe J (2006) Site-specific replacement of Y356 with 3,4-dihydroxyphenylalanine in the beta2 subunit of E. coli ribonucleotide reductase. J Am Chem Soc 128(8):2522–2523. https://doi.org/10.1021/ja057776q

Seyedsayamdost MR, Chan CTY, Mugnaini V, Stubbe J, Bennati M (2007) PELDOR spectroscopy with DOPA-beta2 and NH2Y-alpha2s: distance measurements between residues involved in the radical propagation pathway of E. coli ribonucleotide reductase. J Am Chem Soc 129(51): 15748–15749. https://doi.org/10.1021/ja076459b

Sharma A, Shakeel T, Gupta M, Rajacharya GH, Yazdani SS (2021) Biophysical and structural studies reveal marginal stability of a crucial hydrocarbon biosynthetic enzyme acyl ACP reductase. Sci Rep 11(1):12045. https://doi.org/10.1038/s41598-021-91232-0

Shu L, Nesheim JC, Kauffmann K, Münck E, Lipscomb JD, Que L Jr (1997) An FeIV2O2 diamond core structure for the key intermediate Q of methane monooxygenase. Science 275(5299): 515–518

Singh JS, Singh DP (2017) Methanotrophs: an emerging bioremediation tool with unique broad spectrum methane monooxygenase (MMO) enzyme. In: Agro-environmental sustainability, pp 1–18

Sjöberg BM, Reichard P, Gräslund A, Ehrenberg A (1978) The tyrosine free radical in ribonucleotide reductase from Escherichia coli. J Biol Chem 253(19):6863–6865

Srinivas V, Lebrette H, Lundin D, Kutin Y, Sahlin M, Lerche M, Eirich J, Branca RMM, Cox N, Sjöberg B-M, Högbom M (2018) Metal-free ribonucleotide reduction powered by a DOPA radical in Mycoplasma pathogens. Nature 563(7731):416–420. https://doi.org/10.1038/s41586-018-0653-6

Srinivas V, Banerjee R, Lebrette H, Jones JC, Aurelius O, Kim I-S, Pham CC, Gul S, Sutherlin KD, Bhowmick A, John J, Bozkurt E, Fransson T, Aller P, Butryn A, Bogacz I, Simon P, Keable S, Britz A, Tono K, Kim KS, Park S-Y, Lee SJ, Park J, Alonso-Mori R, Fuller FD, Batyuk A, Brewster AS, Bergmann U, Sauter NK, Orville AM, Yachandra VK, Yano J, Lipscomb JD, Kern J, Högbom M (2020) High-resolution XFEL structure of the soluble methane monooxygenase hydroxylase complex with its regulatory component at ambient temperature in two oxidation states. J Am Chem Soc 142(33):14249–14266. https://doi.org/10.1021/jacs.0c05613

Stillman TJ, Hempstead PD, Artymiuk PJ, Andrews SC, Hudson AJ, Treffry A, Guest JR, Harrison PM (2001) The high-resolution X-ray crystallographic structure of the ferritin (EcFtnA) of Escherichia coli; comparison with human H ferritin (HuHF) and the structures of the Fe(3+) and Zn(2+) derivatives. J Mol Biol 307(2):587–603. https://doi.org/10.1006/jmbi.2001.4475

Theil EC (2011) Ferritin protein nanocages use ion channels, catalytic sites, and nucleation channels to manage iron/oxygen chemistry. Curr Opin Chem Biol 15(2):304–311. https://doi.org/10.1016/j.cbpa.2011.01.004

Theil EC (2012) Ferritin protein nanocages-the story. Nanotechnol Percept 8(1):7–16. https://doi.org/10.4024/n03th12a.ntp.08.01

Thomas WC, Brooks FP, Burnim AA, Bacik J-P, Stubbe J, Kaelber JT, Chen JZ, Ando N (2019) Convergent allostery in ribonucleotide reductase. Nat Commun 10(1):2653. https://doi.org/10.1038/s41467-019-10568-4

Tinberg CE, Lippard SJ (2011) Dioxygen activation in soluble methane monooxygenase. Acc Chem Res 44(4):280–288

Tomter AB, Zoppellaro G, Andersen NH, Hersleth H-P, Hammerstad M, Røhr ÅK, Sandvik GK, Strand KR, Nilsson GE, Bell CB, Barra A-L, Blasco E, Le Pape L, Solomon EI, Andersson KK (2013) Ribonucleotide reductase class I with different radical generating clusters. Coord Chem Rev 257(1):3–26. https://doi.org/10.1016/j.ccr.2012.05.021

Torrents E (2014) Ribonucleotide reductases: essential enzymes for bacterial life. Front Cell Infect Microbiol 4. https://doi.org/10.3389/fcimb.2014.00052

Trehoux A, Mahy J-P, Avenier F (2016) A growing family of O2 activating dinuclear iron enzymes with key catalytic diiron(III)-peroxo intermediates: Biological systems and chemical models. Coord Chem Rev 322:142–158. https://doi.org/10.1016/j.ccr.2016.05.014

Uhlin U, Eklund H (1994) Structure of ribonucleotide reductase protein R1. Nature 370(6490): 533–539. https://doi.org/10.1038/370533a0

Uppsten M, Färnegårdh M, Domkin V, Uhlin U (2006) The first holocomplex structure of ribonucleotide reductase gives new insight into its mechanism of action. J Mol Biol 359(2): 365–377. https://doi.org/10.1016/j.jmb.2006.03.035

Wallar BJ, Lipscomb JD (1996) Dioxygen activation by enzymes containing binuclear non-heme iron clusters. Chem Rev 96(7):2625–2657

Wallar BJ, Lipscomb JD (2001) Methane monooxygenase component B mutants alter the kinetics of steps throughout the catalytic cycle. Biochemistry 40(7):2220–2233

Wang W, Lippard SJ (2014) Diiron oxidation state control of substrate access to the active site of soluble methane monooxygenase mediated by the regulatory component. J Am Chem Soc 136(6):2244–2247. https://doi.org/10.1021/ja412351b

Wang Q, Huang X, Zhang J, Lu X, Li S, Li J-J (2014a) Engineering self-sufficient aldehyde deformylating oxygenases fused to alternative electron transfer systems for efficient conversion of aldehydes into alkanes. Chem Commun Camb Engl 50(33):4299–4301. https://doi.org/10.1039/c4cc00591k

Wang W, Iacob RE, Luoh RP, Engen JR, Lippard SJ (2014b) Electron transfer control in soluble methane monooxygenase. J Am Chem Soc 136(27):9754–9762. https://doi.org/10.1021/ja504688z

Wang W, Liang AD, Lippard SJ (2015) Coupling oxygen consumption with hydrocarbon oxidation in bacterial multicomponent monooxygenases. Acc Chem Res 48(9):2632–2639. https://doi.org/10.1021/acs.accounts.5b00312

Wang C, Zhao C, Hu L, Chen H (2016) Calculated mechanism of cyanobacterial aldehyde-deformylating oxygenase: asymmetric aldehyde activation by a symmetric diiron cofactor. J Phys Chem Lett 7(21):4427–4432. https://doi.org/10.1021/acs.jpclett.6b02061

Wang Q, Bao L, Jia C, Li M, Li J-J, Lu X (2017a) Identification of residues important for the activity of aldehyde-deformylating oxygenase through investigation into the structure-activity relationship. BMC Biotechnol 17(1):31. https://doi.org/10.1186/s12896-017-0351-8

Wang VC-C, Maji S, Chen PP-Y, Lee HK, Yu SS-F, Chan SI (2017b) Alkane oxidation: methane monooxygenases, related enzymes, and their biomimetics. Chem Rev 117(13):8574–8621. https://doi.org/10.1021/acs.chemrev.6b00624

Wang W, Zhang Y, Zhao G, Wang H (2021) Ferritin with atypical ferroxidase centers takes B-channels as the pathway for Fe^{2+} uptake from *Mycoplasma*. Inorg Chem 60(10): 7207–7216. https://doi.org/10.1021/acs.inorgchem.1c00265

Warui DM, Li N, Nørgaard H, Krebs C, Bollinger JM, Booker SJ (2011) Detection of formate, rather than carbon monoxide, as the stoichiometric coproduct in conversion of fatty aldehydes to alkanes by a cyanobacterial aldehyde decarbonylase. J Am Chem Soc 133(10):3316–3319. https://doi.org/10.1021/ja111607x

Warui DM, Pandelia M-E, Rajakovich LJ, Krebs C, Bollinger JM, Booker SJ (2015) Efficient delivery of long-chain fatty aldehydes from the *Nostoc punctiforme* Acyl–Acyl carrier protein reductase to its cognate aldehyde-deformylating oxygenase. Biochemistry 54(4):1006–1015. https://doi.org/10.1021/bi500847u

Yokoyama K, Uhlin U, Stubbe J (2010) A hot oxidant, 3-NO2Y122 radical, unmasks conformational gating in ribonucleotide reductase. J Am Chem Soc 132(43):15368–15379. https://doi.org/10.1021/ja1069344

Yu F, Cangelosi VM, Zastrow ML, Tegoni M, Plegaria JS, Tebo AG, Mocny CS, Ruckthong L, Qayyum H, Pecoraro VL (2014) Protein design: toward functional metalloenzymes. Chem Rev 114(7):3495–3578. https://doi.org/10.1021/cr400458x

Zhang J, Wallar BJ, Popescu CV, Renner DB, Thomas DD, Lipscomb JD (2006) Methane monooxygenase hydroxylase and B component interactions. Biochemistry 45(9):2913–2926

Zhang J, Lu X, Li J-J (2013) Conversion of fatty aldehydes into alk (a/e)nes by in vitroreconstituted cyanobacterial aldehyde-deformylating oxygenase with the cognate electron transfer system. Biotechnol Biofuels 6(1):86. https://doi.org/10.1186/1754-6834-6-86

Zheng H, Lipscomb JD (2006) Regulation of methane monooxygenase catalysis based on size exclusion and quantum tunneling. Biochemistry 45(6):1685–1692

Zhou YJ, Kerkhoven EJ, Nielsen J (2018) Barriers and opportunities in bio-based production of hydrocarbons. Nat Energy 3(11):925–935. https://doi.org/10.1038/s41560-018-0197-x

Chapter 5
Still no Rest for the Reductases: Ribonucleotide Reductase (RNR) Structure and Function: An Update

Marcus J. C. Long, Phillippe Ly, and Yimon Aye

Abstract Herein we present a multidisciplinary discussion of ribonucleotide reductase (RNR), the essential enzyme uniquely responsible for conversion of ribonucleotides to deoxyribonucleotides. This chapter primarily presents an overview of this multifaceted and complex enzyme, covering RNR's role in enzymology, biochemistry, medicinal chemistry, and cell biology. It further focuses on RNR from mammals, whose interesting and often conflicting roles in health and disease are coming more into focus. We present pitfalls that we think have not always been dealt with by researchers in each area and further seek to unite some of the field-specific observations surrounding this enzyme. Our work is thus not intended to cover any one topic in extreme detail, but rather give what we consider to be the necessary broad grounding to understand this critical enzyme holistically. Although this is an approach we have advocated in many different areas of scientific research, there is arguably no other single enzyme that embodies the need for such broad study than RNR. Thus, we submit that RNR itself is a paradigm of interdisciplinary research that is of interest from the perspective of the generalist and the specialist alike. We hope that the discussions herein will thus be helpful to not only those wanting to tackle RNR-specific problems, but also those working on similar interdisciplinary projects centering around other enzymes.

Keywords Ribonucleotide reductases · Moonlighting functions · Deoxynucleotide pools · DNA replication and repair · Genome maintenance

M. J. C. Long
University of Lausanne (UNIL), Lausanne, Switzerland

Department of Biochemistry, UNIL, Epalinges, Switzerland
e-mail: marcusjohncurtis.long@unil.ch

P. Ly · Y. Aye (✉)
Swiss Federal Institute of Technology Lausanne (EPFL), Lausanne, Switzerland

EPFL SB ISIC LEAGO, Lausanne, Switzerland
e-mail: phillippe.ly@epfl.ch; yimon.aye@epfl.ch

J. R. Harris, J. Marles-Wright (eds.), *Macromolecular Protein Complexes IV*,
Subcellular Biochemistry 99, https://doi.org/10.1007/978-3-031-00793-4_5

Introduction

Life as we know it likely started with a ribonucleic acid-based information storage device, in what is now called the RNA world. This code was multipurpose, as it could replicate, store information, replicate, and perform biological functions, the latter two aspects, at least, similar to the ribozymes, riboswitches, etc. that we know today. However, owing principally to the 2′-hydroxyl group vicinal to the 3-′-phosphate group in RNA, this structure is relatively unstable and far from apposite in making a robust coded message lasting for many years, or perhaps millennia (Peris et al. 2020). To prolong genetically encoded information, this intrinsic hydrolysis pathway in RNA needed to be suppressed. Of course, this could have been achieved in several ways: numerous proteins can bind to and stabilize RNA (Maori et al. 2019). Furthermore, it transpires that various chemical units can form a backbone to scaffold nucleobases in a defined order, such as peptide nucleic acid and hexose nucleic acid (Pitsch et al. 1993), which are at least as stable as RNA. However, life chose a deceptively simple strategy, reduction of the C2' hydroxymethine group within RNA to a methylene group.

Chemically, hydroxymethine reduction can be performed in multiple steps via classic named chemical reactions such as "Barton-McCombie reduction," a radical-mediated process requiring tributyltin hydride and proceeding through a xanthate ester intermediate, neither of which is at all common in biology, despite tin likely being a trace element (Mertz 1974), or chemoselective oxidation, Wolff-Kirchner reduction, a process requiring, at least in the chemistry laboratory, formation of a hydrazone and then strong base and reflux, liberating nitrogen (Fig. 5.1). Life evolved to reduce NTPs to dNTPs using a single catalyst, albeit a heterodimeric molecule, that functions at ambient temperatures. Similar to the Barton-McCombie reaction, this enzyme complex uses radical chemistry, although the pathway is distinct from those that were known at the time of the discovery of the enzyme in 1961 (Reichard et al. 1961). Perhaps more surprisingly, this single complex reduces each NTP to the corresponding dNTP through an exquisitely allosterically controlled process. The entity responsible for this catalysis in humans and higher eukaryotes is the two subunits of ribonucleotide reductase, α and β, working in tandem, minimally as a dimer of dimers. These subunits form a self-regulatory loop whose allosteric regulation of chemical activity is particularly intricate. Indeed, all DNA-based life has at least one form of RNR; some even have numerous forms, such as *E. coli*. We will focus mainly on work from higher eukaryotes and yeast, which only contain a class of RNR referred to as class Ia. It happens that *E. coli* also possesses a class Ia RNR (in addition to a class Ib system). Due to its stability and ready preparation relative to mammalian counterparts, *E. coli* class Ia RNR has been used as a paradigm for the chemistry undertaken by class Ia RNRs.

As the chemistry performed by RNR is essential for DNA-based life, as far as we are aware, we must assume that RNR, or an equivalent dNTP-forming process (Poole et al. 2014) in some form, is at least as old as DNA-based life (Lundin et al. 2015). Thus, to study RNR is to investigate one of the oldest enzymes that is

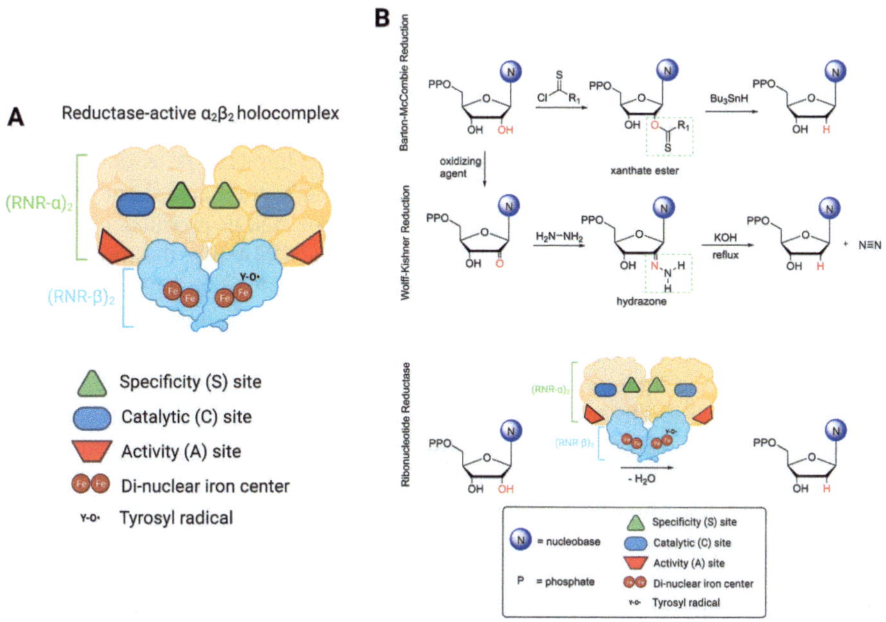

Fig. 5.1 (**a**) The minimal active state of class Ia RNR (to which RNRs from *E. coli*, yeast, mouse, and human belong) is composed of two dimeric subunits that come together to form the reductase-active holocomplex. The small subunit (β) contains the di-nuclear iron center responsible for the generation of the tyrosyl radical. Through a proton-coupled electron transfer (PCET) mechanism, the larger subunit (α) is able to reduce ribose to deoxyribose in its catalytic center (blue oval). (**b**) The same transformation can be performed through synthetic methods such as the Barton-McCombie or Wolff-Kishner method. Note that these processes are illustrative and are not meant to reflect known transformations in the literature on a specific substrate

perhaps most fundamental to, and certainly most emblematic of, DNA-based life. Unsurprisingly, this protein, like many ancient enzymes, has developed new tricks during evolution (Huberts and van der Klei 2010; Jeffery 2020), leading to a rich, diverse, and often contradicting biology, involving several moonlighting functions. We will broach these topics in the context of in vitro biochemistry, roles in DNA damage repair, medicinal chemistry, and non-catalytic roles of RNR.

Thus, here we aim to instil readers with a multivalent understanding of the chemistry, biochemistry, and biology of RNR. We stress that, at least in our opinion, developing a detailed understanding of *all* these facets is key to appreciating the nuanced and often contrasting roles played by RNR in biology and medicinal chemistry. We will focus on data accrued since our previous in-depth RNR review of 2015 (Aye et al. 2015), to which readers are pointed should they wish to read up on earlier work.

RNR Chemistry

Large tomes have been written on RNR chemistry. For this review, we will deal only with the salient points to understand the broader biology. RNR is responsible for the rate-limiting step in the reduction of all nucleotides in humans. RNR does this by taking NDPs and converting them to dNDPs. Crucially, this chemistry *does not* create dNTPs directly, requiring kinases to convert the dNDP products to dNTPs. Furthermore, harking back to RNR bridging RNA and DNA life, the formation of TTP is achieved via RNR-catalyzed conversion of UDP to dUDP that is then converted to TTP, via thymidylate synthase, and thymidylate kinase. It has been proposed that defects in UDP conversion to TTP may be responsible for low TTP pools (which has a knock-on effect on dGTP pools) in terminally differentiated myotubes, leading to DNA damage phenotypes observed in cell cycle-reactivated myotubes (Pajalunga et al. 2017). This observation highlights the biological importance of steps that occur post RNR activity, and how they can impact more than one dNTP. Thymidylate kinase deficiency can also elevate dUTP concentrations that can lead to elevated dU incorporation into DNA (Hu et al. 2012).

Of the two subunits making up RNR, the larger subunit, RNR-α, also called R1, or RRM1 (typically around 90 kDa in molecular weight in mammals), is essentially a classical protein component that has three defined nucleotide binding sites, one of which is the catalytic, or C-, site, which binds NDPs, and the other two sites are allosteric sites that bind (d)NTPs. The purposes of the allosteric sites will be explained later. The smaller subunit, RNR-β, also called RR2, R2, RR2M, or RRM2, is usually around 40 kDa in size and possesses no known nucleotide binding sites. The two subunits of RNR also have species-specific names in yeasts and *E. coli*, which we will not dwell on here, but for completeness they are listed in Table 5.1. Humans and many other vertebrates possess another RNR-β paralog, p53R2. This alternative subunit (~80% homologous to human RNR-β and with very similar specific activity) functions chemically in a similar way to RNR-β (Zhou et al. 2005), in the sense that it performs the same function and is similarly sensitive to the same drugs. We will thus discuss this subunit mostly in terms of its biological behaviors, some of which are distinct from RNR-β.

NDP reduction occurs in the large subunit C-site via a transiently formed cysteine radical. However, the large subunit lacks the ability to form the cysteine radical and hence perform reductive chemistry on its own. Association with the small RNR subunit, or RNR-β, is required to activate RNR-α. RNR-β contains a diferric center that is responsible for forming a permanent tyrosyl radical ($OFe^{III}{}_2$-Y) at RNR-β-Y176 (human numbering, which will be used throughout, with reference

Table 5.1 Nomenclature of RNR subunits in humans, yeast, and *E. coli*

Organism	Large subunit name (RNR-α)	Small subunit name (RNR-β)
H. sapiens	R1, RRM1	RR2, R2, RR2M, RRM2
S. cerevisiae	RNR1, RNR3	RNR2, RNR4
E. coli	nrdA	nrdB

to other species if necessary). Formation of this radical requires oxidizing equivalents (most likely oxygen), as the iron cofactor is assembled in vitro and in cells from Fe^{II}, a metal low on the Irving-Williams series, allowing facile ligand exchange (Cotruvo and Stubbe 2011). It is likely that OFe^{III}_2-Y can catalyze approximately 30 turnovers prior to needing replenishing by another oxidizing equivalent (Chimploy et al. 2000). Because of the relative instability of RNR-β and the need for reoxidation, hypoxia is prone to compromising RNR activity through inhibition of RNR-β. p53R2 appears to be able to retain its activity during hypoxia better than RNR-β; this stability appears to be linked to a better oxidation scavenging ability of p53R2 (likely lower $K_m(O_2)$), which has been ascribed to several specific mutations that exist between the two enzymes (Foskolou et al. 2017). In crucian carp, whose cells can still divide despite experiencing hypoxia for many days at low temperature, a specific isoform of p53R2 may assist dNTP formation due to its inherent longevity (Sandvik et al. 2012) (no decay in radical over 24 h at 0°C, or for 1 h at room temperature). Mammalian RNR-β has a half-life of 10–20 min at 37°C (Aye et al. 2012). In RNR-β's active or holo form, the radical is observable as a doublet by EPR ($g = 4.3$). A similar radical is also visible in the whole-cell EPR of some cell lines, e.g., K562 cells, as well as in cell lines engineered to overexpress RNR-β, typically through induced resistance during elevated exposure to RNR-β-targeting drugs (Aye et al. 2012); one such line is derived from TA3 cells. RNR-β can also exist in the apo (no iron, no radical) and met (iron loaded, but no radical) forms. These have no activity and do not show a tyrosyl radical.

In order for the tyrosyl radical on RNR β to oxidize the cysteine on RNR-α, a long-range proton-coupled electron transfer process is necessary (Fig. 5.2). Transfer occurs through several conserved aromatic residues that form a 35 Å molecular wire, allowing reversible transfer from the stable radical site on RNR-β to the active cysteine in the C-site on RNR-α, at C439. In this way, the transient radical on C439 can abstract a hydrogen radical from the 3′-position on the pentose ring of a

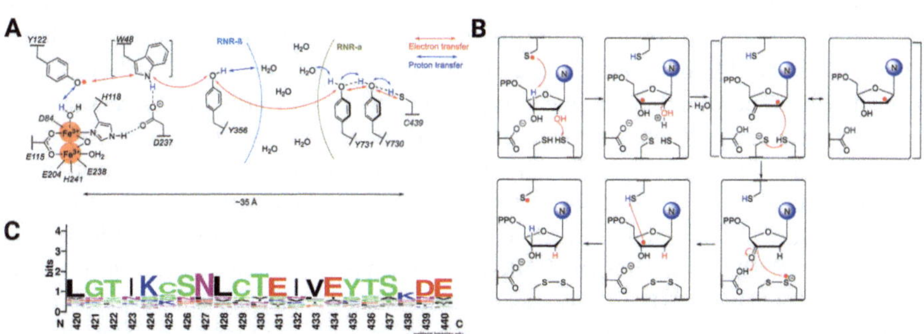

Fig. 5.2 (**a**) A closer picture of the residues known to be involved in the 35-Å transfer of the radical from Tyr122 of RNR-β to C439 in the active site of *E. coli* RNR-α. (**b**) General mechanism of the reduction of ribose to deoxyribose in RNR-α where the radical electron (red dot) is involved in the abstraction of the 3′ hydrogen from ribose. (**c**) Sequence logo of the region around Cys429 (active site residue in human), showing high conservation for this residue across RNR-α from 855 species

bound substrate molecule, triggering ejection of the C2′-OH unit, to form a secondary cation at C2′. This radical is directly reduced by hydride transfer from a cysteine pair distinct from C439, generating a disulfide in the process. This reduction step leaves the initially generated C3′-radical that is resonance stabilized by the 3-′-OH. This radical can abstract the hydrogen from the C439, giving the product. To ensure turnover, the disulfide on RNR-α formed in this process must be re-reduced. Re-reduction can be catalyzed in vitro by small-molecule reducing agents, such as DTT. Thioredoxin or glutaredoxin (in the presence of a large amount of glutathione (Sengupta 2014)) are able to reactivate oxidized RNR-α (complexed to RNR-β or p53R2 (Domkin and Chabes 2014; Sengupta et al. 2019)) through an *in trans* transfer system that leads to migration of the disulfide from the C-site to a mobile C-terminal tail that is accessible to enzymatic reductants. These enzymatic reduction pathways are the only re-reduction processes that occur in cells, and these reducing agents can also be used in vitro. In vitro, the C-terminal cysteine tail is dispensable when using DTT as this reductant can directly reduce disulfides in the C-site; this is not possible with phosphine reductants (Domkin and Chabes 2014). Unsurprisingly, overexpression of thioredoxin stimulates RNR activity in cells; combined inhibition of RNR and thioredoxin can lead to synergistic anticancer effects (Lou et al. 2017). In *S. pombe*, loss of the two RNR reducing enzymes (thioredoxin and glutaredoxin) is lethal, unless peroxiredoxin is deleted; these data were interpreted to be due to rerouting of reducing equivalents to reactivate RNR-α (Boronat et al. 2017).

It is critical to note that most of our chemical understanding of RNR was unravelled using the more stable, more active, and much easier to handle *E. coli*, class Ia, enzymes (Hogbom et al. 2003). *E. coli* RNR has proven to be a paradigm for RNR mechanistic analysis and appears to be a faithful model for the chemistry of most class Ia RNRs (Stubbe et al. 2003). Based on studies on the *E. coli* enzyme, it is assumed that conformational changes during catalysis and not chemistry are limiting (Ge et al. 2003). Although the parallels between human and *E. coli* enzymes are readily apparent (see Fig. 5.3b, overlap of the two structures), there are marked differences at several levels. In terms of stabilities, it transpires that the diferric center is less well bound in the human RNR-β than that in the *E. coli* ortholog. This leads to differences in the way the two proteins behave towards iron chelators (Aye et al. 2012), such as the approved anticancer drug hydroxyurea (HU), which in *E. coli* behaves as a reductant, forming the met form of the protein. Reduction of the tyrosyl radical has been proposed to occur via a reduction-assisted proton wire across a hydrogen bonding network leading from the surface to the $Y^{\cdot}$ (Offenbacher and Barry 2020). In the human RNR-β, which has a water-accessible iron center (Kauppi et al. 1996), HU behaves as a chelator removing iron from the active site with concomitant loss of the radical. For these reasons, the met form of the human RNR-β was believed to be inaccessible, although it was later discovered that other RNR-β inhibitors such a triapine, which has not yet been approved, can form the met form of human RNR-β, at least transiently in vitro and in cells (Aye et al. 2012).

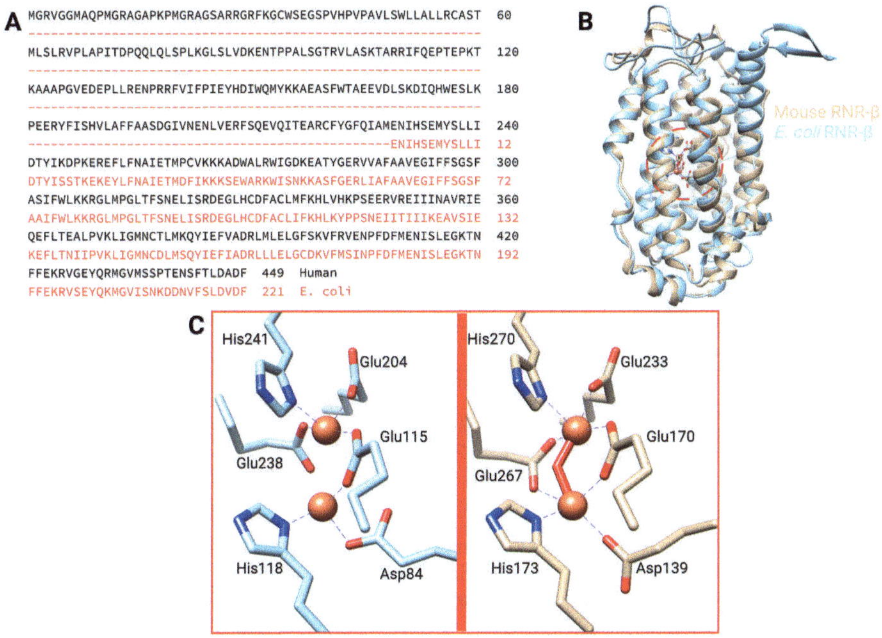

Fig. 5.3 (**a**) Sequence alignment of RNR-β from human and its homologous equivalent in *E. coli*. (**b**) Structural overlap of crystal structures of mouse RNR-β (tan; PDB 1 W68) complexed with Fe_2O and *E. coli* RNR-β in complex with Fe^{3+} (light blue; PDB 1MXR). (**c**) Using the previously mentioned structures, the specific residues that interact with the ferric ions are highlighted

Thus, it is always important when evaluating RNR chemistry/drug mechanisms to understand which enzyme the data were derived from and what evidence models proposed use. Although we have little basis to doubt any of the basic models of chemistry of RNR, it is typically better to use human components wherever necessary.

To complicate matters, although association between RNR-α and RNR-β is essential in numerous species (Xie et al. 2020), the affinity of RNR-α to RNR-β is quite low, and hence, it has proven difficult to study the complex structurally (Fig. 5.4). Nevertheless, our understanding of this process has grown steadily over the years, mostly because of work on the model *E. coli* enzyme, which is of course a simplification. Based on work mapping the pathway the electron takes between *E. coli* RNR-α and RNR-β (Minnihan et al. 2013), a doubly substituted RNR-β was recently used in complex with an active RNR-α to trap the $\alpha_2\beta_2$ complex and observe it by cryo-EM at 3.6-Å resolution (Kang et al. 2020). Although such experiments are indeed very informative and help improve our overall understanding of RNR, it is noteworthy that many of these have been performed using mutants. Hence, they must be viewed with even more caution, even though they paint an overall consistent picture.

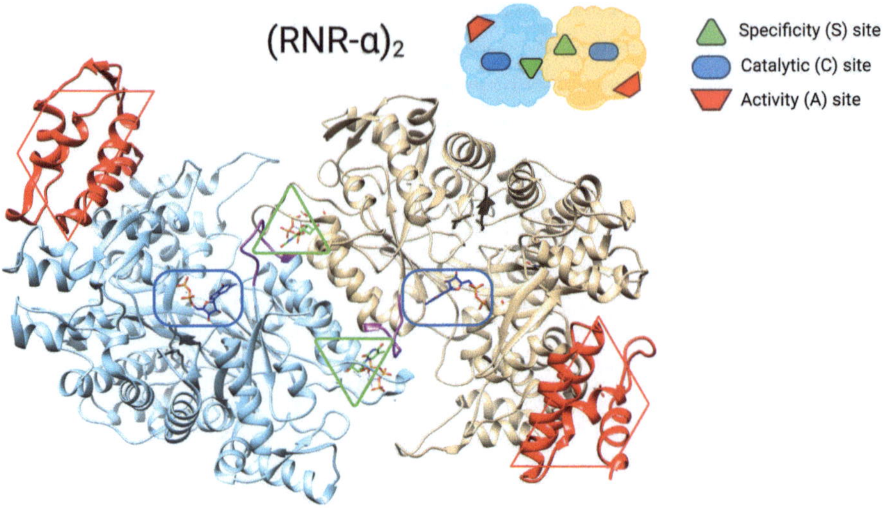

Fig. 5.4 Crystal structure of human RNR-α dimer (PDB 3HND) showing the A-site residues (red), GDP bound in the C-site (blue rounded rectangle), TTP bound in the S-site (green triangle), and the strand of amino acids that comprise loop 2 (purple)

RNR Subunit Biochemistry

RNR biochemistry extends the chemical observations above to achieve a more holistic understanding of how substrate and activity preferences link to the RNR essential role of sustaining and balancing dNTPs at the structural level. Most of the biochemical regulation occurs on RNR-α, on which we will focus here. As described above, the α-subunit of RNR has two additional (d)NTP binding sites beyond its catalytic site. These two allosteric sites are responsible for the overall activity (the activity, or "A"-site) and the substrate preference (or specificity, "S"-site). Distinct from the C-site, these allosteric sites bind (d)NTPs.

Binding of ATP to the A-site stimulates activity: dATP binding to this site suppresses activity. For this reason, typically RNR activity assays are conducted in the presence of saturating ATP. It is intriguing that among the five (d)NTPs (A, T, C, G, and U), only adenosine metabolites are used as a metric for dNTP/NTP ratios. Perhaps the large cellular concentrations of ATP, likely ubiquitous distribution, and awesome power of the cell to create ATP render this molecule an ideal metric, or possibly the use of ATP as an energy source means that its concentrations are more likely to fluctuate than that of other NTPs (GTP's role as an energy currency is less general than that of ATP (Sumita et al. 2016, p. 61)). These factors couple together to make ATP/dATP ratios largest in both dividing and quiescent cells relative to other NTP/dNTPs (Kong et al. 2018). It also transpires that dNTP hydrolysis enzymes also sense ATP in a complementary way, as we will discuss below; such an observation does not explain why ATP is a universal yardstick, but it implies a coregulatory loop.

The S-site tunes binding of substrates in the C-site. This relationship obviously acts to balance relative dNTP pools to obtain optimal conditions for faithful replication. In class Ia RNRs (like the human enzyme), dATP or ATP docking at the S-site induces CDP/UDP reduction, whereas dGTP and dTTP binding promote ADP and GDP reduction, respectively. Thus, we have described the A-site as a break/gas pedal, whereas the S-site is a gear stick, fine-tuning the activity of RNR, dependent on the situation. The mode of allosteric regulation through the S-site has been investigated in the *E. coli* enzyme using X-ray crystallography. This work uncovered a series of residues in an overall weakly conserved nine amino acid region of the protein (referred to as loop 2) that interacts with both the bound diphosphate and the S-site-bound triphosphate. In this way, substrate binding can be tuned by S-site occupancy (Zimanyi et al. 2016). Aspects of individual dNTPs orchestrating repositioning of loop 2 have also been proposed (Knappenberger et al. 2016), and correlated changes in specific residues (N291 and P294) across different species affecting allosteric regulation have also been released (Knappenberger et al. 2017).

It is critical to appreciate the nature of the specific site binders and tuning reaction conditions when designing RNR activity assays. It is also critical to consider all sites and relevant affinities where known. For instance, typical assay conditions, which use high concentrations of ATP to stimulate activity, are not conducive for detecting A-site competitive inhibitors, and hence where possible, activity in the absence of ATP (which *is* measurable, but significantly less than that in the presence of ATP) should always be considered. Such competitive relationships can also give important insights into site selectivity. As a further consequence of ATP being used as a stimulator of activity and ATP binding to the S-site, typically RNR activity assays are run measuring CDP reduction. But of course, inhibitors may have matched or mismatched S-site binders, meaning again it is easy to miss good inhibitors in vitro when performing single-substrate RNR activity assays. A large number of different assays have been released to measure RNR activity and they all have their benefits and weaknesses; some assays measure all four activities simultaneously (Chimploy and Mathews 2001). However, this laboratory has favored the use of radioactive CDP as a substrate. This assay is powerful enough to allow extraction of complex, time-dependent inhibition constants (Aye et al. 2012; Wisitpitthaya et al. 2016). A final consideration when planning RNR assays is the nature of the re-reductant. The use of enzymatic re-reduction systems may be the most biologically relevant, although enzymatic re-reduction processes are potentially subject to inhibition by compounds being screened. Repeating inhibition assays (particularly for molecules that inhibit) in the presence of small-molecule reductants is a good way to rule out this process, at least in vitro.

Allostery and Oligomerization, the Basis of RNR Activity Regulation

Structurally, the A-site exists in an approximately 100 amino acid N-terminal cone region that is not always present in class I RNRs and may indeed be duplicated in some RNRs from some species, such as *P. aeruginosa* (Jonna et al. 2015). The human RNR-α N-terminal cone region can be purified independently of the catalytic body. Indeed, the N-terminal cone in the human enzyme can be transposed onto the *E. coli* catalytic body, generating a catalytically active hybrid that is functional with *E. coli* RNR-β selectivity (the corresponding *E. coli* catalytic body without an N-terminal cone is inactive) (Fu et al. 2013). This implies that the N-terminal cone regions of *E. coli* and human enzymes can partially compensate for each other. However, structurally, the *E. coli* and eukaryotic forms of RNR are regulated by ATP differently. dATP binding to human RNR-α creates a hexameric state on the order of 1–2 min post dATP addition to the mainly monomeric fully apo protein (Fu et al. 2014; Aye et al. 2012; Aye and Stubbe 2011). A similar hexameric state has been shown in the yeast enzyme, for which the first crystal structure of a dATP-bound RNR-α hexamer was released (Fairman et al. 2011), and this is overall in agreement with a human RNR-α_6 EM structure and a human RNR-α_6 low-resolution crystal structure (Ando et al. 2016), all of which show a ring of RNR-α dimer-linked inter-dimer through contacts between N-terminal ATP binding cones (Brignole et al. 2018). There has been some debate about the orientation of the active sites in these structures (two different arrangements were observed in the yeast structure). Based on the more recent human structures, it seems that there is increasing belief that the active sites point to the center of the RNR-α_6 ring, at least for dATP-induced hexamers; the system is, however, quite nuanced, as we will discuss later. It is very clear that dATP-induced mammalian RNR-α_6 can be formed in the absence of RNR-β. In the human RNR-α, using a mixture of randomly fluorophore-labelled proteins, dATP-induced oligomerization proceeds via an activated monomeric state, which dimerizes and then proceeds rapidly to the hexameric state (Fu et al. 2014). SAXS experiments, on native RNR-α_6, are overall in agreement with this model (Ando et al. 2016). The RNR-α_6 state is unstable in the absence of dATP, and owing to the rapid reversibility of dATP binding, observation of RNR-α_6 requires that dATP be present throughout experiments (Aye and Stubbe 2011; Aye et al. 2012). There is confusion in the literature about how well RNR-β can bind RNR-α_6. It is postulated that if RNR-β can bind to the dATP-induced hexameric state, it is bound in a conformation that is unable to undergo productive electron transfer.

Intriguingly, ATP (low mM) may also usher formation of *similar* RNR-α hexamers in the absence of RNR-β, although this structure is not observed under all conditions by all groups. We believe that this confusion likely reflects that the ATP-induced RNR-α hexamers are less stable than those induced by dATP. Increasing ATP concentrations (10 mM) may form filamentous structures. Seeing as cellular ATP levels are in the region of the concentration at which filament formation has been observed, and as such filaments are apparently not observed in cells, such observations could be due to artifacts. RNR-β has been proposed to interact with the dATP and ATP-induced RNR-α hexamers differently. These differences are proposed to lead to changes in conformation in ATP-induced RNR-α_6, allowing activity. However, the overall intricacies of these differences are not established.

The *E. coli* class I enzyme, on the other hand, creates $\alpha_4\beta_4$ tetramers in the presence of dATP, existing as a ring of two repeating $\alpha_2\beta_2$ units. This state is believed to have doubled the distance between the tyrosyl radical and the active site cysteine, inhibiting activity (Greene et al. 2018). Hexameric states of *E. coli* RNR-α are not apparently formed, although $(\alpha_4\beta_4)_2$ structures, in which the two $\alpha_4\beta_4$ rings interlock (Zimanyi et al. 2012), have been reported. From these discussions, it is clear that one crucial difference between the mammalian RNRs and *E. coli* RNRs is that in *E. coli*, higher-order oligomerization requires RNR-β. Another take-home is that in many instances, *E. coli* RNR-α is an overall poor biochemical model of its human ortholog, particularly in terms of oligomerization. From a cell biology perspective, the picture is also unclear. In *E. coli* RNR, mutations that disrupt the α,β-interface in the $\alpha_4\beta_4$ state cause loss of inhibition of the *E. coli* enzyme by dATP (Chen et al. 2018). This result has been interpreted to imply that oligomerization is the final driving force for inhibition, although this assertion is far from proven. Nevertheless, some of these dimer-disturbing mutations were linked to a previously reported mutator phenotype in *E. coli*, which is consistent with the need for RNR to self-regulate to achieve balanced dNTP pools. A mutation in the ATP binding cone of mammalian RNR-α (D57N) renders RNR-α immune to inhibition through dATP binding (Aye and Stubbe 2011; Fu et al. 2014), echoing these data. Using fluorescent labelling and size-exclusion chromatography (SEC), the loss of activity regulation in the mammalian RNR-α (D57N) mutant appears to reflect an inability of the protein to form inactive RNR-α hexamers. Human (Caras and Martin 1988) and yeast (Fleck et al. 2013) cells expressing this mutant show severely elevated dNTP levels and show elevated mutation rates, consistent with similar data from *E. coli*.

Some forms of RNR-α are believed to have other allosteric sites (Thomas et al. 2019). Novel ligandable binding sites have also been reported for the human enzyme (Ahmad et al. 2015). The relation of these novel sites to allosteric regulation is not well established. It is thus possible that allostery has more surprises in stall for the future. Finally, it should be noted that, in line with the movement away from classical enzymology to a more structurally orientated approach, investigations on RNR have begun to gravitate to X-ray and EM analysis. However, careful analysis of kinetic constants, affinities, and other simple parameters often remains unknown or poorly defined, especially for the human enzyme. Furthermore, investigations into RNR have been undertaken oftentimes using mutants that may cause unexpected artifacts. Techniques that can give insights into changes in the wild-type human enzymes during cycling/ligand binding would be most welcome. Several techniques using relatively simple physical methods have been disclosed (Offenbacher and Barry 2020; Watson et al. 2021), although these may lack the penetrance and residue specificity to provide significant new insights.

RNR-Targeting Drugs and Their Links to Allostery and Chemistry

There are numerous RNR-targeting drugs approved for clinical use, and many are in trials or reported from preclinical studies. These can be broken down to RNR-β targeters and RNR-α targeters. RNR-β-targeting drugs affect the radical/iron active site and are often chelators. Compounds targeting RNR-α typically bind one of the known binding sites, either through hijacking radical chemistry, directly blocking the active site (C-site binding), or through allosteric regulation (A- or S-site binding). RNR-α inhibitors often look like nucleotides. Thus, in many ways, the properties of these drugs reflect what is known about the individual RNR subunits (Fig. 5.5).

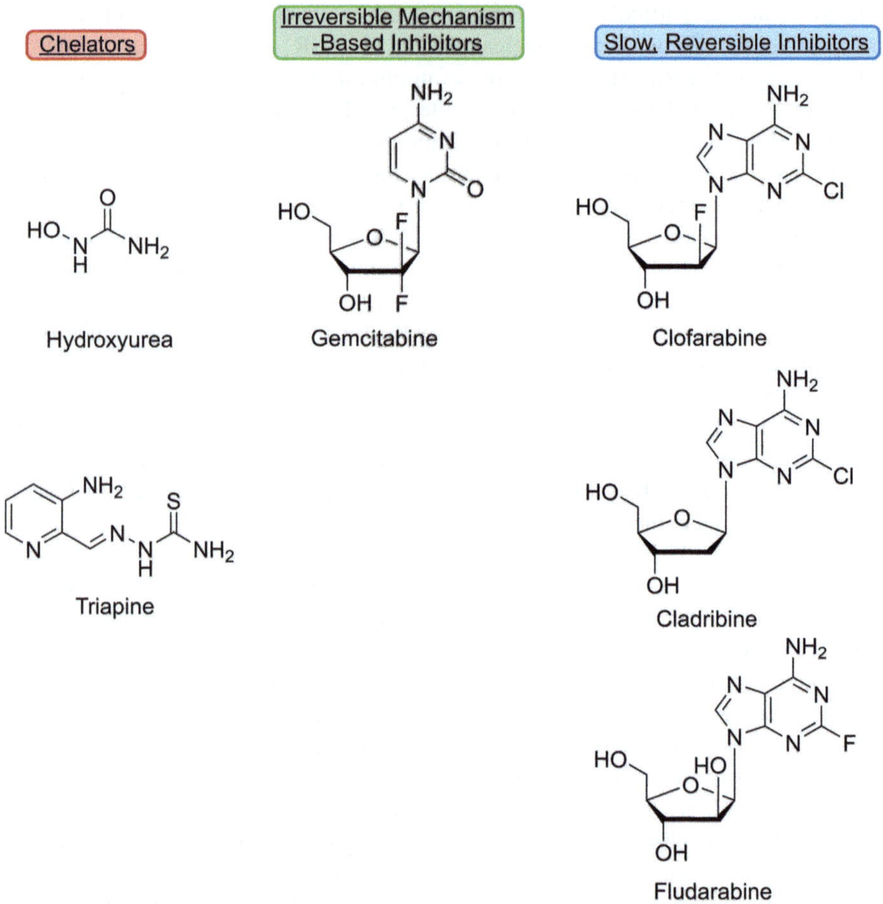

Fig. 5.5 Structures of approved drugs, categorized by mode of action. The chelating inhibitors target RNR-β by disrupting the necessary tyrosyl radical and di-iron nuclear center, while the nucleotide mimics target RNR-α by either alkylation or binding to the A-site to promote the formation of hexamers

We will discuss the data below that lead to the conclusion in vitro and in cells that some approved drugs and some new molecules target a specific subunit of RNR. We have not discussed the large number of molecules reported that simply inhibit RNR in vitro, or are generically cytotoxic/perturb dNTP pools as part of their pharmaceutical program. We believe such results are not, in of themselves, a firm basis for discussion on RNR-inhibiting drug candidates as the cellular target remains to be elucidated carefully. However, it should also be noted that many of the approved drugs, including hydroxyurea, and many of the nucleotides (Çelik et al. 2018) are pleiotropic, i.e., they target many proteins, at least in vitro. Especially in the case of nucleotides, these RNR inhibitors can also serve as substrates for polymerases (Prakasha Gowda et al. 2010) and affect nucleotide pool regulation through other mechanisms, some of which may be dependent on RNR inhibition (Wisitpitthaya et al. 2016) (Fig. 5.6).

As target validation is critical for evaluating drug data, and this is often poorly demonstrated in the literature and overall poorly understood in several research communities, it is relevant to briefly discuss methods to validate a protein as a

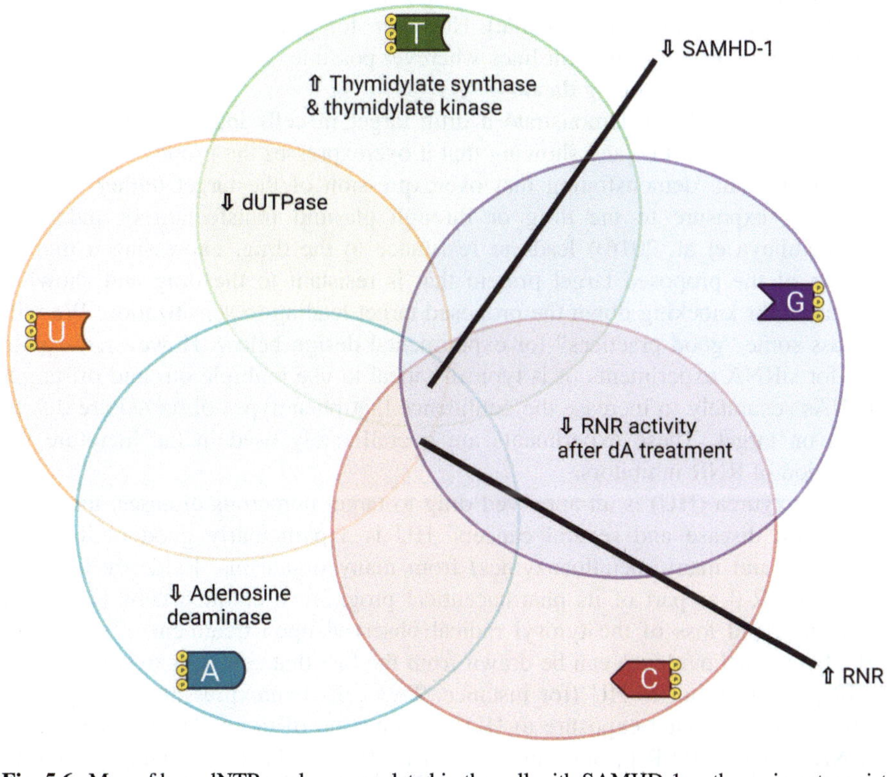

Fig. 5.6 Map of how dNTP pools are regulated in the cell with SAMHD-1 as the main antagonist that opposes RNR reductase activity by hydrolyzing dNTPs into their respective deoxynucleotides and tripolyphosphates. Arrows indicate if the effector increases (up arrow) or decreases (down arrow) the pools, respectively

relevant target of a drug in cells. Similar ideas are applicable to organisms, although they may be more difficult to perform. Demonstrating in vitro inhibition of a purified protein at a specific concentration of an inhibitor does not necessitate that a specific protein be a target in cells at the concentrations used in the in vitro assay. Furthermore, as many of the compounds used in RNR inhibition are administered as the nucleoside (which needs conversion to the nucleotide to be active) or can act as chelators, whose mode of action could be different from the purified drug in vitro, the active concentration of the drug and even the active form of the drug in cells are often debated. Measuring residual target enzyme activity in cell lysates post drug treatment, observing oligomerization changes due to drug treatment, or identifying a drug-target complex through mass spectrometric analysis are encouraging evidence that the proposed target is engaged in cells. However, these methods *do not prove* that *sufficient* inhibition occurs in cells to create a phenotype and hence on their own are insufficient to prove a cellular mechanism. Finally, we would point out that it is critical to research specific cell lines that were used and culture conditions that are intended to be used in assays. All these can change drug behavior, or be used to create lines that should be sensitized or resistant to specific mechanisms proposed. Generally, it is best to choose lines that best mimic the disease you are studying (or primary tumor lines, if available). However, for general experimentation, it is better to use a panel of different lines wherever possible, as mutations are common that can lead to confounding data.

Classic methods to demonstrate a drug target in cells include the following: evolving a resistant line and showing that it overexpresses the proposed target or a resistant mutant, demonstrating that overexpression of the target (either through escalating exposure to the drug or through plasmid transfection or induction (Wisitpitthaya et al. 2016)) leads to resistance to the drug, expressing a mutant version of the proposed target protein that is resistant to the drug and showing resistance, or knocking down the proposed target leading to sensitization. We will discuss some "good practices" for experimental design below. However, we note that for siRNA experiments, it is typically ideal to use multiple on- and off-target siRNAs separately to increase the confidence that phenotypes observed are due to being on target. These experiments are overall rarely used in the literature for validation of RNR inhibitors.

Hydroxyurea (HU) is an approved drug to target numerous diseases, including sickle cell disease and several cancers. HU is a particularly good inhibitor of RNR-β's (and many metalloenzymes) from many organisms. Evidence that HU targets RNR-β as part of its pharmaceutical program in cultured cells is derived from the rapid loss of the tyrosyl radical observed upon treatment of cells with HU. Functional evidence can be drawn from the fact that cell lines overexpressing RNR-β are resistant to HU (for instance, TA3 cells overexpressing RNR-β were derived from elevated exposure to HU as a function of time). Treating cells with siRNA targeting RNR-β, but with residual RNR-β levels still allowing growth to occur, sensitize cells to HU treatment in A549 and HCT-116 cells. siRNA targeting RNR-α does not have this effect (Reid et al. 2009). Other chelators, such as triapine, a drug formerly used in clinical trials for numerous cancers, including renal (Knox

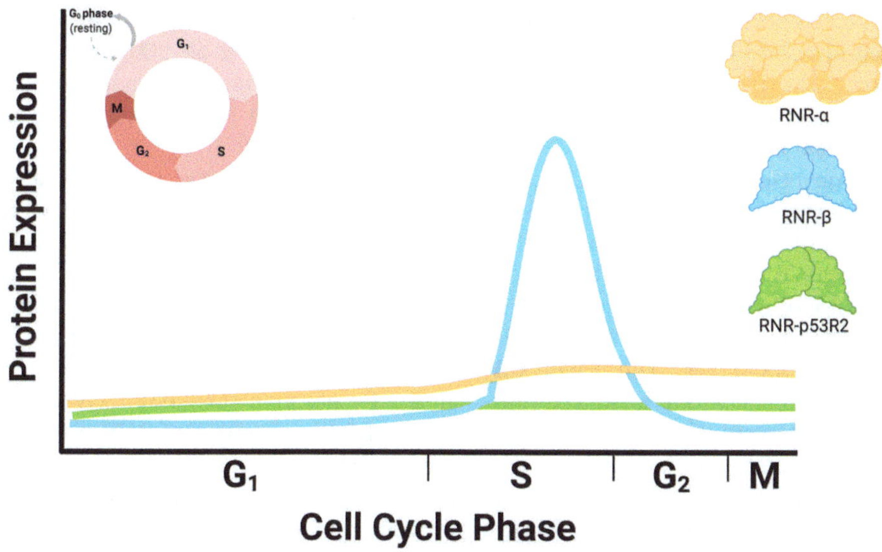

Fig. 5.7 Relative RNR subunit expression across the cell cycle. RNR activity is increased during the S-phase, indicating expression of RNR-α (yellow)and RNR-β (blue); however, only the latter, smaller subunit is rapidly degraded by ubiquitin ligases. On the other hand, RNR-p53β (green) is stably expressed throughout, maintaining a minimal level of RNR reductase activity

et al. 2007) and uterine cancers (Kunos et al. 2019), are not sufficiently strong metal chelators and/or are too large to remove iron from the active site of human RNR-β and instead complex to iron and reduce the tyrosine radical site to form a met human enzyme (Aye et al. 2012) (Fig. 5.7).

The four approved RNR-α targeting nucleoside drugs break down into two groups, irreversible and slow binding, reversible inhibitors. Based on historical precedent, the first group considered is irreversible inhibition, which occurs through hijacking RNR's chemical mechanism. The only approved inhibitor in this class is gemcitabine, which is an approved treatment for pancreatic cancer. In this molecule, radical abstraction by RNR-α creates a situation where a potent electrophile is generated within the RNR-α active site through loss of, in the case of bacterial enzymes, both fluorines and, in human and bacterial enzymes, the nucleobase. This species is then able to alkylate RNR, leading to inhibition and formation of a tight α,β-complex, which is believed to differ between *E. coli* and human proteins. In cells, a two- to threefold overexpression of RNR-α is sufficient to grant significant (approximately threefold) resistance to HEK293T-REX cells treated with gemcitabine (Wisitpitthaya et al. 2016). These experiments were carried out using a line that upregulates RNR-α expression in the presence of tetracycline from a specifically defined locus. Thus, the same cells that were sensitive to gemcitabine in the absence of tetracycline were desensitized to gemcitabine in the presence of tetracycline. Such experimental conditions/design helps to avoid artifacts of transfection and further gives better control expression of the resistance-inducing gene.

Partial knockdown of RNR-α, under conditions that still allowed growth to occur, sensitized A549 and HCT-116 cells to gemcitabine (Reid et al. 2009). Perhaps consistent with the formation of the tight RNR-α/RNR-β complex, conditions that lead to overexpression of RNR-β have been reported to lead to resistance to gemcitabine (Tu et al. 2017).

Conversely, clofarabine (ClF), cladribine (ClA), and fludarabine (FlU; typically administered in patients and cells as the monophosphate, FlUMP, form) were, over a period of several years, shown to target the A-site of RNR-α. All these drugs all appear to function as dATP analogs, as they all form inactive RNR-α hexamers in the absence of RNR-β in vitro. The triphosphate of ClF is established as the active form (Gandhi and Plunkett 2006) in cells and patients. Cells expressing ectopic His_6-tagged RNR-α were treated with ClF and the resulting hexameric complexes post 30-min treatment were isolated and analyzed by negative stain EM (Aye et al. 2012). This result proves target engagement and further shows that RNR-α oligomeric regulation occurs in cells, which has not been possible with dA treatment. ClF-induced hexamers of endogenous RNR-α were also trapped through cross-linking. The validation of the A-site as the principal target of these molecules was shown by the use of RNR-α-D57N; a mutant that (as described above) is unable to oligomerize in the presence of dATP, and hence is not inhibited by dATP, is also not inhibited by ClFTP, ClATP, and FlUTP in vitro (Wisitpitthaya et al. 2016). In cells, expression of RNR-α-D57N (two- to threefold above wt-RNR-α) is completely protective for ClF and ClA treatment (experiments again performed using a tetracycline-inducible system, which expressed an amount of the induced RNR-α similar to that in the wild type). Expression of RNR-α-D57N (which is as sensitive to gemcitabine as wt RNR-α in vitro), two- to threefold above wt-RNR-α, was only as protective against gemcitabine toxicity as wild-type expression. Surprisingly, RNR-α-D57N expression was not protective against FlUMP toxicity (Wisitpitthaya et al. 2016). Thus, RNR-α could not be assigned confidently as a target of FlUTP, despite this molecule's ability to inhibit RNR-α in vitro. It should be noted that the affinity of FlUTP for RNR-α is much lower than for ClA and ClF, and hence, the action of FlUMP could be different due to FlUMP having an unknown binding partner that it targets preferentially over RNR-α.

What is all the more remarkable about these data are that the diphosphates of ClF, ClA, and FlU are also inhibitors of RNR-α. However, these species also are unable to inhibit RNR-α-D57N. As these diphosphates are most likely C-site binders, this implies that C-site binding can affect oligomerization and further explains why RNR-α-D57N expression can transfer almost complete resistance to cells treated with ClF or ClA (Wisitpitthaya et al. 2016). These data overall are in accordance with data from *E. coli* RNR-α that implied that it is oligomerization that leads to inactive states of RNR, not ligand binding (Fu et al. 2013; Ando et al. 2016).

Finally, a series of non-nucleotide reversible, competitive inhibitors of RNR-α have been released. These molecules, despite having relative low affinity ($IC_{50} \sim 5\ \mu M$), were shown to synergize in cells with siRNA targeting RNR-α (Huff et al. 2018). It will be interesting to see the effect of these molecules, while continuing to investigate the on-target effects of the resulting compounds correctly.

Nevertheless, this is a promising start. Certainly, identifying inhibitors with better on-target activity against RNR-α and better therapeutic indices is a very worthwhile process (Misko et al. 2019). Strategies based on RNAi, such as the RNR-β-targeting oligonucleotide, GTI-2040, that showed an acceptable safety profile in Phase I clinical trials (Malik et al. 2018), would also fit into this category. Such a strategy may help suppress genotoxic/mutagenic effects caused by current RNR-inhibiting drugs. It should be noted that many genotoxic anticancer agents cause mutagenesis in treated patients, and some of these may be worse than RNR-targeting drugs: reports in chicken DT-40 cells showed that hydroxyurea and gemcitabine are minimally mutagenic as compared to that of approved drugs like cisplatin (Szikriszt et al. 2016; Ahmad et al. 2017), even though all show a positive result in the AMES test for mutagenesis in bacteria. Other studies have found chromosomal aberrations in gemcitabine-treated human lymphocytes and in bone marrow cells extracted from mice treated with gemcitabine (Aly et al. 2003; Aydemir et al. 2005). Furthermore, some of the off-target effects ascribed to RNR inhibition may well be intrinsic to RNR inhibition either through misregulating dNTP pools or through interfering with RNR's complex biology. Thus, alternative approaches to harness RNR's moonlighting roles in drug design are also welcomed (see below).

Drug-Induced RNR-α Hexamers

For the above reasons, RNR-α hexamerization has become an important aspect in drug design. In fact, we now have access to at least seven molecules that can induce RNR-α inactive hexamers (dATP, ClFTP, ClATP, FlUTP, ClFDP, ClADP, FlUDP), ostensibly through binding to two different sites. Obviously, Occam's razor interpretation is that hexamers induced by the different compounds are all similar and that their common structure in some way differs from the hexamer that may be formed when ATP binds to the A-site. Data countermanding this single inhibited hexameric RNR-α hypothesis, which was already present from early yeast RNR-α_6 crystal structures (as discussed above), started to arise from investigations into ClFTP. dATP-induced hexamers collapse rapidly post drug departure; for instance, to observe dATP-induced RNR-α hexamers through size-exclusion chromatography (SEC), dATP must be present in the SEC running buffer. However, ClFTP-induced hexamers remained stable post "complete" release of the drug, even when run on SEC in the absence of ClFTP, or during lysis, and protein purification from cultured mammalian cells treated with ClF (Aye and Stubbe 2011; Aye et al. 2012). Such stability could reflect the ability of the hexamer to sustain binding at fractional ClFTP occupancy, assuming that such fractional occupancy of ClFTP would not have been detectable. As more dATP analogs were investigated, it became clearer that there is heterogeneity across all RNR-α hexameric states. For instance, the different hexameric states of RNR-α in the presence of different drugs, although indistinguishable by SEC, are differentially cleaved by a protease under saturating concentrations of each drug; the protease was not inhibited by any of the molecules used (Wisitpitthaya et al. 2016). Negative stain EM also showed different structural arrangements among the hexamers induced by different hexamerizers (Fu et al. 2018). Critically, even within hexamers caused by the *same drug*, a large degree of heterogeneity was observed. Although these data imply a very intricate regulatory process, it should be remembered that the S-site is reorganized at the atomic level to optimize the active site to accommodate specific nucleotides. It is thus not surprising that such subtle allostery could occur in RNR-α. The plasticity of RNR-α hexamers is particularly intriguing and opens doors to a huge scope of sensing and signalling functions that we discuss later. As these hexamers are caused by RNR-α-inhibiting drugs, the role of these hexamers in drug action is an ongoing question that we will address below as well. Finally, although not necessarily completely diagnostic, RNR-α_6 formation could readily form the basis of an inhibitor screen, which would bypass the need for much more complex activity assays (Figs. 5.8 and 5.9).

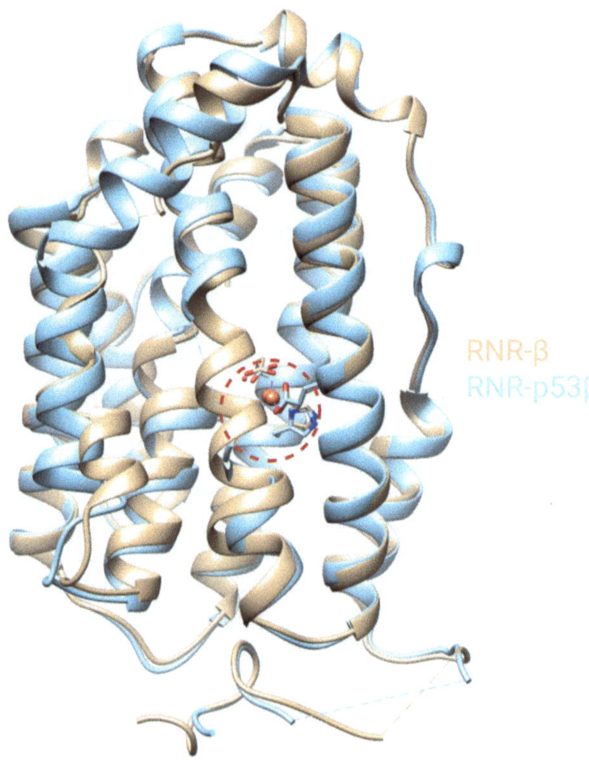

Fig. 5.8 Overlay of crystal structures of human RNR-β (tan; PDB 2UW2) and RNR-p53β (light blue; PDB 2VUX) which have around 80% homology. The nuclear iron center can be seen

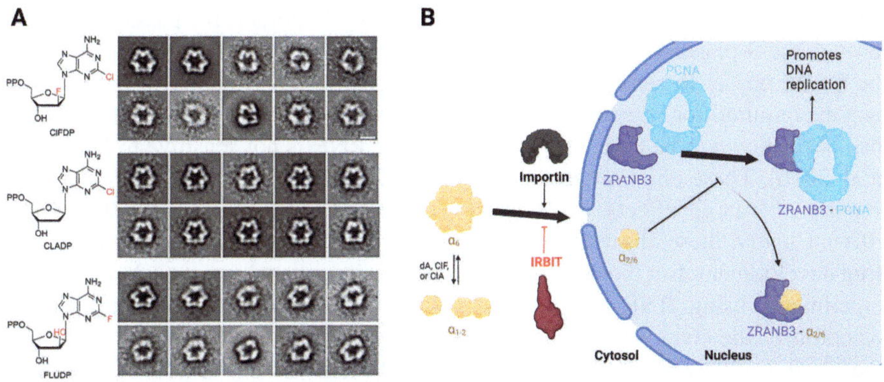

Fig. 5.9 (**a**) RNR-α forms hexamers in the presence of dA, ClF, ClA, or FLU that inhibits its reductase ability. (**b**) The RNR-α oligomer is imported into the nucleus by importin or prevented by IRBIT. Inside the nucleus, it is able to bind to ZRANB3 and thus hinders DNA replication by reducing ZRANB3 binding to PCNA

RNR-Inhibited dNTP Pools and the Cell Cycle: Implications for Studies with RNR Inhibitors

One classic manifestation of RNR inhibition in cells is reduction of dNTP pools. In HEK293T cells treated with dA for 30 min, almost complete loss of dCTP and dGTP is observed. Curiously, TTP levels are not affected. Similar data have been reported for cells treated with ClF (Daly et al. 2016) in several lines (Parker et al. 1991). Loss of dNTP pools is rapidly followed by suppression of DNA synthesis, observable by a suppression in the incorporation of BrdU or EdU into DNA using whole-cell imaging of flow cytometry, or fiber staining (Fu et al. 2018). Loss of DNA synthesis occurs on a similar timescale to loss of dNTPs, in HEK293T cells treated with either dA or ClF. A similar rate of suppression of DNA synthesis is found in U2OS cells treated with dA or ClF on a similar timescale. These data agree with other reports for ClF treatment of HL-60 and KG1 cells, although DNA synthesis in some other lines, such as MOLM-13, responds slower to ClF treatment (Seedhouse et al. 2009). These observations also agree with data postulating that nucleotide levels are limiting for DNA synthesis (Stodola and Burgers 2016). Suppression of DNA synthesis post RNR inhibition is reflected by a suppression of movement through the S-phase, which is manifest in K562 cells treated with ClF initially by a loss in G2/M cells that is apparent 3 h posttreatment (approximately half the time taken to go through the G2/M phase (Cooper 2000)). Similar S-phase progression suppression is observed in other cell lines (Çelik et al. 2018) treated with ClF or ClA. There is some evidence that this slowing of S-phase progression is mediated by Chk1 Ser345 phosphorylation (Mesa et al. 2005). Over time, ClF-treated K562 cells typically lock in the early S-phase, or the G1/S border. HU treatment of cells is also known to lock cells in the G1/S or early S-phase in various organisms (Borel et al. 2002; Apraiz et al. 2017).

These observations all point to RNR activity being important for passage into and through the S-phase. It appears that RNR inhibition does not affect passage through the G2/M phase. Recently, it was shown in yeast that a mutation to an essential enzyme required for ergosterol sensitizes *S. pombe* to HU (Xu et al. 2016). HU has also been shown to be toxic to organisms whose RNR is not sensitive to HU (Liew et al. 2016). These observations serve as reminders that many RNR inhibitors have important off-target effects, which may only manifest in sensitized lines (Zhang et al. 2020). Clearly, these observations have implications for personalized medicine and drug development. But these data also serve as an important message when planning experiments using RNR inhibitors to measure cell cycle-specific effects, or to synchronize the cell cycle: measure effects as a function of time, assess cell cycle at each timepoint, and try to assess the effects of drugs as early as possible (as DNA synthesis and cell cycle changes happen rapidly). When using RNR inhibitors to block cell cycle progression, remember that RNR inhibitors are typically pleiotropic and could cause artifacts that may readily be confounding. Indeed, it is best to use various methods to block the cell cycle through different mechanisms to try to ensure fairness.

Other dNTP Regulation Pathways

RNR is typically marketed as a self-regulatory machine that is key for balancing dNTP pool levels. However, there are other mechanisms that act to regulate dNTP pool levels. We have already discussed kinases and TTP metabolic enzymes that are needed downstream of RNR to create dNTP pools and their role in health and disease and how these are cell cycle regulated. Another key adjuvant to RNR activity is adenosine deaminase, which is responsible for converting adenosine metabolites to inosine. In this way this enzyme prevents RNR inhibition due to toxic levels of dATP accumulating in cells. Adenosine deaminase appears to be particularly important in the immune system. Adenosine deaminase deficiency is associated with severe combined immunodeficiency.

There are also enzymes that oppose RNR activity (Rampazzo et al. 2010). These enzymes are also important for genetic stability. The principal opposing force to RNR is the sterile-α motif and histidine-aspartate containing protein-1 (SAMHD-1). This is a triphosphatase that converts dNTPs to deoxynucleosides and tripolyphosphate. It appears to be active throughout the cell cycle, although it undergoes cell cycle-dependent phosphorylation, which is poorly understood (Tramentozzi et al. 2018). This enzyme has two allosteric sites: an overall activity site (allosteric site 1) and a second allosteric site. GTP and dGTP ($K_D \sim 0.2$ μM) (Miazzi et al. 2014) stimulate activity by binding to allosteric site 1. In this state, binding of any dNTP to allosteric site 2 gives an active tetramer (Ji et al. 2014) (K_D nucleotide 1–20 μM) (Miazzi et al. 2014). Thus, SAMHD-1 lacks the intricate allosteric regulation of RNR. Nevertheless, dATP has the weakest affinity for the active site, and in the presence of dNTPs, the rate of dATP hydrolysis is slowest by

SAMHD-1. This finding may further explain why RNR-α uses dATP levels as a metric for dNTP pool levels. It has been proposed that inhibition of SAMHD-1 via lowering dNTP pool levels through inhibition of RNR can aid the mode of action of some drugs (Rudd et al. 2020) that require phosphorylation. Consistent with a role antagonistic to RNR, knockdown of SAMHD-1 enlarges dNTP pools (Tramentozzi et al. 2018), stimulates proliferation, and promotes passage through the cell cycle, in THP-1 cells (Bonifati et al. 2016). Intriguingly, not all SAMHD-1's roles are antagonistic to RNR, as like RNR (Burkhalter et al. 2009), SAMHD-1 promotes DNA damage repair by homologous recombination. This function of SAMHD-1 appears to be independent of its canonical activity and functions through recruitment to DNA double-stranded breaks, promoting end resection (Daddacha et al. 2017).

dUTPase is a particularly important protein acting downstream of RNR. This enzyme hydrolyzes dUTP that can be formed especially upon upregulation of RNR activity (Chen et al. 2016). dUTPase acts as a tumor suppressor countermanding the incorporation of dU into DNA. Its low expression in conjunction with high RNR-β expression is a particularly poor prognosis marker for cancer patients. These data implicate that measurement of dUTP should routinely be made when measuring dNTP pool levels upon RNR overexpression and in cancer profiling. Of course, RNR expression level data should ideally be interpreted in relation to the expression of these key regulators. Since some of these relationships have only recently become known, it is possible that older expression level data need to be reevaluated to build up a complete picture of how RNR activity functions in disease.

Regulation of RNR Activity in Cells

The elaborate mechanisms discussed above fit into a neat package that explains many salient observations in vitro and further help to explain how individual proteins work synergistically to balance dNTP pools in cells. However, in cells there are several additional control points that directly regulate RNR activity. These include cell cycle control (through RNR-β ubiquitination and other posttranslational modifications), subcellular localization, checkpoint control, and regulation by associating proteins.

Cell cycle control is perhaps the most important aspect of cellular RNR regulation. In cells that are not synthesizing genomic DNA, i.e., those not in the S-phase, dNTP pools are maintained low. Given the significant increase in demand for dNTPs in cells during the S-phase, it is not surprising that RNR activity peaks in the S-phase and is rapidly diminished out of the S-phase. Both principal subunits of RNR are transcribed as the cell starts to enter the S-phase, with mRNA levels peaking in the early S-phase. For RNR-α, in unstressed cells, this regulation has minimal effect as RNR-α protein has a long half-life; like many large, long-lived proteins, RNR-α is believed to be degraded by autophagy (Roberts and Deretic 2008; Dyavaiah et al.

2011) relatively slowly. RNR-β stability, on the other hand, is regulated by two ubiquitin ligases that prime RNR-β for degradation via the proteasome out of the S-phase: the Skp1/Cullin/F-box (SCF) that promotes RNR-β degradation in the G2-phase and the anaphase-promoting complex (APC) that promotes degradation in the M and G0 phases. APC-mediated degradation is mediated by recognition of a KEN box on the RNR-β N-terminus (recognized by Cdh1-APC). Thus, RNR-β has a short half-life, and its protein levels (and hence RNR activity) peak in the S-phase. Indeed, this high demand for dNTPs in the S-phase explains why RNR inhibitors block passage through the S-phase, but do not affect the G2/M exit (*note:* this is true for many compounds so this observation is in no way diagnostic for RNR inhibition). There is some requirement of dNTPs out of the S-phase for mitochondrial DNA synthesis and repair. These are provided by p53R2. Consistent with this role, defects in p53R2 lead to mitochondrial disease (Lim et al. 1993). p53R2 can perform this function as it lacks a KEN box, making it stably expressed throughout the cell cycle. It should also be noted that thymidylate synthase and other TTP-producing enzymes downstream of the RNR also peak in the S-phase (Ke and Chang 2004; Le François et al. 2007; Lee et al. 2014), meaning that dNTP synthesis in and out of the S-phase may well function differently.

It has also recently been reported that RNR-α activity might be increased by phosphorylation at S559. This phosphorylation was enacted by CDK2/cyclin A during the S- and G2-phases of the cell cycle (Shu et al. 2020). Expression of an RNR-α-S559D mutant, mimicking phosphorylation, in an RNR-α-knockdown background caused elevated lysate RNR activity, increased dNTPs, and increased DNA synthesis rates. Loss of this phosphorylation, through rescue of the same RNR-α knockdown lines with RNR-α-S559A, had the opposite effect and also increased γH2AX and sensitized colony formation to the ATR inhibitor VE822. Aside from having potentially important ramifications for dNTP pool maintenance that in our opinion still need further elucidation, these mutants could directly prove very useful in further cellular experiments on moonlighting and activity-associated roles of RNR. However, prior to that, careful in vitro experimentation using the recombinant enzymes will be necessary. Should the RNR-α-S559D prove to be intrinsically more active than that of the wt-RNR-α in vitro, this mutant could also overcome some of the issues with in vitro kinetics of RNR-α. On the other hand, RNR-β acetylation at K95 by KAT7 suppresses RNR-β homodimer formation, leading to suppression of RNR activity (Chen et al. 2019a, b). K95 acetylation is removed by Sirt2, whose levels peak in the S-phase. Knockdown of KAT7 showed an increase in all dNTP pools: knockdown of Sirt2 weakly decreased all dNTPs, reminiscent of the behavior of dNTP pools upon RNR inhibitor treatments. P53R2 has been reported to be selectively cleaved by caspase 8, a protease immediately upstream of the executioner caspase, caspase 3, in the apoptosis pathway. This cleavage inactivates the enzyme (Tebbi et al. 2015).

Subcellular Localization

Although some reports suggest otherwise, when careful and controlled confocal immunofluorescence using different antibodies in conjunction with multiple on- versus off-target siRNAs to validate on-target binding of the antibodies, RNR-α and RNR-β are predominantly, although not entirely, cytosolic in numerous cell lines. There are some reports that p53R2 is more nuclear-localized than both canonical RNR subunits, although this is debatable; subcellular localization of canonical RNR subunits is also likely context dependent, so the cell lines compared need to be controlled and validated carefully. Based on the majority of data for the canonical RNR complex, it is typically believed that dNTPs are created by RNR in the cytosol and can enter the nucleus, likely through passive diffusion, leading to homogeneity between the nucleus and the cytosol. Although quite feasible, it should be noted that there is evidence that dNTPs are retained within the nucleus post perforation of the cytoplasm using mild detergent. Using this approach, cell-cycle-induced fluctuations of nuclear dNTPs were shown to be less acute than cytosolic fluctuations (Leeds et al. 1985). This result implies that the nuclear membrane acts in some way as a second regulator of dNTP pool levels. There also appear to be special mechanisms to control distribution of dNTP pools between the cytosol and nucleus (Mun and Mathews 1991). Such data and general concerns highlight that we are overall very poorly informed on local dNTP pools in cells and how these levels change.

RNR has been reported to translocate to the nucleus to provide dNTPs at DNA damage sites, particularly in the G1 phase, where dNTP pools are lowest. A similar localization mechanism bringing RNR to sites of DNA replication has been suggested even in *E. coli* (Guzmán et al. 2002). In mammalian cells, this RNR nuclear translocation is believed to occur through interaction with the histone acetyltransferase, TIP60 (Niida et al. 2010). It should be noted that *all* the components of dNTP synthesis have not been shown to translocate on mass to DNA damage sites, although there is a report of recruitment of RNR-α, RNR-β, and thymidylate kinase to DNA damage events (Hu et al. 2012). This report further showed that loss of thymidylate kinase elevated dU incorporation into DNA, which could be explained by RNR localizing to DNA damage sites causing elevated dUTP levels that could not be converted to TTP. dU incorporation could be rescued by suppressing RNR-α recruitment to DNA damage events, through overexpression of the C-terminus of RNR-α fused to NLS-YFP, which is proposed to block the interaction of full-length RNR-α with TIP60. Although an interesting hypothesis, it should also be noted that in G1, DNA repair occurs principally through nonhomologous end joining, which requires minimal dNTPs (Mao et al. 2008) and is not affected by HU treatment (Burkhalter et al. 2009), possibly at least in part due to the lack of dNTPs present (Burkhalter et al. 2009; Zhao et al. 2017). It remains poorly understood how the constructs and mutants used affect the choice of DNA damage repair in cells. RNR-β expression in G1/G0 is also minimal, raising the question of what phase of the cell cycle was being measured in the cells in which RNR-β/thymidylate kinase colocalization (referred to above) was observed.

Contrary to the TIP60 mechanism, RNR-β nuclear shuttling, catalyzed by Dif1 in budding yeast, inhibits RNR activity (Wu and Huang 2008). This observation serves as another reminder that translocation of only part of the RNR complex is counter-productive to RNR activity. Of course, for this inhibition mechanism to function, a significant proportion of RNR must translocate into the nucleus, and this has not been convincingly demonstrated in mammalian cells.

RNR Binders That Regulate RNR Activity

The example of Dif1 brings us onto direct RNR activity regulators. RNR binders regulating reductase activity were first reported in yeast. These yeast regulators can be broken down into two camps. The first is direct inhibitors of RNR activity. The principal protagonist in this class is the budding yeast protein Sml1. This protein is evolutionarily related to Dif1, but serves as an inhibitor of yeast RNR-α (Nestoras et al. 2010), inhibiting re-reduction of the active site post nucleotide reduction (Chabes and Stillman 2007). There is a loosely similar protein in fission yeast, Spd1, although this protein also appears to be linked to nuclear import, performing a function similar to Dif1 (above). Spd1 may also directly inhibit RNR-α activity, and it should be noted that promoting nuclear entry (of the smaller RNR subunit) and activity inhibition via interaction with RNR-α is not necessarily mutually exclusive. Wtm1 is believed to serve as the nuclear anchor of the small RNR subunits in budding yeast (Lee and Elledge 2006). Deletion of many of these RNR regulatory proteins, such as Sml1 or Wtm1, are permissive for lethal mutations/deletions in yeast, such as Mec1 and Rad53 deletions (Zhao et al. 1998). Intriguingly, Mec1 and Rad53 function as important tumor suppressors in humans (where they are known as ATR and Chk1), which are bypassed, lost, or altered in some cancers (Lam et al. 2004; Hoyer et al. 2012). Thus, identification of RNR activity regulators in mammalian cells could readily open new avenues for understanding how cancer cells can retain viability when ostensibly essential DNA damage regulatory genes are lost.

The search for higher eukaryotic RNR activity regulators has been long and has really only reached its infancy. It was clear from genomic studies that there is no direct sequence-related Sml1 analogue in humans. It was later reported that IRBIT, a protein involved in regulating ER calcium levels, suppresses RNR activity (Arnaoutov and Dasso 2014). dNTP pools in HeLa cells are sensitive to IRBIT levels, and some evidence for IRBIT controlling passage through the cell cycle has been provided. Mechanistically, IRBIT behaves dissimilarly to Sml1, in the sense that it does not directly inhibit RNR-α, but rather acts as an adjuvant stabilizing the dATP-induced RNR-α_6 (Fu et al. 2018). It was shown that the interaction of human RNR-α with Sml1 has little dATP dependence, although it is known that the human RNR-α does not interact with Sml1 in the same way as Sml1 interacts with yeast RNR-α (Chabes et al. 1999). Nevertheless, Sml1 inhibits yeast RNR-α via an oligomerization-independent mechanism, further implying that the links between Sml1 and IRBIT are minimal (Misko et al. 2016). Recently, it was shown that IRBIT functions through RNR to coordinate proper intestinal development in *Drosophila* (Arnaoutov et al. 2020). Although the precise mechanism of this developmental process was not elucidated, these data add further weight to the in vivo relevance of the RNR-α-IRBIT interaction (Arnaoutov et al. 2020). However, such results do not necessarily implicate a dNTP-specific role. Indeed, the hexamerization dependence of the IRBIT-RNR-α interaction implies to us that RNR activity regulation is not the principal role of this interaction. We elaborate on this below.

Role of dNTPs in Health and Disease

dNTP pool imbalances are linked to several diseases, including synucleinopathies (which were proposed to function in part through elevated RNR-α degradation (Sampaio-Marques et al. 2019) through autophagy stimulation), progeria (Hämäläinen et al. 2019), viral evolution (Deforche et al. 2007), and, of course, cancer. Overall, the data paint a picture of the need to meticulously balance dNTP pools for S-phase entry, correct DNA synthesis/S-phase progression, and likely complete the S-phase. Although elevated dNTP pools are achievable in cells through many different means (external dosing of nucleosides, knockdown of SAMDH1, and overexpression of RNR/RNR hyperactive mutants, as discussed above), it should be noted that in mammalian cells DNA damage induction does not show a large elevation in dNTP pools (Håkansson et al. 2006) (unlike bacteria, yeast (Elledge and Davis 1990), and likely *Drosophila* (Akdemir et al. 2007)). It is unknown if this observation reflects a true divergence in the way DNA damage is martialled, or if it reflects lack of nuance in the way pools are measured. Intriguingly, elevated dNTP pool levels found in bacteria during DNA damage are mutagenic (Gon et al. 2011), so resistance to dNTP pool elevation could reflect an evolved change in martialling of DNA damage response in mammalian cells. Indeed, the principal effect of elevated dNTP pools through any mechanism is mutagenesis. This is because minor dNTP pool elevation can promote intrinsic errors in DNA synthesis through impairing proofreading. These errors are similar for both the leading and lagging strands (Buckland et al. 2014). However, elevated dNTP pool levels can also impinge on genome integrity in other ways, such as through affecting origin usage and choice of repair mechanisms, for instance (Burkhalter et al. 2009). Of course, an even relatively minor dNTP pool elevation can also suppress mutations/defects that may otherwise stimulate apoptosis, trigger damage checkpoints (Chabes and Stillman 2007), or martial other stress responses. In multicellular organisms, such papering over the cracks may not be an optimal strategy, as it can suppresses genetic defects incurred by mitotic recombination and mutation that could lead to malignancy. Finally, in mammalian cells, hyperelevated dNTP pool levels suppress S-phase entry (Franzolin et al. 2013); similar growth suppression effects are observed when dNTP pools are severely elevated in yeast (Chabes and Stillman 2007). This sets a ceiling on how wildly dNTP pools can be imbalanced.

RNR Activity During Cancer Development

RNR activity is unsurprisingly linked to tumor formation: elevated dNTP pools can also help promote oncogenesis by providing an environment that can overcome oncogene-induced senescence, stimulate DNA damage repair, and compensate for loss of essential checkpoint genes (e.g., Mec1/ATR (Lopez-Contreras et al. 2015), or Rad53/Chk1 (Gan et al. 2017)) which are often lost or mutated in cancer. Elevated or imbalanced dNTP pools can also help promote a mutagenic environment by impairing proofreading and slippage of the replisome (through the next nucleotide effect). dNTP pool levels may also affect the choice of DNA damage repair pathway and, in systems prone to hyperrecombination, may suppress recombination (Fasullo et al. 2010), perhaps by promoting synthesis near damage sites (Sabouri et al. 2008). These choices are also linked to the cell cycle and how such choices would affect cancer progression is context dependent.

Recently, a screen in yeast uncovered several mutations to RNR that lead to severely altered dNTP pool levels, in otherwise unperturbed backgrounds (Schmidt et al. 2019). From 23 identified mutator RNR alleles, 11 were located near the S-site, two were at the A-site, and two were at the C-site. The result of the dNTP pool changes caused by expressing these mutants was elevated mispairing and frameshifting, apparently driven by the next-nucleotide effect; however, this was observed for the majority, although not all, of mutants only when Exo1, an exonuclease involved in mismatch repair and recombination (Fiorentini et al. 1997), was deleted. It was found that the most severe phenotypes occurred when three of the four dNTPs were elevated. This state, particularly when dGTP elevation occurred, was particularly harmful. Several of the RNR mutations identified in this screen are annotated in cancer databases, implying that RNR misregulation could be linked to cancer etiology.

However, the overall picture is nuanced: RNR activity is a self-balancing and autoregulatory process that overall makes dNTPs, which are required for healthy proliferation. RNR is also a suppressor of replication stress by providing dNTPs which aid resolution of damage events. Of course, such behavior (either through choice of polymerase, promoting recombination, or the next-nucleotide effect, as discussed above) can lead to errors in replication. Nevertheless, in fish, exogenous nucleoside treatment can rescue defects associated with replication stress (Danilova et al. 2014). Elevating dNTP pool levels, through elevated expression of RNR-β, can also suppress breakage at fragile sites (areas of late replication that commonly undergo DNA damage and deletion under replication stress and are linked to cancer progression (Glover et al. 2017)) in ATR-deficient cells and likely in mice (Lopez-Contreras et al. 2015). Instability at fragile site loci can functionally contribute to cancer (Richards 2001), at least in some instances. Thus, RNR activity elevation could be linked to tumor suppression. Intriguingly, common fragile sites are different in different cell types, pointing to a nuanced picture, that serves to further muddy the waters and could help explain some of the tissue-specific variations in the correlation of RNR and cancer.

Thus, RNR activity and cancer development is context dependent, including the duration of replication stress, localization of stress, genotype, and mode of stress, among others. Data from yeast all the way to mice clearly demonstrate that RNR activity can aid cells surmounting certain stresses and even losses of important replication stress proteins. Of course, when misregulated, such behavior may be considered as overriding or modulating (Lamb et al. 2020) arguably more important checkpoints, or promoting conditions that are overall deleterious/mutagenic. These questions are particularly complex as there is a general problem in the cancer and related fields that we often seek to retrofit genotypes of mature cancer to explain etiology, or do vice versa in disease models that may not be entirely faithful to the origins and progression of a specific disease. Nevertheless, for many proteins that confer cellular protection against malignancies, damage and stress does not simply loose that function in cancer cells; it is certainly not inherently contradictory to assume that some of the functions afforded by cytoprotective proteins would be even more necessary to cancers, due to accelerated growth requirements and overall elevated stress levels therein. For instance, loss of glutathione-S-transferase (GST) isoforms is a marker for increased cancer risk, indicative of a tumor-suppressive function (Okat 2018). However, many cancers are addicted to GST activity (or the activity of its transcription factor Nrf2 (Nagel et al. 2016)), and GST overexpression in cancer can confer protection against several anticancer drugs (Singh 2015).

RNR Activity in Cancer Treatment

The strong link between RNR activity and mature cancer cells is reinforced by several observations, including the elevation of RNR activity in cancer cells and the use of RNR inhibitors in cancer therapy (see above). Many other RNR-targeting drugs have been or are currently in clinical trials. The success and continued interest in RNR inhibitors underscore the critical importance of RNR in cancer biology and, consequentially, the *relatively* minor role played by this enzyme in differentiated cells. This link is perhaps more strengthened in the light that targeting such an essential enzyme is a somewhat nonideal strategy, as it targets all rapidly growing cells and can impinge on senescent cells as well.

Nevertheless, given the elevated replication stress, elevated S-phase, etc. shown by cancer cells, RNR inhibition is particularly damaging for cancer cells over non-transformed cells. Highly consistent with cellular data, tumors treated with an RNR-targeting drug have been shown in the vast majority of studies to be more resistant to that specific drug if they overexpress the RNR subunit targeted by the drug. This has been taken by many authors to mean that, consequentially, the overexpressed subunit is somehow acting as a tumor promoter. However, it is, in fact, a consequence of the overexpression of the drug target and does not *necessarily* have anything to do with a tumor-promoting behavior of the RNR subunit in question. As discussed above, these data can be readily replicated in cultured human cells whose proliferation does not depend on excess RNR activity. Nevertheless, these data do suggest a means to profile tumors for RNR-targeting drug sensitivity, which is practically useful for clinicians.

We will proceed to discuss the role of individual RNR subunits in cancer. We will deliberately not engage drug-related data, due to the confusion these data can cause.

RNR-β and p53R2 in Cancer

An objective view of the literature uncovers some interesting traits, which also need to be considered carefully when interpreting RNR-β's link to oncogenesis. RNR-β is agreed by most authors to be strongly elevated in cancer cells; RNR-β levels are also strongly correlated with the severity of cancer (Zhang et al. 2014) and are inversely correlated with patient survival (Mah et al. 2015; Chen et al. 2019a, b), possibly in a gender-specific manner in some cases; knockdown of RNR-β suppresses these behaviors in cultured cells. RNR-β is considered to be responsible for resistance to many non-RNR-targeting cancer drugs (Zhan et al. 2021). On the face of it, these data provide evidence that RNR-β is a tumor promoter. However, it is not always clear how much of the association of RNR-β with cancer is due to elevation of the S-phase that tends to occur in cancers versus an addiction to RNR activity that must occur due to S-phase elevation (and hence RNR-β stimulation) in all rapidly growing cells (Song et al. 2017). Indeed, there is a strong correlation between malignancy and growth rate that has been known for almost 100 years (Mottram 1934). Furthermore, studies in which RNR-β are knocked down in cancer cells show G1/early S-phase arrest, similar to what is expected if RNR-β were performing its canonical function (Wang et al. 2018). Although this is not a disproof of the hypothesis, it does not indicate that RNR-β is behaving unexpectedly. Of course, better insights into how RNR-β affects tumors could be gleaned by evaluating how RNR-β expression changes across the cell cycle in tumor cells and attenuated lines. Such experiments are rarely performed. In lieu of such data, fiber staining, which measures individual DNA synthesis events (obligatorily in the S-phase; i.e., in a[n] [approximately] cell cycle-*independent* manner), is occasionally provided as a metric for the ability of cells to synthesize DNA in the S-phase, which can be correlated to global RNR-β levels. However, there are many factors affecting replication fork speed, and hence, these data must be interpreted with caution.

However, specific mechanisms of promoting RNR-β expression in cancers have been elucidated; these overall point to a specific promotion of RNR-β by oncogenes and oncogenic effects conferred by RNR-β expression. E2F1, a transcription elongation factor, is linked to RNR-β overexpression that was shown to drive growth and invasion (Fang et al. 2015). Similar observations have been made for other cancer-associated proteins, including human papillomavirus E7 (Wang et al. 2014), CREB1

(Fang et al. 2016), BRAC1 (Rasmussen et al. 2016), JUN (Tu et al. 2017), and FOXM1, a critical driver of prostate cancer (Mazzu et al. 2019). Conversely, several tumor suppressors reduce RNR-β expression (Osako et al. 2019; Goss et al. 2021). Furthermore, RNR-β appears to promote the transformation of several cells. RNR-β also cooperates with oncogenes, such as MYCN, to promote tumor formation in several animal models of neuroblastoma and cancer patients (Nunes et al. 2020). RNR-β is also linked to a host of cancer phenotypes, including promoting angiogenesis, through suppression of TSP-1 and promotion of VEGF (Zhang et al. 2009), and increasing invasion and migration, likely in part through elevation of matrix metalloprotease expression (Zhong et al. 2016) and through stimulating Akt activity (Ma et al. 2020). Mice significantly overexpressing RNR-β appear to develop lung tumors (Xu et al. 2008) and adenocarcinoma. However, mice weakly overexpressing RNR-β show no increase in tumorigenicity (Lopez-Contreras et al. 2015). RNR-β expression has also been linked to resistance to anticancer drugs, such as tamoxifen (Teicher et al. 2015). Unsurprisingly, RNR-β overexpression in cancer is strongly linked to low survival and a poor outcome (Zou et al. 2019).

In some ways, p53R2 can be used as a comparison. As this enzyme is not cell cycle regulated, but contains a similar cofactor and performs similar chemical functions to RNR-β (Cho and Yen 2016), the comparison is fair, if somewhat naïve. p53R2 downregulation is associated with metastasis in hepatocarcinoma, regardless of p53 status: overexpression of p53R2 countermands these effects in cultured cells and in mice (Tian et al. 2014). This effect was linked to suppression of the Akt pathway, which, as mentioned above, is believed to be stimulated by RNR-β (Cho and Yen 2016; Sun et al. 2019). Akt stimulation was observed in wild-type and mutant p53 cancer cells upon p53R2 knockdown. These effects were proposed to proceed through reducing p21, a downstream effector of p53 (Tarangelo and Dixon 2018). Overexpression of p53R2 (albeit a GFP fusion) behaved oppositely and lowered S-phase component (Zhang et al. 2011). p53R2 expression has also been linked to promoting FOXO3 activity, which was proposed to serve a proliferation inhibition role (Cho et al. 2014). However, in other cancers, such as cervical cancer (Bo et al. 2017), p53R2 has been proposed to promote Akt activity and promote metastasis, similarly to what has been proposed for RNR-β (Cho and Yen 2016). In some cancers, such as nasopharyngeal (Chen et al. 2017) and colorectal cancers (Liu et al. 2011), there is an inverse relationship between overall survival and p53R2 expression, although this does not extend to all cancers, and survival and p53R2 are positively correlated in other cancers (Chen et al. 2017; Ding et al. 2019), such as small cell lung cancer (Uramoto et al. 2006) and esophageal (Okumura et al. 2006) cancers. Significant overexpression of p53R2 in mice also leads to lung neoplasms, similarly to expression of RNR-β. Finally, it has been reported that RNR-α expression correlates positively with RNR-β expression in 31 different types of cancer. This trend was not observed with p53R2 (Ding et al. 2019).

Thus, on the whole, p53R2 seems to be significantly less linked to cancer development than RNR-β and certainly has more links to tumor suppression than RNR-β. Although not impossible to rationalize, it is difficult to understand how two proteins with the same activity and 80% homology play such vastly different roles, pointing to pleiotropic behaviors of both proteins. Of course, many of the studies used different cell lines, or expression systems, or patient data, all of which can lead to contrasting conclusions due to background differences nothing to do with the different RNRs used. Studies into RNR-β's role in cancer should therefore examine the roles of both RNR-β paralogs, possibly incorporating KEN box RNR-β mutants to see if there are genuinely differences between the proteins, or if the differences reflect cell-type intrinsic defects. Either way, studies would be rendered much more insightful.

RNR-α: A Tumor Suppressor

Reports of the tumor suppressor-like behavior of RNR-α have circulated for numerous years. It has been known in *Drosophila* that RNR-α knockdown suppresses UV-induced apoptosis and caspase activity upregulation (Akdemir et al. 2007); (d)ATP is involved in the formation of the apoptosome (Bao and Shi 2007), so this may not be particularly surprising. This link is not established in mammalian cells, although the RNR-α associator IRBIT is a controller of apoptosis (Bonneau et al. 2016), which may be a connection worth investigating. RNR-α overexpression suppresses the transformation potential by oncogenic Ras in NIH3t3 cells (Fan et al. 1997). Critically, knockdown of RNR-α led to elevation of transformation by oncogenic Ras. On the other hand, cells with low expression of *both* subunits of RNR and thymidylate synthase undergo oncogene senescence in a similar assay (re-overexpression of these three genes was sufficient to rescue senescence (Mannava et al. 2013)). RNR-α also appears to be downregulated in tumors more frequently than RNR-β and p53R2 (Ding et al. 2019). Overall, these data are indicative of tumor suppression behavior. It should further be noted that although RNR-α is essential, mice with a "single functional allele" of RNR-α (wt-RNR-α/RNR-α-W688G) are fully *via*ble, even though homozygous RNR-α-W688G mice are not. This finding indicates that catalytically active RNR-α is in excess (Specks et al. 2015), although it should be noted that the activity of the RNR-α-W688G mutant was not formally calculated, and hence, this protein may retain some activity. Further, consistent with RNR-α being a tumor suppressor, RNR-α lies in a region commonly lost in cancer. Consistent with these data, patients with "high" RNR-α expression show better overall pancreatic and non-small lung cancer survival rates than in those with "low" expression (Bepler et al. 2004; Zheng et al. 2007; Akita et al. 2009). These data further agree with data from mice overexpressing RNR-α, which showed overall reduced tumor propensity. It should be noted that these clinical reports are not necessarily well considered in the field, although nonetheless, they have never been disproven. Indeed, other studies have failed to find a strong link between RNR-α expression and survival (Grossi et al. 2015), and others have found an anticorrelation, albeit in later-stage cancers (Souglakos et al. 2008; Lee et al. 2010), or cancers different to those studied above (Jiang et al. 2017; Sagawa et al. 2017; Ma et al. 2017; Ding et al. 2019). There are numerous examples of

RNR-α overexpression in cancer, which have been reported through a vast number of mechanisms, including suppression of miRNAs (Fan et al. 2016). Other interesting data, which are contrary to data from those for RNR-β, show that RNR-α overexpression suppresses cancer invasion (Zhou et al. 1997; Fan et al. 1997; Gautam et al. 2003; Cao et al. 2003), migration, and metastasis.

Modes of RNR-α Tumor Suppression and Links to Approved Drug Mechanisms

One of the first links of RNR-α to tumor suppressor behavior was through PTEN, one of the tumor suppressors most commonly mutated/silenced (Chalhoub and Baker 2009) in cancers. PTEN is responsible for converting proliferation-stimulating phosphoinositol-3-phosphate signals to phosphoinositol-2-phosphate on the plasma membrane (although it may have other functions). In this way, PTEN serves as a negative regulator of PI3K/Akt signaling and other pro-proliferation signaling pathways. In colorectal cancer cells, RNR-α-mediated stimulation of PTEN was observed in the cell expression of both active and inactive RNR-α (Qi et al. 2015). These data agree with earlier data linking RNR-α to PTEN expression (Gautam et al. 2003) and stimulation. However, there has been little investigation into the correlation of RNR-α (Qi et al. 2015) and PTEN expression/mutation across numerous cancers, and furthermore, there has been little investigation into how RNR-α expression behaves in PTEN mutant cell lines (although lung cancer, where this link is arguably best studied, overall has relatively low PTEN mutation rates (Milella et al. 2015)). The mechanism of action is overall not well elucidated. As noted above, RNR-β is also linked to the upregulation of Akt activity, a major downstream effector negatively regulated by PTEN. This link has not been extensively investigated either.

Results of a yeast two-hybrid screening showed that RNR-α is an interactor with the nuclear protein ZRANB3 (Fu et al. 2018). ZRANB3 is best known as a regulator of the response to DNA damage during replication. ZRANB3 performs this role by interacting with K63-linked polyubiquitinated PCNA, a state of PCNA that occurs only on DNA damage. ZRANB3 interacts with ubiquitin K63-linked PCNA through two zinc finger domains and two functionally coupled PCNA interacting domains, the PIP box and a C-terminal APIM domain. Obviously, PCNA ubiquitination strengthens the ZRANB3 interaction, and by virtue of the PIP and APIM domains on ZRANB3, ZRANB3 can interact with non-ubiquitinated PCNA, which it does particularly in the later S-phase, likely in replication factories. Although ZRANB3 is not an essential gene, knockdown of ZRANB3 partially suppresses DNA synthesis in several cell lines. Cells lacking ZRANB3 are refractive to Myc and Ras-induced transformation and show reduced invasion, making ZRANB3 a prime example of non-oncogene addiction (Puccetti et al. 2019; Long et al. 2020).

Nuclear RNR-α interacts with ZRANB3 competitively with respect to PCNA, likely through the interaction with the APIM domain. Nuclear RNR-α (regardless of activity) also shows suppression of DNA synthesis, which is lost in ZRANB3 knockdown cells. Expression of nuclear RNR-α (either catalytically active or inactive) is the active agent in the suppression of RasG12V-induced transformation of NIH3t3 cells. However, this function of nuclear RNR-α only occurs when ZRANB3 is expressed. Thus, RNR-α is an inhibitor of ZRANB3's cancer-related functions (Long et al. 2020).

Hexamerization of RNR-α in cells (through treatment of cells with dA, ClF, or ClA) causes selective RNR-α nuclear translocation in numerous cells in less than 30 min. Neither RNR-β nor p53R2 translocates in this time period. This translocation is inhibited by IRBIT and does not occur in RNR-α-D57N-expressing cells, although this translocation is not impeded when nucleotide levels are elevated and does not depend on TIP60. In cells in which ZRANB3 has been knocked down or knocked out, the suppressive effects of 30-min ClF or dA treatment on DNA synthesis are negligible compared to cells that express ZRANB3. These data are consistent with the ZRANB3-PCNA complex being an early target of ClF or ClA through promoting RNR-α nuclear translocation. Such results also overall indicate that RNR-α_6 could readily be able to be mined for its gain of function properties and RNR activity regulatory behaviors. Conversely, compounds that block the ZRANB3-PCNA interaction could also readily be mined for drugs.

Outlook

We hope that we have outlined how interesting the RNR field is and how it is a bridge between so many different "worlds," from the RNA to the DNA world, the in vitro world to the cell, and indeed the world of screening to the clinic. RNR biology is certainly not easy: the chemistry is sui generis, the allostery is exquisite, and its role in multiple intersection moonlighting roles in cells is a particularly challenging conundrum. However, the field as a whole is beginning to bring forward tools to understand RNRs better, through the use of novel physical methods, as well as through functional mutants, siRNA, inducible expression, etc. It is critical to note that many of these mutants are either derived from, or understood on the back of excellent in vitro work that has ramifications to cellular and disease models. Thus, RNR makes for a particularly satisfying arena where basic science plays a huge role in guiding studies of disease. Throughout ongoing studies, we must continue to apply the collective knowledge that has been built up to study these complex enzymes to the best of our abilities. It is further important that we approach each problem from as many angles as possible: these days too often mutants are identified whose activity is not well characterized; pathways are proposed whose intersections are not investigated through rigorous genetic methods, as a function of cell cycle changes and for all RNR units; inhibitors are created that are not investigated thoroughly enough to prove their target. This requires RNR researchers to be multifaceted and to embrace complexity, not just through their own comfort zones, but also through using assays with which they may not be familiar.

References

Ahmad MF, Huff SE, Pink J et al (2015) Identification of non-nucleoside human ribonucleotide reductase modulators. J Med Chem 58:9498–9509. https://doi.org/10.1021/acs.jmedchem.5b00929

Ahmad MF, Alam I, Huff SE et al (2017) Potent competitive inhibition of human ribonucleotide reductase by a nonnucleoside small molecule. Proc Natl Acad Sci U S A 114:8241–8246. https://doi.org/10.1073/pnas.1620220114

Akdemir F, Christich A, Sogame N et al (2007) p53 directs focused genomic responses in Drosophila. Oncogene 26:5184–5193. https://doi.org/10.1038/sj.onc.1210328

Akita H, Zheng Z, Takeda Y et al (2009) Significance of RRM1 and ERCC1 expression in resectable pancreatic adenocarcinoma. Oncogene 28:2903–2909. https://doi.org/10.1038/onc.2009.158

Aly MS, Ashour MB, El Nahas SM, Abo Zeid MAF (2003) Genotoxicity and cytotoxicity of the anticancer drugs gemcitabine and cisplatin, separately and in combination: in vivo studies. J Biol Sci 3:961–972. https://doi.org/10.3923/jbs.2003.961.972

Ando N, Li H, Brignole EJ et al (2016) Allosteric inhibition of human ribonucleotide reductase by dATP entails the stabilization of a hexamer. Biochemistry 55:373–381. https://doi.org/10.1021/acs.biochem.5b01207

Apraiz A, Mitxelena J, Zubiaga A (2017) Studying cell cycle-regulated gene expression by two complementary cell synchronization protocols. JoVE. https://doi.org/10.3791/55745

Arnaoutov A, Dasso M (2014) IRBIT is a novel regulator of ribonucleotide reductase in higher eukaryotes. Science 345:1512–1515. https://doi.org/10.1126/science.1251550

Arnaoutov A, Lee H, Plevock Haase K et al (2020) IRBIT directs differentiation of intestinal stem cell progeny to maintain tissue homeostasis. iScience 23:100954. https://doi.org/10.1016/j.isci.2020.100954

Aydemir N, Çelikler S, Bilaloğlu R (2005) In vitro genotoxic effects of the anticancer drug gemcitabine in human lymphocytes. Mutat Res 582:35–41. https://doi.org/10.1016/j.mrgentox.2004.12.013

Aye Y, Brignole EJ, Long MJC et al (2012) Clofarabine targets the large subunit (α) of human ribonucleotide reductase in live cells by assembly into persistent hexamers. Chem Biol 19:799–805. https://doi.org/10.1016/j.chembiol.2012.05.015

Aye Y, Li M, Long MJC, Weiss RS (2015) Ribonucleotide reductase and cancer: biological mechanisms and targeted therapies. Oncogene 34:2011–2021. https://doi.org/10.1038/onc.2014.155

Aye Y, Stubbe J (2011) Clofarabine 5′-di and -triphosphates inhibit human ribonucleotide reductase by altering the quaternary structure of its large subunit. Proc Natl Acad Sci USA 108:9815–9820. https://doi.org/10.1073/pnas.1013274108

Bao Q, Shi Y (2007) Apoptosome: a platform for the activation of initiator caspases. Cell Death Differ 14:56–65. https://doi.org/10.1038/sj.cdd.4402028

Bepler G, Sharma S, Cantor A, et al (2004) *RRM1* and *PTEN* as prognostic parameters for overall and disease-free survival in patients with non–small-cell lung cancer. JCO 22:1878–1885. https://doi.org/10.1200/JCO.2004.12.002

Bo H, Weiyang C, Zhang Y (2017) p53R2 is oncogenic in human cervical cancer. Cell Cycle 16:1323–1324. https://doi.org/10.1080/15384101.2017.1337978

Bonifati S, Daly MB, St. Gelais C et al (2016) SAMHD1 controls cell cycle status, apoptosis and HIV-1 infection in monocytic THP-1 cells. Virology 495:92–100. https://doi.org/10.1016/j.virol.2016.05.002

Bonneau B, Ando H, Kawaai K et al (2016) IRBIT controls apoptosis by interacting with the Bcl-2 homolog, Bcl2l10, and by promoting ER-mitochondria contact. eLife 5:e19896. https://doi.org/10.7554/eLife.19896

Borel F, Lacroix FB, Margolis RL (2002) Prolonged arrest of mammalian cells at the G1/S boundary results in permanent S phase stasis. J Cell Sci 115:2829–2838. https://doi.org/10.1242/jcs.115.14.2829

Boronat S, Domènech A, Carmona M et al (2017) Lack of a peroxiredoxin suppresses the lethality of cells devoid of electron donors by channelling electrons to oxidized ribonucleotide reductase. PLoS Genet 13:e1006858. https://doi.org/10.1371/journal.pgen.1006858

Brignole EJ, Tsai K-L, Chittuluru J et al (2018) 3.3-Å resolution cryo-EM structure of human ribonucleotide reductase with substrate and allosteric regulators bound. eLife 7:e31502. https://doi.org/10.7554/eLife.31502

Buckland RJ, Watt DL, Chittoor B et al (2014) Increased and imbalanced dNTP pools symmetrically promote both leading and lagging strand replication infidelity. PLoS Genet 10:e1004846. https://doi.org/10.1371/journal.pgen.1004846

Burkhalter MD, Roberts SA, Havener JM, Ramsden DA (2009) Activity of ribonucleotide reductase helps determine how cells repair DNA double strand breaks. DNA Repair 8:1258–1263. https://doi.org/10.1016/j.dnarep.2009.07.009

Cao M-Y, Lee Y, Feng N-P et al (2003) Adenovirus-mediated ribonucleotide reductase R1 gene therapy of human colon adenocarcinoma. Clin Cancer Res 9:4553–4561

Caras IW, Martin DW (1988) Molecular cloning of the cDNA for a mutant mouse ribonucleotide reductase M1 that produces a dominant mutator phenotype in mammalian cells. Mol Cell Biol 8:2698–2704. https://doi.org/10.1128/mcb.8.7.2698-2704.1988

Çelik H, Sciandra M, Flashner B et al (2018) Clofarabine inhibits Ewing sarcoma growth through a novel molecular mechanism involving direct binding to CD99. Oncogene 37:2181–2196. https://doi.org/10.1038/s41388-017-0080-4

Chabes A, Stillman B (2007) Constitutively high dNTP concentration inhibits cell cycle progression and the DNA damage checkpoint in yeast Saccharomyces cerevisiae. Proc Natl Acad Sci 104:1183–1188. https://doi.org/10.1073/pnas.0610585104

Chabes A, Domkin V, Thelander L (1999) Yeast Sml1, a protein inhibitor of ribonucleotide reductase. J Biol Chem 274:36679–36683. https://doi.org/10.1074/jbc.274.51.36679

Chalhoub N, Baker SJ (2009) PTEN and the PI3-kinase pathway in cancer. Annu Rev Pathol Mech Dis 4:127–150. https://doi.org/10.1146/annurev.pathol.4.110807.092311

Chen C-W, Tsao N, Huang L-Y et al (2016) The impact of dUTPase on ribonucleotide reductase-induced genome instability in cancer cells. Cell Rep 16:1287–1299. https://doi.org/10.1016/j.celrep.2016.06.094

Chen J, Li S, Xiao Y et al (2017) p53R2 as a novel prognostic biomarker in nasopharyngeal carcinoma. BMC Cancer 17:846. https://doi.org/10.1186/s12885-017-3858-4

Chen PY-T, Funk MA, Brignole EJ, Drennan CL (2018) Disruption of an oligomeric interface prevents allosteric inhibition of Escherichia coli class Ia ribonucleotide reductase. J Biol Chem 293:10404–10412. https://doi.org/10.1074/jbc.RA118.002569

Chen G, Luo Y, Warncke K et al (2019a) Acetylation regulates ribonucleotide reductase activity and cancer cell growth. Nat Commun 10:3213. https://doi.org/10.1038/s41467-019-11214-9

Chen W, Yang L, Xu L et al (2019b) Bioinformatics analysis revealing prognostic significance of *RRM2* gene in breast cancer. Biosci Rep 39:BSR20182062. https://doi.org/10.1042/BSR20182062

Chimploy K, Mathews CK (2001) Mouse ribonucleotide reductase control. J Biol Chem 276:7093–7100. https://doi.org/10.1074/jbc.M006232200

Chimploy K, Tassotto ML, Mathews CK (2000) Ribonucleotide reductase, a possible agent in deoxyribonucleotide pool asymmetries induced by hypoxia. J Biol Chem 275:39267–39271. https://doi.org/10.1074/jbc.M006233200

Cho EC, Yen Y (2016) Novel regulators and molecular mechanisms of p53R2 and its disease relevance. Biochimie 123:81–84. https://doi.org/10.1016/j.biochi.2016.01.008

Cho E-C, Kuo M-L, Liu X et al (2014) Tumor suppressor FOXO3 regulates ribonucleotide reductase subunit RRM2B and impacts on survival of cancer patients. Oncotarget 5:4834–4844. https://doi.org/10.18632/oncotarget.2044

Cooper GM (2000) The eukaryotic cell cycle. In: The cell: a molecular approach, 2nd edn. Sinauer Associates, Sunderland, MA

Cotruvo JA, Stubbe J (2011) Class I ribonucleotide reductases: metallocofactor assembly and repair in vitro and in vivo. Annu Rev Biochem 80:733–767. https://doi.org/10.1146/annurev-biochem-061408-095817

Daddacha W, Koyen AE, Bastien AJ et al (2017) SAMHD1 promotes DNA end resection to facilitate DNA repair by homologous recombination. Cell Rep 20:1921–1935. https://doi.org/10.1016/j.celrep.2017.08.008

Daly MB, Roth ME, Bonnac L et al (2016) Dual anti-HIV mechanism of clofarabine. Retrovirology 13:20. https://doi.org/10.1186/s12977-016-0254-0

Danilova N, Bibikova E, Covey TM et al (2014) The role of DNA damage response in zebrafish and cellular models of diamond Blackfan anemia. Dis Models Mech. https://doi.org/10.1242/dmm.015495

Deforche K, Camacho R, Laethem KV et al (2007) Estimating the relative contribution of dNTP Pool imbalance and APOBEC3G/3F editing to HIV evolution in vivo. J Comput Biol 14:1105–1114. https://doi.org/10.1089/cmb.2007.0073

Ding Y, Zhong T, Wang M et al (2019) Integrative analysis reveals across-cancer expression patterns and clinical relevance of ribonucleotide reductase in human cancers. Front Oncol 9:956. https://doi.org/10.3389/fonc.2019.00956

Domkin V, Chabes A (2014) Phosphines are ribonucleotide reductase reductants that act *via* C-terminal cysteines similar to thioredoxins and glutaredoxins. Sci Rep 4:5539. https://doi.org/10.1038/srep05539

Dyavaiah M, Rooney JP, Chittur SV et al (2011) Autophagy-dependent regulation of the DNA damage response protein ribonucleotide reductase 1. Mol Cancer Res 9:462–475. https://doi.org/10.1158/1541-7786.MCR-10-0473

Elledge SJ, Davis RW (1990) Two genes differentially regulated in the cell cycle and by DNA-damaging agents encode alternative regulatory subunits of ribonucleotide reductase. Genes Dev 4:740–751. https://doi.org/10.1101/gad.4.5.740

Fairman JW, Wijerathna SR, Ahmad MF et al (2011) Structural basis for allosteric regulation of human ribonucleotide reductase by nucleotide-induced oligomerization. Nat Struct Mol Biol 18:316–322. https://doi.org/10.1038/nsmb.2007

Fan H, Huang A, Villegas C, Wright JA (1997) The R1 component of mammalian ribonucleotide reductase has malignancy-suppressing activity as demonstrated by gene transfer experiments. Proc Natl Acad Sci 94:13181–13186. https://doi.org/10.1073/pnas.94.24.13181

Fan P, Liu L, Yin Y et al (2016) MicroRNA-101-3p reverses gemcitabine resistance by inhibition of ribonucleotide reductase M1 in pancreatic cancer. Cancer Lett 373:130–137. https://doi.org/10.1016/j.canlet.2016.01.038

Fang Z, Gong C, Liu H et al (2015) E2F1 promote the aggressiveness of human colorectal cancer by activating the ribonucleotide reductase small subunit M2. Biochem Biophys Res Commun 464:407–415. https://doi.org/10.1016/j.bbrc.2015.06.103

Fang Z, Lin A, Chen J et al (2016) CREB1 directly activates the transcription of ribonucleotide reductase small subunit M2 and promotes the aggressiveness of human colorectal cancer. Oncotarget 7:78055–78068. https://doi.org/10.18632/oncotarget.12938

Fasullo M, Tsaponina O, Sun M, Chabes A (2010) Elevated dNTP levels suppress hyper-recombination in Saccharomyces cerevisiae S-phase checkpoint mutants. Nucleic Acids Res 38:1195–1203. https://doi.org/10.1093/nar/gkp1064

Fiorentini P, Huang KN, Tishkoff DX et al (1997) Exonuclease I of Saccharomyces cerevisiae functions in mitotic recombination in vivo and in vitro. Mol Cell Biol 17:2764–2773. https://doi.org/10.1128/MCB.17.5.2764

Fleck O, Vejrup-Hansen R, Watson A et al (2013) Spd1 accumulation causes genome instability independently of ribonucleotide reduction but functions to protect the genome when deoxynucleotide pools are elevated. J Cell Sci. https://doi.org/10.1242/jcs.132837

Foskolou IP, Jorgensen C, Leszczynska KB et al (2017) Ribonucleotide reductase requires subunit switching in hypoxia to maintain DNA replication. Mol Cell 66:206–220.e9. https://doi.org/10.1016/j.molcel.2017.03.005

Franzolin E, Pontarin G, Rampazzo C et al (2013) The deoxynucleotide triphosphohydrolase SAMHD1 is a major regulator of DNA precursor pools in mammalian cells. Proc Natl Acad Sci 110:14272–14277. https://doi.org/10.1073/pnas.1312033110

Fu Y, Long MJC, Rigney M et al (2013) Uncoupling of allosteric and oligomeric regulation in a functional hybrid enzyme constructed from Escherichia coli and human ribonucleotide reductase. Biochemistry 52:7050–7059. https://doi.org/10.1021/bi400781z

Fu Y, Lin H-Y, Wisitpitthaya S et al (2014) A fluorimetric readout reporting the kinetics of nucleotide-induced human ribonucleotide reductase oligomerization. ChemBioChem 15:2598–2604. https://doi.org/10.1002/cbic.201402368

Fu Y, Long MJC, Wisitpitthaya S et al (2018) Nuclear RNR-α antagonizes cell proliferation by directly inhibiting ZRANB3. Nat Chem Biol 14:943–954. https://doi.org/10.1038/s41589-018-0113-5

Gan H, Yu C, Devbhandari S et al (2017) Checkpoint kinase Rad53 couples leading- and lagging-strand DNA synthesis under replication stress. Mol Cell 68:446–455.e3. https://doi.org/10.1016/j.molcel.2017.09.018

Gandhi V, Plunkett W (2006) Clofarabine and nelarabine: two new purine nucleoside analogs. Curr Opin Oncol 18:584–590. https://doi.org/10.1097/01.cco.0000245326.65152.af

Gautam A, Li Z-R, Bepler G (2003) RRM1-induced metastasis suppression through PTEN-regulated pathways. Oncogene 22:2135–2142. https://doi.org/10.1038/sj.onc.1206232

Ge J, Yu G, Ator MA, Stubbe J (2003) Pre-steady-state and steady-state kinetic analysis of *E. coli* class I ribonucleotide reductase. Biochemistry 42:10071–10083. https://doi.org/10.1021/bi034374r

Glover TW, Wilson TE, Arlt MF (2017) Fragile sites in cancer: more than meets the eye. Nat Rev Cancer 17:489–501. https://doi.org/10.1038/nrc.2017.52

Gon S, Napolitano R, Rocha W et al (2011) Increase in dNTP pool size during the DNA damage response plays a key role in spontaneous and induced-mutagenesis in Escherichia coli. Proc Natl Acad Sci 108:19311–19316. https://doi.org/10.1073/pnas.1113664108

Goss KL, Koppenhafer SL, Waters T et al (2021) The translational repressor 4E-BP1 regulates RRM2 levels and functions as a tumor suppressor in Ewing sarcoma tumors. Oncogene 40:564–577. https://doi.org/10.1038/s41388-020-01552-0

Greene BL, Nocera DG, Stubbe J (2018) Basis of dATP inhibition of RNRs. J Biol Chem 293:10413–10414. https://doi.org/10.1074/jbc.H118.003717

Grossi F, Dal Bello MG, Salvi S et al (2015) Expression of ribonucleotide reductase subunit-2 and thymidylate synthase correlates with poor prognosis in patients with resected stages I–III non-small cell lung cancer. Dis Markers 2015:1–18. https://doi.org/10.1155/2015/302649

Guzmán EC, Caballero JL, Jiménez-Sánchez A (2002) Ribonucleoside diphosphate reductase is a component of the replication hyperstructure in Escherichia coli: ribonucleoside diphosphate reductase in the replication complex. Mol Microbiol 43:487–495. https://doi.org/10.1046/j.1365-2958.2002.02761.x

Håkansson P, Hofer A, Thelander L (2006) Regulation of mammalian ribonucleotide reduction and dNTP pools after DNA damage and in resting cells. J Biol Chem 281:7834–7841. https://doi.org/10.1074/jbc.M512894200

Hämäläinen RH, Landoni JC, Ahlqvist KJ et al (2019) Defects in mtDNA replication challenge nuclear genome stability through nucleotide depletion and provide a unifying mechanism for mouse progerias. Nat Metab 1:958–965. https://doi.org/10.1038/s42255-019-0120-1

Hogbom M, Galander M, Andersson M et al (2003) Displacement of the tyrosyl radical cofactor in ribonucleotide reductase obtained by single-crystal high-field EPR and 1.4-A x-ray data. Proc Natl Acad Sci 100:3209–3214. https://doi.org/10.1073/pnas.0536684100

Hoyer J, Ekici AB, Endele S et al (2012) Haploinsufficiency of ARID1B, a member of the SWI/SNF-A chromatin-remodeling complex, is a frequent cause of intellectual disability. Am J Hum Genet 90:565–572. https://doi.org/10.1016/j.ajhg.2012.02.007

Hu C-M, Yeh M-T, Tsao N et al (2012) Tumor cells require thymidylate kinase to prevent dUTP incorporation during DNA repair. Cancer Cell 22:36–50. https://doi.org/10.1016/j.ccr.2012.04.038

Huberts DHEW, van der Klei IJ (2010) Moonlighting proteins: an intriguing mode of multitasking. Biochim Biophys Acta 1803:520–525. https://doi.org/10.1016/j.bbamcr.2010.01.022

Huff SE, Mohammed FA, Yang M et al (2018) Structure-guided synthesis and mechanistic studies reveal sweetspots on naphthyl salicyl hydrazone scaffold as non-nucleosidic competitive, reversible inhibitors of human ribonucleotide reductase. J Med Chem 61:666–680. https://doi.org/10.1021/acs.jmedchem.7b00530

Jeffery CJ (2020) Enzymes, pseudoenzymes, and moonlighting proteins: diversity of function in protein superfamilies. FEBS J 287:4141–4149. https://doi.org/10.1111/febs.15446

Ji X, Tang C, Zhao Q et al (2014) Structural basis of cellular dNTP regulation by SAMHD1. Proc Natl Acad Sci 111:E4305–E4314. https://doi.org/10.1073/pnas.1412289111

Jiang K, Zhi T, Xu W et al (2017) MicroRNA-1468-5p inhibits glioma cell proliferation and induces cell cycle arrest by targeting RRM1. Am J Cancer Res 7:784–800

Jonna VR, Crona M, Rofougaran R et al (2015) Diversity in overall activity regulation of ribonucleotide reductase. J Biol Chem 290:17339–17348. https://doi.org/10.1074/jbc.M115.649624

Kang G, Taguchi AT, Stubbe J, Drennan CL (2020) Structure of a trapped radical transfer pathway within a ribonucleotide reductase holocomplex. Science 368:424–427. https://doi.org/10.1126/science.aba6794

Kauppi B, Nielsen BB, Ramaswamy S et al (1996) The three-dimensional structure of mammalian ribonucleotide reductase protein R2 reveals a more-accessible iron-radical site than Escherichia coliR2. J Mol Biol 262:706–720. https://doi.org/10.1006/jmbi.1996.0546

Ke P-Y, Chang Z-F (2004) Mitotic degradation of human thymidine kinase 1 is dependent on the anaphase-promoting complex/cyclosome-Cdh1-mediated pathway. Mol Cell Biol 24:514–526. https://doi.org/10.1128/MCB.24.2.514-526.2004

Knappenberger AJ, Ahmad MF, Viswanathan R et al (2016) Nucleoside analogue triphosphates allosterically regulate human ribonucleotide reductase and identify chemical determinants that drive substrate specificity. Biochemistry 55:5884–5896. https://doi.org/10.1021/acs.biochem.6b00594

Knappenberger AJ, Grandhi S, Sheth R et al (2017) Phylogenetic sequence analysis and functional studies reveal compensatory amino acid substitutions in loop 2 of human ribonucleotide reductase. J Biol Chem 292:16463–16476. https://doi.org/10.1074/jbc.M117.798769

Knox JJ, Hotte SJ, Kollmannsberger C et al (2007) Phase II study of Triapine® in patients with metastatic renal cell carcinoma: a trial of the National Cancer Institute of Canada clinical trials group (NCIC IND.161). Investig New Drugs 25:471–477. https://doi.org/10.1007/s10637-007-9044-9

Kong Z, Jia S, Chabes AL et al (2018) Simultaneous determination of ribonucleoside and deoxyribonucleoside triphosphates in biological samples by hydrophilic interaction liquid chromatography coupled with tandem mass spectrometry. Nucleic Acids Res 46:e66–e66. https://doi.org/10.1093/nar/gky203

Kunos CA, Andrews SJ, Moore KN et al (2019) Randomized phase II trial of triapine-cisplatin-radiotherapy for locally advanced stage uterine cervix or vaginal cancers. Front Oncol 9:1067. https://doi.org/10.3389/fonc.2019.01067

Lam MH, Liu Q, Elledge SJ, Rosen JM (2004) Chk1 is haploinsufficient for multiple functions critical to tumor suppression. Cancer Cell 6:45–59. https://doi.org/10.1016/j.ccr.2004.06.015

Lamb NA, Bard J, Buck MJ, Surtees JA (2020) A targeted next generation sequencing approach to develop robust, genotype-specific mutation profiles and uncover novel variants in *Saccharomyces cerevisiae*. Genetics

Le François BG, Maroun JA, Birnboim HC (2007) Expression of thymidylate synthase in human cells is an early G1 event regulated by CDK4 and p16INK4A but not E2F. Br J Cancer 97:1242–1250. https://doi.org/10.1038/sj.bjc.6604020

Lee YD, Elledge SJ (2006) Control of ribonucleotide reductase localization through an anchoring mechanism involving Wtm1. Genes Dev 20:334–344. https://doi.org/10.1101/gad.1380506

Lee JJ, Maeng CH, Baek SK et al (2010) The immunohistochemical overexpression of ribonucleotide reductase regulatory subunit M1 (RRM1) protein is a predictor of shorter survival to gemcitabine-based chemotherapy in advanced non-small cell lung cancer (NSCLC). Lung Cancer 70:205–210. https://doi.org/10.1016/j.lungcan.2010.02.005

Lee M-H, Wang L, Chang Z-F (2014) The contribution of mitochondrial thymidylate synthesis in preventing the nuclear genome stress. Nucleic Acids Res 42:4972–4984. https://doi.org/10.1093/nar/gku152

Leeds JM, Slabaugh MB, Mathews CK (1985) DNA precursor pools and ribonucleotide reductase activity: distribution between the nucleus and cytoplasm of mammalian cells. Mol Cell Biol 5: 3443–3450. https://doi.org/10.1128/MCB.5.12.3443

Liew LP, Lim ZY, Cohen M et al (2016) Hydroxyurea-mediated cytotoxicity without inhibition of ribonucleotide reductase. Cell Rep 17:1657–1670. https://doi.org/10.1016/j.celrep.2016.10.024

Lim AZ, McFarland R, Taylor RW, Gorman GS (1993) RRM2B mitochondrial DNA maintenance defects. In: Adam MP, Ardinger HH, Pagon RA et al (eds) GeneReviews®. University of Washington, Seattle, WA

Liu X, Lai L, Wang X et al (2011) Ribonucleotide reductase small subunit M2B prognoses better survival in colorectal cancer. Cancer Res 71:3202–3213. https://doi.org/10.1158/0008-5472.CAN-11-0054

Long MJC, Zhao Y, Aye Y (2020) Clofarabine commandeers the RNR-α-ZRANB3 nuclear signaling axis. Cell Chem Biol 27:122–133.e5. https://doi.org/10.1016/j.chembiol.2019.11.012

Lopez-Contreras AJ, Specks J, Barlow JH et al (2015) Increased *Rrm2* gene dosage reduces fragile site breakage and prolongs survival of ATR mutant mice. Genes Dev 29:690–695. https://doi.org/10.1101/gad.256958.114

Lou M, Liu Q, Ren G et al (2017) Physical interaction between human ribonucleotide reductase large subunit and thioredoxin increases colorectal cancer malignancy. J Biol Chem 292:9136–9149. https://doi.org/10.1074/jbc.M117.783365

Lundin D, Berggren G, Logan D, Sjöberg B-M (2015) The origin and evolution of ribonucleotide reduction. Life 5:604–636. https://doi.org/10.3390/life5010604

Ma CX, Suman V, Goetz MP et al (2017) A phase II trial of neoadjuvant MK-2206, an AKT inhibitor, with anastrozole in clinical stage II or III *PIK3CA*-mutant ER-positive and HER2-negative breast cancer. Clin Cancer Res 23:6823–6832. https://doi.org/10.1158/1078-0432.CCR-17-1260

Ma C, Luo H, Cao J et al (2020) Independent prognostic implications of *RRM2* in lung adenocarcinoma. J Cancer 11:7009–7022. https://doi.org/10.7150/jca.47895

Mah V, Alavi M, Márquez-Garbán DC et al (2015) Ribonucleotide reductase subunit M2 predicts survival in subgroups of patients with non-small cell lung carcinoma: effects of gender and smoking status. PLoS One 10:e0127600. https://doi.org/10.1371/journal.pone.0127600

Malik L, Zwiebel A, Cooper J (2018) A phase I pharmacokinetic and pharmacodynamic study of GTI-2040 in combination with gemcitabine in patients with solid tumors. Cancer Chemother Pharmacol 82:533–539. https://doi.org/10.1007/s00280-018-3647-z

Mannava S, Moparthy KC, Wheeler LJ et al (2013) Depletion of Deoxyribonucleotide pools is an endogenous source of DNA damage in cells undergoing oncogene-induced senescence. Am J Pathol 182:142–151. https://doi.org/10.1016/j.ajpath.2012.09.011

Mao Z, Bozzella M, Seluanov A, Gorbunova V (2008) DNA repair by nonhomologous end joining and homologous recombination during cell cycle in human cells. Cell Cycle 7:2902–2906. https://doi.org/10.4161/cc.7.18.6679

Maori E, Navarro IC, Boncristiani H et al (2019) A secreted RNA binding protein forms RNA-stabilizing granules in the Honeybee Royal Jelly. Mol Cell 74:598–608.e6. https://doi.org/10.1016/j.molcel.2019.03.010

Mazzu YZ, Armenia J, Chakraborty G et al (2019) A novel mechanism driving poor-prognosis prostate cancer: overexpression of the DNA repair gene, ribonucleotide reductase small subunit M2 (RRM2). Clin Cancer Res 25:4480–4492. https://doi.org/10.1158/1078-0432.CCR-18-4046

Mertz W (1974) The newer essential trace elements, chromium, tin, vanadium, nickel and silicon. Proc Nutr Soc 33:307–313. https://doi.org/10.1079/PNS19740054

Mesa RA, Loegering D, Powell HL et al (2005) Heat shock protein 90 inhibition sensitizes acute myelogenous leukemia cells to cytarabine. Blood 106:318–327. https://doi.org/10.1182/blood-2004-09-3523

Miazzi C, Ferraro P, Pontarin G et al (2014) Allosteric regulation of the human and mouse deoxyribonucleotide triphosphohydrolase sterile α-motif/histidine-aspartate domain-containing protein 1 (SAMHD1). J Biol Chem 289:18339–18346. https://doi.org/10.1074/jbc.M114.571091

Milella M, Falcone I, Conciatori F et al (2015) PTEN: multiple functions in human malignant tumors. Front Oncol 5. https://doi.org/10.3389/fonc.2015.00024

Minnihan EC, Nocera DG, Stubbe J (2013) Reversible, long-range radical transfer in E. coli class Ia ribonucleotide reductase. Acc Chem Res 46:2524–2535. https://doi.org/10.1021/ar4000407

Misko TA, Wijerathna SR, Radivoyevitch T et al (2016) Inhibition of yeast ribonucleotide reductase by Sml1 depends on the allosteric state of the enzyme. FEBS Lett 590:1704–1712. https://doi.org/10.1002/1873-3468.12207

Misko TA, Liu Y-T, Harris ME et al (2019) Structure-guided design of anti-cancer ribonucleotide reductase inhibitors. J Enzyme Inhib Med Chem 34:438–450. https://doi.org/10.1080/14756366.2018.1545226

Mottram JC (1934) On the correlation between malignancy and the rate of growth of tar warts in mice. Am J Cancer 22:801–830. https://doi.org/10.1158/ajc.1934.801

Mun BJ, Mathews CK (1991) Cell cycle-dependent variations in deoxyribonucleotide metabolism among Chinese hamster cell lines bearing the Thy-mutator phenotype. Mol Cell Biol 11:20–26. https://doi.org/10.1128/MCB.11.1.20

Nagel R, Semenova EA, Berns A (2016) Drugging the addict: non-oncogene addiction as a target for cancer therapy. EMBO Rep 17(1516–1531):10.15252/embr.201643030

Nestoras K, Mohammed AH, Schreurs A-S et al (2010) Regulation of ribonucleotide reductase by Spd1 involves multiple mechanisms. Genes Dev 24:1145–1159. https://doi.org/10.1101/gad.561910

Niida H, Katsuno Y, Sengoku M et al (2010) Essential role of Tip60-dependent recruitment of ribonucleotide reductase at DNA damage sites in DNA repair during G1 phase. Genes Dev 24: 333–338. https://doi.org/10.1101/gad.1863810

Nunes C, Depestel L, Mus L, et al (2020) RRM2 is a target for synthetic lethal interactions with replication stress checkpoint addiction in high-risk neuroblastoma. Cancer Biol

Offenbacher AR, Barry BA (2020) A proton wire mediates proton coupled electron transfer from hydroxyurea and other hydroxamic acids to tyrosyl radical in class Ia ribonucleotide reductase. J Phys Chem B 124:345–354. https://doi.org/10.1021/acs.jpcb.9b08587

Okat Z (2018) Clinical importance of glutathione-s-transferase enzyme polymorphisms in cancer. IPMRJ 3:10.15406/ipmrj.2018.03.00153

Okumura H, Natsugoe S, Yokomakura N et al (2006) Expression of p53R2 is related to prognosis in patients with esophageal squamous cell carcinoma. Clin Cancer Res 12:3740–3745. https://doi.org/10.1158/1078-0432.CCR-05-2416

Osako Y, Yoshino H, Sakaguchi T et al (2019) Potential tumor-suppressive role of microRNA-99a-3p in sunitinib-resistant renal cell carcinoma cells through the regulation of RRM2. Int J Oncol. https://doi.org/10.3892/ijo.2019.4736

Pajalunga D, Franzolin E, Stevanoni M et al (2017) A defective dNTP pool hinders DNA replication in cell cycle-reactivated terminally differentiated muscle cells. Cell Death Differ 24:774–784. https://doi.org/10.1038/cdd.2017.4

Parker WB, Shaddix SC, Chang CH et al (1991) Effects of 2-chloro-9-(2-deoxy-2-fluoro-beta-D-arabinofuranosyl)adenine on K562 cellular metabolism and the inhibition of human ribonucleotide reductase and DNA polymerases by its 5′-triphosphate. Cancer Res 51:2386–2394

Peris D, Janssen K, Barthel HJ et al (2020) DNA from resin-embedded organisms: past, present and future. PLoS One 15:e0239521. https://doi.org/10.1371/journal.pone.0239521

Pitsch S, Wendeborn S, Jaun B, Eschenmoser A (1993) Why pentose- and not hexose-nucleic acids? Part VII. Pyranosyl-RNA ('P-RNA'). Preliminary communication. Helv Chim Acta 76: 2161–2183. https://doi.org/10.1002/hlca.19930760602

Poole AM, Horinouchi N, Catchpole RJ et al (2014) The case for an early biological origin of DNA. J Mol Evol 79:204–212. https://doi.org/10.1007/s00239-014-9656-6

Prakasha Gowda AS, Polizzi JM, Eckert KA, Spratt TE (2010) Incorporation of gemcitabine and cytarabine into DNA by DNA polymerase β and ligase III/XRCC1. Biochemistry 49:4833–4840. https://doi.org/10.1021/bi100200c

Puccetti MV, Adams CM, Kushinsky S, Eischen CM (2019) Smarcal1 and Zranb3 protect replication forks from Myc-induced DNA replication stress. Cancer Res 79:1612–1623. https://doi.org/10.1158/0008-5472.CAN-18-2705

Qi H, Lou M, Chen Y et al (2015) Non-enzymatic action of RRM1 protein upregulates PTEN leading to inhibition of colorectal cancer metastasis. Tumor Biol 36:4833–4842. https://doi.org/10.1007/s13277-015-3137-4

Rampazzo C, Miazzi C, Franzolin E et al (2010) Regulation by degradation, a cellular defense against deoxyribonucleotide pool imbalances. Mutat Res 703:2–10. https://doi.org/10.1016/j.mrgentox.2010.06.002

Rasmussen RD, Gajjar MK, Tuckova L et al (2016) BRCA1-regulated RRM2 expression protects glioblastoma cells from endogenous replication stress and promotes tumorigenicity. Nat Commun 7:13398. https://doi.org/10.1038/ncomms13398

Reichard P, Baldesten A, Rutberg L (1961) Formation of deoxycytidine phosphates from cytidine phosphates in extracts from Escherichia coli. J Biol Chem 236:1150–1157. https://doi.org/10.1016/S0021-9258(18)64258-9

Reid G, Wallant NC, Patel R et al (2009) Potent subunit-specific effects on cell growth and drug sensitivity from optimised siRNA-mediated silencing of ribonucleotide reductase. J RNAi Gene Silencing 5:321–330

Richards RI (2001) Fragile and unstable chromosomes in cancer: causes and consequences. Trends Genet 17:339–345. https://doi.org/10.1016/S0168-9525(01)02303-4

Roberts EA, Deretic V (2008) Autophagic proteolysis of Long-lived proteins in nonliver cells. In: Deretic V (ed) Autophagosome and phagosome. Humana Press, Totowa, NJ, pp 111–117

Rudd SG, Tsesmetzis N, Sanjiv K et al (2020) Ribonucleotide reductase inhibitors suppress SAMHD 1 ara-CTP ase activity enhancing cytarabine efficacy. EMBO Mol Med 12: 10.15252/emmm.201910419

Sabouri N, Viberg J, Goyal DK et al (2008) Evidence for lesion bypass by yeast replicative DNA polymerases during DNA damage. Nucleic Acids Res 36:5660–5667. https://doi.org/10.1093/nar/gkn555

Sagawa M, Ohguchi H, Harada T et al (2017) Ribonucleotide reductase catalytic subunit M1 (RRM1) as a novel therapeutic target in multiple myeloma. Clin Cancer Res 23:5225–5237. https://doi.org/10.1158/1078-0432.CCR-17-0263

Sampaio-Marques B, Guedes A, Vasilevskiy I et al (2019) α-Synuclein toxicity in yeast and human cells is caused by cell cycle re-entry and autophagy degradation of ribonucleotide reductase 1. Aging Cell 18. https://doi.org/10.1111/acel.12922

Sandvik GK, Tomter AB, Bergan J et al (2012) Studies of ribonucleotide reductase in crucian carp—an oxygen dependent enzyme in an anoxia tolerant vertebrate. PLoS One 7:e42784. https://doi.org/10.1371/journal.pone.0042784

Schmidt TT, Sharma S, Reyes GX et al (2019) A genetic screen pinpoints ribonucleotide reductase residues that sustain dNTP homeostasis and specifies a highly mutagenic type of dNTP imbalance. Nucleic Acids Res 47:237–252. https://doi.org/10.1093/nar/gky1154

Seedhouse C, Grundy M, Shang S et al (2009) Impaired S-phase arrest in acute myeloid leukemia cells with a FLT3 internal tandem duplication treated with clofarabine. Clin Cancer Res 15: 7291–7298. https://doi.org/10.1158/1078-0432.CCR-09-1222

Sengupta R (2014) Thioredoxin and glutaredoxin-mediated redox regulation of ribonucleotide reductase. WJBC 5:68. https://doi.org/10.4331/wjbc.v5.i1.68

Sengupta R, Coppo L, Mishra P, Holmgren A (2019) Glutathione-glutaredoxin is an efficient electron donor system for mammalian p53R2–R1-dependent ribonucleotide reductase. J Biol Chem 294:12708–12716. https://doi.org/10.1074/jbc.RA119.008752

Shu Z, Li Z, Huang H et al (2020) Cell-cycle-dependent phosphorylation of RRM1 ensures efficient DNA replication and regulates cancer vulnerability to ATR inhibition. Oncogene 39:5721–5733. https://doi.org/10.1038/s41388-020-01403-y

Singh S (2015) Cytoprotective and regulatory functions of glutathione S-transferases in cancer cell proliferation and cell death. Cancer Chemother Pharmacol 75:1–15. https://doi.org/10.1007/s00280-014-2566-x

Song Y, Marmion RA, Park JO et al (2017) Dynamic control of dNTP synthesis in early embryos. Dev Cell 42:301–308.e3. https://doi.org/10.1016/j.devcel.2017.06.013

Souglakos J, Boukovinas I, Taron M et al (2008) Ribonucleotide reductase subunits M1 and M2 mRNA expression levels and clinical outcome of lung adenocarcinoma patients treated with docetaxel/gemcitabine. Br J Cancer 98:1710–1715. https://doi.org/10.1038/sj.bjc.6604344

Specks J, Lecona E, Lopez-Contreras AJ, Fernandez-Capetillo O (2015) A single conserved residue mediates binding of the ribonucleotide reductase catalytic subunit RRM1 to RRM2 and is essential for mouse development. Mol Cell Biol 35:2910–2917. https://doi.org/10.1128/MCB.00475-15

Stodola JL, Burgers PM (2016) Resolving individual steps of Okazaki-fragment maturation at a millisecond timescale. Nat Struct Mol Biol 23:402–408. https://doi.org/10.1038/nsmb.3207

Stubbe J, Nocera DG, Yee CS, Chang MCY (2003) Radical initiation in the class I ribonucleotide reductase: long-range proton-coupled electron transfer? Chem Rev 103:2167–2202. https://doi.org/10.1021/cr020421u

Sumita K, Lo Y-H, Takeuchi K et al (2016) The lipid kinase PI5P4Kβ is an intracellular GTP sensor for metabolism and tumorigenesis. Mol Cell 61:187–198. https://doi.org/10.1016/j.molcel.2015.12.011

Sun H, Yang B, Zhang H et al (2019) RRM2 is a potential prognostic biomarker with functional significance in glioma. Int J Biol Sci 15:533–543. https://doi.org/10.7150/ijbs.30114

Szikriszt B, Póti Á, Pipek O et al (2016) A comprehensive survey of the mutagenic impact of common cancer cytotoxics. Genome Biol 17:99. https://doi.org/10.1186/s13059-016-0963-7

Tarangelo A, Dixon S (2018) The p53-p21 pathway inhibits ferroptosis during metabolic stress. Oncotarget 9(24572–24573):10.18632/oncotarget.25362

Tebbi A, Guittet O, Tuphile K et al (2015) Caspase-dependent proteolysis of human ribonucleotide reductase small subunits R2 and p53R2 during apoptosis. J Biol Chem 290:14077–14090. https://doi.org/10.1074/jbc.M115.649640

Teicher BA, Polley E, Kunkel M et al (2015) Sarcoma cell line screen of oncology drugs and investigational agents identifies patterns associated with gene and microRNA expression. Mol Cancer Ther 14:2452–2462. https://doi.org/10.1158/1535-7163.MCT-15-0074

Thomas WC, Brooks FP, Burnim AA et al (2019) Convergent allostery in ribonucleotide reductase. Nat Commun 10:2653. https://doi.org/10.1038/s41467-019-10568-4

Tian H, Ge C, Li H et al (2014) Ribonucleotide reductase M2B inhibits cell migration and spreading by early growth response protein 1-mediated phosphatase and tensin homolog/Akt1 pathway in hepatocellular carcinoma. Hepatology 59:1459–1470. https://doi.org/10.1002/hep.26929

Tramentozzi E, Ferraro P, Hossain M et al (2018) The dNTP triphosphohydrolase activity of SAMHD1 persists during S-phase when the enzyme is phosphorylated at T592. Cell Cycle 17:1102–1114. https://doi.org/10.1080/15384101.2018.1480216

Tu M, Li H, Lv N et al (2017) Vasohibin 2 reduces chemosensitivity to gemcitabine in pancreatic cancer cells *via* Jun proto-oncogene dependent transactivation of ribonucleotide reductase regulatory subunit M2. Mol Cancer 16:66. https://doi.org/10.1186/s12943-017-0619-6

Uramoto H, Sugio K, Oyama T et al (2006) P53R2, p53 inducible ribonucleotide reductase gene, correlated with tumor progression of non-small cell lung cancer. Anticancer Res 26:983–988

Wang N, Zhan T, Ke T et al (2014) Increased expression of RRM2 by human papillomavirus E7 oncoprotein promotes angiogenesis in cervical cancer. Br J Cancer 110:1034–1044. https://doi.org/10.1038/bjc.2013.817

Wang N, Li Y, Zhou J (2018) Downregulation of ribonucleotide reductase subunits M2 induces apoptosis and G1 arrest of cervical cancer cells. Oncol Lett. https://doi.org/10.3892/ol.2018.7806

Watson RA, Offenbacher AR, Barry BA (2021) Detection of catalytically linked conformational changes in wild-type class Ia ribonucleotide reductase using reaction-induced FTIR spectroscopy. J Phys Chem B 125:8362–8372. https://doi.org/10.1021/acs.jpcb.1c03038

Wisitpitthaya S, Zhao Y, Long MJC et al (2016) Cladribine and fludarabine nucleotides induce distinct hexamers defining a common mode of reversible RNR inhibition. ACS Chem Biol 11:2021–2032. https://doi.org/10.1021/acschembio.6b00303

Wu X, Huang M (2008) Dif1 controls subcellular localization of ribonucleotide reductase by mediating nuclear import of the R2 subunit. Mol Cell Biol 28:7156–7167. https://doi.org/10.1128/MCB.01388-08

Xie S, Luo H, Huang Y et al (2020) A missense mutation in a large subunit of ribonucleotide reductase confers temperature-gated tassel formation. Plant Physiol 184:1979–1997. https://doi.org/10.1104/pp.20.00219

Xu X, Page JL, Surtees JA et al (2008) Broad overexpression of ribonucleotide reductase genes in mice specifically induces lung neoplasms. Cancer Res 68:2652–2660. https://doi.org/10.1158/0008-5472.CAN-07-5873

Xu Y, Singh A, Alter GM (2016) Hydroxyurea induces cytokinesis arrest in cells expressing a mutated sterol-14α-demethylase in the ergosterol biosynthesis pathway. Genetics 204:959–973. https://doi.org/10.1534/genetics.116.191536

Zhan Y, Jiang L, Jin X et al (2021) Inhibiting RRM2 to enhance the anticancer activity of chemotherapy. Biomed Pharmacother 133:110996. https://doi.org/10.1016/j.biopha.2020.110996

Zhang K, Hu S, Wu J et al (2009) Overexpression of RRM2 decreases thrombspondin-1 and increases VEGF production in human cancer cells in vitro and in vivo: implication of RRM2 in angiogenesis. Mol Cancer 8:11. https://doi.org/10.1186/1476-4598-8-11

Zhang K, Wu J, Wu X et al (2011) p53R2 inhibits the proliferation of human cancer cells in association with cell-cycle arrest. Mol Cancer Ther 10:269–278. https://doi.org/10.1158/1535-7163.MCT-10-0728

Zhang H, Liu X, Warden CD et al (2014) Prognostic and therapeutic significance of ribonucleotide reductase small subunit M2 in estrogen-negative breast cancers. BMC Cancer 14:664. https://doi.org/10.1186/1471-2407-14-664

Zhang J, Hou D, Annis J et al (2020) Inactivating mutations in exonuclease and polymerase domains in DNA polymerase delta alter sensitivities to inhibitors of dNTP synthesis. DNA Cell Biol 39:50–56. https://doi.org/10.1089/dna.2019.5125

Zhao X, Muller EGD, Rothstein R (1998) A suppressor of two essential checkpoint genes identifies a novel protein that negatively affects dNTP pools. Mol Cell 2:329–340. https://doi.org/10.1016/S1097-2765(00)80277-4

Zhao X, Wei C, Li J et al (2017) Cell cycle-dependent control of homologous recombination. Acta Biochim Biophys Sin 49:655–668. https://doi.org/10.1093/abbs/gmx055

Zheng Z, Chen T, Li X et al (2007) DNA synthesis and repair genes *RRM1* and *ERCC1* in lung cancer. N Engl J Med 356:800–808. https://doi.org/10.1056/NEJMoa065411

Zhong Z, Cao Y, Yang S, Zhang S (2016) Overexpression of RRM2 in gastric cancer cell promotes their invasiveness *via* AKT/NF-κB signaling pathway. Pharmazie 71:280–284

Zhou B-S, Tsai P, Ker R et al (1997) Overexpression of transfected human ribonucleotide reductase M2 subunit in human cancer cells enhances their invasive potential. Clin Exp Metastasis 16:43–49. https://doi.org/10.1023/A:1006559901771

Zhou B, Shao J, Su L et al (2005) A dityrosyl-diiron radical cofactor center is essential for human ribonucleotide reductases. Mol Cancer Ther 4:1830–1836. https://doi.org/10.1158/1535-7163.MCT-05-0273

Zimanyi CM, Ando N, Brignole EJ et al (2012) Tangled up in knots: structures of inactivated forms of E. coli class Ia ribonucleotide reductase. Structure 20:1374–1383. https://doi.org/10.1016/j.str.2012.05.009

Zimanyi CM, Chen PY-T, Kang G, et al (2016) Molecular basis for allosteric specificity regulation in class Ia ribonucleotide reductase from Escherichia coli. eLife 5:e07141. https://doi.org/10.7554/eLife.07141

Zou Y, Zhou J, Xu B et al (2019) Ribonucleotide reductase subunit M2 as a novel target for clear-cell renal cell carcinoma. Onco Targets 12:3267–3275. https://doi.org/10.2147/OTT.S196347

[illegible]

Chapter 6
Structure and Dynamics of the Human Multi-tRNA Synthetase Complex

Myung Hee Kim and Beom Sik Kang

Abstract Aminoacyl-tRNA synthetases (ARSs) are essential enzymes that ligate amino acids to their cognate tRNAs during protein synthesis. A growing body of scientific evidence acknowledges that ubiquitously expressed ARSs act as crossover mediators of biological processes, such as immunity and metabolism, beyond translation. In particular, a cytoplasmic multi-tRNA synthetase complex (MSC), which consists of eight ARSs and three ARS-interacting multifunctional proteins in humans, is recognized to be a central player that controls the complexity of biological systems. Although the role of the MSC in biological processes including protein synthesis is still unclear, maintaining the structural integrity of MSC is essential for life. This chapter deals with current knowledge on the structural aspects of the human MSC and its protein components. The main focus is on the regulatory functions of MSC beyond its catalytic activity.

Keywords Aminoacyl-tRNA synthetase · Multi-tRNA synthetase complex (MSC) · Protein synthesis · Non-canonical function · MSC structure

Introduction

Aminoacyl-tRNA synthetases (ARSs) are evolutionary conserved enzymes. They are found in a range of organisms (i.e., from prokaryotic bacteria to higher eukaryotes), and they decode genetic information for protein synthesis by catalyzing the ligation of amino acids to their cognate tRNAs (Schimmel 1987). These essential enzymes are expressed ubiquitously and, generally speaking, there are 20 specific ARSs that correspond to the 20 different amino acids. ARS catalyzes a two-step

M. H. Kim (✉)
Infection and Immunity Research Laboratory, Metabolic Regulation Research Center, Korea Research Institute of Bioscience and Biotechnology (KRIBB), Daejeon, South Korea
e-mail: mhk8n@kribb.re.kr

B. S. Kang (✉)
School of Life Sciences, Kyungpook National University, Daegu, South Korea
e-mail: bskang2@knu.ac.kr

J. R. Harris, J. Marles-Wright (eds.), *Macromolecular Protein Complexes IV*, Subcellular Biochemistry 99, https://doi.org/10.1007/978-3-031-00793-4_6

transesterification reaction. In the first step, ARS recognizes its cognate amino acid in the presence of adenosine triphosphate (ATP) and forms an enzyme–aminoacyl adenylate (amino acid–adenosine monophosphate; AMP) intermediate complex by releasing pyrophosphate. In the second step, the cognate ARS tRNA binds to the complex via an anticodon-binding domain (ABD) and forms a covalent bond with an amino acid by releasing AMP (McClain 1993). The resulting aminoacylated tRNA enters the ribosome to decode mRNA.

In eukaryotic systems, there are two categories of ARS enzymes. The first includes the cytoplasmic ARSs, which are responsible for supplying aminoacyl-tRNA conjugates for general protein translation. The second category comprises the mitochondrial ARSs. These ARSs are imported into the mitochondrial matrix for charging amino acids to their mitochondrial genome-encoded tRNA molecules. ARSs are abbreviated using the single-letter code for their corresponding substrate amino acid followed by "ARS1" for cytoplasmic enzymes and by "ARS2" for their mitochondrial counterparts. This chapter will focus on cytoplasmic ARSs.

All ARSs have a catalytic core, where the ligation reaction occurs, and an ABD (Fig. 6.1a–d). The ARSs are grouped into two classes, namely, class I and II. The class to which an ARS belongs to depends on its catalytic core structure (Fig. 6.1e) (Burbaum and Schimmel 1991; Cusack et al. 1990; Eriani et al. 1990; Ribas de Pouplana and Schimmel 2001). Class I ARS, for example, forms a Rossmann fold containing a parallel β-sheet with the HIGH and KMSKS motifs, which are involved in substrate binding (Fig. 6.1a, c). By contrast, the catalytic core structure of class II ARS consists of an antiparallel β-sheet flanked by α-helices containing three conserved motifs called motif 1, 2, and 3 (Fig. 6.1b, d). There is no functional difference between class I and class II enzymes. However, class I enzymes bind to the minor groove of the tRNA acceptor stem and attach an amino acid to the 2'-OH of the tRNA terminal nucleotide. Class II enzymes bind to the major groove of the acceptor stem and attach the amino acid to the 3'-OH of the terminal nucleotide. The ARSs belonging to class I exist mostly as monomers, while class II ARSs are typically dimers (Rubio Gomez and Ibba 2020). For example, methionyl-tRNA synthetase 1 (MARS1; Fig. 6.1a, c), a class I ARS, and aspartyl-tRNA synthetase 1 (DARS1; Fig. 6.1b, d), a class II ARS, function as a monomer and dimer, respectively.

Eukaryotic ARSs evolved additional motifs and domains that provide eukaryotic-specific biological functions beyond translation (Guo et al. 2010b; Hyeon et al. 2019; Jeong et al. 2019; Kim et al. 2011; Kwon et al. 2019; Schimmel 2018). The protein components of eukaryotic ARSs assemble into the multi-tRNA synthetase complex (MSC) in different combinations, primarily through these motifs and domains (Fig. 6.2). In yeast, MARS1 and glutamyl-tRNA synthetase 1(EARS1) form a triple complex with the aminoacyl-tRNA synthetase cofactor 1 protein (Arc1p) (Kim and Kim 2020; Simader et al. 2006). The MSC of the parasite protozoan *Toxoplasma gondii* consists of MARS1, EARS1, tyrosyl-tRNA synthetase 1 (YARS1), glutaminyl-tRNA synthetase 1 (QARS1), and the nonenzymatic protein Tg-p43 (Kim and Kim 2020; van Rooyen et al. 2014). In the MSC of *Trypanosoma brucei*, prolyl-tRNA synthetase 1 (PARS1), QARS1, and MARS1 are associated with three non-ARS proteins, namely, the multiple ARS complex-associated proteins (MCP)-1, MCP-2, and MCP-3 (Cestari et al. 2013).

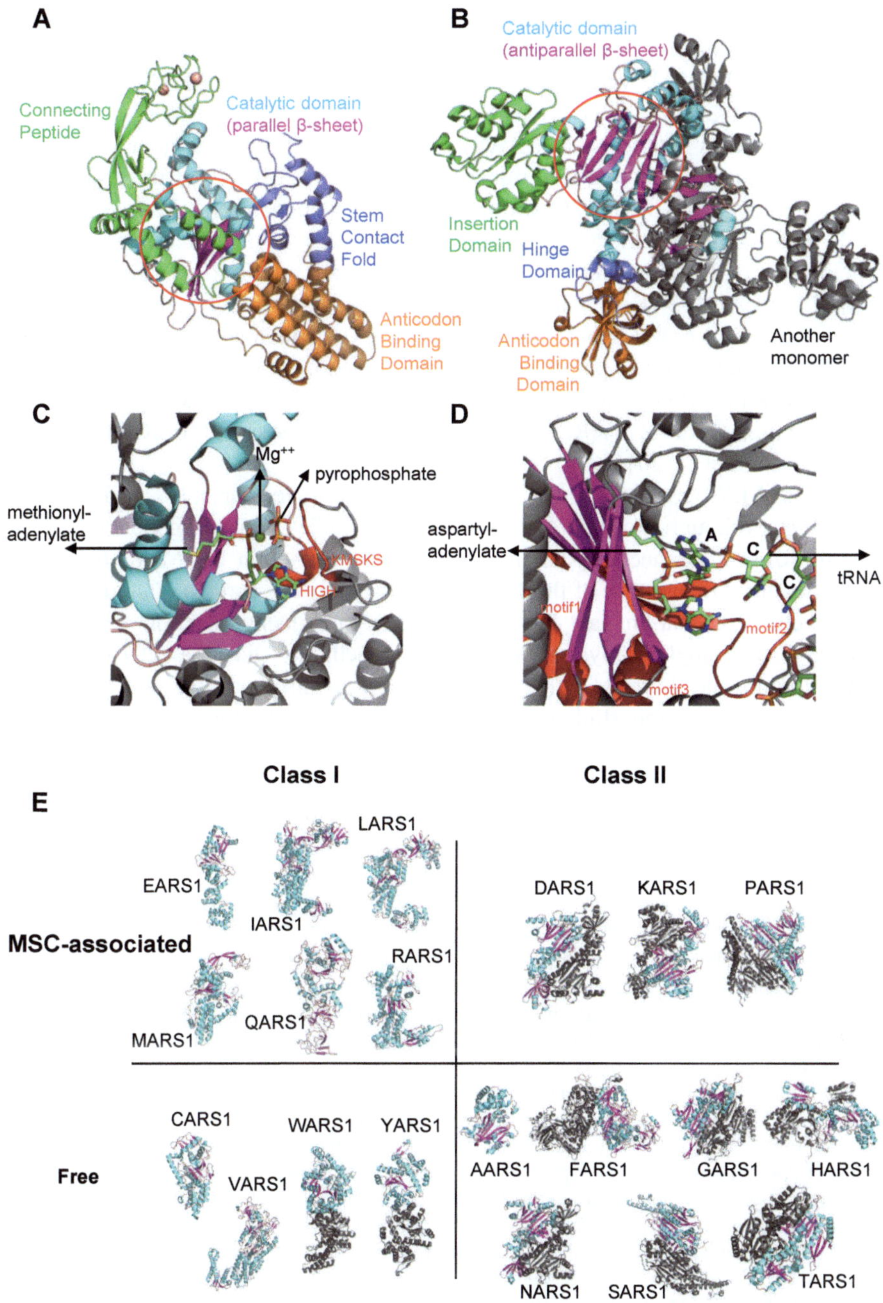

Fig. 6.1 Structure of class I and II ARSs. Ribbon diagrams of human MARS1 (**a**) and DARS1 (**b**), which belong to class I and class II, respectively. MARS1 consists of a catalytic domain (cyan and magenta), an anticodon-binding domain (orange), a connecting peptide (green), and a stem contact fold (slate purple). Figure modified from Kim et al. (2021a). DARS1 forms a dimer. One DARS1

In higher eukaryotes, more ARSs are part of the MSC (Fig. 6.2). The nematode *Caenorhabditis elegans* forms an MSC with eight ARSs, including isoleucyl-tRNA synthetase 1 (IARS1), valyl-tRNA synthetase 1 (VARS1), EARS1, lysyl-tRNA synthetase 1 (KARS1), leucyl-tRNA synthetase 1 (LARS1), MARS1, QARS1, and arginyl-tRNA synthetase 1 (RARS1), and the core scaffold protein aminoacyl-tRNA synthetase-interacting multifunctional protein (AIMP) 2. At the amino acid level, *C. elegans* AIMP2 is 33% similar to its human counterpart (Havrylenko et al. 2011; Kim and Kim 2020). Eukaryotes from flies to mammals form MSCs containing eight ARSs and the three nonenzymatic scaffold proteins AIMP1, AIMP2, and AIMP3 (Kim and Kim 2020). These MSCs harbor the largest ARS, namely, glutamyl-prolyl-tRNA synthetase 1(EPRS1), which possesses two catalytic activities within a single polypeptide. In addition to the canonical ARSs, recent studies show that more proteins can associate with the MSC. For example, threonyl-tRNA synthetase like-2 (TARSL2) is found to be associated with MSC in mammalian cells (Park et al. 2015).

Human MSC consists of 24 proteins with a molecular mass estimated at ~1.5 MDa. Although biochemical and molecular studies propose models that illustrate the overall connections among the 24 proteins within the MSC, the three-dimensional (3D) structure of the MSC remains unknown. Structural studies of the 3D structure of mammalian MSC were performed using electron microscopy (EM) and small-angle X-ray scattering (SAXS) methods (Dias et al. 2013; Norcum 1991; Norcum and Warrington 1998). However, large discrepancies exist among the results of the structural studies at low resolution, which suggest that MSC has conformational flexibility.

Genetic depletion of the core scaffold protein, AIMP2, from the MSC is lethal in mice (Kim et al. 2002). This observation indicates that the structural integrity of the MSC is essential for life. However, the mechanisms underlying MSC

Fig. 6.1 (continued) monomer consists of a catalytic domain (cyan and magenta), an anticodon-binding domain (orange), an insertion domain (green), and a hinge domain (slate purple). The other monomer is in grey. Figure modified from Kim et al. (2013). Red circles in the MARS1 and DARS1 structures highlight the catalytic domains. Ribbon diagrams of the catalytic cores of MARS1 (**c**) and DARS1 (**d**). The catalytic core of MARS1, a class I ARS, has a Rossmann fold consisting of a parallel β-sheet (magenta) with α-helices. HIGH and KMSKS motifs are colored in red. Methionyl-adenylate and pyrophosphate, the reaction intermediates of *Leishmania major* MARS1, are shown as sticks. Figure modified from Larson et al. (2011). The catalytic core of *Escherichia coli* DARS1, a class II ARS, has an antiparallel β-sheet (magenta). Three conserved motifs are colored in red. Aspartyl-adenylate and nucleotides at the 3′ end of the tRNA are shown as sticks. Figure modified from Moulinier et al. (2001). (**e**) ARSs for the 20 amino acids are grouped into class I and II. Six class I ARSs and three class II ARSs assemble to form the MSC. PDB IDs of the ARS structures used for the ribbon diagrams are 4XEM (AARS1), 1LI5 (CARS1), 4J15 (DARS1), 1N78 (EARS1), 3L4G (FARS1), 5E6M (GARS1), 4G85 (HARS1), 7D5C (IARS1), 4YCU (KARS1), 6LPF (LARS1), 5GL7 (MARS1), 5XIX (NARS1), 4HVC (PARS1), 4R3Z (QARS1), 4Q2T (RARS1), 3VBB (SARS1), 4P3N (TARS1), 1GAX (VARS1), 1R6T (WARS1), and 5THH (YARS1). Helices and ribbons in a monomer are colored in cyan and magenta, respectively. The other molecules in dimeric structures are colored in grey

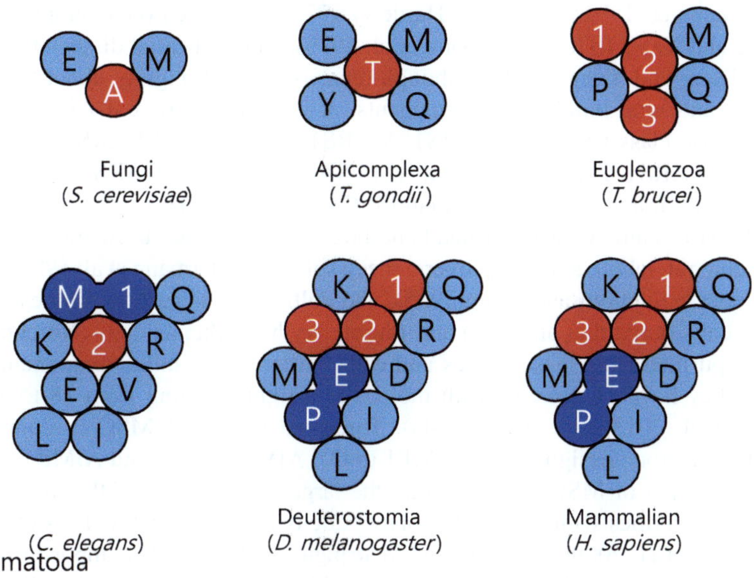

Fig. 6.2 Aminoacyl-tRNA synthetase complexes from various organisms. *Saccharomyces cerevisiae* has a complex consisting of EARS1 (E), MARS1 (M), and Arc1p (A). The apicomplexan, *Toxoplasma gondii*, has EARS1 (E), MARS1 (M), YARS1 (Y), QARS1 (Q), and TARS1 (T) as components of its complex. The complex of the euglenozoa, *Trypanosoma brucei*, consists of MARS1 (M), PARS1 (P), QARS1 (Q), MCP1(1), MCP2 (2), and MCP3 (3). Nematoda *Caenorhabditis elegans* has a complex consisting of EARS1 (E), IARS1 (I), KARS1 (K), LARS1 (L), MARS1 (M), QARS1 (Q), RARS1 (R), VARS1 (V), and AIMP2 (2). AIMP1 (1) is connected to MARS1 as its C-terminal domain. *Drosophila melanogaster*, which belongs to Deuterostomia, and humans (*Homo sapiens*) have a complex consisting of nine ARSs (D, E, I, K, L, M, P, Q, and R) and three nonenzymatic proteins (1, 2, and 3). In *D. melanogaster* and *H. sapiens*, genes encoding EARS1 (E) and PARS1 (R) are fused to form a bifunctional protein

depletion-induced lethality remain unknown. Accumulation of information on the overall shape of the MSC, together with detailed structural data on several of its protein components revealed by X-ray crystallography, provides insights into the 3D architecture and dynamic nature of the MSC. In this chapter, we focus on the human MSC and present current knowledge on how MSC structure is integrated by its various components. Finally, we discuss how the MSC functions in a broad range of biological processes beyond protein synthesis.

Mammalian MSC

Several cytoplasmic ARSs in higher eukaryotic organisms are part of a high-molecular-weight complex, called the MSC. Evidence of physical associations among these ARSs comes from conventional co-purification (Denney 1977;

Kellermann et al. 1979; Som and Hardesty 1975), affinity chromatographic analysis (Dang and Yang 1978; Kellermann et al. 1979), and sequential dissociation studies (Van Dang and Yang 1979). Although there is some ambiguity about the stoichiometry and total number of components, at least nine different ARSs, including monomeric class I (IARS1, LARS1, MARS1, QARS1, and RARS1) and dimeric class II (KARS1, PARS1, and aspartyl-tRNA synthetase 1 [DARS1]) enzymes, are consistently found in the mammalian MSC (Lee et al. 2004a).

The MSC complex was purified to homogeneity from various mammalian tissues, including rat, rabbit, and sheep livers, sheep spleen (Kellermann et al. 1982; Mirande et al. 1982; Siddiqui and Yang 1985), and cultured cells such as Chinese hamster ovary cells (Lazard et al. 1987) and murine erythroleukemia cells (Norcum 1989). MSCs purified from these tissues and cells harbor very similar components. Typically, they have 11 proteins with molecular weights ranging from 18 to 150 kDa (Dias et al. 2013; Lee et al. 2004a). Three out of the 11 MSC components are auxiliary factors, designated as AIMP1 (p43), AIMP2 (p38), and AIMP3 (p18).

Purification of MSCs revealed that the largest component of the MSC has two catalytic activities within a single polypeptide (Cerini et al. 1991; Dias et al. 2013; Kaminska et al. 2009b; Norcum and Warrington 1998). MSC components EARS1 and PARS1 exist as a bifunctional enzyme within a single polypeptide called EPRS1 (Cerini et al. 1991).

Like other mammalian MSCs, human MSC consists of 11 components, including eight ARS proteins and the three scaffold proteins AIMP1, AIMP2, and AIMP3 (Fig. 6.2). This chapter will focus on the structure of human MSC.

Appended Domains in Human MSC

In general, eukaryotic ARSs are bigger than their prokaryotic counterparts due to acquisition of additional domains and motifs during evolution, which confer eukaryote-specific functions beyond protein synthesis (Guo et al. 2010b). The occurrence of the structural additions is linked to functional versatility and correlates with the increased biological complexity of higher organisms (Guo et al. 2010b; Kim and Kim 2020). The domains and motifs appended to the N- and C-terminus of the catalytic core domain play an important role in the assembly of the MSC structure by mediating intermolecular interactions (Fig. 6.3).

Several ARSs have their own unique sequence motif (UNE), while others harbor common structural domains such as leucine heptad repeats and glutathione transferase (GST)-homology domains, which are also found in other non-ARS-proteins (Fig. 6.3). QARS1 has a unique sequence motif (UNE-Q) at the N-terminus, while IARS1 and LARS1 retain UNE-I and UNE-L at their C-terminus, respectively. Particularly, IARS1 contains two tandemly arranged UNEs at the C-terminus, namely, UNE-I1 and UNE-I2. UNE-I1 is found in all eukaryotes from yeast to humans, while UNE-I2 exists only in vertebrates. Amino acid sequence analysis suggests that UNE-I2 contains two repeated motifs consisting of about 90 amino

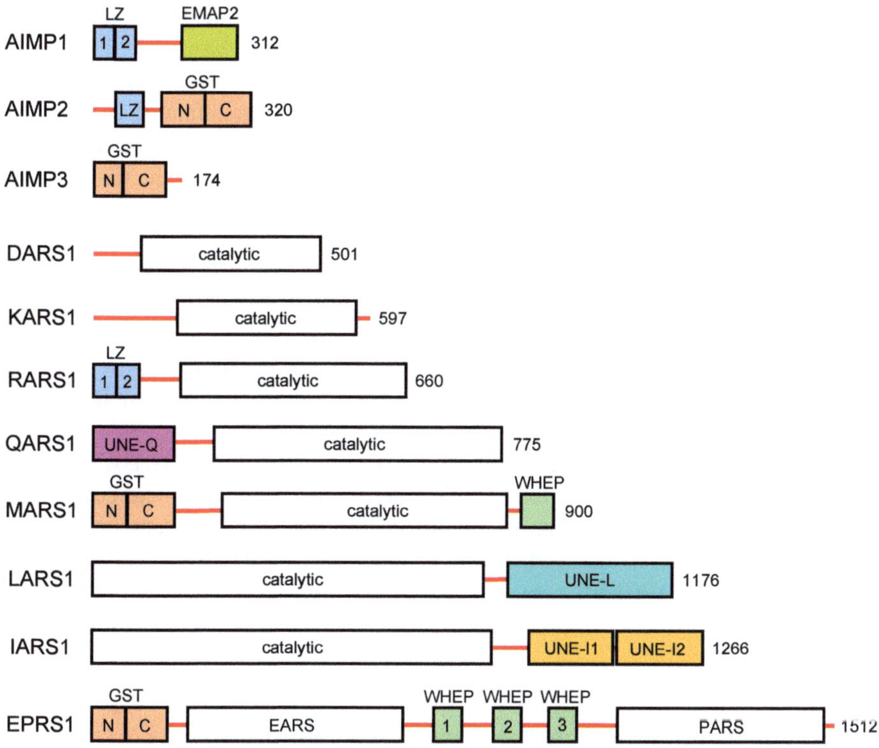

Fig. 6.3 Domain architectures of the proteins in the human multi-tRNA synthetase complex. Defined or predicted structural elements are presented as unshaded squares and rectangles, and disordered regions are shown as red solid lines. AIMP1, AIMP2, and RARS1 have leucine zipper (LZ) heptad repeats. AIMP2, AIMP3, MARS1, and EPRS1 have a GST domain consisting of GST-N and GST-C subdomains. MARS1 and EPRS1 have one and three WHEP domains, respectively. ARSs contain a catalytic domain that includes the tRNA-binding domain, except for EPRS1, which has catalytic domains for EARS1 and PARS1. QARS1, IARS1, and LARS1 have motifs, UNE-Q, UNE-I, and UNE-L, respectively, that are unique to each component

acids for each repeat. These motifs interact with EPRS1 in the MSC (Rho et al. 1996, 1999).

A unique variant domain, called WHEP, is found in EPRS1. The name WHEP comes from its presence in tryptophanyl-tRNA synthetase 1 (WARS1), histidyl-tRNA synthetase 1(HARS1), and EPRS1. In EPRS1, EARS1 and PARS1 are connected by a linker containing triple repeats of the 50–amino acid WHEP domain, adopting a helix-turn-helix structure. EPRS1 is the only bifunctional tRNA synthetase with two distinct aminoacylation activities, and its WHEP domains in the linker mediate a noncanonical translation regulatory function. WHEP domains are present in variable numbers within the EPRS proteins of diverse bilaterian taxa. For example, human EPRS1 contains three WHEP domains, while its fly counterpart has six WHEP domains (Kim and Kim 2020). MARS1 also has a WHEP domain at its C-terminus. The structure of the C-terminal part of MARS1 was determined by

NMR spectroscopy (PDB ID 2DJV). The structure reveals a WHEP domain comprising a helix-turn-helix motif and lysine-rich region.

Among the MSC components, MARS1, EPRS1, AIMP2, and AIMP3 contain a GST-like domain. These GST-like domains are critical for maintaining MSC structural integrity (Cho et al. 2019; Hahn et al. 2019). The GST-like domains are found in proteins such as the EF1B gamma subunit, which is involved in the translational machinery (Koonin et al. 1994). VARS1 also contains a GST-like domain at its N-terminus (Guo et al. 2010a).

Leucine zipper (LZ) motifs are found in RARS1, AIMP1, and AIMP2. The LZ is a common protein motif with a heptad leucine repeat that forms a coiled-coil structure. RARS1 is composed of two tandem LZ motifs at the N-terminus. AIMP1 contains two LZ motifs in a long helix at the N-terminus, followed by the C-terminal EMAPII-like domain, while AIMP2 carries the LZ motif in the N-terminal region, followed by a C-terminal GST domain.

Although they do not contain structured appendages, KARS1, DARS1, and AIMP2 have a long peptide region at their N-terminus. These unstructured appendages of KARS1 and DARS1 are not involved in intermolecular interactions in the MSC (Cho et al. 2019; Hahn et al. 2019; Ofir-Birin et al. 2013). However, the N-terminal peptide of AIMP2 is responsible for the interaction with dimeric KARS1 (Ofir-Birin et al. 2013).

MSC Assembly Through the Appended Domains

Human MSC proteins have additional regions adjacent to the domains that function in enzymatic reactions for protein synthesis. The special appendage regions such as UNE, GST, and LZ either are unstructured or form domains that function in intermolecular interactions within the MSC. EMAPII-like and WHEP domains participate in important biological processes, such as immune regulation (Lee et al. 2016; Shalak et al. 2001; Yao et al. 2012). However, the role of these domains in integrating the structure of MSC is unclear.

Leucine Zipper–Mediated Assembly

Among the human MSC components, RARS1, AIMP1, and AIMP2 contain leucine heptad repeat sequence motifs (Fig. 6.3). RARS1 and AIMP1 have two LZ motifs (LZ1 and LZ2) in the N-terminus, which appear as two segmented regions consisting of about 30 amino acids for each region (Fu et al. 2014). By contrast, AIMP2 has an LZ motif between the N-terminal peptide (responsible for KARS1 interaction) and a C-terminal GST domain (Ahn et al. 2003; Kim and Kim 2020; Ofir-Birin et al. 2013). The crystal structure of the trimeric subcomplex consisting of QARS1, RARS1, and AIMP1 shows a coiled-coil interaction between RARS1 LZ2, and AIMP1 LZ2 (Fu et al. 2014).

Twofold crystallographic symmetry based on a SAXS analysis showed that the trimeric subcomplex structure forms a hexameric complex consisting of two trimeric (QARS1–RARS1–AIMP1) subcomplexes (Fig. 6.4a) (Fu et al. 2014). In the hexameric structure, two N-terminal AIMP1 helices arrange into a twofold symmetrical X-shaped structure, which facilitates coiled-coil interactions between RARS1 LZ1–AIMP1 LZ1 and RARS1 LZ2–AIMP1 LZ2.

AIMP2 interacts with AIMP1 through its LZ domain (Fu et al. 2014). The LZ1 of AIMP1 is responsible for the interaction with the AIMP2 LZ. Within the dimeric coiled-coil interactions, AIMP2 is unable to associate with the tetrameric complex in a 2:2 stoichiometry in the RARS1–AIMP1 complex. AIMP1 LZ1 forms a coiled-coil structure with AIMP2 LZ instead of the RARS1 LZ1 in the presence of AIMP2. This is because LZ1 interaction between AIMP1 and RARS1 is not essential for the interaction between AIMP1 and RARS1. A hypothetical interaction model can be considered, which involves the QARS1–RARS1–AIMP1–AIMP2 tetrameric and the heterotrimeric QARS1–RARS1–AIMP1 complexes and the symmetry-related RARS1 (Fig. 6.4b). In the tetrameric complex, typical coiled-coil structures form between AIMP1 LZ2 and RARS1 LZ2 and between AIMP1 LZ1 and AIMP2 LZ. The RARS1 LZ1 would then be free to bind to another partner (Fig. 6.4c). The threonyl-tRNA synthetase-like protein-2 (TARSL2), which has an LZ motif, associates with MSC (Park et al. 2015). The LZ1 motif of RARS1 could be a partner of TARSL2.

Formation of a coiled-coil structure by two LZ motifs occurs through hydrophobic interactions. Assembly of hydrophobic surfaces in the helical structures of LZ motifs can form not only a coiled-coil structure, but also three-strand coiled-coil interactions and a four-helix coiled-coil structure (Fig. 6.4d) (Durr and Bosshard 2000; Lupas et al. 2017). Several interaction modes with the LZ motifs from RARS1, AIMP1, and AIMP2 within the MSC are possible. One possibility is formation of a RARS1–AIMP1–AIMP2 complex in a 2:2:2 stoichiometry based on the tetrameric RARS1–AIMP1 complex with the three-strand coiled-coil structures (Fig. 6.4e). Because AIMP2 forms a dimer, AIMP2 LZ may be involved in the formation of a four-helix coiled-coil structure with AIMP1 LZ1 in a 2:2 stoichiometry, yielding a RARS1–AIMP1–AIMP2 complex in a 2:2:2 stoichiometry (Fig. 6.4e).

GST-Like Domain-Mediated Assembly

GST is an enzyme for detoxification. Its detoxification function is accomplished by conjugating the sulfhydryl group of glutathione to xenobiotic substrates (Allocati et al. 2018). Throughout evolution, GST-like structures have spread into different proteins, including ARSs and other translational factors, to assume nonenzymatic functions (Guo and Yang 2014; Koonin et al. 1994; Oakley 2005). Human MSC contains GST-like domains in MARS1, EPRS1, AIMP2, and AIMP3. Although several functions are attributed to these MSC-embedded GSTs, accumulating biochemical and structural evidence indicates that they play crucial roles in protein assembly and folding.

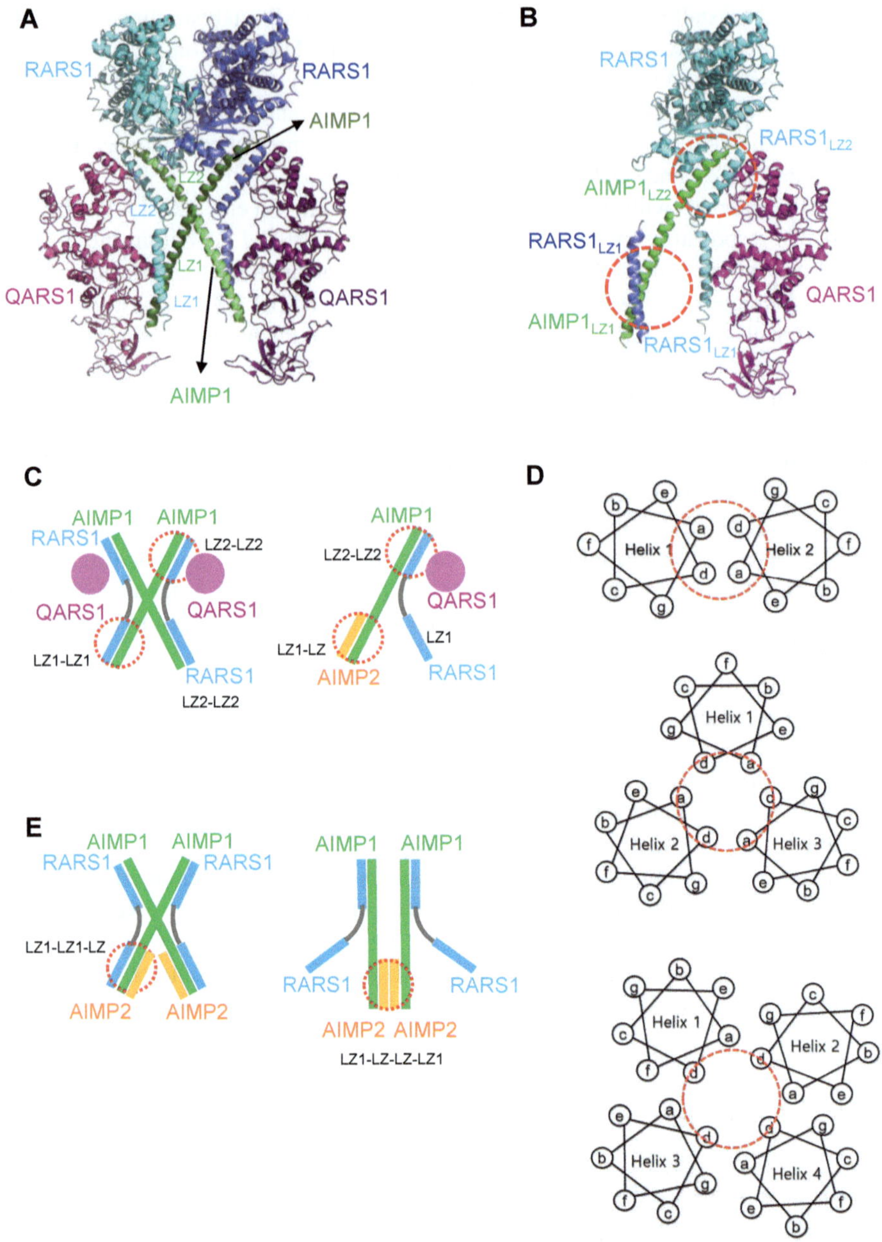

Fig. 6.4 Interactions through the leucine zipper motifs in human MSC. (**a**) Ribbon diagram of the hexameric QARS1–RARS1–AIMP1 complex shows coiled-coil interactions between RARS1 (cyan) and AIMP1 (green). Figure modified from Fu et al. (2014). (**b**) The first leucine zipper motif (LZ1) of RARS1 resides on the trimeric QARS1–RARS1–AIMP1 complex. The first (LZ1) and second (LZ2) motifs in a long helix of AIMP1 (green) have coiled-coil interactions with the first (slate purple) and second (cyan) motifs from two separate RARS1, respectively. (**c**) Schematic

Because functional GST enzymes exist as homodimers, MSC proteins containing a GST domain are predicted to be homodimers. However, unlike canonical GST, the GST domains of MARS1, EPRS1, AIMP2, and AIMP3 do not form a homodimer. Instead, they exist as monomeric proteins, but can form heterodimers in a specific manner. For example, MARS1 GST interacts with AIMP3 GST, while EPRS1 GST binds to AIMP2 GST (Fig. 6.5).

In the MSC, GST assumes a dimeric conformation with other GST domains via two types of binding interface. The GST domain consists of two subdomains, namely, GST-N and GST-C. The canonical interaction of enzymatic GST as a homodimer is induced through an interface that encompasses the GST-N and GST-C subdomains. This region of GST interaction is called binding interface 1. The other type of binding interface, called binding interface 2, forms through the GST-C subdomains. In the MSC, the MARS1 and EPRS1 GST domains interact with the AIMP3 and AIMP2 GST domains, respectively, by way of the canonical binding interface 1. Binding interface 1 consists of the last helix of the GST-N subdomain and the first helix of the GST-C subdomain. The GST domains of EPRS1 and AIMP3 can also interact with each other through binding interface 2, which consists of a helix and a loop in the GST-C subdomains. Consequently, the GST domain in MSC binds two GSTs simultaneously through binding interfaces 1 and 2. For instance, EPRS1 binds to AIMP2 and AIMP3, while AIMP3 binds to MARS1 and EPRS1. Specific interaction between these GST domains yields a tetrameric GST complex in the order MARS1–AIMP3–EPRS1–AIMP2 (Fig. 6.5). GST domains from four component proteins of the MSC assemble linearly. Another interface on the AIMP2 GST-N subdomain interacts with DARS1.

UNE Domain-Mediated Assembly

IARS1 and LARS1 contain the UNE regions UNE-I and UNE-L, respectively, at their C-terminus. IARS1 is further subdivided into UNE-I1 and UNE-I2. The C-terminal region of LARS1 is indispensable for its localization within the MSC. This is assuming that LARS1 UNE-L is involved in the interaction with an MSC member. EPRS1 interacts with IARS1 via UNE-I, which links to LARS1. Thus, IARS1 and LARS1 localize in the MSC through EPRS1.

Fig. 6.4 (continued) diagrams for coiled-coil interactions in the hexameric QARS1–RARS1–AIMP1 complex (left) and in the model of trimeric QARS1–RARS1–AIMP1 complex with AIMP2 (right). (**d**) Leucine zippers containing heptad repeats form variable coiled-coil structures of α-helices. Hydrophobic cores with residues at the a and d positions in the helical wheels are highlighted with a red dashed circle in the two (top), three (middle), and four (bottom)-helix coiled-coil structures. (**e**) Schematic diagrams for three-helix (left) and four-helix (right) coiled-coil interactions in a hypothetical RARS1–AIMP1–AIMP2 complex

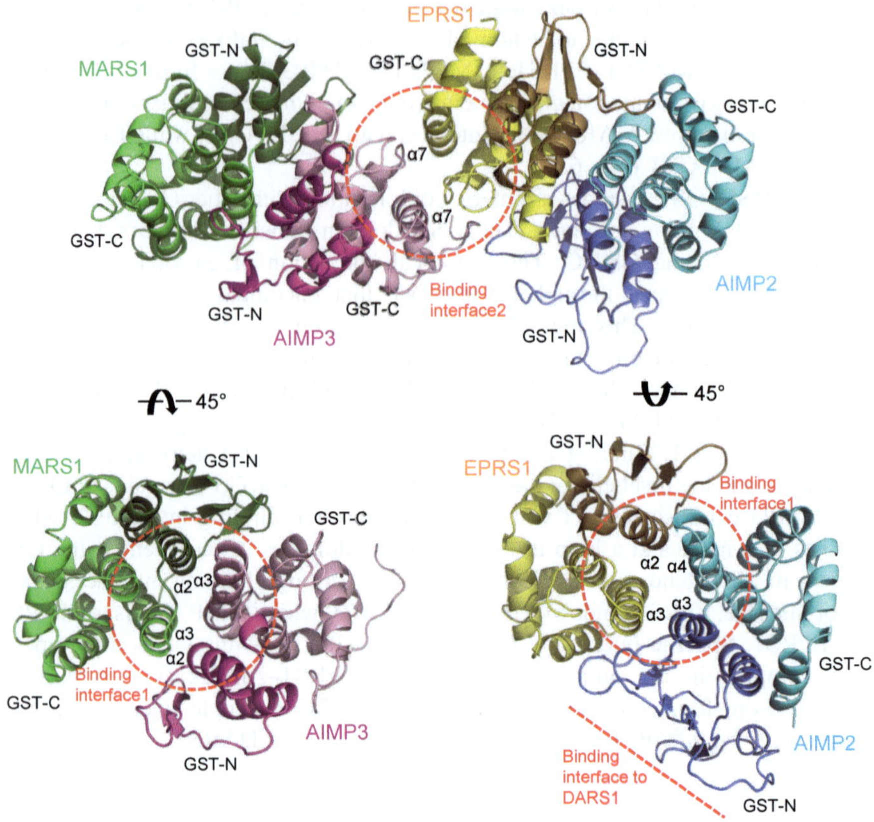

Fig. 6.5 Interaction among MSC protein components through glutathione transferase homology domains. Ribbon diagram of the tetrameric complex consisting of the GST-homology domains from MARS1 (green), AIMP3 (magenta), EPRS1 (yellow), and AIMP2 (cyan), derived from the crystal structure of the MARS1–AIMP3–EPRS1–AIMP2–DARS1 pentameric complex (PDB ID 5Y6L). Four GST domains consist of GST-N (dark color) and GST-C (bright color) subdomains. Interaction between AIMP3 and EPRS1 occurs via binding interface 2 at their GST-C subdomain. Interaction between MARS1 and AIMP3, and between EPRS1 and AIMP2, occurs through binding interface 1, which involves two helices from GST-N and GST-C. AIMP2 has a binding interface at its β-sheet in the GST-N subdomain, which interacts with DARS1. The binding interfaces 1 and 2 are circled in red. The binding interface of AIMP2 for DARS1 is indicated by a dashed line

Architecture of the Human MSC

The unicellular organism *Saccharomyces cerevisiae* forms a primitive eukaryotic MSC. It consists of MARS1, EARS1, and an auxiliary protein, Arc1p. Arc1p interacts simultaneously with MARS1 and EARS1, resulting in a MARS1–Arc1p–EARS1 complex with no direct connection between the two ARSs. Human AIMP3 links MARS1 to EPRS1 in a manner similar to the yeast Arc1p. In general, the mammalian MSC is composed of 11 proteins that include three nonenzymatic scaffold proteins. The interactions among these scaffold proteins within the MSC are complex. The functional versatility of MSC scaffold proteins correlates with increasing biological complexity. The complex structure of tRNA synthetases is of interest not only for understanding the protein biosynthetic machinery, but also as a model system to study protein assembly beyond the quaternary structure.

Three Subcomplexes in MSC

Biochemical studies provide insights into interactions among components within the mammalian MSC. For example, cross-linked complexes from rabbit identified neighboring protein pairs within the MSC (Norcum and Warrington 1998). Furthermore, systematic depletion analysis to determine the structural proximity and turnover of MSC components in HeLa cells shows that the stability of some components depends on their neighbors (Han et al. 2006). This study reveals that among the other MSC components, KARS1 is the most independent for specifying its own stability. KARS1 is also essential for the stability of other MSC components. A structural biological study revealed that KARS1 interacts exclusively with the N-terminal region of AIMP2 (Ofir-Birin et al. 2013). The two dimeric proteins KARS1 and DARS1 bind tightly to AIMP2 (Robinson et al. 2000). Assembly analysis of the MSC purified from HeLa cells produced a human MSC connectivity map showing that AIMP3 links MARS1 to the complex. AIMP1 recruits QARS1 and RARS1 to the MSC, while AIMP2 connects the two subcomplexes I (MARS1, AIMP3, EPRS1, IARS1, and LARS1) and II (AIMP1, QARS1, and RARS1) (Kaminska et al. 2009a). Recent structural studies show that the AIMP2 GST domain interacts tightly with DARS1 and EPRS1 (Cho et al. 2019; Hahn et al. 2019). These observations suggest that AIMP2 functions as a core scaffold protein to maintain the structural integrity of the MSC.

Taking the above studies together, a two-dimensional linkage map of human MSC components is proposed (Fig. 6.6). Among the MSC components, AIMP2 has the largest number of binding proteins, while EPRS1 has only three binding partners. Based on the interactions of AIMP2 and EPRS1 with their binding proteins, human MSC is classified into three subcomplexes (Fig. 6.6). The N-terminal region of AIMP2-mediated subcomplex I consists of QARS1, RARS1, AMIP1, and KARS1. Subcomplex II, which is mediated by the GST domains of AIMP2 and EPRS1, is composed of MARS1, AIMP3, and DARS1. Subcomplex III contains the EPRS1-mediated interaction proteins IARS1 and LARS1.

Details of Interactions Within Subcomplexes

MSC subcomplexes based on the relationships among component proteins are displayed as linear interactions (Fig. 6.7). AIMP2 binds subcomplexes I (KARS1

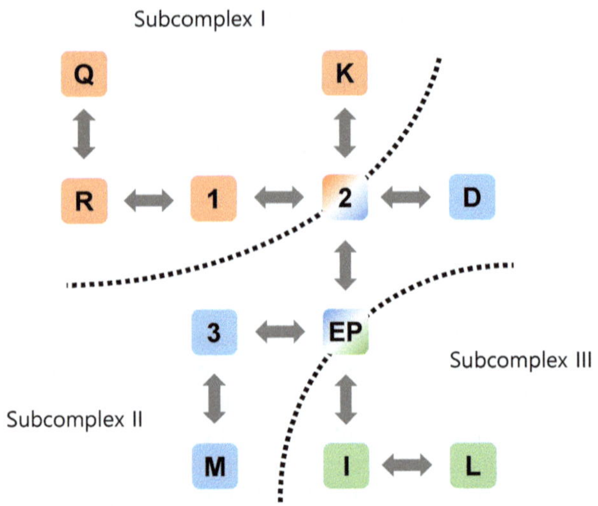

Fig. 6.6 Binary interactions among MSC components. AIMP2 (2) interacts with KARS1 (K), DARS1 (D), EPRS1 (EP), and AIMP1 (1), while EPRS1 binds to AIMP2, AIMP3 (3), and IARS1 (I). IARS1 binds LARS1 (L), and AIMP3 interacts with MARS1 (M). AIMP1, AIMP3, RARS1 (R), and IARS1 each interact with two proteins. AIMP1 binds to RARS1 and AIMP2, while AIMP3 interacts with EPRS1 and MARS1. RARS1 binds to QARS1 (Q) and AIMP1, while IARS1 binds to EPRS1 and LARS1. The MSC can be divided into subcomplexes I, II, and III by dividing AIMP3 and EPRS1. In each subcomplex, the component proteins are linked linearly

and AIMP1) and II (EPRS1 and DARS1) via its N-terminal region and C-terminal GST domain, respectively. EPRS1 binds subcomplexes II (AIMP2 and AIMP3) and III (IARS1) via its N-terminal GST domain and C-terminal region, respectively. Subcomplex I forms through interactions between LZ motifs, while subcomplex II forms through interactions among GST domains. Subcomplex III is built by UNE region-mediated interactions.

Formation of subcomplex I is mediated by the N-terminal peptide and follows the AIMP2 LZ motif (Fig. 6.7a). A previous study revealed that the ternary complex of QARS1, RARS1, and AIMP1 anchors to AIMP2 through interactions with AIMP1 (Fu et al. 2014), leading to the hypothesis that AIMP1 and AIMP2 form a coiled-coil structure through their LZ motifs. RARS1 also contains LZ motifs. RARS1 LZ2 interacts with AIMP1 LZ2. The N-terminal domain of human QARS1 interacts with RARS1 (Rho et al. et al. 1999). QARS1 anchors to the MSC by interacting with RARS1 LZ2 through its C-terminal core (Fu et al. 2014). This interaction interface forms via hydrophobic residues and two ionic pairs. Thus, the interaction between AIMP1 LZ2 and AIMP2 LZ, together with the interaction between the N-terminal peptide of AIMP2 and KARS1, completes formation of a linear QARS1–RARS1–AIMP1–AIMP2–KARS1 subcomplex.

Subcomplex II forms mainly through interactions among the GST domains of MARS1, EPRS1, AIMP2, and AIMP3 (Fig. 6.7b). The four GST domains play a central role in the formation of the subcomplex in the order MARS1–AIMP3–

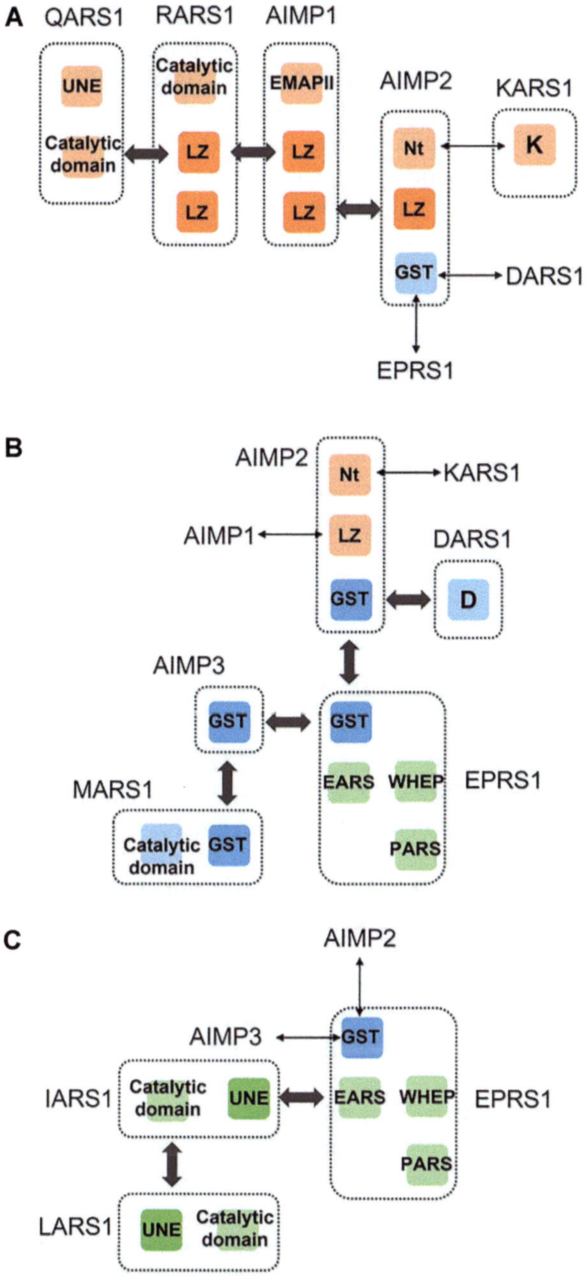

Fig. 6.7 Interactions between the component proteins of the three subcomplexes in human MSC. (**a**) Subcomplex I is a QARS1–RARS1–AIMP1–AIMP2–KARS1 pentameric complex. It assembles through the leucine zipper (LZ) motifs. RARS1, AIMP1, and AIMP2 have LZ motifs, and they assemble via coiled-coil interactions. QARS1 binds to the LZ of RARS1. KARS1 binds to the N-terminal peptide of AIMP2. Linking of AIMP2 to EPRS1 and DARS1 of subcomplex II occurs

EPRS1–AIMP2. This tetrameric GST complex functions as a scaffold for other MSC components. The interaction between the AIMP2 GST domain and DARS1 completes formation of the pentameric MARS1–AIMP3–EPRS1–AIMP2–DARS1 complex. The C-terminal tail of AIMP3 enhances the interaction between DARS1 and the tetrameric GST complex (Cho et al. 2019).

Subcomplex III is a trimeric complex consisting of EPRS1, IARS1, and LARS1 (Fig. 6.7c). A previous study shows that IARS1 interacts with LARS1 and EPRS1 (Rho et al. 1999). Furthermore, the same study reports that the C-terminal UNE region of IARS1 interacts with the EPRS1 WHEP region and the C-terminal LARS1. Another study reveals that the EARS1 domain, instead of the WHEP domain region, is required for the interaction with IARS1 (Halawani et al. 2018). It is likely that the EARS1 domain in EPRS1 interacts with the IARS1 UNE region, and the LARS1 UNE region mediates the interaction between LARS and IARS. However, a role for the EPRS WHEP region in mediating associations between EPRS1 and IARS1 cannot be ruled out. These observations highlight the critical role of EPRS1 in integrating the MSC structure via its N-terminal GST domain in subcomplex II. The other EPRS1 regions, such as EARS1 and WHEP, are essential for subcomplex III formation.

Symmetry of Subcomplexes in MSC

Most components comprising human MSCs are class I monomeric ARSs. Class II dimeric ARSs, including DARS1, KARS1, and EPRS1 via PARS1, duplicate their binding partners, suggesting that MSC is maintained as a symmetrical structure.

The three class II ARSs are located in each subcomplex. In subcomplex I, the N-terminal end (residues 1–31) of AIMP2 binds to the KARS1 dimer (Ofir-Birin et al. 2013). KARS1 interaction with AIMP2 leads to formation of a QARS1/RARS1/AIMP1/AIMP2/KARS1 complex in a 1:1:1:1:2 stoichiometry. This complex is duplicated by dimeric DARS1 in subcomplex II. Thus, the subcomplex I forms in a 2:2:2:2:4 stoichiometry in a linear arrangement of the order QARS1–RARS1–AIMP1–AIMP2–(KARS1/KARS1)–(KARS1/KARS1)–AIMP2–AIMP1–RARS1–QARS1.

The DARS1 dimer plays two key roles in MSC. First, it mediates formation of the symmetrical subcomplex I structure. Second, it holds two tetrameric GST complexes

Fig. 6.7 (continued) through its GST domain. (**b**) The MARS1–AIMP3–EPRS1–AIMP2–DARS1 pentameric complex is subcomplex II. It assembles through the GST-like domains. Four GST domains are linked linearly in the order MARS1–AIMP3–EPRS1–AIMP2. DARS1 binds to the GST domain of AIMP2. (**c**) The EPRS1–IARS1–LARS1 trimeric complex is subcomplex III. LARS1 binds to IARS1 via its UNE-L domain, while IARS1 binds to EPRS1 via its UNE-I domain. In EPRS1, the EARS1 domain is responsible for the interaction with IARS1. AIMP2 and EPRS1 are responsible for interaction with subcomplexes I and III, respectively in subcomplex II to achieve a 2:2:2:2:2 stoichiometry. The linear arrangement of subcomplex II follows the order MARS1–AIMP3–EPRS1(GST)–AIMP2–DARS1–DARS1–AIMP2–EPRS1(GST)–AIMP3–MARS1. Structural models of the decameric subcomplex II generated by SAXS are consistent with this linear arrangement (Cho et al. 2019).

In subcomplex III, EPRS1 plays a structural role as a PARS1-mediated dimer. As described, subcomplex III contains three ARSs in the order LARS1–IARS1–EPRS1 in a 2:2:2 stoichiometry. Thus, the linear arrangement of ARSs in subcomplex III in a two-dimensional layout is LARS1–IARS1–(EARS1–PARS1)–(PARS1–EARS1)–IARS1–LARS1.

In summary, the KARS1, DARS1, and PARS1 domains in EPRS1 reside at the twofold symmetrical center of each subcomplex. Subcomplexes I and II are linked through AIMP2, while subcomplexes II and III are linked through EPRS1 (Fig. 6.8a). These stoichiometric and topological analyses provide a twofold symmetrical, two-dimensional model of an MSC containing 24 molecules (Fig. 6.8b).

Spatial Arrangement of MSC Components

Experimental evidence obtained thus far indicates that MSC plays a central role in controlling biological system complexity to maintain life (Kim et al. 2002; Kim and Kim 2020; Kwon et al. 2019; Schimmel 2018). Although the role of MSCs beyond protein synthesis is still unclear, the MSC structure (based on studies of its protein components) correlates with increasing biological complexity from yeast to humans. MSC was first discovered several decades ago. However, it remains unknown how MSC is assembled and maintained and how its functions are coordinated with other biological processes.

EM studies (Norcum and Boisset 2002) and SAXS (Dias et al. 2013) using MSCs isolated from rabbit shows low-resolution 3D structures of holo-MSC (Fig. 6.9). A model of the 3D MSC structure was constructed from EM images of negatively stained and frozen hydrated samples using three independent angular assignment methods (Fig. 6.9a). EM analysis reveals a globular structure that is approximately $19 \times 16 \times 10$ nm in size. The 3D volumes show an asymmetric triangular arrangement of protein domains around a deep central cavity (Norcum and Boisset 2002).

Results from six independent ab initio shape determinations reveal an extended and multiarmed shape of rabbit MSC with a radius of gyration and maximal diameter of 13.9 and 52 nm, respectively (Fig. 6.9b). These results suggest an MSC structure with a non-globular elongated shape consisting of broadly distributed protein components that differ from previous EM 3D reconstructions (Dias et al. 2013).

Structural images are snapshots of the dynamic conformations adopted by the MSC, and many represent the most abundant shapes and sizes under specific experimental conditions. As described in the section on "Symmetry of Subcomplexes in MSC," human holo-MSC contains 24 protein molecules, and its

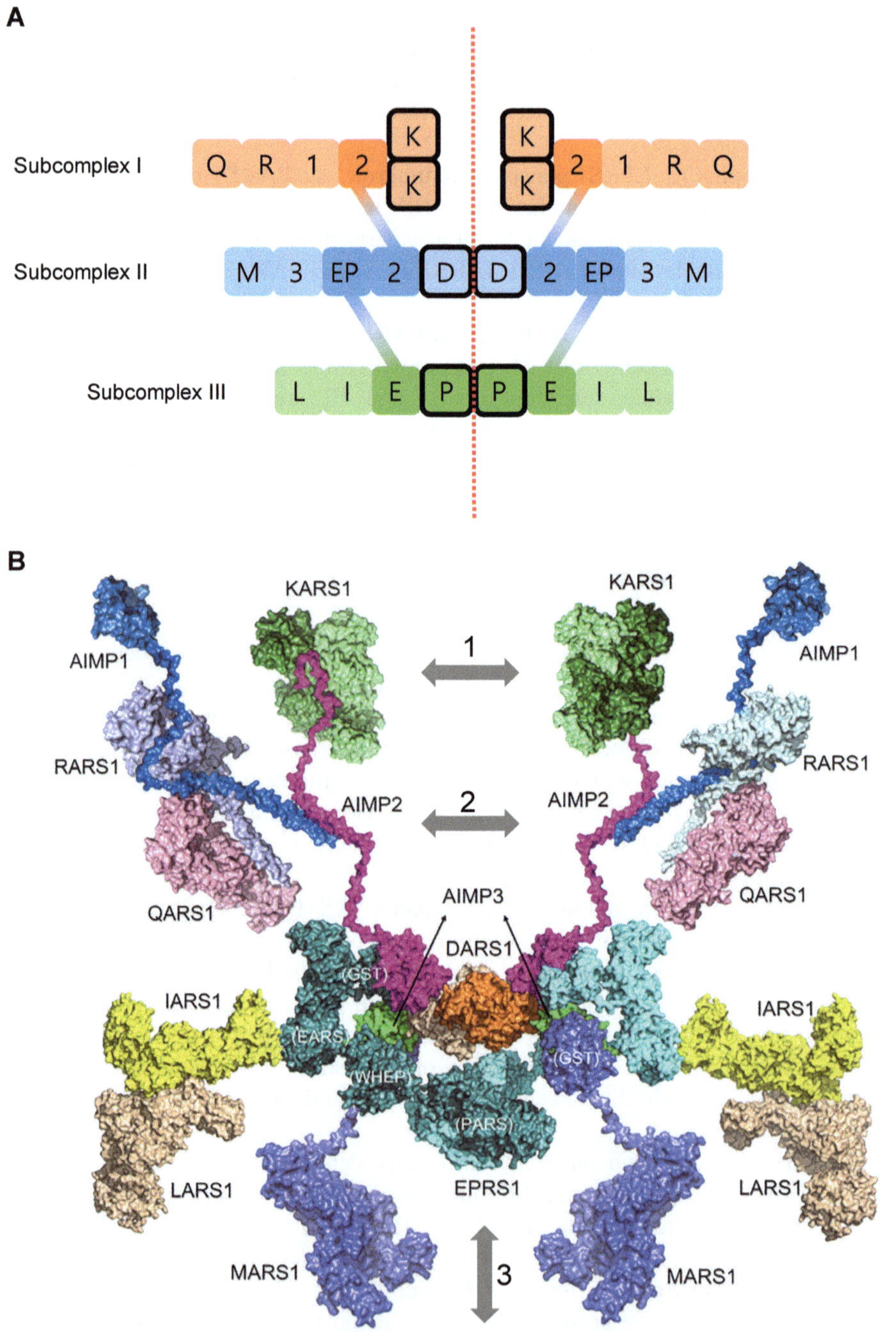

Fig. 6.8 Overall connections in the MSC. (**a**) Duplication of MSC subcomplexes. Among the MSC components, KARS1, DARS1, and PARS1 belong to class II ARS and form homodimers. Homodimerization of EPRS1 through PARS1 enables duplication of subcomplex III. A DARS1

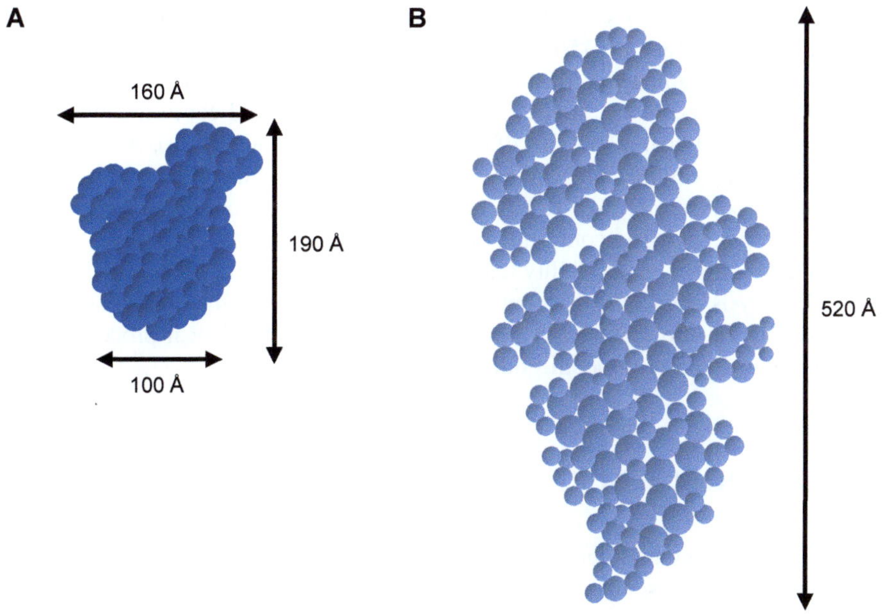

Fig. 6.9 Size estimation of an MSC supramolecular complex. (**a**) The overall shape of rabbit MSC built by SAXS. Figure modified from Dias et al. (2013). (**b**) A reconstructed EM image of rabbit MSC. Figure modified from Norcum and Boisset (2002)

molecular weight is estimated to be ~1.5 MDa. From this value, the volume of the MSC is ~1.85×10^6 Å^3, based on the specific volume of a protein (~0.74 cm^3/g) (Brown et al. 2011; Kim et al. 2021a). Assuming that MSC protein components are crystallographically packed in solution, the volume of the MSC, including solvent molecules in the void space between components, is calculated to be 3.75×10^6 Å^3

Fig. 6.8 (continued) dimer linking two tetrameric GST complexes is responsible for the duplication of subcomplex II. In subcomplex I, AIMP2 binds to a KARS1 dimer. Two AIMP2s in the duplicated subcomplex II hold two copies of subcomplex I. The twofold symmetric conformation of DARS1 and PARS1 dimers allows a twofold symmetric arrangement of MSC components. The symmetric axis is indicated by a red dotted line. (**b**) Spatial arrangement of 24 protein molecules in a twofold symmetrical MSC model. A DARS1 dimer holds two AIMP2s, and each AIMP2 links to EPRS1 through its GST domains. Two EPRS1s hold each other using the PARS domain, and two IARS1–LARS1 complexes are attached through each EARS1 domain of an EPRS1 dimer. MARS1 links to each EPRS1 through AIMP3. The QARS1–RARS1–AIMP1 complex aligns to the LZ of AIMP2, while the KARS1 dimer is linked to the N-terminal peptide of AIMP2. Grey arrows indicate possible movements in the arrangement of component proteins. Two KARS1 dimers may form a tetrameric complex with two antiparallel N-terminal peptides from two AIMP2 molecules (arrow 1). Coiled-coil interactions between multiple helices enable additional interactions among the complexes of leucine zipper motifs (arrow 2). The PARS1 dimer moves away from the DARS1 dimer by extension of EPRS1 linker peptides (arrow 3)

by applying a Matthew's coefficient of 2.5 per Dalton. This represents a cubic volume with a length of 150 Å. By comparison, the approximate volumes of hexahedrons containing the MSC measured by EM and SAXS are 3.0×10^6 and 5.0×10^6 Å^3, respectively, suggesting compact and extended conformations of the MSC caused by dynamic motions of components (Kim et al. 2021a). As described, the MSC plays central roles in protein synthesis and functions in a series of biological processes through tRNA charging, as well as interactions with other cellular components. The conformational flexibility of the MSC correlates with its multifunctionality. The possibility that some of MSC components are missing from the EM analysis cannot be discounted. As such, uncovering the precise structure–function relationship among MSC components and the biological processes they regulate requires more in-depth studies.

Molecular Root of MSC Dynamics

Disordered peptides between the structural domains of ARSs and AIMPs provide flexibility during assembly of the MSC. A cross-linking mass spectrometry (XL-MS) approach using HEK293T cell–derived MSC reveals structural information on the flexibility of appended proteins, which is characteristic of nearly all MSC components (Khan et al. 2020). Moreover, a split-NanoLuc complementation system enables a better understanding of the flexible framework of the mammalian MSC under physiological conditions (Kong and Kim 2020). Thus, all of these analyses and observations lead to the conclusion that the human MSC is an intrinsically dynamic complex that controls multiple cellular functions and protein synthesis simultaneously.

KARS1

KARS1 integrates MSC components from *C. elegans* to humans (Kim and Kim 2020). As a class II ARS, KARS1 exists as a dimer. One dimeric KARS1 interacts with one of the AIMP2 N-terminal end peptides in human MSC (Fig. 6.10a). This allows two KARS1 dimers and two AIMP2s to assemble into a V-form within the MSC structure (Ofir-Birin et al. 2013). The two AIMP2 molecules bind to homodimeric DARS1 via their GST domains, and these interactions hold two KARS1 dimers in the MSC (Fig. 6.10b). In this conformation, two KARS1 dimers are not able to directly interact. As such, each KARS1 dimer could have independent flexibility in the MSC to exert both canonical and noncanonical functions.

Another crystal structure of the KARS1 and AIMP2 complex reveals that two KARS1 dimers are sequestered by two AIMP2 molecules into a tight X-form assembly (Fig. 6.10c, d) (Hei et al. 2019). In this form, the two AIMP2 N-terminal end peptides form an antiparallel scaffold and hold two KARS1 dimers through four binding motifs and additional interactions with other MSC components

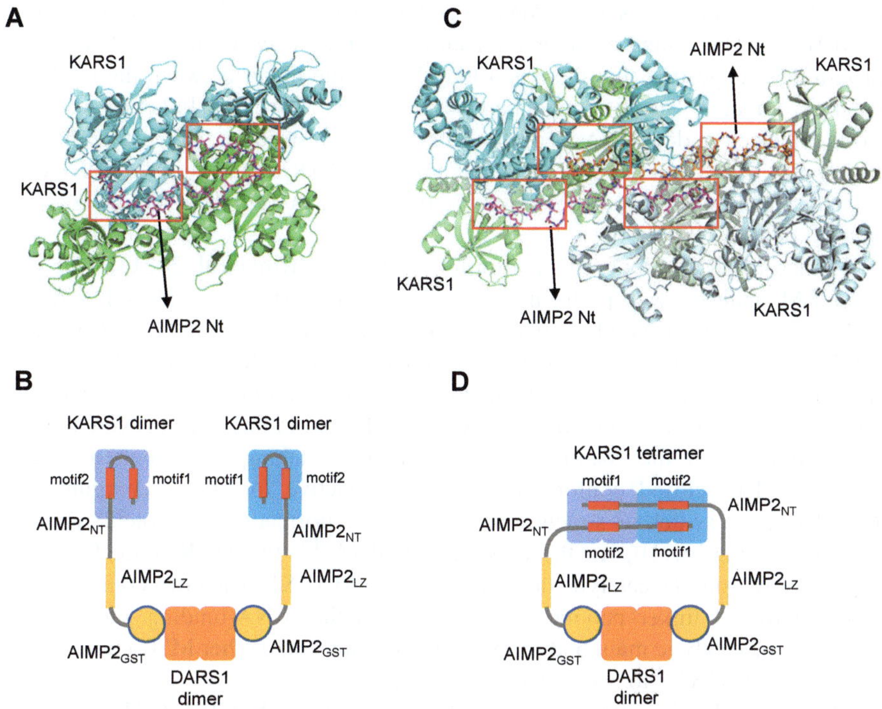

Fig. 6.10 Conformational changes in the MSC due to alternative assembly of KARS1 and AIMP2. (**a**) The N-terminal peptide of an AIMP2 molecule binds to a KARS1 dimer. There are two KARS1-binding motifs in the N-terminal peptide of AIMP2. Each motif binds to each binding pocket in twofold symmetrical positions in a KARS1 dimer. Red rectangles mark the binding pockets of KARS1. (**b**) A schematic diagram of two AIMP2–KARS1 complexes connected by DARS1. The two complexes are separated from each other (V-form). (**c**) The two N-terminal peptides from two AIMP2 molecules are arranged in an antiparallel orientation. Each peptide contains two motifs, which bind two KARS1 dimers simultaneously. (**d**) A schematic diagram of the interactions between two AIMP2s complexed with DARS1 to form a tetrameric KARS1 (X-form)

(Fig. 6.10c, d). Four catalytic subunits of KARS1 in this tightly assembled complex are all accessible for tRNA recognition (Hei et al. 2019). In addition, human disease–associated mutations (L350H and P390R) in KARS1 (Verrigni et al. 2017) disturb both X-form complex assembly and enzyme activity (Hei et al. 2019). These observations suggest that the V and X forms of KARS1 coexist dependently under cellular physiological conditions.

MARS1

Human MARS1 is a 900-residue protein. It has two appended GST and WHEP domains at the N- and C-terminal regions of the main protein body, respectively. The

GST domain complexes with AIMP3 and is encompassed by residues Met1 to Lys191. The first strand of the Rossmann fold within its catalytic main body starts at Asn265, suggesting that there is a long 73-residue linker between the GST and catalytic domains. The N-terminal part (residues 211–220) of this linker region is not visible in electron density maps of the MARS1 crystal structure, suggesting that it is highly flexible (Kim et al. 2021a). Unlike human MARS1, the yeast MARS1 GST domain attaches directly to the catalytic domain, with very few linking residues (Galani et al. 2001).

Human MARS1 adopts two conformational states (Kim et al. 2021a). Negative stain EM analysis was performed using purified MARS1 complexed with the AIMP3, GST, and EARS1 domains of EPRS1 and with AIMP2 GST (Cho et al. 2015). The 3D reconstruction model of the complex shows that the MARS1 catalytic main body associates closely with its GST domain and other protein components (Fig. 6.11a). In this conformation, the MARS1 tRNA-binding site is covered by the GST domain and the N-terminal linker peptide, making it difficult to accept $tRNA^{Met}$. This closed conformation may be inactive for tRNA charging, and MARS1 has to switch its structure into an open conformation for protein synthesis to proceed (Fig. 6.11b). In the presence of $tRNA^{Met}$, the MARS1 GST domain moves away from the catalytic domain to convert MARS1 to an open conformation. The N-terminal linker peptide dissociates from the tRNA-binding site, thereby relieving the catalytic main body from its GST domain and other MSC components. This allows $tRNA^{Met}$ to access the active site.

Because AIMP3 contacts DARS1 directly, and the MARS1 GST domain is adjacent to DARS1 (Cho et al. 2019; Hahn et al. 2019), the MARS1 catalytic domain positions itself around DARS1. The 3D architecture of MSC determined using CL-MS reveals a cross-link between Lys451 of DARS1 and Lys729 of MARS1 (Khan et al. 2020). In the MARS1–AIMP3–EPRS1–AIMP2–DARS1 model, the distances from A211 at the MARS1 GST domain to the two K451 residues in a DARS1 dimer are 59 and 62 Å. The space between A221 and K729 at the MARS1 catalytic core is 31 Å. Thus, the N-terminal linker (A211–A221) in MARS1 is extended as an open conformation to allow cross-linking between DARS1 Lys451 and MARS1 Lys729 (Fig. 6.11b). In the MSC, MARS1 is likely present in a dynamic equilibrium between the open and closed conformations.

EPRS1

Human EPRS1 consists of GST, EARS1, three WHEP, and PARS1 domains (Fig. 6.11c). Small-angle neutron scattering (SANS) experiments provide models for GST-EARS1 with two WHEP domains and half of the third WHEP domain (WHEP5/6). The size estimate of GST-EARS1 is 140 × 69 Å, while that of GST-EARS1–WHEP5/6 is 170 × 79 Å (Fig. 6.11d). The difference (30 Å) in the longitudinal direction is caused by WHEP5/6. The WHEP domain, which is found

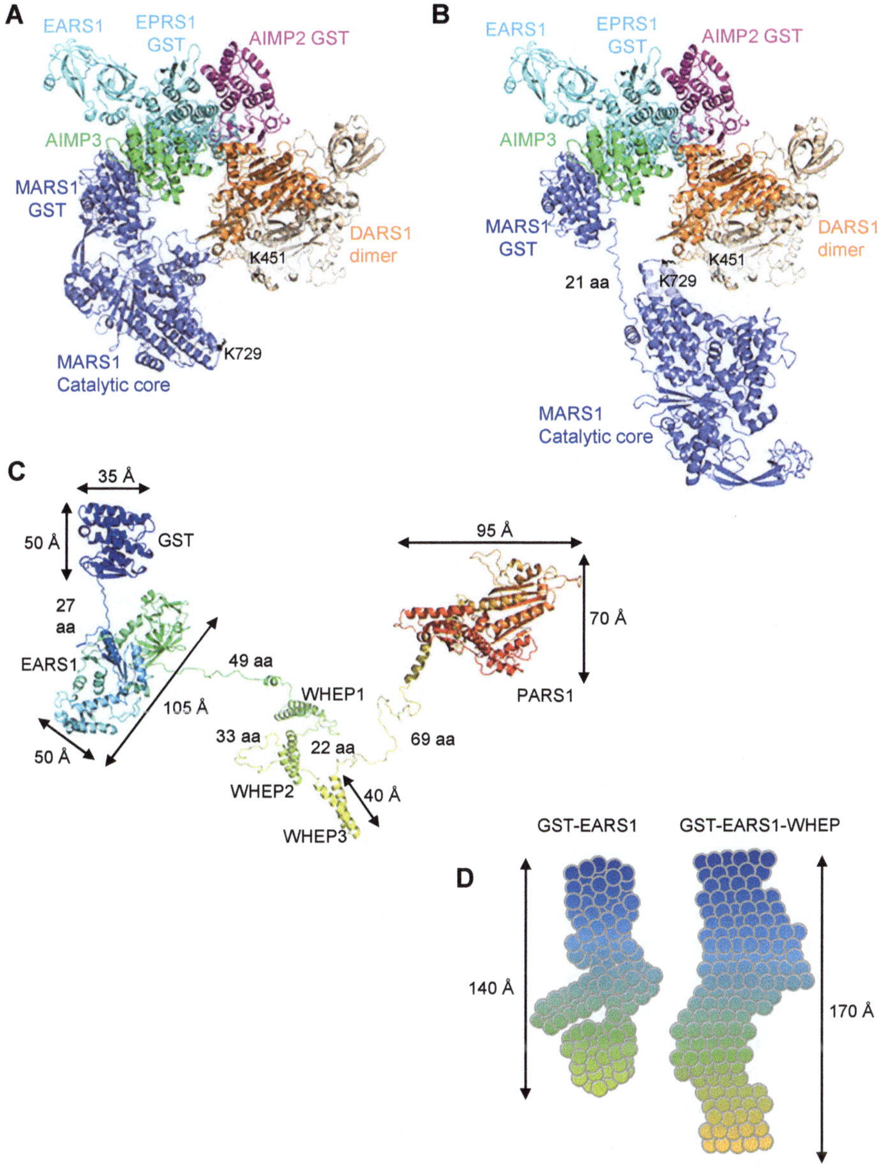

Fig. 6.11 Local conformational changes in MSC components. Closed (**a**) and open (**b**) conformations of MARS1. Ribbon diagrams show that the complex consists of MARS1 (blue), AIMP3 (green), EPRS1 (GST and EARS1) (cyan), AIMP2 (GST) (magenta), and a DARS1 dimer (orange/wheat). The linker peptide, which consists of 21 residues between the GST and catalytic domains of MARS1, can expand to allow relatively free positioning of the MARS1 catalytic domain against other MSC component proteins in the open conformation. K729 of MARS1 is located close to K451 of DARS1. (**c**) Domain sizes of EPRS1. Linker peptides connect the domains of EPRS1. A peptide with 27 residues lies between the N-terminal GST and EARS1 domains. The EARS1 and PARS1 domains are connected to three WHEP domains by peptides comprising 49 and 69 residues,

only in eukaryotic ARSs, is a 46–amino acid helix-turn-helix that is 40 Å in length. The linker between EARS1 and WHEP1 has 49 residues, while two linkers within the WHEP domain region consist of 22 and 33 residues (Fig. 6.11c). The size of EPRS1 estimated by SANS indicate that EARS1 and WHEP domains associate closely.

Size-exclusion chromatographic analysis of GST-EARS1–WHEP5/6 shows that its apparent molecular mass is larger than expected. The SAXS-derived structure of the MSC exhibits an elongated shape with multiple arms, which is about 510 Å long. To achieve this size, the linker peptide between the WHEP and PARS1 domains in EPRS1 should extend because extensible regions within subcomplexes I and II of the MSC are limited.

Flexibility of MSC Components via Linker Peptides

Human ARSs have additional peptides and domains adjacent to the catalytic core domains. MSC component proteins are multidomain proteins that contain linker peptides among their structured domains (see Fig. 6.3). These linker regions provide flexibility not only for specifying individual protein conformations, but also for the overall shape of the MSC.

Protein complex assembly through interactions among the compact domains and motifs of MSC components provide rigidity to the MSC body. For example, GST domains from four proteins produce a rigid body, while LZ motifs from three proteins form a stem body. The structural stability of the domain complexes is increased via multiple interactions. However, the linker parts of MSC component proteins also contribute to structural disorder with respect to the orientation of rigid bodies.

Structural disorders are expected wherever linker peptides are present (Fig. 6.3). QARS1 has a linker between UNE-Q and the catalytic core domains. EPRS1 has five regions between GST, EARS1, three WHEP, and PARS1 domains (as shown in Fig. 6.11c). MARS1 has two regions, between GST and the catalytic domains and between ABD and WHEP. LARS1 and IARS1 have linkers between their catalytic cores and UNE-L or UNE-I, respectively. RARS1 has peptides between LZ and the catalytic core. AIMP1 has a linker between LZ and EMAPII, while AIMP2 has flexible regions between the N-terminal KARS1-binding motif and LZ and between LZ and GST.

Fig. 6.11 (continued) respectively. There are 33 residues and 22 residues between WHEP1 and WHEP2 and between WHEP2 and WHEP3, respectively. (**d**) SANS analyses reveal the relatively compact shape of EPRS1. The envelopes for GST-EPRS1 (left) and GST-EARS1-WHEP5/6 (right) are 170 and 140 Å in length, respectively. The overall shape and the size of the envelopes indicate that the GST, EARS1, and WHEP domains are packed compactly. The envelopes are colored as in (**c**). Figure modified from Halawani et al. (2018)

Structural disorder in a protein can be limited. AIMP1 utilizes the N-terminal LZ to anchor to the MSC. The linker peptide confers freedom in the orientation of the C-terminal EMAPII against the MSC. However, the flexibility of a linker peptide can alter the orientation of the subcomplexes. AIMP2 has a linker peptide connecting the GST domain and the LZ motif, which are involved in the assembly of subcomplexes I and II, respectively. The flexibility of the linker results in a dramatic conformational change in the MSC. Similarly, a small rotation in the linker position between the GST and EARS1 domains in EPRS1 causes a huge conformational alteration in the MSC because EPRS1 links subcomplexes II and III.

MSC Dynamics–Based Function Repertoire

Genetic depletion of the core MSC scaffold protein, AIMP2, causes lethality in mice without affecting global protein synthesis and cell viability. This demonstrates that the structural integrity of the MSC is essential at the systems and organismal level, but not necessarily at the cellular level (Kim et al. 2002). Human MSC is recognized as a depot system from which its components are released via posttranslational modifications to control cell signaling networks in a stimulus-dependent manner (Hyeon et al. 2019; Kim and Kim 2020; Lee et al. 2004a) (Fig. 6.12). These activities correlate with structurally appended region-mediated flexible assembly of the MSC. How the MSC provides a broad range of biological functions beyond translation is a fundamental biological question that needs to be addressed. Examples of MSC-driven biological functions are discussed below.

KARS1

KARS1 activates the immune response by mast cells (Yannay-Cohen et al. 2009). The immune system is activated by KARS1 through microphthalmia-associated transcription factor (MITF)-dependent transcriptional control, which induces production of allergic mediators by the cells. MITF activity is suppressed in quiescent mast cells through interaction with the histidine triad nucleotide-binding protein (Hint) repressor in the nucleus. Transcriptional activation requires disruption of MITF/Hint interaction (Lee et al. 2004b). Under normal physiological conditions, KARS1 localizes to cytoplasmic MSC through interaction with AIMP2. Upon immune activation, KARS1 is phosphorylated at Ser207, which is located at the junction of the catalytic and tRNA-binding domains (Ofir-Birin et al. 2013). Phosphorylation causes a structural change, leading to disruption of its interaction with AIMP2, and dissociation of KARS1 from the MSC (Ofir-Birin et al. 2013). Dissociated KARS1 then translocates from the cytoplasm to the nucleus, where it binds to the MITF/Hint complex. KARS1-mediated disruption of the MITF/Hint complex triggers production of the second messenger, diadenosine tetraphosphate, which

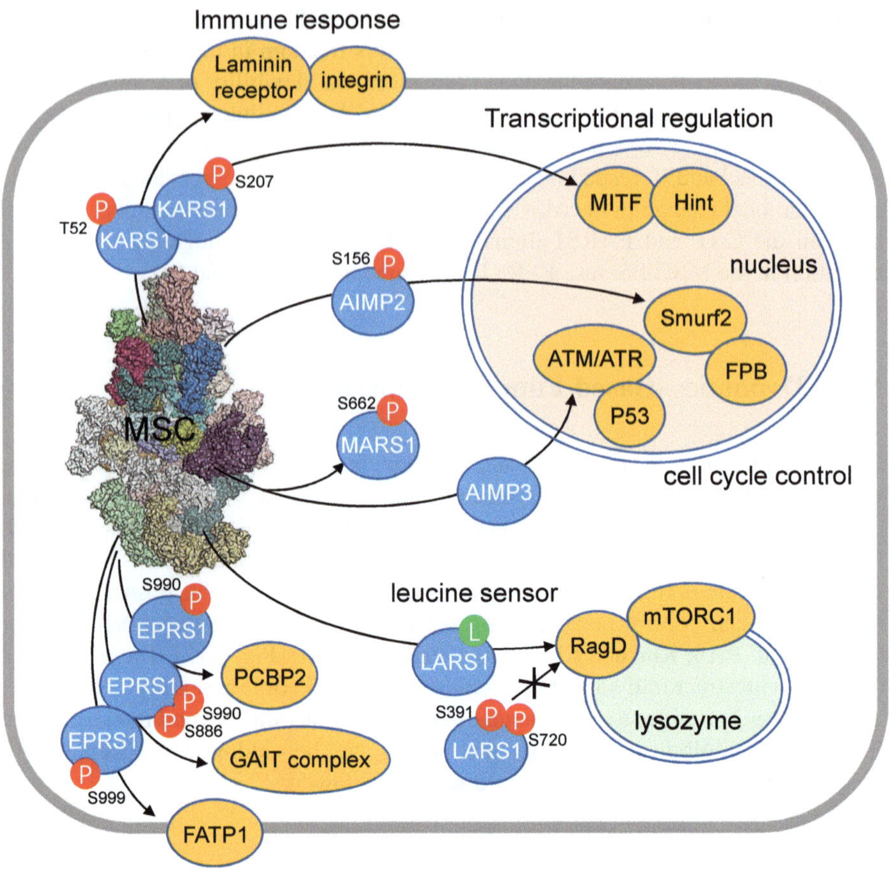

Fig. 6.12 Proposed functions of MSC components released from the human MSC. Among the component proteins, EPRS1, LARS1, KARS1, MARS1, AIMP2, and AIMP3 are shown in the figure. Released EPRS1 associates with PCBP2, the GAIT complex, or FATP1, after phosphorylation of serine residues. LARS1 with leucine translocates to the lysozyme and modulates RagD, which regulates mTORC1. Binding of leucine to LARS1 is blocked by phosphorylation at residues S391 and S720. Phosphorylated MARS1 at S662 releases AIMP3, which then translocates to the nucleus. In the nucleus, AIMP3 interacts with ATM/ATR. Phosphorylated AIMP2 at S156 translocates to the nucleus and binds to smurf2. Phosphorylated KARS1 at T52 interacts with a laminin receptor, while KARS1 phosphorylated at S207 translocates to the nucleus and binds to MITF

transcriptionally activates MITF-targeted genes (Yannay-Cohen et al. 2009). These studies demonstrate that a single conformational change triggered by phosphorylation of the MSC switches KARS1 from a translation to a transcription function.

KARS1 is highly expressed by various cancer cells and tissues. This enzyme is phosphorylated at Thr52 in a laminin-dependent manner (Kim et al. 2012, 2014). The modification-mediated dissociation of KARS1 from the MSC leads to its translocation to the plasma membrane and subsequent association with the 67-kDa laminin receptor (67LR). This event prevents 67LR from neural precursor cell–expressed developmentally downregulated protein 4 (Nedd4)-mediated ubiquitination and increases 67LR stability in the plasma membrane. This in turn enhances laminin-induced cell migration, a process associated with cancer metastasis. These observations suggest a specific way to control KARS1–67LR interaction-mediated metastasis (Kim et al. 2014).

LARS1

LARS1 is a leucine sensor that regulates mTORC1 activity. The mTORC1 complex functions in translation, metabolism, autophagy, and cell growth (Han et al. 2012; Kim et al. 2017; Lee et al. 2018b; Yoon et al. 2020). Although it is not known how LARS1 is released from its home MSC, a structural study shows two distinct functional states of LARS1 for mTORC1 activation (Kim et al. 2021b). Leucine binding to the catalytic domain leads to a conformational change in its C-terminal domain, leading to an association with RagD. RagD is the GTPase-activating protein required for mTORC1 activation. The work of Kim et al. (2021b) unveils the molecular basis for leucine-dependent long-range communication between the catalytic and RagD-binding domains of LARS1 for mTORC1 activation.

Upon its translocation to the lysosome, LARS1 interacts directly with, and facilitates GTP hydrolysis of, RagD in a leucine-dependent manner (Han et al. 2012). Kinetic analysis of the RagD–RagB GTPase cycle during leucine signaling further demonstrates that LARS1 serves as an initiating switch via GTP hydrolysis of RagD. GTP hydrolysis regulates the entire Rag GTPase cycle and mTORC1 activation, whereas Sestrin2 acts as a terminating switch by controlling GTP hydrolysis of RagB in the Rag GTPase–mTORC1 axis (Lee et al. 2018b).

LARS1 also regulates translation through the glucose-dependent control of leucine usage (Yoon et al. 2020). Upon sensing leucine in the presence of glucose, LARS1 functions in protein synthesis by leucylating the corresponding tRNA and activating mTORC1. Moreover, LARS1 inhibits mTORC1 activation under low glucose conditions. Under low glucose, no leucine-dependent interaction between LARS1 and RagD occurs. The absence of LARS1–RagD interaction under low glucose is driven by the sequential phosphorylation of LARS1 at Ser391 and Ser720 by Unc-51-like autophagy activating kinase 1. This event saves ATP consumption and allows leucine to be used for ATP generation, leading to tolerance against glucose starvation by rebalancing leucine metabolism (Yoon et al. 2020). These results reveal the coupling between an amino acid and glucose to mediate LARS1 function during mTORC1 activation.

EPRS1

In addition to its function in protein synthesis, EPRS1 plays multiple roles in regulating immunity and metabolism via stimulus-dependent phosphorylation at the Ser886, Ser990, or Ser999 residues (Arif et al. 2009, 2012, 2017; Kim and Kim 2020; Lee et al. 2016, 2018a; Yao et al. 2012). In mouse and human monocytes and macrophages, interferon (IFN)-gamma treatment releases EPRS1 from the MSC when it is phosphorylated at S886 and Ser999. These residues are located in the linker containing the WHEP domains (Arif et al. 2009, 2012). Phosphorylated EPRS then associates with other proteins to form the IFN-gamma-activated inhibitor of translation (GAIT) complex. The GAIT complex binds to the 3′-untranslated region containing GAIT elements in inflammation-related mRNAs, leading to the translational inhibition of selected transcripts (Yao et al. 2012).

In adipocytes, the insulin-activated mTORC1–S6K1 axis phosphorylates EPRS1 at Ser999. Phosphorylation induces the release of EPRS1 from the MSC. Free EPRS1 then interacts with and transports fatty acid transport-1 (FATP1) to the plasma membrane for long-chain fatty acid uptake and triglyceride synthesis (Arif et al. 2017).

Upon RNA virus infection, the MSC senses the virus and activates EPRS1 to promote antiviral immunity via infection-specific phosphorylation of Ser990 (Lee et al. 2016, 2018a). Infection-specific phosphorylation of EPRS induces its dissociation from the MSC. Upon its release from the MSC, EPRS1 interacts with poly (rC) binding protein 2 (PCBP2), a negative regulator of mitochondrial antiviral signaling protein (MAVS), which is critical for antiviral immunity. Interaction of EPRS1 with PCBP2 blocks PCBP2-mediated ubiquitination of MAVS and suppresses viral replication.

The bifunctional EPRS1 resides at the exterior of the MSC, and the phosphorylated serine residues are located in the WHEP domain region. The use of multiple and selective phosphorylation sites in an accessible region of EPRS1 presents a highly efficient system of switching EPRS1 function in response to different stimuli.

AIMP2

AIMP2 is a core scaffold protein that integrates MSC structure (Cho et al. 2019; Hahn et al. 2019; Han et al. 2008). Upon stimulation of HeLa cells with transforming growth factor-beta (TGF-beta), AIMP2 is phosphorylated at Ser156, triggering its detachment from DARS1 and release from the MSC (Kim et al. 2016). Free AIMP2 translocates to the nucleus, where it interacts with SMAD-specific E3 ubiquitin protein ligase 2 (Smurf2), which enhances ubiquitination of the far upstream element (FUSE)-binding protein (FBP). FBP is a transcriptional activator of c-myc, which functions to suppress tumorigenesis (Kim et al. 2003).

Upon DNA damage, AIMP2 is also phosphorylated and released from the MSC. Released AIMP2 translocates to the nucleus, where it acts as a proapoptotic mediator by interacting directly with the p53 tumor suppressor to protect p53 from MDM2-mediated ubiquitination and degradation (Han et al. 2008).

AIMP3

AIMP3 is released from the MSC in a MARS1-dependent manner to function in the nucleus. It interacts tightly with the MARS1 GST domain. Upon DNA damage, MARS1 is phosphorylated at Ser662 by the general control non-repressed 2 kinase (Kwon et al. 2011). Phosphorylation of Ser662 induces a conformational change in MARS1 and disrupts its interaction with AIMP3, thereby releasing AIMP3 from the MSC. The released AIMP3 translocates to the nucleus, where it interacts directly with ataxia–telangiectasia mutated (ATM)/ATM- and RAD3-related (ATR) kinases to activate p53. The activation of p53 by the ATM/ATR–AIMP3 pathway is a potent tumor suppressor (Park et al. 2005).

Perspectives

There is a clear tendency for the addition of motifs and domains to eukaryotic ARSs during evolution. The emergence of such structural domains of ARSs is linked to their functional versatility. Experimental evidence has uncovered noncanonical functions of ARSs, in addition to their conventional roles in protein synthesis. The functional diversification of eukaryotic ARSs driven by the evolution of new motifs and domains likely reflects the increasing biological complexity of higher organisms. These noncanonical functions are evident not only in the cytoplasm, but also in the nucleus and outside the cell (Guo and Schimmel 2013). The non-translational functions of ARSs can also occur through the actions of alternative splice variants or proteolytic fragments (Lo et al. 2014).

Most intriguing is that the activities of many ARSs in higher eukaryotes are regulated by their presence in a cytoplasmic depot system called the MSC, which assembles in most cases via appendage of its individual protein components. Although the role of the MSC in cells is unclear, there are proposals that it facilitates the channeling of amino-acylated tRNAs to the ribosome during protein synthesis (Negrutskii and Deutscher 1991, 1992; Sang Lee et al. 2002; Sivaram and Deutscher 1990; Stapulionis and Deutscher 1995). The MSC also serves as a cellular reservoir of components with various noncanonical cellular functions. Under stress conditions, for example, protein components are released from the MSC via posttranslational modifications (mainly phosphorylation). Upon release, some of these protein components participate in non-translational events, such as immunity and cell metabolism (Arif et al. 2017; Lee et al. 2016; Yao et al. 2012; Yoon et al. 2020).

Although the MSC was first discovered several decades ago, and knowledge of ARSs has increased, several aspects about their roles in modulating cellular processes remain unknown. For instance, it is not clear why MSC components exist as a complex in the cytoplasm and how ARSs regulate protein synthesis when they are within the MSC complex. Furthermore, the mechanisms by which the MSC assembles, and how this assembly coordinates the specific release of its individual components to initiate noncanonical cellular functions, remain unresolved. The size and number of MSC components appear to be related to the complexity of life. However, little is known about the correlation between MSC composition and the complexity of biological systems. No single approach can answer these questions because of the dynamic nature and functional complexity of the MSC. Most studies to date have only investigated the behavior of one or a few MSC components. Therefore, fundamental questions regarding the mechanisms that coordinate protein synthesis and cellular signaling have yet to be addressed at the system and organism levels. These questions are among the most challenging in biology.

In conclusion, domain addition and functional expansion are not limited to the ARSs in the MSC. The emergence of new domains and structures is observed in almost all ARSs. Research progress on ARSs and MSCs raises new questions in biology that could advance the fields of protein translation and cell signaling.

Acknowledgements This work is supported by the National Research Foundation grant, funded by the Ministry of Science and ICT of Korea (2020R1A2C1003541 to B.S.K) and by a Samsung Science and Technology Foundation grant (SSTF-BA1902-04 to M.H.K.).

References

Ahn HC, Kim S, Lee BJ (2003) Solution structure and p43 binding of the p38 leucine zipper motif: coiled-coil interactions mediate the association between p38 and p43. FEBS Lett 542:119–124

Allocati N, Masulli M, Di Ilio C, Federici L (2018) Glutathione transferases: substrates, inihibitors and pro-drugs in cancer and neurodegenerative diseases. Oncogenesis 7:8

Arif A, Jia J, Mukhopadhyay R, Willard B, Kinter M, Fox PL (2009) Two-site phosphorylation of EPRS coordinates multimodal regulation of noncanonical translational control activity. Mol Cell 35:164–180

Arif A, Chatterjee P, Moodt RA, Fox PL (2012) Heterotrimeric GAIT complex drives transcript-selective translation inhibition in murine macrophages. Mol Cell Biol 32:5046–5055

Arif A, Terenzi F, Potdar AA, Jia J, Sacks J, China A, Halawani D, Vasu K, Li X, Brown JM et al (2017) EPRS is a critical mTORC1-S6K1 effector that influences adiposity in mice. Nature 542: 357–361

Brown PH, Balbo A, Zhao H, Ebel C, Schuck P (2011) Density contrast sedimentation velocity for the determination of protein partial-specific volumes. PLoS One 6:e26221

Burbaum JJ, Schimmel P (1991) Structural relationships and the classification of aminoacyl-tRNA synthetases. J Biol Chem 266:16965–16968

Cerini C, Kerjan P, Astier M, Gratecos D, Mirande M, Semeriva M (1991) A component of the multisynthetase complex is a multifunctional aminoacyl-tRNA synthetase. EMBO J 10:4267–4277

Cestari I, Kalidas S, Monnerat S, Anupama A, Phillips MA, Stuart K (2013) A multiple aminoacyl-tRNA synthetase complex that enhances tRNA-aminoacylation in African trypanosomes. Mol Cell Biol 33:4872–4888

Cho HY, Maeng SJ, Cho HJ, Choi YS, Chung JM, Lee S, Kim HK, Kim JH, Eom CY, Kim YG et al (2015) Assembly of multi-tRNA synthetase complex via heterotetrameric glutathione transferase-homology domains. J Biol Chem 290:29313–29328

Cho HY, Lee HJ, Choi YS, Kim DK, Jin KS, Kim S, Kang BS (2019) Symmetric assembly of a decameric subcomplex in human multi-tRNA synthetase complex via interactions between glutathione transferase-homology domains and aspartyl-tRNA synthetase. J Mol Biol 431: 4475–4496

Cusack S, Berthet-Colominas C, Hartlein M, Nassar N, Leberman R (1990) A second class of synthetase structure revealed by X-ray analysis of Escherichia coli seryl-tRNA synthetase at 2.5 A. Nature 347:249–255

Dang CV, Yang DC (1978) Affinity chromatography of rat liver aminoacyl-tRNA synthetase complex. Biochem Biophys Res Commun 80:709–714

Denney RM (1977) Detection and partial purification of rapidly sedimenting forms of aminoacyl-transfer ribonucleic acid synthetases from human placenta. Arch Biochem Biophys 183:156–167

Dias J, Renault L, Perez J, Mirande M (2013) Small-angle X-ray solution scattering study of the multi-aminoacyl-tRNA synthetase complex reveals an elongated and multi-armed particle. J Biol Chem 288:23979–23989

Durr E, Bosshard HR (2000) Folding of a three-stranded coiled coil. Protein Sci 9:1410–1415

Eriani G, Delarue M, Poch O, Gangloff J, Moras D (1990) Partition of tRNA synthetases into two classes based on mutually exclusive sets of sequence motifs. Nature 347:203–206

Fu Y, Kim Y, Jin KS, Kim HS, Kim JH, Wang D, Park M, Jo CH, Kwon NH, Kim D et al (2014) Structure of the ArgRS-GlnRS-AIMP1 complex and its implications for mammalian translation. Proc Natl Acad Sci USA 111:15084–15089

Galani K, Grosshans H, Deinert K, Hurt EC, Simos G (2001) The intracellular location of two aminoacyl-tRNA synthetases depends on complex formation with Arc1p. EMBO J 20:6889–6898

Guo M, Schimmel P (2013) Essential nontranslational functions of tRNA synthetases. Nat Chem Biol 9:145–153

Guo M, Yang XL (2014) Architecture and metamorphosis. Top Curr Chem 344:89–118

Guo M, Schimmel P, Yang XL (2010a) Functional expansion of human tRNA synthetases achieved by structural inventions. FEBS Lett 584:434–442

Guo M, Yang XL, Schimmel P (2010b) New functions of aminoacyl-tRNA synthetases beyond translation. Nat Rev Mol Cell Biol 11:668–674

Hahn H, Park SH, Kim HJ, Kim S, Han BW (2019) The DRS-AIMP2-EPRS subcomplex acts as a pivot in the multi-tRNA synthetase complex. IUCrJ 6:958–967

Halawani D, Gogonea V, DiDonato JA, Pipich V, Yao P, China A, Topbas C, Vasu K, Arif A, Hazen SL et al (2018) Structural control of caspase-generated glutamyl-tRNA synthetase by appended noncatalytic WHEP domains. J Biol Chem 293:8843–8860

Han JM, Lee MJ, Park SG, Lee SH, Razin E, Choi EC, Kim S (2006) Hierarchical network between the components of the multi-tRNA synthetase complex: implications for complex formation. J Biol Chem 281:38663–38667

Han JM, Park BJ, Park SG, Oh YS, Choi SJ, Lee SW, Hwang SK, Chang SH, Cho MH, Kim S (2008) AIMP2/p38, the scaffold for the multi-tRNA synthetase complex, responds to genotoxic stresses via p53. Proc Natl Acad Sci USA 105:11206–11211

Han JM, Jeong SJ, Park MC, Kim G, Kwon NH, Kim HK, Ha SH, Ryu SH, Kim S (2012) Leucyl-tRNA synthetase is an intracellular leucine sensor for the mTORC1-signaling pathway. Cell 149:410–424

Havrylenko S, Legouis R, Negrutskii B, Mirande M (2011) Caenorhabditis elegans evolves a new architecture for the multi-aminoacyl-tRNA synthetase complex. J Biol Chem 286:28476–28487

Hei Z, Wu S, Liu Z, Wang J, Fang P (2019) Retractile lysyl-tRNA synthetase-AIMP2 assembly in the human multi-aminoacyl-tRNA synthetase complex. J Biol Chem 294:4775–4783

Hyeon DY, Kim JH, Ahn TJ, Cho Y, Hwang D, Kim S (2019) Evolution of the multi-tRNA synthetase complex and its role in cancer. J Biol Chem 294:5340–5351

Jeong SJ, Park S, Nguyen LT, Hwang J, Lee EY, Giong HK, Lee JS, Yoon I, Lee JH, Kim JH et al (2019) A threonyl-tRNA synthetase-mediated translation initiation machinery. Nat Commun 10:1357

Kaminska M, Havrylenko S, Decottignies P, Gillet S, Le Marechal P, Negrutskii B, Mirande M (2009a) Dissection of the structural organization of the aminoacyl-tRNA synthetase complex. J Biol Chem 284:6053–6060

Kaminska M, Havrylenko S, Decottignies P, Le Marechal P, Negrutskii B, Mirande M (2009b) Dynamic organization of aminoacyl-tRNA synthetase complexes in the cytoplasm of human cells. J Biol Chem 284:13746–13754

Kellermann O, Brevet A, Tonetti H, Waller JP (1979) Macromolecular complexes of aminoacyl-tRNA synthetases from eukaryotes. 1. Extensive purification and characterization of the high-molecular-weight complex(es) of seven aminoacyl-tRNA synthetases from sheep liver. Eur J Biochem 99:541–550

Kellermann O, Tonetti H, Brevet A, Mirande M, Pailliez JP, Waller JP (1982) Macromolecular complexes from sheep and rabbit containing seven aminoacyl-tRNA synthetases. I. Species specificity of the polypeptide composition. J Biol Chem 257:11041–11048

Khan K, Baleanu-Gogonea C, Willard B, Gogonea V, Fox PL (2020) 3-dimensional architecture of the human multi-tRNA synthetase complex. Nucleic Acids Res 48:8740–8754

Kim MH, Kim S (2020) Structures and functions of multi-tRNA synthetase complexes. Enzyme 48: 149–173

Kim JY, Kang YS, Lee JW, Kim HJ, Ahn YH, Park H, Ko YG, Kim S (2002) p38 is essential for the assembly and stability of macromolecular tRNA synthetase complex: implications for its physiological significance. Proc Natl Acad Sci USA 99:7912–7916

Kim MJ, Park BJ, Kang YS, Kim HJ, Park JH, Kang JW, Lee SW, Han JM, Lee HW, Kim S (2003) Downregulation of FUSE-binding protein and c-myc by tRNA synthetase cofactor p38 is required for lung cell differentiation. Nat Genet 34:330–336

Kim S, You S, Hwang D (2011) Aminoacyl-tRNA synthetases and tumorigenesis: more than housekeeping. Nat Rev Cancer 11:708–718

Kim DG, Choi JW, Lee JY, Kim H, Oh YS, Lee JW, Tak YK, Song JM, Razin E, Yun SH et al (2012) Interaction of two translational components, lysyl-tRNA synthetase and p40/37LRP, in plasma membrane promotes laminin-dependent cell migration. FASEB J 26:4142–4159

Kim KR, Park SH, Kim HS, Rhee KH, Kim BG, Kim DG, Park MS, Kim HJ, Kim S, Han BW (2013) Crystal structure of human cytosolic aspartyl-tRNA synthetase, a component of multi-tRNA synthetase complex. Proteins 81:1840–1846

Kim DG, Lee JY, Kwon NH, Fang P, Zhang Q, Wang J, Young NL, Guo M, Cho HY, Mushtaq AU et al (2014) Chemical inhibition of prometastatic lysyl-tRNA synthetase-laminin receptor interaction. Nat Chem Biol 10:29–34

Kim DG, Lee JY, Lee JH, Cho HY, Kang BS, Jang SY, Kim MH, Guo M, Han JM, Kim SJ et al (2016) Oncogenic mutation of AIMP2/p38 inhibits its tumor-suppressive interaction with Smurf2. Cancer Res 76:3422–3436

Kim JH, Lee C, Lee M, Wang H, Kim K, Park SJ, Yoon I, Jang J, Zhao H, Kim HK et al (2017) Control of leucine-dependent mTORC1 pathway through chemical intervention of leucyl-tRNA synthetase and RagD interaction. Nat Commun 8:732

Kim DK, Lee HJ, Kong J, Cho HY, Kim S, Kang BS (2021a) Structural basis for the dynamics of human methionyl-tRNA synthetase in multi-tRNA synthetase complexes. Nucleic Acids Res 49:6549–6568

Kim S, Yoon I, Son J, Park J, Kim K, Lee JH, Park SY, Kang BS, Han JM, Hwang KY et al (2021b) Leucine-sensing mechanism of leucyl-tRNA synthetase 1 for mTORC1 activation. Cell Rep 35: 109031

Kong J, Kim S (2020) Cell-based analysis of pairwise interactions between the components of the multi-tRNA synthetase complex. FASEB J 34:10476–10488

Koonin EV, Mushegian AR, Tatusov RL, Altschul SF, Bryant SH, Bork P, Valencia A (1994) Eukaryotic translation elongation factor 1 gamma contains a glutathione transferase domain--study of a diverse, ancient protein superfamily using motif search and structural modeling. Protein Sci 3:2045–2054

Kwon NH, Kang T, Lee JY, Kim HH, Kim HR, Hong J, Oh YS, Han JM, Ku MJ, Lee SY et al (2011) Dual role of methionyl-tRNA synthetase in the regulation of translation and tumor suppressor activity of aminoacyl-tRNA synthetase-interacting multifunctional protein-3. Proc Natl Acad Sci USA 108:19635–19640

Kwon NH, Fox PL, Kim S (2019) Aminoacyl-tRNA synthetases as therapeutic targets. Nat Rev Drug Discov 18:629–650

Larson ET, Kim JE, Zucker FH, Kelley A, Mueller N, Napuli AJ, Verlinde CL, Fan E, Buckner FS, Van Voorhis WC et al (2011) Structure of Leishmania major methionyl-tRNA synthetase in complex with intermediate products methionyladenylate and pyrophosphate. Biochimie 93: 570–582

Lazard M, Mirande M, Waller JP (1987) Expression of the aminoacyl-tRNA synthetase complex in cultured Chinese hamster ovary cells. Specific depression of the methionyl-tRNA synthetase component upon methionine restriction. J Biol Chem 262:3982–3987

Lee SW, Cho BH, Park SG, Kim S (2004a) Aminoacyl-tRNA synthetase complexes: beyond translation. J Cell Sci 117:3725–3734

Lee YN, Nechushtan H, Figov N, Razin E (2004b) The function of lysyl-tRNA synthetase and Ap4A as signaling regulators of MITF activity in FcepsilonRI-activated mast cells. Immunity 20:145–151

Lee EY, Lee HC, Kim HK, Jang SY, Park SJ, Kim YH, Kim JH, Hwang J, Kim JH, Kim TH et al (2016) Infection-specific phosphorylation of glutamyl-prolyl tRNA synthetase induces antiviral immunity. Nat Immunol 17:1252–1262

Lee EY, Kim S, Kim MH (2018a) Aminoacyl-tRNA synthetases, therapeutic targets for infectious diseases. Biochem Pharmacol 154:424–434

Lee M, Kim JH, Yoon I, Lee C, Fallahi Sichani M, Kang JS, Kang J, Guo M, Lee KY, Han G et al (2018b) Coordination of the leucine-sensing rag GTPase cycle by leucyl-tRNA synthetase in the mTORC1 signaling pathway. Proc Natl Acad Sci USA 115:E5279–E5288

Lo WS, Gardiner E, Xu Z, Lau CF, Wang F, Zhou JJ, Mendlein JD, Nangle LA, Chiang KP, Yang XL et al (2014) Human tRNA synthetase catalytic nulls with diverse functions. Science 345: 328–332

Lupas AN, Bassler J, Dunin-Horkawicz S (2017) The structure and topology of alpha-helical coiled coils. Subcell Biochem 82:95–129

McClain WH (1993) Rules that govern tRNA identity in protein synthesis. J Mol Biol 234:257–280

Mirande M, Cirakoglu B, Waller JP (1982) Macromolecular complexes from sheep and rabbit containing seven aminoacyl-tRNA synthetases. III. Assignment of aminoacyl-tRNA synthetase activities to the polypeptide components of the complexes. J Biol Chem 257:11056–11063

Moulinier L, Eiler S, Eriani G, Gangloff J, Thierry JC, Gabriel K, McClain WH, Moras D (2001) The structure of an AspRS-tRNA(asp) complex reveals a tRNA-dependent control mechanism. EMBO J 20:5290–5301

Negrutskii BS, Deutscher MP (1991) Channeling of aminoacyl-tRNA for protein synthesis in vivo. Proc Natl Acad Sci USA 88:4991–4995

Negrutskii BS, Deutscher MP (1992) A sequestered pool of aminoacyl-tRNA in mammalian cells. Proc Natl Acad Sci USA 89:3601–3604

Norcum MT (1989) Isolation and electron microscopic characterization of the high molecular mass aminoacyl-tRNA synthetase complex from murine erythroleukemia cells. J Biol Chem 264: 15043–15051

Norcum MT (1991) Structural analysis of the high molecular mass aminoacyl-tRNA synthetase complex. Effects of neutral salts and detergents. J Biol Chem 266:15398–15405

Norcum MT, Boisset N (2002) Three-dimensional architecture of the eukaryotic multisynthetase complex determined from negatively stained and cryoelectron micrographs. FEBS Lett 512: 298–302

Norcum MT, Warrington JA (1998) Structural analysis of the multienzyme aminoacyl-tRNA synthetase complex: a three-domain model based on reversible chemical crosslinking. Protein Sci 7:79–87

Oakley AJ (2005) Glutathione transferases: new functions. Curr Opin Struct Biol 15:716–723

Ofir-Birin Y, Fang P, Bennett SP, Zhang HM, Wang J, Rachmin I, Shapiro R, Song J, Dagan A, Pozo J et al (2013) Structural switch of lysyl-tRNA synthetase between translation and transcription. Mol Cell 49:30–42

Park BJ, Kang JW, Lee SW, Choi SJ, Shin YK, Ahn YH, Choi YH, Choi D, Lee KS, Kim S (2005) The haploinsufficient tumor suppressor p18 upregulates p53 via interactions with ATM/ATR. Cell 120:209–221

Park SJ, Ahn HS, Kim JS, Lee C (2015) Evaluation of multi-tRNA Synthetase complex by multiple reaction monitoring mass spectrometry coupled with size exclusion chromatography. PLoS One 10:e0142253

Rho SB, Lee KH, Kim JW, Shiba K, Jo YJ, Kim S (1996) Interaction between human tRNA synthetases involves repeated sequence elements. Proc Natl Acad Sci USA 93:10128–10133

Rho SB, Kim MJ, Lee JS, Seol W, Motegi H, Kim S, Shiba K (1999) Genetic dissection of protein-protein interactions in multi-tRNA synthetase complex. Proc Natl Acad Sci USA 96:4488–4493

Ribas de Pouplana L, Schimmel P (2001) Two classes of tRNA synthetases suggested by sterically compatible dockings on tRNA acceptor stem. Cell 104:191–193

Robinson JC, Kerjan P, Mirande M (2000) Macromolecular assemblage of aminoacyl-tRNA synthetases: quantitative analysis of protein-protein interactions and mechanism of complex assembly. J Mol Biol 304:983–994

Rubio Gomez MA, Ibba M (2020) Aminoacyl-tRNA synthetases. RNA 26:910–936

Sang Lee J, Gyu Park S, Park H, Seol W, Lee S, Kim S (2002) Interaction network of human aminoacyl-tRNA synthetases and subunits of elongation factor 1 complex. Biochem Biophys Res Commun 291:158–164

Schimmel P (1987) Aminoacyl tRNA synthetases: general scheme of structure-function relationships in the polypeptides and recognition of transfer RNAs. Annu Rev Biochem 56:125–158

Schimmel P (2018) The emerging complexity of the tRNA world: mammalian tRNAs beyond protein synthesis. Nat Rev Mol Cell Biol 19:45–58

Shalak V, Kaminska M, Mitnacht-Kraus R, Vandenabeele P, Clauss M, Mirande M (2001) The EMAPII cytokine is released from the mammalian multisynthetase complex after cleavage of its p43/proEMAPII component. J Biol Chem 276:23769–23776

Siddiqui FA, Yang DC (1985) Generation of multiple forms of methionyl-tRNA synthetase from the multi-enzyme complex of mammalian aminoacyl-tRNA synthetases by endogenous proteolysis. Biochim Biophys Acta 828:177–187

Simader H, Hothorn M, Kohler C, Basquin J, Simos G, Suck D (2006) Structural basis of yeast aminoacyl-tRNA synthetase complex formation revealed by crystal structures of two binary sub-complexes. Nucleic Acids Res 34:3968–3979

Sivaram P, Deutscher MP (1990) Existence of two forms of rat liver arginyl-tRNA synthetase suggests channeling of aminoacyl-tRNA for protein synthesis. Proc Natl Acad Sci USA 87: 3665–3669

Som K, Hardesty B (1975) Isolation and partial characterization of an aminoacyl-tRNA synthetase complex from rabbit reticulocytes. Arch Biochem Biophys 166:507–517

Stapulionis R, Deutscher MP (1995) A channeled tRNA cycle during mammalian protein synthesis. Proc Natl Acad Sci USA 92:7158–7161

Van Dang C, Yang DC (1979) Disassembly and gross structure of particulate aminoacyl-tRNA synthetases from rat liver. Isolation and the structural relationship of synthetase complexes. J Biol Chem 254:5350–5356 van Rooyen JM, Murat JB, Hammoudi PM, Kieffer-Jaquinod S, Coute Y, Sharma A, Pelloux H, Belrhali H, Hakimi MA (2014) Assembly of the novel five-component apicomplexan multi-aminoacyl-tRNA synthetase complex is driven by the hybrid scaffold protein Tg-p43. PLoS One 9:e89487

Verrigni D, Diodato D, Di Nottia M, Torraco A, Bellacchio E, Rizza T, Tozzi G, Verardo M, Piemonte F, Tasca G et al (2017) Novel mutations in KARS cause hypertrophic cardiomyopathy and combined mitochondrial respiratory chain defect. Clin Genet 91:918–923

Yannay-Cohen N, Carmi-Levy I, Kay G, Yang CM, Han JM, Kemeny DM, Kim S, Nechushtan H, Razin E (2009) LysRS serves as a key signaling molecule in the immune response by regulating gene expression. Mol Cell 34:603–611

Yao P, Potdar AA, Arif A, Ray PS, Mukhopadhyay R, Willard B, Xu Y, Yan J, Saidel GM, Fox PL (2012) Coding region polyadenylation generates a truncated tRNA synthetase that counters translation repression. Cell 149:88–100

Yoon I, Nam M, Kim HK, Moon HS, Kim S, Jang J, Song JA, Jeong SJ, Kim SB, Cho S et al (2020) Glucose-dependent control of leucine metabolism by leucyl-tRNA synthetase 1. Science 367:205–210

[illegible]

Chapter 7
On the Regulation of Mitosis by the Kinetochore, a Macromolecular Complex and Organising Hub of Eukaryotic Organisms

Victor M. Bolanos-Garcia

Abstract The kinetochore is the multiprotein complex of eukaryotic organisms that is assembled on mitotic or meiotic centromeres to connect centromeric DNA with microtubules. Its function involves the coordinated action of more than 100 different proteins. The kinetochore acts as an organiser hub that establishes physical connections with microtubules and centromere-associated proteins and recruits central protein components of the spindle assembly checkpoint (SAC), an evolutionarily conserved surveillance mechanism of eukaryotic organisms that detects unattached kinetochores and destabilises incorrect kinetochore-microtubule attachments. The molecular communication between the kinetochore and the SAC is highly dynamic and tightly regulated to ensure that cells can progress towards anaphase until each chromosome is properly bi-oriented on the mitotic spindle. This is achieved through an interplay of highly cooperative interactions and concerted phosphorylation/dephosphorylation events that are organised in time and space.

This contribution discusses our current understanding of the function, structure and regulation of the kinetochore, in particular, how its communication with the SAC results in the amplification of specific signals to exquisitely control the eukaryotic cell cycle. This contribution also addresses recent advances in machine learning approaches, cell imaging and proteomics techniques that have enhanced our understanding of the molecular mechanisms that ensure the high fidelity and timely segregation of the genetic material every time a cell divides as well as the current challenges in the study of this fascinating molecular machine.

Keywords Kinetochore · Genome stability · Chromosome segregation · Signal amplification · Cell cycle regulation · Kinetochore-microtubule network · Spindle assembly checkpoint (SAC) · Cancer · Macromolecular assemblies · Aneuploidy

---

V. M. Bolanos-Garcia (✉)
Department of Biological and Medical Sciences, Faculty of Health and Life Sciences, Oxford Brookes University, Oxford, UK
e-mail: vbolanos-garcia@brookes.ac.uk

J. R. Harris, J. Marles-Wright (eds.), *Macromolecular Protein Complexes IV*, Subcellular Biochemistry 99, https://doi.org/10.1007/978-3-031-00793-4_7

Introduction

In eukaryotic organisms, mitosis is defined as the process in which a cell divides to give place to two genetically identical descendant cells. This highly regulated biological process is divided into six successive phases: prophase, prometaphase, metaphase, anaphase, telophase and cytokinesis. Accurate chromosome segregation requires attachment of the kinetochore, a large multiprotein complex assembled on the centromere of each chromosome, to the spindle microtubules. In this way, the kinetochore functions as a macromolecular structure interface for microtubule attachment and enables molecular surveillance mechanisms to monitor and ensure proper microtubule attachment, which is required for the accurate and timely segregation of chromosomes upon cell division (Lara-Gonzalez et al. 2021; Ferreira and Maiato 2021; Yamamoto 2021; Navarro and Cheeseman 2021). The kinetochore protein complexes play a key role in mitosis by connecting spindle microtubules to the centromeric region of chromosomes (Hara and Fukagawa Hara and Fukagawa 2017, 2020; Funabiki and Wynne 2013; Westhorpe and Straight 2013). Undoubtedly, the definition of kinetochore composition and architecture has advanced tremendously since the proteins CENP-A, CENP-B and CENP-C were first identified as kinetochore central components (Earnshaw and Rothfield 1985; Palmer et al. 1987). Indeed, to date more than one hundred proteins have been reported to contribute to the formation of a functional kinetochore in vertebrates, in which functions must be coordinated in time and space (Audett and Maresca 2020; Cheeseman and Desai 2008).

Kinetochore-microtubule attachment can be subdivided into several stages, starting with kinetochore capture along the walls of microtubules, followed by the configuration of end-on attachment to the tips of microtubules. Once all the kinetochores have achieved end-on attachment and biorientation, chromosomes are enabled to segregate (Shrestha et al. 2017). Because the accurate transmission of the genetic material to descendants is crucial for the self-perpetuation of a species, it is not surprising that this stage of the cell division process is tightly controlled. However, this process is prone to gross chromosome segregation errors, which must be corrected to avoid chromosome instability and aneuploidy (Curtis et al. 2020; Sarkar et al. 2021). It has been estimated that one error occurs in every 105 division events, a rate that increases in some types of tumours and results in chromosomal instability (CIN) and aneuploidy, a condition linked to gain or loss of a region of chromosomes (and in extreme cases of whole chromosomes) (Curtis et al. 2020; Sarkar et al. 2021).

Similar to many other cellular signalling systems, elucidation of structure-function relationship has been instrumental to enhance our understanding of the kinetochore. Recent advances in high-resolution light microscopy and structural biology methods including single-particle electron microscopy have revealed mechanistic details of the pulling forces exerted by the microtubule on the kinetochores during kinetochore-microtubule interactions with an unprecedented level of detail

(Blundell and Chaplin 2021; Knossow et al. 2020; Lander and Glaeser 2021; Navarro and Cheeseman 2021; Renda and Khodjakov 2021; Yan et al. 2019).

In this contribution, I discuss how cooperative interactions and an interplay of protein kinase and phosphatase catalytic activities work together to regulate chromosome-microtubule attachment in time and space. I argue that even though the functions and mode of regulation of protein kinases in mitosis, including their roles in the stabilisation of chromosome-microtubule attachment, have been investigated for decades, important mechanistic details remain poorly understood and that even less is known about the precise roles of protein phosphatases in the regulation of the molecular communication between chromosomes and microtubules and the spindle assembly checkpoint (SAC) to ensure the tight regulation of mitosis. Last, the current challenges and foreseeable developments in the study of the kinetochore and its role in the control of cell division in eukaryotic organisms are addressed.

Kinetochore Structural Organisation

The kinetochore structure is organised in three layers: the fibrous corona, which emanates from the outer plate or outer kinetochore and decreases upon kinetochore-microtubule attachment; the outer plate where the physical attachment between the kinetochore and microtubules takes place; and the inner plate or inner kinetochore, a structure built on the centromere to recruit outer kinetochore-microtubule proteins onto the centromere (Hara and Fukagawa 2020; Funabiki and Wynne 2013; Westhorpe and Straight 2013). In simple terms, the centromere is a specific region on the chromatin that contains the histone H3 protein variant (Ferrand et al. 2020), whereas the inner kinetochore includes proteins that interact with DNA and the outer kinetochore comprises proteins that mediate microtubule binding.

Although unlike the centromere, the kinetochore seems to be highly conserved across species, the biochemical and functional characterisation of the inner kinetochore and the microtubule-binding outer kinetochore from diverse origin has revealed interspecific variability in the composition of the kinetochore (Tromer et al. 2019). For example, in live budding yeast and other eukaryotic organisms, the inner and other kinetochore regions exhibit different distributions during metaphase. Such composition variations define diverse architectures of centromere-associated networks and prove that homologous kinetochore protein complexes can mediate different functions in different organisms (Hamilton and Davis 2020).

The formation of a functional kinetochore requires the assembly of the proteins that constitute the centromere-associated network (CCAN) and the KMN network. The latter assembly will be discussed in detail in the next section. As for the CCAN, this is a protein complex on the chromatin region harbouring the centromere-specific histone CENP-A (Fig. 7.1) (Hori et al. 2013; Perpelescu and Fukagawa 2011; Kixmoeller et al. 2020). In budding yeast, the kinetochore comprises a single nucleosome, containing the centromere-specific H3 variant CENP-A^{Cse4}, bound to specific kinetochore proteins that connect to a single microtubule. Despite the

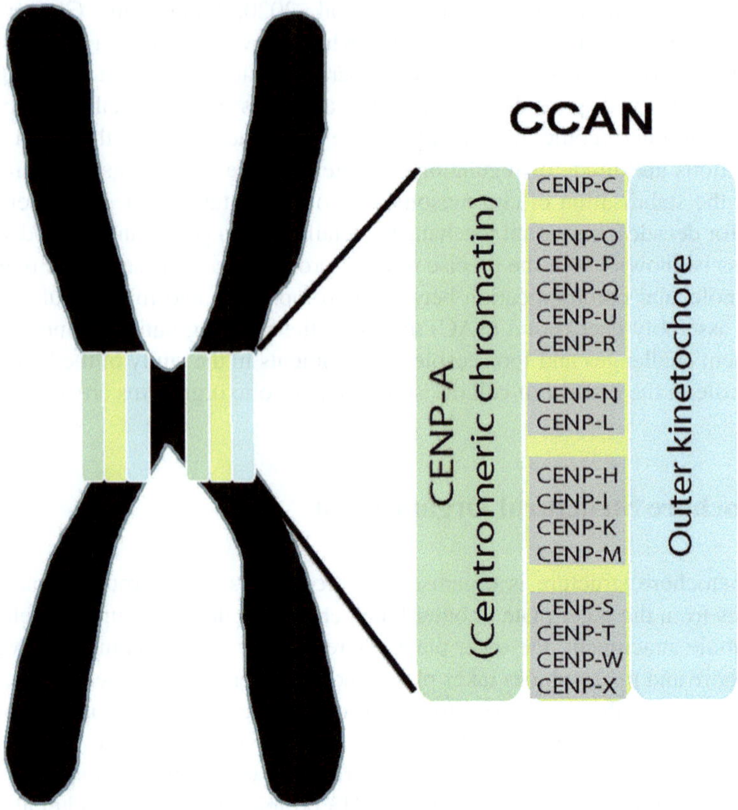

Fig. 7.1 The CCAN complex. The figure shows the relative localisation of the CCAN in the centromere and the 16 proteins that compose this complex

relative simplicity of the *S. cerevisiae* kinetochore, its regulation is complex, involving the integration of regulatory signals triggered by post-translational modifications (notably but not exclusively by phosphorylation and dephosphorylation events) from cell cycle, DNA damage and repair checkpoints, as well as stress-responsive mechanisms and central metabolic pathways. Ultimately, these concerted interactions control chromosome segregation and ensure the maintenance of a stable genome (revised by Klemm et al. 2021).

The CENP-A nucleosome physically associates with the CCAN protein component CENP-C, and the interaction is required to promote kinetochore formation (Watanabe et al. 2019). Given the central function of CENP-C in this process (Przewloka et al. 2011; Screpanti et al. 2011), its association with CENP-A is highly regulated during cell division (Watanabe et al. 2019). The interaction of CENP-C with CENP-A involves the C-terminal region of CENP-C, which harbours a CENP-A-binding motif that is highly conserved across higher organisms. The binding of CENP-C to CENP-A is promoted by Cdk1-dependent phosphorylation of CENP-C, a functional feature that evidences the high plasticity of interactions between centromere and kinetochore proteins throughout the cell cycle (Watanabe et al. 2019; Allu et al. 2019).

The KMN Network

The structural core of the kinetochore is the KMN network, which constitutes a docking platform for the kinetochore recruitment of SAC protein components. The KMN network is located in the outer kinetochore and composed by the single protein Knl1 (a protein also known as CASC5, Blinkin and AF15Q14 in humans, Spc105 in budding yeast and flies and Spc7 in fission yeast) (Kiyomitsu et al. 2007, 2011; Bolanos-Garcia et al. 2009) and the protein complexes Ndc80 and Mis12 (Musacchio and Desai 2017; Varma and Salmon 2012). The Mis12 protein assembly is constituted by the proteins Mis12, Mtw1 and MIND, whereas the Ndc80 complex is made of the proteins Ndc80 and Hec1. As mentioned earlier, although the amino acid sequence in most kinetochore proteins is clearly divergent, including those that define the KMN network, the overall architecture of the kinetochore remains highly conserved across species (van Hooff et al. 2017; Tanaka 2013; Westhorpe and Straight 2013; Przewloka and Glover 2009). The KMN is important for the establishment of proper kinetochore-microtubule attachments and essential for cell viability. In plants, central protein components of the kinetochore and the SAC were first identified in *Arabidopsis thaliana* (Lermontova et al. 2008) and subsequently in other plant species (Komaki and Schnittger 2016, 2017; Su et al. 2021; van Hooff et al. 2017). Further investigations in plants have shown that kinetochore-SAC communication share conserved functions with vertebrate and fungi species. However, the devil is in the details and important species-specific features have been found in a number of key kinetochore and mitotic checkpoint proteins that account for unique kinetochore-SAC features compared to their animals and yeast counterparts, providing compelling evidence of the co-evolution of the SAC and the kinetochore (Kops et al. 2020; Komaki and Schnittger 2016, 2017; Su et al. 2021; van Hooff et al. 2017).

Knl1

Knl1 is a large, predominantly disordered protein of the KMN network that acts as a molecular platform to recruit several other proteins to the kinetochore (Kiyomitsu et al. 2007; Przewloka and Glover 2009; Santaguida and Musacchio 2009; Ghongane et al. 2014). Knl1 importantly contributes to the control of the KMN network assembly/disassembly through its binding to protein phosphatase 1 (PP1), Bub1, BubR1, Bub3, Zwint-1 and the component of the Mis12 complex, Nsl1 (Fig. 7.2). Although the various protein-binding motifs of Knl1 are distributed

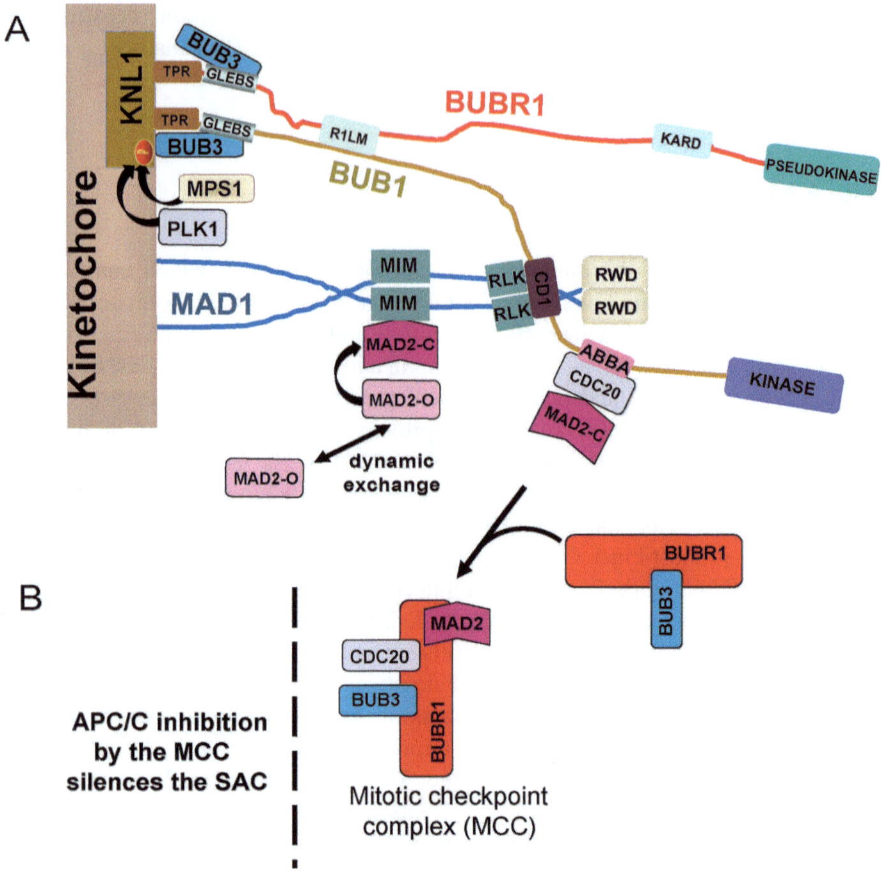

Fig. 7.2 (**a**) Knl1-dependent interactions that mount a SAC response. Phosphorylation of Knl1 by the dual-specificity protein kinase monopolar spindle 1 (Mps1) and polo-like kinase 1 (Plk1) promotes the kinetochore localisation of Bub1-Bub3 (Shepperd et al. 2012; von Schubert et al. 2015; Yamagishi et al. 2012; London et al. 2012; Zhang et al. 2017; Fischer et al. 2021). Bub1 at kinetochores then binds to Mad1 which is associated with Mad2 in its close conformation (c-MAD2) (De Antoni et al. 2005). These complexes recruit cytosolic Mad2 in its open conformation (o-Mad2) and Cdc20. Cdc20 binds Mad2 through a SLiM, commonly referred to as the Mad2-interacting motif (MIM) (Izawa and Pines 2012), inducing a large conformational remodelling of Mad2 from o-Mad2 to c-Mad2 (Luo et al. 2002, 2004; Sironi et al. 2002). The conversion of o-Mad2 to c-Mad2 is the rate-limiting step of MCC assembly and is catalysed at the kinetochore by Bub1 (Piano et al. 2021; Lara-Gonzalez et al. 2021b). (**b**) The subsequent incorporation into the Cdc20-c-Mad2 complex of the BubR1-Bub3 heterodimer completes the assembly of the MCC. The catalytic formation of the MCC is required for inhibition of the APC/Cdc20 by the SAC, thus providing time for the establishment of proper kinetochore-microtubule attachments

throughout the polypeptide chain, most functional motifs are concentrated in the N-terminal region, namely, the linear motifs implicated in PP1 binding, the motif that mediates the binding of the TPR motifs of Bub1 and BubR1 (e.g., the PI motif) (Bolanos-Garcia et al. 2011; Krenn et al. 2012; Rosenberg et al. 2011; London et al.

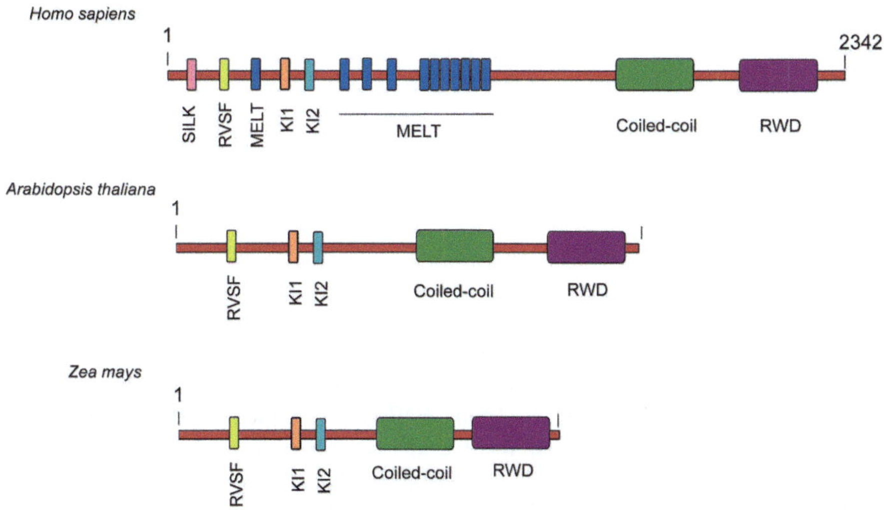

Fig. 7.3 Evolutionary divergence of the kinetochore. This is illustrated using Knl1 from three different species as an example. In humans, phosphorylation of the MELT motifs is required for the kinetochore recruitment of SAC proteins, whereas in *Arabidopsis thaliana* and maize, the MELT motifs are dispensable. In contrast, the KI motifs implicated in Bub1 and BubR1 binding (KI1 and KI2, respectively) as well as the RWD domain implicated in Nsl1 binding are highly conserved in the three species

2012; Shepperd et al. 2012; Funabiki and Wynne 2013; Vleugel et al. 2013) and the MELT motifs that mediate binding of Bub3 upon MELT phosphorylation by Mps1 (Zhang et al. 2014; Primorac et al. 2013). In contrast, the Knl1 C-terminal region harbours one segment that binds directly to Nsl1, thus mediating the kinetochore recruitment of the Mis12 complex (Fig. 7.2; Petrovic et al. 2014). Details of the composition of the Mis12 complex is presented in the section below.

The majority of Knl1 homologues contain an arrangement of motif repeat units, the MELT motif (Tromer et al. 2015; Vleugel et al. 2013; Zhang et al. 2014). However, the number of MELT repeat units and the specific amino acid sequence of the MELT repeats vary widely across species, which initially suggested that these differences may account for species-specific recognition of Knl1 interaction partners (revised in Ghongane et al. 2014). Indeed, recent phylogenetic studies that included *Arabidopsis thaliana* (Tromer et al. 2015) and functional and bioinformatics studies on maize (*Zea mays*) kinetochores (Su et al. 2021) indicate that these plant species lack the MELT motifs as well as the canonical KI motifs (Fig. 7.3). Moreover, in the latter case, the study showed that these motifs are dispensable for Knl1 interaction with SAC components (Su et al. 2021). To what extent this feature is shared by other flowering plants remains to be established. In mammals, depletion and/or suppression of the expression of Knl1 leads to extensive chromosome segregation defects with phenotypes that closely resemble the chromosomal aberrations caused by Bub1 and BubR1 expression suppression (Kiyomitsu et al. 2007; Cheeseman et al. 2006, 2008), whereas in plants, loss of Knl1 function results in aberrant chromosomal congression and segregation and gross defects in cell division during kernel development (Su et al. 2021). Taken together, the studies show the essential role of Knl1 and the KMN network in cell division across species.

Mis12 Complex

Mis12 is a subcomplex composed of four different proteins: Nnf1, Dsn1, Nsl1 and the one that gives the complex its name, Mis12. Through Mis12, the KMN network establishes molecular communication with the inner centromere (Petrovic et al. 2010, Petrovic et al. 2014; Venkei et al. 2011). In humans, the Nnf1 subunit of the Mis12 subcomplex binds to CENP-C and CENP-T, two proteins that form part of the inner kinetochore repertoire and the CCAN complex (Yan et al. 2019). More precisely, the attachment of centromeres to microtubules occurs via CENP-C binding to Knl1 and Mis12, which in turn recruits Ndc80 and the binding of CENP-T to Ndc80 (Rago et al. 2015; McKinley and Cheeseman 2016). The interaction between CCAN and KMN is highly dynamic and modulated through changes in the phosphorylation status of KMN-CCAN components (Hara et al. 2018). We have a good understanding of the molecular interactions that hold the Mis12 complex together because crystal structures of the holocomplex and subcomplexes from different organisms have been solved (Protein Data Bank IDs 5LSK, 5T51, 5WWL, 6MJE, 3VZA). One example of the overall topology of a Mis12 assembly is shown in Fig. 7.4, namely, details of the crystal structure of the human Mis12 complex in association with a CENP-C fragment.

Ndc80 Complex

Ndc80 is a highly conserved complex and a core component of the kinetochore-microtubule attachment machinery in eukaryotic cells (McCleland et al. 2003, 2004; Wigge and Kilmartin 2001; DeLuca and Musacchio 2012). The Ndc80 complex plays a major role in the attachment of assembling/disassembling microtubule ends to the chromosomes (reviewed by Ustinov et al. 2020) and is composed of four subunits: Spc24, Spc25, Nuf2 and Ndc80 (the subunit that gives its name to the entire subcomplex) (Ciferri et al. 2005, 2008; Wei et al. 2005, 2007; Wan et al. 2009). The assembly defines a dumbbell-shaped molecule with Spc24-Spc25 and Nuf2-Ndc80 located in opposite ends of the molecule (Fig. 7.5). Nuf2-Ndc80 association is required to prime Ndc80 complex binding to microtubules, whereas the initial assembly of the Spc24-Spc25 complex drives the binding of the Ndc80 complex to Knl1 and the Mis12 subcomplex (Cheeseman et al. 2006; Kiyomitsu et al. 2007; Wei et al. 2007; Ciferri et al. 2008; Wan et al. 2009; Joglekar and DeLuca 2009).

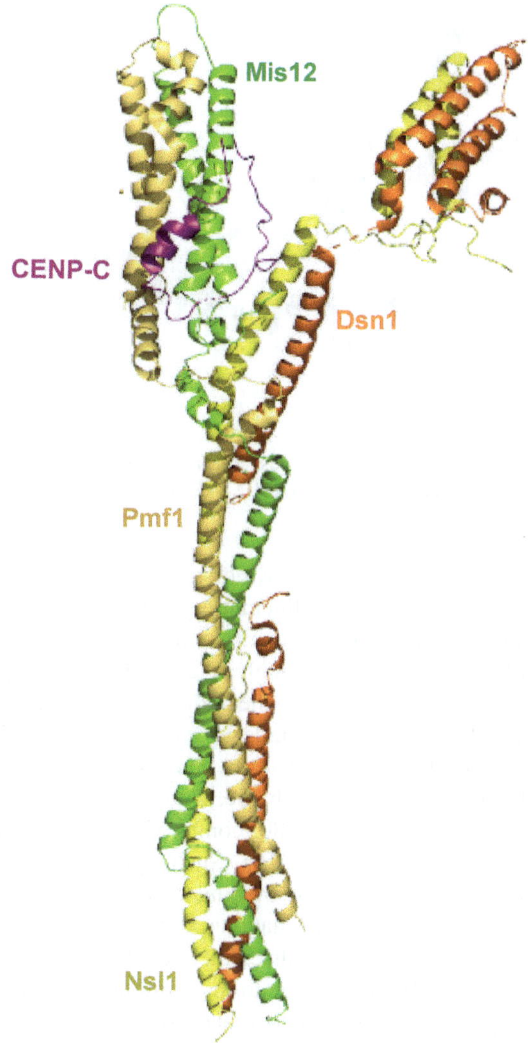

Fig. 7.4 The Mis 12 complex. The figure shows the crystal structure of a CENP-C fragment bound to the Mis12 complex (Protein Data Bank, PDB, ID 5LSK)

It has been established that microtubule attachments must be sufficiently stable to transduce force. At the same time, the strength of these attachments must be tightly regulated to ensure timely, error-free progression through mitosis. How can the fine-tuning sense and regulation of this process be achieved? Insights into this fascinating aspect of kinetochore function are drawn from diverse studies. For instance, the integration of biochemical, structural and cellular biology approaches shows that a small, disordered region within the Hec1 subunit of the Ndc80 complex—the N-terminal "tail" domain—plays an important role in the generation and regulation of kinetochore-microtubule attachment and that tight bending of the Ndc80 complex from budding yeast inhibits its binding to microtubules (revised by Wimbish and

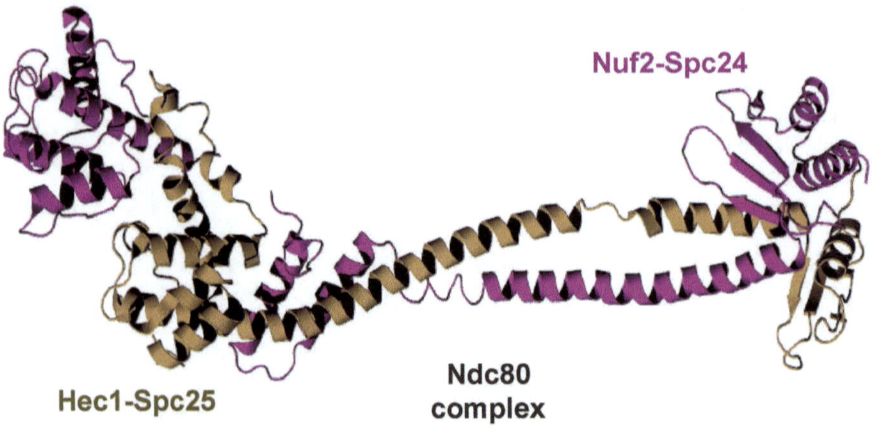

Fig. 7.5 The Ndc80 complex. The figure shows the crystal structure of a bonsai version of the human Ndc80 complex (PDB ID 2VE7)

DeLuca 2020). Moreover, Förster resonance energy transfer (FRET) measurements have indicated large fluctuations between straight and bent forms of the Ndc80 complex and that this complex has a clear preference for binding to straightened microtubules (Scarborough et al. 2019). More precisely, the loop region of the Ndc80 complex enables the adoption of a bent conformation. Its deletion promotes straightening of the complex (Scarborough et al. 2019). The picture emerging from these studies indicate that the Mis12 complex (e.g. known as the MIND complex in the budding yeast orthologue, Dimitrova et al. 2016) enhances microtubule binding by opposing the Ndc80 complex's tightly bent (auto-inhibited) conformation (Scarborough et al. 2019). Such dynamic conformational changes suggest that prior to its assembly at the kinetochore, the Ndc80 complex interchanges between bent and open conformations and that, once it is built, its association with the Mind12 complex stabilises the Ndc80 assembly in a straightened form, resulting in higher affinity binding to microtubules (Scarborough et al. 2019). To what extent this dynamic mode of interactions and dramatic structural reorganisation may be conserved in other species remains to be established.

Now, in order to segregate chromosomes properly, the kinetochore fibre must be dynamic and generate and respond to force. How the kinetochore remodels under force remains poorly understood. This is largely in part due to the impossibility to biochemically reconstitute kinetochore fibres in vitro and because controlled forces exerted to kinetochore fibres in vivo remains an important technical challenge. However, in a recent study on mammalian kinetochore fibres, Long et al. (2020) used microneedles to test the extent a sustained force affects kinetochore fibre dynamics and structure. Interestingly, these authors showed that force lengthens kinetochore fibres through a mechanism that favours plus-end polymerisation rather than by increasing the rate of polymer formation. They also noted that upon break, kinetochore fibres still attach to kinetochores or poles, supporting the notion that different mechanisms of local force dissipation by the k-fibre restrain force transmission in order to preserve the mechanical robustness of the spindle (Long et al. 2020). Future work is needed to establish the mechanical-sensing features that are implicated in kinetochore-microtubule interactions and how exactly these forces are transduced to the SAC.

Other Kinetochore-Associated Subcomplexes

The RZZ Complex

There are additional kinetochore complexes that play important roles in the attachment of chromosomes to spindle microtubules and the SAC such as the RZZ complex. This is an assembly of ca. 800 kDa composed of the proteins Rod-Zwilch-Zw10 (hence the name RZZ). Upon self-assembly to form filaments, the RZZ complex contributes to the formation of the outer corona of the kinetochore, the structure implicated in the regulation of microtubule attachments and the kinetochore recruitment of key SAC protein components (Mosalaganti et al. 2017; Rodriguez-Rodriguez et al. 2018; Pereira et al. 2018; Sacristan et al. 2018; Barbosa et al. 2020; Kops et al. 2005; Wang et al. 2004; Starr et al. 2000). A model of the human RZZ complex that integrated biochemical reconstitution and cross-linking mass spectrometry analyses with single-particle electron cryomicroscopy and 3D structural modelling showed that RZZ is structurally related to coat scaffolds implicated in membrane cargo trafficking, including clathrin, Sec13-Sec31 and αβ'ε-COP (Mosalaganti et al. 2017). The outer kinetochore protein Zwint-1 (human Zw10 interacting protein-1) was initially identified from a search for interacting partners of human Zw10 using the yeast two-hybrid system (Obuse et al. 2004). In humans, Zwint-1 corresponds to a 43-kDa protein predicted to harbour an extended coiled-coil domain. Zwint-1 localises to the kinetochore in prometaphase prior to Zw10 localisation and remains there until late anaphase, suggesting a role in Zw10 kinetochore recruitment (Barbosa et al. 2020). Through its direct binding to Zw10, Zwint-1 interacts with the Rod-Zwilch-Zw10 (RZZ) complex (Kiyomitsu et al. 2011; Barbosa et al. 2020). Zwint-1 also interacts with Knl1, specifically with the C-terminal region encompassing residues 1904-2316 in the human protein (Petrovic et al. 2010). Importantly, the kinetochore localisation of Knl1 depends not only on Zwint-1 but also on the Mis12 complex (Varma et al. 2013), and it is possible that the RZZ and Mis12 complexes act cooperatively to recruit Knl1 to the kinetochore. Moreover, Zwint-1 seems to interact with fully active centromeres, suggesting an important role in correct centromere function. In addition to the aforementioned interactions, it has been suggested that localisation of the RZZ complex to the kinetochore involves its physical interaction with the SAC kinase Bub1 (Zhang et al. 2019). Specifically, with the Bub1 region located downstream of a conserved domain defined by the amino acid residues 485-521 in the human orthologue (Zhang et al. 2019). However, it remains to be firmly established if the interaction between Bub1 and protein components of the RZZ complex is direct, is transient in nature and/or requires additional factors for the establishment of a productive interaction.

Another important partner of the RZZ complex is the dynein adaptor protein Spindly. This is a molecule with important roles in chromosomal segregation during cell division. The Spindly N-terminal region mediates the binding of this protein to the microtubule motor dynein and its activator dynactin, whereas the Spindly C-terminal domain physically interacts with the RZZ complex in the outermost layer of the kinetochore. In human cells, the Spindly C-terminal region directly binds to Rod, while in *C. elegans*, Spindly-C binds to Rod-1 and Zwilch-1. The latter two interactions are predominantly mediated by two disordered regions that are C-terminally adjacent to the short segments of α-helical propensity (Henen et al. 2021). The differential substrate specificity of Spindly in humans and worms represent yet another example of evolutionary divergence of the kinetochore and important functional features that are species-specific (Henen et al. 2021; Kops et al. 2020; Tromer et al. 2019).

The kinetochore localisation of the RZZ complex is affected by inhibition of Aurora B kinase activity and this has no effect on Zwint-1 recruitment. Zwint-1 is phosphorylated by Aurora B. Expression of a Zwint-1 mutant that evades Aurora B phosphorylation impaired kinetochore localisation of RZZ-dependent proteins and induced defects in chromosome movement during prometaphase. Taken together, these observations confirmed Zwint-1 as a bona fide Aurora B substrate and a protein required for kinetochore assembly and SAC silencing at metaphase. It also provides yet another example of the importance of concerted phosphorylation and dephosphorylation events in the regulation of chromosome segregation and the maintenance of a stable genome.

The CENP-O Class

The proteins CENP-O/P/U/Q/R associate to form a stable complex required for proper kinetochore function. Such complex is commonly referred to as CENP-O class. Although the function and structure of a similar complex constituted by the proteins CENP-O/P/U/Q homologues (e.g. the COMA complex) from yeast has been established, the mechanism underpinning the assembly of the human CENP-O class complex appears to be significantly different, with important details still to be established. A recent report showed that the CENP-U regions encompassing the amino acid residues 241-360 and the C-terminal half of CENP-Q play key roles in the establishment of a CENP-U/CENP-Q subcomplex which in turn physically interacts with CENP-O/P at least in vitro (Cao et al. 2021). The same report showed that the kinetochore protein CENP-R establishes a physical interaction with CENP-U and CENP-Q but not with CENP-O/P (Cao et al. 2021).

The CCAN Complex

The constitutive centromere-associated network (CCAN) is a 16-subunit assembly that plays a fundamental role in the organisation of the interface between the kinetochore and the specialised chromosome region known as the centromere. A hallmark of this specialised chromatin region is the presence of the histone H3 variant CENP-A. In essence, the CCAN bridges CENP-A to the spindle-binding moiety of the kinetochore. The structure of the human inner kinetochore CCAN complex consisting of the protein CENP-C N-terminal region (residues 1-544) and the subcomplexes CENP-O/CENP-P/CENP-Q/CENPU/CENP-R (CENP-OPQUR complex), CENP-N/CENP-L (CENP-LN), CENPH/CENP-I57-765/CENP-K/CENP-M (CENP-HIKM) and CENP-T/CENP-W/CENPS/CENP-X (CENP-TWSX) was solved recently through cryoelectron microscopy (Pesenti et al. 2022). The atom coordinates were deposited in the PDB (ID 7QOO) in December 2021 by Vetter et al. (Fig. 7.6a). The structure of the human CCAN complex revealed new structural and mechanistic details of its interaction with CENP-A, providing important clues about the specific recognition of the CENP-A nucleosome by the CCAN (Pesenti et al. 2022. Importantly, the human CCAN complex shows an overall structure similarity to that of CCAN from budding yeast (a CCAN orthologue that in this species is commonly referred to as Ctf), particularly in its general shape. However, a closer inspection of the two complexes reveal important differences that account for the mode of binding of the CCAN complexes to the nucleosome. In particular, the CENP-C residues were implicated in CENP-A binding in each organism. Another important feature of the entire human CCAN complex is the significant structural differences it shows compared to previously reported structures of truncated forms of CENP-C in complex with CENP-A nucleosomes (Fig. 7.6b) (Allu et al. 2019; Chittori et al. 2018; Pentakota et al. 2017; Tian et al. 2018). Future studies are required to define the 3D structure of the CCAN complex bound to DNA or nucleosomes (Pesenti et al. 2022).

Kinetochore-Microtubule Attachments

Incorrect kinetochore-microtubule attachments are destabilised through phosphorylation of protein components of the KMN network by the protein kinases Aurora B and Polo/Plk1, two master regulators of mitosis, which impairs microtubule binding to the KMN network (Welburn et al. 2010; Zhu et al. 2013). More precisely, phosphorylation of the N-terminal region of the KMN protein component Ndc80 by Aurora B kinase leads to the destabilisation of incorrect kinetochore-microtubule attachments (Zhu et al. 2013). The kinetochore communicates to the spindle assembly checkpoint (SAC), a complex signalling system that prevents the onset of anaphase until proper kinetochore-microtubule attachments are achieved (Jia et al. 2013; Foley and Kapoor 2013; Hardwick et al. 2000; Warren et al. 2002; Morrow

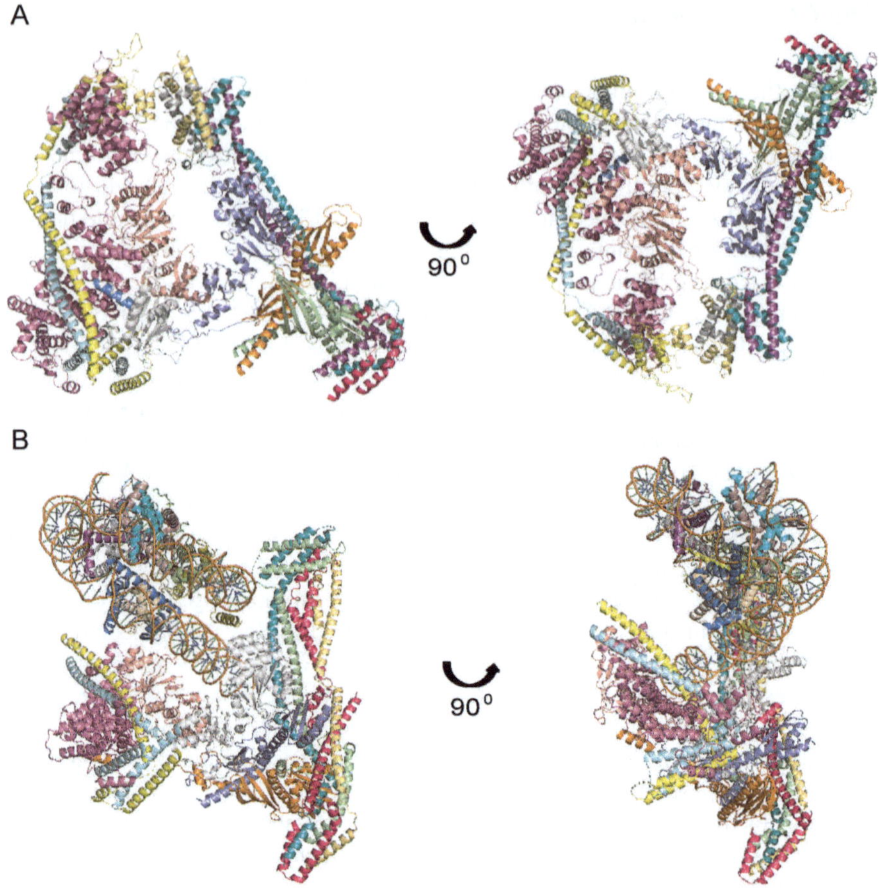

Fig. 7.6 The CCAN complex. (**a**) 3D structure of the entire *Homo sapiens* CCAN in complex with CENP-A (PDB ID 7QOO) (Vetter et al. 2021; Pesenti et al. 2022). (**b**) 3D structure of the *Saccharomyces cerevisiae* CCAN-CENP-A complex (PDB ID 6QLD)

et al. 2005; Yao and Dai 2012). When even one single chromosome is unattached or improperly attached, a mitotic protein kinase of dual specificity, Mps1 (also known as TTK), is recruited to the kinetochore to phosphorylate the kinetochore organising protein Knl1 on multiple Met-Glu-Leu-Thr (MELT) motif repeats that span the Knl1 N-terminal to the middle region (revised in Ghongane et al. 2014). This post-translational modification of Knl1 in turn recruits other SAC proteins to the kinetochore (e.g. Bub1-Bub3 and BubR1-Bub3), where Bub3 physically interacts with the phosphorylated MELT motifs, ultimately delaying anaphase onset until all attachment errors are corrected (Musacchio and Salmon 2007; Lara-Gonzalez et al. 2012). Once all the errors are corrected, proper kinetochore-microtubule attachments are established, subsequently stabilising the chromosomes that at this stage are congregated at the spindle equator and silencing the SAC (Ciferri et al.

2008; DeLuca and Musacchio 2012; Fuller and Stukenberg 2009). How the incorrect attachment to the microtubule is sensed by the SAC is an aspect discussed in the next section below.

Kinetochore-SAC Communication Often Involves Large Conformational Transitions

Generally speaking, multiprotein assemblies that form cooperatively are unlikely to occur by accident (Blundell et al. 2002; Bolanos-Garcia et al. 2012). In the case of kinetochore-SAC communication, the cooperative association of higher-order complexes arising from specific and often low-affinity binary interactions favours signal amplification, thus ensuring an effective response to defects in kinetochore-microtubule attachment. Recent structural studies show that the establishment of kinetochore interaction with the SAC often involve important conformational changes, including disorder-to-order transitions that are important to mount a proper SAC response. The cooperative assembly of the KMN subcomplexes Mis12 and Ndc80 described above serves to illustrate how the establishment of higher-order signalling complexes can operate to modulate SAC signalling. Additional examples of this in the kinetochore-SAC signalling axis are Bub1 and BubR1 (Mad3 in yeast), in which their conserved tetratricopeptide repeat (TPR) motifs mediate the interactions with the KI motifs of Knl1 (Fig. 7.7a). Specific hydrophobic amino acid residues that are mapped onto the Bub1 and BubR1 TPR regions establish extensive hydrophobic interactions with the Bub1- and BubR1-binding motifs present in N-terminal Knl1. For instance, in the case of the BubR1-Knl1 interaction, the Knl1 residues I213, F215, F218 and I219 dock into BubR1 pockets, thus ensuring high specificity and sensitive regulation (Fig. 7.7b).

The multidomain Bub1 and BubR1 paralogues also harbour a conserved stretch of about 40 amino acid residues immediately after the N-terminal TPR domain that harbour a Bub3-binding region commonly referred to as the GLE2p-binding sequence (GLEBS) motif. The crystal structures of two different complexes Mad3 GLEBS motif-Bub3 and Bub1 GLEBS-Bub3 revealed an extensive interface along the top surface of Bub3 upon complex formation (Larsen and Harrison 2004) (Fig. 7.8a) and a mode of binding that is compatible with a drastic disorder-to-order transition. Other examples of this class of rearrangements relevant to the communication between the kinetochore and the SAC are the binding of the WD40 fold protein Bub3 to the Knl1 MELT motifs and the binding of the Knl1 RWD domain to a synthetic peptide that mimics Nsl1 (Fig. 7.8b, c, respectively) (Kiyomitsu et al. 2011; Bolanos-Garcia et al. 2011; Krenn et al. 2012; Primorac et al. 2013; Petrovic et al. 2014; Ghongane et al. 2014).

The high amino acid sequence conservation of the Knl1 MELT and KI motifs from diverse species is shown in Fig. 7.9. In all these Knl1 complexes, the binding of an otherwise predominantly disordered protein fragment containing a small linear

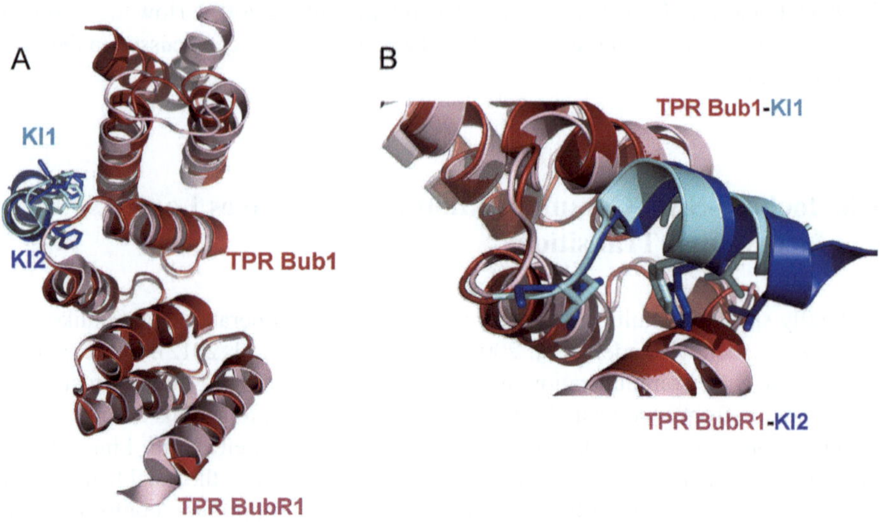

Fig. 7.7 Conformational transitions of KMN SLiMs-I. The N-terminal Bub1 and BubR1 regions harbour a triple tandem of the TPR motif. Superposition of the structures of human TPR Bub1 and human TPR BubR1 in complex with the KI motifs of Knl1 (pdb 4AIG and 3SI5, respectively) revealed a similar mode of binding that is consistent with large disorder-to-order conformational changes

interaction motif (SLiM) to the globular partner involves dramatic conformational transitions in the former and little conformational changes upon complex formation in the latter (e.g. the globular) molecule. Moreover, the formation of several subcomplexes of KMN-SAC proteins falls into the category of non-obligate complexes (Jones and Thornton 1996), which often involve a relatively small buried area, ranging from approximately 500–1700 Å2 (Table 7.1), a feature that accounts for the transient nature of these and other kinetochore complexes. Additional points of contact between the kinetochore and the SAC have been proposed such as the physical interaction of Mps1 with the Ndc80 complex (Ji et al. 2015; Hiruma et al. 2015), in a process that at least in budding yeast seems to involve Ndc80 phosphorylation by Mps1 (Sarangapani et al. 2021). Future studies should aim to dissect more precisely the mechanism by which Mps1 interaction with Ndc80 contributes to the regulation of kinetochore-microtubule attachment and how this signal is transduced to mount a SAC response.

It is possible that the local conformational changes in Knl1 triggered by disorder-to-order transitions upon ligand binding influence the interaction of this kinetochore protein with other interacting partners such as protein kinases and phosphatases to regulate the SAC, thus contributing to the tight regulation of chromosome segregation (Liu et al. 2010; Rosenberg et al. 2011). This is a fundamental aspect of kinetochore function that is discussed in some detail in the next section.

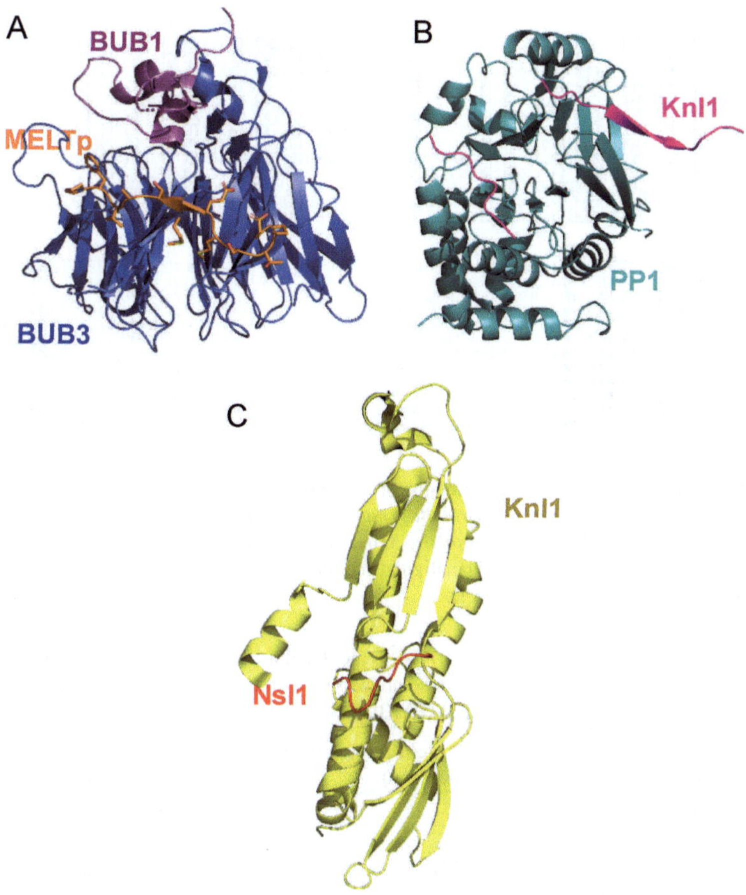

Fig. 7.8 Conformational transitions of KMN SLiMs-II. (**a**) Figure showing the crystal structure of yeast Bub3 in complex with the MELTp motif of yeast Knl1 (commonly referred to as Spc105) and a Bub1 fragment harbouring the Bub3-binding region (PDB ID reference 4BL0 and Primorac et al. 2013). (**b**) Crystal structure of the protein phosphatase 1 (PP1) holoenzyme in complex with a Knl1 fragment (PDB ID 6CZO). (**c**) The RWD domain of Knl1 in complex with Nsl1 (PDB 4NF9). In these four structures, key protein-protein interactions implicated in the formation of the complex are characterised by important disorder-to-order conformational transitions

Regulation of Kinetochore-SAC Communication

The exquisite regulation of mitosis through the communication between the kinetochore and the SAC is a fine example of how a highly dynamic association and dissociation of protein complexes in a temporal and spatial framework can ensure high selectivity and fidelity in cell signalling cascades. The serine/threonine protein kinases Mps1, Bub1, BubR1 and Plk1 play essential roles in the regulation of the kinetochore-SAC signalling axis (Fig. 7.2a). Mps1 is a dual-activity kinase and the

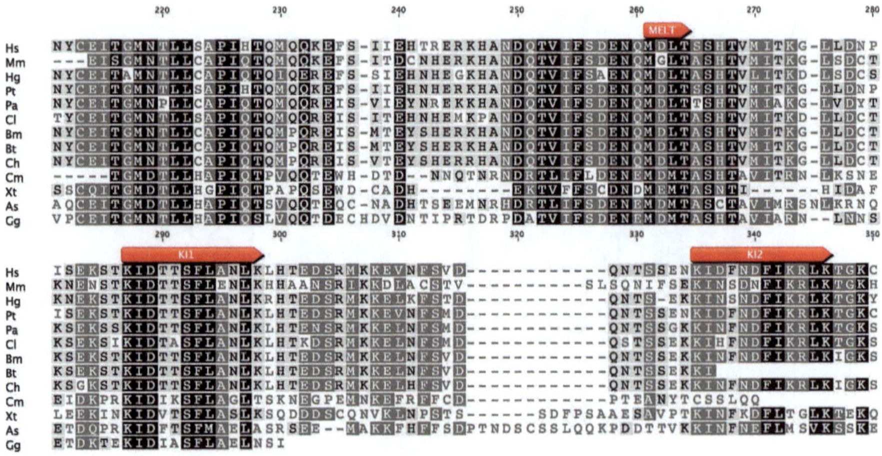

Fig. 7.9 Amino acid sequence alignments of the MELT, KI1 and KI2 motifs from diverse organisms. As, *Alligator sinensis*; Bm, *Bos mutus*; Ch, *Capra hircus*; Cl, *Canis lupus*; Hg, *Heterocephalus glaber*; Hs, *Homo sapiens*; Mm, *Mus musculus*; Pa, *Pteropus alecto*; Pt, *Pan troglodytes*; Xt, *Xenopus tropicalis*. The aligned sequences show that these three functional motifs are highly conserved across these species

Table 7.1 Analysis of the interface area in Å^2, calculated as the difference in the total accessible surface areas of isolated and interfacing structures divided by two, according to the ePISA (Protein Interfaces, Surfaces and Assemblies) tool (EMBL-EBI, UK)

Protein complex	Protein-protein interface area (Å^2)	PDB code	References
Bub3-MELT motif of Knl1 in ternary complex	654	4BL0	Primorac et al. (2013)
Bub1-KI-1 motif of Knl1	527	4A1G	Krenn et al. (2012)
BubR1-KI-2 motif of Knl1	464	3SI5	Bolanos-Garcia et al. (2011)
Nsl1-RWD domain of Knl1	565	4NF9	Petrovic et al. (2014)
Knl1-PP1 holoenzyme	1340	6CZO	Bajaj et al. (2018)
Mis12-CENP-C	1181	5LSK	Petrovic et al. (2016)

most upstream regulator of the SAC (Weiss and Winey 1996; Abrieu et al. 2001; Maciejowski et al. 2010; Tipton et al. 2013). Mps1 has been identified as one of the top 25 genes associated with cancer and therefore recognised as a bona fide target to develop anticancer therapies (Carter et al. 2006; Janssen et al. 2009). Bub1 is recruited early to the kinetochore and required for the kinetochore localisation of other checkpoint proteins in cells that have the checkpoint unsatisfied. It also plays an important role in the assembly of the inner centromere. In humans, BubR1 works in the SAC as a pseudokinase and is required for the establishment of proper kinetochore-microtubule attachment and chromosome alignment. Also, BubR1 associates with the SAC proteins Bub3, Mad2 and Cdc20 to form the mitotic checkpoint complex (MCC), a macromolecular assembly that inhibits the E3 ubiquitin ligase activity of the anaphase-promoting complex (also known as the cyclosome, APC/C) towards its substrates securin and cyclin B1 (Fig. 7.2b) (Tang et al. 2004; Vanoosthuyse and Hardwick 2005; Boyarchuk et al. 2007; Bolanos-Garcia and Blundell 2011; Elowe 2011; Chao et al. 2012 Lara-Gonzalez et al. 2021; Curtis and Bolanos-Garcia 2019), thus delaying mitotic progression until defects in chromosome attachment to microtubules are resolved.

Comparative sequence analysis of Bub1 and BubR1 suggests that in vertebrates, BubR1 kinase activity has been lost as a result of degeneration in several kinase subdomains including the glycine-rich and the activation loop regions. The resulting amino acid substitutions are incompatible with catalysis, thus resulting in a reclassification of BubR1 as a pseudokinase in SAC signalling (Suijkerbuijk et al. 2012; Tromer et al. 2016; Breit et al. 2015). However, it may be possible that in a different context, BubR1 may exhibit enzymatic activity such as in the regulation of cell signalling cascades by certain growth factor receptors or the use of alternative sequences in the kinase domain to support a catalytic function.

On the other hand, a functional role for BubR1 that implicates the kinase domain and the region immediately preceding this domain has been reported. At least in humans, CENP-E kinetochore localisation has been reported to be dependent on the BubR1 kinase domain (Ciossani et al. 2018; Craske and Welburn 2020; Legal et al. 2020; Chan et al. 1999), where the most C-terminal residues of this domain define a helix that mediates electrostatic interactions with an acidic patch on CENP-E (Ciossani et al. 2018; Legal et al. 2020). Although this interaction seems to be required for the docking of CENP-E at attached kinetochores, after a prolonged absence of microtubule attachment (for example, following treatment with the drug nocodazole) the interaction appears to be dispensable, and instead, CENP-E binding becomes dependent on proteins that compose the kinetochore fibrous corona (Legal et al. 2020).

A paralogue of BubR1 is the protein kinase Bub1 and the latter has been reported to associate with the kinetochore protein CENP-F. Such interaction mirrors that of BubR1 with CENP-E, adding support to the notion that the BUB1/R1-CENP-E/F binary interactions have co-evolved as a functional unit of the kinetochore-SAC signalling axis (Ciossani et al. 2018).

Silencing the SAC is required once errors in kinetochore-microtubule attachments have been repaired. Protein phosphatase 1 (PP1) and PP2A-B56 play key roles in silencing the SAC. PP1 kinetochore recruitment is promoted by the PP2A-mediated dephosphorylation of the Knl1 docking motif RVxF/SILK, which is located in the Knl1 N-terminal region, whereas BubR1 association with PP2A-B56 is required for the dephosphorylation of the Knl1 MELT repeats in late prometaphase, which is facilitated by the recruitment of PP1 to Knl1 (Espert et al. 2014; Nijenhuis et al. 2014; Cordeiro et al. 2020). Through direct binding to PP2A-B56, BubR1 can also drive PP2A-B56 specificity towards Cdc20 bound to the MCC (Hein et al. 2021). Because Cdc20 is also dephosphorylated by PP2A-B55, it has been suggested that BubR1 acts as a molecular switch that regulates Cdc20 inhibition/activation in the SAC. In such regulatory model, BubR1-mediated dephosphorylation of Cdc20 bound to the MCC by PP2A-B56 primes Cdc20 for APC/C activation prior to PP2A-B55 activation later in anaphase and mitosis exit (Hein et al. 2021).

BubR1-bound PP2A-B56 also dephosphorylates Bub1 at the Mad1 binding site and the T-loop region of Mps1 (Suijkerbuijk et al. 2012; Ji et al. 2017; Hayward et al. 2019). In addition to its role in SAC silencing, BubR1-bound PP2A has been reported to negatively affect the interaction between Bub1 and Mad1, to have an important role in the regulation of chromosome congression (Xu et al. 2013) and to reverse Aurora B phosphorylation at the outer kinetochore, ultimately resulting in the stabilisation of kinetochore-microtubule attachments (Suijkerbuijk et al. 2012; Nijenhuis et al. 2014).

The recruitment of PP2A-B56 to the kinetochore is mediated via the kinetochore attachment regulatory domain (KARD) of BubR1 (Suijkerbuijk et al. 2012; Vallardi et al. 2019; Foley et al. 2011; Wang et al. 2016a, b), implicating a hydrophobic pocket in the B56 subunit that is conserved in all B56 isoforms and a consensus interacting motif (LxxIxE, residues L669-E674 in the human BubR1) within the KARD (Suijkerbuijk et al. 2012; Hertz et al. 2016; Wang et al. 2016a; b). The LxxIxE motif is another example of a small linear interaction motif (SLiM). BubR1 and PP2A-B56 are sequentially recruited to the kinetochore during early prometaphase resulting in dephosphorylation of Bub1 at residue T461 by PP2A-B56, which in turn disturbs the interaction of Bub1 with Mad1 (Qian et al. 2017; Vallardi et al. 2019; Gama Braga et al. 2020). Interestingly, disruption of the BubR1-PP2A-B56 interaction leads to chronic SAC activation and mitotic cell death.

The protein kinase Aurora B also plays an important role in the regulation of the kinetochore, such as the destabilisation of incorrect kinetochore-microtubule attachments (Zhu et al. 2013; Welburn et al. 2010). However, the roles of Aurora B in mitosis are varied and complex and will not be discussed here. Instead, the reader is invited to consult the excellent reviews on Aurora kinase B that have been published recently (McVey et al. 2021; Broad and DeLuca 2020; Cairo and Lacefield 2020). In addition to Aurora B kinase activity, further interactions seem to be important for the regulation of kinetochore function such as the interaction of Bub1 kinase with Sgo1 to regulate kinetochore biorientation (Fernius and Hardwick 2007; Peplowska et al. 2014). However, the role of a Bub1-Sgo1 signalling axis in the regulation of the kinetochore is an aspect that merits further investigations. Moreover, apart from concerted phosphorylation and dephosphorylation events, additional layers of kinetochore regulation have been identified that significantly influence kinetochore protein stability, turnover, signal reversibility, subcellular localisation and the hierarchical order of protein complex formation and disassembly. Such additional layers of regulation include other types of post-translational modifications such as acetylation, ubiquitylation/deubiquitylation and sumoylation (Pawson and Nash 2003; Kim et al. 2006; Seet et al. 2006; Simorellis and Flynn 2006; Mao et al. 2011; Wan et al. 2012; Jia et al. 2013).

In summary, to avoid defective kinetochore-microtubule attachments during mitosis, eukaryotic organisms rely on multiple kinetochore kinases and fewer phosphatases as well as kinase-independent mechanisms. A surge in protein phosphorylation, which is orchestrated by several protein kinases, and opposing dephosphorylation events that are catalysed by a handful of protein phosphatases represent an elegant and effective strategy to control cell cycle progression through mitosis. In contrast to the roles of protein kinases in cell division regulation, which has been studied for decades, only in recent years has the importance of phosphatases in mitosis become increasingly appreciated. In this regard, an excellent review of mitotic phosphatases, from a description of their functional and structural properties to the current understanding of their mode of regulation, was published recently by Moura and Conde (2019).

The Kinetochore as a Potential Drug Target

The current literature suggests that the kinetochore-microtubule interface may represent an opportunity to develop new drugs to interfere with aberrant cell division in, for example, cancer cells (Matson and Stukenberg 2011). Although a number of drugs that interfere with the dynamics of microtubule polymerisation/depolymerisation are already in use in the clinic, they exhibit important disadvantages. For example, Taxol, also known as paclitaxel or PTX, is one of the most widely used drugs to treat tumours. However, it also suffers from important side effects and the induction of rapid resistance to the drug in the treated cells (Gallego-Jara et al. 2020). PTX induces mitotic arrest as its main mechanism of action. However, it may also cause cell death in tumour cells via the activation of metabolic apoptosis, which in turn may be due to transcription factor p53 activation, a molecule widely associated with tumour progression (Ren et al. 2018; Strobel et al. 1996).

Another drug, docetaxel, is a small-size compound that reversibly binds to microtubules with high affinity (Matesanz et al. 2008), thus impeding microtubule disassembly. This condition leads to microtubule accumulation inside the cell, triggering apoptosis. Both in vitro and in vivo analyses have shown the antineoplastic activity of docetaxel to be effective against a wide range of known cancer cells, to cooperate with other antineoplastic agents and to have greater cytotoxicity than paclitaxel, possibly due to its more rapid intracellular uptake (Gallego-Jara et al. 2020; Lyseng-Williamson and Fenton 2005). Due to these advantages, docetaxel is used in patients with local recurrence of breast cancer, locally advanced breast cancer and secondary breast tumours (Breast Cancer Care 2021). It is also given before (neoadjuvant therapy) and after surgery (adjuvant therapy) and used prior to radiotherapy or hormone therapy (Breast Cancer Care 2021). More recently, another small-size molecule with demonstrated high tumour growth inhibitory activity and selectivity index on human cancer cells was reported. This new molecule is a diarylpentanoid compound termed BP-M345 (Novais et al. 2021). This drug was reported to induce alterations of the mitotic spindles of cancer cells, causing microtubule perturbations that result in irreversible chromosome congression errors during mitosis and ultimately cancer cell death (Novais et al. 2021).

In summary, molecules that target microtubules are some of the most important antimitotic drugs currently in use to treat diverse malignancies, including cancer. Through the perturbation of mitotic spindle assembly, microtubule poisons can trigger the onset of the SAC, leading to mitotic arrest and subsequent apoptosis. However, microtubule poisons present important limitations such as undesired side effects, high toxicity, induction of the rapid acquisition of drug resistance, mitotic slippage and failure to induce apoptosis in cancer cells (Henriques et al. 2019). These important limitations continue to stimulate the search and development of novel, more effective drugs that target microtubules for the treatment of cancer and other chronic diseases.

Recent Technology-Driven Advances to Study the Kinetochore

Recently, a game changer achievement in the field of protein structure was reported, namely, the prediction with high accuracy of three-dimensional protein structure based on the primary structure of the molecule using an artificial intelligence (AI) approach. This breakthrough and AI system named AlphaFold and developed by the company DeepMind initially covered the human proteome and the proteomes of several other organisms (Jumper et al. 2021; Varadi et al. 2022). A second release incorporated the majority of manually curated UniProt entries. The next frontier in this area is the prediction of the 3D structures of protein complexes, which seems far more challenging for a number of reasons, including the induction of large disorder-to-order conformational transitions upon complex formation that occur in multiple cases. It can be anticipated that the application of AI approaches to the study of the kinetochore structure, composition and regulation will be helpful to predict the formation of yet unidentified kinetochore protein complexes and to test structure-guided hypothesis to dissect more precisely the role of the kinetochore in the control of cell division and how defects in the process lead to genome instability and aneuploidy.

The rise of deep learning and machine learning technologies have also allowed researchers to automate image classification, a strategy that has been applied to generate a computational model of the kinetochore (Lawrimore et al. 2019). More specifically, a convolutional neural net (CNN) has been developed to correctly classify fluorescent images of inner and outer kinetochore proteins (Lawrimore et al. 2019).

Deep learning and machine learning have been used to produce a computational model of fission-yeast mitosis, which postulates that progressive restriction of attachment geometry, destabilisation of misaligned attachments and attachment force dependence are required to achieve chromosome biorientation (Edelmaier et al. 2020). The computational model also suggests that these molecular mechanisms contribute to persistent biorientation, driving chromosome segregation to the poles after anaphase onset (Edelmaier et al. 2020). Ultimately, the in silico model provides a theoretical framework for the systematic exploration of the effect of altering key requirements on chromosome biorientation, force generation and spindle length regulation in the mitotic spindle.

During mitotic progression, the kinetochore undergoes dynamic morphological changes that can be monitored by fluorescence and electron microscopy (Hara and Fukagawa 2020). Such dynamics changes provide important clues of the kinetochore functions in the control of cell cycle progression. Expansion microscopy is a novel super-resolution microscopy technique that relies in the physical expansion of fixed specimens to allow features closer than the diffraction limit of light (~250 nm) to be resolvable by confocal microscopy. Indeed, expansion microscopy of the mitotic spindle using tubulin labelling has permitted the visualisation of microtubule subpopulations within the mitotic spindle and the characterisation of the architecture of mitotic spindle microtubules before and after perturbation of the structure of the mitotic spindle (Ponjavić et al. 2021).

Serial block face-scanning electron microscopy (SBF-SEM) is a technique that can be used to visualise the ultrastructure of entire cells, including components of the mitotic spindle such as microtubules, kinetochores, centrosomes and chromosomes, thus enhancing our understanding of how the mitotic spindle operates at the molecular level. Although transmission electron microscopy (TEM) has higher resolution, the reconstruction of large volumes using TEM and tomography is labour intensive, whereas SBF-SEM takes only days to process, image and segment samples (Clarke and Royle 2018). Hence, SBF-SEM fills the resolution gap between light microscopy (LM) and TEM (Clarke and Royle 2018). When combined with LM, SBF-SEM provides a platform where dynamic cellular events can be selected and imaged at high resolution and this is expected to be very useful in the study of the kinetochore (Clarke and Royle 2018).

Another area of fast development is single-cell proteomics, which for the first time has enabled the definition of protein dynamics changes with special resolution (Petrov et al. 2021; Liotta et al. 2021; Gnann et al. 2021; Xing et al. 2020; Trinkle-Mulcahy 2019; Doupé and Perrimon 2016). The application of such approached to the study of the kinetochore-SAC signalling axis should enhance our understanding of the molecular changes that regulate the progression of the cell cycle and the control of cell division in humans and other eukaryotic organisms.

What Lies Ahead?

The recent advances in AI systems, high-resolution live cell imaging methods, single-cell electron microscopy and single-cell proteomics techniques with spatial resolution enable a holistic approach to investigate the function and regulation of the kinetochore with spatial context, including the definition of the dynamics of protein subcellular localisation and translocations, remodelling of interaction networks of the kinetochore individual protein components and kinetochore-SAC subcomplexes, definition of the dynamics of the kinetochore-SAC signalling in a disease microenvironment (e.g. tumours, neurological disorders) and comparative spatial proteomics approaches to study the evolution of the kinetochore-SAC signalling axis.

There is growing consensus that the kinetochore is a dynamic structure whose composition changes as the cell cycle progresses (Hamilton et al. 2019) and the mounting evidence of multiple pathways whereby the microtubule-binding elements of the outer kinetochore may be recruited by inner kinetochore proteins. Considering that the in vitro reconstitution of a functional kinetochore holocomplex is now doable, the structural characterisation of the kinetochore by electron microscopy should shed new light into the molecular interactions responsible of its assembly/disassembly and the identification and characterisation of the large unstructured regions of individual kinetochore protein components that undergo dramatic conformational remodelling upon complex formation. Undoubtedly, elucidation of the 3D structure of the whole kinetochore-microtubule-SAC axis would be a remarkable achievement in the structural biology and cell division regulation fields.

In can also be anticipated that in-depth studies of selected new interactions implicating kinetochore-microtubule-SAC communication using biochemical, biophysical, cellular and structural biology approaches will illuminate important species-specific aspects of chromosome segregation control. Future research should also clarify the true potential of the kinetochore-SAC signalling axis as a source of novel biomarkers of disease and new drug targets and a suitable system to apply innovative proteolysis-targeted therapies to treat human and animal malignancies of poor prognosis. In the years to come, the study of the kinetochore appears as exciting as ever!

Acknowledgments I thank Andrea Musacchio, Stefan Raunser and Ingrid Vetter of the Max-Planck Institute of Molecular Physiology, Dortmund, Germany, for kindly sharing the atom coordinates of the human CCAN complex (PDB ID 7QOO) prior to its publication.

References

Abrieu A, Magnaghi-Jaulin L, Kahana JA, Peter M, Castro A, Vigneron S, Lorca T, Cleveland DW, Labbe JC (2001) Mps1 is a kinetochore-associated kinase essential for the vertebrate mitotic checkpoint. Cell 106:83–93

Allu PK, Dawicki-McKenna JM, Van Eeuwen T, Slavin M, Braitbard M, Xu C, Kalisman N, Murakami K, Black BE (2019) Structure of the human core centromeric nucleosome complex. Curr Biol 29:2625–2639.e5

Audett MR, Maresca TJ (2020) The whole is greater than the sum of its parts: at the intersection of order, disorder, and kinetochore function. Essays Biochem 64:349–358

Bajaj R, Bollen M, Peti W, Page R (2018) KNL1 binding to PP1 and microtubules is mutually exclusive. Structure 26:1327–1336.e4

Barbosa J, Conde C, Sunkel C (2020) RZZ-SPINDLY-DYNEIN: you got to keep 'em separated. Cell Cycle 19:1716–1726

Blundell TL, Chaplin AK (2021) The resolution revolution in X-ray diffraction, Cryo-EM and other technologies. Prog Biophys Mol Biol 160:2–4

Blundell TL, Bolanos-Garcia VM, Chirgadze DY, Harmer NJ, Lo T, Pellegrini L, Sibanda BL (2002) Asymmetry in the multiprotein systems of molecular biology. Struct Chem 13:405–412

Bolanos-Garcia VM, Blundell TL (2011) BUB1 and BUBR1: multifaceted kinases of the cell cycle. Trends Biochem Sci 36:141–150

Bolanos-Garcia VM, Kiyomitsu T, D'Arcy S, Chirgadze DY, Grossmann JG, Matak-Vinkovic D, Venkitaraman AR, Yanagida M, Robinson CV, Blundell TL (2009) The crystal structure of the N-terminal region of BUB1 provides insight into the mechanism of BUB1 recruitment to kinetochores. Structure 17:105–116

Bolanos-Garcia VM, Lischetti T, Matak-Vinkovic D, Cota E, Simpson PJ, Chirgadze DY, Spring DR, Robinson CV, Nilsson J, Blundell TL (2011) Structure of a Blinkin-BUBR1 complex reveals an interaction crucial for kinetochore-mitotic checkpoint regulation via an unanticipated binding site. Structure 19:1691–1700

Bolanos-Garcia VM, Wu Q, Ochi T, Chirgadze DY, Sibanda BL, Blundell TL (2012) Spatial and temporal organisation of multiprotein assemblies: achieving sensitive control in information-rich cell regulatory systems. Philos Trans A Math Phys Eng Sci 370:3023–3039

Boyarchuk Y, Salic A, Dasso M, Arnaoutov A (2007) Bub1 is essential for assembly of the functional inner centromere. J Cell Biol 176:919–928

Breast Cancer Care (2021). https://www.breastcancercare.org.uk/information-support/facing-breast-cancer/going-through-treatment-breast-cancer/chemotherapy/docetaxel-taxotere

Breit C et al (2015) Role of intrinsic and extrinsic factors in the regulation of the mitotic checkpoint kinase Bub1. PLoS One 10:e0144673

Broad AJ, DeLuca JG (2020) The right place at the right time: Aurora B kinase localization to centromeres and kinetochores. Essays Biochem 64:299–311

Cairo G, Lacefield S (2020) Establishing correct kinetochore-microtubule attachments in mitosis and meiosis. Essays Biochem 64:277–287

Cao B, Zhao C, Zhang Y, Wang X, Ye J, Hu L, He X (2021) The novel interaction mode among centromere sub-complex CENP-O/P/U/Q/R. J Mol Recogn 34:e2892

Carter SL, Eklund AC, Kohane IS, Harris LN, Szallasi Z (2006) A signature of chromosomal instability inferred from gene expression profiles predicts clinical outcome in multiple human cancers. Nat Genet 38:1043–1048

Chan GK, Jablonski SA, Sudakin V, Hittle JC, Yen TJ (1999) Human BUBR1 is a mitotic checkpoint kinase that monitors CENP-E functions at kinetochores and binds the cyclosome/APC. J Cell Biol 146:941–954

Chao WC, Kulkarni K, Zhang Z, Kong EH, Barford D (2012) Structure of the mitotic checkpoint complex. Nature 484:208–213

Cheeseman IM, Desai A (2008) Molecular architecture of the kinetochore-microtubule interface. Nat Rev Mol Cell Biol 9:33–46

Cheeseman IM, Chappie JS, Wilson-Kubalek EM, Desai A (2006) The conserved KMN network constitutes the core microtubule-binding site of the kinetochore. Cell 127:983–997

Cheeseman IM, Hori T, Fukagawa T, Desai A (2008) KNL1 and the CENP-H/I/K complex coordinately direct kinetochore assembly in vertebrates. Mol Biol Cell 19:587–594

Chittori S, Hong J, Saunders H, Feng H, Ghirlando R, Kelly AE, Bai Y, Subramaniam S (2018) Structural mechanisms of centromeric nucleosome recognition by the kinetochore protein CENP-N. Science 359:339–343

Ciferri C, De Luca J, Monzani S, Ferrari KJ, Ristic D, Wyman C, Stark H, Kilmartin J, Salmon ED, Musacchio A (2005) Architecture of the human Ndc80-Hec1 complex, a critical constituent of the outer kinetochore. J Biol Chem 280:29088–29095

Ciferri C, Pasqualato S, Screpanti E, Varetti G, Santaguida S, Dos Reis G, Maiolica A, Polka J, De Luca JG, De Wulf P, Salek M, Rappsilber J, Moores CA, Salmon ED, Musacchio A (2008) Implications for kinetochore-microtubule attachment from the structure of an engineered Ndc80 complex. Cell 133:427–439

Ciossani G, Overlack K, Petrovic A, Huis in 't Veld PJ, Koerner C, Wohlgemuth S, Maffini S, Musacchio A (2018) The kinetochore proteins CENP-E and CENP-F directly and specifically interact with distinct BUB mitotic checkpoint Ser/Thr kinases. J Biol Chem 293:10084–10101
Clarke NI, Royle SJ (2018) Correlating light microscopy with serial block face scanning electron microscopy to study mitotic spindle architecture. Methods Cell Biol 145:29–43
Cordeiro MH, Smith RJ, Saurin AT (2020) Kinetochore phosphatases suppress autonomous Polo-like kinase 1 activity to control the mitotic checkpoint. J Cell Biol 219:e202002020
Craske B, Welburn JPI (2020) Leaving no-one behind: how CENP-E facilitates chromosome alignment. Essays Biochem 64:313–324
Curtis NL, Bolanos-Garcia VM (2019) The anaphase promoting complex/cyclosome (APC/C): a versatile E3 ubiquitin ligase. Subcell Biochem 93:539–623
Curtis LN, Ruda GF, Brennan P, Bolanos-Garcia VM (2020) Deregulation of chromosome segregation and cancer. Annu Rev Cancer Biol 4:257–278
De Antoni A, Pearson CG, Cimini D, Canman JC, Sala V, Nezi L, Mapelli M, Sironi L, Faretta M, Salmon ED, Musacchio A (2005) The Mad1/Mad2 complex as a template for Mad2 activation in the spindle assembly checkpoint. Curr Biol 15:214–225
DeLuca JG, Musacchio A (2012) Structural organization of the kinetochore-microtubule interface. Curr Opin Cell Biol 24:48–56
Dimitrova YN, Jenni S, Valverde R, Khin Y, Harrison SC (2016) Structure of the MIND complex defines a regulatory focus for yeast kinetochore assembly. Cell 167:1014–1027.e12
Doupé DP, Perrimon N (2016) Toward a systems understanding of signaling pathway function. Curr Top Dev Biol 117:221–236
Earnshaw WC, Rothfield N (1985) Identification of a family of human centromere proteins using autoimmune sera from patients with scleroderma. Chromosoma 91:313–321
Edelmaier C, Lamson AR, Gergely ZR, Ansari S, Blackwell R, McIntosh JR, Glaser MA, Betterton MD (2020) Mechanisms of chromosome biorientation and bipolar spindle assembly analyzed by computational modeling. Elife 9:e48787
Elowe S (2011) Bub1 and BubR1: at the interface between chromosome attachment and the spindle checkpoint. Mol Cell Biol 31:3085–3093
Espert A, Uluocak P, Bastos RN, Mangat D, Graab P, Gruneberg U (2014) PP2A-B56 opposes Mps1 phosphorylation of Knl1 and thereby promotes spindle assembly checkpoint silencing. J Cell Biol 206:833–842
Fernius J, Hardwick KG (2007) Bub1 kinase targets Sgo1 to ensure efficient chromosome biorientation in budding yeast mitosis. PLoS Genet 3:e213
Ferrand J, Rondinelli B, Polo SE (2020) Histone variants: guardians of genome integrity. Cells 9:2424
Ferreira LT, Maiato H (2021) Prometaphase. Semin Cell Dev Biol 117:52–61
Fischer ES, Yu CWH, Bellini D, McLaughlin SH, Orr CM, Wagner A, Freund SMV, Barford D (2021) Molecular mechanism of Mad1 kinetochore targeting by phosphorylated Bub1. EMBO Rep 22:e52242
Foley EA, Kapoor TM (2013) Microtubule attachment and spindle assembly checkpoint signalling at the kinetochore. Nat Rev Mol Cell Biol 14:25–37
Foley EA, Maldonado M, Kapoor TM (2011) Formation of stable attachments between kinetochores and microtubules depends on the B56-PP2A phosphatase. Nat Cell Biol 13:1265–1271
Fuller BG, Stukenberg PT (2009) Cell division: righting the check. Curr Biol 19:R550–R553
Funabiki H, Wynne DJ (2013) Making an effective switch at the kinetochore by phosphorylation and dephosphorylation. Chromosoma 122:135–158
Gallego-Jara J, Lozano-Terol G, Sola-Martínez RA, Cánovas-Díaz M, de Diego PT (2020) A compressive review about Taxol®: history and future challenges. Molecules 25:5986
Gama Braga L, Cisneros AF, Mathieu MM, Clerc M, Garcia P, Lottin B, Garand C, Thebault P, Landry CR, Elowe S (2020) BUBR1 pseudokinase domain promotes kinetochore PP2A-B56 recruitment, spindle checkpoint silencing, and chromosome alignment. Cell Rep 33:108397

Ghongane P, Kapanidou M, Asghar A, Elowe S, Bolanos-Garcia VM (2014) The dynamic protein KNL1: a kinetochore rendezvous. J Cell Sci 127:3415–3423
Gnann C, Cesnik AJ, Lundberg E (2021) Illuminating non-genetic cellular heterogeneity with imaging-based spatial proteomics. Trends Cancer 7:278–282
Hamilton GE, Davis TN (2020) Biochemical evidence for diverse strategies in the inner kinetochore. Open Biol 10:200284
Hamilton G, Dimitrova Y, Davis TN (2019) Seeing is believing: our evolving view of kinetochore structure, composition, and assembly. Curr Opin Cell Biol 60:44–52
Hara M, Fukagawa T (2017) Critical foundation of the kinetochore: the constitutive centromere-associated network (CCAN). Prog Mol Subcell Biol 56:29–57
Hara M, Fukagawa T (2020) A dynamics of kinetochore structure and its regulations during mitotic progression. Cell Mol Life Sci 77:2981–2995
Hara M, Ariyoshi M, Okumura EI, Hori T, Fukagawa T (2018) Multiple phosphorylations control recruitment of the KMN network onto kinetochores. Nat Cell Biol 20:1378–1388
Hardwick KG, Johnston RC, Smith DL, Murray AW (2000) MAD3 encodes a novel component of the spindle checkpoint which interacts with Bub3p, CDC20p, and MAD2p. J Cell Biol 148: 871–882
Hayward D, Bancroft J, Mangat D, Alfonso-Pérez T, Dugdale S, McCarthy J, Barr FA, Gruneberg U (2019) Checkpoint signaling and error correction require regulation of the MPS1 T-loop by PP2A-B56. J Cell Biol 218:3188–3199
Hein JB, Garvanska DH, Nasa I, Kettenbach AN, Nilsson J (2021) Coupling of Cdc20 inhibition and activation by BubR1. Cell Biol 220:e202012081
Henen MA, Myers W, Schmitt LR, Wade KJ, Born A, Nichols PJ, Vögeli B (2021) The disordered spindly C-terminus interacts with RZZ subunits ROD-1 and ZWL-1 in the kinetochore through the same sites in C. elegans. J Mol Biol 433:166812
Henriques AC, Ribeiro D, Pedrosa J, Sarmento B, Silva PMA, Bousbaa H (2019) Mitosis inhibitors in anticancer therapy: When blocking the exit becomes a solution. Cancer Lett 440-441:64–81
Hertz EPT, Kruse T, Davey NE, López-Méndez B, Sigurðsson JO, Montoya G, Olsen JV, Nilsson J (2016) A conserved motif provides binding specificity to the PP2A-B56 phosphatase. Mol Cell 63:686–695
Hiruma Y, Sacristan C, Pachis ST, Adamopoulos A, Kuijt T, Ubbink M, von Castelmur E, Perrakis A, Kops GJ (2015) Cell division cycle. Competition between MPS1 and microtubules at kinetochores regulates spindle checkpoint signaling. Science 348:1264–1267
Hori T, Shang WH, Takeuchi K, Fukagawa T (2013) The CCAN recruits CENP-A to the centromere and forms the structural core for kinetochore assembly. J Cell Biol 200:45–60
Izawa D, Pines J (2012) Mad2 and the APC/C compete for the same site on Cdc20 to ensure proper chromosome segregation. J Cell Biol 199:27–37
Janssen A, Kops GJ, Medema RH (2009) Elevating the frequency of chromosome missegregation as a strategy to kill tumor cells. Proc Natl Acad Sci USA 106:19108–19113
Ji Z, Gao H, Yu H (2015) Cell division cycle. Kinetochore attachment sensed by competitive Mps1 and microtubule binding to Ndc80C. Science 348:1260–1264
Ji Z, Gao H, Jia L, Li B, Yu H (2017) A sequential multi-target Mps1 phosphorylation cascade promotes spindle checkpoint signaling. Elife 6:e22513
Jia L, Kim S, Yu H (2013) Tracking spindle checkpoint signals from kinetochores to APC/C. Trends Biochem Sci 38:302–311
Joglekar AP, DeLuca JG (2009) Chromosome segregation: Ndc80 can carry the load. Curr Biol 19: R404–R407
Jones S, Thornton JM (1996) Principles of protein-protein interactions. Proc Natl Acad Sci USA 93: 13–20
Jumper J, Evans R, Pritzel A, Green T, Figurnov M, Ronneberger O, Tunyasuvunakool K, Bates R, Žídek A, Potapenko A, Bridgland A, Meyer C, Kohl SAA, Ballard AJ, Cowie A, Romera-Paredes B, Nikolov S, Jain R, Adler J, Back T, Petersen S, Reiman D, Clancy E, Zielinski M, Steinegger M, Pacholska M, Berghammer T, Bodenstein S, Silver D, Vinyals O, Senior AW,

Kavukcuoglu K, Kohli P, Hassabis D (2021) Highly accurate protein structure prediction with AlphaFold. Nature 596:583–589
Kim SC, Sprung R, Chen Y, Xu Y, Ball H, Pei J, Cheng T, Kho Y, Xiao H, Xiao L, Grishin NV, White M, Yang XJ, Zhao Y (2006) Substrate and functional diversity of lysine acetylation revealed by a proteomics survey. Mol Cell 23:607–618
Kixmoeller K, Allu PK, Black BE (2020) The centromere comes into focus: from CENP-A nucleosomes to kinetochore connections with the spindle. Open Biol 10:200051
Kiyomitsu T, Obuse C, Yanagida M (2007) Human Blinkin/AF15q14 is required for chromosome alignment and the mitotic checkpoint through direct interaction with Bub1 and BubR1. Dev Cell 13:663–676
Kiyomitsu T, Murakami H, Yanagida M (2011) Protein interaction domain mapping of human kinetochore protein Blinkin reveals a consensus motif for binding of spindle assembly checkpoint proteins Bub1 and BubR1. Mol Cell Biol 31:998–1011
Klemm C, Thorpe PH, Ólafsson G (2021) Cell-cycle phospho-regulation of the kinetochore. Curr Genet 67:177–193
Knossow M, Campanacci V, Khodja LA, Gigant B (2020) The mechanism of tubulin assembly into microtubules: insights from structural studies. iScience 23:101511
Komaki S, Schnittger A (2016) The spindle checkpoint in plants-a green variation over a conserved theme? Curr Opin Plant Biol 34:84–91
Komaki S, Schnittger A (2017) The spindle assembly checkpoint in Arabidopsis is rapidly shut off during severe stress. Dev Cell 43:172–185.e5
Kops GJ, Kim Y, Weaver BA, Mao Y, McLeod I, Yates JR 3rd, Tagaya M, Cleveland DW (2005) ZW10 links mitotic checkpoint signaling to the structural kinetochore. J Cell Biol 169:49–60
Kops GJPL, Snel B, Tromer EC (2020) Evolutionary dynamics of the spindle assembly checkpoint in eukaryotes. Curr Biol 30:R589–R602
Krenn V, Wehenkel A, Li X, Santaguida S, Musacchio A (2012) Structural analysis reveals features of the spindle checkpoint kinase Bub1-kinetochore subunit Knl1 interaction. J Cell Biol 196: 451–467
Lander GC, Glaeser RM (2021) Conquer by cryo-EM without physically dividing. *Biochem Soc Trans*. 49:2287–2298
Lara-Gonzalez P, Westhorpe FG, Taylor SS (2012) The spindle assembly checkpoint. Curr Biol 22: R966–R980
Lara-Gonzalez P, Pines J, Desai A (2021) Spindle assembly checkpoint activation and silencing at kinetochores. Semin Cell Dev Biol 117:86–98
Lara-Gonzalez P, Kim T, Oegema K, Corbett K, Desai A (2021b) A tripartite mechanism catalyzes Mad2-Cdc20 assembly at unattached kinetochores. Science 371:64–67
Larsen NA, Harrison SC (2004) Crystal structure of the spindle assembly checkpoint protein Bub3. J Mol Biol 344:885–892
Lawrimore J, Doshi A, Walker B, Bloom K (2019) AI-assisted forward modeling of biological structures. Front Cell Dev Biol 7:279
Legal T, Hayward D, Gluszek-Kustusz A, Blackburn EA, Spanos C, Rappsilber J, Gruneberg U, Welburn JPI (2020) The C-terminal helix of BubR1 is essential for CENP-E-dependent chromosome alignment. J Cell Sci 133:jcs246025
Lermontova I, Fuchs J, Schubert I (2008) The Arabidopsis checkpoint protein Bub3.1 is essential for gametophyte development. Front Biosci 13:5202–5211
Liotta LA, Pappalardo PA, Carpino A, Haymond A, Howard M, Espina V, Wulfkuhle J, Petricoin E (2021) Laser capture proteomics: spatial tissue molecular profiling from the bench to personalized medicine. Expert Rev Proteomics 18:845–861
Liu D, Vleugel M, Backer CB, Hori T, Fukagawa T, Cheeseman IM, Lampson LA (2010) Regulated targeting of protein phosphatase 1 to the outer kinetochore by KNL1 opposes Aurora B kinase. J Cell Biol 188:809–920
London N, Ceto S, Ranish JA, Biggins S (2012) Phosphoregulation of Spc105 by Mps1 and PP1 regulates Bub1 localization to kinetochores. Curr Biol 22:900–906

Long AF, Suresh P, Dumont S (2020) Individual kinetochore-fibers locally dissipate force to maintain robust mammalian spindle structure. J Cell Biol 219:e201911090
Luo X, Tang Z, Rizo J, Yu H (2002) The MAD2 spindle checkpoint protein undergoes similar major conformational changes upon binding to either MAD1 or CDC20. Mol Cell 9:59–71
Luo X, Tang Z, Xia G, Wassmann K, Matsumoto T, Rizo J, Yu H (2004) The Mad2 spindle checkpoint protein has two distinct natively folded states. Nat Struct Mol Biol 11:338–345
Lyseng-Williamson KA, Fenton C (2005) Docetaxel: a review of its use in metastatic breast cancer. Drugs 65:2513–2531
Maciejowski J, George KA, Terret ME, Zhang C, Shokat KM, Jallepalli PV (2010) Mps1 directs the assembly of Cdc20 inhibitory complexes during interphase and mitosis to control M phase timing and spindle checkpoint signaling. J Cell Biol 190:89–100
Mao YS, Zhang B, Spector DL (2011) Biogenesis and function of nuclear bodies. Trends Genet 27: 295–306
Matesanz R, Barasoain I, Yang CG, Wang L, Li X, de Inés C, Coderch C, Gago F, Barbero JJ, Andreu JM, Fang WS, Díaz JF (2008) Optimization of taxane binding to microtubules: binding affinity dissection and incremental construction of a high-affinity analog of paclitaxel. Chem Biol 15:573–585
Matson DR, Stukenberg PT (2011) Spindle poisons and cell fate: a tale of two pathways. Mol Interven 11:141–150
McCleland ML, Gardner RD, Kallio MJ, Daum JR, Gorbsky GJ, Burke DJ, Stukenberg PT (2003) The highly conserved Ndc80 complex is required for kinetochore assembly, chromosome congression, and spindle checkpoint activity. Genes Dev 17:101–114
McCleland ML, Kallio MJ, Barrett-Wilt GA, Kestner CA, Shabanowitz J, Hunt DF, Gorbsky GJ, Stukenberg PT (2004) The vertebrate Ndc80 complex contains Spc24 and Spc25 homologs, which are required to establish and maintain kinetochore-microtubule attachment. Curr Biol 14: 131–137
McKinley KL, Cheeseman IM (2016) The molecular basis for centromere identity and function. Nat Rev Mol Cell Biol 17:16–29
McVey SL, Cosby JK, Nannas NJ (2021) Aurora B tension sensing mechanisms in the kinetochore ensure accurate chromosome segregation. Int J Mol Sci 22:8818
Morrow CJ, Tighe A, Johnson VL, Scott MI, Ditchfield C, Taylor SS (2005) BUB1 and aurora B cooperate to maintain BubR1-mediated inhibition of APC/CCDC20. J Cell Sci 118:3639–3652
Mosalaganti S, Keller J, Altenfeld A, Winzker M, Rombaut P, Saur M, Petrovic A, Wehenkel A, Wohlgemuth S, Müller F, Maffini S, Bange T, Herzog F, Waldmann H, Raunser S, Musacchio A (2017) Structure of the RZZ complex and molecular basis of its interaction with Spindly. J Cell Biol 216:961–981
Moura M, Conde C (2019) Phosphatases in mitosis: roles and regulation. Biomolecules 9:55
Musacchio A, Desai A (2017) A molecular view of kinetochore assembly and function. Biology 6:5
Musacchio A, Salmon ED (2007) The spindle-assembly checkpoint in space and time. Nat Rev Mol Cell Biol 8:379–393
Navarro AP, Cheeseman IM (2021) Kinetochore assembly throughout the cell cycle. Semin Cell Dev Biol 117:62–74
Nijenhuis W, Vallardi G, Teixeira A, Kops GJ, Saurin AT (2014) Negative feedback at kinetochores underlies a responsive spindle checkpoint signal. Nat Cell Biol 16:1257–1264
Novais P, Silva PMA, Moreira J, Palmeira A, Amorim I, Pinto M, Cidade H, Bousbaa H (2021) BP-M345, a new diarylpentanoid with promising antimitotic activity. Molecules 26:7139
Obuse C, Iwasaki O, Kiyomitsu T, Goshima G, Toyoda Y, Yanagida M (2004) A conserved Mis12 centromere complex is linked to heterochromatic HP1 and outer kinetochore protein Zwint-1. Nat Cell Biol 6:1135–1141
Palmer DK, O'Day K, Wener MH, Andrews BS, Margolis RL (1987) A 17-kD centromere protein (CENP-A) copurifies with nucleosome core particles and with histones. J Cell Biol 104:805–815

Pawson T, Nash P (2003) Assembly of cell regulatory systems through protein interaction domains. Science 300:445–452

Pentakota S, Zhou K, Smith C, Maffini S, Petrovic A, Morgan GP, Weir JR, Vetter IR, Musacchio A, Luger K (2017) Decoding the centromeric nucleosome through CENP-N. Elife 6:e33442

Peplowska K, Wallek AU, Storchova Z (2014) Sgo1 regulates both condensin and Ipl1/Aurora B to promote chromosome biorientation. PLoS Genet 10:e1004411

Pereira C, Reis RM, Gama JB, Celestino R, Cheerambathur DK, Carvalho AX, Gassmann R (2018) Self-assembly of the RZZ complex into filaments drives kinetochore expansion in the absence of microtubule attachment. Curr Biol 28:3408–3421.e8

Perpelescu M, Fukagawa T (2011) The ABCs of CENPs. *Chromosoma* 120:425–446

Pesenti ME, Raisch T, Conti D, Hoffmann I, Vogt D, Prumbaum D, Vetter IR, Raunser S, Musacchio A (2022) Structure of the human inner kinetochore CCAN complex and its significance for human centromere organization. Mol Cell S1097-2765:00390-2.

Petrov PN, Müller H, Glaeser RM (2021) Perspective: emerging strategies for determining atomic-resolution structures of macromolecular complexes within cells. J Struct Biol 214:107827

Petrovic A, Pasqualato S, Dube P, Krenn V, Santaguida S, Cittaro D, Monzani S, Massimiliano L, Keller J, Tarricone A, Maiolica A, Stark H, Musacchio A (2010) The MIS12 complex is a protein interaction hub for outer kinetochore assembly. J Cell Biol 190:835–852

Petrovic A, Mosalaganti S, Keller J, Mattiuzzo M, Overlack K, Krenn V, De Antoni A, Wohlgemuth S, Cecatiello V, Pasqualato S, Raunser S, Musacchio A (2014) Modular assembly of RWD domains on the Mis12 complex underlies outer kinetochore organization. Mol Cell 53: 591–605

Petrovic A, Keller J, Liu Y, Overlack K, John J, Dimitrova YN, Jenni S, van Gerwen S, Stege P, Wohlgemuth S, Rombaut P, Herzog F, Harrison SC, Vetter IR, Musacchio A (2016) Structure of the MIS12 Complex and molecular basis of its interaction with CENP-C at human kinetochores. Cell 167:1028–1040.e15

Piano V, Alex A, Stege P, Maffini S, Stoppiello GA, Huis In 't Veld PJ, Vetter IR, Musacchio A (2021) CDC20 assists its catalytic incorporation in the mitotic checkpoint complex. Science 371:67–71

Ponjavić I, Vukušić K, Tolić IM (2021) Expansion microscopy of the mitotic spindle. Methods Cell Biol 161:247–274

Primorac I, Weir JR, Chiroli E, Gross F, Hoffmann I, van Gerwen S, Ciliberto A, Musacchio A (2013) Bub3 reads phosphorylated MELT repeats to promote spindle assembly checkpoint signaling. Elife 2:e01030

Przewloka MR, Glover DM (2009) The kinetochore and the centromere: a working long distance relationship. Annu Rev Genet 43:439–465

Przewloka MR, Venkei Z, Bolanos-Garcia VM, Debski J, Dadlez M, Glover DM (2011) CENP-C is a structural platform for kinetochore assembly. Curr Biol 21:399–405

Qian J, García-Gimeno MA, Beullens M, Manzione MG, Van der Hoeven G, Igual JC, Heredia M, Sanz P, Gelens L, Bollen M (2017) An attachment-independent biochemical timer of the spindle assembly checkpoint. Mol Cell 68:715–730.e5

Rago F, Gascoigne KE, Cheeseman IM (2015) Distinct organization and regulation of the outer kinetochore KMN network downstream of CENP-C and CENP-T. Curr Biol 25:671–677

Ren X, Zhao B, Chang H, Xiao M, Wu Y, Liu Y (2018) Paclitaxel suppresses proliferation and induces apoptosis through regulation of ROS and the AKT/MAPK signaling pathway in canine mammary gland tumor cells. Mol Med Rep 17:8289–8299

Renda F, Khodjakov A (2021) Role of spatial patterns and kinetochore architecture in spindle morphogenesis. Semin Cell Dev Biol 117:75–85

Rodriguez-Rodriguez JA, Lewis C, McKinley KL, Sikirzhytski V, Corona J, Maciejowski J, Khodjakov A, Cheeseman IM, Jallepalli PV (2018) Distinct roles of RZZ and Bub1-KNL1 in mitotic checkpoint signaling and kinetochore expansion. Curr Biol 28:3422–3429.e5

Rosenberg JS, Cross FR, Funabiki H (2011) KNL1/Spc105 recruits PP1 to silence the spindle assembly checkpoint. Curr Biol 21:942–947

Sacristan C, Ahmad MUD, Keller J, Fermie J, Groenewold V, Tromer E, Fish A, Melero R, Carazo JM, Klumperman J, Musacchio A, Perrakis A, Kops GJ (2018) Dynamic kinetochore size regulation promotes microtubule capture and chromosome biorientation in mitosis. Nat Cell Biol 20:800–810

Santaguida S, Musacchio A (2009) The life and miracles of kinetochores. EMBO J 28:2511–2531

Sarangapani KK, Koch LB, Nelson CR, Asbury CL, Biggins S (2021) Kinetochore-bound Mps1 regulates kinetochore-microtubule attachments via Ndc80 phosphorylation. J Cell Biol 220: e202106130

Sarkar S, Sahoo PK, Mahata S, Pal R, Ghosh D, Mistry T, Ghosh S, Bera T, Nasare VD (2021) Mitotic checkpoint defects: en route to cancer and drug resistance. Chromosome Res 29:131–144

Scarborough EA, Davis TN, Asbury CL (2019) Tight bending of the Ndc80 complex provides intrinsic regulation of its binding to microtubules. Elife 8:e44489

Screpanti E, De Antoni A, Alushin GM, Petrovic A, Melis T, Nogales E, Musacchio A (2011) Direct binding of Cenp-C to the Mis12 complex joins the inner and outer kinetochore. Curr Biol 20:391–398

Seet BT, Dikic I, Zhou MM, Pawson T (2006) Reading protein modifications with interaction domains. Nat Rev Mol Cell Biol 7:473–483

Shepperd LA, Meadows JC, Sochaj AM, Lancaster TC, Zou J, Buttrick GJ, Rappsilber J, Hardwick KG, Millar JB (2012) Phosphodependent recruitment of Bub1 and Bub3 to Spc7/KNL1 by Mph1 kinase maintains the spindle checkpoint. Curr Biol 22:891–899

Shrestha RL, Conti D, Tamura N, Braun D, Ramalingam RA, Cieslinski K, Ries J, Draviam VM (2017) Aurora-B kinase pathway controls the lateral to end-on conversion of kinetochore-microtubule attachments in human cells. Nat Commun 8:150

Simorellis AK, Flynn PF (2006) Fast local backbone dynamics of encapsulated ubiquitin. J Am Chem Soc 128:9580–9581

Sironi L, Mapelli M, Knapp S, De Antoni A, Jeang KT, Musacchio A (2002) Crystal structure of the tetrameric Mad1-Mad2 core complex: implications of a 'safety belt' binding mechanism for the spindle checkpoint. EMBO J 21:2496–2506

Starr DA, Saffery R, Li Z, Simpson AE, Choo KH, Yen TJ, Goldberg ML (2000) HZwint-1, a novel human kinetochore component that interacts with HZW10. J Cell Sci 113:1939–1950

Strobel T, Swanson L, Korsmeyer S, Cannistra SA (1996) BAX enhances paclitaxel-induced apoptosis through a P53-independent pathway. Proc Natl Acad Sci USA 93:14094–14099

Su H, Liu Y, Wang C, Liu Y, Feng C, Sun Y, Yuan J, Birchler JA, Han F (2021) Knl1 participates in spindle assembly checkpoint signaling in maize. Proc Natl Acad Sci USA 118:e2022357118

Suijkerbuijk SJ, van Dam TJ, Karagöz GE, von Castelmur E, Hubner NC, Duarte AM, Vleugel M, Perrakis A, Rüdiger SG, Snel B, Kops GJ (2012) The vertebrate mitotic checkpoint protein BUBR1 is an unusual pseudokinase. Dev Cell 22:1321–1329

Tanaka K (2013) Regulatory mechanisms of kinetochore-microtubule interaction in mitosis. Cell Mol Life Sci 70:559–579

Tang Z, Shu H, Oncel D, Chen S, Yu H (2004) Phosphorylation of CDC20 by Bub1 provides a catalytic mechanism for APC/C inhibition by the spindle checkpoint. Mol Cell 16:387–397

Tian T, Li X, Liu Y, Wang C, Liu X, Bi G, Zhang X, Yao X, Zhou ZH, Zang J (2018) Molecular basis for CENP-N recognition of CENP-A nucleosome on the human kinetochore. Cell Res 28: 374–378

Tipton AR, Ji W, Sturt-Gillespie B, Bekier ME 2nd, Wang K, Taylor WR, Liu ST (2013) Monopolar spindle 1 (MPS1) kinase promotes production of closed MAD2 (C-MAD2) conformer and assembly of the mitotic checkpoint complex. J Biol Chem 288:35149–35158

Trinkle-Mulcahy L (2019) Recent advances in proximity-based labeling methods for interactome mapping. F1000Res 8:F1000

Tromer E, Snel B, Kops GJPL (2015) Widespread recurrent patterns of rapid repeat evolution in the kinetochore scaffold KNL1. Genome Biol Evol 7:2383–2393

Tromer E, Bade D, Snel B, Kops GJ (2016) Phylogenomics guided discovery of a novel conserved cassette of short linear motifs in BubR1 essential for the spindle checkpoint. Open Biol 6: 160315

Tromer EC, van Hooff JJE, Kops GJPL, Snel B (2019) Mosaic origin of the eukaryotic kinetochore. Proc Natl Acad Sci USA 116:12873–12882

Ustinov NB, Korshunova AV, Gudimchuk NB (2020) Protein complex NDC80: properties, functions, and possible role in pathophysiology of cell division. Biochemistry (Mosc) 85: 448–462

Vallardi G, Allan LA, Crozier L, Saurin AT (2019) Division of labour between PP2A-B56 isoforms at the centromere and kinetochore. Elife 8:e42619

van Hooff JJ, Tromer E, van Wijk LM, Snel B, Kops GJ (2017) Evolutionary dynamics of the kinetochore network in eukaryotes as revealed by comparative genomics. EMBO Rep 18:1559–1571

Vanoosthuyse V, Hardwick KG (2005) Bub1 and the multilayered inhibition of CDC20-APC/C in mitosis. Trends Cell Biol 15:231–233

Varadi M, Anyango S, Deshpande M, Nair S, Natassia C, Yordanova G, Yuan D, Stroe O, Wood G, Laydon A, Žídek A, Green T, Tunyasuvunakool K, Petersen S, Jumper J, Clancy E, Green R, Vora A, Lutfi M, Figurnov M, Cowie A, Hobbs N, Kohli P, Kleywegt G, Birney E, Hassabis D, Velankar S (2022) AlphaFold protein structure database: massively expanding the structural coverage of protein-sequence space with high-accuracy models. Nucleic Acids Res 50:D439–D444

Varma D, Salmon ED (2012) The KMN protein network – chief conductors of the kinetochore orchestra. J Cell Sci 125:5927–5936

Varma D, Wan X, Cheerambathur D, Gassmann R, Suzuki A, Lawrimore J, Desai A, Salmon ED (2013) Spindle assembly checkpoint proteins are positioned close to core microtubule attachment sites at kinetochores. J Cell Biol 202:735–746

Venkei Z, Przewloka MR, Glover DM (2011) Drosophila Mis12 complex acts as a single functional unit essential for anaphase chromosome movement and a robust spindle assembly checkpoint. Genetics 187:131–140

Vetter IR, Pesenti M, Raisch T. (2021) Structure of the human inner kinetochore CCAN complex. Protein Data Bank ID 7QOO

Vleugel M, Tromer E, Omerzu M, Groenewold V, Nijenhuis W, Snel B, Kops GJ (2013) Arrayed BUB recruitment modules in the kinetochore scaffold KNL1 promote accurate chromosome segregation. J Cell Biol 203:943–955

von Schubert C, Cubizolles F, Bracher JM, Sliedrecht T, Kops GJPL, Nigg EA (2015) Plk1 and Mps1 cooperatively regulate the spindle assembly checkpoint in human cells. Cell Rep 12:66–78

Wan X, O'Quinn RP, Pierce HL, Joglekar AP, Gall WE, DeLuca JG, Carroll CW, Liu ST, Yen TJ, McEwen BF, Stukenberg PT, Desai A, Salmon ED (2009) Protein architecture of the human kinetochore microtubule attachment site. Cell 137:672–684

Wan J, Subramonian D, Zhang XD (2012) SUMOylation in control of accurate chromosome segregation during mitosis. Curr Protein Pept Sci 13:467–481

Wang H, Hu X, Ding X, Dou Z, Yang Z, Shaw AW, Teng M, Cleveland DW, Goldberg ML, Niu L, Yao X (2004) Human Zwint-1 specifies localization of Zeste White 10 to kinetochores and is essential for mitotic checkpoint signaling. J Biol Chem 279:54590–54598

Wang X, Bajaj R, Bollen M, Peti W, Page R (2016a) Expanding the PP2A interactome by defining a B56-specific SLiM. Structure 24:2174–2181

Wang J, Wang Z, Yu T, Yang H, Virshup DM, Kops GJ, Lee SH, Zhou W, Li X, Xu W, Rao Z (2016b) Crystal structure of a PP2A B56-BubR1 complex and its implications for PP2A substrate recruitment and localization. Protein Cell 7:516–526

Warren CD, Brady DM, Johnston RC, Hanna JS, Hardwick KG, Spencer FA (2002) Distinct chromosome segregation roles for spindle checkpoint proteins. Mol Biol Cell 13:3029–3041

Watanabe R, Hara M, Okumura EI, Hervé S, Fachinetti D, Ariyoshi M, Fukagawa T (2019) CDK1-mediated CENP-C phosphorylation modulates CENP-A binding and mitotic kinetochore localization. J Cell Biol 218:4042–4062

Wei RR, Sorger PK, Harrison SC (2005) Molecular organization of the Ndc80 complex, an essential kinetochore component. Proc Natl Acad Sci USA 102:5363–5367

Wei RR, Al-Bassam J, Harrison SC (2007) The Ndc80/HEC1 complex is a contact point for kinetochore-microtubule attachment. Nat Struct Mol Biol 14:54–55

Weiss E, Winey M (1996) The *Saccharomyces cerevisiae* spindle pole body duplication gene *MPS1* is part of a mitotic checkpoint. J Cell Biol 132:111–123

Welburn JP, Vleugel M, Liu D, Yates JR 3rd, Lampson MA, Fukagawa T, Cheeseman IM (2010) Aurora B phosphorylates spatially distinct targets to differentially regulate the kinetochore-microtubule interface. Mol Cell 38:383–392

Westhorpe FG, Straight AF (2013) Functions of the centromere and kinetochore in chromosome segregation. Curr Opin Cell Biol 25:334–340

Wigge PA, Kilmartin JV (2001) The Ndc80p complex from Saccharomyces cerevisiae contains conserved centromere components and has a function in chromosome segregation. J Cell Biol 152:349–360

Wimbish RT, DeLuca JG (2020) Hec1/Ndc80 tail domain function at the kinetochore-microtubule interface. Front Cell Dev Biol 8:43

Xing QR, Cipta NO, Hamashima K, Liou YC, Koh CG, Loh YH (2020) Unraveling heterogeneity in transcriptome and its regulation through single-cell multi-omics technologies. Front Genet 11:662

Xu P, Raetz EA, Kitagawa M, Virshup DM, Lee SH (2013) BUBR1 recruits PP2A via the B56 family of targeting subunits to promote chromosome congression. Biol Open 2:479–486

Yamagishi Y, Yang CH, Tanno Y, Watanabe Y (2012) MPS1/Mph1 phosphorylates the kinetochore protein KNL1/Spc7 to recruit SAC components. Nat Cell Biol 14:746–752

Yamamoto A (2021) Shake it off: the elimination of erroneous kinetochore-microtubule attachments and chromosome oscillation. Int J Mol Sci 22:3174

Yan K, Yang J, Zhang Z, McLaughlin SH, Chang L, Fasci D, Ehrenhofer-Murray AE, Heck AJR, Barford D (2019) Structure of the inner kinetochore CCAN complex assembled onto a centromeric nucleosome. Nature 574:278–282

Yao Y, Dai W (2012) Mitotic checkpoint control and chromatin remodeling. Front Biosci 17:976–983

Zhang G, Lischetti T, Nilsson J (2014) A minimal number of MELT repeats supports all functions of KNL1 in chromosome segregation. J Cell Sci 127:871–884

Zhang G, Kruse T, López-Méndez B, Sylvestersen KB, Garvanska DH, Schopper S, Nielsen ML, Nilsson J (2017) Bub1 positions Mad1 close to KNL1 MELT repeats to promote checkpoint signalling. Nat Commun 8:15822

Zhang G, Kruse T, Guasch Boldú C, Garvanska DH, Coscia F, Mann M, Barisic M, Nilsson J (2019) Efficient mitotic checkpoint signaling depends on integrated activities of Bub1 and the RZZ complex. EMBO J 38:e100977

Zhu T, Dou Z, Qin B, Jin C, Wang X, Xu L, Wang Z, Zhu L, Liu F, Gao X, Ke Y, Wang Z, Aikhionbare F, Fu C, Ding X, Yao X (2013) Phosphorylation of microtubule-binding protein Hec1 by mitotic kinase Aurora B specifies spindle checkpoint kinase Mps1 signaling at the kinetochore. J Biol Chem 288:36149–36159

Part I
Membrane Protein Complexes

Chapter 8
The Conformational Dynamics of Heterotrimeric G Proteins During GPCR-Mediated Activation

Donghoon Ahn and Ka Young Chung

Abstract Heterotrimeric G proteins (G proteins) are essential cellular signaling proteins that mediate extracellular signals to achieve various cellular functions. G-protein-coupled receptors (GPCRs) are the major guanine nucleotide exchange factors (GEFs) that induce G proteins to release guanosine diphosphate and rapidly bind to guanosine triphosphate, resulting in G protein activation. G proteins undergo dynamic conformational changes during the activation/inactivation process, and the precise structural mechanism of GPCR-mediated G protein activation is of great interest. Over the last decade, a number of GPCR-G protein complex structures have been identified, yet an understanding of the mechanisms underlying allosteric conformational changes during receptor-mediated G protein activation and GPCR-G protein coupling selectivity is only now emerging. This review discusses recent studies on the dynamic conformational changes of G proteins and provides insight into the structural mechanism of GPCR-mediated G protein activation.

Keywords GPCR · G protein · Conformational dynamics · Signaling selectivity

Introduction

Heterotrimeric G proteins (G proteins) are important cellular signaling proteins whose activation status is regulated by the binding of guanosine diphosphate (GDP) or guanosine triphosphate (GTP) (Milligan and Kostenis 2006). G proteins are composed of Gα, Gβ, and Gγ subunits; the Gα subunit is the GDP-/GTP-binding subunit. In the basal state, GDP binds to the Gα subunit, which forms a heterotrimer with the Gβγ subunits (Fig. 8.1a). A guanine nucleotide exchange factor (GEF), such as a G-protein-coupled receptor (GPCR), induces conformational changes in the G protein, which results in the release of GDP (Fig. 8.1a). Under physiological cellular conditions, GTP rapidly binds to the empty Gα subunit, which dissociates from the

D. Ahn · K. Y. Chung (✉)
School of Pharmacy, Sungkyunkwan University, Suwon, Republic of Korea
e-mail: kychung2@skku.edu

J. R. Harris, J. Marles-Wright (eds.), *Macromolecular Protein Complexes IV*,
Subcellular Biochemistry 99, https://doi.org/10.1007/978-3-031-00793-4_8

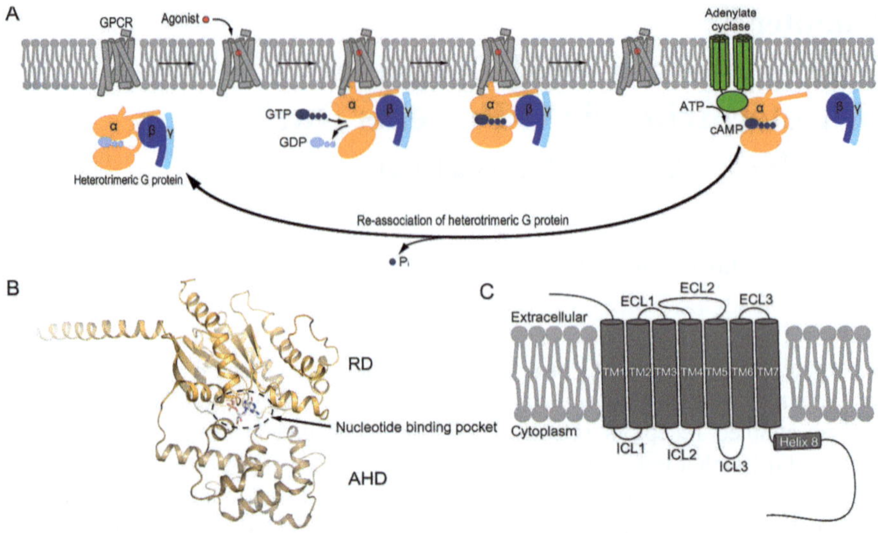

Fig. 8.1 Overall features of G protein signaling and structure. (**a**) The G protein activation/inactivation pathway regulated by a G protein-coupled receptor (GPCR). (**b**) G protein α-subunit structure (RD, orange; AHD, light brown; Protein Data Bank [PDB], 6EG8) with GDP. The Gα subunit has the nucleotide-binding pocket between the Ras-like domain (RD) and the α-helical domain (AHD). (**c**) General structure of a GPCR. A GPCR has seven transmembrane domains (TM), intracellular loops (ICLs), and extracellular loops (ECLs)

receptor and Gβγ subunits (Fig. 8.1a). The dissociated GTP-bound Gα subunit and free Gβγ subunits are then fully activated and can regulate downstream signaling molecules, such as adenylate cyclase. The intrinsic GTPase activity of the Gα subunit hydrolyzes GTP into GDP, which leads to the reassociation of the Gα and Gβγ subunits (Fig. 8.1a) (Milligan and Kostenis 2006).

G proteins are categorized into four subtypes (Gs, Gi/o, Gq/11, and G12/13) based on the sequence and function of the Gα subunit, and there are 16 Gα subunit genes in humans (Syrovatkina et al. 2016). All Gα subunits have a similar structure with two domains: the Ras-like domain (RD) and the α-helical domain (AHD) (Fig. 8.1b). GDP or GTP localizes within the nucleotide-binding pocket between the RD and AHD (Fig. 8.1b). The RD contains intrinsic GTPase activity and provides interfaces for GEFs and effector proteins.

GPCRs are the major proteins with GEF activity that interact with G proteins. In humans, there are approximately 800 GPCRs (Syrovatkina et al. 2016). GPCRs are the largest family of membrane receptors and mediate a variety of pathophysiological functions (Sriram and Insel 2018). Because of their critical physiological and pathological roles, GPCRs are the target of approximately one-third of currently developed medicines (Sriram and Insel 2018). They share a seven-transmembrane domain structure (Fig. 8.1c), with the cytosolic part of a receptor as the G-protein-binding site, while the ligand-binding sites vary depending on the receptor. GPCRs can couple with one or more G proteins, and the coupling efficiencies often differ depending on the G protein subtypes (Inoue et al. 2019). Primary GPCR-G protein coupling often shows higher efficacy and/or faster interactions, whereas secondary coupling shows lower efficacy and/or slower interactions.

Because of the importance of G proteins in fundamental cellular signaling and drug development, the structural mechanism of G protein activation by GPCRs is of great interest to researchers. It has been discovered that G proteins undergo dynamic conformational changes during GPCR-mediated activation. In this review, we will discuss the conformational dynamics of GPCR-mediated G protein activation.

GPCR-G Protein Complex Structures

Over the last decade, the structures of various GPCR-G protein complexes have been revealed owing to advancements in biophysical technologies such as cryogenic electron microscopy (cryo-EM) and X-ray crystallography (Garcia-Nafria and Tate 2019; Glukhova et al. 2018). The first GPCR-G protein complex structure, the X-ray crystal structure of the beta-2 adrenergic receptor (β_2AR)-Gs complex, was reported in 2011 (Fig. 8.2a) (Rasmussen et al. 2011). Since then, numerous GPCR-G protein complex structures have been revealed by X-ray crystallography and cryo-EM, and the structures are well summarized in a GPCR-specialized database called GPCRdb (GPCRdb.org) (Fig. 8.2). There were 82 GPCR-Gs, 46 GPCR-Gi, and 4 GPCR-Gq structures deposited in the Protein Data Bank (PDB) (rcsb.org) at the time this review was written.

In vertebrates, GPCRs are categorized into six classes depending on their sequence and function, including class A (rhodopsin-like receptor family), class B

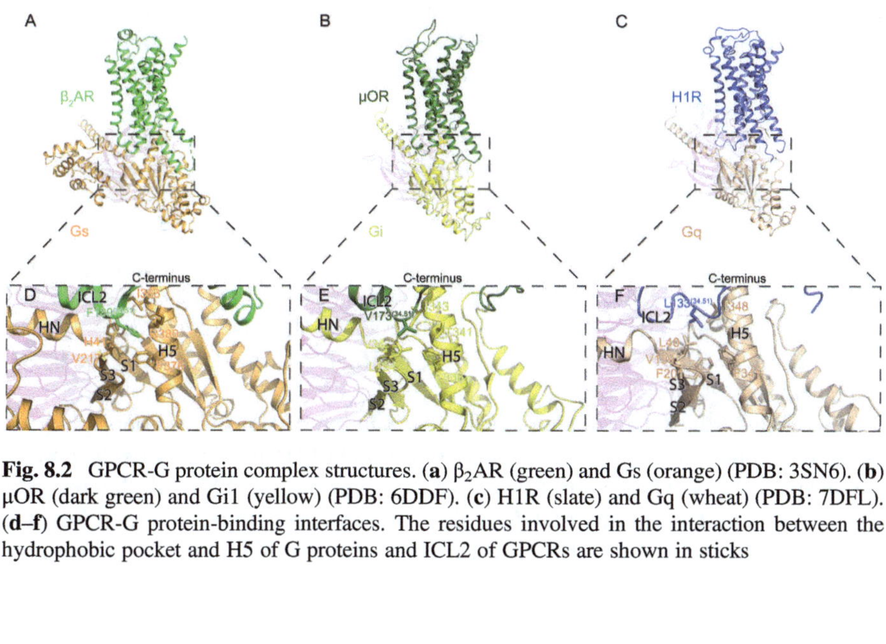

Fig. 8.2 GPCR-G protein complex structures. (**a**) β_2AR (green) and Gs (orange) (PDB: 3SN6). (**b**) μOR (dark green) and Gi1 (yellow) (PDB: 6DDF). (**c**) H1R (slate) and Gq (wheat) (PDB: 7DFL). (**d–f**) GPCR-G protein-binding interfaces. The residues involved in the interaction between the hydrophobic pocket and H5 of G proteins and ICL2 of GPCRs are shown in sticks

(secretin receptor family), class B2 (adhesion receptor family), class C (metabotropic glutamate receptor family), and class F (frizzled/smoothened family) (Bjarnadottir et al. 2006). Among these classes, class A is the largest, accounting for nearly 85% of the human GPCR genes. Class A GPCRs share sequence homology at conserved residues, whereas other classes are not homologous at these sites. Therefore, in this review, we focused on the class A GPCR-mediated dynamic structural changes in G proteins.

As shown in Fig. 8.2, the GPCR-G protein complex structures have been revealed in the Gs, Gi/o, and Gq/11 subfamilies. These structures have shown similar GPCR-G protein-binding interfaces. The C-terminal part of helix 5 (H5) of Gα is the major site that binds the cytosolic core of the receptor, which is composed of transmembrane (TM) domains and intracellular loops (ICLs) (Fig. 8.2). Another major binding interface exists between the ICL2 of the receptor and the hydrophobic pocket formed by the HN/S1 hinge, S2, S3, and H5 (Fig. 8.2). However, the interaction with ICL2 is less prominent in GPCR-Gi/o coupling than in GPCR-Gs or GPCR-Gq coupling (Fig. 8.2d–f), and a few studies have suggested that this interaction may give rise to the selectivity of GPCR-G protein coupling (Ma et al. 2020; Flock et al. 2017; Kim et al. 2020). For example, hydrogen/deuterium exchange mass spectrometry (HDX-MS) and mutation studies have shown that the interaction with the ICL2 of a receptor is crucial for GPCR-Gs coupling but not for GPCR-Gi/o coupling (Du et al. 2019; Kim et al. 2020).

The relatively similar binding interfaces of the GPCR-Gs, GPCR-Gi/o, and GPCR-Gq/11 complexes suggest that there is a common structural mechanism for GPCR-induced G protein activation. Figure 8.3 compares the structures of the G protein and GPCR before and after complex formation. The GDP-bound G protein is stabilized by interactions between the p-loop of Gα and β-phosphate of GDP and between the S5/HG and S6/H5 loops of Gα and guanine ring of GDP (Fig. 8.3a). In all the reported GPCR-G protein complex structures except those that used a mini-G protein (a G protein in which Gα AHD is truncated), GDP is missing; therefore, these loops and the neighboring helices or β-strands were more disordered and flexible (Fig. 8.3a) (Chung et al. 2011).

Another dramatic structural difference within the G protein before and after interaction with a receptor is the insertion of H5 into the receptor cytosolic core, which is accompanied by the rotation and translation of H5 (Fig. 8.3a). Along with this movement, the interaction between H1 and H5, which has been suggested to stabilize the GDP- or GTP-bound state (Flock et al. 2015), is disturbed (Fig. 8.3a). The rotation and translation of H5 and perturbation of the interaction between H1 and H5 have been suggested as the key structural factors that induce GPCR-mediated GDP release (Kaya et al. 2014, 2016; Flock et al. 2015; Ghosh and Garcia-Marcos 2020).

Nonetheless, the rotation and degree of translation of H5 differed depending on the GPCR-G protein pairs (Fig. 8.3b). The insertion of H5 into the cytosolic core of the receptor is possible with the outward movement of TM6 of the receptor (Fig. 8.3c). The C-terminus of H5 of Gα has long been suggested as a key site rendering selectivity of GPCR-G protein coupling, and the outward movement of

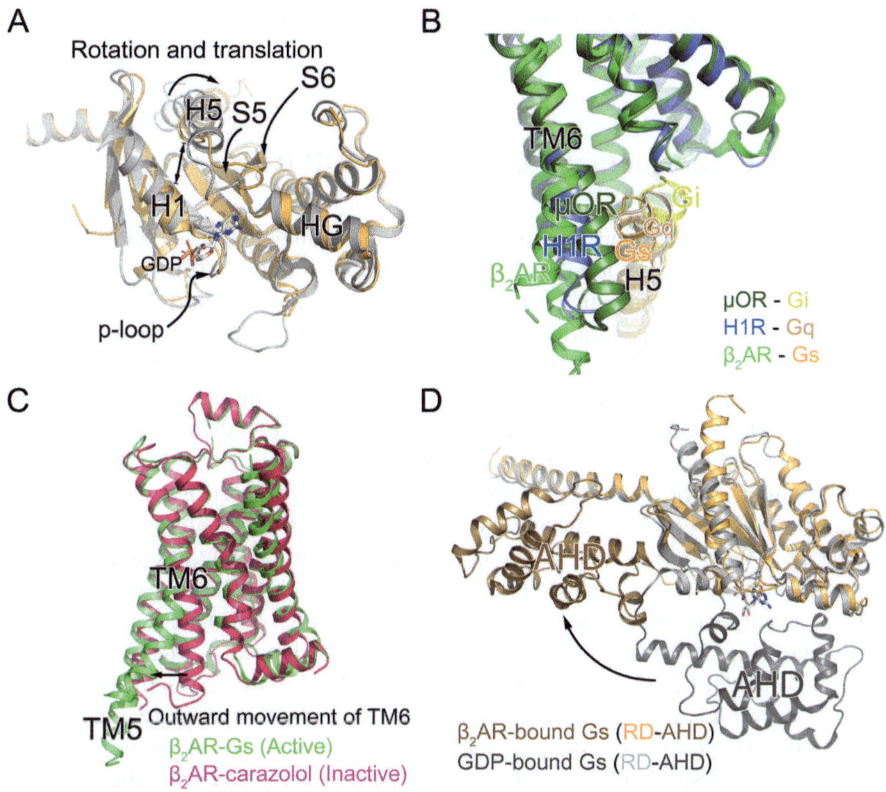

Fig. 8.3 Comparison of inactive and active G proteins and GPCRs. (**a**) Structural comparison between GDP-bound G protein α-subunit (gray, PDB: 6EG8) and GPCR-bound G protein α-subunit (orange, PDB: 3SN6). (**b**) Structural comparison of the different GPCR-G protein complex structures. β_2AR (green) and Gs (orange) (PDB: 3SN6). μOR (dark green) and Gi1 (yellow) (PDB: 6DDF). H1R (slate) and Gq (wheat) (PDB: 7DFL). (**c**) Structural comparison between the active β_2AR (green, PDB: 3SN6) and the inactive β_2AR (pink, PDB: 2RH1). (**d**) Structural comparison between GDP-bound G protein α-subunit (RD: gray, AHD: dark gray) and GPCR-bound G protein α-subunit (RD: orange, AHD: light brown) focused on dramatically displaced AHD

TM6 is often smaller in Gi-coupled receptors than Gs-coupled receptors (Fig. 8.3b), resulting in a narrower cavity. The C-terminus is bulkier in Gαs than that in Gαi, and therefore, the C-terminus of Gαs cannot be inserted into the cytosolic core of the Gi-coupled receptors (Rose et al. 2014; Semack et al. 2016; Van Eps et al. 2018; Sandhu et al. 2019; Flock et al. 2017).

In addition to the conformational changes at the RD, the AHD was so dramatically displaced that the nucleotide-binding pocket was exposed to the solvent (Fig. 8.3d). These details will be discussed below in the section "Conformational Dynamics Within Gα AHD." Moreover, the sequence of events regarding structure during GPCR-mediated GDP release has not been fully elucidated, and we will discuss current proposals in the section "Dynamic Conformational Changes for GDP Release."

Dynamic Conformational Changes in GDP Release

The reported structures provide insight into the binding interfaces and structural changes of the receptor and G protein upon complex formation (Figs. 8.2 and 8.3). Scientists in this field agree that disturbance of the p-loop and/or S6/H5 loop, the two key GDP-contacting sites, induce GDP release (Sun et al. 2018; Rose et al. 2015; Ham et al. 2021; Calebiro et al. 2021). However, the precise allosteric structural mechanism by which these loops are altered by receptor binding at a site remote from the GDP-binding pocket remains unclear. Nonetheless, there have been some efforts to reveal the sequence of events that lead to GPCR-mediated GDP release.

Although most of the reported GPCR-G protein complex structures show similar conformations (Fig. 8.2), recent cryo-EM studies of the neurotensin receptor 1 (NTSR1)-Gi complex and vasopressin V2 receptor (V2R)-Gs complex show more than one distinct structure (Fig. 8.4a–d). For the NTSR1-Gi1 complex, cryo-EM revealed two structures, canonical (C state) (Fig. 8.4a) and noncanonical (NC state) (Fig. 8.4b) (Kato et al. 2019). The C state showed a structure similar to other reported GPCR-G protein complexes (Fig. 8.4a), but the Gi1 in the NC state is rotated approximately 45° relative to the receptor as compared to the C state (Fig. 8.4b). This rotation in the NC state results in a loss of the interaction between the ICL2 of NTSR1 and H5 of Gαi (Fig. 8.4b, insert). However, the NC state establishes new interactions between the H4/S6 of Gαi and H8 of the receptor, H3/S5 and H2/S4 of Gαi and ICL1 of the receptor, and Gβ and ICL1 and ICL2 of the receptor (Fig. 8.4b, insert).

Cryo-EM of the V2R-Gs complex revealed three different structures: loose (L), tight 1 (T1), and tight 2 (T2) states with a population distribution of 16%, 48%, and 36%, respectively (Bous et al. 2021). The authors defined structures for the L and T1 states (Fig. 8.4c, d). The T1 state showed a tighter interaction between the C-terminal part of H5 and the receptor than that of the L state, with different orientations between Gs and the receptor (Fig. 8.4c, d). The authors suggested that the reported L and T1 states represent the average of states that are part of a much larger conformational ensemble (Bous et al. 2021). The V2R-Gs complex structures also showed that both L and T states slightly differ from other GPCR-G protein complexes (Fig. 8.4e). In the V2R-Gs complex structures, there are many direct contacts between ICL1 of the receptor and Gβ or HN of Gαs; the position of H5 of Gαs is different in the V2R-Gs complex compared to other GPCR-Gs structures in that the C-terminus of H5 of Gαs forms distinct contacts with V2R. Taken together, this suggests that the structure of the GPCR-G protein complex is very dynamic and can adopt different conformations.

What is striking in these distinct GPCR-G protein complex structures is that none of them contain GDP (Figs. 8.2 and 8.4a–e), suggesting that these structures are

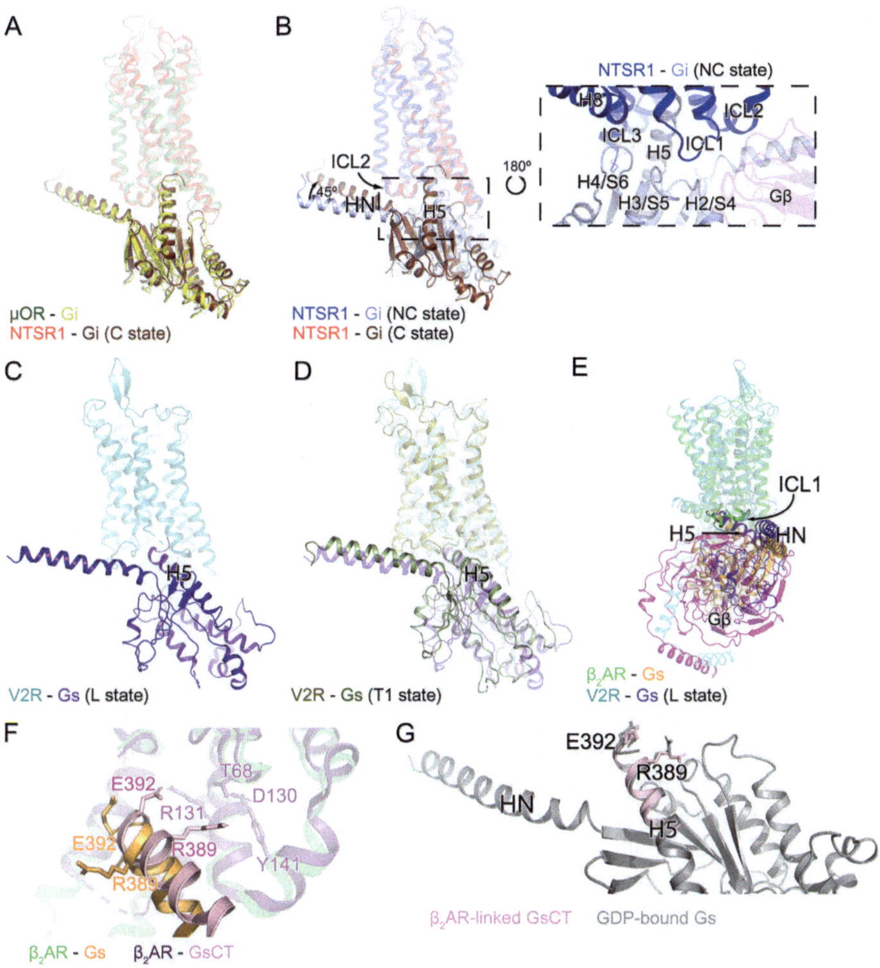

Fig. 8.4 Variable structures of GPCR-G protein complexes. (**a**) Structural comparison between μOR (dark green)-Gi1 (yellow) (PDB: 6DDF) and C state NTSR1 (red)-Gi1 (brown) (PDB: 6OS9). (**b**) Structural comparison between C state NTSR1 (red)-Gi1 (brown) (PDB: 6OS9) and NC state NTSR1 (blue)-Gi1 (light blue) (PDB: 6OSA). (**c**) Structure of the loose state (L state) of V2R (deep teal) and Gs (purple-blue) complex (PDB: 7BB6). (**d**) Structure of the tight 1 state (T state) V2R (deep olive) and Gs (smudge) complex (PDB: 7BB7). (**e**) Structural comparison between the L state V2R (deep teal)-Gs (α, purple-blue; β, magenta; γ, cyan) complex and β_2AR (green)-Gs (orange) complex (PDB: 3SN6). (**f**) Structural comparison between β_2AR (green)-Gs (orange) and β_2AR (deep purple)-C terminal peptide of H5 (GsCT, light pink) (PDB: 6E67). (**g**) Structural comparison between GsCT (light pink) and GDP-bound Gs (gray, PDB: 6EG8)

captured after GDP has been released from the G protein. To date, no GPCR-G protein complex structure has been reported with GDP, which makes it difficult to study the early stages of GPCR-G protein interaction. Nevertheless, recent studies using various biochemical and biophysical techniques have provided insight into the stepwise allosteric conformational changes during GPCR-mediated GDP release (Liu et al. 2019; Du et al. 2019; Sun et al. 2018). Recently, an X-ray crystal structure of the β_2AR fused with the C-terminal peptide of H5 of Gαs was reported (Fig. 8.4f) (Liu et al. 2019). Unexpectedly, the interacting residues between the β_2AR and the C-terminal part of H5 were different between the β_2AR-Gs complex and β_2AR fused with the C-terminal peptide of H5 of Gαs (Fig. 8.4f). The positions of residues R389 and E392 of Gαs are similar in the GDP-bound Gs structure and in the peptide-fused β_2AR structure but not in the β_2AR-Gs structure (Fig. 8.4g). Interestingly, these residues face outward in the GDP-bound Gs structure (Fig. 8.4g) and contact β_2AR in the peptide-fused β_2AR structure (Fig. 8.4f), but are rotated in the β_2AR-Gs structure (Fig. 8.4f). These structures show that the C-terminal part of H5 of GDP-bound Gs structure is more similar to that in the peptide-fused β_2AR structure than that in the β_2AR-Gs structure. Comparison of these structures and other biochemical and molecular dynamic simulation studies suggest that the peptide-fused β_2AR structure might represent an earlier stage of activation (i.e., before GDP release) than that of the β_2AR-Gs structure (Liu et al. 2019).

The question of what key structural factors lead to GDP release remains. A rhodopsin construct with a mutation in ICL2 has reportedly failed to activate transducin (Gt), although it can interact with Gt (Franke et al. 1990). This study suggests that the binding and release of GDP do not occur simultaneously and that there are stepwise structural events during GPCR-mediated G protein activation. Recently, molecular dynamics (MD) simulations and biophysical studies have hinted at the sequence of structural events during GPCR-mediated G protein activation (Du et al. 2019; Sun et al. 2018). An MD simulation study by the Bowman group revealed that tilting rather than upward movement of H5 is the rate-limiting step for GDP release (Sun et al. 2018). The authors also suggested that H1 and β-sheets of S1–S3 are allosteric communication hubs.

In another biophysical study, HDX-MS and hydroxyl radical footprinting mass spectrometry (HRF-MS) were adopted to study the sequential events during β_2AR-mediated GDP release from Gs (Du et al. 2019). In this study, stepwise conformational changes in Gαs during β_2AR-mediated GDP release were proposed. The first proposed step is binding of the C-terminus of Gαs to the receptor. The next step is interaction of the conserved large hydrophobic phenylalanine residue in ICL2 (residue 34.51, according to the GPCR numbering system provided by gpcrdb.org) with the hydrophobic pocket of Gαs. This was suggested to be the key step for GDP release because mutation of phenylalanine residue 34.51 to alanine failed to induce GDP release from Gs, although the mutant receptor could still interact with Gs. After GDP release, further conformational changes in the C-terminal part of H5 of Gαs have been observed (Du et al. 2019), which is consistent with the MD simulation data that suggested that the tilting of H5 is the rate-limiting step for GDP release and upward movement occurs after GDP release (Sun et al. 2018). The prolonged conformational change after GDP release is also related to the diversity of structures observed in the NTSR1-Gi1 complex (Fig. 8.4a–c and NC states) and the V2R-Gs complex (Fig. 8.4c, d, l, and T states).

Although not a class A GPCR, the structure of the $GABA_B$-Gi complex showed that the C-terminus of Gαi is barely inserted into the receptor core, and GDP can still be released (Mao et al. 2020). This observation implies that the C-terminus of Gαi does not need to be inserted into the receptor core to trigger GDP release. An HDX-MS study with β_2AR and Gi also revealed that the HDX profile at the C-terminal part of H5 did not change during β_2AR-mediated GDP release from Gi (Kim et al. 2020), suggesting that insertion of the C-terminal part of H5 into the cytosolic core of the receptor does not occur and is not necessary for β_2AR-mediated GDP release from Gi. Moreover, unlike the β_2AR-Gs coupling, where deletion of the four C-terminal residues of Gαs resulted in loss of coupling with β_2AR (Du et al. 2019), deletion of the C-terminal four residues of Gαi did not affect coupling with β_2AR (Kim et al. 2020), suggesting that the structural involvement of the C-terminus of Gα differs between GPCR-Gs and GPCR-Gi coupling. In summary, these recent biochemical and biophysical studies suggest that there are dynamic conformational changes during and after GDP release and that G protein subtypes adopt different dynamic conformations.

Conformational Dynamics Within Gα AHD

Among all GTPase proteins, the AHD is not present in the small GTPase proteins but only in the α-subunit of the heterotrimeric G proteins. The RD is involved in all canonical G protein functions, such as GTPase activity, Gβγ binding, GEF binding, regulator of G protein signaling (RGS) binding, and effector binding. However, an understanding of the function of the AHD in G protein signaling is slowly developing (Dohlman and Jones 2012).

Although the RD is the major binding site for other proteins such as GEF, RGS, and effector proteins, it has also been suggested to be part of the binding site for other signaling proteins such as PLCβ (Kwan et al. 2015), GIRK (Mase et al. 2012), and RGS (Nance et al. 2013; Soundararajan et al. 2008). Importantly, when the Gα subunit sequences were aligned (Fig. 8.5), the AHD showed weaker sequence homology than that of the RD. Therefore, the AHD may provide protein-binding specificity for these binding partners (Dohlman and Jones 2012).

As shown in Fig. 8.3d, the AHD was displaced in the GPCR-G protein complex structure. The position of the AHD in the GDP-released state is considered not static, but dynamically changing (Westfield et al. 2011). Moreover, upon GDP release, the Gα subunit becomes structurally dynamic (Abdulaev et al. 2006) and the α-helical content of the Gα subunit is decreased (Tanaka et al. 1998), implying that the AHD becomes more flexible. However, whether the displacement of the AHD occurs as a cause or result of GDP release has not been fully elucidated. A previous study revealed that GDP can be released even without domain opening and that GDP remains in the RD even after AHD displacement (Dror et al. 2015), which implies that AHD displacement is not a necessary factor for GDP release.

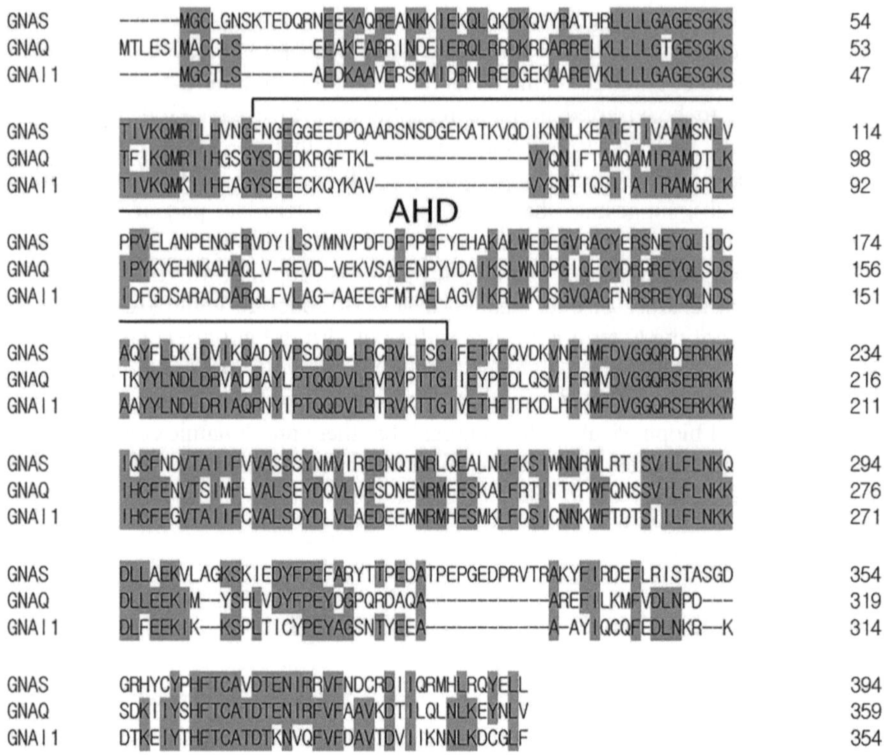

Fig. 8.5 Sequence alignment of G proteins, Gs, Gi1, and Gq. Identical residues are shaded. The α-helical domain (AHD) has fewer similarities than that of the Ras-like domain

Although domain opening is not required for GDP release, there are a few studies showing that interdomain interaction differences between different G protein subtypes are related to the different GDP/GTP exchange rates (Zhang et al. 2004; Marin et al. 2001). It has also been reported that phosphorylation of AHD at the interdomain interface by receptor tyrosine kinases activates Gi due to destabilization of the interdomain interaction (Kalogriopoulos et al. 2020). Moreover, studies with AtGPA1 (a self-activating G protein from the plant *Arabidopsis thaliana*) have confirmed that the AHD controls GDP/GTP exchange (Jones et al. 2011, 2012). Crystallographic and biophysical studies have shown that the AHD of AtGPA1 is intrinsically dynamic and disordered (Jones et al. 2011). A swapping mutant containing the Gαi RD and AtGPA1 AHD showed an increased GDP/GTP exchange rate compared to wild-type Gαi (Jones et al. 2011), and further swapping mutants with smaller segments of AHD suggested that HA and HB are the major sites for regulating G protein activation by AHD (Jones et al. 2012). In summary, the structural dynamics of the AHD affect the G protein activation rate and binding selectivity of G protein-binding partners.

Perspectives

G proteins are structurally very dynamic, as we have discussed so far. They adopt various conformations including a GDP-bound basal state, GTP-bound fully active state, and various conformations of the receptor-bound state. G proteins undergo dynamic stepwise conformational changes during GPCR-mediated activation. Although our understanding of the conformational dynamics of G proteins has advanced, there are still unanswered questions. The most interesting question is the early-stage structure during GPCR-G protein complex formation. Currently available high-resolution structures of GPCR-G protein complexes have failed to provide the structural mechanism of GPCR-G protein selectivity, potentially because selectivity is determined at the early stage of interaction rather than the final stage after GDP release. Furthermore, although not discussed in this review, G proteins have been reported to pre-couple with a basal state receptor (Huang et al. 2021; Andressen et al. 2018; Damian et al. 2015; Qin et al. 2011). Structural studies with the early-stage or pre-coupled GPCR-G protein complex would provide valuable insights into the GPCR-G protein coupling selectivity mechanism.

Another question related to the conformational dynamics of G proteins concerns the AHD. The AHD shows the weakest sequence similarities between different Gα subtypes and undergoes the most dramatic conformational changes during GPCR-G protein coupling. The functional significance of the AHD has recently emerged, and thus, it is tempting to suggest it may have a novel function. Taken together, we are still in the process of understanding the dynamic nature of heterotrimeric G proteins, with more to be discovered in the future.

Acknowledgments This work was supported by the National Research Foundation of Korea and funded by the Ministry of Science and ICT of Korea (NRF-2021R1A2C3003518) and under the framework of the international cooperation program managed by the National Research Foundation of Korea (NRF-2021K2A9A2A08000144).

Declarations of Interest The authors declare no conflict of interest.

References

Abdulaev NG, Ngo T, Ramon E, Brabazon DM, Marino JP, Ridge KD (2006) The receptor-bound "empty pocket" state of the heterotrimeric G-protein alpha-subunit is conformationally dynamic. Biochemistry 45(43):12986–12997. https://doi.org/10.1021/bi061088h

Andressen KW, Ulsund AH, Krobert KA, Lohse MJ, Bunemann M, Levy FO (2018) Related GPCRs couple differently to Gs: preassociation between G protein and 5-HT7 serotonin receptor reveals movement of Galphas upon receptor activation. FASEB J 32(2):1059–1069. https://doi.org/10.1096/fj.201700486R

Bjarnadottir TK, Gloriam DE, Hellstrand SH, Kristiansson H, Fredriksson R, Schioth HB (2006) Comprehensive repertoire and phylogenetic analysis of the G protein-coupled receptors in human and mouse. Genomics 88(3):263–273. https://doi.org/10.1016/j.ygeno.2006.04.001

Bous J, Orcel H, Floquet N, Leyrat C, Lai-Kee-Him J, Gaibelet G, Ancelin A, Saint-Paul J, Trapani S, Louet M, Sounier R, Demene H, Granier S, Bron P, Mouillac B (2021) Cryo-electron microscopy structure of the antidiuretic hormone arginine-vasopressin V2 receptor signaling complex. Sci Adv 7(21). https://doi.org/10.1126/sciadv.abg5628

Calebiro D, Koszegi Z, Lanoiselee Y, Miljus T, O'Brien S (2021) G protein-coupled receptor-G protein interactions: a single-molecule perspective. Physiol Rev 101(3):857–906. https://doi.org/10.1152/physrev.00021.2020

Chung KY, Rasmussen SG, Liu T, Li S, DeVree BT, Chae PS, Calinski D, Kobilka BK, Woods VL Jr, Sunahara RK (2011) Conformational changes in the G protein Gs induced by the beta2 adrenergic receptor. Nature 477(7366):611–615. https://doi.org/10.1038/nature10488

Damian M, Mary S, Maingot M, M'Kadmi C, Gagne D, Leyris JP, Denoyelle S, Gaibelet G, Gavara L, de Souza G, Costa M, Perahia D, Trinquet E, Mouillac B, Galandrin S, Gales C, Fehrentz JA, Floquet N, Martinez J, Marie J, Baneres JL (2015) Ghrelin receptor conformational dynamics regulate the transition from a preassembled to an active receptor: Gq complex. Proc Natl Acad Sci USA 112(5):1601–1606. https://doi.org/10.1073/pnas.1414618112

Dohlman HG, Jones JC (2012) Signal activation and inactivation by the Gα helical domain: a long-neglected partner in G protein signaling. Sci Signal 5(226):re2. https://doi.org/10.1126/scisignal.2003013

Dror RO, Mildorf TJ, Hilger D, Manglik A, Borhani DW, Arlow DH, Philippsen A, Villanueva N, Yang Z, Lerch MT, Hubbell WL, Kobilka BK, Sunahara RK, Shaw DE (2015) Signal transduction. Structural basis for nucleotide exchange in heterotrimeric G proteins. Science 348(6241):1361–1365. https://doi.org/10.1126/science.aaa5264

Du Y, Duc NM, Rasmussen SGF, Hilger D, Kubiak X, Wang L, Bohon J, Kim HR, Wegrecki M, Asuru A, Jeong KM, Lee J, Chance MR, Lodowski DT, Kobilka BK, Chung KY (2019) Assembly of a GPCR-G protein complex. Cell 177(5):1232–1242. https://doi.org/10.1016/j.cell.2019.04.022

Flock T, Ravarani CNJ, Sun D, Venkatakrishnan AJ, Kayikci M, Tate CG, Veprintsev DB, Babu MM (2015) Universal allosteric mechanism for Gα activation by GPCRs. Nature 524(7564): 173–179. https://doi.org/10.1038/nature14663

Flock T, Hauser AS, Lund N, Gloriam DE, Balaji S, Babu MM (2017) Selectivity determinants of GPCR-G-protein binding. Nature 545(7654):317–322. https://doi.org/10.1038/nature22070

Franke RR, Konig B, Sakmar TP, Khorana HG, Hofmann KP (1990) Rhodopsin mutants that bind but fail to activate transducin. Science 250(4977):123–125. https://doi.org/10.1126/science.2218504

Garcia-Nafria J, Tate CG (2019) Cryo-EM structures of GPCRs coupled to Gs, Gi and Go. Mol Cell Endocrinol 488:1–13. https://doi.org/10.1016/j.mce.2019.02.006

Ghosh P, Garcia-Marcos M (2020) Do all roads Lead to Rome in G-protein activation? Trends Biochem Sci 45(3):182–184. https://doi.org/10.1016/j.tibs.2019.10.010

Glukhova A, Draper-Joyce CJ, Sunahara RK, Christopoulos A, Wootten D, Sexton PM (2018) Rules of engagement: GPCRs and G proteins. ACS Pharmacol Transl Sci 1(2):73–83. https://doi.org/10.1021/acsptsci.8b00026

Ham D, Ahn D, Ashim J, Cho Y, Kim HR, Yu W, Chung KY (2021) Conformational switch that induces GDP release from Gi. J Struct Biol 213(1):107694. https://doi.org/10.1016/j.jsb.2020.107694

Huang SK, Pandey A, Tran DP, Villanueva NL, Kitao A, Sunahara RK, Sljoka A, Prosser RS (2021) Delineating the conformational landscape of the adenosine A2A receptor during G protein coupling. Cell 184(7):1884–1894. https://doi.org/10.1016/j.cell.2021.02.041

Inoue A, Raimondi F, Kadji FMN, Singh G, Kishi T, Uwamizu A, Ono Y, Shinjo Y, Ishida S, Arang N, Kawakami K, Gutkind JS, Aoki J, Russell RB (2019) Illuminating G-protein-coupling selectivity of GPCRs. Cell 177(7):1933–1947. https://doi.org/10.1016/j.cell.2019.04.044

Jones JC, Duffy JW, Machius M, Temple BR, Dohlman HG, Jones AM (2011) The crystal structure of a self-activating G protein alpha subunit reveals its distinct mechanism of signal initiation. Sci Signal 4. https://doi.org/10.1126/scisignal.2001446

Jones JC, Jones AM, Temple BR, Dohlman HG (2012) Differences in intradomain and interdomain motion confer distinct activation properties to structurally similar Gα proteins. Proc Natl Acad Sci USA 109(19):7275–7279. https://doi.org/10.1073/pnas.1202943109

Kalogriopoulos NA, Lopez-Sanchez I, Lin C, Ngo T, Midde KK, Roy S, Aznar N, Murray F, Garcia-Marcos M, Kufareva I, Ghassemian M, Ghosh P (2020) Receptor tyrosine kinases activate heterotrimeric G proteins via phosphorylation within the interdomain cleft of Gαi. Proc Natl Acad Sci USA 117(46):28763–28774. https://doi.org/10.1073/pnas.2004699117

Kato HE, Zhang Y, Hu H, Suomivuori CM, Kadji FMN, Aoki J, Krishna Kumar K, Fonseca R, Hilger D, Huang W, Latorraca NR, Inoue A, Dror RO, Kobilka BK, Skiniotis G (2019) Conformational transitions of a neurotensin receptor 1-Gi1 complex. Nature 572(7767): 80–85. https://doi.org/10.1038/s41586-019-1337-6

Kaya AI, Lokits AD, Gilbert JA, Iverson TM, Meiler J, Hamm HE (2014) A conserved phenylalanine as a relay between the alpha5 helix and the GDP binding region of heterotrimeric Gi protein alpha subunit. J Biol Chem 289(35):24475–24487. https://doi.org/10.1074/jbc.M114.572875

Kaya AI, Lokits AD, Gilbert JA, Iverson TM, Meiler J, Hamm HE (2016) A conserved hydrophobic core in Gαi1 regulates G protein activation and release from activated receptor. J Biol Chem 291(37):19674–19686. https://doi.org/10.1074/jbc.M116.745513

Kim HR, Xu J, Maeda S, Duc NM, Ahn D, Du Y, Chung KY (2020) Structural mechanism underlying primary and secondary coupling between GPCRs and the Gi/o family. Nat Commun 11(1):3160. https://doi.org/10.1038/s41467-020-16975-2

Kwan DH, Wong KM, Chan AS, Yung LY, Wong YH (2015) An intact helical domain is required for Gα14 to stimulate phospholipase Cβ. BMC Struct Biol 15:18. https://doi.org/10.1186/s12900-015-0043-3

Liu X, Xu X, Hilger D, Aschauer P, Tiemann JKS, Du Y, Liu H, Hirata K, Sun X, Guixa-Gonzalez R, Mathiesen JM, Hildebrand PW, Kobilka BK (2019) Structural insights into the process of GPCR-G protein complex formation. Cell 177(5):1243–1251. https://doi.org/10.1016/j.cell.2019.04.021

Ma X, Hu Y, Batebi H, Heng J, Xu J, Liu X, Niu X, Li H, Hildebrand PW, Jin C, Kobilka BK (2020) Analysis of β2AR-Gs and β2AR-Gi complex formation by NMR spectroscopy. Proc Natl Acad Sci USA. https://doi.org/10.1073/pnas.2009786117

Mao C, Shen C, Li C, Shen DD, Xu C, Zhang S, Zhou R, Shen Q, Chen LN, Jiang Z, Liu J, Zhang Y (2020) Cryo-EM structures of inactive and active GABAB receptor. Cell Res 30(7):564–573. https://doi.org/10.1038/s41422-020-0350-5

Marin EP, Krishna AG, Archambault V, Simuni E, Fu WY, Sakmar TP (2001) The function of interdomain interactions in controlling nucleotide exchange rates in transducin. J Biol Chem 276(26):23873–23880. https://doi.org/10.1074/jbc.M101197200

Mase Y, Yokogawa M, Osawa M, Shimada I (2012) Structural basis for modulation of gating property of G protein-gated inwardly rectifying potassium ion channel (GIRK) by i/o-family G protein alpha subunit (Gαi/o). J Biol Chem 287(23):19537–19549. https://doi.org/10.1074/jbc.M112.353888

Milligan G, Kostenis E (2006) Heterotrimeric G-proteins: a short history. Br J Pharmacol 147(Suppl 1):S46–S55. https://doi.org/10.1038/sj.bjp.0706405

Nance MR, Kreutz B, Tesmer VM, Sterne-Marr R, Kozasa T, Tesmer JJ (2013) Structural and functional analysis of the regulator of G protein signaling 2-gαq complex. Structure 21(3): 438–448. https://doi.org/10.1016/j.str.2012.12.016

Qin K, Dong C, Wu G, Lambert NA (2011) Inactive-state preassembly of G(q)-coupled receptors and G(q) heterotrimers. Nat Chem Biol 7(10):740–747. https://doi.org/10.1038/nchembio.642

Rasmussen SG, DeVree BT, Zou Y, Kruse AC, Chung KY, Kobilka TS, Thian FS, Chae PS, Pardon E, Calinski D, Mathiesen JM, Shah ST, Lyons JA, Caffrey M, Gellman SH, Steyaert J, Skiniotis G, Weis WI, Sunahara RK, Kobilka BK (2011) Crystal structure of the β2 adrenergic receptor-Gs protein complex. Nature 477(7366):549–555. https://doi.org/10.1038/nature10361

Rose AS, Elgeti M, Zachariae U, Grubmuller H, Hofmann KP, Scheerer P, Hildebrand PW (2014) Position of transmembrane helix 6 determines receptor G protein coupling specificity. J Am Chem Soc 136(32):11244–11247. https://doi.org/10.1021/ja5055109

Rose AS, Zachariae U, Grubmuller H, Hofmann KP, Scheerer P, Hildebrand PW (2015) Role of structural dynamics at the receptor G protein Interface for signal transduction. PLoS One 10(11): e0143399. https://doi.org/10.1371/journal.pone.0143399

Sandhu M, Touma AM, Dysthe M, Sadler F, Sivaramakrishnan S, Vaidehi N (2019) Conformational plasticity of the intracellular cavity of GPCR-G-protein complexes leads to G-protein promiscuity and selectivity. Proc Natl Acad Sci USA 116(24):11956–11965. https://doi.org/10.1073/pnas.1820944116

Semack A, Sandhu M, Malik RU, Vaidehi N, Sivaramakrishnan S (2016) Structural elements in the Gαs and Gαq C termini that mediate selective G protein-coupled receptor (GPCR) signaling. J Biol Chem 291(34):17929–17940. https://doi.org/10.1074/jbc.M116.735720

Soundararajan M, Willard FS, Kimple AJ, Turnbull AP, Ball LJ, Schoch GA, Gileadi C, Fedorov OY, Dowler EF, Higman VA, Hutsell SQ, Sundstrom M, Doyle DA, Siderovski DP (2008) Structural diversity in the RGS domain and its interaction with heterotrimeric G protein α-subunits. Proc Natl Acad Sci USA 105(17):6457–6462. https://doi.org/10.1073/pnas.0801508105

Sriram K, Insel PA (2018) G protein-coupled receptors as targets for approved drugs: how many targets and how many drugs? Mol Pharmacol 93(4):251–258. https://doi.org/10.1124/mol.117.111062

Sun X, Singh S, Blumer KJ, Bowman GR (2018) Simulation of spontaneous G protein activation reveals a new intermediate driving GDP unbinding. eLife 7. https://doi.org/10.7554/eLife.38465

Syrovatkina V, Alegre KO, Dey R, Huang XY (2016) Regulation, signaling, and physiological functions of G-proteins. J Mol Biol 428(19):3850–3868. https://doi.org/10.1016/j.jmb.2016.08.002

Tanaka T, Kohno T, Kinoshita S, Mukai H, Itoh H, Ohya M, Miyazawa T, Higashijima T, Wakamatsu K (1998) α helix content of G protein alpha subunit is decreased upon activation by receptor mimetics. J Biol Chem 273(6):3247–3252. https://doi.org/10.1074/jbc.273.6.3247

Van Eps N, Altenbach C, Caro LN, Latorraca NR, Hollingsworth SA, Dror RO, Ernst OP, Hubbell WL (2018) Gi- and Gs-coupled GPCRs show different modes of G-protein binding. Proc Natl Acad Sci USA 115(10):2383–2388. https://doi.org/10.1073/pnas.1721896115

Westfield GH, Rasmussen SG, Su M, Dutta S, DeVree BT, Chung KY, Calinski D, Velez-Ruiz G, Oleskie AN, Pardon E, Chae PS, Liu T, Li S, Woods VL Jr, Steyaert J, Kobilka BK, Sunahara RK, Skiniotis G (2011) Structural flexibility of the G alpha s alpha-helical domain in the β2-adrenoceptor Gs complex. Proc Natl Acad Sci USA 108(38):16086–16091. https://doi.org/10.1073/pnas.1113645108

Zhang Q, Dickson A, Doupnik CA (2004) Gbetagamma-activated inwardly rectifying K(+) (GIRK) channel activation kinetics via Galphai and Galphao-coupled receptors are determined by Galpha-specific interdomain interactions that affect GDP release rates. J Biol Chem 279(28): 29787–29796. https://doi.org/10.1074/jbc.M403359200

Chapter 9
Regulation of Lytic Machineries by the FtsEX Complex in the Bacterial Divisome

Martín Alcorlo, Siseth Martínez-Caballero, Rafael Molina, and Juan A. Hermoso

Abstract The essential membrane complex FtsE/FtsX (FtsEX), belonging to the ABC transporter superfamily and widespread among bacteria, plays a relevant function in some crucial cell wall remodeling processes such as cell division, elongation, or sporulation. FtsEX plays a double role by recruiting proteins to the divisome apparatus and by regulating lytic activity of the cell wall hydrolases required for daughter cell separation. Interestingly, FtsEX does not act as a transporter but uses the ATPase activity of FtsE to mechanically transmit a signal from the cytosol, through the membrane, to the periplasm that activates the attached hydrolases. While the complete molecular details of such mechanism are not yet known, evidence has been recently reported that clarify essential aspects of this complex system. In this chapter we will present recent structural advances on this topic. The three-dimensional structure of FtsE, FtsX, and some of the lytic enzymes or their cognate regulators revealed an unexpected scenario in which a delicate set of intermolecular interactions, conserved among different bacterial genera, could be at the core of this regulatory mechanism providing exquisite control in both space and time of this central process to assist bacterial survival.

Keywords Bacterial division · Divisome · FtsE · FtsX · PcsB · Lytic regulation · SPOR · 3D structure · Mechano-transmission · ABC transporters

M. Alcorlo · R. Molina · J. A. Hermoso (✉)
Department of Crystallography and Structural Biology, Institute of Physical Chemistry "Rocasolano", CSIC, Madrid, Spain
e-mail: malcorlo@iqfr.csic.es; rmolina@iqfr.csic.es; xjuan@iqfr.csic.es

S. Martínez-Caballero
Department of Crystallography and Structural Biology, Institute of Physical Chemistry "Rocasolano", CSIC, Madrid, Spain

Department of Chemistry of Biomacromolecules, Universidade Nacional Autonoma de Mexico, Ciudad de México, Mexico
e-mail: siseth.martinez@iquimica.unam.mx

J. R. Harris, J. Marles-Wright (eds.), *Macromolecular Protein Complexes IV*, Subcellular Biochemistry 99, https://doi.org/10.1007/978-3-031-00793-4_9

Introduction

Bacterial division is an essential cellular process that is pivotal for its adaptive strategy. During cell division, exquisite orchestration of both cell wall biosynthetic and lytic machineries is required to produce the daughter cells. Understanding this process at the molecular level is not only fundamental for cell biology, but also for providing new therapeutic targets in the fight against antimicrobial resistance. Most bacterial cells divide by a binary fission system, which is the splitting of a bacterial cell into two progeny cells that are approximately equal in size. However, this relatively simple mechanism has many variations, and the number of newly described organisms in which noncanonical division occurs is increasing, reflecting a vast diversity of bacterial strategies used for the duplication of cells (i.e., polar budding (Brown et al. 2012), longitudinal division (Leisch et al. 2012), and sporulation, among others). In all cases, bacterial division requires synthesis of a new septal cell wall, which must be split down the middle by one or more murein hydrolases to separate the resulting daughter cells. The peptidoglycan (PG) is the major component of the bacterial cell wall and is composed of linear alternating units of N-acetylglucosamine (NAG) and N-acetylmuramic acid (NAM) residues linked by β-1,4 bonds and cross-linked by peptide stems attached to the NAM unit. The core structure of the attached non-cross-linked stem peptide can have some variations. Generally, in G(−) bacteria, the stem peptide is composed of L-Ala-γ-D-Glu-m-DAP-D-Ala-D-Ala, while L-Ala-γ-D-Glu-L-Lys-D-Ala-D-Ala is generally present in G(+) bacteria (Alcorlo et al. 2017). This common scheme can be modified constitutively or in response to different environmental stresses at the glycan part (e.g., *O*-acetylation, *N*-deacetylacion, or amidation) or at the peptide stem (e.g., incorporation of Gly residues or noncanonical D-amino acids) (Cava and de Pedro 2014).

The cell fission process is coordinated by a dynamic macromolecular complex of proteins, referred to as the divisome (Bi and Lutkenhaus 1991), that creates and closes the septum to separate the dividing cells (Fig. 9.1). In several model bacteria, the divisome has been proposed to assemble in temporally distinct stages. The most studied system is that of the G(−) bacterium *Escherichia coli*, but with some exceptions; homologs of the *E. coli* divisome proteins are also present in G(+) bacteria such as *Streptococcus pneumoniae* (Fig. 9.1), and it seems likely that the pneumococcal divisome assembles in the same order (recently reviewed in Briggs et al. (2021)). Thus, first, FtsZ undergoes GTP-dependent polymerization into filaments that form a defined ringlike structure (known as the Z ring) at the site of cytokinesis. The exact composition of FtsZ-ring divisomes, the mechanism of timing and triggering of septal PG synthesis, and the involvement of PG remodeling by hydrolases are only partly understood and vary widely among different bacterial species (Gamba et al. 2009; Yao et al. 2017; Du et al. 2019). Then, membrane-associated FtsZ-interacting proteins, such as FtsA and ZipA in *E. coli*, tether the Z ring to the cell envelope, which results in an initial complex called the proto-ring (Fig. 9.1c). In *S. pneumoniae*, cell division also starts with the localization of the

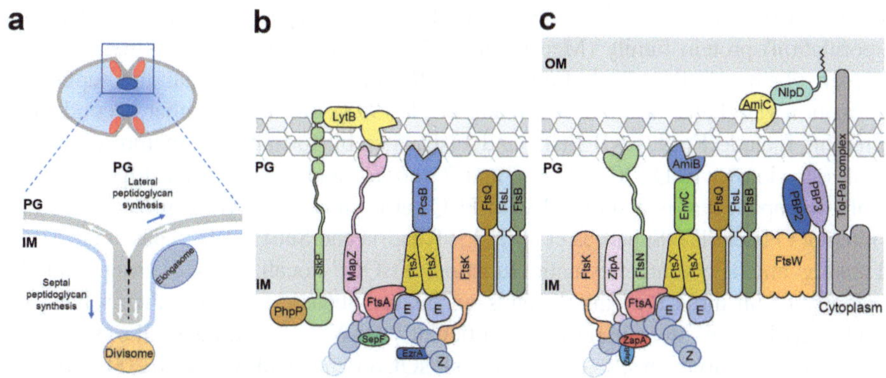

Fig. 9.1 The bacterial divisome. (**a**) Schematic cartoon showing the localization of the divisome and elongasome in *S. pneumoniae*. Peptidoglycan synthesis is indicated with arrows. PG, peptidoglycan; IM, inner membrane. (**b**) Main components of the pneumococcal divisome. Essential lytic enzyme PcsB is directly regulated by the FtsEX system, while septal cell wall thickness and final daughter cell separation performed by LytB are controlled by StkP. (**c**) Main components of the *E. coli* divisome. FtsEX-mediated ATP hydrolysis stimulates amidase activity (AmiB in this figure) through the effector protein EnvC. Amidase AmiC is controlled by the NlpD effector protein that is attached to the outer membrane (OM). See text for details

tubulin-like FtsZ and the actin-like FtsA cell division proteins (Lara et al. 2005) that assemble at mid-cell to form the Z-ring required for the recruitment of the other components to complete cytokinesis (Fig. 9.1b).

The Z ring is targeted to the site of division by related but distinct mechanisms (reviewed in Haeusser and Margolin 2016; den Blaauwen et al. (2017)). In the rod-shaped bacteria such as *E. coli* and *Bacillus subtilis*, a bipolar gradient that is formed by the Min system (oscillating in *E. coli* and fixed in *B. subtilis*) and nucleoid occlusion inhibits the assembly of FtsZ. Thus, the Z ring is restricted to the lowest region of negative regulation at mid-cell (Monahan et al. 2014; Rowlett and Margolin 2015). Partitioning of FtsZ filaments/bundles into daughter cells occurs in ovoid-shaped (ovococcus) bacteria, such as *S. pneumoniae*, by a mechanism that is fundamentally different from the Min and nucleoid occlusion systems present in rod-shaped bacteria (Wu and Errington 2011; Rowlett and Margolin 2015). Early in division, MapZ (also known as LocZ) forms a ring at mid-cell, before FtsZ recruitment, where it serves to drive the assembly of the division machinery. MapZ is not absolutely required for the formation of the Z ring, but is important for the correct positioning at the division site as shown in cells without MapZ where they are severely deformed (Fleurie et al. 2014). EzrA and SepF proteins are also involved in anchoring pneumococcal FtsZ filaments and bundles to cell membranes. In addition, EzrA is required for pneumococcal FtsZ ring formation (Perez et al. 2021) (Fig. 9.1c).

During a later stage, the proto-ring recruits a second group of divisome proteins that are involved in the synthesis of the septal cell wall (Aarsman et al. 2005; Gamba et al. 2009). These include the penicillin-binding proteins (PBPs), such as PBP3 (also known as FtsI) in *E. coli* or PBP2x in *S. pneumoniae*, which perform transpeptidation of the PG strands, and FtsW, a member of the SEDS (shape, elongation, division, and sporulation) protein family (Meeske et al. 2016), involved in the polymerization of glycan chains of PG. Both FtsW and PBP3/PBP2x are essential for septal PG formation (Pogliano et al. 1997; Wang et al. 1998). In *E. coli*, other second-stage proteins include FtsN, FtsB FtsL, and FtsQ, thereby forming the mature divisome. FtsB (also known as DivIC in *E. coli* and *B. subtilis*) and FtsL associate together into a stable complex that is associated with FtsQ (also known as DivIB in *B. subtilis* and other G(+) bacteria) (Saparoea et al. 2013; Khadria and Senes 2013; Glas et al. 2015). Lacking any apparent enzymatic function, FtsQLB localizes as a group and the stability of all three proteins depends on direct interactions among themselves and with other late cell division proteins, particularly FtsW, FtsI, and FtsN, all of which are recruited to the division site subsequent to FtsQLB (Chen and Beckwith 2001; Robson et al. 2002; Buddelmeijer and Beckwith 2004; Daniel et al. 2006). Despite their critical role in cytokinesis, recruitment of FtsI and FtsW to the nascent division site is insufficient to drive cytokinesis. Instead, a growing body of evidence supports a role for FtsN as the trigger for cytokinesis by both direct interactions with FtsI and FtsW, but also through its role as an activator of FtsA-mediated recruitment of FtsQLB (reviewed in den Blaauwen et al. (2017)). Significantly, although FtsA, FtsQLB, FtsI, and FtsW are widely conserved, FtsN is limited to G(−) organisms, suggesting that other bacteria, such as *S. pneumoniae*, utilize different mechanisms to activate division. Consistent with this idea, cell division proteins are phosphorylated in several bacterial species, including *B. subtilis*, *S. pneumoniae*, *Staphylococcus aureus*, and *Mycobacteria*—but not in *E. coli*—suggesting a potential additional route for activation (for a review, see Manuse et al. (2016)). Similar to other G(+) bacteria, *S. pneumoniae* possesses a conserved signaling system consisting of a Ser/Thr protein kinase, StkP, and a cognate PP2C-type phosphatase, PhpP (Echenique et al. 2004; Nováková et al. 2005). StkP belongs to the subfamily of eukaryotic-type Ser/Thr kinases, eSTKs, and is composed of a cytoplasmic kinase domain, a transmembrane region, and an extracellular C-terminal region constituted by four PASTA (penicillin-binding protein and serine/threonine kinase-associated) domains. Several StkP substrates play a role in cell division, including MapZ (reviewed in Vollmer et al. (2019)). Both StkP and its cognate PhpP localize to the division site with active PG synthesis. PASTA domains of StkP are required for its septal localization, indicating that they bind to newly synthesized and still uncross-linked PG chains in vivo. Interestingly, PASTA domains in StkP have been reported to regulate the septal cell wall thickness and control final cell separation through their interaction with the lytic enzyme LytB at the very late stage of cell division (Zucchini et al. 2017). It is worth mentioning that PASTA domains are also present in PBPs from G(+) bacteria but only in those PBPs involved in cell division such as PBP2x (*S. pneumoniae*), PBP2B (*B. subtilis*), or PBP1 (*S. aureus*). While the role of PASTA domains in PBPs is not fully understood, it has been reported that they play a role in protein–protein recognition in *B. subtilis* by strengthening the interaction between PBP2B and DivIB (Angeles et al. 2020), to promote PBP2x catalysis in *S. pneumoniae* by recognition of nascent PG chains (Bernardo-García et al. 2018) or to recognize PG fragments and regulate interactions with other divisome proteins in *S. aureus* (Martínez-Caballero et al. 2021).

Once the divisome complex is assembled, PBPs that are required for septal growth must couple their enzymatic activity to membrane constriction, thereby localizing cell wall hydrolysis. In G(−) bacteria, this regulation needs to extend to the outer membrane. The Tol–Pal (peptidoglycan-associated lipoprotein) complex, which physically connects the cytoplasmic membrane and the outer membrane in *E. coli*, is crucial for coordinating the constriction of the outer membrane and the synthesis of peptidoglycan (Gerding et al. 2007). At the heart of this system is CpoB (also known as YbgF), which regulates PBP1B activity in response to the Tol–Pal complex in the outer membrane (Gray et al. 2015).

Following septation and membrane constriction, hydrolases separate bacterial daughter cells. In *E. coli*, amidases that localize at mid-cell (Bernhardt and De Boer 2003) are involved in daughter cell separation. In *S. pneumoniae*, the endopeptidase PcsB (from protein required for cell wall separation of group B streptococcus) plays the same role as *E. coli* amidases. These hydrolases may signal the Z ring through a conserved pair of proteins, FtsE and its membrane partner FtsX, thereby coordinating ring constriction with septum growth and accompanying hydrolysis of the cell wall (Corbin et al. 2007; Yang et al. 2011). In addition, *S. pneumoniae* and its relatives have an FtsEX-independent hydrolase, LytB (an endo-β-1,4-N-acetylglucosaminidase), required for the terminal separation of daughter cells at the end of cell division. Hence, the FtsEX complex plays a central role in coordinating the different divisome proteins and, specially, in regulating the activity of the potentially dangerous lytic machineries required for cell division.

In this chapter we will describe the present knowledge on both the regulatory role of the FtsEX complex and associated lytic and regulatory proteins as well as their three-dimensional structure. The main emphasis will be placed on the pneumococcal case for which many pieces of this complex puzzle have been reported recently.

FtsEX, a Type VII Member of the ABC Transporter Superfamily, Is Involved in the Regulation of Bacterial Cell Division

The general organization of FtsEX complexes resembles that of ABC transporters (Schmidt et al. 2004; Arends et al. 2009; Mir et al. 2015). The ABC transporter family is widespread in all domains of life and belongs to the ATP-binding cassette (ABC) superfamily, which uses the hydrolysis of ATP to fuel diverse biological processes (Saurin et al. 1999), such as development and growth, nutrient import, cellular detoxification, lipid homeostasis, signal transduction, antiviral defense, and antigen presentation (Locher 2016). All ABC transporters share a basic architecture comprising at least two intracellular nucleotide-binding domains (NBDs) and two transmembrane domains (TMDs). The NBDs supply energy via ATP binding and hydrolysis, while the TMDs form the transmembrane permeation pathway. In the FtsEX complex, an FtsE cytoplasmic dimer forms the NBDs and an FtsX dimer in the inner membrane forms the TMDs (Pichoff et al. 2019). The NBD dimer can adopt open or closed conformations. NBDs are coupled to the TMDs through the so-called coupling helix at the NBD–TMD interface (reviewed in reference (Locher 2016)). FtsEX is a member of a small subfamily (type VII) of the ABC superfamily that includes the G(−) transporters MacB and LolCDE, neither of which are transporters but instead use a mechanotransmission mechanism to perform work in the periplasm (Crow et al. 2017; Greene et al. 2018). This mechanism is tightly wed to the structural architecture of type VII ABC transporters and it is anticipated that similar conformational changes are conserved throughout this superfamily (Crow et al. 2017; Greene et al. 2018; Pichoff et al. 2019). In the particular case of FtsEX, it is believed to transduce a conformational signal through the coupling helices in the TMD; this requires ATP hydrolysis and leads to the activation of PG hydrolases.

The essential FtsEX complex is widely conserved across bacteria and plays a central role by recruiting proteins to the divisome apparatus and by regulating periplasmic muralytic activity from the cytosol. Concerning recruitment, it has been reported that, in *E. coli*, FtsE is one of the earliest proteins that assembles along with FtsX at the mid-cell site during cell division. FtsEX directly participates in this critical process by promoting septal-ring assembly in a low-osmolarity medium (Schmidt et al. 2004). Recent studies revealed that FtsEX acts on FtsA to promote divisome assembly, and continual ATP hydrolysis by FtsE is needed for the divisome to synthesize septal PG (Du et al. 2016). In the divisome of G(−) bacteria, FtsE also interacts with the conserved tail of FtsZ (Du et al. 2019), linking the Z ring to the transmembrane protein FtsX, therefore suggesting that FtsEX may function as a membrane anchor for FtsZ by using ATP binding and hydrolysis to regulate Z-ring constriction (Schmidt et al. 2004; Corbin et al. 2007). In addition, FtsEX directly recruits periplasmic division PG hydrolases to the septum.

Concerning the regulation of bacterial division by PG hydrolases, a principle that has emerged from recent studies is that the division PG hydrolases are usually autoinhibited (Vollmer et al. 2008; Yang et al. 2012b; Vollmer 2012; Mavrici et al. 2014) and require interaction by regulatory proteins, such as EnvC and NlpD in *E. coli*, to activate synchronized cleavage of the PG cell wall. PG hydrolases vary among different bacterial species (Sham et al. 2011; Yang et al. 2011; Meisner et al. 2013; Mavrici et al. 2014), and their lytic activity is thus regulated, spatially and temporally, by coupling their activation to the assembly of the cytokinetic ring (Uehara et al. 2010) through the FtsEX complex. In *E. coli*, *Neisseria gonorrhoeae*, *Aeromonas hydrophila*, and *Flavobacterium johnsoniae*, *ftsE* and/or *ftsX* mutants exhibit division defects, indicating that FtsEX's function in cell division is conserved in these organisms (Salmond et al. 1984; Kempf and McBride 2000; Merino et al. 2001; Ramirez-Arcos et al. 2001). In *S. pneumoniae*, FtsX interacts with the PG hydrolase PcsB (Sham et al. 2011, 2013; Massidda et al. 2013; Bartual et al. 2014). FtsE, FtsX, and PcsB are essential for growth in some serotype strains of *S. pneumoniae* (Ng et al. 2003, 2004; Sham et al. 2013), and their absence causes severely diminished growth and cell morphology defects in others (Giefing et al. 2008; Giefing-Kroll et al. 2011). In contrast, the FtsEX of *B. subtilis* has no obvious role in cell division but instead controls the peptidoglycan hydrolase CwlO, which

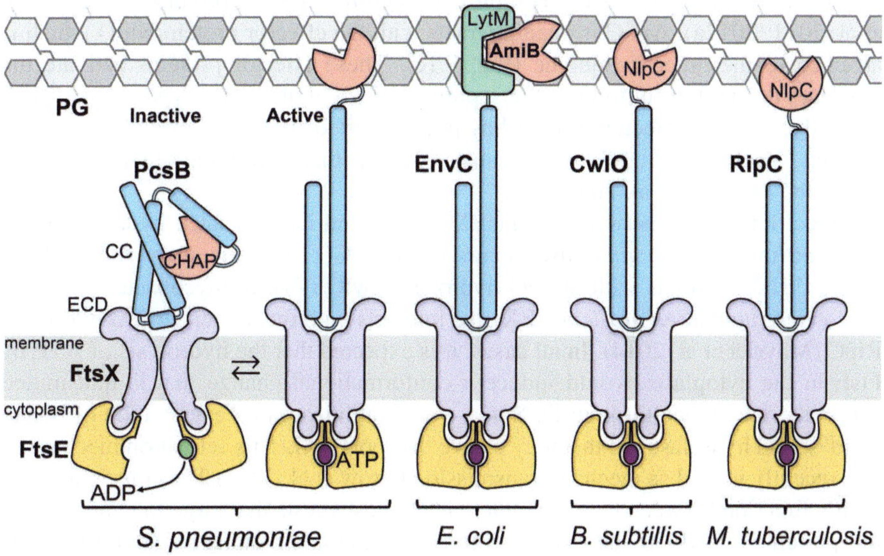

Fig. 9.2 Regulatory role of lytic enzymes by the FtsEX in different bacteria. Schematic models showing the activation mechanism of PcsB promoted by the FtsEX complex (left panel) and comparison with the *E. coli* FtsEX–EnvC–amidase complex, the *B. subtilis* FtsEX–CwlO elongation hydrolase complex, and the *M. tuberculosis* FtsEX–RipC hydrolase complex

plays an important role in cell wall expansion during cell elongation (Domínguez-Cuevas et al. 2013; Meisner et al. 2013; Brunet et al. 2019) and regulates entry into sporulation (Garti-Levi et al. 2008). In addition, two uncharacterized proteins (SweD and SweC) were recently identified in *B. subtilis* as essential cofactors of FtsEX in controlling PG hydrolase activity along the lateral cell wall. Homologs of SweD and SweC are found in a subset of firmicutes and it was proposed that these and analogous factors enable FtsEX to function outside the divisome to control cell wall elongation hydrolases (Brunet et al. 2019). FtsEX was recently reported to be involved in the control of both the early and late stages of cytokinesis in the alphaproteobacterium *Caulobacter crescentus* (Meier et al. 2017) and in the control of the activity of autolysin CwlO to regulate biofilm formation in *Bacillus velezensis* strain SQR9, a plant-beneficial rhizobacterium (Li et al. 2018).

Therefore, the FtsEX complex has emerged as a broadly conserved regulatory module in controlling cell wall hydrolases in diverse bacterial species. In most characterized examples, FtsEX is integrated into the division machinery (or in cell wall elongation in *B. subtilis*) where it controls cell wall cleavage during cytokinesis. However, their arrangement in G(+) and G(−) bacteria is different; while in G(+) bacteria, the FtsEX complex directly interacts with the PG hydrolase, in G(−) bacteria, an effector protein mediates the interaction between FtsEX and the corresponding lytic enzyme (Fig. 9.2). Thus, in *E. coli*, the amidases AmiA and AmiB, which play an important role in releasing daughter cells (Heidrich et al.

2001), interact with the FtsEX complex through the effector proteins EnvC (Yang et al. 2011, 2012a). AmiC instead interacts with the effector protein NlpD which is attached to the outer membrane (Fig. 9.1c). These effector proteins activate the amidases during the splitting of the septum (Uehara et al. 2010; Typas et al. 2011). Recently, another effector protein has been reported in *E. coli*, an NlpD paralog (termed "YgeR" or "ActS"), which can activate the three amidases with a preference for AmiC (Serrano et al. 2021).

Direct interaction between the FtsEX system and its associated hydrolase PcsB observed in *S. pneumoniae* also occurs in other G(+) bacteria such as *B. subtilis* where FtsEX interacts with the PG hydrolase CwlO (74). In mycobacteria such as *Mycobacterium tuberculosis*, FtsEX also interacts with the associated PG hydrolase RipC (Mavrici et al. 2014). In all cases, it is expected that the hydrolysis of ATP by FtsE in the cytoplasm would induce a conformational change that is transmitted across the membrane through FtsX and would culminate in the activation of bound amidase/PG hydrolases so that they cleave the septal PG. This activation mechanism was recently named as mechanotransmission (Crow et al. 2017).The conservation of FtsEX function between G(−) and G(+) bacteria suggests that targeting and activating proteins outside of the cytoplasmic membrane is an ancient function of ABC transporters (Zeng and Charkowski 2021).

Role of Endopeptidases and Amidases in Bacterial Cell Division

PG, the largest bacterial macromolecule, is a single cross-linked entity that encases the entire bacterium, and consequently, murein hydrolase activity is required to separate the resulting daughter cells. The general structure of PG is a backbone of NAG–NAM disaccharides with peptide stems appended to the NAM units (Fig. 9.3a). Endopeptidases and amidases perform this lytic activity by cleaving at specific positions of the peptide stems (Fig. 9.3a). The endopeptidases cleave the amide bonds within the stem peptide itself, while the amidases are *N*-acetylmuramoyl-L-alanine amidases that break down peptide cross-links by cleaving bonds connecting stem peptides to the *N*-acetylmuramic acid component of the glycan strands (Fig. 9.3a) (Alcorlo et al. 2017). Both types of enzymes can be divided into different subfamilies according to the Pfam (Mistry et al. 2021) database. In G(+) bacteria, essential lytic enzymes in cell division usually have endopeptidase activity through a CHAP (from carboxy terminal cysteine, histidine-dependent amidohydrolase/peptidase, Pfam: PF05257) domain or through an NlpC/P60 domain (Pfam: PF00877), both papain-like domains. The CHAP domain is characterized by an α plus β fold containing two invariant residues, a cysteine and a histidine, where the Cys is the catalytic residue in a nucleophilic-attack mechanism (Bateman and Rawlings 2003). The NlpC/P60 domain has a conserved active site

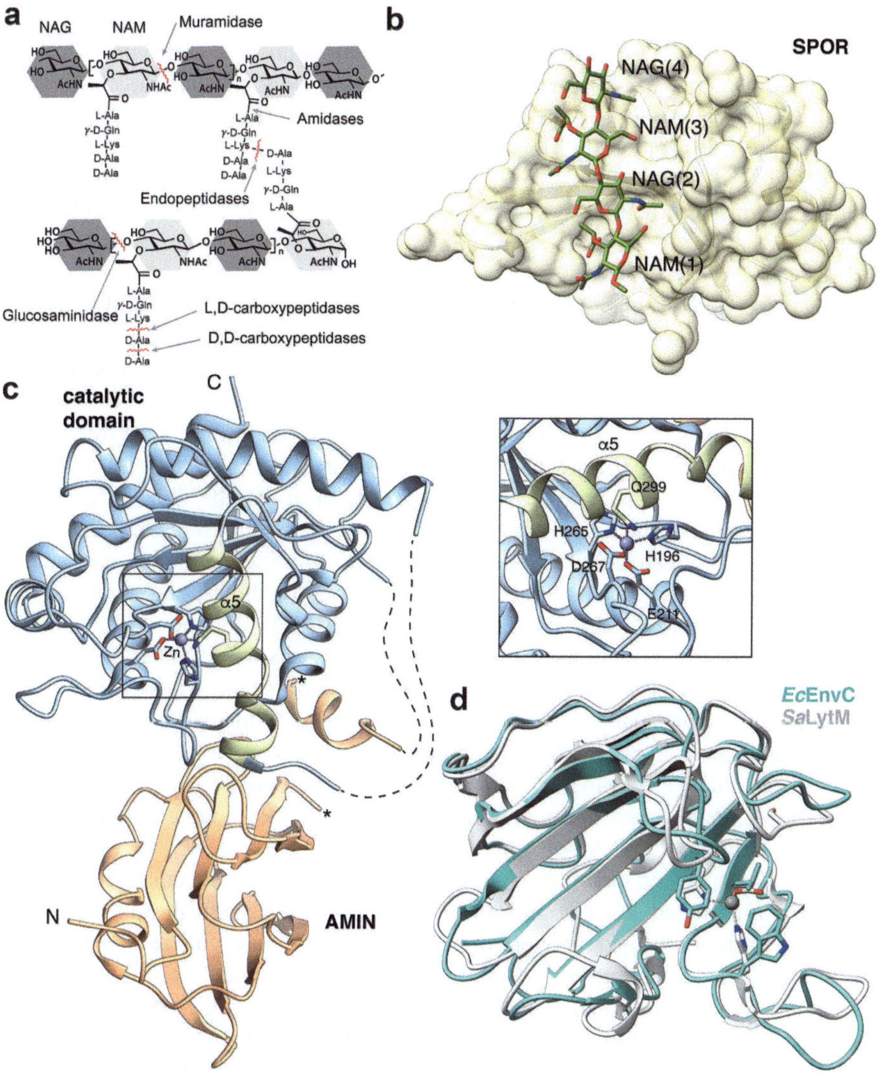

Fig. 9.3 Three-dimensional structures of SPOR domains, amidases, and LytM proteins. (**a**) General scheme of PG in G(+) and the different bacterial peptidoglycan hydrolases involved in cell wall processing. (**b**) SPOR domain from *P. aeruginosa* in complex with a denuded PG (green sticks) (PDB code 6I09) (Alcorlo et al. 2019). The protein's ribbon diagram and molecular surface are shown. (**c**) The 3D structure of AmiC (PDB code 4BIN) (Rocaboy et al. 2013). The AMIN domain is colored in salmon, the amidase domain in blue, and the blocking α5-helix in yellow. A zoomed view of the catalytic site is boxed showing the residues involved in Zn^{+2} coordination (in sticks) including the Q299 from α5 that blocks the active site. (**d**) Structural comparison between the LytM–EnvC domain of *E. coli* (colored in cyan, PDB code 4BH5) (Peters 2013) and the LytM of *S. aureus* (colored in gray, PDB code 2B13) (Firczuk et al. 2005). The residues involved in Zn^{+2} coordination and catalysis are depicted in sticks

comprising a cysteine and a histidine residue and a third variable residue that can be either asparagine or less frequently histidine (Anantharaman and Aravind 2003).

In G(−) bacteria, the two more relevant groups of lytic proteins in cell division are the *N*-acetylmuramoyl-L-alanine amidases of type 3 (Pfam: PF01520) and the LytM peptidases from family M23 (Pfam: PF01551). In *E. coli*, the three amidases related with cell division (AmiA, AmiB, and AmiC) are Zn^{+2}-dependent metalloenzymes that belong to the amidase_3 family (PF01520). AmiB and AmiC are localized at the division site, while AmiA is dispersed in dividing cells (Rocaboy 2013).

As a direct consequence of the action of amidases, denuded glycans (PG devoid of peptide stems) are transiently formed at the division site (Yahashiri et al. 2017; Alcorlo et al. 2019). The sporulation-related repeat (SPOR) domains sense this modified PG at the septum to direct proteins to the bacterial division site. The relevance of the action exerted by amidases is reflected by the fact that SPOR domains (Pfam: PF05036) are ubiquitous in prokaryotes, in both G(−) and G(+) bacteria (13,422 proteins from over 4388 species, as of October 2021). In *E. coli*, four SPOR-domain-containing proteins have been identified (DamX, DedD, FtsN, and RlpA), all of which are involved in cell division, with FtsN evaluated as essential (Gerding et al. 2009; Ryan Arends et al. 2010). Recently, the crystal structures of a SPOR domain (from *Pseudomonas aeruginosa* RlpA) in complex with denuded glycans have been reported, unveiling the molecular basis for denuded PG recognition that delineate a conserved pattern in other SPOR domains (Alcorlo et al. 2019) (Fig. 9.3b). Structural analysis revealed that recognition of the carboxylate moieties of NAM units at positions 1 and 3 is crucial for complex formation (Fig. 9.3b). These carboxylate moieties become available only after the removal of peptide stems by the action of the abovementioned amidases. If peptide stems were included in the structures of the reported SPOR–PG complexes, the strong steric clashes with the SPOR structure would preclude binding by the PG. The key conserved features identified in denuded PG recognition by SPOR domains were the presence of a conserved basic patch and a network of polar interactions involving residues shaping the SPOR domain binding site. Many of the protein–ligand interactions involve main-chain atoms rather than side-chain moieties (Alcorlo et al. 2019). This feature helps explain why all SPOR domains possess the requirements for recognition of denuded PG despite their low sequence conservation. Thus, binding occurs through a combination of hydrogen-bond networks to main-chain atoms, mediated by water molecules or by polar and salt-bridge interactions provided by a small number of critical residues. Interestingly, while peptide stem recognition is not allowed, SPOR domains can recognize some other PG modifications. Binding of the end of the PG strands through the 1,6-anhNAM, as a consequence of the action of lytic transglycosylases, or *O*-acetylation on NAM units is also allowed, while *N*-deacetylation on NAG units is not (Alcorlo et al. 2019). Thus, action of amidases is not only important for its own lytic activity, which ultimately leads to the separation of daughter cells, but also as a way to provide a precise spatiotemporal regimen for recruitment of the individual proteins in the divisome.

Three-Dimensional Structures of Amidase AmiC from *E. coli* and Regulatory LytM Domains

As observed in the Pfam database, 12 different crystal structures of amidase_3 family members have been solved so far from different bacterial, mycobacterial, or phage species, most of them presenting just the catalytic domain. Concerning *E. coli*, only the 3D structure of AmiC has been reported (Rocaboy et al. 2013). Interestingly, this structure shows, for the first time, the N-terminal AMIN domain that interacts with the PG and the C-terminal amidase domain (Fig. 9.3c). The AMIN domain is folded in a β-sandwich of two symmetrical four-stranded β-sheets exposing highly conserved motifs on the two outer faces. The catalytic domain of AmiC (Fig. 9.3c) presents the Zn^{+2}-dependent catalytic center in which the Zn^{+2} is coordinated by two His residues (H196 and H265) and a Glu residue (E211). These residues are strictly conserved in the amidase_3 family. Interestingly, AmiC presents an α-helix (α5) that occludes the active site (Fig. 9.3c). Structural comparison with the amidase_3 members that possess accessible active sites revealed that the inhibitory helix also induces conformational changes in a nearby loop provoking reorientation of the side chain of D267 to coordinate catalytic ion (Rocaboy et al. 2013). Residue Q299 from α5 also chelates the metal ion, thus blocking the active site center (Fig. 9.3c).

The activation of amidase AmiC is regulated by the interaction with NlpD that would displace the α5 unmasking the active site of AmiC (Rocaboy et al. 2013). Although the three amidases of *E. coli* have the same catalytic activity, they present different domain arrangements. Thus, while AmiB and AmiC both present one AMIN domain in the N-terminal followed by the catalytic amidase domain, AmiB also contains a LysM domain for PG recognition. On the contrary, AmiA presents only the catalytic domains without any other cell-wall-binding domains.

Amidases related with cell division in G(−) bacterial species present an insertion of approximately 50 residues that is absent in amidases from phages (Korndörfer et al. 2006) or those involved in the sporulation process such as LytC of *B. subtilis* (Smith and Foster 1995). This insertion corresponds to the region containing the helix blocking the active site as observed in the AmiC of *E. coli* (Rocaboy et al. 2013) or in the AmiB from *Bartonella henselae* (Yang et al. 2012b). Therefore, it is expected that the regulatory process suggested for AmiC will be similar in its cognate AmiA and AmiB of *E. coli* as well as in other G(−) bacteria.

Activation of the amidases involved in cell division requires the intervention of a special class of LytM domains lacking catalytic activity. LytM-domain-containing proteins are LAS peptidases (from lysostaphin-type enzymes, D-Ala-D-Ala metallopeptidases, and sonic hedgehog) that play different roles during bacterial cell cycle through hydrolysis of bacterial PG (Figueroa-Cuilan et al. 2021). LytM proteins with PG hydrolase activity present a Zn^{+2}-dependent active site and have been identified in G(−) (Möll et al. 2010; Sycuro et al. 2010) and G(+) bacteria such as *S. aureus* (Firczuk et al. 2005). However, the LytM domain of EnvC in *E. coli*, which shows a similar 3D structure than its catalytic LytM homologues, is degenerated and does not have the residues required to coordinate the catalytic metal ion (Fig. 9.3d) (Uehara et al. 2010; Peters 2013). Interestingly, mutation studies in the EnvC degenerated zone indicated that this region is critical for amidase activation both in vivo and in vitro through binding the autoinhibition helix of AmiB (Peters 2013). This interaction would promote a reversible conformational change in the amidase allowing the release of autoinhibitory α-helix and leaving the active site available (Yang et al. 2012a). This mechanism of EnvC regulation of amidases is expected to work for the other degenerated LytM factors although no experimental evidence has yet been reported for NlpD or YgeR.

Recently, the crystal structure of full-length EnvC in complex with the extracellular domain (ECD) of FtsX, $FtsX_{ECD}$, has been reported (Cook et al. 2020). EnvC presents three functional domains: a coiled-coil (CC) domain, a central regulatory domain, and a degenerated LytM domain (Fig. 9.6a). Interestingly, the EnvC structure showed the binding site of the LytM domain blocked by a regulatory domain or "restraining arm" composed of a 50-residue α-helix mimicking the autoinhibitory helix of AmiC. Considering that (*i*) FtsEX directly recruits EnvC through the $FtsX_{ECD}$ and (*ii*) ATPase-defective variants of FtsEX still recruit EnvC to the septum but fail to promote cell separation (Yang et al. 2011), a conformational change via the regulation of FtsEX is required to liberate the LytM domain and to allow interaction with its cognate amidases. As detailed below, the structure of EnvC with the ECD of FtsX (Cook et al. 2020) provides a very interesting model to understand how this activation process could take place at the molecular level.

The Three-Dimensional Structure of Pneumococcal Endopeptidase PcsB

PcsB is the essential PG hydrolase required for the separation of daughter cells in *S. pneumoniae*. Crystal structure determination of PcsB (Bartual et al. 2014) revealed the presence of three distinct regions: a coiled-coil (CC) domain, $PcsB_{CC}$, containing leucine zipper motifs (residues 41–266), a linker region (267–278), and a CHAP domain, $PcsB_{CHAP}$ (279–392) (Fig. 9.4a, b). The structure can be roughly encased in an acute triangle with two long sides (~112 and ~118 Å) and a short side of ~66 Å. The $PcsB_{CC}$ domain is arranged in five helices, three long bent helices and two short ones conferring the overall V-shaped architecture of PcsB. The linker is an alanine-rich region that connects the last helix from the CC domain with the globular catalytic CHAP domain. This CHAP domain displays a classical α + β papain-like peptidase fold where the reactive cysteine, Cys292, is positioned (Fig. 9.4c). Interestingly, the dimeric arrangement of PcsB molecules found in the crystal structure was confirmed in solution by performing small-angle X-ray scattering (SAXS) and analytical ultracentrifugation (AUC) at high protein concentrations. This dimer is formed by 3D domain swapping (Liu and Eisenberg 2002), in which both monomers exchange their catalytic domains. The dimer places both catalytic domains in the

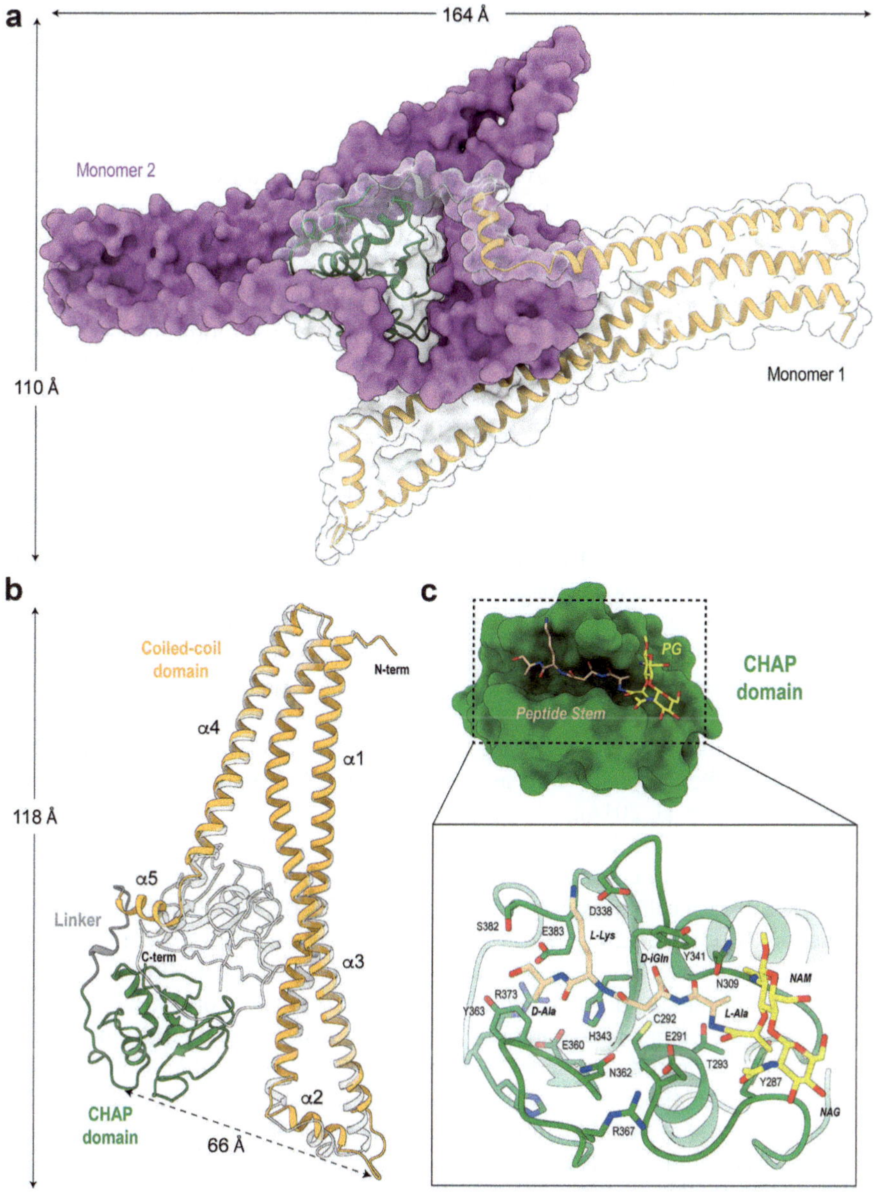

Fig. 9.4 3D structure of PcsB. (**a**) Dimeric arrangement of PcsB in the crystal structure (PDB code 4CGK) (Bartual et al. 2014). (**b**) Modular disposition of PcsB in the context of its monomer crystal structure superimposed into the monomeric model predicted by AlphaFold 2 (grey) (Jumper et al. 2021). (**c**) The $PcsB_{CHAP}$ with a PG fragment docked into the catalytic site. Boxed area shows a detailed view of the active site with catalytic and substrate binding residues represented as capped sticks

middle with their active sites oriented in opposite directions and with the CC domains extending outwards on the same axis in a 164-Å long dimer. The V-shaped CC domain of the monomer clamps the CHAP domain of the other in a narrow cavity. At lower protein concentrations, SAXS and AUC experiments also suggested an inactive module arrangement of PcsB for the monomer species wherein its CHAP domain is inserted into the internal cavity formed by its own CC domain (Bartual et al. 2014). This monomeric model is nowadays also supported by the protein prediction folding calculations performed with AlphaFold 2 (Jumper et al. 2021) (Fig. 9.4b).

To gain insights into PcsB substrate recognition, Bartual et al. (2014) also performed docking calculations with GOLD (Verdonk et al. 2003) using a peptide stem fragment, L-Ala-D-isoGln-L-Lys-D-Ala, present in pneumococcal PG. Top scoring positions nicely placed the scissile bond between isoGln and L-Lys close to the catalytic Cys292 and His343. In addition, the ligand was strongly stabilized by many hydrogen bonds and salt links with residues in the active site providing a feasible model of PcsB–peptide stem interaction (Fig. 9.4c). Thus, the active site of the $PcsB_{CHAP}$ is fully functional but substrate access to the reactive center would be impeded by the CC domain from the dimeric partner or by its own CC domain in the monomer. In either monomer or dimer, PcsB presents an inactive conformation in which the CC helices act as "molecular tweezers" clamping the catalytic domain and blocking access to the active site center. Activation of PcsB would require release of the CHAP domain by the FtsEX complex. Besides, experiments with the CHAP domain alone or with LytF-$PcsB_{CHAP}$ (a chimera in which the $LytF_{CHAP}$ domain of the LytF fratricine was replaced by the CHAP domain of PcsB) proved the muralytic activity of the $PcsB_{CHAP}$ as well as the role of the CC domain in the regulatory process (Bartual et al. 2014).

Previous studies involving temperature-sensitive (Ts) mutations and suppressor mutants of PcsB and FtsX, respectively, revealed mutations in the large (ECL1) and small (ECL2) extracellular loops of FtsX and in the $PcsB_{CC}$ affecting the FtsX–PcsB interaction (Sham et al. 2011, 2013). The Ts mutations in $PcsB_{CC}$ were located just in the middle of α1- and α3-helices and at the base of the α4-helix (Fig. 9.4b), while the connecting regions in FtsX involved the distal portion of the ECL1 and ECL2 loops. Thus, taking in consideration these previous experimental results and considering that the estimated thickness of the cross wall was compatible with the overall dimensions of the dimer (Fig. 9.4a), a mechanism of action was proposed to explain how its inactive state could be turned into an active state by the action of the FtsEX complex (Bartual et al. 2014). Briefly, during division, each daughter cell would present a PcsB monomer to adopt a dimeric state at the septum in which the V-shaped CC domain of each monomer acts as a pair of molecular tweezers locking the catalytic CHAP domain of each partner in the inactive configuration. This would imply that the PcsB dimer extends across the cross wall and interacts with FtsEX complexes at both ends. The release of the catalytic domains would be assisted by an ATP-driven conformational change in the transmembrane FtsEX complex, ensuring that the enzyme digests the cell wall only at the right time in the right place (Bartual et al. 2014). The recent structural results on the EnvC–$FtsX_{ECD}$ system in *E. coli*

(Cook et al. 2020) offers a new interpretation for the activation of monomeric PcsB in *S. pneumoniae* that will be discussed later.

Structural Information on the FtsX Membrane Protein

FtsX is an integral membrane protein with cytoplasmic amino and carboxyl termini, four transmembrane segments (TM1–TM4), and one large and one small extracellular loops denoted ECL1 and ECL2, respectively (Fig. 9.5a). ECL1 is an approximately 100-residue-long folded domain, hereafter referred to as extracellular domain (ECD). These two regions (ECD and ECL2) interact in the periplasmic space with either cell wall hydrolases or effector proteins, which results in the activation of PG hydrolysis via a yet unknown mechanism (Sham et al. 2011, 2013; Yang et al. 2011, 2012a; Meisner et al. 2013; Bartual et al. 2014; Mavrici et al. 2014; Meier et al. 2017). In the cytoplasm, FtsX interacts with FtsE through a region located between TM2 and TM3 (Fig. 9.5a). This region is predicted to be a helix that is roughly parallel to the membrane plane and constitutes the coupling helix that would be inserted in a hydrophobic cavity present in FtsE. In addition, the FtsX N-terminus in the gram-negative *E. coli* and *P. aeruginosa* bacteria presents an additional long region (~50 residues) that is absent in gram-positive organisms.

Although the complete structure of FtsX still remains unknown, 3D structures of its ECD ($FtsX_{ECD}$), connecting the first two transmembrane helices (TM1 and TM2), have been reported for different bacteria including *M. tuberculosis* (PDB code 4N8N (Mavrici et al. 2014)), *S. pneumoniae* (i.e., PDB code 6HFX (Rued et al. 2019)), and *E. coli* (PDB code 6TPI (Cook et al. 2020)) (Fig. 9.5). They all have a similar fold arranged into a bilobal structure (Fig. 9.5b). The upper lobe presents a central core that comprises two βαβ secondary structure motifs that interlock to form a central four-membered β-sheet with two flanking helices. This domain is homologous to the so-called Porter subdomain found in *E. coli* MacB and LolC, which is a conserved feature of all type VII ABC transporters and likely an intrinsic part of the transmission apparatus (Crow et al. 2017; Greene et al. 2018). The lower lobe, termed the X-lobe (Pichoff et al. 2019), is α-helical and comprises a conserved approximately 10-residue-long α-helix and a shorter α-helical region that may adopt different conformations, reflecting its dynamic behavior and conformational plasticity. It is worth mentioning that in all reported pneumococcal crystal structures (PDB codes 6HEH, 6HFX, and 6HEE (Rued et al. 2019)), a molecule of detergent was found inserted into the X-lobe. The presence of detergents was critical, as in their absence, crystals diffracted at very low resolution (<4 Å) indicating that their presence was required to stabilize the mobility in this protein region (Rued et al. 2019). In agreement with this observation, the NMR solution structure of pneumococcal ECD (PDB code 6MK7) further corroborated the mobility of this region (Rued et al. 2019). The main differences among the ECD available structures are the presence of an extended β-hairpin in the pneumococcal case (a hallmark from streptococcal species (Rued et al. 2019)) and a disulfide bond present in the

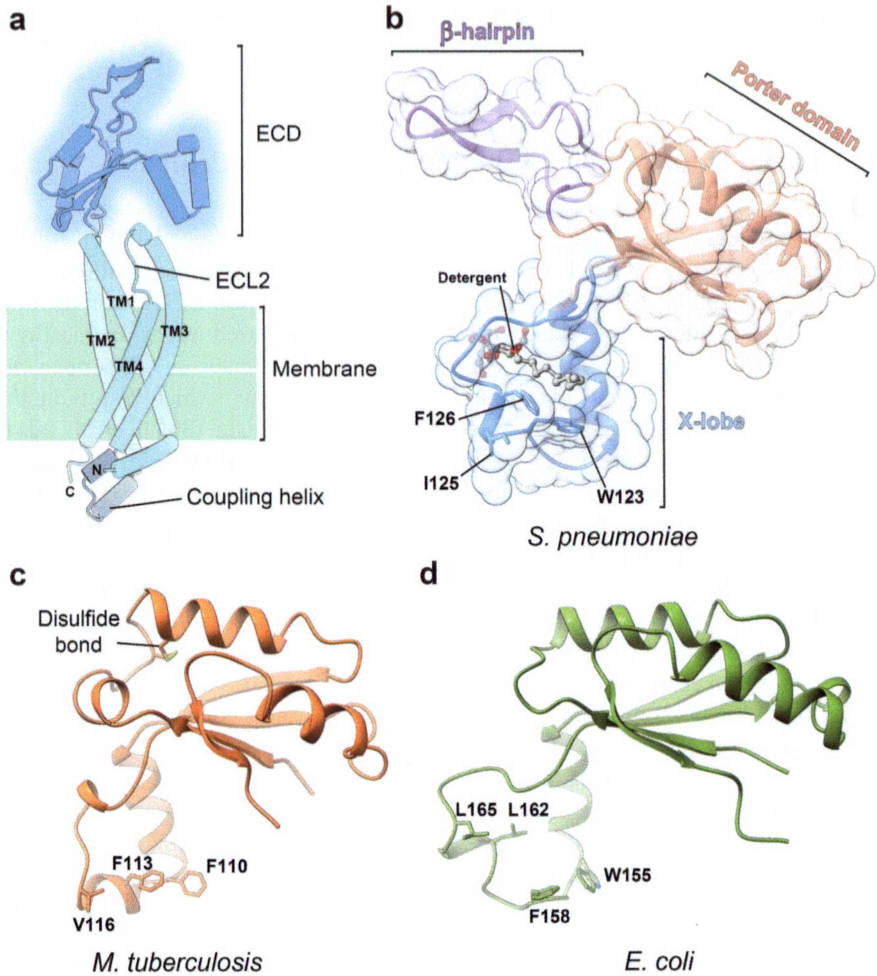

Fig. 9.5 3D structure of FtsX. (**a**) Schematic cartoon showing the topology of the FtsX monomer from *S. pneumoniae*. Transmembrane regions (colored in cyan) are numbered from 1 to 4. ECD is colored in dark blue and the coupling helix in gray. (**b–d**) Comparison of $FtsX_{ECD}$ domains displayed in cartoon. Structures are derived from PDB entries 6HFX (*S. pneumoniae* FtsX), 4N8N (*M. tuberculosis* FtsX, colored in orange), and 6TPI (*E. coli* FtsX, colored in green) (Mavrici et al. 2014; Rued et al. 2019; Cook et al. 2020). The molecular surface of pneumococcal ECD on panel B is shown including the upper lobe that comprises the Porter domain (colored in salmon), the lower X-lobe (colored in cyan), and the β-hairpin insertion (colored in violet). A detergent molecule (n-decyl-β-D-maltoside) inserted in the X-lobe is displayed in ball and sticks. Some of the key hydrophobic residues located in the X-lobe that are relevant in the interaction with the CC of their corresponding binding partners are numbered and depicted in sticks

M. tuberculosis ECD, just at the place of the β-hairpin insertion of the pneumococcal ECD. It is important to highlight that that the X-lobe, strictly conserved in FtsX, is absent in other type VII ABC transporters pointing to its importance for FtsX-specific functions. In agreement, the X-lobe has been reported to be directly implicated in PG hydrolase binding (Yang et al. 2011; Du et al. 2016; Rued et al. 2019; Cook et al. 2020). In all reported 3D structures of the ECD, the upper and lower lobes define a large hydrophobic cleft previously suggested as a strong candidate for the interaction surface between FtsX and its specific PG hydrolase (Mavrici et al. 2014). All of them present hydrophobic residues (mainly Phe and Trp residues) in the distal part of the X-lobe (Fig. 9.5b–d) that would be involved in the interaction with the binding partner. In addition, characterization of *S. pneumoniae* D39-derived strains harboring mutations in the X-lobe showed that this subdomain is essential for cell viability and required for proper cell division of the pneumococcus (Rued et al. 2019).

The Three-Dimensional Structure of Pneumococcal FtsE

Despite the relevant biological roles exerted by the FtsEX complex, no structural information was available for any FtsE until recently, when the high-resolution structure of the pneumococcal FtsE bound to different nucleotides was reported (Alcorlo et al. 2020). Structural analysis revealed that FtsE contains all the conserved structural motifs and functional critical regions observed in NBDs of the ABC transporters that are associated with ATPase activity (reviewed in Locher (2016) and Rees et al. (2009)) (Fig. 9.6a), that is, (*i*) the Walker A motif (also known as the phosphate-binding loop or P-loop), which binds the α- and β-phosphates of the nucleotide; (*ii*) a Mg ion-binding site or the Walker B motif, which includes the catalytic glutamate; (*iii*) the A-loop, which provides an aromatic side chain that packs against the purine ring; (*iv*) the switch region, which contains a histidine residue that stabilizes the transition-state geometry during ATP hydrolysis; (*v*) the α-helical subdomain that contains the signature LSGGE motif, which pins and orients ATP of the dimeric partner during hydrolysis and is the hallmark of the ABC transporter superfamily; (*vi*) the Q-loop (between the Walker A and the signature motifs), named for a conserved glutamine, which interacts with the γ-phosphate through a water bond and provides contacts to the TMD; and finally, (*vii*) the dimerization or D-loop, which is involved in dimerization and helps coupling hydrolysis to transport (reviewed in Rees et al. (2009)). The Walker A, Walker B, A-loop, Q-loop, and switch region form the nucleotide-binding site (Higgins 1992, 2001; Schneider and Hunke 1998).

Overall, the FtsE fold comprises two distinct thick lobes (lobe I and II) grouped together resembling an "L" shape with convex and concave sides, as previously reported for other NBDs belonging to the ABC transporter family (103) (Fig. 9.6a). FtsE lobe I consists of a combination of α- and β-structures and contains the Walker A and Walker B motifs and the key residues for ATP hydrolysis and/or binding

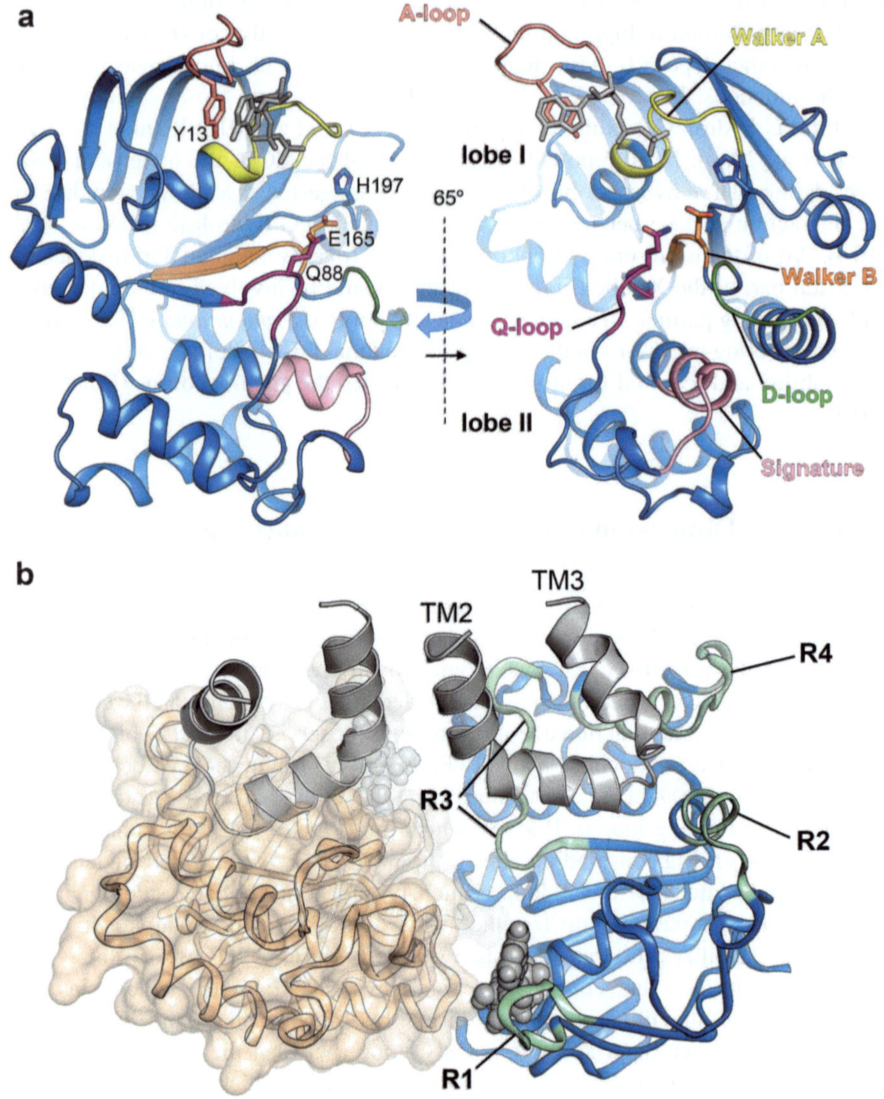

Fig. 9.6 3D structure of pneumococcal FtsE in complex with ADP. (**a**) Ribbon representation of the FtsE monomer structure (PDB 6Z4W). The locations of the conserved and functionally critical motifs present in the FtsE structure are highlighted, including Walker A/B, A-loop, D-loop, Q-loop, and signature motif (colored in yellow/orange, salmon, green, magenta, and pink, respectively). ADP (colored in gray), the catalytic glutamate (E165), the conserved glutamine (Q88) from the Q-loop, the aromatic side chains (Y13) from the A-loop, and switch histidine (H197) are shown as sticks. (**b**) Model for the FtsE dimer in the ADP-bound form including the coupling helices from FtsX (gray carton). FtsE monomers are displayed as cartoon and colored in pale brown and blue, respectively. The molecular surface of the left FtsE monomer is shown. Several regions on the right FtsE monomer, in which the main structural variability is located, are indicated highlighted in pale green (R1, R2, R3, and R4; see main text for details). TM1 and TM2: FtsX transmembrane region 1 and 2, respectively. ADP is depicted in gray spheres

which are located in the P-loop. The ATP-binding pocket is located at the boundary of lobe I. The perpendicular lobe II is only α-helical and contains the signature motif. The hinge between lobe I and lobe II contains both the switch region and the Q-loop, contacting the γ-phosphate of the ATP molecule.

Interestingly, FtsE structures were obtained in two different crystal forms ($P2_1$ and P1, PDB codes 6Z63 and 6Z4W, respectively) (Alcorlo et al. 2020). Their comparison showed that the distance and angle between the two lobes fluctuate, revealing an intramolecular breathing motion within the FtsE monomer. In addition, the FtsE structure presented an unexpectedly high conformational plasticity clustered in four regions (R1–R4) (Fig. 9.6b); three of them (R2–R4) shape a hydrophobic cavity in which the cytosolic region (the coupling helix) of the FtsX partner is inserted (98). These regions are believed to play a relevant role in the transmission from the cytosol through the membrane in order to regulate lytic activity in the periplasm. The residues corresponding to the FtsX coupling helix, responsible for contacting FtsE, were also identified and validated by in vitro binding assays and in vivo mutagenesis studies showing that this interaction is essential for cell growth and proper morphology (Alcorlo et al. 2020). The FtsX coupling helix presents a significant degree of conservation among the FtsX proteins from different organisms and likely corresponds to the unique cytoplasmic loop present in FtsX. This cytoplasmic loop has also been pinpointed as an FtsE-binding site in *E. coli* (Du et al. 2019).

FtsE performs its catalytic activity in vivo as a dimer (Mir et al. 2015). The availability of similar NBD structures in complex with their corresponding TMDs (Okada et al. 2017; Crow et al. 2017) allowed the authors to generate a model for the pneumococcal FtsE dimer in both the ADP- and ATP-bound conformations in complex with the FtsX coupling helix (Alcorlo et al. 2020). In the dimer, each FtsE monomer is predicted to orient its conserved motifs toward the other monomer, thus generating two ATP-binding sites between the P-loops of one monomer and the signature motif of the other. This analysis provided evidence on how the difference between the ATP-/ADP-bound states in FtsE would dramatically alter the interaction of FtsEX with the PG hydrolase PcsB in pneumococcal division. Analysis of the ADP and ATP structures for FtsE revealed that the disposition and conformation of R2–R4 regions was different in each case, thus altering the global shape of the FtsX-binding cavity. Structural changes observed in the backbone or in the side-chain conformation of some residues (such as Y90) in these regions would correlate with the changes between the ADP and ATP conformations in FtsE (Alcorlo et al. 2020). Overall, regions R2 to R4 would play a relevant role in the transmission from the cytosol through the membrane in order to regulate lytic activity in the periplasm.

In the same study, a composite model was presented for the FtsEX complex by superposition of the Porter subdomains in the pneumococcal ECD and in MacB, in both the ATP- and ADP-bound states. Interestingly, depending on the ADP- or ATP-bound state, the helical lobe of ECD changes its orientation completely (Alcorlo et al. 2020). In the model, just a rotation of ECD propelled by the scissorlike action of the transmembrane domains driven by the dimerization of the FtsE would be needed to provide the driving force necessary to regulate the lytic activity in the pneumococcal division (Alcorlo et al. 2020).

Mechanotransmission Model for the Activation of PcsB by the FtsEX Complex

FtsEX is a member of a small subfamily (type VII) of the ABC superfamily that includes the MacB efflux pump and LolCDE lipoprotein trafficking machinery (Crow et al. 2017; Kaplan et al. 2018). These two ABC transporters are closely related to FtsEX, as they have the same transmembrane topology and an ECD located between the first and second membrane-spanning helices. As a result, they have four TM segments and assemble into ABC transporters as dimers with a total of eight TM segments (Dassa and Bouige 2001; Kobayashi et al. 2003; Daley et al. 2005). MacB and LolCDE are not transporters but instead couple cytoplasmic ATP hydrolysis with periplasmic conformational changes to perform their function in the periplasm. This process has been recently referred to as mechanotransmission (Crow et al. 2017; Greene et al. 2018). In the case of FtsEX, the mechanism by which FtsEX's mechanotransmission regulates the activity of its corresponding binding partners remained unknown. However, some recently reported structures have shed light on the molecular mechanism underlying this process.

The ADP- and ATP-bound conformations of MacB of *E. coli* (PDB codes 5GKO and 5LJ6, respectively) have provided relevant insights into the molecular mechanism for the mechanotransmission type VII family of the ABC transporters (Crow et al. 2017). MacB, like FtsX, presents a Porter subdomain on its ECD which is directly connected to the raised helices between TM1 and TM2 in the same way found in FtsX. Comparison of the MacB structures with and without ATP reveals long-range conformational changes in the TM helices, stalk helices, and ECD (Crow et al. 2017). ATP binding induces dimerization of the NBD domains causing the coupling helices to push upwards on the TM2 helices. This push into the rigid periplasmic helical bundle on top of TM1 and TM2 helices causes significant changes in the ECD (Crow et al. 2017). Based on the structural analysis of the structures available for MacB, FtsEX is thought to function by a similar mechanotransmission mechanism. This would require regulation of conformational changes by the ATPase cycle of FtsE in the ECDs of FtsX, which in turn triggers PG hydrolase activity by releasing the catalytic/activator domain from the claws of its regulatory CC domain (Alcorlo et al. 2020).

Recently, the structure of the full-length EnvC of *E. coli* in complex with $FtsX_{ECD}$ has been reported (PDB code 6TPI, Fig. 9.7a) (Cook et al. 2020). The structure of full-length EnvC revealed its different domains: the N-terminal coiled-coil ($EnvC_{CC}$) domain (residues 35–220), a central regulatory domain (the so-called restraining arm, 221–272), and a globular C-terminal LytM domain ($EnvC_{LytM}$, residues 318–413) (Fig. 9.6a, left panel) (Cook et al. 2020). $EnvC_{LytM}$ is degenerated and

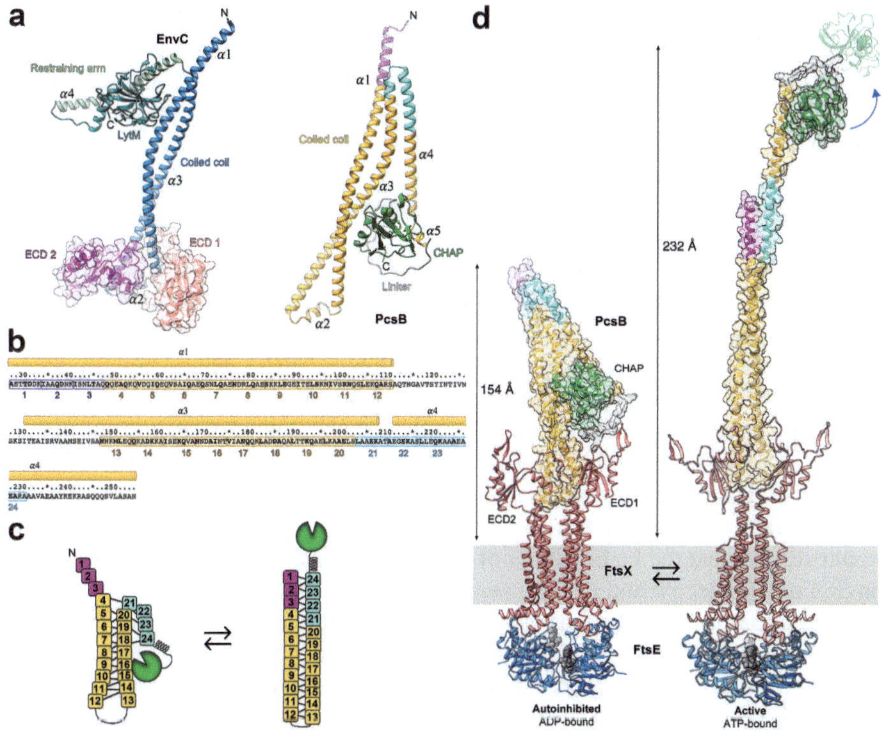

Fig. 9.7 Mechanotransmission model for the activation of PcsB by the FtsEX complex. (**a**) Left, 3D structure EnvC (ribbon) bound to $FtsX_{ECD}$ domains (represented by their molecular surface). Right, model of the inactive PcsB monomer as predicted by AlphaFold 2 (Jumper et al. 2021). The different structural elements are labeled. The structure of PcsB is colored to emphasize the location of unpaired heptads at both N- (purple) and C-termini (cyan). The catalytic Cys residue in PcsB is displayed in spheres. In PcsB, the α4 would play a similar role as that exerted by the restraining arm in EnvC. (**b**) Identification of heptad repeats in the $PcsB_{CC}$ domain. The panel follows the same color scheme of the PcsB structure presented in panel A. The location of 24 heptads is marked with boxes and numbered. Paired heptads are shown with yellow boxes, while those that remain unpaired are marked with purple (N-terminus) and cyan (C-terminus) boxes. (**c**) Schematic representations of the PcsB structure in the autoinhibited and active forms. Heptad repeats are shown as numbered boxes; the color code for the PcsB regions is the same as the one used in panels A and B. (**d**) Hypothetical models for the autoinhibited and active forms of the FtsEX–PcsB complex. The molecular surface of PcsB is shown. The interaction between heptads 1 to 3 and 22 to 24 is shown in the PcsB activated state, according to the previously proposed mechanism by Cook et al. (2020) and the hypothesized compression of the $PcsB_{CC}$ due to FtsEX mechanotransmission. The catalytic Cys in the CHAP domain is shown in spheres. Up to 18 residues from the CHAP and the Ala-rich linker (colored in gray) could move upon activation, providing extra plasticity to reach PG regions far away from the FtsEX–PcsB complex

lacks lytic activity but instead serves as an activator of amidases AmiA and AmiB (Uehara et al. 2010). The 3D structure of $EnvC_{LytM}$ alone was reported previously (PDB code 4bh5) and revealed a groove in its surface in which AmiA and AmiB are likely bound. Residues in this binding site were shown to be critical for amidase activation in vivo and in vitro (Peters 2013). Binding of ATP to FtsE in the cytoplasm would induce mechanotransmission by FtsEX, leading to a conformational change, through the $EnvC_{CC}$, and towards the LytM domain and restraining arm. This conformational change is supposed to cause the release of the LytM domain from the restraining arm, allowing it to bind and activate AmiA and AmiB by displacing amidase's own blocking helix (Yang et al. 2012a). ATP hydrolysis eventually leads the system to reset, returning the LytM domain to the restraining the arm and releasing the amidase.

Surprisingly, the crystal structure of the EnvC–$FtsX_{ECD}$ complex revealed a 1:2 stoichiometry with two distinct FtsX-binding sites located within the antiparallel CC domain of EnvC (Fig. 9.7a). The two interfaces are mostly hydrophobic and dominated by interactions arising from the X-lobe of ECD which partially wraps around each of the two long helices that make up the $EnvC_{CC}$. The same FtsX molecular surface recognizes two structurally distinct target sites, being an example of dual recognition. For this, the flexibility in the X-lobe, a feature observed in all reported ECD structures, is crucial since it must adopt different conformations at each interface to accommodate two different binding surfaces. Thus, this structure would suggest that the FtsEX dimer of dimers (2FtsE+2FtsX) would interact with a single EnvC molecule and not two as might be expected.

In addition, the $EnvC_{CC}$ has a hairpin-like structure arranged in two long antiparallel α-helices that are joined by a short one, sharing the same architecture as the one found in the $PcsB_{CC}$. The amino acid sequence of $EnvC_{CC}$ shows 25 seven-residue "heptad" repeats. The first residue of each heptad is typically a leucine, isoleucine, or valine, and the fourth and fifth positions are most often occupied by either leucine or glutamine (Cook et al. 2020). The heptad repeats fall within the two long α-helices and are important to stabilize, mainly by hydrophobic interactions, the extended conformation of the CC. Noticeably, $EnvC_{CC}$ presents several unpaired heptads found within the N-terminus and the restraining arm. This particular architecture of the EnvC molecule and the preexisting structures of the homologous ABC transporter, MacB, allowed the authors to propose an elegant molecular model explaining the conformational changes at the heart of amidase regulation. In the proposed model, in the ADP-bound or nucleotide-free form of the FtsEX–EnvC complex, EnvC is locked in an inactive resting state in which the LytM domain is tightly bound to the restraining arm, preventing its interaction with PG amidases. Despite a lack of information regarding the precise details on how the arrangement of the $FtsX_{ECD}$ domains provokes release of the restraining arm from the LytM domain, the authors proposed a hypothetical model in which the $FtsX_{ECD}$ domains are squeezed together by a mechanotransmission mechanism similar to that of MacB (Crow et al. 2017). This would provoke the restraining arm (and the preceding linker) to form an extension of the existing CC by packing against the exposed EnvC N-terminus that would allow the release of the LytM domain for amidase binding. In this situation, unpaired heptad repeats within the EnvC N-terminus and within the restraining arm would reunite upon the formation of an extended conformation of the $EnvC_{CC}$ (Cook et al. 2020).

In the pneumococcal case, as described before, the crystal structure of PcsB presented a CC domain arranged in five helices, three long bent helices (α1, α3, and α4) and two short ones (α2 and α5), the alanine-rich (266-AAAPVRAKVRP-279) linker region, and the globular catalytic CHAP domain (Fig. 9.7a, right). The $PcsB_{CC}$ domain acts as long molecular tweezers that hold the globular catalytic CHAP domain in an inactive state (Bartual et al. 2014) with the α4 playing a role similar to that of the restraining arm in the EnvC case (Fig. 9.7a). Interestingly, we identify in $PcsB_{CC}$ a pattern of heptad repeats (Fig. 9.7b) similar to that found in $EnvC_{CC}$. In addition, the initial residues at the beginning of the α1, which were not visible in the crystal structure but present in the construct, also present a heptad pattern that would correspond to three unpaired heptads (heptads 1 to 3). The predicted structure for this N-terminal region by AlphaFold 2 (Jumper et al. 2021) would add 18 residues to the long α1-helix (Fig. 9.7a). The last three heptads present in the $PcsB_{CC}$ are heptads 22 to 24 and are located in the beginning of the α4-helix. In the resting inactive state observed in the PcsB crystal structure, heptads 22 to 24 are packed against heptads 19 to 21 from the C-terminus of the CC (Fig. 9.7c). When PcsB turns into an activated state, by analogy with the mechanism proposed by Cook et al. for EnvC (Cook et al. 2020), the region corresponding to heptads 22 to 24 might dramatically alter its location by reuniting with heptads 1 to 3 (Fig. 9.7c). This would result in an extended conformation for the $PcsB_{CC}$ that would provoke the release of the CHAP domain more than 200 Å away from the FtsEX complex (Fig. 9.7d). If PcsBs were to engage FtsEX as a monomer, as EnvC does, then the α4-helix of PcsB could perform an equivalent function as the restraining arm in the gating access to the C-terminal domain. The CHAP domain could have extra mobility since it is linked to the resulting extended helix through an 18-residue-long linker. Measurements from electron micrographs showed that the thickness of the pneumococcal cross wall is in the range of 170–250 Å (Straume et al. 2020). Consequently, the estimated thickness of the cross wall is consistent with this activation model.

Conclusions and Future Perspectives

The abovementioned mechanism could, in principle, be extrapolated to other microorganisms since many recent studies have extended the role of FtsEX as a regulator of PG hydrolysis in organisms such as *C. crescentus* (Meier et al. 2017; Zielińska et al. 2017), *B. anthracis* (Margulieux et al. 2017), *M. tuberculosis* (Mavrici et al. 2014), *B. subtilis* (Domínguez-Cuevas et al. 2013; Meisner et al. 2013), *Corynebacterium glutamicum* (Lim et al. 2019), and *Fusobacterium nucleatum* (Wu et al. 2018). In most bacteria, including *E. coli*, *N. gonorrhoeae*, *A. hydrophila*, and *F. johnsoniae*, the regulatory role of FtsEX in cell wall hydrolysis is performed at the septum, thus indicating a conserved function in cell division. In these systems, mechanotransmission by FtsEX would drive direct release of the enzymatic domain from an internal autoinhibitory element giving access to cell wall substrates.

The $PcsB_{CC}$ and $EnvC_{CC}$, for which 3D structures are known, share a common core architecture arranged into three long bent helices (α1, α3, and α4), with the first two linked by a short helix (α2) that, in both PcsB and EnvC cases, present high B factor values. Interestingly, while the α1- and α3-helices establish strong hydrophobic interactions through paired heptads, most of them containing bulky hydrophobic residues (Leu or Ile), α4, which also contains heptads in its sequence, does not establish a similar strong interaction with α3. This is due to both (*i*) the presence of a globular domain at the end of α4 (LytM domain in EnvC and CHAP domain in PcsB) and (*ii*) the reduction in the number of bulky hydrophobic residues in α4-heptads that is accompanied by a large increase in the number of Ala residues, thus providing extra plasticity for this region. It is noteworthy that α-helical refolding of Ala-rich regions has been observed in the activation mechanisms of other enzymes such as lipases (Carrasco-López et al. 2009).

In the resting inactivated state of the PG hydrolase (or the adaptor in the case of EnvC), the CC is slightly overwound in comparison to a its ideal geometry, suggesting that it may be under strain (Cook et al. 2020). Thus, when the $FtsX_{ECD}$ domains are squeezed together by the FtsEX mechanotransmission mechanism, they would compress the CC of its corresponding binding partner forcing it to adopt an extended geometry by reuniting the unpaired heptads present in this region. Based on our structural predictions, we speculate that this same CC architecture can be extrapolated to RipC and CwlO whose structures are still unknown. However, the number and presence of unpaired heptads at both the N- and C-termini of the CC core and the composition of the linker region connecting the CC to the specific C-terminal domain depends on the system under consideration. In RipC, the predicted signal peptide (SP, 33 residues long) is followed by an N-terminal CC domain (residues 34–217) and a unique proline-rich linker (compared to EnvC, ClwO, and PcsB) (residues 218–270) connecting CC and the NlpC/P60 peptidase domain. This linker is predicted to include a low complexity region and could be involved in binding to a potential regulatory partner. In addition, RipC has no predicted N-terminal unpaired heptads, which suggests a different regulatory mechanism as compared with the other PG hydrolases discussed above in which the formation of an extended CC is based on the reunion of N- and C-terminal unpaired heptads upon activation.

Also based on our structural predictions, ClwO very likely offers an architecture similar to that presented by EnvC. This would include (*i*) an α4-component with a similar length to the α4/restraining arm of EnvC and (*ii*) the presence of four unpaired N-terminal heptads that may reunite, upon activation, with the unpaired heptads also present in the restraining arm, in the same way as proposed for EnvC (Cook et al. 2020). However, CwlO presents a singular serine-rich linker of approximately 50 residues connecting the CC domain to the NlpC/P60 catalytic domain that is not predicted to be structured. Its role is currently unknown, but the presence of a low-complexity region in this linker suggests that it may act as a potential binding site for a regulatory partner.

How mechanotransmission enables FtsEX to regulate PG hydrolases is a central question whose molecular details still remain unanswered. Much work still needs to be done to understand how the ATPase cycle is triggered and how it is controlled in space and time for precise control of cell wall degradation. In addition, other pending questions are what triggers FtsE's ATPase activity in the cytoplasm, what is the role of the FtsA–FtsX interaction, and how it is produced. It is interesting to note that in some organisms, FtsEX is not essential or only partially essential (Ec or Mt's FtsEX are essential in low-osmolarity media, but cells are less affected in higher-osmolarity conditions).

Finally, considering the conservation among bacterial species and the relevance of this system in cell viability, a deep understanding of cell division regulation by FtsEX could serve as an important base to design new classes of antibiotics targeting this essential machinery.

Acknowledgements We are grateful to Dr. Douglas Laurents for critical reading of the text. This study was funded by the Spanish Ministry of Science and Innovation, which is part of Agencia Estatal de Investigación (AEI), through the project PID2020-115331GB-100.

References

Aarsman MEGG, Piette A, Fraipont C et al (2005) Maturation of the *Escherichia coli* divisome occurs in two steps. Mol Microbiol 55:1631–1645. https://doi.org/10.1111/j.1365-2958.2005.04502.x

Alcorlo M, Martínez-Caballero S, Molina R, Hermoso JA (2017) Carbohydrate recognition and lysis by bacterial peptidoglycan hydrolases. Curr Opin Struct Biol 44:87–100

Alcorlo M, Dik DA, De Benedetti S et al (2019) Structural basis of denuded glycan recognition by SPOR domains in bacterial cell division. Nat Commun 101(10):1–13. https://doi.org/10.1038/s41467-019-13354-4

Alcorlo M, Straume D, Lutkenhaus J et al (2020) Structural characterization of the essential cell division protein Ftse and its interaction with Ftsx in *Streptococcus pneumoniae*. MBio 11:1–20. https://doi.org/10.1128/mBio.01488-20

Anantharaman V, Aravind L (2003) Evolutionary history, structural features and biochemical diversity of the NlpC/P60 superfamily of enzymes. Genome Biol 4:R11. https://doi.org/10.1186/GB-2003-4-2-R11

Angeles DM, Macia-Valero A, Bohorquez LC, Scheffers DJ (2020) The PASTA domains of *Bacillus subtilis* PBP2B strengthen the interaction of PBP2B with DIVIB. Microbiology (United Kingdom) 166:826–836. https://doi.org/10.1099/mic.0.000957

Arends SJR, Kustusch RJ, Weiss DS (2009) ATP-binding site lesions in FtsE impair cell division. J Bacteriol 191:3772–3784. https://doi.org/10.1128/JB.00179-09

Bartual SG, Straume D, Stamsas GA et al (2014) Structural basis of PcsB-mediated cell separation in *Streptococcus pneumoniae*. Nat Commun 5:3842. https://doi.org/10.1038/ncomms4842

Bateman A, Rawlings ND (2003) The CHAP domain: a large family of amidases including GSP amidase and peptidoglycan hydrolases. Trends Biochem Sci 28:234–237

Bernardo-García N, Mahasenan KV, Batuecas MT et al (2018) Allostery, recognition of nascent peptidoglycan, and cross-linking of the cell wall by the essential penicillin-binding protein 2x of *Streptococcus pneumoniae*. ACS Chem Biol 13. https://doi.org/10.1021/acschembio.7b00817

Bernhardt TG, De Boer PAJJ (2003) The *Escherichia coli* amidase AmiC is a periplasmic septal ring component exported via the twin-arginine transport pathway. Mol Microbiol 48:1171–1182. https://doi.org/10.1046/j.1365-2958.2003.03511.x

Bi E, Lutkenhaus J (1991) FtsZ ring structure associated with division in *Escherichia coli*. Nature 354:161–164. https://doi.org/10.1038/354161a0

Briggs NS, Bruce KE, Naskar S et al (2021) The pneumococcal divisome: dynamic control of *Streptococcus pneumoniae* cell division. Front Microbiol 12:2998. https://doi.org/10.3389/fmicb.2021.737396

Brown PJB, de Pedro MA, Kysela DT et al (2012) Polar growth in the alphaproteobacterial order Rhizobiales. Proc Natl Acad Sci 109:1697–1701. https://doi.org/10.1073/PNAS.1114476109

Brunet YR, Wang X, Rudner DZ (2019) SweC and SweD are essential co-factors of the FtsEX-CwlO cell wall hydrolase complex in *Bacillus subtilis*. PLoS Genet 15:e1008296. https://doi.org/10.1371/journal.pgen.1008296

Buddelmeijer N, Beckwith J (2004) A complex of the *Escherichia coli* cell division proteins FtsL, FtsB and FtsQ forms independently of its localization to the septal region. Mol Microbiol 52: 1315–1327. https://doi.org/10.1111/j.1365-2958.2004.04044.x

Carrasco-López C, Godoy C, de las Rivas B, et al (2009) Activation of bacterial thermo alkalophilic lipases is spurred by dramatic structural rearrangements. J Biol Chem 284. https://doi.org/10.1074/jbc.M808268200

Cava F, de Pedro MA (2014) Peptidoglycan plasticity in bacteria: emerging variability of the murein sacculus and their associated biological functions. Curr Opin Microbiol 18:46–53

Chen JC, Beckwith J (2001) FtsQ, FtsL and FtsI require FtsK, but not FtsN, for co-localization with FtsZ during *Escherichia coli* cell division. Mol Microbiol 42:395–413. https://doi.org/10.1046/j.1365-2958.2001.02640.x

Cook J, Baverstock TC, McAndrew MBL et al (2020) Insights into bacterial cell division from a structure of EnvC bound to the FtsX periplasmic domain. Proc Natl Acad Sci USA 117:28355–28365. https://doi.org/10.1073/pnas.2017134117

Corbin BD, Wang Y, Beuria TK, Margolin W (2007) Interaction between cell division proteins FtsE and FtsZ. J Bacteriol 189:3026–3035. https://doi.org/10.1128/JB.01581-06

Crow A, Greene NP, Kaplan E, Koronakis V (2017) Structure and mechanotransmission mechanism of the MacB ABC transporter superfamily. Proc Natl Acad Sci USA 114:12572–12577. https://doi.org/10.1073/pnas.1712153114

Daley DO, Rapp M, Granseth E et al (2005) Biochemistry: global topology analysis of the *Escherichia coli* inner membrane proteome. Science 308:1321–1323. https://doi.org/10.1126/science.1109730

Daniel RA, Noirot-Gros MF, Noirot P, Errington J (2006) Multiple interactions between the transmembrane division proteins of *Bacillus subtilis* and the role of FtsL instability in divisome assembly. J Bacteriol 188:7396–7404. https://doi.org/10.1128/JB.01031-06

Dassa E, Bouige P (2001) The ABC of ABCs: a phylogenetic and functional classification of ABC systems in living organisms. Res Microbiol 152:211–229. https://doi.org/10.1016/S0923-2508(01)01194-9

den Blaauwen T, Hamoen LW, Levin PA (2017) The divisome at 25: the road ahead. Curr Opin Microbiol 36:85–94

Domínguez-Cuevas P, Porcelli I, Daniel RA, Errington J (2013) Differentiated roles for MreB-actin isologues and autolytic enzymes in *Bacillus subtilis* morphogenesis. Mol Microbiol 89:1084–1098. https://doi.org/10.1111/mmi.12335

Du S, Pichoff S, Lutkenhaus J (2016) FtsEX acts on FtsA to regulate divisome assembly and activity. Proc Natl Acad Sci USA 113:E5052–E5061. https://doi.org/10.1073/pnas.1606656113

Du S, Henke W, Pichoff S, Lutkenhaus J (2019) How FtsEX localizes to the Z ring and interacts with FtsA to regulate cell division. Mol Microbiol 112:881–895. https://doi.org/10.1111/mmi.14324

Echenique J, Kadioglu A, Romao S et al (2004) Protein serine/threonine kinase StkP positively controls virulence and competence in *Streptococcus pneumoniae*. Infect Immun 72:2434–2437. https://doi.org/10.1128/IAI.72.4.2434-2437.2004

Figueroa-Cuilan WM, Randich AM, Dunn CM et al (2021) Diversification of LytM protein functions in polar elongation and cell division of *Agrobacterium tumefaciens*. Front Microbiol 0:2341. https://doi.org/10.3389/FMICB.2021.729307

Firczuk M, Mucha A, Bochtler M (2005) Crystal structures of active LytM. J Mol Biol 354:578–590. https://doi.org/10.1016/j.jmb.2005.09.082

Fleurie A, Lesterlin C, Manuse S et al (2014) MapZ marks the division sites and positions FtsZ rings in *Streptococcus pneumoniae*. Nature 5167530(516):259–262. https://doi.org/10.1038/nature13966

Gamba P, Veening JW, Saunders NJ et al (2009) Two-step assembly dynamics of the *Bacillus subtilis* divisome. J Bacteriol 191:4186–4194. https://doi.org/10.1128/JB.01758-08

Garti-Levi S, Hazan R, Kain J et al (2008) The FtsEX ABC transporter directs cellular differentiation in *Bacillus subtilis*. Mol Microbiol 69:1018–1028. https://doi.org/10.1111/j.1365-2958.2008.06340.x

Gerding MA, Ogata Y, Pecora ND et al (2007) The trans-envelope Tol-Pal complex is part of the cell division machinery and required for proper outer-membrane invagination during cell constriction in *E. coli*. Mol Microbiol 63:1008–1025. https://doi.org/10.1111/j.1365-2958.2006.05571.x

Gerding MA, Liu B, Bendezú FO et al (2009) Self-enhanced accumulation of FtsN at division sites and roles for other proteins with a SPOR domain (DamX, DedD, and RlpA) in *Escherichia coli* cell constriction. J Bacteriol 191:7383. https://doi.org/10.1128/JB.00811-09

Giefing C, Meinke AL, Hanner M et al (2008) Discovery of a novel class of highly conserved vaccine antigens using genomic scale antigenic fingerprinting of pneumococcus with human antibodies. J Exp Med 205:117–131. https://doi.org/10.1084/jem.20071168

Giefing-Kroll C, Jelencsics KE, Reipert S et al (2011) Absence of pneumococcal PcsB is associated with overexpression of LysM domain-containing proteins. Microbiology 157:1897–1909. https://doi.org/10.1099/mic.0.045211-0

Glas M, Van Den Berg B, Van Saparoea H, McLaughlin SH et al (2015) The soluble periplasmic domains of *Escherichia coli* cell division proteins FtsQ/FtsB/FtsL form a trimeric complex with submicromolar affinity. J Biol Chem 290:21498–21509. https://doi.org/10.1074/jbc.M115.654756

Gray AN, Egan AJF, van't Veer IL et al (2015) Coordination of peptidoglycan synthesis and outer membrane constriction during *Escherichia coli* cell division. Elife 4. https://doi.org/10.7554/ELIFE.07118

Greene NP, Kaplan E, Crow A, Koronakis V (2018) Antibiotic resistance mediated by the MacB ABC transporter family: a structural and functional perspective. Front Microbiol 9:950. https://doi.org/10.3389/fmicb.2018.00950

Haeusser DP, Margolin W (2016) Splitsville: structural and functional insights into the dynamic bacterial Z ring. Nat Rev Microbiol 14:305–319

Heidrich C, Templin MF, Ursinus A et al (2001) Involvement of N-acetylmuramyl-L-alanine amidases in cell separation and antibiotic-induced autolysis of *Escherichia coli*. Mol Microbiol 41:167–178. https://doi.org/10.1046/j.1365-2958.2001.02499.x

Higgins CF (1992) ABC transporters: from microorganisms to man. Annu Rev Cell Biol 8:67–113. https://doi.org/10.1146/ANNUREV.CB.08.110192.000435

Higgins CF (2001) ABC transporters: physiology, structure and mechanism – an overview. Res Microbiol 152:205–210. https://doi.org/10.1016/S0923-2508(01)01193-7

Jumper J, Evans R, Pritzel A et al (2021) Highly accurate protein structure prediction with AlphaFold. Nature 596:583–589. https://doi.org/10.1038/s41586-021-03819-2

Kaplan E, Greene NP, Crow A, Koronakis V (2018) Insights into bacterial lipoprotein trafficking from a structure of LolA bound to the LolC periplasmic domain. Proc Natl Acad Sci 115:E7389–E7397. https://doi.org/10.1073/PNAS.1806822115

Kempf MJ, McBride MJ (2000) Transposon insertions in the *Flavobacterium johnsoniae ftsX* gene disrupt gliding motility and cell division. J Bacteriol 182:1671–1679. https://doi.org/10.1128/jb.182.6.1671-1679.2000

Khadria AS, Senes A (2013) The transmembrane domains of the bacterial cell division proteins FtsB and FtsL form a stable high-order oligomer. Biochemistry 52:7542–7550. https://doi.org/10.1021/BI4009837

Kobayashi N, Nishino K, Hirata T, Yamaguchi A (2003) Membrane topology of ABC-type macrolide antibiotic exporter MacB in *Escherichia coli*. FEBS Lett 546:241–246. https://doi.org/10.1016/S0014-5793(03)00579-9

Korndörfer IP, Danzer J, Schmelcher M et al (2006) The crystal structure of the bacteriophage PSA endolysin reveals a unique fold responsible for specific recognition of *Listeria* cell walls. J Mol Biol 364:678–689. https://doi.org/10.1016/j.jmb.2006.08.069

Lara B, Rico AI, Petruzzelli S et al (2005) Cell division in cocci: localization and properties of the *Streptococcus pneumoniae* FtsA protein. Mol Microbiol 55:699–711. https://doi.org/10.1111/J.1365-2958.2004.04432.X

Leisch N, Verheul J, Heindl NR et al (2012) Growth in width and FtsZ ring longitudinal positioning in a gammaproteobacterial symbiont. Curr Biol 22:R831–R832

Li Q, Li Z, Li X et al (2018) FtsEX-CwlO regulates biofilm formation by a plant-beneficial rhizobacterium *Bacillus velezensis* SQR9. Res Microbiol 169:166–176. https://doi.org/10.1016/j.resmic.2018.01.004

Lim HC, Sher JW, Rodriguez-Rivera FP et al (2019) Identification of new components of the RipC-FtsEX cell separation pathway of Corynebacterineae. PLoS Genet 15:e1008284. https://doi.org/10.1371/journal.pgen.1008284

Liu Y, Eisenberg D (2002) 3D domain swapping: as domains continue to swap. Protein Sci 11:1285–1299. https://doi.org/10.1110/ps.0201402

Locher KP (2016) Mechanistic diversity in ATP-binding cassette (ABC) transporters. Nat Struct Mol Biol 23:487–493. https://doi.org/10.1038/nsmb.3216

Manuse S, Fleurie A, Zucchini L et al (2016) Role of eukaryotic-like serine/threonine kinases in bacterial cell division and morphogenesis. FEMS Microbiol Rev 40:41–56. https://doi.org/10.1093/FEMSRE/FUV041

Margulieux KR, Liebov BK, Tirumala VSKKS et al (2017) Bacillus anthracis peptidoglycan integrity is disrupted by the chemokine CXCL10 through the FtsE/X complex. Front Microbiol 8. https://doi.org/10.3389/fmicb.2017.00740

Martínez-Caballero S, Mahasenan KV, Kim C et al (2021) Integrative structural biology of the penicillin-binding protein-1 from *Staphylococcus aureus*, an essential component of the divisome machinery. Comput Struct Biotechnol J 19:5392–5405. https://doi.org/10.1016/J.CSBJ.2021.09.018

Massidda O, Novakova L, Vollmer W et al (2013) From models to pathogens: how much have we learned about *Streptococcus pneumoniae* cell division? Environ Microbiol 15:3133–3157. https://doi.org/10.1111/1462-2920.12189

Mavrici D, Marakalala MJ, Holton JM, et al (2014) *Mycobacterium tuberculosis* FtsX extracellular domain activates the peptidoglycan hydrolase, RipC. Proc Natl Acad Sci USA 111:8037–8042. https://doi.org/10.1073/pnas.1321812111

Meeske AJ, Riley EP, Robins WP et al (2016) SEDS proteins are a widespread family of bacterial cell wall polymerases. Nature 5377622(537):634–638. https://doi.org/10.1038/nature19331

Meier EL, Daitch AK, Yao Q et al (2017) FtsEX-mediated regulation of the final stages of cell division reveals morphogenetic plasticity in *Caulobacter crescentus*. PLoS Genet 13:e1006999. https://doi.org/10.1371/journal.pgen.1006999

Meisner J, Montero Llopis P, Sham LT et al (2013) FtsEX is required for CwlO peptidoglycan hydrolase activity during cell wall elongation in *Bacillus subtilis*. Mol Microbiol 89:1069–1083. https://doi.org/10.1111/mmi.12330

Merino S, Altarriba M, Gavin R et al (2001) The cell division genes (*ftsE* and *X*) of *Aeromonas hydrophila* and their relationship with opsonophagocytosis. FEMS Microbiol Lett 198:183–188. https://doi.org/10.1111/j.1574-6968.2001.tb10640.x

Mir MA, Arumugam M, Mondal S et al (2015) *Mycobacterium tuberculosis* cell division protein, FtsE, is an ATPase in dimeric form. Protein J 34:35–47. https://doi.org/10.1007/s10930-014-9593-7

Mistry J, Chuguransky S, Williams L et al (2021) Pfam: the protein families database in 2021. Nucleic Acids Res 49:D412–D419. https://doi.org/10.1093/NAR/GKAA913

Möll A, Schlimpert S, Briegel A et al (2010) DipM, a new factor required for peptidoglycan remodeling during cell division in *Caulobacter crescentus*. Mol Microbiol 77:90. https://doi.org/10.1111/J.1365-2958.2010.07224.X

Monahan LG, Liew ATF, Bottomley AL, Harry EJ (2014) Division site positioning in bacteria: one size does not fit all. Front Microbiol 5:19

Ng WL, Robertson GT, Kazmierczak KM et al (2003) Constitutive expression of PcsB suppresses the requirement for the essential VicR (YycF) response regulator in *Streptococcus pneumoniae* R6. Mol Microbiol 50:1647–1663. https://doi.org/10.1046/J.1365-2958.2003.03806.X

Ng WL, Kazmierczak KM, Winkler ME (2004) Defective cell wall synthesis in *Streptococcus pneumoniae* R6 depleted for the essential PcsB putative murein hydrolase or the VicR (YycF) response regulator. Mol Microbiol 53:1161–1175. https://doi.org/10.1111/j.1365-2958.2004.04196.x

Nováková L, Sasková L, Pallová P et al (2005) Characterization of a eukaryotic type serine/threonine protein kinase and protein phosphatase of *Streptococcus pneumoniae* and identification of kinase substrates. FEBS J 272:1243–1254. https://doi.org/10.1111/j.1742-4658.2005.04560.x

Okada U, Yamashita E, Neuberger A et al (2017) Crystal structure of tripartite-type ABC transporter MacB from *Acinetobacter baumannii*. Nat Commun 8:1336. https://doi.org/10.1038/s41467-017-01399-2

Perez AJ, Bazan Villicana J, Tsui H-CT et al (2021) FtsZ-ring regulation and cell division are mediated by essential EzrA and accessory proteins ZapA and ZapJ in *Streptococcus pneumoniae*. Front Microbiol. https://doi.org/10.3389/FMICB.2021.780864

Peters NT (2013) Structure-function analysis of the LytM domain of EnvC, an activator of cell wall remodelling at the *Escherichia coli* division site. Mol Microbiol 89:690–701. https://doi.org/10.1111/mmi.12304

Pichoff S, Du S, Lutkenhaus J (2019) Roles of FtsEX in cell division. Res Microbiol 170:374–380. https://doi.org/10.1016/j.resmic.2019.07.003

Pogliano J, Pogliano K, Weiss DS et al (1997) Inactivation of FtsI inhibits constriction of the FtsZ cytokinetic ring and delays the assembly of FtsZ rings at potential division sites. Proc Natl Acad Sci 94:559–564. https://doi.org/10.1073/PNAS.94.2.559

Ramirez-Arcos S, Salimnia H, Bergevin I et al (2001) Expression of *Neisseria gonorrhoeae* cell division genes *ftsZ*, *ftsE* and *minD* is influenced by environmental conditions. Res Microbiol 152:781–791. https://doi.org/10.1016/S0923-2508(01)01261-X

Rees DC, Johnson E, Lewinson O (2009) ABC transporters: the power to change. Nat Rev Mol Cell Biol 10:218–227. https://doi.org/10.1038/nrm2646

Robson SA, Michie KA, Mackay JP et al (2002) The *Bacillus subtilis* cell division proteins FtsL and DivIC are intrinsically unstable and do not interact with one another in the absence of other septasomal components. Mol Microbiol 44:663–674. https://doi.org/10.1046/j.1365-2958.2002.02920.x

Rocaboy M (2013) Structural and biochemical study of the proteins AmiC, NlpD and FtsW involved in the bacterial cell division. Thesis

Rocaboy M, Herman R, Sauvage E et al (2013) The crystal structure of the cell division amidase amic reveals the fold of the AMIN domain, a new peptidoglycan binding domain. Mol Microbiol 90:267–277. https://doi.org/10.1111/mmi.12361

Rowlett VW, Margolin W (2015) The Min system and other nucleoid-independent regulators of Z ring positioning. Front Microbiol 6. https://doi.org/10.3389/FMICB.2015.00478

Rued BE, Alcorlo M, Edmonds KA et al (2019) Structure of the large extracellular loop of FtsX and its interaction with the essential peptidoglycan hydrolase PcsB in *Streptococcus pneumoniae*. MBio 10. https://doi.org/10.1128/mBio.02622-18

Ryan Arends SJ, Williams K, Scott RJ et al (2010) Discovery and characterization of three new *Escherichia coli* septal ring proteins that contain a SPOR domain: DamX, DedD, and RlpA. J Bacteriol 192:242–255. https://doi.org/10.1128/JB.01244-09

Salmond GPC, Plakidou S et al (1984) Genetic analysis of essential genes in the *ftsE* region of the *Escherichia coli* genetic map and identification of a new cell division gene, *ftsS*. Mol Gen Genet 197:304–308. https://doi.org/10.1007/BF00330978

Saurin W, Hofnung M, Dassa E (1999) Getting in or out: early segregation between importers and exporters in the evolution of ATP-binding cassette (ABC) transporters. J Mol Evol 48:22–41. https://doi.org/10.1007/PL00006442

Schmidt KL, Peterson ND, Kustusch RJ et al (2004) A predicted ABC transporter, FtsEX, is needed for cell division in *Escherichia coli*. J Bacteriol 186:785–793

Schneider E, Hunke S (1998) ATP-binding-cassette (ABC) transport systems: functional and structural aspects of the ATP-hydrolyzing subunits/domains. FEMS Microbiol Rev 22:1–20. https://doi.org/10.1111/j.1574-6976.1998.tb00358.x

Serrano CKG, Winkle M, Martorana AM et al (2021) ActS activates peptidoglycan amidases during outer membrane stress in *Escherichia coli*. Mol Microbiol 116:329–342. https://doi.org/10.1111/MMI.14712

Sham LT, Barendt SM, Kopecky KE, Winkler ME (2011) Essential PcsB putative peptidoglycan hydrolase interacts with the essential FtsXSpn cell division protein in *Streptococcus pneumoniae* D39. Proc Natl Acad Sci USA 108:E1061–E1069. https://doi.org/10.1073/pnas.1108323108

Sham LTT, Jensen KR, Bruce KE, Winkler ME (2013) Involvement of FtsE ATPase and FtsX extracellular loops 1 and 2 in FtsEX-PcsB complex function in cell division of *Streptococcus pneumoniae* D39. MBio 4:1–9. https://doi.org/10.1128/mBio.00431-13

Smith TJ, Foster SJ (1995) Characterization of the involvement of two compensatory autolysins in mother cell lysis during sporulation of *Bacillus subtilis* 168. J Bacteriol 177:3855–3862. https://doi.org/10.1128/JB.177.13.3855-3862.1995

Straume D, Piechowiak KW, Olsen S et al (2020) Class A PBPs have a distinct and unique role in the construction of the pneumococcal cell wall. Proc Natl Acad Sci USA 117:6129–6138. https://doi.org/10.1073/pnas.1917820117

Sycuro LK, Pincus Z, Gutierrez KD et al (2010) Peptidoglycan crosslinking relaxation promotes *Helicobacter pylori*'s helical shape and stomach colonization. Cell 141:822–833. https://doi.org/10.1016/j.cell.2010.03.046

Typas A, Banzhaf M, Gross CA, Vollmer W (2011) From the regulation of peptidoglycan synthesis to bacterial growth and morphology. Nat Rev Microbiol 102(10):123–136. https://doi.org/10.1038/nrmicro2677

Uehara T, Parzych KR, Dinh T, Bernhardt TG (2010) Daughter cell separation is controlled by cytokinetic ring-activated cell wall hydrolysis. EMBO J 29:1412–1422. https://doi.org/10.1038/EMBOJ.2010.36

van den Saparoea HB, Glas M, IGWH V et al (2013) Fine-mapping the contact sites of the *Escherichia coli* cell division proteins FtsB and FtsL on the FtsQ protein. American Society for Biochemistry and Molecular Biology, Rockville

Verdonk ML, Cole JC, Hartshorn MJ et al (2003) Improved protein–ligand docking using GOLD. Proteins 52:609–623. https://doi.org/10.1002/prot.10465

Vollmer W (2012) Bacterial growth does require peptidoglycan hydrolases. Mol Microbiol 86:1031–1035. https://doi.org/10.1111/mmi.12059

Vollmer W, Joris B, Charlier P, Foster S (2008) Bacterial peptidoglycan (murein) hydrolases. FEMS Microbiol Rev 32:259–286. https://doi.org/10.1111/j.1574-6976.2007.00099.x

Vollmer W, Massidda O, Tomasz A (2019) The cell wall of *Streptococcus pneumoniae*. Microbiol Spectr 7. https://doi.org/10.1128/microbiolspec.gpp3-0018-2018

Wang L, Khattar MK, Donachie WD, Lutkenhaus J (1998) FtsI and FtsW are localized to the septum in *Escherichia coli*. J Bacteriol 180:2810–2816. https://doi.org/10.1128/jb.180.11.2810-2816.1998

Wu LJ, Errington J (2011) Nucleoid occlusion and bacterial cell division. Nat Rev Microbiol 101(10):8–12. https://doi.org/10.1038/nrmicro2671

Wu C, Al Mamun AAM, Luong TT et al (2018) Forward genetic dissection of biofilm development by *Fusobacterium nucleatum*: novel functions of cell division proteins FtsX and EnvC. MBio 9. https://doi.org/10.1128/MBIO.00360-18

Yahashiri A, Jorgenson MA, Weiss DS (2017) The SPOR domain, a widely conserved peptidoglycan binding domain that targets proteins to the site of cell division. J Bacteriol 199:e00118–e00117. https://doi.org/10.1128/JB.00118-17

Yang DC, Peters NT, Parzych KR et al (2011) An ATP-binding cassette transporter-like complex governs cell-wall hydrolysis at the bacterial cytokinetic ring. Proc Natl Acad Sci USA 108: E1052–E1060. https://doi.org/10.1073/pnas.1107780108

Yang DC, Tan K, Joachimiak A, Bernhardt TG (2012a) A conformational switch controls cell wall-remodelling enzymes required for bacterial cell division. Mol Microbiol 85:768–781. https://doi.org/10.1111/j.1365-2958.2012.08138.x

Yang DC, Tan K, Joachimiak A, Bernhardt TG (2012b) A conformational switch controls cell wall remodeling enzymes required for bacterial cell division. Mol Microbiol 85:768. https://doi.org/10.1111/J.1365-2958.2012.08138.X

Yao Q, Jewett AI, Chang Y et al (2017) Short FtsZ filaments can drive asymmetric cell envelope constriction at the onset of bacterial cytokinesis. EMBO J 36(1577–1589):10.15252/embj.201696235

Zeng Y, Charkowski AO (2021) The role of ATP-binding cassette transporters in bacterial phytopathogenesis. Phytopathology 111:600–610. https://doi.org/10.1094/PHYTO-06-20-0212-RVW

Zielińska A, Billini M, Möll A et al (2017) LytM factors affect the recruitment of autolysins to the cell division site in *Caulobacter crescentus*. Mol Microbiol 106:419–438. https://doi.org/10.1111/MMI.13775

Zucchini L, Mercy C, Garcia PS et al (2017) PASTA repeats of the protein kinase StkP interconnect cell constriction and separation of *Streptococcus pneumoniae*. Nat Microbiol 32(3):197–209. https://doi.org/10.1038/s41564-017-0069-3

Chapter 10
Structure, Function, and Regulation of the Kainate Receptor

Surbhi Dhingra, Juhi Yadav, and Janesh Kumar

Abstract Neural communication and modulation are complex processes. Ionotropic glutamate receptors (iGluRs) significantly contribute to mediating the fast-excitatory branch of neurotransmission in the mammalian brain. Kainate receptors (KARs), a subfamily of the iGluRs, act as modulators of the neuronal circuitry by playing important roles at both the post- and presynaptic sites of specific neurons. The functional tetrameric receptors are formed by two different gene families, low agonist affinity (GluK1-GluK3) and high agonist affinity (GluK4-GluK5) subunits. These receptors garnered attention in the past three decades, and since then, much work has been done to understand their localization, interactome, physiological functions, and regulation. Cloning of the receptor subunits (GluK1-GluK5) in the early 1990s led to recombinant expression of kainate receptors in heterologous systems. This facilitated understanding of the functional differences between subunit combinations, splice variants, trafficking, and drug discovery. Structural studies of individual domains and recent full-length homomeric and heteromeric kainate receptors have revealed unique functional mechanisms, which have answered several long-standing questions in the field of kainate receptor biology. In this chapter, we review the current understanding of kainate receptors and associated disorders.

Keywords Kainate · Electrophysiology · Cryo-EM · Glutamate · Synaptic transmission · Synaptic plasticity

---

Surbhi Dhingra and Juhi Yadav contributed equally with all other contributors.

S. Dhingra · J. Yadav · J. Kumar (✉)
Laboratory of Membrane Protein Biology, National Centre for Cell Science, NCCS Complex, S. P. Pune University, Pune, Maharashtra, India
e-mail: janesh@nccs.res.in

J. R. Harris, J. Marles-Wright (eds.), *Macromolecular Protein Complexes IV*,
Subcellular Biochemistry 99, https://doi.org/10.1007/978-3-031-00793-4_10

Introduction

Glutamate is the major excitatory neurotransmitter in the vertebrate brain. Its effects are mediated through ionotropic and metabotropic glutamate receptors (iGluRs and mGluRs, respectively). iGluRs form cation-selective pore that allow the passage of ions like Na^+, K^+, and Ca^{2+} resulting in rapid depolarization of postsynaptic membranes leading to signaling events necessary for memory formation, synaptic plasticity, and neuronal development (Fig. 10.1a). The iGluRs are divided into four subfamilies based on sequence similarities and preferred pharmacological agonists, namely, NMDA (N-methyl-D-aspartate), AMPA (α-amino-3-hydroxy-5-methyl-4-isoxazolepropionate), KA (kainate), and delta receptors. A typical receptor subunit is formed of modular domains, viz., the amino-terminal domain (ATD), the ligand-binding domain (LBD), the transmembrane domain (TMD), and the C-terminal domain (CTD) (Traynelis et al. 2010) (Fig. 10.1b). Functional iGluRs assemble as tetramers in homomeric or heteromeric configuration with subunits from within subfamilies.

Of all the iGluRs, the study of kainate receptors (KARs) lagged behind AMPARs and NMDARs due to the lack of specific pharmacological tools that isolated them from AMPARs. The discovery of GYKI53655, which selectively antagonizes AMPARs, aided in isolating KAR responses in the regions where their expression overlapped, facilitating study of KAR functions (Fig. 10.1c) (Paternain et al. 1995; Wilding and Huettner 1995). KARs play the role of modulators of neurotransmission and neuronal development in the mammalian brain as they are present on both the pre- and postsynaptic sites on the neurons. They perform the classical role of ligand-gated ion channels on the postsynaptic neuronal compartments. In addition, they are also known to interact via the non-canonical G-protein coupled metabotropic pathway on the presynaptic sites and fine-tune the secretion of excitatory and inhibitory neurotransmitters [as reviewed in (Lerma 2003; Huettner 2003; Pinheiro and Mulle 2006)]. This modulatory behavior of neuronal circuits makes KARs an excellent choice for therapeutic targeting and has been used extensively to study antagonists to reduce deteriorating effects of several neuropathies like epilepsy, migraine, and pain (Jane et al. 2009; Lerma and Marques 2013). In this chapter, we review the advances in the field that led to a better understanding of the physiology of kainate receptors.

Genomic Localization and Gene Expression Profiling of KARs

Kainate receptors are formed by two subunit families based on their affinities for the ligand, low-affinity (GluK1-GluK3) and high-affinity (GluK4-GluK5). The low-affinity subunits can form homomeric or facultative heteromeric channels with the low- or high-affinity subunits and reach the synapse. In contrast, the high-affinity subunits are obligate heteromers that require low-affinity subunits to reach the

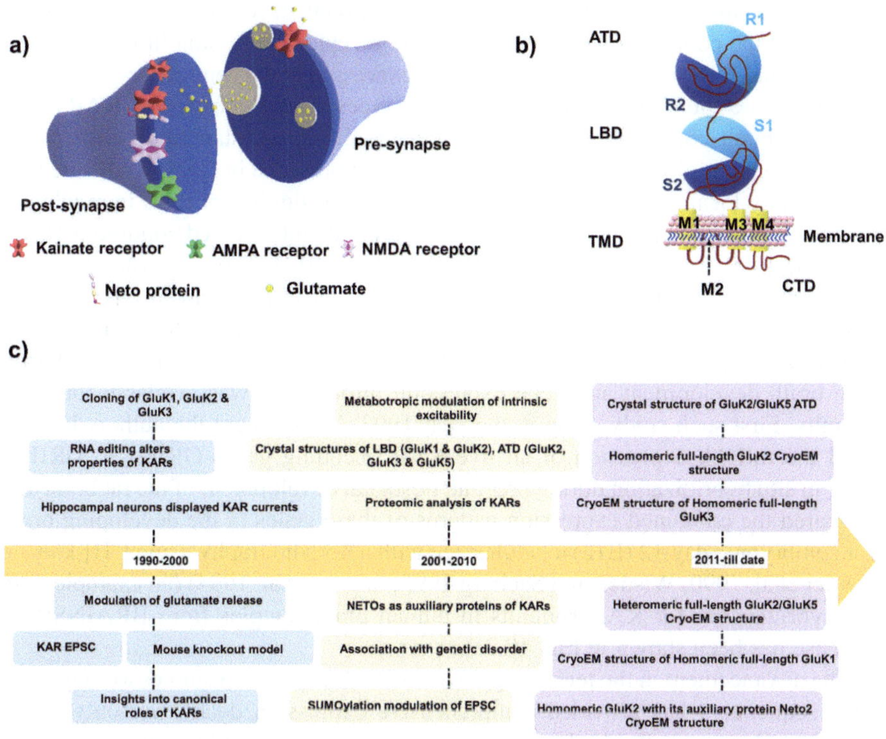

Fig. 10.1 History of KARs. (**a**) Cartoon representation of localization of KARs, pre- and postsynapse. As depicted by cartoon, KARs are present at both post- and presynaptic neurons; (**b**) membrane topology of KARs illustrating the three transmembrane domains (M1, M3, and M4) and the pore loop (M2), the extracellular amino-terminal domain (ATD), ligand-binding domain (LBD), and the intracellular carboxy-terminus domain (CTD). R1 and R2 represent sub-domains in the ATD, whereas S1 and S2 are modular lobes present in the LBD. (**c**) Timeline of discoveries in KAR neurobiology, starting from the cloning of the first KAR in the 1990s to the discovery of heteromeric KARs and homomeric GluK2 with its auxiliary protein structure in 2021, have given great insights into the receptors' biophysical and physiological properties (Bettler et al. 1990; Sommer et al. 1991; Lerma et al. 1993; Chittajallu et al. 1996; Castillo et al. 1997; Vignes and Collingridge 1997; Mulle et al. 1998; Bortolotto et al. 1999; Contractor et al. 2001; Melyan et al. 2002; Delorme et al. 2004; Nanao et al. 2005; Naur et al. 2005; Mayer 2005; Coussen et al. 2005; Martin et al. 2007; Dutta et al. 2007; Zhang et al. 2009; Kumar et al. 2009, 2011; Kumar and Mayer 2010; Meyerson et al. 2014; Kumari et al. 2019; Khanra et al. 2021; Selvakumar et al. 2021; He et al. 2021)

neuronal surface. In the early 1990s, based on Southern analysis of DNAs from different origin and fluorescence in situ hybridization, human GRIK1-GRIK5 were mapped on chromosomes 21q21.3 (Eubanks et al. 1993; Barbon and Barlati 2000), 6q16.3 (Paschen et al. 1994; Barbon et al. 2001), 1p34.3 (Puranam et al. 1993), 11q23.3, and 19q13.2 (Szpirer et al. 1994), respectively (corresponding rat gene locations: 11q11, 20q13, 5q36, 8q22, and 1q21, respectively).

Over the last few decades, a tremendous amount of work has been done to understand the control of ionotropic glutamate receptor expression in neuronal and non-neuronal cells. KARs are predominately expressed in the central nervous system (CNS) and are found in various neuronal cell types. The expression of KARs with various subunit composition and splice variants differs substantially between cell types and brain regions. Apart from their prevalent presence in CNS, KARs are also differentially localized in other body tissues and play significant roles in regulating the body's physiological activities in various normal and diseased conditions (Kidd and Isaac 1999; Tashiro et al. 2003; Stepulak et al. 2014). The expression levels of KARs are integrated through numerous environmental stimuli, and specific subunits expressed on particular cells are strong determinants of synaptic phenotypes.

The Bettler group demonstrated that GRIK1 shows a regulated expression during mice brain development and gene expression information was available for other subunits. First such study came in the year 1993–1994, where the authors investigated the mRNA expression of all five KAR encoding genes-GRIK1 to GRIK5, using in situ hybridization during pre- and postnatal development of the rat brain and compared the combined expression patterns of these genes in the developing brain from embryonic day 12 (E12) to adult stage with autoradiography using [^{3}H]-kainate (Bettler et al. 1990; Wisden and Seeburg 1993; Szpirer et al. 1994) (an example of in situ hybridization for KAR subunits in human brain samples from BRAINSPAN database has been shown in Fig. 10.2a).

The advancement in the technology from mRNA expression studies to single-cell RNA sequencing has provided comprehensive expression details not only in the brain but also in all the body tissues. A compilation of publicly available RNA sequencing (RNA-seq) data generated by GTEx consortium of KARs in different tissues and cell types of the human body and transcriptomics data available on BRAINSPAN consortium is shown in Fig. 10.2b and c. GTEx RNA-seq data revealed the abundance of GRIK2 and GRIK5 in the cerebellar hemisphere and cerebellum regions of the brain, which is also supported by various in vitro and in vivo experiments (Bahn et al. 1994). A high level of GRIK5 expression was observed in different organs like the testis, pituitary, colon, sigmoid, thyroid, tibial artery, uterus, etc., whereas GRIK1, GRIK3, and GRIK4 show prevalent expression in the brain tissues (Fig. 10.2b). Analysis of BRAINSPAN transcriptomics data gives an insight into the expression of various KAR subtypes in different regions of the brain with respect to age (Fig. 10.2c). Recently, various databases were analyzed to provide an exhaustive compilation of splice variants and RNA-edited versions of all the iGluRs (including KARs) present in the human brain. The study also identified the presence of two new splice variants in KARs: GRIK4–2 and GRIK5-b (Herbrechter et al. 2021), the functional importance of which remains to be determined. A schematic of all the splice variants of KARs is shown in Fig. 10.3a.

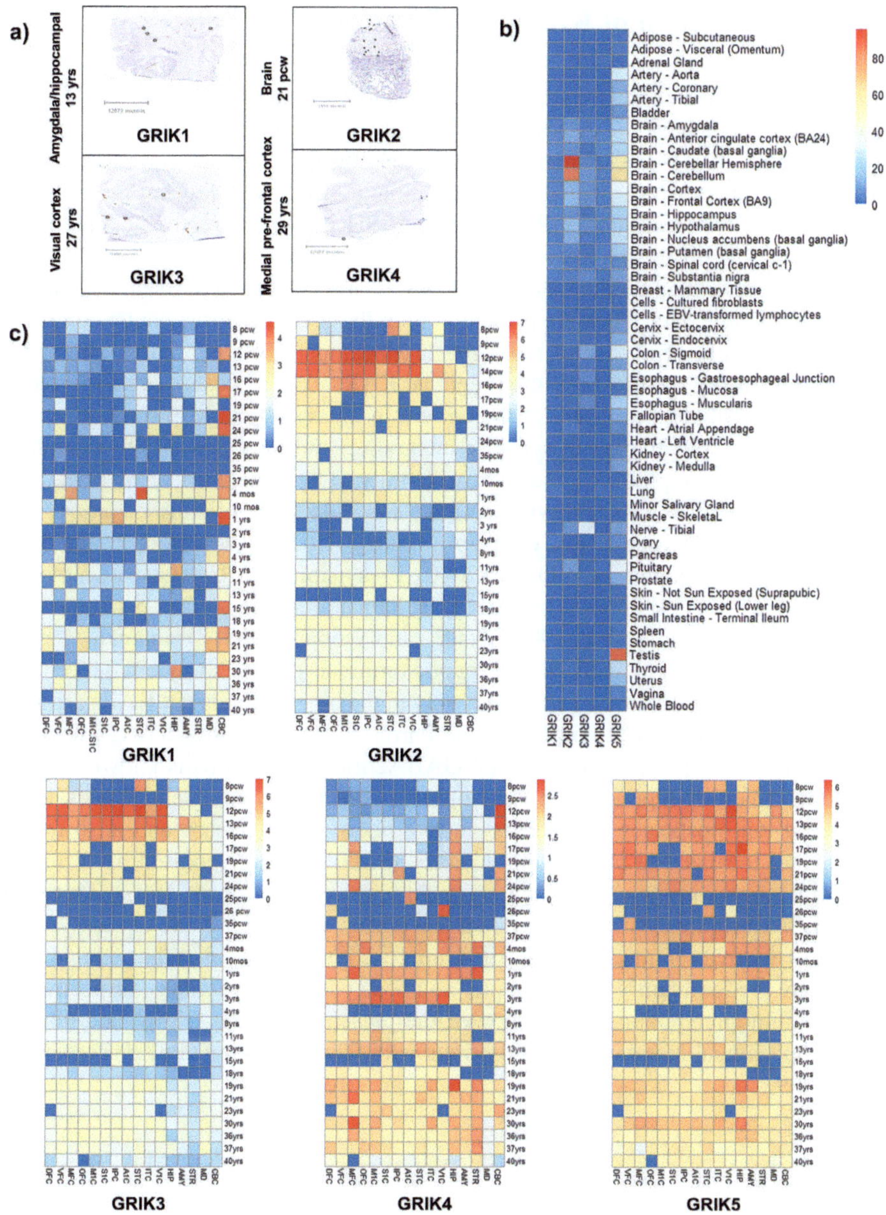

Fig. 10.2 Regional gene-level expression of KARs. (**a**) In situ hybridization images of KARs in different regions of the brain obtained from the BRAINSPAN database. (**b**) Publicly available data from GTEx was taken and plotted as a heatmap depicting the abundance of KAR subunits (GRIK1-GRIK5) in different body tissues. Data revealed the abundance of GRIK5 throughout the body. (**c**) Open access data available at BRAINSPAN was used to map KARs expression in various brain regions at different developmental stages of humans. The regions of brain are abbreviated as the following: DFC (dorsolateral prefrontal cortex), VFC (ventrolateral prefrontal cortex), MFC (anterior [rostral] cingulate [medial prefrontal] cortex), OFC (orbital frontal cortex), M1C (primary motor cortex area M1, area 4), S1C (primary somatosensory cortex (area S1, areas 3,1,2), IPC

Spatial Expression of KARs

Although RNA-seq expression data provides sensitive, quantitative transcript information, these data have the disadvantage of analyzing a mixture of cell types. In situ hybridization (ISH) and immunofluorescence studies have an advantage over RNA-seq as it provides spatial expression on a single-cell level. Figure 10.2a highlights the localization of GRIK1–4 using ISH in various regions of the brain from samples of different age groups.

KAR expression is ubiquitous in the CNS as observed by RNA-seq analysis of publicly available data. They are present at both the pre- and postsynaptic compartments (Chittajallu et al. 1996), with the hippocampal CA3 pyramidal and CA1 interneurons being the most explored regions. KARs exist as postsynaptic receptors and mediate a small component of the current in mossy fibers (MF)-CA3 synapses that make their role evident in synaptic integration and their capacity to modulate synaptic activity, emphasizing that the excitability of brain networks may be tightly controlled through KARs. Presynaptic roles for KARs are established in several regions of the brain via synaptic electrophysiology (Rodríguez-Moreno et al. 1997; Clarke et al. 1997; Kamiya and Ozawa 2000). Presynaptic KARs presence on mossy fiber terminals was established by early [^{3}H] kainate-binding studies performed on rat hippocampus (Bettler et al. 1990; Wisden and Seeburg 1993; Szpirer et al. 1994).

KARs are targeted to axons at presynaptic terminals onto CA3 pyramidal cells in the hippocampus and other neurons. Not only in the hippocampus but KARs are also found in thalamocortical terminals in the barrel cortex and amygdala (Daw et al. 2007). Also, their presence in neonatal GABAergic interneurons is known to play important roles in hippocampal development. Electrophysiological studies have suggested that postsynaptic KARs at mossy fiber-CA3 synapses and at different regions are localized at postsynaptic density (PSD); this is supported by different biochemical and immunogold labeling techniques (Garcia et al. 1998). Presence of GluK2/GluK5 heteromer is reported at the postsynaptic membrane of mossy fiber synapses, whereas immunogold electron microscopy has revealed that GluK4 and GluK5 are predominantly located at pre- and postsynaptic membranes, respectively. A co-immunoprecipitation study showed the presence of both GluK4 and GluK5 from hippocampal membrane extracts along with GluK2, but not GluK3 (Fernandes et al. 2009).

Fig. 10.2 (continued) (posteroventral [inferior] parietal cortex), A1C (primary auditory cortex core), STC ([posterior caudal] superior temporal cortex, area 22c), ITC (inferolateral temporal cortex, area TEv, area 20), V1C (primary visual cortex, striate cortex, area V1/17), HIP (hippocampus), AMY (amygdaloid complex), STR (striatum), MD (mediodorsal nucleus of thalamus), CBC (cerebellar cortex)

a)

Gene	GRIK1	GRIK2	GRIK3	GRIK4	GRIK5
Locus (human)	21q21.3	6q16.3	1p34.3	11q23.3	19q13.2
Exons	17	16	16	21	20

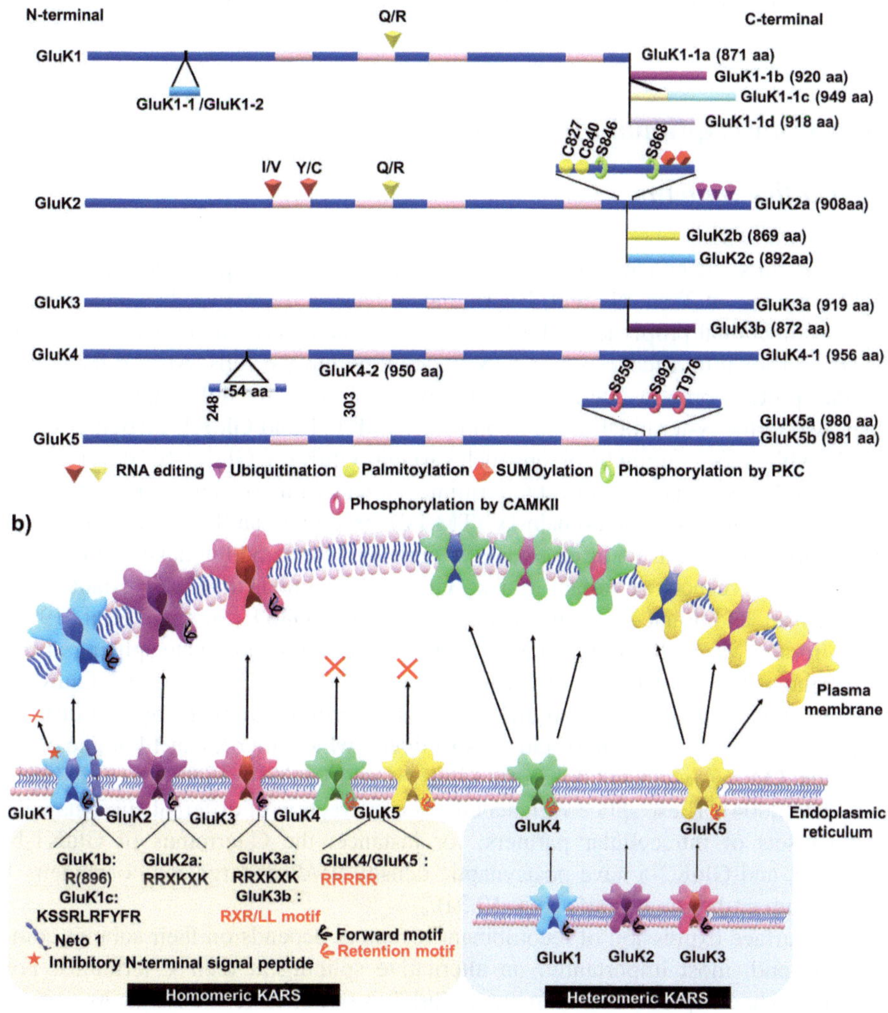

Fig. 10.3 Diversity of KARs subunit and their surface expression. (**a**) Chromosomal position of kainate receptor subunit genes in humans has been tabulated. Splice variants, RNA editing, and post-translational modifications of KARs are depicted. GluK1 has two splice variants at ATD, GluK1–1, GluK1–2, and 4 CTD splice variants: GluK1-a, GluK1-b, GluK1-c, and GluK1-d, respectively. GluK2 has three splice variants: GluK2-a, GluK2-b and GluK2-c. It has multiple phosphorylation sites at its C-terminal. GluK3-a and GluK3-b are the two C-terminal splice variants of GluK3. Recently, a splice resulted due to deletion of 54 amino acids at the ATD of GluK4–2 is depicted, whereas a single amino acid addition in GluK5 gave rise to the GluK5-b variant. GluK1 and GluK2 also undergo RNA editing and different post-translational modifications –phosphorylation, ubiquitination, palmitoylation, and SUMOylation that impart functional differences to the subunits. (**b**) Surface expression of KARs is regulated by alternative splicing and different

Pre- and Post-Translational Modifications of KARs

The physiological properties of KARs are modulated by post-transcriptional mechanisms like RNA splicing and editing, which leads to the production of functionally distinct proteins from a single gene by modifying the transcribed RNA.

Post-Transcriptional Modifications

Alternative Splicing

KAR subunits reflect a great functional diversity as they undergo distinct alternative RNA splicing at their N- and C-terminal domains; these modifications impart various functional properties to KARs, especially the C-termini alternative splicing has been shown to affect receptor's localization within the cells. Also, it facilitates distinct protein–protein interactions. Six splice variants of GluK1 are reported to date, two at the N-terminal domain, namely, GluK1–1 and GluK1–2 (Bettler et al. 1990), while its C-termini have four such variants, GluK1-a, GluK1-b, GluK1-c, and GluK1-d. GluK1-a is the shortest C-terminus variant with 16 amino acid residues at the end of transmembrane domain IV (TM IV), producing an 871 residue polypeptide. GluK1-b contains an insert after Q869 that leads to an additional 49 amino acid residues with respect to variant a; also, GluK1-c is similar to splice variant b but contains an additional 29 amino acid cassette after Q869 (Sommer et al. 1992; Pinheiro and Mulle 2006). Other KAR subunits also show alternate splicing at the C terminus. For instance, GluK2 can produce GluK2-a, GluK2-b, and GluK2-c variants, GluK2-a and GluK2-b differ in their C-terminal domains by two distinct sequences of 54 and 15 amino acids, respectively, which are present 14 amino acids after the TM IV, whereas GluK3 has GluK3-a and GluK3-b (Schiffer et al. 1997; Jaskolski 2004). These splice variants infer the interaction of different subunits with distinct sets of intracellular partners; for instance, the C-terminus of GluK1-b, GluK1-c, and GluK2-a have postsynaptic density-95/discs large/zona occludens-1 (PDZ) domain-binding motifs (Fig. 10.3a).

The surface expression of recombinant receptors depends on their subunit composition and, most importantly, on alternative splicing at their C-terminus. For instance, when expressed as homomer, GluK5 is confined to the ER as its surface delivery is restricted by the presence of the retention motif RXR. This retention signal gets occluded when it is expressed as a heteromer that comprises the GluK2a subunit splice variant (Jaskolski 2004). A detailed description of receptor trafficking is given in the section below.

Fig. 10.3 (continued) molecular determinants. In addition, forward and retention motifs present at the CTD and ATD also regulate the surface trafficking and ER retention of the receptors

RNA Editing

RNA editing is known to modulate the kinetic and ionic properties of glutamate receptors, and it is defined as the "transcriptional or post-transcriptional modification of an RNA molecule that changes its coding specificity by a process other than splicing". Adenosine-to-inosine (A-to-I) is the most abundant editing found in mammals; the process is catalyzed by adenosine deaminase acting on RNA (ADAR) I and II enzymes (Orlandi et al. 2012). ADARs modify adenosines to inosines in specific coding regions. These inosines are read as guanosines by the translational machinery, leading to alteration in the codon of the edited nucleotide and resulting in an amino acid change in the expressed protein. The functional diversity of the kainate receptor is extended via RNA editing of two subunits, namely, GluK1 and GluK2. They undergo single editing at the Q/R636 (GluK1) and Q/R621 (GluK2) positioned at the channel forming segment—the transmembrane domain II (TM2) where glutamine gets substituted with arginine. GluK2 has two additional RNA editing sites, the I/V567 and the Y/C571 located in the transmembrane domain I (TM1) where isoleucine can be substituted with valine and tyrosine can be replaced by cysteine, respectively (Egebjerg and Heinemann 1993; Köhler et al. 1993). Editing at Q/R in GluK1 has various functional consequences, as it results in the formation of receptor channels with markedly reduced permeability for divalent cations (Ca^{2+}). Also, it has been reported that this change leads to inhibition of receptors by membrane fatty acids and affects the neuronal maturation in the spinal horn. Editing at Q/R, I/V, and Y/C in GluK2 imposes finer modification of ionic properties and affects ER retention (Burnashev et al. 1995).

Post-Translational Modifications

N-Glycosylation

N-glycosylation is a vital post-translational modification as it confers potential structural diversity and is related to fundamental steps in the receptor biogenesis, such as efficient protein folding and its transportation from the ER (Everts et al. 1999). The process involves the conjugation of oligosaccharides to the asparagine moieties in the extracellular domains of the receptor subunit and their processing. There are nearly four to twelve N-glycosylation consensus motifs present on individual receptor subunits, and these oligosaccharide additions can constitute almost 10% or more of the mass of the mature receptors (Hughes 1994; Parker et al. 2013). Proteomic analysis of glycans and glycopeptides revealed discrete sites of N-glycosylation in AMPARs, KARs, and NMDARs (Parker et al. 2013). These sites are positioned within the extracellular linker domain connecting the amino-terminal and the ligand-binding domains involved in the channel gating (Vernon et al. 2017). N-glycosylation also confers unique functional properties to the receptor, as the molecular composition of glycans substantially alters KAR biophysical properties like desensitization, recovery from desensitization, and deactivation (Vernon et al. 2017; Kumari et al. 2019).

Phosphorylation

Considerable advances have been made in the past few years to understand the dynamically regulated post-translational phosphorylation sites present in iGluRs. Ongoing advancements in the field, such as using a strategic combination of specific phosphoprotein antibodies, site-directed mutagenesis, and chromophore-tagged receptors, have made it possible to associate the functional importance of phosphorylation.

Protein Kinase C (PKC)-Mediated Phosphorylation of GluK2

PKC phosphorylation of GluK2 at residues S846 and S868 exerts multiple levels of control over KAR trafficking. It promotes SUMOylation and KAR endocytosis. The phosphomimetic serine-to-aspartate mutations of residues 846 and 868 in GluK2 impede KAR trafficking to the cell surface (Nasu-Nishimura et al. 2010; Chamberlain et al. 2012). Furthermore, these PKC phosphorylation sites are involved in the ER exit of KARs, and preventing PKC phosphorylation by mutating S846 and S868 to alanine increases basal rates of secretory pathway trafficking. PKC-mediated phosphorylation in KARs can differentially regulate the KAR-mediated synaptic transmission and is also involved in LTD in DRG neurons, the perirhinal cortex and the hippocampus (Lerma 2003).

CaMKII Phosphorylation of GluK5

It has been shown that GluK5 functional properties are also controlled via phosphorylation; CAMKII phosphorylates GluK5 at its C-terminal domain. Its phosphorylation enhances the receptor's surface expression but reduces neuronal synaptic localization. Moreover, it reduces GluK5 and PSD95 interaction. It has been shown that direct phosphorylation of GluK5 is required for KAR LTD at MF-CA3 synapses as the expression of non-phosphorylatable GluK5 did not restore KAR LTD (Carta et al. 2013).

SUMOylation

SUMOylation is a three-step process in which small ubiquitin-like modifier (SUMO) family member(s) are attached covalently to lysine residues present in the target proteins; this process occurs via an enzymatic cascade (Wilkinson and Henley 2010).

GluK2 is shown to undergo SUMOylation, which is required for agonist-dependent endocytosis of GluK2 (Konopacki et al. 2011). It has also been reported that PKC-mediated phosphorylation of GluK2 at serine 868 enhances GluK2 SUMOylation, leading to receptor endocytosis (Wilkinson and Henley 2010). It has been found that both phosphorylation and SUMOylation are necessary for PKC-dependent LTD of KARs at mossy fiber-CA3 synapses in the hippocampus.

Palmitoylation

Palmitoylation is a post-translational modification in which a C16 fatty acid acyl chain is covalently bound to a serine, threonine, or cysteine residue of a nascent protein via an ester or thioester bond. Among KARs, only GluK2 has been reported to undergo palmitoylation at residues C827 and C840 in the C-terminal domain (Pickering et al. 1995). A comparative study of the wild-type and non-palmitoylated mutant (C827A, C840A) revealed that palmitoylation might not modulate kainate channel function directly but instead affect channel functions indirectly by regulating the phosphorylation state of the receptor. The kainate-gated currents produced by the non-palmitoylated mutant receptor were indistinguishable from those of the wild-type because non-palmitoylated mutant acts as a better substrate for protein kinase C than the wild-type GluK2 receptor (Pickering et al. 1995).

Ubiquitination

Ubiquitination is an enzymatic process carried out by a set of three enzymes: ubiquitin-activating enzyme (E1), ubiquitin-conjugating enzyme (E2), and ubiquitin-protein ligase (E3). Briefly, ubiquitin is activated by E1; this activated ubiquitin is then transferred to E2; in the final step, glycine at the C-terminal of ubiquitin recognizes and binds to lysine in the target protein via E3. Repeated addition of ubiquitin to the target protein leads to its degradation by proteasomal machinery. Ubiquitination is reported to be an important regulator of synaptic plasticity. In KARs, only GluK2 is known to undergo ubiquitination, which results in the decrease of its surface expression (Salinas et al. 2006).

Trafficking/Retention Motifs

The controlled trafficking of KARs to specific pre-, post-, and extrasynaptic locations that have retention in ER is fundamental to many aspects of neuronal functions. ER-sorting mechanisms and membrane delivery of KARs depend primarily on the subunit composition and splice variants. Additionally, trafficking/retention motifs are added post-translationally, which regulate KARs release or retention in ER (Fig. 10.3b).

GluK1

Two C-terminal splice variants, GluK1-a and GluK1-b, are predominantly present at the plasma membrane, whereas most GluK1 variants are retained within the ER; for instance, GluK1-c has an RXR motif within the CTD and is known to be highly confined to the ER (Jaskolski 2004). The retention of GluK1–2b in ER is due to the presence of a motif that consists of positively charged arginine (R896) and surrounding amino acids located within the alternatively spliced C-terminal region as well as two basic amino acids (R900 and K901) further downstream (Ren et al. 2003b). Also, the trafficking of GluK1–2b receptors could possibly be phosphorylation status-dependent as the introduction of a phosphorylation-mimetic mutation (T898E) near the ER retention motif promotes their surface delivery (Ren et al. 2003b). Additionally, Duan et al. revealed that the N-terminal cleaved signal peptide of GluK1 acts as an inhibitor of forward trafficking of these receptor by binding with the ATD (Duan et al. 2018).

GluK2

Although GluK2-a and GluK2-b are known to be targeted to the plasma membrane, GluK2-a surface targeting is more robust due to the presence of a four amino acids forward trafficking motif [873 RRXKXK 878] (Jaskolski 2004); also PKC phosphorylation of GluK2 at residues S846 and S868 restricts its forward trafficking (Nasu-Nishimura et al. 2010).

GluK3

GluK3 has two splice forms: GluK3-a and GluK3-b, as described earlier. The strong surface expression of GluK3-a is promoted by a forward trafficking motif similar to GluK2-a, while GluK3-b is retained in the ER. Unlike GluK1-c and GluK5,

GluK3-b ER retention is not controlled by the RXR motif, however, rapid endocytosis and degradation of surface-expressed GluK3-b are regulated by a dileucine motif within the CTD (Huyghe et al. 2011). Also, it has been suggested that mutation at the dileucine motif reduces its EndoH sensitivity, perhaps increasing ER retention (Huyghe et al. 2011).

GluK4

Not much is known about GluK4 forward trafficking and ER retention motifs. However, there are five arginine residues within the CTD of the receptor, which are known for the ER retention of the receptor (Ren et al. 2003a).

GluK5

A poly-arginine motif (RRRRR) similar to GluK4 within the CTD of homomeric GluK5 receptors is known for its ER retention; removing this motif increases its surface expression (Ren et al. 2003a). Additionally, the RXR motif in the intracellular loop preceding the TM II has been identified to mediate ER retention (Nasu-Nishimura et al. 2006). Furthermore, it has been demonstrated that this poly arginine motif interacts with the coat protein complex I (COPI), which is thought to mediate ER-retrieval of misfolded or unassembled proteins from the Golgi to ER (Vivithanaporn et al. 2006).

Functional Characterization of KARs

The study of functional roles of KARs was a difficult task in the initial days due to the lack of specific antagonists that could isolate KAR currents from AMPARs. The use of GYKI 53655 as AMPAR-specific antagonist opened up the field of KAR physiological studies (Paternain et al. 1995; Wilding and Huettner 1995). It was found that there was a stark difference between the kinetics of AMPARs and KARs, with the excitatory postsynaptic currents (EPSCs) of KARs having a longer rise and slower decay time with respect to AMPARs (Castillo et al. 1997; Kidd and Isaac 2001). Cloning of individual subunits, their expression in heterologous systems (Xenopus oocytes, HEK cells), and whole-cell/outside-out patch electrophysiological studies have helped understand the differences in the kinetics of each KAR subtype. Recombinant GluK2 (Q) is the most studied receptor because of its high expression in heterologous systems, large and stable currents. Depending on the subtypes that form the homomeric or heteromeric KAR receptor, the sensitivity toward different ligands, desensitization, deactivation, and recovery kinetics are

Table 10.1 Electrophysiological assays for recombinant KARs and their auxiliary proteins NETO1/NETO2 in a heterologous system (Native represents properties of receptor alone; Neto1 and Neto2 panels show receptor properties in presence of either of the auxiliary proteins; ? represents unavailability of quantifiable data; * represents the different calculation method used by Straub et al. that considers both the fast and slow components of the desensitization)

		KAR Subtype						
Properties		GluK1	GluK2	GluK3	GluK2/GluK4	GluK1/GluK5	GluK2/GluK5	GluK3/GluK5
Desensitization rate (ms)	Native	67.8 ± 12.9	13.7 ± 1.2	12.6 ± 0.7	?	0.7 ± 0.1	5.9 ± 0.6/3.5 ± 0.2*	?
	Neto1	35.4 ± 7.4	17.9 ± 2.6	9.1 ± 0.5	?	?	6.2 ± 0.6 (Steady state current=8.2 ± 0.2%)	?
	Neto2	439.1 ± 71.5	30.4 ± 6.8	18.7 ± 1.4	?	3.3 ± 0.5	11.4 ± 1.2*	?
Peak amplitude (pA)	Native	?	?	?	447 ± 92	200 ± 40	391 ± 26	55 ± 6
	Neto1	282 ± 65% from native	143 ± 13% from native	303 ± 92% from native	561 ± 118	335 ± 180	467 ± 179	194 ± 35
	Neto2	573 ± 63% from native	146 ± 18% from native	1348 ± 535% from native	?		?	?
Deactivation (ms)	Native	1.7 ± 0.4	2	1.0 ± 0.1	?	Mono-exponential decay	Slightly slower than K2 homomer	
	Neto1	?	?	?	?	?	?	?
	Neto2	3.9 ± 0.7	Tf = 2.3 ms; Ts=7	?	?	Bi-exponential decay	~3 fold slower than K2 homomer	
Recovery from Desensitization	Native	5.1 ± 0.5s	1.7 ± 0.2s	3.4 ± 0.3s	?	3.8s	1.74 ± 0.19s	?
	Neto1	0.9 ± 0.1s	0.6 ± 0.1s	1.2 ± 0.1s	?	?	0.25 ± 0.05s	?
	Neto2	4.9 ± 0.3s	1.1 ± 0.2s	4.2 ± 0.4s	?	Tf = 450ms; Ts = similar to native	TF = 61 ms; Ts = similar to native	?
Agonist affinity (Kd, μM)	Native	485 ± 74	603 ± 29	EC_{50} + 10.6 ± 0.2 mM	EC_{50} = 43.2 ± 7	EC_{50} = 7.8 ± 1.3	EC_{50} = 6 ± 1.2	EC_{50} = 6 ± 0.7
	Neto1	14 ± 7	208 ± 81	<Native	EC_{50} = 9.1 ± 1.2	EC_{50} = 5.2 ± 0.2	EC_{50} = 5 ± 0.3	EC_{50} = 4.8 ± 0.5
	Neto2	64 ± 28	139 ± 62	<Native	?	?	?	?

Rectification	Native	?	0.07 ± 0.02	?	?	?	?	?
	Neto1	?	0.66 ± 0.03	?	?	?	?	?
	Neto2	?	0.48 ± 0.05	?	?	?	?	?
Surface expression/Synaptic targeting	Native	1	1	1	?	?	?	?
	Neto1	~1.2 fold	~1.1 fold	~0.7 fold	?	?	?	?
	Neto2	~1.5 fold	~0.8 fold	~1.5 fold	?	?	?	?
References		Copits et al. (2011)	Fisher and Mott (2012)	Perrais et al. (2009)	Fisher and Mott (2012)	Fisher and Mott (2012)	Straub et al. (2011)	Fisher and Mott (2012)
		Straub et al. (2011)	Fisher and Mott (2012)	Palacios-Filardo et al. (2016)			Straub et al. (2011)	
		Palacios-Filardo et al. (2016)	Palacios-Filardo et al. (2016)				Fisher and Mott (2012)	

summarized in Table 10.1. For example, GluK1 and GluK2 show variable kinetics to pharmacological agonists like kainate and domoate (Swanson et al. 1997; Swanson and Heinemann 1998). GluK3 shows least affinity for glutamate in all the homomeric KAR subunits (Perrais et al. 2009). The presence of GluK4 or GluK5 changes the affinity for glutamate of heteromeric receptors as well as slows the deactivation and rate of desensitization at low agonist concentrations (Fisher and Fisher 2014). Moreover, Q/R editing in the TM II region of the GluK1 and GluK2 was shown to change the conductance-voltage relationship of the homomeric receptors, which also affects the permeability toward calcium as well as the expression on the surface (Egebjerg and Heinemann 1993; Swanson et al. 1996; Gurung et al. 2018). KARs are also affected by extracellular ions like Na^+ and Cl^- that stabilize the ligand-binding domain and affect the kinetic properties of the receptors, thus acting as allosteric modulators of the KAR functioning (Plested and Mayer 2007; Plested et al. 2008). It was observed that there were discrepancies in the kinetics of recombinant receptors in heterologous systems vs. neuronal slices (Straub et al. 2011; Lerma 2011). NETO1 and NETO2 were found to interact with KARs as their auxiliary partners and were shown to affect the pharmacological properties of these receptors. The effects of NETOs are described in detail in the next section.

Regulatory Partners of KARs

The regulatory processes involved in the constitutive and activity-dependent trafficking of KARs are subtle and complex. KARs surface delivery, residence time, internalization, and recycling or degradation are orchestrated by the intricate combination of protein-protein interaction and post-translational modifications. It has been reported that genetic omission of KARs in the mice shows no kainate receptor activity, whereas their overexpression does not enhance kainate-receptor-mediated excitatory postsynaptic current (EPSC), suggesting that native kainate receptors may contain additional modulatory proteins. Several cytoplasmic proteins have been identified as kainate receptor interactors, namely, SAP90/PSD95, the cadherin/catenin complex, KRIP6, Actinfillin, PICK1, Syntenin, GRIP, NETO1, and NETO2 (Garcia et al. 1998; Coussen et al. 2002; Hirbec et al. 2003; Salinas et al. 2006; Laezza et al. 2007).

Additionally, it was observed that postsynaptic KAR functions were modulated by C1q-like proteins (C1ql2 and C1ql3) along with presynaptic neurexin-3 via interactions with GluK2 and GluK4 ATDs, which was confirmed in neurons and heterologous system, HEK293 cell lines (Matsuda et al. 2016). This study also highlighted the role of C1q proteins and neurexins in pathological conditions associated with KARs.

Identification of NETOs as a Kainate Receptor-Binding Protein

Neuropilin and tolloid-like (NETO) proteins were cloned and characterized as a novel member of CUB and LDLa containing family in genetic screens for transcripts expressed in the brain and retina. However, the functional relevance of these proteins was unidentified at that time (Michishita et al. 2003, 2004). The first study to identify NETOs interaction with KARs came in the year 2009, when Zhang et al. co-purified NETO2 with KARs (GluK2 and GluK3) from the mammalian brain and overruled the assumption that NETO2 interacts with AMPARs and NMDARs (Zhang et al. 2009). Concurrently NETO1-a structural homolog was proposed to act as an interacting protein for NMDARs (Ng et al. 2009). A substantial reduction (~50%) and faster decay kinetics of KAR-mediated EPSCs at hippocampal MF-CA3 synapses in Neto1-null and Neto1/2-double-null mice provided the key evidence that NETO1 is an auxiliary protein for KARs (Tang et al. 2011). Following the initial identification of NETO2 as a protein associated with the GluK2 and GluK3, both NETO1 and NETO2 were shown to interact with other recombinant kainate receptors.

Synaptic Targeting and Functional Modulation

NETO proteins play an essential role in functional modulation and synaptic targeting of kainate receptors; their co-assembly differentially impact all aspects of the receptor functions, which depend on both the receptor subtype and NETO isoform. The synaptic targeting of KARs by NETO proteins is still a contentious issue. The first study to report the role of NETO proteins in the trafficking and synaptic targeting of kainate receptors highlighted that NETO2 promotes the plasma membrane localization in both heterologous cells and hippocampal neurons, whereas NETO1 reduced the neuronal membrane localization (Copits et al. 2011). NETO2 was found to slow down the deactivation and desensitization of glutamate-gated currents from homomeric GluK2 receptors by Zhang W et al. in 2009; subsequently, the same effect was observed on many other homomeric and heteromeric subunits (Copits et al. 2011; Straub et al. 2011). NETO1 has a bidirectional effect on receptor kinetics. On one hand, it increases the desensitization of GluK1 receptors, on the other hand, it slows down the deactivation and desensitization of GluK2/GluK5 heteromeric receptors (Copits et al. 2011; Straub et al. 2011). NETO2 has been shown to increase open probability and open time of KARs without affecting their single-channel conductance (Zhang et al. 2009). Furthermore, NETO2 consistently decreases the deactivation rates by ~two-fold, whereas it can show a larger magnitude of increase in desensitization rates (Copits et al. 2011).

NETO Subunits and Synaptic Current Modulation

NETO1 and NETO2 are composed of two discrete structurally conserved, ~110 amino acid protein interaction CUB (complement C1r/C1s, Uegf, and Bmp1) domains, followed by an extracellular juxtamembrane low-density lipoprotein class A domain, which are considered to form a complex with calcium and adopt a rigid secondary structure that restricts conformational freedom. The intracellular C-terminal domains possess an AP2 adaptor protein endocytic motif and, in the case of NETO1, a type I PDZ ligand at the extreme C terminus. The interaction between NETO proteins and KARs is primarily governed through the second extracellular CUB domain and is absent when both CUB domains are deleted (Tang et al. 2011). The intracellular C-terminal domain also influences the KAR functioning as the deletion of intracellular C-terminal domain or neutralization of a series of three adjacent positively charged residues located near TM domain in NETO1/2 resulted in a decrease in the rectification (Fisher and Mott 2012). Recent studies have shown that ATD of KARs play an important role in the association with its auxiliary proteins (Li et al. 2019; Vinnakota et al. 2021). Furthermore, the interaction of GluK2 receptor with Neto is state dependent and the apparant affinity of Neto is lower for the desensitized receptor compared to those trappeed in closed state (Vinnakota et al. 2021). These findings are supported by the latest cryo-EM structures of GluK2-Neto2 complex captured in different states (He et al. 2021).

Structural Characterization of KARs

Cloning of different subunits of KARs in the early 1990s allowed analysis of sequences and common structural motifs present in all the subunits of this family and other iGluRs. This analysis showed that each subunit that comprises an ATD, homologous to leucine/isoleucine/valine (LIV)-binding proteins (O'Hara et al. 1993; Ayalon and Stern-Bach 2001; Kumar et al. 2009), was required for assembly of the receptors. LBDs, homologous to bacterial periplasmic proteins bound to glutamate and the two lobes of this domain, was found to be interrupted with the ion channel (Nakanishi et al. 1990; Stern-Bach et al. 1994). The pore region was similar to an inverted K^+ channel (Wo and Oswald 1995; Kuner et al. 2003), and the cytoplasmic domain contained multiple interaction sites like PDZ binding motifs, KRIP6-binding motif, etc. (Garcia et al. 1998; Hirbec et al. 2003; Coussen et al. 2005; Laezza et al. 2007).

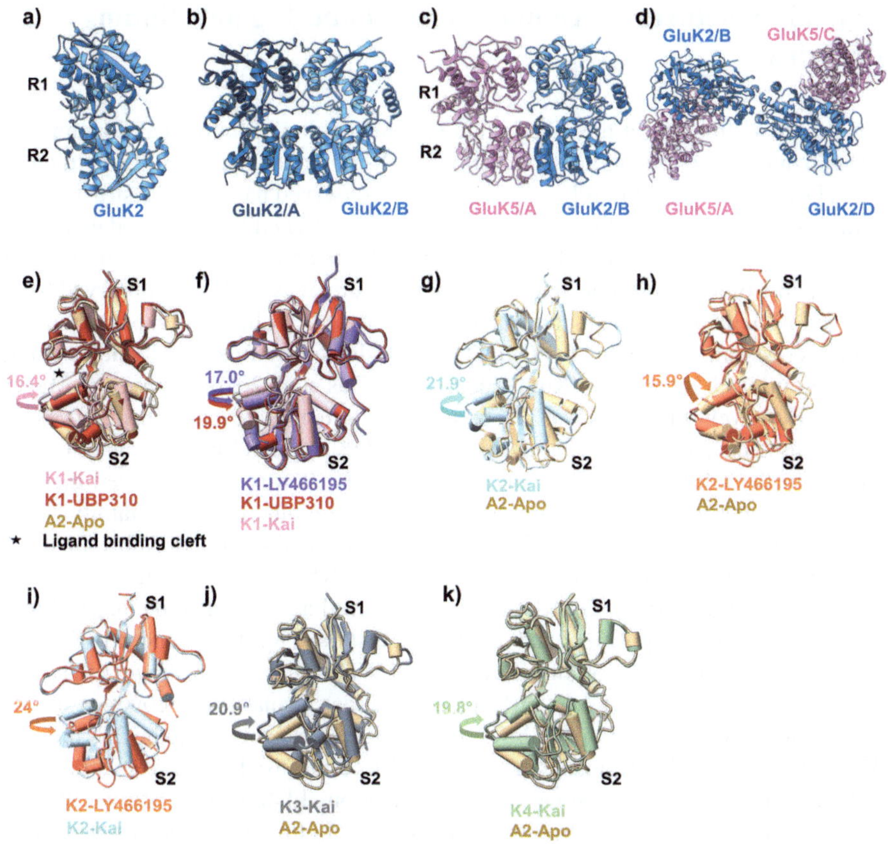

Fig. 10.4 Assembly of kainate receptor amino-terminal and ligand-binding domains. Panels (**a**) and (**b**) show the side views of the crystal structure of GluK2 (PDB: 3H6G) that crystallized as a dimer. The clamshell structure observed in the monomer (a) is formed from R1 and R2 sub-domains. The light and dark colors in dimer (**b**) are shown to distinguish between the two monomers. Panels (**c**) and (**d**) show the ATD arrangement for heteromeric receptor, GluK2/GluK5 (PDB: 3QLV) that crystallized as a classical N-shaped tetramer with GluK2 subunits in the proximal positions (B/D) and GluK5 in the distal (A/C). The position of subunits was confirmed biochemically using cysteine mutants by Kumar et al. 2011 and the cryo-EM structure of full-length GluK2/GluK5 by Khanra et al. 2021. (**e**) shows cleft closure on the binding of agonist (kainate; PDB:3C32) or antagonist (UBP310; PDB:2F34) to GluK1 LBD with respect to apo state (PDB:1FTO) of GluA2 LBD. S1 lobes of GluK1 were superimposed with respect to the apo state (gold) to depict the LBD conformational changes on the binding of kainate (pink) and UBP310 (brick). (**f**) Conformational changes in the LBD of GluK1 in different states were compared using the angle of rotation of domains. Closed or antagonist bound states (UBP310; PDB:2F34, brick; LY466195; PDB:2QS4, purple) were compared to the desensitized state (kainate; PDB:3C32, pink). (**g**) and (**h**) show the LBD cleft closures for kainate and LY466195 bound GluK2 (PDB:1TT1, light blue; PDB:5CMK, orange, respectively) w.r.t. the apo state. (**i**) demonstrates that the degree of closure in the antagonist bound form of GluK2 is lesser (24°) than the agonist bound form. (**j**) and (**k**) show agonist bound states of GluK3 (PDB:4E0W, grey) and GluK4 (PDB:5IKB, sea green) crystallized in desensitized states using kainate w.r.t. the apo state of GluA2

Crystal Structures of Amino-Terminal and Ligand-Binding Domains

The modular structure of the iGluRs allowed the cloning of individual domains for crystallization and structure determination via X-ray crystallography. The structure of ATD across the KAR family was found to be conserved based on the structures of GluK2, GluK3 that crystallized as ATD dimers. The ATD dimers showed extensive contacts between the R1 and R2 domains (Fig. 10.4a and b). These structures were also capable of forming stable tetramers (dimer of dimer assembly) as observed from interactions with the adjacent unit cell and were proved biochemically using cysteine mutant cross-linking studies for GluK2 ATD followed by functional studies using outside-out patch recordings (Kumar et al. 2009; Das et al. 2010; Kumar and Mayer 2010). However, the GluK5 ATD homodimer structure revealed differences in dimeric assembly with respect to the GluK2 or GluK3. The R1 interface was loosely packed in the GluK5 ATD crystallographic dimers. It was also observed that the tetrameric interface residues preserved in the homomeric KAR family were substituted in GluK5, which likely prevent the dimer-of-dimer assembly of these receptors. Hence, this might lead to GluK4-GluK5 being obligate heteromers (Kumar and Mayer 2010). GluK2/GluK5 crystallized as a dimer of the heterodimer (GluK2 in proximal position and GluK5 in distal) (Fig. 10.4c and d). Further, analytical ultracentrifugation of GluK2/GluK5 ATDs in solution showed that they have very high affinities (K_d 11 nM) for heterodimer formation. In contrast, the K_d values for homodimerization of GluK2 and GluK5 were estimated to be 250 nM and 350 μM, respectively (Kumar et al. 2011). Zhao et al. used fluorescence-detected sedimentation velocity analytical ultracentrifugation (FDS-SV) to estimate the K_d values more precisely for heterodimeric assemblies. In this study, they used purified proteins tagged with DyLight 488 and subjected them to FDS-SV to understand the association of homomeric or heteromeric assembly for each of the AMPAR and KAR subunit ATDs. They correlated the estimated Gibbs free energy to the stability of each complex. They showed that heteromeric ATDs of KARs (especially, GluK2/GluK5) and AMPARs are thermodynamically more stable than their homomeric counterparts. Also, AMPAR ATDs formed tighter homomeric/heteromeric ATD complexes with respect to KARs (Zhao et al. 2017). To date, GluK1 and GluK4 ATD structures are not reported.

The ligand-binding domain (LBD) is the most dynamic part of the receptor, and different conformational states of LBDs have been captured using constructs where the S1 and S2 sub-domains of LBD of each receptor were joined by GT linker(s). The S1S2-LBDs were crystallized in different conformations, namely, apo (without ligand), desensitized and closed states with the help of agonists and antagonists. GluK1 in the presence of glutamate was the first KAR LBD that was crystallized (Naur et al. 2005; Mayer 2005). Physiologically, the LBDs assemble as back-to-back dimers with interactions at the S1 sub-domain. As shown in Fig. 10.4(e–f), GluK1 LBD adopts a closed conformation (upward movement of S2 to close the binding cleft) on the binding of the agonist, kainate with respect to the apo state structure of

GluA2 LBD (Armstrong and Gouaux 2000). On the other hand, binding of an inhibitor like UBP310 (Mayer et al. 2006) or LY466195 (Alushin et al. 2011; Meyerson et al. 2016) leads to an extended conformation of the ligand-binding cleft similar to the apo state. In GluK2 (Fig. 10.4g–i), even binding of the antagonist LY466195 leads to changes in rotation angle of S2, although it does not cause cleft closure like the binding of the agonist, kainate. Figure 10.4e–k shows conservation in the overall LBD structure of all the subunits (homomeric, GluK1–3 and heteromeric, GluK4). However, all of them close incompletely even in the presence of high-affinity agonist kainate, suggesting that the channel kinetics is much more complicated and required studies of full-length receptor structure to understand the channel physiology. Ahmed et al. (2011 and 2013) revealed in AMPAR LBDs using cysteine cross-linking and hydrogen-deuterium exchange experiments that even partial agonists like kainate can cause full domain closure. However, the stability of the closed conformation determines the efficacy of the ligand (Ahmed et al. 2011, 2013). Nevertheless, the crystal structures of LBDs with different additives and mutants helped understand the mechanisms of ligand binding between different receptors, facilitating the designing of a wide range of antagonists that could be further optimized and used as drug candidates. Multiple reviews have summarized the functional insights gained from the LBD structures (Kumar and Mayer 2013; Mayer 2016, 2021; Møllerud et al. 2017).

Cryo-EM Structures of Full-Length KARs

Full-length structures of kainate receptors are needed to understand the molecular details of the receptor trapped in different states to decipher the mechanisms of action, but obtaining full-length crystal structures of these receptors has been a challenging task as they have poor expression, low stability in solution, and several flexible regions that lead to poor yields. The first full-length structure of a KAR was that of homotetrameric GluK2 receptor trapped in the desensitized state, solved using cryo-electron microscopy (Cryo-EM) (Meyerson et al. 2014). This showed a unique arrangement at the ATD and LBD layers, which differed from the published AMPAR full-length crystal structure trapped in a closed-state (Sobolevsky et al. 2009). Figure 10.5a shows the first structure of GluK3 (Kumari et al. 2019) captured in the desensitized state. Both the GluK2 and GluK3 receptors trapped in desensitized state showed a similar three-layered arrangement of ATD, LBD, and TMD domains. The ATD layer shows an intact dimer-of-dimer assembly with twofold symmetry, whereas the LBDs are arranged in a quasi-four-fold symmetry matching the fourfold symmetry of the TM domains forming the pore. The desensitized conformation seems to be conserved throughout the KAR family, as this is also observed in the heteromeric GluK2/GluK5 (Khanra et al. 2021) receptors, as shown in Fig. 10.5(b, d, and f). The domain swapping arrangement at the ATD and the LBD layer was also conserved similar to that observed in AMPA and NMDA receptors (Sobolevsky et al. 2009; Wang and Furukawa 2019). KAR

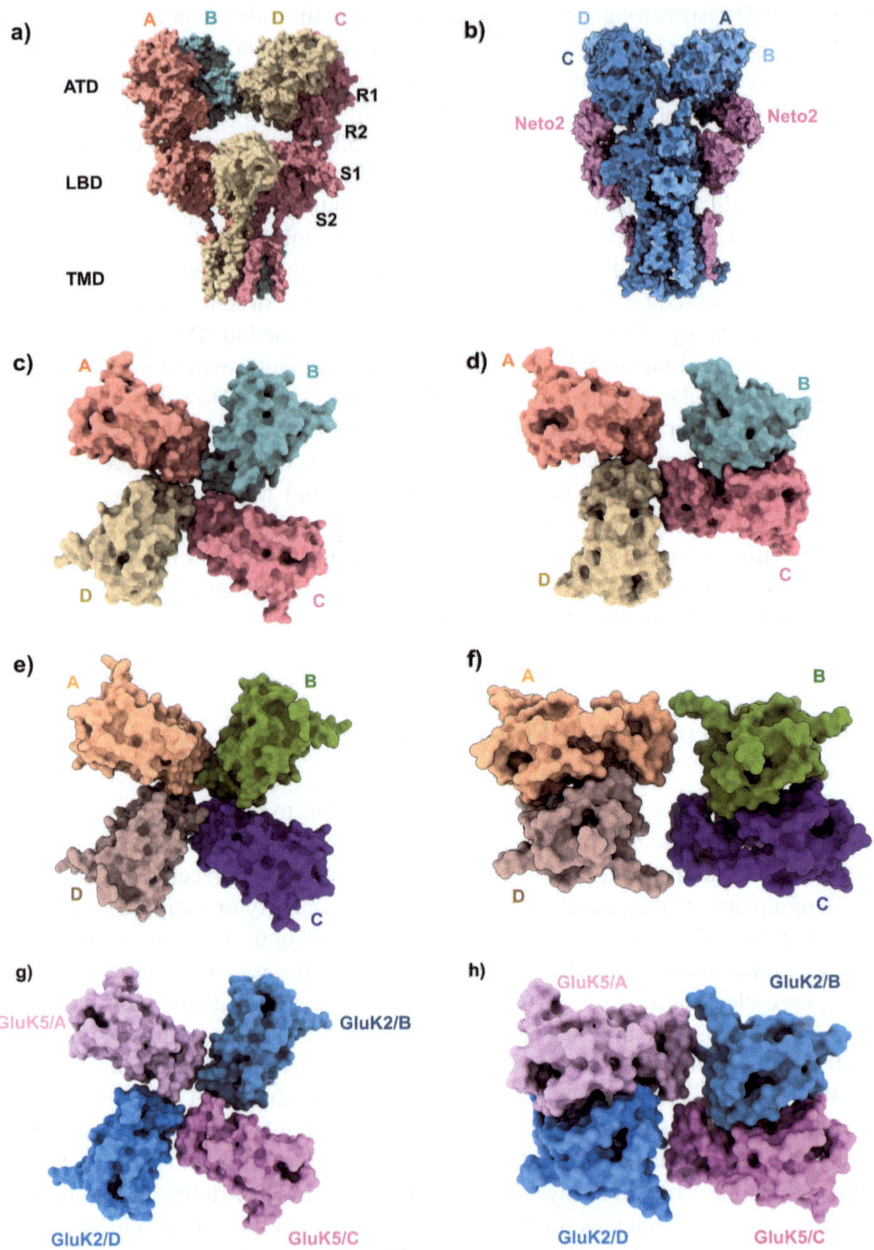

Fig. 10.5 The full-length structures of KARs solved using cryo-EM. (**a**) Side view of the full-length cryo-EM structure of GluK3-SYM (PDB:6JFY) captured the receptor in the desensitized state, the overall Y-shaped structure is conserved throughout the KAR family. Different colors denote the four subunits of the receptor (A—coral, B—cadel blue, C—violet red, and D—tan). (**b**) shows the side view of the full-length GluK2 (in closed state, PDB:7F5A) in complex with its auxiliary protein Neto2 (receptor—blue, auxiliary subunit—pink) (**c**) shows the arrangement of GluK3 LBD in desensitized state wherein all four subunits show a quasi-four-fold symmetry. (**d**) displays the closed state (PDB: 6JFZ) of the GluK3 receptor LBD, wherein the LBD dimerization is

structures captured in antagonist bound form are also available. However, unlike the desensitized state, some differences have been observed in the arrangement of the LBDs in the closed state. Figure 10.5(c, e, and g) shows the differences at the LBD layer between GluK3, GluK2, and GluK2/GluK5 trapped in the closed state, respectively. While the LBD dimers are intact for GluK2 and GluK2/GluK5, it is not the case in GluK3. In GluK3, only two LBD domains adopt the dimeric arrangement necessary for receptor activation. The other two remain in a desensitized-like state, which is a deviation from the classical closed-state assembly (Kumari et al. 2019, 2020). The presence of unique residue Asp759 at the LBD dimer interface, shown to constitute a Zn^{2+}-binding site (Veran et al. 2012), influences the stability of the LBD dimers and hence, could be responsible for the observed nonclassical closed-state conformation in the GluK3 receptors. The lower LBD dimer stability in GluK3 is consistent with the faster desensitization and low glutamate sensitivity of these receptors. GluK2/GluK5 full-length structure also demonstrated the role of TM3-S2 linker of GluK2 at B/D position in accommodating large-scale movements of the LBD between the desensitized and closed states of the heteromeric receptors. This structure also proved that GluK5 used even at ten times the concentration of GluK2 occupied only the distal positions (A/C) in the heteromeric receptor. The tetramer was formed of dimers of heterodimers, which was previously shown by the crystal structure of GluK2/GluK5 ATDs, cysteine cross-linking experiments (Kumar et al. 2011), and smFRET studies (Litwin et al. 2020). The recent structure of GluK1-2 homomeric receptor shows conservation of desensitized state among the homomeric KAR family pointing towards a conserved mechanism (Selvakumar et al. 2021). To date, no structures are known for heteromeric receptors containing GluK4 subunits or heteromers consisting of only the low affinity subunits.

KARS in Genetic Disorders

As already discussed above, kainate receptors are important modulators of neuronal circuits and plasticity during brain development. KARs have been associated with a range of CNS disorders like neurodegenerative (epilepsy, pain, Huntington's), mood disorders (depression, bipolar), anxiety (fear memory, obsessive-compulsive disorder), and developmental (Schizophrenia, autism) as summarized in Table 10.2. There have been contrasting studies about KARs' connection with brain disorders

Fig. 10.5 (continued) affected at the A/D subunit due to the presence of residue D759 at the interaction interface that might destabilize the LBD dimer assembly as observed by Kumari et al. (2020). (**e**) and (**f**) show LBD arrangement in desensitized (PDB: 5KUF) and closed states (PDB:5KUH) of full-length GluK2 homomeric KARs. (**g**) and (**h**) demonstrate LBD arrangement in desensitized (PDB:7KS3) and closed state (PDB:7KS0) of full-length GluK2/GluK5 heteromeric KARs

Table 10.2 Kainate receptors association with nervous system disorders

Sr. No.	KAR subunit	Disorder (abnormality)	Reference(s)
1	GluK1	Epilepsy (tetranucleotide repeat polymorphism, increased expression)	(Sander et al. 1997; Izzi et al. 2002; Lucarini et al. 2007; Li et al. 2010)
		Neuropathic pain	(Palecek et al. 2004; Ko et al. 2005)
		Bipolar disorder	(Woo et al. 2007)
		Fear memory	(Ko et al. 2005; Zhuo 2017)
		Schizophrenia (rs469472)	(Woo et al. 2007; Hirata et al. 2012)
2	GluK2	Epilepsy	(Peret et al. 2014)
		Neuropathic pain	(Ruscheweyh and Sandkühler 2002; Ko et al. 2005)
		Huntington's (linkage disequilibrium)	(Rubinsztein et al. 1997; MacDonald et al. 1999; Lee et al. 2012)
		Bipolar disorder	(Shaltiel et al. 2008)
		Obsessive-compulsive disorder, OCD (rs1556995)	(Delorme et al. 2004; Sampaio et al. 2011)
		Fear memory	(Ko et al. 2005)
		Autism (M867I)	(Jamain et al. 2002; Shuang et al. 2004; Dutta et al. 2007)
		Schizophrenia (maternal transmission disequilibrium)	(Shibata et al. 2002; Bah et al. 2004)
3	GluK3	Bipolar disorder	(Wilson et al. 2006)
		Depression (T928G or S310A)	(Wilson et al. 2006; Schiffer and Heinemann 2007)
		Schizophrenia (S310A)	(Begni et al. 2002; Ahmad et al. 2009; Kilic et al. 2010; Lowry et al. 2013)
4	GluK4	Epilepsy	(Grigorenko et al. 1997; Das et al. 2012)
		Bipolar disorder (indel 3' UTR mutation)	(Pickard et al. 2008; Lowry et al. 2013)
		Depression	(Paddock et al. 2007)
		Schizophrenia	(Pickard et al. 2006; Koromina et al. 2019)
5	GluK5	Epilepsy	(Porter et al. 2006; Das et al. 2012)

which suggest that other pathways of glutamate signal transmission are also involved.

Kainate has been known for decades to cause epileptogenic seizures, and most of the studies have been done using mouse models for temporal lobe epilepsy (Falcón-Moya et al. 2018). Similarly, overexpression of KARs could lead to sensitivity of neurons to various insults; for example, this can be observed in Down's syndrome where due to trisomy of Chromosome 21, GluK1 protein levels (GRIK is present on 21q22.1) would increase (Altuna et al. 2021). Similarly, an increase in transcript levels for GRIK2 and GRIK4 in the hippocampus and temporal lobe cortex was observed in the patients with epilepsy (Grigorenko et al. 1997). Additionally, it was observed that GluK1 antagonists were able to alleviate the symptoms of epilepsy

(Figueiredo et al. 2010; Matute 2011; Grosenbaugh et al. 2018). GluK2 knockout in mice showed less sensitivity to kainate-induced seizures or accompanying cell death (Mulle et al. 1998). Similarly, studies have shown the involvement of KAR subunits in other neurodegenerative disorders like pain and Huntington's, as listed in Table 10.2. KARs (GluK1-GluK4) are also implicated in mood disorders, along with a range of other factors. For example, GluK3 isolated from patients was shown to be a specific allelic variant with the S310A variant. However, it does not change the receptor's structure but might change its stability and hence protein levels (Schiffer and Heinemann 2007). Interestingly, 2-SNP haplotypes (rs2282586 and rs1944522) of GRIK4 exhibited a protective role against bipolar disorder (Pickard et al. 2008). Among all the KAR subunits actively involved in amygdala functions, only GluK2 is associated with anxiety disorders like fear memory and OCD (Delorme et al. 2004; Ko et al. 2005; Sampaio et al. 2011). Downregulated expression of KAR mRNA levels has been profoundly established for developmental disorders like schizophrenia. Moreover, intronic as well as exonic SNPs from patient samples indicate that the regulation at RNA rather than only at protein level can lead to developmental disorders. An interesting case of GluK4 overexpression due to a similar molecular anomaly in both the mouse model and human subjects (rs58575285 polymorphism) leads to an increased synaptic transmission that results in a reduction in social interaction (hallmark symptom of autism) (Whalley et al. 2009; Aller et al. 2015; Arora et al. 2018). KARs' physiology and disorders are extensively covered in the following reviews (Lerma 2003, 2006; Pinheiro and Mulle 2006; Bowie 2008; Lerma and Marques 2013; Yuan et al. 2015; Valbuena and Lerma 2021). Recently, a GluK5 knockout generated in mice for understanding the role of GluK5 in psychiatric disorders displayed for the first time that it is involved in depressive-like behavior and contextual memory (Iida et al. 2021).

Future Perspectives

Although research on KARs has progressed rapidly in the past three decades in every aspect, through physiological, pharmacological, and structural studies, several important questions still await an answer. In the last few years, immense work has been done to characterize different receptor subunits and their splice forms; still, we lack high-affinity binders (antibodies or nanobodies) to distinguish each of these variants in native regions of the brain. Furthermore, while we recently gained structural insights into heteromeric receptors, the details of KAR interaction in the native system and how different environmental cues affect KARs remains unknown. We now have huge datasets depicting the temporal and spatial localization of KARs throughout the body but how this selective expression is orchestrated is the biggest mystery. Very few studies report the role of GluK5 in CNS disorders, although it is known that the expression of GluK5 is ubiquitous in the brain. Hence, it would be important to understand the part played by this subunit in neuropathies. We expect future studies that answer these questions will strengthen our knowledge on KARs role in development, plasticity, and diseases.

References

Ahmad Y, Bhatia MS, Mediratta PK et al (2009) Association between the ionotropic glutamate receptor kainate3 (GRIK3) Ser310Ala polymorphism and schizophrenia in the Indian population. World J Biol Psychiatry 10:330–333. https://doi.org/10.3109/15622970802688044

Ahmed AH, Wang S, Chuang H-H, Oswald RE (2011) Mechanism of AMPA receptor activation by partial agonists: disulfide trapping of closed lobe conformations*. J Biol Chem 286:35257–35266. https://doi.org/10.1074/jbc.M111.269001

Ahmed AH, Ptak CP, Fenwick MK et al (2013) Dynamics of cleft closure of the GluA2 ligand-binding domain in the presence of full and partial agonists revealed by hydrogen-deuterium exchange *. J Biol Chem 288:27658–27666. https://doi.org/10.1074/jbc.M113.495564

Aller MI, Pecoraro V, Paternain AV et al (2015) Increased dosage of high-affinity Kainate receptor gene grik4 alters synaptic transmission and reproduces autism spectrum disorders features. J Neurosci 35:13619–13628. https://doi.org/10.1523/JNEUROSCI.2217-15.2015

Altuna M, Giménez S, Fortea J (2021) Epilepsy in down syndrome: a highly prevalent comorbidity. J Clin Med 10:2776. https://doi.org/10.3390/jcm10132776

Alushin GM, Jane D, Mayer ML (2011) Binding site and ligand flexibility revealed by high resolution crystal structures of GluK1 competitive antagonists. Neuropharmacology 60:126–134. https://doi.org/10.1016/j.neuropharm.2010.06.002

Armstrong N, Gouaux E (2000) Mechanisms for activation and antagonism of an AMPA-sensitive glutamate receptor: crystal structures of the GluR2 ligand binding core. Neuron 28:165–181. https://doi.org/10.1016/s0896-6273(00)00094-5

Arora V, Pecoraro V, Aller MI et al (2018) Increased Grik4 gene dosage causes imbalanced circuit output and human disease-related behaviors. Cell Rep 23:3827–3838. https://doi.org/10.1016/j.celrep.2018.05.086

Ayalon G, Stern-Bach Y (2001) Functional assembly of AMPA and kainate receptors is mediated by several discrete protein-protein interactions. Neuron 31:103–113. https://doi.org/10.1016/s0896-6273(01)00333-6

Bah J, Quach H, Ebstein RP et al (2004) Maternal transmission disequilibrium of the glutamate receptor GRIK2 in schizophrenia. Neuroreport 15:1987–1991. https://doi.org/10.1097/00001756-200408260-00031

Bahn S, Volk B, Wisden W (1994) Kainate receptor gene expression in the developing rat brain. J Neurosci 14:5525–5547

Barbon A, Barlati S (2000) Genomic organization, proposed alternative splicing mechanisms, and RNA editing structure of GRIK1. Cytogenet Cell Genet 88:236–239. https://doi.org/10.1159/000015558

Barbon A, Vallini I, Barlati S (2001) Genomic organization of the human GRIK2 gene and evidence for multiple splicing variants. Gene 274:187–197. https://doi.org/10.1016/s0378-1119(01)00611-4

Begni S, Popoli M, Moraschi S et al (2002) Association between the ionotropic glutamate receptor kainate 3 (GRIK3) ser310ala polymorphism and schizophrenia. Mol Psychiatry 7:416–418. https://doi.org/10.1038/sj.mp.4000987

Bettler B, Boulter J, Hermans-Borgmeyer I et al (1990) Cloning of a novel glutamate receptor subunit, GluR5: expression in the nervous system during development. Neuron 5:583–595. https://doi.org/10.1016/0896-6273(90)90213-y

Bortolotto ZA, Clarke VR, Delany CM et al (1999) Kainate receptors are involved in synaptic plasticity. Nature 402:297–301. https://doi.org/10.1038/46290

Bowie D (2008) Ionotropic glutamate receptors & CNS disorders. CNS Neurol Disord Drug Targets 7:129–143

Burnashev N, Zhou Z, Neher E, Sakmann B (1995) Fractional calcium currents through recombinant GluR channels of the NMDA, AMPA and kainate receptor subtypes. J Physiol 485(Pt 2): 403–418. https://doi.org/10.1113/jphysiol.1995.sp020738

Carta M, Opazo P, Veran J et al (2013) CaMKII-dependent phosphorylation of GluK5 mediates plasticity of kainate receptors. EMBO J 32:496–510. https://doi.org/10.1038/emboj.2012.334

Castillo PE, Malenka RC, Nicoll RA (1997) Kainate receptors mediate a slow postsynaptic current in hippocampal CA3 neurons. Nature 388:182–186. https://doi.org/10.1038/40645

Chamberlain SEL, González-González IM, Wilkinson KA et al (2012) SUMOylation and phosphorylation of GluK2 regulate kainate receptor trafficking and synaptic plasticity. Nat Neurosci 15:845–852. https://doi.org/10.1038/nn.3089

Chittajallu R, Vignes M, Dev KK et al (1996) Regulation of glutamate release by presynaptic kainate receptors in the hippocampus. Nature 379:78–81. https://doi.org/10.1038/379078a0

Clarke VR, Ballyk BA, Hoo KH et al (1997) A hippocampal GluR5 kainate receptor regulating inhibitory synaptic transmission. Nature 389:599–603. https://doi.org/10.1038/39315

Contractor A, Swanson G, Heinemann SF (2001) Kainate receptors are involved in short- and long-term plasticity at mossy fiber synapses in the hippocampus. Neuron 29:209–216. https://doi.org/10.1016/s0896-6273(01)00191-x

Copits BA, Robbins JS, Frausto S, Swanson GT (2011) Synaptic targeting and functional modulation of GluK1 kainate receptors by the auxiliary neuropilin and tolloid-like (NETO) proteins. J Neurosci 31:7334–7340. https://doi.org/10.1523/JNEUROSCI.0100-11.2011

Coussen F, Normand E, Marchal C et al (2002) Recruitment of the kainate receptor subunit glutamate receptor 6 by cadherin/catenin complexes. J Neurosci Off 22:6426–6436

Coussen F, Perrais D, Jaskolski F et al (2005) Co-assembly of two GluR6 kainate receptor splice variants within a functional protein complex. Neuron 47:555–566. https://doi.org/10.1016/j.neuron.2005.06.033

Das U, Kumar J, Mayer ML, Plested AJR (2010) Domain organization and function in GluK2 subtype kainate receptors. Proc Natl Acad Sci U S A 107:8463–8468. https://doi.org/10.1073/pnas.1000838107

Das A, Wallace GC, Holmes C et al (2012) Hippocampal tissue of patients with refractory temporal lobe epilepsy is associated with astrocyte activation, inflammation, and altered expression of channels and receptors. Neuroscience 220:237–246. https://doi.org/10.1016/j.neuroscience.2012.06.002

Daw MI, Scott HL, Isaac JTR (2007) Developmental synaptic plasticity at the Thalamocortical input to barrel cortex: mechanisms and roles. Mol Cell Neurosci 34:493–502. https://doi.org/10.1016/j.mcn.2007.01.001

Delorme R, Krebs M-O, Chabane N et al (2004) Frequency and transmission of glutamate receptors GRIK2 and GRIK3 polymorphisms in patients with obsessive compulsive disorder. Neuroreport 15:699–702. https://doi.org/10.1097/00001756-200403220-00025

Duan G-F, Ye Y, Xu S et al (2018) Signal peptide represses GluK1 surface and synaptic trafficking through binding to amino-terminal domain. Nat Commun 9:4879. https://doi.org/10.1038/s41467-018-07403-7

Dutta S, Das S, Guhathakurta S et al (2007) Glutamate receptor 6 gene (GluR6 or GRIK2) polymorphisms in the Indian population: a genetic association study on autism spectrum disorder. Cell Mol Neurobiol 27:1035–1047. https://doi.org/10.1007/s10571-007-9193-6

Egebjerg J, Heinemann SF (1993) Ca2+ permeability of unedited and edited versions of the kainate selective glutamate receptor GluR6. Proc Natl Acad Sci U S A 90:755–759

Eubanks JH, Puranam RS, Kleckner NW et al (1993) The gene encoding the glutamate receptor subunit GluR5 is located on human chromosome 21q21.1-22. 1 in the vicinity of the gene for familial amyotrophic lateral sclerosis. Proc Natl Acad Sci U S A 90:178–182. https://doi.org/10.1073/pnas.90.1.178

Everts I, Petroski R, Kizelsztein P et al (1999) Lectin-induced inhibition of desensitization of the kainate receptor GluR6 depends on the activation state and can be mediated by a single native or ectopic N-linked carbohydrate side chain. J Neurosci 19:916–927

Falcón-Moya R, Sihra TS, Rodríguez-Moreno A (2018) Kainate receptors: role in epilepsy. Front Mol Neurosci 11:217. https://doi.org/10.3389/fnmol.2018.00217

Fernandes HB, Catches JS, Petralia RS et al (2009) High affinity kainate receptor subunits are necessary for ionotropic but not metabotropic signaling. Neuron 63:818–829. https://doi.org/10.1016/j.neuron.2009.08.010

Figueiredo TH, Qashu F, Apland JP et al (2010) The GluK1 (GluR5) kainate/AMPA receptor antagonist LY293558 reduces soman-induced seizures and neuropathology. J Pharmacol Exp Ther 336(2):303–312. https://doi.org/10.1124/jpet.110.171835

Fisher MT, Fisher JL (2014) Contributions of different kainate receptor subunits to the properties of recombinant homomeric and heteromeric receptors. Neuroscience 278:70–80. https://doi.org/10.1016/j.neuroscience.2014.08.009

Fisher JL, Mott DD (2012) The auxiliary subunits Neto 1 and Neto2 reduce voltage-dependent inhibition of recombinant Kainate receptors. J Neurosci 32:12928–12933. https://doi.org/10.1523/JNEUROSCI.2211-12.2012

Garcia EP, Mehta S, Blair LA et al (1998) SAP90 binds and clusters kainate receptors causing incomplete desensitization. Neuron 21:727–739. https://doi.org/10.1016/s0896-6273(00)80590-5

Grigorenko E, Glazier S, Bell W et al (1997) Changes in glutamate receptor subunit composition in hippocampus and cortex in patients with refractory epilepsy. J Neurol Sci 153:35–45. https://doi.org/10.1016/s0022-510x(97)00180-9

Grosenbaugh DK, Ross BM, Wagley P, Zanelli SA (2018) The role of Kainate receptors in the pathophysiology of hypoxia-induced seizures in the neonatal mouse. Sci Rep 8:7035. https://doi.org/10.1038/s41598-018-24722-3

Gurung S, Evans AJ, Wilkinson KA, Henley JM (2018) ADAR2-mediated Q/R editing of GluK2 regulates kainate receptor upscaling in response to suppression of synaptic activity. J Cell Sci 131:jcs222273. https://doi.org/10.1242/jcs.222273

He L, Sun J, Gao Y et al (2021) Kainate receptor modulation by NETO2. Nature 599:325–329. https://doi.org/10.1038/s41586-021-03936-y

Herbrechter R, Hube N, Buchholz R, Reiner A (2021) Splicing and editing of ionotropic glutamate receptors: a comprehensive analysis based on human RNA-Seq data. Cell Mol Life Sci 78:5605–5630. https://doi.org/10.1007/s00018-021-03865-z

Hirata Y, Zai CC, Souza RP et al (2012) Association study of GRIK1 gene polymorphisms in schizophrenia: case-control and family-based studies. Hum Psychopharmacol 27:345–351. https://doi.org/10.1002/hup.2233

Hirbec H, Francis JC, Lauri SE et al (2003) Rapid and differential regulation of AMPA and kainate receptors at hippocampal mossy fibre synapses by PICK1 and GRIP. Neuron 37:625–638. https://doi.org/10.1016/s0896-6273(02)01191-1

Huettner JE (2003) Kainate receptors and synaptic transmission. Prog Neurobiol 70:387–407. https://doi.org/10.1016/s0301-0082(03)00122-9

Hughes TE (1994) Transmembrane topology of the glutamate receptors: a tale of novel twists and turns. J Mol Neurosci 5:211–217. https://doi.org/10.1007/BF02736722

Huyghe D, Veran J, Labrousse VF et al (2011) Endocytosis of the glutamate receptor subunit GluK3 controls polarized trafficking. J Neurosci 31:11645–11654. https://doi.org/10.1523/JNEUROSCI.2206-11.2011

Iida I, Konno K, Natsume R et al (2021) A comparative analysis of kainate receptor GluK2 and GluK5 knockout mice in a pure genetic background. Behav Brain Res 405:113194. https://doi.org/10.1016/j.bbr.2021.113194

Izzi C, Barbon A, Kretz R et al (2002) Sequencing of the GRIK1 gene in patients with juvenile absence epilepsy does not reveal mutations affecting receptor structure. Am J Med Genet 114:354–359. https://doi.org/10.1002/ajmg.10254

Jamain S, Betancur C, Quach H et al (2002) Linkage and association of the glutamate receptor 6 gene with autism. Mol Psychiatry 7:302–310. https://doi.org/10.1038/sj.mp.4000979

Jane DE, Lodge D, Collingridge GL (2009) Kainate receptors: pharmacology, function and therapeutic potential. Neuropharmacology 56:90–113. https://doi.org/10.1016/j.neuropharm.2008.08.023

Jaskolski F (2004) Subunit composition and alternative splicing regulate membrane delivery of Kainate receptors. J Neurosci 24:2506–2515. https://doi.org/10.1523/JNEUROSCI.5116-03.2004

Kamiya H, Ozawa S (2000) Kainate receptor-mediated presynaptic inhibition at the mouse hippocampal mossy fibre synapse. J Physiol 523:653–665. https://doi.org/10.1111/j.1469-7793.2000.t01-1-00653.x

Khanra N, Brown PM, Perozzo AM et al (2021) Architecture and structural dynamics of the heteromeric GluK2/K5 kainate receptor. eLife 10:e66097. https://doi.org/10.7554/eLife.66097

Kidd FL, Isaac JT (1999) Developmental and activity-dependent regulation of kainate receptors at thalamocortical synapses. Nature 400:569–573. https://doi.org/10.1038/23040

Kidd FL, Isaac JT (2001) Kinetics and activation of postsynaptic kainate receptors at thalamocortical synapses: role of glutamate clearance. J Neurophysiol 86:1139–1148. https://doi.org/10.1152/jn.2001.86.3.1139

Kilic G, Ismail Kucukali C, Orhan N et al (2010) Are GRIK3 (T928G) gene variants in schizophrenia patients different from those in their first-degree relatives? Psychiatry Res 175:43–46. https://doi.org/10.1016/j.psychres.2008.10.001

Ko S, Zhao M-G, Toyoda H et al (2005) Altered behavioral responses to noxious stimuli and fear in glutamate receptor 5 (GluR5)- or GluR6-deficient mice. J Neurosci 25:977–984. https://doi.org/10.1523/JNEUROSCI.4059-04.2005

Köhler M, Burnashev N, Sakmann B, Seeburg PH (1993) Determinants of Ca2+ permeability in both TM1 and TM2 of high affinity kainate receptor channels: diversity by RNA editing. Neuron 10:491–500. https://doi.org/10.1016/0896-6273(93)90336-p

Konopacki FA, Jaafari N, Rocca DL et al (2011) Agonist-induced PKC phosphorylation regulates GluK2 SUMOylation and kainate receptor endocytosis. Proc Natl Acad Sci U S A 108:19772–19777. https://doi.org/10.1073/pnas.1111575108

Koromina M, Flitton M, Blockley A et al (2019) Damaging coding variants within kainate receptor channel genes are enriched in individuals with schizophrenia, autism and intellectual disabilities. Sci Rep 9:19215. https://doi.org/10.1038/s41598-019-55635-4

Kumar J, Mayer ML (2010) Crystal structures of the glutamate receptor ion channel GluK3 and GluK5 amino-terminal domains. J Mol Biol 404:680–696. https://doi.org/10.1016/j.jmb.2010.10.006

Kumar J, Mayer ML (2013) Functional insights from glutamate receptor Ion Channel structures. Annu Rev Physiol 75:313–337. https://doi.org/10.1146/annurev-physiol-030212-183711

Kumar J, Schuck P, Jin R, Mayer ML (2009) The N-terminal domain of GluR6-subtype glutamate receptor ion channels. Nat Struct Mol Biol 16:631–638. https://doi.org/10.1038/nsmb.1613

Kumar J, Schuck P, Mayer ML (2011) Structure and assembly mechanism for heteromeric kainate receptors. Neuron 71:319–331. https://doi.org/10.1016/j.neuron.2011.05.038

Kumari J, Vinnakota R, Kumar J (2019) Structural and functional insights into GluK3-kainate receptor desensitization and recovery. Sci Rep 9:10254. https://doi.org/10.1038/s41598-019-46770-z

Kumari J, Bendre AD, Bhosale S et al (2020) Structural dynamics of the GluK3-kainate receptor neurotransmitter binding domains revealed by cryo-EM. Int J Biol Macromol 149:1051–1058. https://doi.org/10.1016/j.ijbiomac.2020.01.282

Kuner T, Seeburg PH, Guy HR (2003) A common architecture for K+ channels and ionotropic glutamate receptors? Trends Neurosci 26:27–32. https://doi.org/10.1016/s0166-2236(02)00010-3

Laezza F, Wilding TJ, Sequeira S et al (2007) KRIP6: a novel BTB/kelch protein regulating function of kainate receptors. Mol Cell Neurosci 34:539–550. https://doi.org/10.1016/j.mcn.2006.12.003

Lee J-H, Lee J-M, Ramos EM et al (2012) TAA repeat variation in the GRIK2 gene does not influence age at onset in Huntington's disease. Biochem Biophys Res Commun 424:404–408. https://doi.org/10.1016/j.bbrc.2012.06.120

Lerma J (2003) Roles and rules of kainate receptors in synaptic transmission. Nat Rev Neurosci 4: 481–495. https://doi.org/10.1038/nrn1118

Lerma J (2006) Kainate receptor physiology. Curr Opin Pharmacol 6:89–97. https://doi.org/10.1016/j.coph.2005.08.004

Lerma J (2011) Net (o) excitement for kainate receptors. Nat Neurosci 14:808–810. https://doi.org/10.1038/nn.2864

Lerma J, Marques JM (2013) Kainate receptors in health and disease. Neuron 80:292–311. https://doi.org/10.1016/j.neuron.2013.09.045

Lerma J, Paternain AV, Naranjo JR, Mellström B (1993) Functional kainate-selective glutamate receptors in cultured hippocampal neurons. Proc Natl Acad Sci U S A 90:11688–11692. https://doi.org/10.1073/pnas.90.24.11688

Li YJ, Duan GF, Sun JH, Wu D, Ye C, Zang YY, Chen GQ, Shi YY, Wang J, Zhang W, Shi YS (2019) Neto proteins regulate gating of the kainate-type glutamate receptor GluK2 through two binding sites. J Biol Chem 294(47):17889–17902. https://doi.org/10.1074/jbc.RA119.008631

Li J-M, Zeng Y-J, Peng F et al (2010) Aberrant glutamate receptor 5 expression in temporal lobe epilepsy lesions. Brain Res 1311:166–174. https://doi.org/10.1016/j.brainres.2009.11.024

Litwin DB, Paudyal N, Carrillo E et al (2020) The structural arrangement and dynamics of the heteromeric GluK2/GluK5 kainate receptor as determined by sm FRET. Biochim Biophys Acta Biomembr 1862:183001. https://doi.org/10.1016/j.bbamem.2019.05.023

Lowry ER, Kruyer A, Norris EH et al (2013) The GluK4 Kainate receptor subunit regulates memory, mood, and Excitotoxic neurodegeneration. Neuroscience 235:215–225. https://doi.org/10.1016/j.neuroscience.2013.01.029

Lucarini N, Verrotti A, Napolioni V et al (2007) Genetic polymorphisms and idiopathic generalized epilepsies. Pediatr Neurol 37:157–164. https://doi.org/10.1016/j.pediatrneurol.2007.06.001

MacDonald ME, Vonsattel JP, Shrinidhi J et al (1999) Evidence for the GluR6 gene associated with younger onset age of Huntington's disease. Neurology 53:1330–1332. https://doi.org/10.1212/wnl.53.6.1330

Martin S, Nishimune A, Mellor JR, Henley JM (2007) SUMOylation regulates kainate-receptor-mediated synaptic transmission. Nature 447:321–325. https://doi.org/10.1038/nature05736

Matsuda K, Budisantoso T, Mitakidis N et al (2016) Transsynaptic modulation of Kainate receptor functions by C1q-like proteins. Neuron 90:752–767. https://doi.org/10.1016/j.neuron.2016.04.001

Matute C (2011) Therapeutic potential of kainate receptors. CNS Neurosci Ther 17:661–669. https://doi.org/10.1111/j.1755-5949.2010.00204.x

Mayer ML (2005) Crystal structures of the GluR5 and GluR6 ligand binding cores: molecular mechanisms underlying kainate receptor selectivity. Neuron 45:539–552. https://doi.org/10.1016/j.neuron.2005.01.031

Mayer ML (2016) Structural biology of glutamate receptor ion channel complexes. Curr Opin Struct Biol 41:119–127. https://doi.org/10.1016/j.sbi.2016.07.002

Mayer ML (2021) Structural biology of kainate receptors. Neuropharmacology 190:108511. https://doi.org/10.1016/j.neuropharm.2021.108511

Mayer ML, Ghosal A, Dolman NP, Jane DE (2006) Crystal structures of the kainate receptor GluR5 ligand binding core dimer with novel GluR5-selective antagonists. J Neurosci 26:2852–2861. https://doi.org/10.1523/JNEUROSCI.0123-06.2005

Melyan Z, Wheal HV, Lancaster B (2002) Metabotropic-mediated kainate receptor regulation of IsAHP and excitability in pyramidal cells. Neuron 34:107–114. https://doi.org/10.1016/s0896-6273(02)00624-4

Meyerson JR, Kumar J, Chittori S et al (2014) Structural mechanism of glutamate receptor activation and desensitization. Nature 514:328–334. https://doi.org/10.1038/nature13603

Meyerson JR, Chittori S, Merk A et al (2016) Structural basis of kainate subtype glutamate receptor desensitization. Nature 537:567–571. https://doi.org/10.1038/nature19352

Michishita M, Ikeda T, Nakashiba T et al (2003) A novel gene, Btcl 1, encoding CUB and LDLa domains is expressed in restricted areas of mouse brain. Biochem Biophys Res Commun 306: 680–686. https://doi.org/10.1016/s0006-291x(03)01035-0

Michishita M, Ikeda T, Nakashiba T et al (2004) Expression of Btcl2, a novel member of Btcl gene family, during development of the central nervous system. Brain Res Dev Brain Res 153:135–142. https://doi.org/10.1016/j.devbrainres.2004.06.012

Møllerud S, Frydenvang K, Pickering DS, Kastrup JS (2017) Lessons from crystal structures of kainate receptors. Neuropharmacology 112:16–28. https://doi.org/10.1016/j.neuropharm.2016.05.014

Mulle C, Sailer A, Pérez-Otaño I et al (1998) Altered synaptic physiology and reduced susceptibility to kainate-induced seizures in GluR6-deficient mice. Nature 392:601–605. https://doi.org/10.1038/33408

Nakanishi N, Shneider NA, Axel R (1990) A family of glutamate receptor genes: evidence for the formation of heteromultimeric receptors with distinct channel properties. Neuron 5:569–581. https://doi.org/10.1016/0896-6273(90)90212-x

Nanao MH, Green T, Stern-Bach Y et al (2005) Structure of the kainate receptor subunit GluR6 agonist-binding domain complexed with domoic acid. Proc Natl Acad Sci U S A 102:1708–1713. https://doi.org/10.1073/pnas.0409573102

Nasu-Nishimura Y, Hurtado D, Braud S et al (2006) Identification of an endoplasmic reticulum-retention motif in an intracellular loop of the kainate receptor subunit KA2. J Neurosci 26:7014–7021. https://doi.org/10.1523/JNEUROSCI.0573-06.2006

Nasu-Nishimura Y, Jaffe H, Isaac JTR, Roche KW (2010) Differential regulation of kainate receptor trafficking by phosphorylation of distinct sites on GluR6. J Biol Chem 285:2847–2856. https://doi.org/10.1074/jbc.M109.081141

Naur P, Vestergaard B, Skov LK et al (2005) Crystal structure of the kainate receptor GluR5 ligand-binding core in complex with (S)-glutamate. FEBS Lett 579:1154–1160. https://doi.org/10.1016/j.febslet.2005.01.012

Ng D, Pitcher GM, Szilard RK et al (2009) Neto 1 is a novel CUB-domain NMDA receptor-interacting protein required for synaptic plasticity and learning. PLoS Biol 7:e41. https://doi.org/10.1371/journal.pbio.1000041

O'Hara PJ, Sheppard PO, Thøgersen H et al (1993) The ligand-binding domain in metabotropic glutamate receptors is related to bacterial periplasmic binding proteins. Neuron 11:41–52. https://doi.org/10.1016/0896-6273(93)90269-w

Orlandi C, Barbon A, Barlati S (2012) Activity regulation of adenosine deaminases acting on RNA (ADARs). Mol Neurobiol 45:61–75. https://doi.org/10.1007/s12035-011-8220-2

Paddock S, Laje G, Charney D et al (2007) Association of GRIK4 with outcome of antidepressant treatment in the STAR*D cohort. Am J Psychiatry 164:1181–1188. https://doi.org/10.1176/appi.ajp.2007.06111790

Palacios-Filardo J, Aller MI, Lerma J (2016) Synaptic targeting of kainate receptors. Cereb Cortex 26(4):1464–1472. https://doi.org/10.1093/cercor/bhu244

Palecek J, Neugebauer V, Carlton SM et al (2004) The effect of a kainate GluR5 receptor antagonist on responses of spinothalamic tract neurons in a model of peripheral neuropathy in primates. Pain 111:151–161. https://doi.org/10.1016/j.pain.2004.06.012

Parker BL, Thaysen-Andersen M, Solis N et al (2013) Site-specific glycan-peptide analysis for determination of N-glycoproteome heterogeneity. J Proteome Res 12:5791–5800. https://doi.org/10.1021/pr400783j

Paschen W, Blackstone CD, Huganir RL, Ross CA (1994) Human GluR6 kainate receptor (GRIK2): molecular cloning, expression, polymorphism, and chromosomal assignment. Genomics 20:435–440. https://doi.org/10.1006/geno.1994.1198

Paternain AV, Morales M, Lerma J (1995) Selective antagonism of AMPA receptors unmasks kainate receptor-mediated responses in hippocampal neurons. Neuron 14:185–189. https://doi.org/10.1016/0896-6273(95)90253-8
Peret A, Christie LA, Ouedraogo DW et al (2014) Contribution of aberrant GluK2-containing kainate receptors to chronic seizures in temporal lobe epilepsy. Cell Rep 8:347–354. https://doi.org/10.1016/j.celrep.2014.06.032
Perrais D, Coussen F, Mulle C (2009) Atypical functional properties of GluK3-containing Kainate receptors. J Neurosci 29:15499–15510. https://doi.org/10.1523/JNEUROSCI.2724-09.2009
Pickard BS, Malloy MP, Christoforou A et al (2006) Cytogenetic and genetic evidence supports a role for the kainate-type glutamate receptor gene, GRIK4, in schizophrenia and bipolar disorder. Mol Psychiatry 11:847–857. https://doi.org/10.1038/sj.mp.4001867
Pickard BS, Knight HM, Hamilton RS et al (2008) A common variant in the 3′UTR of the GRIK4 glutamate receptor gene affects transcript abundance and protects against bipolar disorder. Proc Natl Acad Sci 105:14940–14945. https://doi.org/10.1073/pnas.0800643105
Pickering DS, Taverna FA, Salter MW, Hampson DR (1995) Palmitoylation of the GluR6 kainate receptor. Proc Natl Acad Sci U S A 92:12090–12094. https://doi.org/10.1073/pnas.92.26.12090
Pinheiro P, Mulle C (2006) Kainate receptors. Cell Tissue Res 326:457–482. https://doi.org/10.1007/s00441-006-0265-6
Plested AJR, Mayer ML (2007) Structure and mechanism of kainate receptor modulation by anions. Neuron 53:829–841. https://doi.org/10.1016/j.neuron.2007.02.025
Plested AJR, Vijayan R, Biggin PC, Mayer ML (2008) Molecular basis of kainate receptor modulation by sodium. Neuron 58:720–735. https://doi.org/10.1016/j.neuron.2008.04.001
Porter BE, Cui X-N, Brooks-Kayal AR (2006) Status epilepticus differentially alters AMPA and kainate receptor subunit expression in mature and immature dentate granule neurons. Eur J Neurosci 23:2857–2863. https://doi.org/10.1111/j.1460-9568.2006.04839.x
Puranam RS, Eubanks JH, Heinemann SF, McNamara JO (1993) Chromosomal localization of gene for human glutamate receptor subunit-7. Somat Cell Mol Genet 19:581–588. https://doi.org/10.1007/BF01233385
Ren Z, Riley NJ, Garcia EP et al (2003a) Multiple trafficking signals regulate kainate receptor KA2 subunit surface expression. J Neurosci 23:6608–6616
Ren Z, Riley NJ, Needleman LA et al (2003b) Cell surface expression of GluR5 kainate receptors is regulated by an endoplasmic reticulum retention signal. J Biol Chem 278:52700–52709. https://doi.org/10.1074/jbc.M309585200
Rodríguez-Moreno A, Herreras O, Lerma J (1997) Kainate receptors presynaptically downregulate GABAergic inhibition in the rat hippocampus. Neuron 19:893–901. https://doi.org/10.1016/s0896-6273(00)80970-8
Rubinsztein DC, Leggo J, Chiano M et al (1997) Genotypes at the GluR6 kainate receptor locus are associated with variation in the age of onset of Huntington disease. Proc Natl Acad Sci U S A 94:3872–3876. https://doi.org/10.1073/pnas.94.8.3872
Ruscheweyh R, Sandkühler J (2002) Role of kainate receptors in nociception. Brain Res Brain Res Rev 40:215–222. https://doi.org/10.1016/s0165-0173(02)00203-5
Salinas GD, Blair LAC, Needleman LA et al (2006) Actinfilin is a Cul 3 substrate adaptor, linking GluR6 Kainate receptor subunits to the ubiquitin-proteasome pathway*. J Biol Chem 281: 40164–40173. https://doi.org/10.1074/jbc.M608194200
Sampaio AS, Fagerness J, Crane J et al (2011) Association between polymorphisms in GRIK2 gene and obsessive-compulsive disorder: a family-based study. CNS Neurosci Ther 17:141–147. https://doi.org/10.1111/j.1755-5949.2009.00130.x
Sander T, Hildmann T, Kretz R et al (1997) Allelic association of juvenile absence epilepsy with a GluR5 kainate receptor gene (GRIK1) polymorphism. Am J Med Genet 74:416–421
Schiffer HH, Heinemann SF (2007) Association of the human kainate receptor GluR7 gene (GRIK3) with recurrent major depressive disorder. Am J Med Genet Part B Neuropsychiatr Genet Off Publ Int Soc Psychiatr Genet 144B:20–26. https://doi.org/10.1002/ajmg.b.30374

Schiffer HH, Swanson GT, Heinemann SF (1997) Rat GluR7 and a carboxy-terminal splice variant, GluR7b, are functional kainate receptor subunits with a low sensitivity to glutamate. Neuron 19: 1141–1146. https://doi.org/10.1016/s0896-6273(00)80404-3

Selvakumar P, Lee J, Khanra N, He C, Munguba H, Kiese L, Broichhagen J, Reiner A, Levitz J, Meyerson JR (2021) Structural and compositional diversity in the kainate receptor family. Cell Rep 37(4):109891. https://doi.org/10.1016/j.celrep.2021.109891

Shaltiel G, Maeng S, Malkesman O et al (2008) Evidence for the involvement of the kainate receptor subunit GluR6 (GRIK2) in mediating behavioral displays related to behavioral symptoms of mania. Mol Psychiatry 13:858–872. https://doi.org/10.1038/mp.2008.20

Shibata H, Shibata A, Ninomiya H et al (2002) Association study of polymorphisms in the GluR6 kainate receptor gene (GRIK2) with schizophrenia. Psychiatry Res 113:59–67. https://doi.org/10.1016/s0165-1781(02)00231-7

Shuang M, Liu J, Jia MX et al (2004) Family-based association study between autism and glutamate receptor 6 gene in Chinese Han trios. Am J Med Genet Part B Neuropsychiatr Genet Off Publ Int Soc Psychiatr Genet 131B:48–50. https://doi.org/10.1002/ajmg.b.30025

Sobolevsky AI, Rosconi MP, Gouaux E (2009) X-ray structure, symmetry and mechanism of an AMPA-subtype glutamate receptor. Nature 462:745–756. https://doi.org/10.1038/nature08624

Sommer B, Köhler M, Sprengel R, Seeburg PH (1991) RNA editing in brain controls a determinant of ion flow in glutamate-gated channels. Cell 67:11–19. https://doi.org/10.1016/0092-8674(91)90568-j

Sommer B, Burnashev N, Verdoorn TA et al (1992) A glutamate receptor channel with high affinity for domoate and kainate. EMBO J 11:1651–1656

Stepulak A, Rola R, Polberg K, Ikonomidou C (2014) Glutamate and its receptors in cancer. J Neural Transm 121:933–944. https://doi.org/10.1007/s00702-014-1182-6

Stern-Bach Y, Bettler B, Hartley M et al (1994) Agonist selectivity of glutamate receptors is specified by two domains structurally related to bacterial amino acid-binding proteins. Neuron 13:1345–1357. https://doi.org/10.1016/0896-6273(94)90420-0

Straub C, Hunt DL, Yamasaki M et al (2011) Distinct functions of kainate receptors in the brain are determined by the auxiliary subunit Neto 1. Nat Neurosci 14:866–873. https://doi.org/10.1038/nn.2837

Swanson GT, Heinemann SF (1998) Heterogeneity of homomeric GluR5 kainate receptor desensitization expressed in HEK293 cells. J Physiol 513:639–646. https://doi.org/10.1111/j.1469-7793.1998.639ba.x

Swanson GT, Feldmeyer D, Kaneda M, Cull-Candy SG (1996) Effect of RNA editing and subunit co-assembly single-channel properties of recombinant kainate receptors. J Physiol 492(Pt 1): 129–142. https://doi.org/10.1113/jphysiol.1996.sp021295

Swanson GT, Gereau RW, Green T, Heinemann SF (1997) Identification of amino acid residues that control functional behavior in GluR5 and GluR6 Kainate receptors. Neuron 19:913–926. https://doi.org/10.1016/S0896-6273(00)80972-1

Szpirer C, Molné M, Antonacci R et al (1994) The genes encoding the glutamate receptor subunits KA1 and KA2 (GRIK4 and GRIK5) are located on separate chromosomes in human, mouse, and rat. Proc Natl Acad Sci U S A 91:11849–11853. https://doi.org/10.1073/pnas.91.25.11849

Tang M, Pelkey KA, Ng D et al (2011) Neto1 is an auxiliary subunit of native synaptic kainate receptors. J Neurosci 31:10009–10018. https://doi.org/10.1523/JNEUROSCI.6617-10.2011

Tashiro A, Dunaevsky A, Blazeski R et al (2003) Bidirectional regulation of hippocampal mossy fiber filopodial motility by Kainate receptors: a two-step model of synaptogenesis. Neuron 38: 773–784. https://doi.org/10.1016/S0896-6273(03)00299-X

Traynelis SF, Wollmuth LP, McBain CJ et al (2010) Glutamate receptor ion channels: structure, regulation, and function. Pharmacol Rev 62:405–496. https://doi.org/10.1124/pr.109.002451

Valbuena S, Lerma J (2021) Losing balance: Kainate receptors and psychiatric disorders comorbidities. Neuropharmacology 191:108558. https://doi.org/10.1016/j.neuropharm.2021.108558

Veran J, Kumar J, Pinheiro PS et al (2012) Zinc potentiates GluK3 glutamate receptor function by stabilizing the ligand binding domain dimer interface. Neuron 76:565–578. https://doi.org/10.1016/j.neuron.2012.08.027

Vernon CG, Copits BA, Stolz JR et al (2017) N-glycan content modulates kainate receptor functional properties. J Physiol 595:5913–5930. https://doi.org/10.1113/JP274790

Vignes M, Collingridge GL (1997) The synaptic activation of kainate receptors. Nature 388:179–182. https://doi.org/10.1038/40639

Vinnakota R, Dhingra S, Kumari J, Ansari MY, Shukla E, Nerkar MD, Kumar J (2021) Role of Neto1 extracellular domain in modulation of kainate receptors. Int J Biol Macromol 192:525–536. https://doi.org/10.1016/j.ijbiomac.2021.10.001

Vivithanaporn P, Yan S, Swanson GT (2006) Intracellular trafficking of KA2 kainate receptors mediated by interactions with coatomer protein complex I (COPI) and 14-3-3 chaperone systems. J Biol Chem 281:15475–15484. https://doi.org/10.1074/jbc.M512098200

Wang JX, Furukawa H (2019) Dissecting diverse functions of NMDA receptors by structural biology. Curr Opin Struct Biol 54:34–42. https://doi.org/10.1016/j.sbi.2018.12.009

Whalley HC, Pickard BS, McIntosh AM et al (2009) Modulation of hippocampal activation by genetic variation in the GRIK4 gene. Mol Psychiatry 14:465. https://doi.org/10.1038/mp.2009.31

Wilding TJ, Huettner JE (1995) Differential antagonism of alpha-amino-3-hydroxy-5-methyl-4-isoxazolepropionic acid-preferring and kainate-preferring receptors by 2,3-benzodiazepines. Mol Pharmacol 47:582–587

Wilkinson KA, Henley JM (2010) Mechanisms, regulation and consequences of protein SUMOylation. Biochem J 428:133–145. https://doi.org/10.1042/BJ20100158

Wilson GM, Flibotte S, Chopra V et al (2006) DNA copy-number analysis in bipolar disorder and schizophrenia reveals aberrations in genes involved in glutamate signaling. Hum Mol Genet 15:743–749. https://doi.org/10.1093/hmg/ddi489

Wisden W, Seeburg PH (1993) A complex mosaic of high-affinity kainate receptors in rat brain. J Neurosci 13:3582–3598

Wo ZG, Oswald RE (1995) Unraveling the modular design of glutamate-gated ion channels. Trends Neurosci 18:161–168. https://doi.org/10.1016/0166-2236(95)93895-5

Woo T-UW, Shrestha K, Amstrong C et al (2007) Differential alterations of kainate receptor subunits in inhibitory interneurons in the anterior cingulate cortex in schizophrenia and bipolar disorder. Schizophr Res 96:46–61. https://doi.org/10.1016/j.schres.2007.06.023

Yuan H, Low C-M, Moody OA et al (2015) Ionotropic GABA and glutamate receptor mutations and human neurologic diseases. Mol Pharmacol 88:203–217. https://doi.org/10.1124/mol.115.097998

Zhang W, St-Gelais F, Grabner CP et al (2009) A transmembrane accessory subunit that modulates kainate-type glutamate receptors. Neuron 61:385–396. https://doi.org/10.1016/j.neuron.2008.12.014

Zhao H, Lomash S, Chittori S et al (2017) Preferential assembly of heteromeric kainate and AMPA receptor amino terminal domains. eLife 6:e32056. https://doi.org/10.7554/eLife.32056

Zhuo M (2017) Cortical kainate receptors and behavioral anxiety. Mol Brain 10:16. https://doi.org/10.1186/s13041-017-0297-8

Chapter 11
Structure, Function, and Variations of the Photosystem I-Antenna Supercomplex from Different Photosynthetic Organisms

Jian-Ren Shen

Abstract Photosystem I (PSI) is a protein complex functioning in light-induced charge separation, electron transfer, and reduction reactions of ferredoxin in photosynthesis, which finally results in the reduction of $NAD(P)^-$ to NAD(P)H required for the fixation of carbon dioxide. In eukaryotic algae, PSI is associated with light-harvesting complex I (LHCI) subunits, forming a PSI-LHCI supercomplex. LHCI harvests and transfers light energy to the PSI core, where charge separation and electron transfer reactions occur. During the course of evolution, the number and sequences of protein subunits and the pigments they bind in LHCI change dramatically depending on the species of organisms, which is a result of adaptation of organisms to various light environments. In this chapter, I will describe the structure of various PSI-LHCI supercomplexes from different organisms solved so far either by X-ray crystallography or by cryo-electron microscopy, with emphasis on the differences in the number, structures, and association patterns of LHCI subunits associated with the PSI core found in different organisms.

Keywords Photosystem I · Light-harvesting complex · X-ray crystallography · Cryo-electron microscopy · Energy harvesting · Energy transfer

Introduction

The photosystem I (PSI)-light harvesting complex I (LHCI) is one of the two supercomplexes involved in photosynthetic light energy harvesting and electron transfer reactions of various photosynthetic organisms. It uses light energy to initiate

J.-R. Shen (✉)
Research Institute for Interdisciplinary Science, and Graduate School of Natural Science and Technology, Okayama University, Okayama, Japan

Key Laboratory of Photobiology, Institute of Botany, Chinese Academy of Sciences, Beijing, China
e-mail: shen@cc.okayama-u.ac.jp

J. R. Harris, J. Marles-Wright (eds.), *Macromolecular Protein Complexes IV*, Subcellular Biochemistry 99, https://doi.org/10.1007/978-3-031-00793-4_11

the charge separation in its reaction center called P_{700}, which is composed of a chlorophyll (Chl) dimer (Fromme et al. 2001; Jordan et al. 2001; Golbeck 2006; Nelson and Junge 2015). The oxidized P_{700} accepts an electron transferred from water via photosystem II (PSII) (Umena et al. 2011; Shen 2015), which is donated to P_{700} via plastocyanin or cytochrome (cyt) c_6 at the electron donor side of PSI. On the other hand, the electron ejected from P_{700} is transferred to several electron acceptors including Chls, phylloquinone, and iron-sulfur centers, and finally to ferredoxin (Fd), which is then used either to reduce NAD(P) via Fd-NAD(P) oxidoreductase or come back to PSI again via cyclic electron transfer (Blankenship 2021; Golbeck 2006; Nelson and Junge 2015). The NAD(P)H generated through the PSI-mediated electron transfer provides the reducing power to reduce CO_2 into sugars.

The abovementioned electron transfer reactions are performed at the "core" of the PSI complex located mostly at the thylakoid membranes of various photosynthetic organisms. Only the PSI core of a primitive cyanobacterium *Gloeobacter* (Nelissen et al. 1995) is located at the cytoplasmic membrane, as this cyanobacterium does not possess thylakoid membrane (Guglielmi et al. 1981). The PSI core complex consists of 9–11 transmembrane subunits, three extrinsic subunits, ~95 Chls, a number of carotenoids, lipids, iron-sulfur centers, phylloquinones, etc., with a molecular weight of around 315–330 kDa for a monomer (Jordan et al. 2001; Qin et al. 2015; Mazor et al. 2017). The composition of this core complex is relatively conserved from prokaryotic organism cyanobacteria to eukaryotes including various algae and higher plants. However, minor changes are observed in some small protein subunits and peripheral Chls, carotenoids, etc., among different organisms (Suga and Shen 2020). In addition, the oligomerization state of the core complex varies among prokaryotes and eukaryotes. In the prokaryotic cyanobacteria, PSI exists either as a trimer (Jordan et al. 2001) or tetramer (Li et al. 2019; Zheng et al. 2019; Kato et al. 2019), whereas in the eukaryotic algae and higher plants, it is exclusively a monomer (Qin et al. 2015; Mazor et al. 2017; Pi et al. 2018; Antoshvili et al. 2019; Qin et al. 2019; Su et al. 2019; Suga et al. 2019; Wang et al. 2021). These changes are not directly related to the electron transfer reactions but are a result of evolutionary changes that occurred during the evolution from prokaryotic cyanobacteria to eukaryotes and are indispensable for the adaptation of PSI to various living environments in which the intensity and quality of light changes dramatically.

In the prokaryotic cyanobacteria, there are no transmembrane light-harvesting subunits under normal growth conditions. The light-harvesting function is partially fulfilled by phycobilisomes which are extrinsic protein subunits associated at the stromal side of the thylakoid membranes (Liu et al. 2013; Watanabe et al. 2014; Zhang et al. 2017; Adir et al. 2020; Ma et al. 2020). Under iron-deficient conditions, an IisA protein will be expressed which is a transmembrane protein and has a sequence similarity with that of CP43, a Chl-binding protein of the PSII core (Laudenbach and Straus 1988; Burnap et al. 1993; Toporik et al. 2019; Cao et al. 2020; Akita et al. 2020). In eukaryotic algae and higher plants, the PSI core complex is surrounded by transmembrane subunits of LHCI, forming a PSI-LHCI supercomplex (Qin et al. 2015; Suga et al. 2016; Mazor et al. 2017; Caspy and Nelson 2018; Netzer-El et al. 2019; Pi et al. 2018; Antoshvili et al. 2019; Qin et al.

2019; Su et al. 2019; Suga et al. 2019; Nagao et al. 2020b; Xu et al. 2020; Wang et al. 2021). LHCI belongs to the Lhc super gene family (Durnford et al. 1999; Green 2003; Hoffman et al. 2011; Dall'Osto et al. 2015), but the numbers and sequences of LHCI associated with the PSI core from different organisms, as well as the pigments that each LHCI subunit binds, differ significantly, which is one of the reasons for the adaptation of various organisms to their individual light environment (Suga and Shen 2020). In this chapter, I will first describe the common electron transfer chain of PSI, followed by descriptions on the composition and organization of protein subunits and pigments in the PSI-LHCI supercomplex from various organisms.

Electron Transfer Chain of PSI

The electron transfer chain (ETC) of PSI is organized into two pseudosymmetric branches called A- and B-branches. Each branch consists of three Chls and one phylloquinone (Q_K) and is coordinated by PsaA and PsaB subunits (see below), respectively, in most organisms (Fig. 11.1a). The electron transfer was thought to start at the reaction center Chls P700, which has a midpoint redox potential (E_m) of 470–490 mV (Kobayashi et al. 2013; Blankenship 2021). P700 is a Chl *a′*/Chl *a* heterodimer (P_A/P_B) in most organisms, where Chl *a′* is the $C13^2$ epimer of Chl

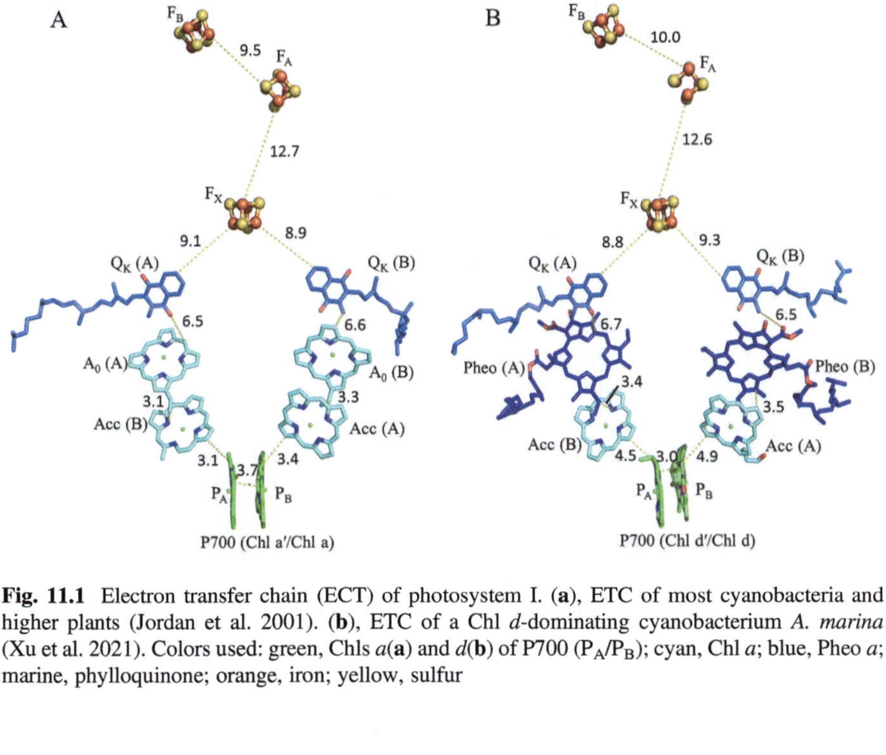

Fig. 11.1 Electron transfer chain (ECT) of photosystem I. (**a**), ETC of most cyanobacteria and higher plants (Jordan et al. 2001). (**b**), ETC of a Chl *d*-dominating cyanobacterium *A. marina* (Xu et al. 2021). Colors used: green, Chls *a*(**a**) and *d*(**b**) of P700 (P_A/P_B); cyan, Chl *a*; blue, Pheo *a*; marine, phylloquinone; orange, iron; yellow, sulfur

a (Jordan et al. 2001; Kobayashi et al. 2005). These two Chls are arranged in parallel, with their interplane distances in the range of 3.0–4.0 Å and Mg-Mg distances of 6–7 Å (Jordan et al. 2001; Qin et al. 2015; Mazor et al. 2017; Wang et al. 2021). Upon absorption of light or acceptance of energy from peripheral antennas, P700 ejects an electron which is accepted by an adjacent Chl *a* acceptor called "accessory Chl" or A_{CC} located at edge-to-edge distances of 3.1–3.5 Å and Mg-Mg distances of 11.5–12.0 Å from P700 (Fig. 11.1a). However, evidence has been presented to suggest that the primary charge separation takes place at one of the two accessory Chls (Acc) and the positive hole then rapidly migrates to P700 and is partially stabilized there (Holzwarth et al. 2006; Muller et al. 2010). These two mechanisms lead to the same charge-separated state, namely, $P700^{+}A_0^{-}$, where the electron of P700 is transferred to A_0, the first stable electron acceptor of PSI which is a pair of Chl *a*, in ~1.5 ps (Blankenship 2021; Golbeck 2006). Thus, these two mechanisms are difficult to distinguish by experiments. The electron is subsequently transferred to one of the two phylloquinone acceptor Q_X called A_1 at edge-to-edge distances of 6.5–6.6 Å in ~25 ps (Fig. 11.1). From phylloquinone, the electron is further transferred to three iron-sulfur centers F_X, F_A, and F_B at the stromal side and then finally to Fd. Among these three iron-sulfur centers, F_X is coordinated at the interface of PsaA and PsaB at the stromal surface and forms the merging point for the A and B branches, whereas F_A and F_B are coordinated by PsaC at the stromal side. E_m of F_B is ~580 mV, which is somewhat more negative than that of F_A (~520 mV), yielding an uphill electron transfer from F_A to F_B (Golbeck 1999). This uphill electron transfer is possible given that the donor and acceptor are in close proximity and that the uphill electron transfer is followed by downhill transfers that make the overall process energetically favorable. The electron is finally transferred from F_B to Fd, yielding reduced Fd.

$P700^{+}$ is re-reduced by either a heme protein cyt c_6 or a blue copper protein plastocyanin (Blankenship 2021; Golbeck 2006), both of which accept electrons from cyt *f* of the cyt b_6/f complex, which come finally from water by the water oxidation reaction of PSII. Cyt c_6 and plastocyanin exist in different organisms: Red and brown algae have cyt c_6 only, higher plants have plastocyanin only, whereas cyanobacteria and green algae have both cyt c_6 and plastocyanin (Sandmann et al. 1986; Sandmann 1986; De la Rosa et al. 2002; Hervás et al. 2003). These two proteins are expressed depending on the growth environment. Under normal growth conditions where iron is available, cyanobacteria and brown algae usually use cyt c_6 as the mobile electron donor to PSI, whereas green algae usually use plastocyanin as the electron donor. Under iron-limiting conditions, the expression of plastocyanin is enhanced, making it a dominating donor. In higher plants, no cyt c_6 is present (proteins homologous to cyt c_6 are found, but they do not fulfill the function of donating electrons to P700, see Slater et al. (2021)), and plastocyanin is the only electron donor (Weigel et al. 2003).

The abovementioned ETC is common to almost all oxygenic photosynthetic organisms ranging from most cyanobacteria to higher plants. However, in Chl *d*-dominating cyanobacteria represented by *Acaryochloris marina*, the reaction center P700 (also called P740 due to the peak wavelength of 740 nm in the light-induced redox difference spectrum of PSI) is found to be a Chl *d′/d* heterodimer, instead of a Chl *a′/a* heterodimer, and the second electron acceptor A_0 is a pair of pheophytin *a* (Pheo *a*) instead of Chl *a* (Fig. 11.1b) (Hu et al. 1995; Akiyama et al. 2001; Gisriel et al. 2021; Hamaguchi et al. 2021; Xu et al. 2021). These alterations likely result in changes in the redox potentials and/or electron transfer dynamics in the PSI core. Indeed, E_m of P740 was found to be 439 mV (Itoh et al. 2007; Bailleul et al. 2008; Schenderlein et al. 2008; Tomo et al. 2008), which is slightly lower than that of P700 as mentioned above. The far-red light of 740 nm has a slightly lower energy of 1.68 eV compared with that of 700 nm (1.77 eV). This may make the A_0 a Pheo *a* molecule, which has a lower E_m than that of Chl *a* or Chl *d* (Kobayashi et al. 2013; Hamaguchi et al. 2021; Xu et al. 2021). These changes may enable the Chl *d*-dominating cyanobacteria survival efficiently under low light conditions. However, the detailed effects of these alterations on the PSI electron transfer are yet to be determined.

One important feature of the PSI-ETC is that electrons can be transferred through both A- and B-branches (Joliot and Joliot 1999; Guergova-Kuras et al. 2001; Muhiuddin et al. 2001; Ramesh et al. 2004), which is different from PSII and the purple photosynthetic bacterial reaction center where electrons are transferred preferentially through one branch. The bidirectional electron transfer has been shown by time-resolved spectroscopy and mutagenesis studies. However, the electron transfer occurs primarily along the A-branch in PSI (Yang et al. 1998; Zech et al. 2000; Guergova-Kuras et al. 2001; Xu et al. 2003; Makita and Hastings 2015), and analyses of the transient absorption changes suggested that the ratios of the A- and B-branch electron transfers are 95:5 at 77 K and 77:23 at 298 K (Makita and Hastings 2015). According to a recent theoretical study, this asymmetry in the electron transfer originates from P_A as a Chl epimer (Mitsuhashi et al. 2021). It is not clear at present how this asymmetry will impact the electron transfer activity and how the epimerized Chl is specifically incorporated into one of the Chl pair of P700.

The PSI-ETC mentioned above takes place at the PSI core, which is associated with different types of light-harvesting antennas depending on the species of the organisms. In cyanobacteria, the PSI core exists alone or is associated with a small amount of phycobiliproteins (Liu et al. 2013; Watanabe et al. 2014), which makes the ratio of PSI/PSII as high as 9:0 to 9.5:0.5 on the Chl basis for the proper growth of the cells (Rogner et al. 1990; Rakhimberdieva et al. 2001). On the other hand, in eukaryotic algae and higher plants, the hydrophilic antenna phycobiliproteins disappear, and the PSI core is surrounded by various numbers of LHCI (FCPI) subunits. In the following, I will describe the structures of the PSI core and its complex with the light-harvesting antennas from various organisms.

Structure of the Cyanobacterial PSI Core

Cyanobacterial PSI does not have a transmembrane light-harvesting system, and they exist predominately in a trimer. However, PSI tetramers are also found in heterocyst-forming cyanobacteria. Under iron-deficient conditions, a subunit called

IsiA is expressed and encircles the PSI core. These three types of structures are described below.

Structure of the PSI Trimer

As stated above, the composition of PSI core is largely conserved from cyanobacteria to higher plants (Table 11.1). However, differences do exist in the small molecular weight subunits among the PSI core of different organisms. The cyanobacterial PSI core contains 12 subunits (PsaA, B, C, D, E, F, I, J, K, L, M, and PsaX). Among them, ten are common to all PSI cores from cyanobacteria to higher plants. PsaK is not found in the diatom PSI but found in all other PSIs, whereas PsaX is present in the cyanobacterial PSI only and absent in all PSIs from other organisms (Table 11.1).

The structure of the PSI core can exist in either trimeric or tetrameric forms. While the trimer is found in most cyanobacteria, tetramer is found in heterocyst-forming cyanobacteria. The structure of a PSI trimer was first solved from a thermophilic cyanobacterium *Thermosynechococcus elongatus* by X-ray

Table 11.1 Subunit compositions of the PSI cores and PSI-LHCI supercomplexes from various organisms

Subunits	Cyanobacteria	Red algae	Diatoms	Green algae	Higher plants
PSI core					
PsaA	+	+	+	+	+
PsaB	+	+	+	+	+
PsaC	+	+	+	+	+
PsaD	+	+	+	+	+
PsaE	+	+	+	+	+
PsaF	+	+	+	+	+
PsaG	–	–	–	+	+
PsaH	–	–	–	+	+
PsaI	+	+	+	+	+
PsaJ	+	+	+	+	+
PsaK	+	+	–	+	+
PsaL	+	+	+	+	+
PsaM	+	+	+	–	–
PsaN	–	–	–	+	+
PsaO	–	+	–	+	+
PsaR	–	–	+	–	–
PsaS	–	–	+	–	–
PsaX	+	–	–	–	–
Light-harvesting antennas					
	Phycobilisomes	3 or 5 Lhcr	16–24 Lhca, Lhcr, Lhcf	8–10 Lhca	4 Lhca

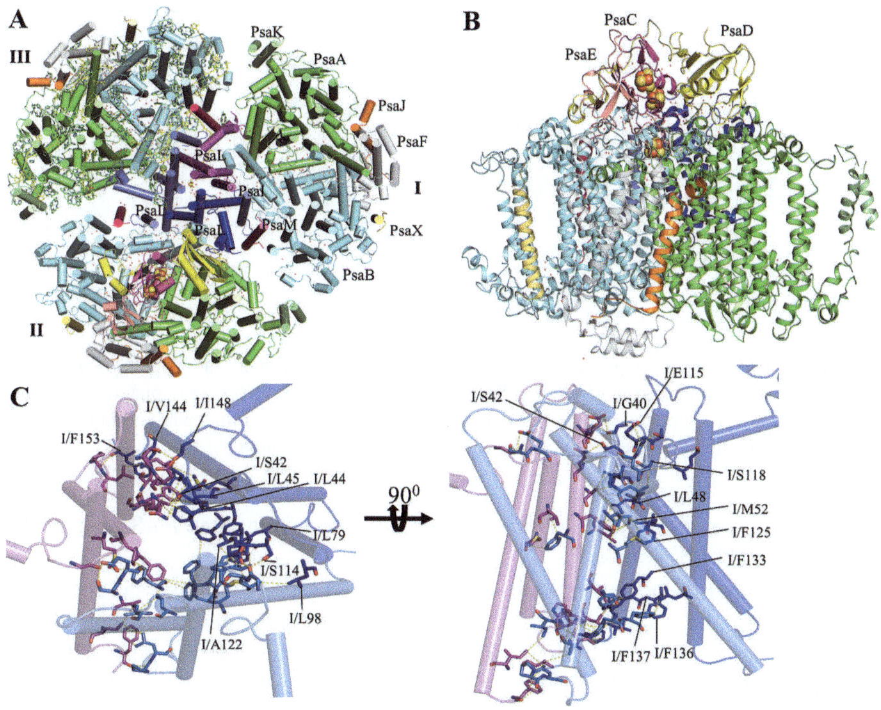

Fig. 11.2 Structure of the PSI core trimer from *Thermosynechococcus elongatus* (Jordan et al. 2001). (**a**), Structure of the PSI trimer with a top view along the membrane normal from the stromal side. Different structural elements are depicted in each of the three (I, II, III) monomers. I, Arrangement of the transmembrane subunits, with the three stromal side extrinsic subunits and all cofactors (except waters, which are shown in small red dots) omitted. Different subunits are colored differently and labeled. II, Same as I, with the three stromal side extrinsic subunits and three Fe-S centers depicted. III, Same as I, with Chls and β-carotenes depicted in green and yellow, respectively. In all the three monomers, subunits are colored the same as in I, with only PsaL colored differently. Water is depicted in dots. (**b**) Side view of the structure of one monomer of PSI (colored as in **A**), including the three stromal side subunits PsaC, PsaD, and PsaE and three Fe-S centers. (**c**) Interface of monomers I, II, and III in A, showing interactions between the three PsaL subunits

crystallography (Fig. 11.2) (Jordan et al. 2001). This structure showed that each PSI monomer is composed of 12 protein subunits and 96 Chls *a*, 22 carotenoids, 2 phylloquinones, 3 Fe_4S_4 clusters, 4 lipids, and around 200 water molecules. The center of each monomer is occupied by two large subunits PsaA/PsaB, which have 11 transmembrane helices each and bind the A- and B-branch ETC components as well as most of the Chls and β-carotenes. In the peripheral region of PsaA/PsaB, four small subunits PsaF, PsaK, PsaJ, and PsaX are located, among which, PsaF providing the docking site for cyt c_6 or plastocyanin in the lumenal side. In the stromal surface, three extrinsic subunits PsaC, PsaD, and PsaE are bound. PsaC are located

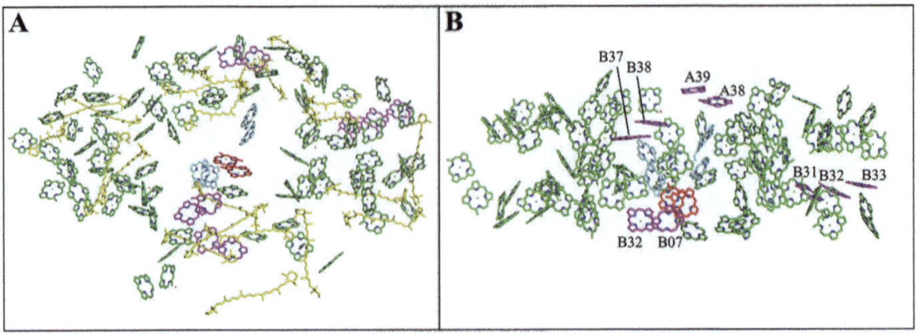

Fig. 11.3 Distribution of pigments in a cyanobacterial PSI monomer. (**a**) Top view along the membrane normal to the stromal side. (**b**) Side view. In both panels, P700 are depicted in red, four Chls of the ETC are depicted in cyan, possible red Chls are depicted in magenta and labeled in B, other Chls are depicted in green, and β-carotenes are depicted in yellow

in the middle between PsaD and PasE and binds two of the three Fe-S centers (F_A and F_B) (Fig. 11.2b).

PsaL is found in the center of the trimer, and two other subunits PsaI and PsaM are located in the vicinity of PsaL. These subunits, together with the PsaA/B subunits, participate in the trimer formation. In particular, PsaL from the three monomers play a central role in the trimer formation (Fig. 11.2c) (Chitnis and Chitnis 1993; Jordan et al. 2001; Kato et al. 2019). A number of resides of PsaL, especially hydrophobic residues, in both the stromal and lumenal sides as well as the transmembrane region, have distances of 3.5–4.2 Å to the nearby residues from PsaL of another monomer; they are thus important for the trimer formation. In particular, comparison of the sequences between trimeric PSI and tetrameric PSI from heterocyst-forming cyanobacteria showed that residues corresponding to Ser42, Ile44, Leu45, Ser118, Ala122 of *T. elongatus* are conserved in the heterocyst-forming cyanobacteria but not in other cyanobacteria (Kato et al. 2019; Li et al. 2019). These residues may therefore be important in determining if the PSI forms trimer or tetramer.

Among the 96 Chls found in a PSI core monomer, 90 serve as core antennas to harvest light energy and transfer them to the reaction center. Most of them, together with most of the 22 β-carotenes, are bound to the PsaA/B subunits (Fig. 11.3). They are distributed over the entire PSI core and surround the PSI reaction center, which facilitate energy harvesting from all directions. Among these Chls, there are several clusters that contain two or three Chls with their interplane distances in the range of 3.5–4.0 Å. These Chl clusters will have a lower energy level than that of the monomeric Chls because of the energy distribution over more than one Chl and thus are called “red Chls.” There is one Chl cluster that has three Chls coupled together (B31, B32, B33; the numbering of Chls are based on Jordan et al. 2001) and three two-Chl clusters in the structure of each PSI core monomer (Fig. 11.3); these Chl clusters may thus serve as red Chls to transfer the light energy to the reaction center as well as to dissipate the excess energy absorbed under strong illumination conditions.

Structure of the PSI Tetramer

The PSI tetramer is found in heterocyst-forming cyanobacteria and is organized in a dimer-of-dimers fashion (Kato et al. 2019; Li et al. 2019; Zheng et al. 2019; Chen et al. 2021). The structure of the PSI core tetramer from *Anabaena* sp. PCC 7120 is shown in Fig. 11.4, which are composed of monomer-A, monomer-A′, monomer-B, and monomer-B′. The structure of each PSI core monomer of *Anabaena* sp. PCC 7210 is essentially the same as that of *T. elongatus*. The interactions among the four monomers can be divided into two types, type 1 and type 2 (Fig. 11.4b, c). Type 1 interactions depict interactions between monomer-A and monomer-B, which is also found in the PSI trimer. As can be seen from Fig. 11.4b, type 1 interactions involve a number of hydrophobic as well as several hydrophilic interactions between two PsaL subunits from monomer-A and monomer-B. Residues of PsaL-The58, Arg61, Asn140, and Ser144 (corresponding to S42, L45, S118 and A122 of *T. elongatus*, Kato et al. 2019) are found to be conserved only in the tetramer-forming cyanobacteria. These residues have large side chains and are found to hamper the formation of the trimer (Kato et al. 2019; Li et al. 2019; Zheng et al. 2019; Chen et al. 2021). Most of these residues participate directly in the type 1 interactions.

Type 2 interactions refer to interactions between the two dimers within the PSI tetramer and are formed between monomers-A and B′ (Fig. 11.4b). This type of interaction is formed between PsaL and PsaA of monomer-A and PsaB of monomer-B′, and most of them are formed between by hydrophilic residues. This type of interaction is found in the PSI tetramer only. Based on these interactions, formation of the different PSI oligomerization states can be explained as follows. In the case that PsaL allows formation of the trimer, the type 1 interface allows three monomers to assemble to a trimer. In the case that PsaL does not allow formation of the trimer, type I interface first allows two monomers to assemble into a dimer. Subsequently, two dimers are joined together by interactions of the type 2 interface, resulting in the formation of the tetramer (Kato et al. 2019).

Structure of the PSI-IsiA Supercomplex

Under iron-deficient conditions, cyanobacteria express an IsiA protein (iron stress-induced protein A), which is a protein homologous to CP43, a Chl-binding subunit in PSII core (Laudenbach and Straus 1988; Burnap et al. 1993; Umena et al. 2011; Shen 2015), and thus is also called CP43′. This protein associates with the PSI core, forming a PSI-IsiA supercomplex. A typical structure of the PSI-IsiA supercomplex

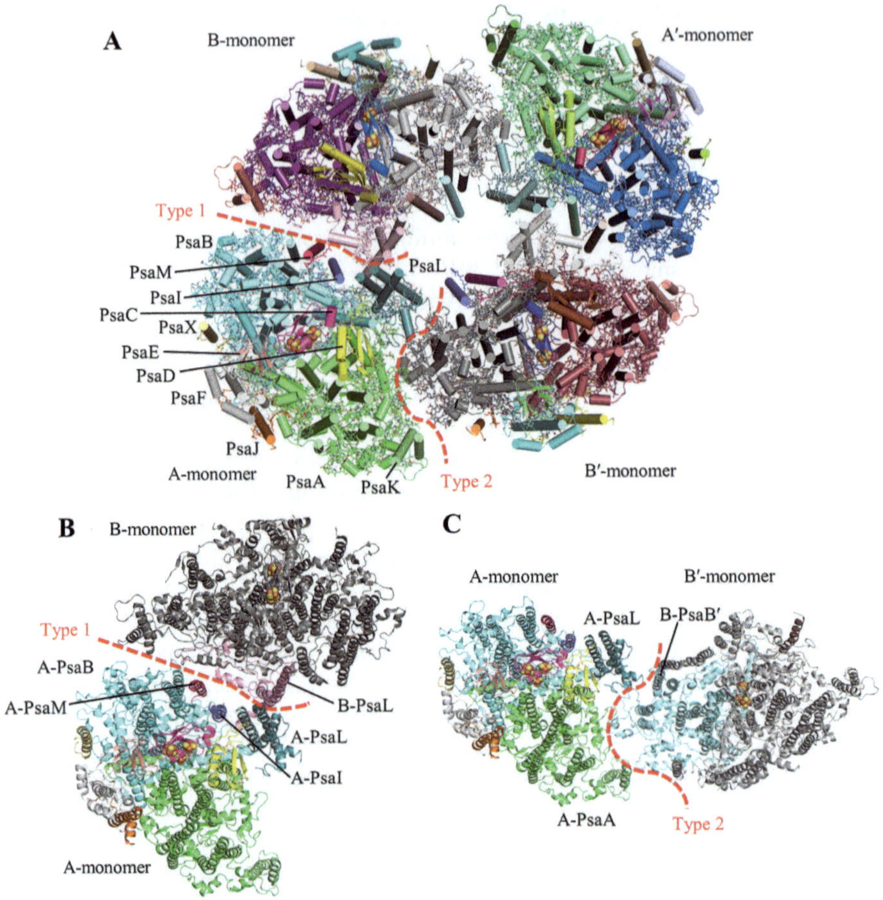

Fig. 11.4 Structure of the PSI tetramer from *Anabaena* sp. PCC 7120 (Kato et al. 2019). (**a**) Structure of the PSI tetramer with a top view from the stromal side. (**b**) Type 1 interactions. (**c**) Type 2 interactions

is shown in Fig. 11.5 (Toporik et al. 2019; Akita et al. 2020; Cao et al. 2020), in which 18 copies of the IsiA protein surround the trimeric core of PSI, forming a closed ring structure. Each PSI core is surrounded by 6 copies of IsaA; thus, the whole structure depicts a threefold symmetry with the PSI core +6 copies of IsaA as the basic unit.

The structure of each Isi-A monomer is essentially the same among the 18 copies. However, it has some flexibility so that they can move either inward or outward the PSI core relative to its adjacent monomer, depending on its position in the PSI-IsiA supercomplex (Toporik et al. 2019). This enables the IsiAs to form a closed ring, and also makes it interact with the PSI core differentially. In the lumenal side, three areas are found to have protein-protein interactions between the IsiA and PSI core; they are A1, A3, and A4. Also in the lumenal side, each IsiA has a stretch extending from

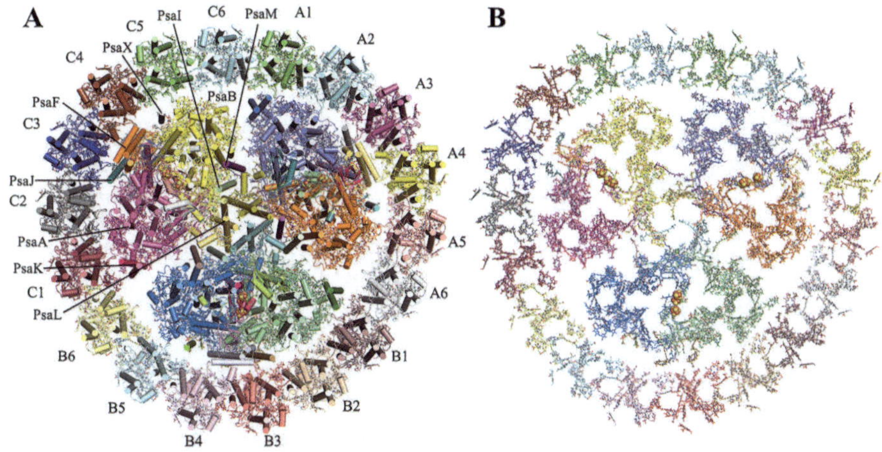

Fig. 11.5 Structure of the PSI-IsiA supercomplex (**a**) and the distribution of pigments (Chls and β-carotenes) in the PSI-IsiA supercomplex (**b**), with a top view from the stromal side (Akita et al. 2020). A1-C6 denotes the 18 copies of the IsiA subunit. The middle of panel A is the PSI core trimer

loop E between transmembrane helices V and VI; this stretch interacts with the PSI core in all the IsiAs (Toporik et al. 2019; Akita et al. 2020). There are also pigment-pigment interactions between IsiA and the PSI core, facilitating the binding of IsiA to the PSI core (Fig. 11.5b).

As mentioned above, the structure of IsiA is similar to that of CP43. However, there are important differences between the two structures. First, the N- and C-terminals of the two subunits are located in different regions in the stromal side. Second, the E loop between helices V and VI in the lumenal side is significantly different (Fig. 11.6a). The E loop of CP43 is much larger and forms an important region for maintaining the function and stability of the Mn_4CaO_5 cluster that catalyzes the water-splitting reaction (Bricker and Frankel 2002; Umena et al. 2011; Shen 2015). This large E loop is much shortened in IsiA, and a part of it interacts with PSI core subunits in the lumenal side as mentioned above. In addition, differences are also found in the pigments bound to the two subunits (Fig. 11.6b). CP43 binds 13 Chls and 3 β-carotenoids, whereas IsiA binds 17 Chls (Fig. 11.6b) and 4 β-carotenoids. Among the extra Chls found in IsiA, three are located in the periphery of the subunit and interact with either the adjacent IsiA subunit or the PSI core, facilitating the energy transfer among the IsiAs as well as between IsiAs and the PSI core. The last extra Chl is found at the stromal side of IsiA, which may facilitate energy transfer from IsiAs to the PSI core or energy dissipation from the PSI core.

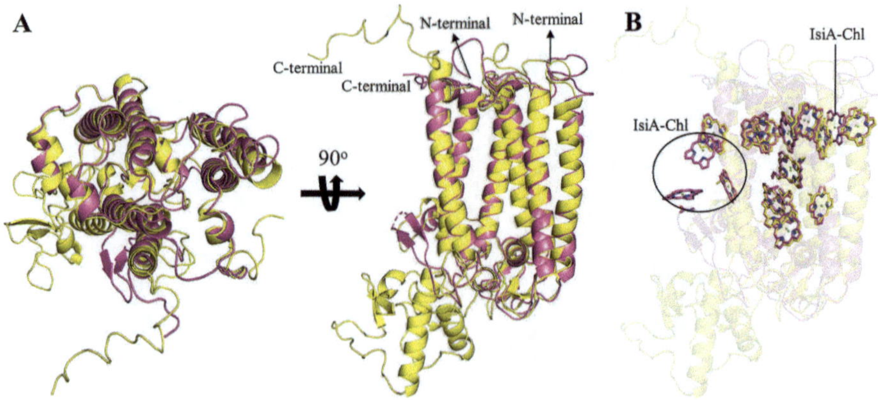

Fig. 11.6 Comparison of the structures between IsiA and CP43 (Umena et al. 2011; Akita et al. 2020). (**a**) Comparison of the protein structures of the two subunits, with a top view from the stromal side (left side) and a side view (right side). (**b**) Comparison of the Chls in the two subunits. The Chls indicated by a black circle and a line show Chls that are present only in the IsiA subunit. Color: light magenta, IsiA; yellow, CP43

Structures of Eukaryotic PSI-LHCI Supercomplexes

LHCI appears in eukaryotes, starting from red algae. Because of the presence of LHCI and also because of the differences in some PSI core subunits between prokaryotes and eukaryotes (Table 11.1), the PSI core of eukaryotes is a monomer. There are also remarkable differences in the number, structure of LHCI bound to the PSI core, as well as the pigments that each LHCI binds. In the following section, I will discuss the structure of the PSI-LHCI supercomplex from each representative eukaryotic organism, focusing on the different LHCI subunits that they bind.

Red Algal PSI-LHCI Supercomplexes

Red algae are one of the most primitive eukaryotic algae, and their photosystems bear both features of cyanobacteria and higher plants. PSII of red algae resembles that of cyanobacteria in that both of them bind phycobilisomes as their light-harvesting antenna (Zhang et al. 2017; Ma et al. 2020; Sui 2021) and both have extrinsic proteins of PsbU and PsbV (Shen and Inoue 1993; Enami et al. 1995; Enami et al. 2008) that are not found in green algal and higher plant PSII. Red algae also resemble cyanobacteria in that they do not have Chl *b*, as the enzyme required for the synthesis of Chl *b* has not appeared in red algae. However, PSI of red algae more resembles green algae and higher plants in that it binds LHCI proteins as their light-harvesting antenna (Pi et al. 2018; Antoshvili et al. 2019). In fact, red algae

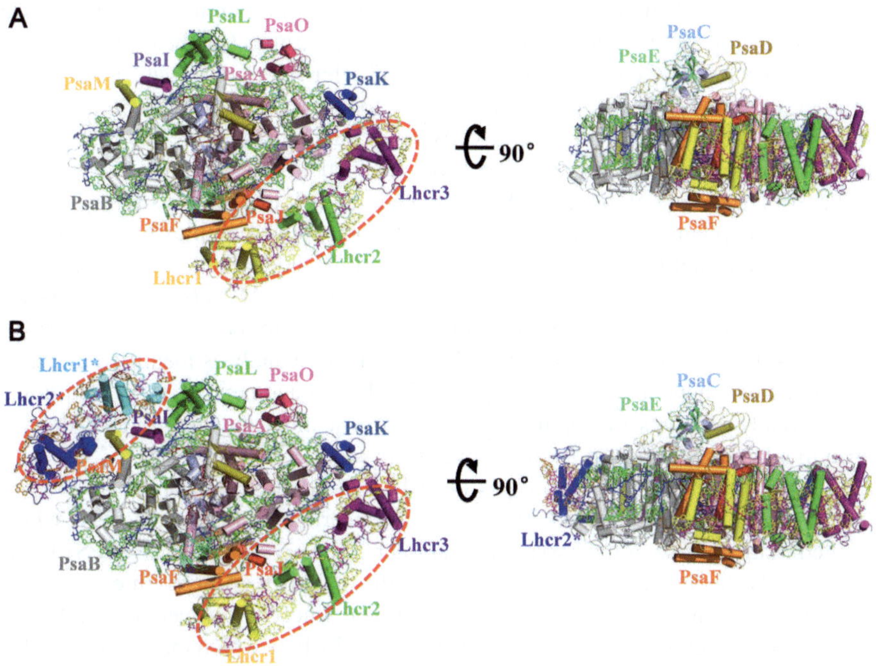

Fig. 11.7 Structure of the red algal PSI-LHCI supercomplex (Pi et al. 2018). (**a**) PSI binding three LHCI subunits. (**b**) PSI binding five LHCI subunits. Left side, top view from the stromal side. Right side, side view of the membrane

may be the most primitive organisms that bind LHCI subunits (Wolfe et al. 1994; Durnford et al. 1999).

The PSI core of red algae is highly conserved from cyanobacteria, but it lacks PsaX that is present in the cyanobacterial PSI, whereas contains PsaO that is absent in the cyanobacterial PSI (Table 11.1). In cyanobacterial PSI, PsaX is located between PsaB and PsaF (Fig. 11.2), which is approximately the opposite side of the place where PsaO is located in the red algal PSI (Fig. 11.7). The function of PsaO is not clear in red algae, but it provides an association point for LHCII in state transition in green algal and higher plant PSI (see below). PsaO binds two Chls, which may enhance the light-harvesting capability of PSI. In the trimeric and tetrameric structures of cyanobacterial PSI, the position of PsaO is occupied by residues from the adjacent monomer, so it is not possible for PsaO to be present in the cyanobacterial PSI. On the other hand, red algal PSI is a monomer; hence, the position of PsaO is free to be occupied.

The reason for a monomeric organization of PSI in red algae is twofold. First, the appearance of PsaO and sequence/structural changes of PsaL may make it unable to form a trimer. Indeed, the sequence and structure of the red algal PsaL is homologous to that in higher plants, in the C-terminal region in that they lack a short helix in the lumenal side, which is found in the cyanobacterial PsaL and participates in the trimer formation (Pi et al. 2018). Second, the attachment of some LHCI subunits at the trimerization site may hinder the formation of the trimer (see below).

The genes for LHCI are termed *lhcr*, with "r" referring to red algae. Three genes for Lhcr are found in the red alga *Cyanidioschyzon merolae* (Matsuzaki et al. 2004). In the cryo-EM structure, however, two types of supercomplex are found. One has three Lhcr subunits forming a single layer attached to one side of PSI, which is similar to that seen in higher plant PSI, although the belt of LHCI is slighted shifted relative to the plant PSI-LHCI (see below). The other supercomplex has five Lhcrs, with two additional Lhcrs associated at the PsaI/L/M side which is the interface of monomer-monomer in the cyanobacterial PSI trimer (Fig. 11.7) (Pi et al. 2018). As there are only three genes, the two additional Lhcrs are products of copies of one or two genes that constitute the main LHCI belt. Structural analysis identified that they are copies of Lhcr1 and Lhcr2, respectively. The attachment of these additional Lhcrs greatly enlarges the light-harvesting cross-section and may enable the red algae to adapt to weak light environment under water.

It is still not clear if the two types of PSI-Lhcrs found in red algae are physiologically relevant to the cells in vivo or they are an artifact due to purification which resulted in the release of the two additional Lhcrs. The crystal structure of PSI-Lhcrs showed only the association of three Lhcrs in the main belt (Antoshvili et al. 2019), suggesting the easy release of the two additional Lhcrs. However, native PAGE analysis of the membranes showed two bands of PSI-Lhcrs which may correspond to the PSI-3Lhcrs and PSI-5Lhcrs (Tian et al. 2017), respectively, and the intensities of these bands are changeable depending on the light intensity under which the cells are grown (Chang et al. 2020). Under low light condition, the amount of the presumable PSI-5Lhcrs is increased, whereas under high light condition, that of the PSI-3Lhcrs is increased concomitant with the decrease in the amount of the PSI-5Lhcrs. These results are in agreement with the presence of two types of PSI-Lhcrs in vivo.

Diatom PSI-FCPI Supercomplexes

Diatoms are diverse eukaryotic algae living in both fresh and sea water across the world and are an important group of photosynthetic organisms, because they contribute to around 25% of the global primary productivity (Smetacek 1999; Falkowski et al. 2004; Levitan et al. 2014). The plastids of diatoms are believed to arise from multiple endosymbiosis events of a primitive red algal cell (Falkowski et al. 2004), and the light-harvesting proteins of its photosystems are unique in that they bind Chl *c* and fucoxanthin (Fx) in addition to Chl *a*. They are thus termed FCPs (fucoxanthin Chl *c*-binding proteins). Diatoms belong to red lineage organisms, because of their red-brown color and the binding of Chl *c* and Fx.

The cryo-EM structures of diatom PSI-FCPI supercomplex have been resolved at 2.40 Å (Nagao et al. 2020b) and 2.38 Å resolutions (Xu et al. 2020), respectively, which showed the binding of different numbers of FCPI subunits (see below). Again, the diatom PSI core is largely conserved, but there are differences with the

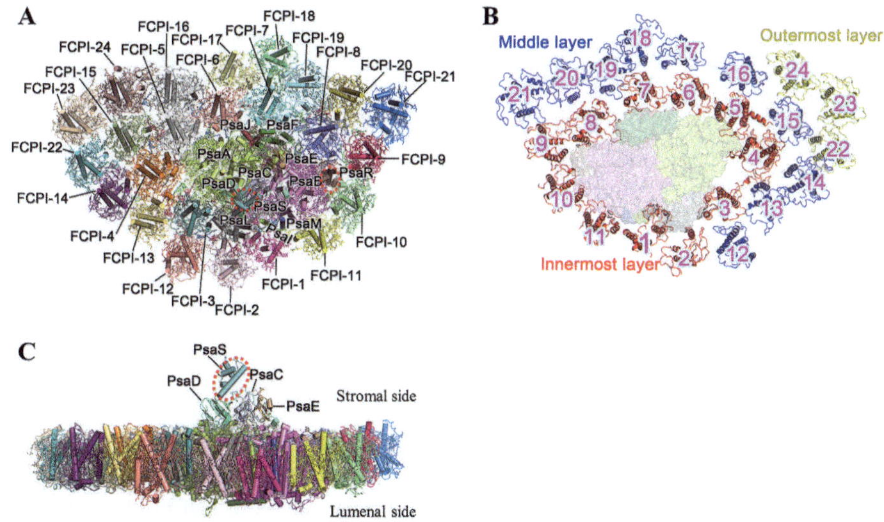

Fig. 11.8 Structure of the PSI-FCPI supercomplex from a diatom *Chaetoceros gracilis* solved by cryo-EM at 2.38 Å resolution (Xu et al. 2020). (**a**) Top view from the stromal side. The names of each subunit are labeled. The areas circled by red dashed lines depict two novel subunits PsaR and PsaS. (**b**) Top view from the lumenal side, showing the arrangement of the 24 FCPIs subunits in three layers. (**c**) Side view of the membrane

cyanobacterial and red algal PSI core. While the diatom PSI core lacks PsaX as that of red algae, it also lacks PsaO and PsaK; the former is found in all eukaryotic organisms except diatoms, whereas the latter is found in all other organisms except diatoms (Table 11.1). Moreover, diatoms possess two novel subunits, termed PsaR and PsaS in one study (Xu et al. 2020) (Fig. 11.8) and Psa28 and Unknown1 in another study (Nagao et al. 2020b) (Fig. 11.9). PsaR/Psa28 is a transmembrane subunit located in the periphery of PsaB, whose location is somewhat similar to PsaG in green algal and higher plant PSI, but they are not exactly the same, and they do not share sequence homology (Nagao et al. 2020b; Xu et al. 2020). PsaR/Psa28 binds one Chl *a* and one Fx, and is surrounded by FCPI subunits on its outside. This suggests a role of PsaR/Psa28 in the binding of FCPI subunits as well as light-harvesting and energy transfer. On the other hand, PsaS/Unknown1 is an extrinsic protein located at the top of the stromal side, covering the PsaC/D/E region in the stromal side. This suggests that PsaS/Unknown1 may be needed to stabilize the stromal side region of the diatom PSI.

The two studies revealed binding of different FCPI subunits in each PSI core. The 2.38 Å structure revealed 24 FCPI subunits, which are arranged in three layers surrounding the PSI core (Fig. 11.8) (Xu et al. 2020). The inner layer contains 11 FCPI subunits and forms a completely closed ring surrounding PSI core, the middle layer contains 10 subunits, forming a semi-closed ring, whereas the outmost layer contains three subunits and is attached to the FCPI-14/15/16 side. On the other hand, the other study revealed the presence of 16 FCPI subunits in each PSI-FCPI

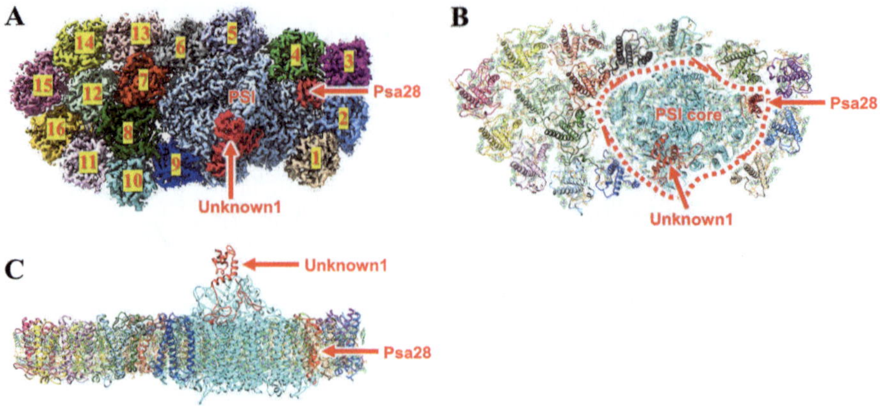

Fig. 11.9 Structure of the PSI-FCPI supercomplex from a diatom *Chaetoceros gracilis* solved by cryo-EM at 2.40 Å resolution (Nagao et al. 2020). (**a**) Top view from the stromal side of the electron potential map, showing the distribution of different subunits. The names of FCPI subunits are labeled in 1–18, and two subunits labeled Psa28 and Unknown1 depict two novel subunits. (**b**) Top view from the stromal side, showing the arrangement of the PSI core and FCPI subunits. (**c**) Side view of the membrane

supercomplex, and they form unclosed rings around the PSI core (Fig. 11.9) (Nagao et al. 2020b). Whether the latter is an artifact due to the loss of some FCPI subunits during purification, or it represents one of the PSI-FCPI supercomplexes in vivo is not clear at present. However, it has been reported that the diatom antenna systems may vary depending on growth conditions such as light or CO_2 concentration (Nagao et al. 2018; Nagao et al. 2019; Akimoto et al. 2020; Nagao et al. 2020a). Thus, it is reasonable to assume that diatoms possess different types of PSI-FCPI supercomplexes with different numbers of FCPI subunits associated in vivo and the number of FCPI subunits associated with each PSI core are regulated by growth conditions, especially that of light intensity and quality. This resembles the PSI-Lhcr supercomplex of red algae, in which the number of Lhcrs associated with each PSI core is also variable (Pi et al. 2018; Chang et al. 2020).

The overall structures of each FCPI are largely similar; however, there are differences in the N-terminal, C-terminal regions, as well as loop regions, which result in different interactions between adjacent FCPI subunits as well as with the PSI core subunits (Nagao et al. 2020b; Xu et al. 2020). The pigments that each FCPI subunit binds also differ slightly, which may be necessary to harvest and transfer the energy in an optimum way. In total, FCPI complexes contain 232 Chl *a*, 34 Chl *c*, and 100 Fx in the PSI-24FCPI structure (Xu et al. 2020) and 128 Chl *a*, 54 Chl *c*, and 54 Fx in the PSI-16FCPI structure (Nagao et al. 2020b). These pigments form a very complicated network that functions to harvest and dissipate the light energy. There are numerous pathways from the Chls/Fxs of FCPI to the PSI core (Fig. 11.10), enabling efficient energy harvesting and transfer from the FCPI to the PSI core. One unique feature of FCPI is its high content of Fxs relative to Chls (100 Fxs + 33 Ddxs:263 Chls in PSI-24FCPI and 54 Fxs + 31 Ddxs:182 Chls in

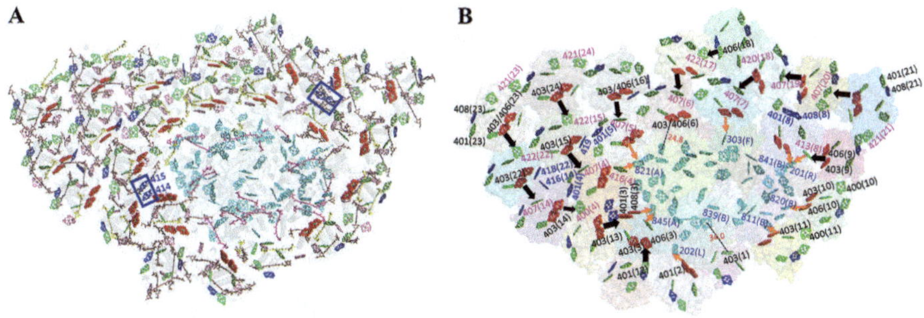

Fig. 11.10 Distribution of Chls and possible energy transfer pathways in the PSI-24FCPI supercomplex (Xu et al. 2020). (**a**) Distribution of all pigments in the PSI-24FCPI supercomplex with a top view from the stromal side. Chl *a* in FCPIs, Chl *a* in core, Chl *c*, Fx, Ddx and Bcr are shown in sticks and colored in green, cyan, blue, darksalmon, yellow and magenta, respectively. Special Chl *a-a* pairs are colored in red; two extra Chl *c-c* pairs are labeled in blue frame. (**b**) Energy transfer pathways from FCPIs to the PSI core with a top view from the stromal side. Orange arrows indicate EET pathways from inner FCPIs to the PSI core and black arrows indicate EET pathways from outer antennae to the inner

PSI-18 FCPI vs. 33 β-carotenoids:143 Chls in green algae, see below) (Wang et al. 2019). This high content of β-carotenoids may not only enable efficient energy harvesting under water where the blue-green region of light is more available but also facilitate energy dissipation under constantly changing light conditions under water.

Green Algal PSI-LHCI Supercomplexes

Green algae are another important group of eukaryotic algae and are the immediate ancestor of land plants. Green algal LHC contains Chl *b* instead of Chl *c* and evolutionarily belongs to the green lineage of photosynthetic organisms (they are also termed Lhcas, since they attach to PSI). The structure of PSI-LHCI from a green alga *Chlamydomonas reinhardtii* is shown in Fig. 11.11 (Suga et al. 2019; Su et al. 2019; Perez-Boerema et al. 2020). The PSI core lacks PsaM but contains PsaG/H/N/O that is commonly found in the green lineage organisms (Table 11.1) (Yu et al. 2008; Suga and Shen 2020). PsaG is homologous to PsaK and is located at its opposite side. It is believed that PsaG arises from gene duplication of PsaK in the green lineage organisms (Okkels et al. 1992; Kjaerulff et al. 1993). Both PsaK and PsaG provide an attachment site for Lhcas. PasH is located between PsaL and PsaI and provides another attaching point of the additional Lhcas (see below). On the other hand, the green algal PSI-LHCI structure does not contain PsaN/O (Qin et al. 2019; Su et al. 2019; Suga et al. 2019; Qin 2021), one of which (PsaO) is found in the PSI-LHCI-LHCII supercomplex after state transition (see below).

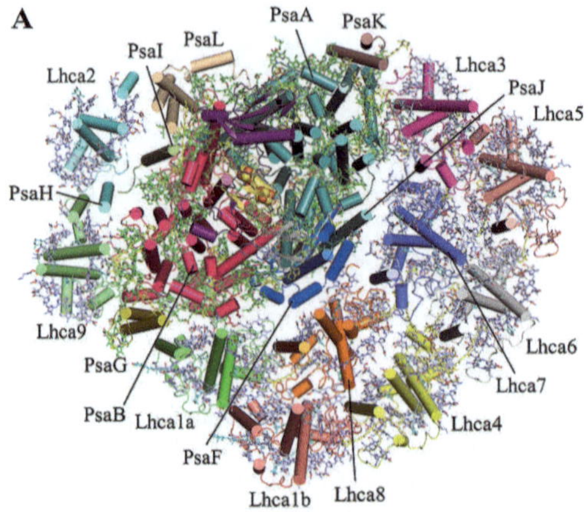

Fig. 11.11 Structure of the PSI-LHCI supercomplex from a green alga *Chlamydomonas reinhardtii*, with a view from the stromal side. (**a**) PSI containing 10 Lhcas. (**b**) PS containing 8 Lhcas. Chls in the PSI core are depicted in green, Chls in LHCIs are depicted in white blue, β-carotenes are depicted in yellow, and luteins are depicted in cyan

The green algae also show two types of the PSI-LHCI structure, one contains 8 Lhcas and the other one contains 10 Lhcas (Fig. 11.11). The PSI-8Lhcas structure has its Lhcas arranged in two layers, with each layer containing four Lhca subunits, in the PsaG/B/F/J/A/K side similar to that of the higher plant PSI-LHCI, whereas the PSI-10Lhcas has two additional Lhcas attached in the opposite side (PsaH/I/L side). This side is the trimerization region in the cyanobacterial PSI, and the attachment of the additional Lhcas makes PSI a monomer in the green algal organisms. Whether these two types of PSI-LHCI exist in vivo or the PSI-8Lhca type is an artifact due to the loss of two additional Lhcas during purification is a matter of debate, and some reports have suggested that the two additional Lhcas bind to the PSI core weakly and hence are easily lost during purification (Tokutsu et al. 2004; Ozawa et al. 2018). However, it is also possible that these two types of PSI-Lhcas are inherently present in the green algal cells, and the relative amount of them may be changeable under different light conditions, reminiscent of the PSI-LHCI supercomplexes in red algae and diatoms.

As the green algae contain only nine Lhca genes (Tokutsu et al. 2004; Ozawa et al. 2018), one of the Lhca subunits must be a copy of these genes. The structure showed that the main LHCI belt consists of Lhca1a/1b/3/4/5/6/7/8 whereas the additional LHCI contains Lhca2/9. Thus, Lhca1 has two copies located in the inner and outer layer of the main LHCI belt. Interestingly, while all Lhca subunits except Lhca2 consist of three transmembrane helices similar to the major LHC protein, Lhca2 has four transmembrane helices which is reminiscent of the structure of PsbS, a subunit functioning in the photoprotection of PSII (Li et al. 2002; Fan et al. 2015). As this Lhca is attached at the outmost region of the PSI-Lhca supercomplex, it may suggest a role in energy dissipation.

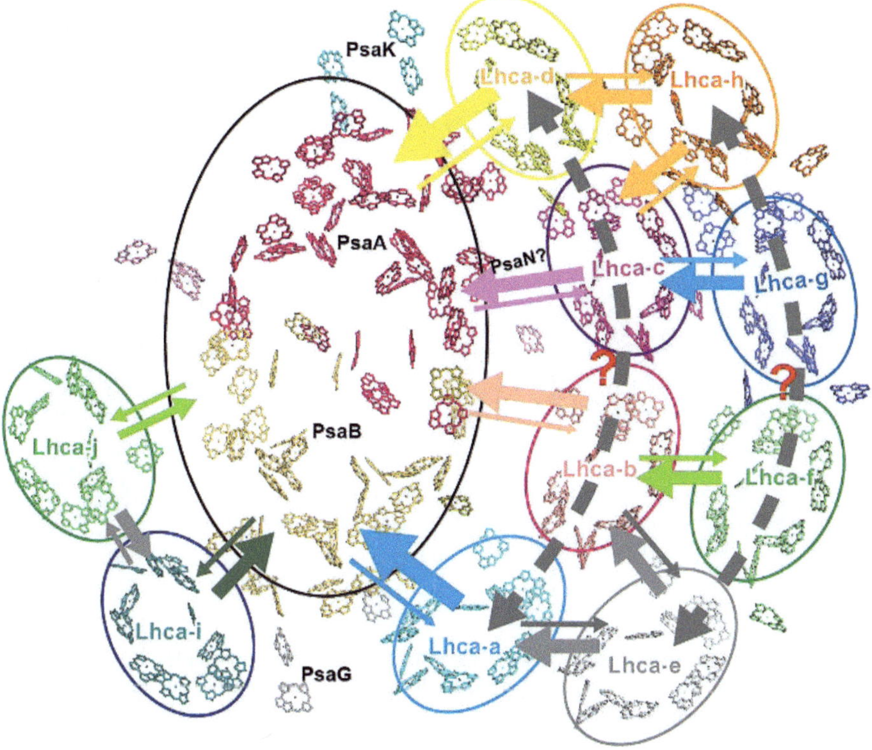

Fig. 11.12 Possible energy transfer pathways from Lhcas to PSI core, and among Lhcas, in the PSI-Lhca supercomplex from a green alga *Bryopsis corticulans* (Qin et al. 2019). Viewed from the stromal side down to the membrane plane

The structures and pigments these Lhcas bind differ slightly among different Lhcas, enabling each subunit occupying a unique position and interacting with its neighboring protein subunits (Fig. 11.11). Each Lhca protein binds 11–17 Chls and 2–5 carotenoids, giving rise to a total of 143 Chls and 33 β-carotenoids. Among these pigments, 20 Chls are identified as Chl *b* and 22 β-carotenoids are identified as lutein (Su et al. 2019; Suga et al. 2019). The distribution of pigments and possible energy transfer pathways in green algae *Bryopsis corticulans* are shown in Fig. 11.12. It is interesting to note that, in this green alga, the β-carotenoids are changed to α-carotene, siphonaxanthin, siphonein, neoxanthin, and violaxanthin (Qin et al. 2019). As this blue-green algae lives in intertidal zones where the light environment may undergo dramatic changes from a very strong light intensity to a very low intensity in a short time, these carotenoids may be necessary for the organism to adapt to such a dramatically changeable environment.

Under light conditions that preferentially excite PSII (shorter wavelength illumination), PSII may be overexcited, and the photosynthetic membranes undergo a remodeling process, in which a part of LHCII moves to PSI to balance the excitation

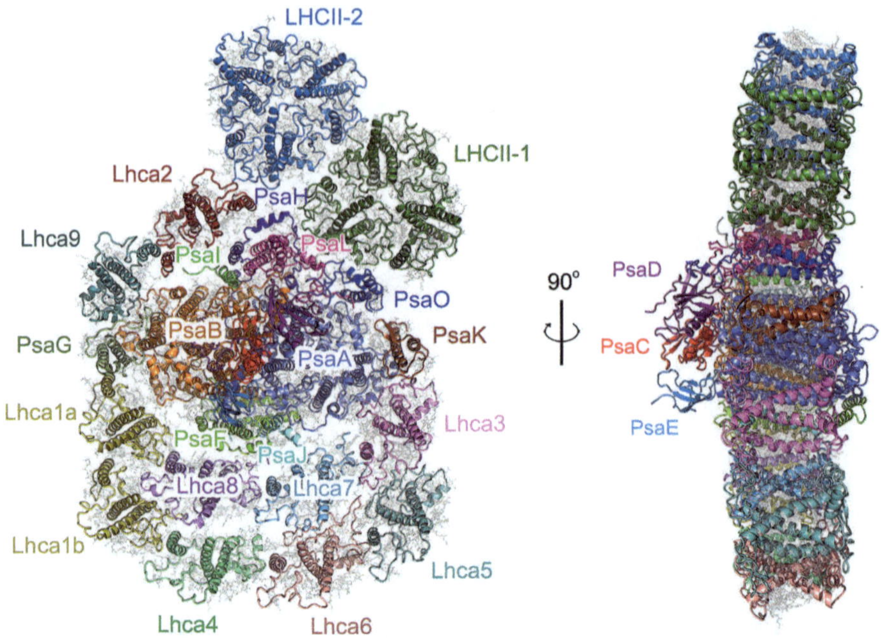

Fig. 11.13 Overall structure the *C. reinhardtii* PSI-LHCI-LHCII supercomplex viewed along the membrane normal from the stromal side (left side) and along the membrane plane (right side) (Huang et al. 2021). Protein subunits are shown in cartoon models and colored differently as shown in the figure

energy between the two photosystems. This process is called state transition (Murata 1969; Bonaventura and Myers 1969; Murata 2009; Minagawa 2011). The movement of LHCII is triggered by phosphorylation of a threonine residue of the N-terminal region of a LHCII subunit (Depege et al. 2003; Lemeille and Rochaix 2010), and this process has been extensively studied in the green alga *C. reinhardtii*, since it has a rather large amplitude of the state transition grown under different light conditions. The structure of the PSI-LHCI-LHCII supercomplex obtained under state 2 has been analyzed from both a higher plant (Pan et al. 2018) and *C. reinhardtii* (Huang et al. 2021) by cryo-EM, and the structure of *C. reinhardtii* is shown in Fig. 11.13. In this structure, two LHCII trimers are moved to PSI and associate at the PsaH/L/O side, increasing the light absorption cross-section of PSI. PsaO is visible because it is covered by one of the LHCII trimers, making it associates with the PSI core rather tightly (Huang et al. 2021; Wu et al. 2021). This is in contrast to that seen in the PSI-LHCI supercomplex, where PsaO is often lost and not visible.

Among the two LHCII trimers, LHCII-1 has direct interactions with the PSI core subunits of PsaH/L/O and can transfer the energy directly to the core. On the other hand, LHCII-2 does not have direct contact with the PSI core but interacts with Lhca2 directly, through which the energy is transferred to the PSI core. The higher plant PSI-LHCI-LHCII structure contains LHCII-1 trimer only and lacks LHCII-2, suggesting that the green algal PSI-LHCI-LHCII binds more Chls under state 2 to increase its absorption cross-section under weak light conditions under water.

Higher Plant PSI-LHCI Supercomplex

Higher plants grow on land where light is usually enough and sometimes even too strong to damage the photosystems. Thus, higher plant PSI-LHCI loses several LHCI subunits compared with various algae grown under water in order to decrease the light energy absorption cross-section. A typical higher plant PSI-LHCI contains four LHCI subunits in one layer located at the interface of PsaF/G/J/K (Qin et al. 2015; Mazor et al. 2017; Wang et al. 2021; Yan et al. 2021), which is similar to the interface where the major layer of LHCI binds in green algae (Fig. 11.14). The core of higher plant PSI is very similar to that of green algae which binds both PsaG and PsaK, and PsaO is often lost without the binding of LHCII. The four LHCI subunits

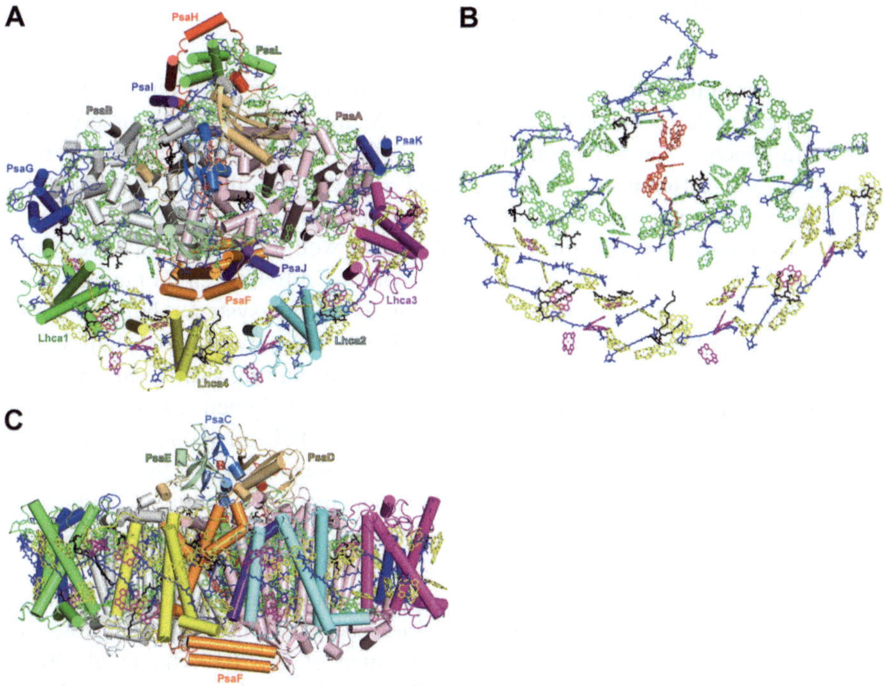

Fig. 11.14 The crystal structure of PSI-LHCI supercomplex from *Pisum sativum*. (**a**) Viewed along the membrane normal from the stromal side. (**b**) The arrangement of pigments (Chls and carotenoids) and other cofactors (lipids, Fe4S4 clusters, and phylloquinones) in the PSI-LHCI supercomplex, with the view direction same as in panel (**a**). For clarity, the protein backbones and phytol chains of Chls were omitted. (**c**) Side view of the structure of the PSI-LHCI supercomplex from the LHCI side, with the phytol chains of Chls omitted

also bind to similar positions with the four LHCI subunits in the inner layer of LHCIs of the green algae, and they form two dimers, the Lhca1/Lhca4 dimer and Lhca2/ Lhca3 dimer. Numerous excitation energy pathways can be found from Lhcas to the PSI core (Qin et al. 2015; Mazor et al. 2017; Wang et al. 2021, and they are distributed in both the stromal layer and lumenal layer, with more pathways located in the stromal layer.

Like the green algae, PsaO is lost in the PSI-LHCI structure but is found in the PSI-LHCI-LHCII structure in which, a LHCII trimer is moved from PSII to PSI under state 2 conditions (Pan et al. 2018). However, the plant PSI-LHCI-LHCII contains only one LHCII trimer, which is slightly shifted toward the PsaK side relative to that seen in the green algal PSI-LHCI-LHCII (Pan et al. 2018; Huang et al. 2021). These results again suggest that light harvesting antenna system is decreased in higher plants compared with green algae and other algae grown under water, probably due to the strong light irradiation conditions on land compared with those under water.

Concluding Marks and Perspectives

The structures of the PSI core and PSI-LHCI supercomplexes from various algae and higher plants have been analyzed by either X-ray crystallography or cryo-EM, providing much information on the composition, arrangement, energy transfer pathways, and differences of proteins and pigments in the PSI core and the PSI-LHCI supercomplexes among different organisms. These studies reveal a surprising diversity in the structures of PSI-LHCI as well as the pigments they bind, which is likely a result of evolution under various different environments. As a general rule, algae grown under water have larger antenna system relative to higher plants grown on land, due to the relative weaker and rapidly changing light conditions under water. The energy transfer pathways are accordingly different among the different organisms. This chapter outlines the protein and pigment arrangements, and some possible energy transfer pathways in the PSI-LHCI supercomplexes revealed by the X-ray and cryo-EM structural analyses, whereas a precise treatment of the energy transfer pathways needs theoretical calculations combined with time-resolved spectroscopy (Croce and van Amerongen 2020). There are also several lineages of organisms whose PSI-LHCI structure have not been solved, such as brown algae, Charophyceae, etc. Even within the same linage, such as green algae, there are differences in the pigment compositions between different organisms, such as β-carotene and lutein in *C. reinhardtii* (Suga et al. 2019; Su et al. 2019), α-carotene, siphonaxanthin, siphonein in *Bryopsis corticulans* (Qin et al. 2019), and β-carotene, diatoxanthin and diadinoxanthin in Euglena (Heelis et al. 1979), and the structures of PSI-LHCI from many species within the same lineage of organisms have not been solved. Finally, most of the cryo-EM structures solved so far have a relatively lower resolution, which hampers our discussion and understanding on the energy transfer pathways in detail, so these resolutions need to be improved in order to fully understand the molecular mechanisms of light energy harvesting and transfer reactions in this supercomplex from various organisms and their adaptations to the different light environment.

References

Adir N, Bar-Zvi S, Harris D (2020) The amazing phycobilisome. Biochim Biophys Acta Bioenerg 1861:148047

Akimoto S, Ueno Y, Yokono M, Shen JR et al (2020) Adaptation of light-harvesting and energy-transfer processes of a diatom *Chaetoceros gracilis* to different light qualities. Photosynth Res 146(1–3):87–93

Akita F, Nagao R, Kato K et al (2020) Structure of a cyanobacterial photosystem I surrounded by octadecameric IsiA antenna proteins. Commun Biol 3:232

Akiyama M, Miyashita H, Kise H et al (2001) Detection of chlorophyll *d'* and pheophytin *a* in a chlorophyll *d*-dominating oxygenic photosynthetic prokaryote *Acaryochloris marina*. Anal Sci 17(1):205–208

Antoshvili M, Caspy I, Hippler M et al (2019) Structure and function of photosystem I in *Cyanidioschyzon merolae*. Photosynth Res 139:499–508

Bailleul B, Johnson X, Finazzi G et al (2008) The thermodynamics and kinetics of electron transfer between cytochrome b_6f and photosystem I in the chlorophyll *d*-dominated cyanobacterium, *Acaryochloris marina*. J Biol Chem 283:25218–22526

Blankenship R (2021) Molecular mechanisms of photosynthesis, 3rd edn. Wiley, Hoboken

Bonaventura C, Myers J (1969) Fluorescence and oxygen evolution from *Chlorella pyrenoidosa*. Biochim Biophys Acta 189:366–383

Bricker TM, Frankel LK (2002) The structure and function of CP47 and CP43 in photosystem II. Photosynth Res 72:131–146

Burnap RL, Troyan T, Sherman LA (1993) The highly abundant chlorophyll-protein complex of iron-deficient *Synechococcus* sp. PCC7942 (CP43') is encoded by the *isiA* gene. Plant Physiol 103:893–902

Cao P, Cao D, Si L et al (2020) Structural basis for energy and electron transfer of the photosystem I-IsiA-flavodoxin supercomplex. Nat Plants 6:167–176

Caspy I, Nelson N (2018) Structure of the plant photosystem I. Biochem Soc Trans 46:285–294

Chang L, Tian L, Ma F et al (2020) Regulation of photosystem I-light-harvesting complex I from a red alga *Cyanidioschyzon merolae* in response to light intensities. Photosynth Res 146:287–297

Chen M, Perez-Boerema A, Zhang L (2021) Distinct structural modulation of photosystem I and lipid environment stabilizes its tetrameric assembly. Nat Plants 6:314–320

Chitnis VP, Chitnis PR (1993) PsaL subunit is required for the formation of photosystem I trimers in the cyanobacterium *Synechocystis* sp. PCC 6803. FEBS Lett 336:330–334

Croce R, van Amerongen H (2020) Light harvesting in oxygenic photosynthesis: structural biology meets spectroscopy. Science 369:eaay2058

Dall'Osto L, Bressan M, Bassi R (2015) Biogenesis of light harvesting proteins. Biochim Biophys Acta 1847:861–871

De la Rosa MA, Navarro JA, Díaz-Quintana A et al (2002) An evolutionary analysis of the reaction mechanisms of photosystem I reduction by cytochrome *c(6)* and plastocyanin. Bioelectrochemistry 55:41–45

Depege N, Bellafiore S, Rochaix JD (2003) Role of chloroplast protein kinase Stt7 in LHCII phosphorylation and state transition in *Chlamydomonas*. Science 299:1572–1575

Durnford DG, Deane JA, Tan S (1999) A phylogenetic assessment of the eukaryotic light-harvesting antenna proteins, with implications for plastid evolution. J Mol Evol 48:59–68

Enami I, Murayama H, Ohta H et al (1995) Isolation and characterization of a photosystem II complex from the red alga *Cyanidium caldarium*: association of cytochrome *c*-550 and a 12 kDa protein with the complex. Biochim Biophys Acta 1232:208–216

Enami I, Okumura A, Nagao R et al (2008) Structures and functions of the extrinsic proteins of photosystem II from different species. Photosynth Res 98:349–363

Falkowski PG, Katz ME, Knoll AH et al (2004) The evolution of modern eukaryotic phytoplankton. Science 305:354–360

Fan M, Li M, Liu Z et al (2015) Crystal structures of the PsbS protein essential for photoprotection in plants. Nat Struct Mol Biol 22:729–735

Fromme P, Jordan P, Krauss N (2001) Structure of photosystem I. Biochim Biophys Acta 1507:5–31

Gisriel CJ, Shen G, Ho MY et al (2021) Structure of a monomeric photosystem II core complex from a cyanobacterium acclimated to far-red light reveals the functions of chlorophylls *d* and *f*. J Biol Chem 298:101424

Golbeck JH (1999) A comparative analysis of the spin state distribution of in vitro and *in vivo* mutants of PsaC. A biochemical argument for the sequence of electron transfer in photosystem I as $F_X \rightarrow F_A \rightarrow F_B \rightarrow$ ferredoxin/flavodoxin. Photosynth Res 61:107–144

Golbeck JH (2006) Photosystem I: the light-driven Plastocyanin: ferredoxin oxidoreductase. Springer, Dordrecht

Green BR (2003) The evolution of light-harvesting antennas. In: Green BR, Parson WW (eds) Light-harvesting antennas in photosynthesis. Kluwer Academic Publishers, Boston, MA, pp 129–168

Guergova-Kuras M, Boudreaux B, Joliot A et al (2001) Evidence for two active branches for electron transfer in photosystem I. Proc Natl Acad Sci U S A 98:4437–4442

Guglielmi G, Cohen-Bazire G, Bryant DA (1981) The structure of *Gloeobacter violaceus* and its phycobilisomes. Arch Microbiol 129:181–189

Hamaguchi T, Kawakami K, Shinzawa-Itoh K et al (2021) Structure of the far-red light utilizing photosystem I of *Acaryochloris marina*. Nat Commun 12:2333

Heelis DV, Kernick W, Phillips GO et al (1979) Separation and identification of the carotenoid pigments of stigmata isolated from light-grown cells of Euglena gracilis strain Z. Arch Microbiol 121:207–211

Hervás M, Navarro JA, De La Rosa MA (2003) Electron transfer between membrane complexes and soluble proteins in photosynthesis. Acc Chem Res 36:798–805

Hoffman GE, Sanchez Puerta MV, Delwiche CF (2011) Evolution of light-harvesting complex proteins from Chl *c*-containing algae. BMC Evol Biol 11:101

Holzwarth AR, Muller MG, Niklas J et al (2006) Ultrafast transient absorption studies on photosystem I reaction centers from *Chlamydomonas reinhardtii*. 2: mutations near the P700 reaction center chlorophylls provide new insight into the nature of the primary electron donor. Biophys J 90:552–565

Hu Q, Miyashita H, Iwasaki I (1995) A photosystem I reaction center driven by chlorophyll *d* in oxygenic photosynthesis. Proc Natl Acad Sci U S A 95:13319–13323

Huang ZSL, Wang W, Mao Z et al (2021) Structure of photosystem I-LHCI-LHCII from the green alga *Chlamydomonas reinhardtii* in state 2. Nat Commun 12:1100

Itoh S, Mino H, Itoh K et al (2007) Function of chlorophyll *d* in reaction centers of photosystems I and II of the oxygenic photosynthesis of *Acaryochloris marina*. Biochemistry 46:12473–12481

Joliot P, Joliot A (1999) *In vivo* analysis of the electron transfer within photosystem I: are the two phylloquinones involved? Biochemistry 38:11130–11136

Jordan P, Fromme P, Witt HT et al (2001) Three-dimensional structure of cyanobacterial photosystem I at 2.5 Å resolution. Nature 411:909–917

Kato K, Nagao R, Jiang TY et al (2019) Structure of a cyanobacterial photosystem I tetramer revealed by cryo-electron microscopy. Nat Commun 10:4929

Kjaerulff S, Andersen B, Nielsen VS et al (1993) The PSI-K subunit of photosystem I from barley (Hordeum vulgare L.). evidence for a gene duplication of an ancestral PSI-G/K gene. J Biol Chem 268:18912–18916

Kobayashi M, Watanabe S, Gotoh T et al (2005) Minor but key chlorophylls in photosystem II. Photosynth Res 84:201–207

Kobayashi M, Akutsu S, Fujinuma D et al (2013) In: Dubinsky Z (ed) Physicochemical properties of chlorophylls in oxygenic photosynthesis — Succession of co-factors from anoxygenic to oxygenic photosynthesis. In Tech, London, pp 47–90

Laudenbach DE, Straus NA (1988) Characterization of a cyanobacterial iron stress-induced gene similar to *psb C*. J Bacteriol 170:5018–5026

Lemeille S, Rochaix JD (2010) State transitions at the crossroad of thylakoid signalling pathways. Photosynth Res 106:33–46

Levitan O, Dinamarca J, Hochman G et al (2014) Diatoms: a fossil fuel of the future. Trends Biotechnol 32:117–124

Li XP, Muller-Moule P, Gilmore AM et al (2002) PsbS-dependent enhancement of feedback de-excitation protects photosystem II from photoinhibition. Proc Natl Acad Sci U S A 99: 15222–15227

Li M, Calteau A, Semchonok DA et al (2019) Physiological and evolutionary implications of tetrameric photosystem I in cyanobacteria. Nat Plants 5:1309–1319

Liu H, Zhang H, Niedzwiedzki DM et al (2013) Phycobilisomes supply excitations to both photosystems in a megacomplex in cyanobacteria. Science 342:1104–1107

Ma J, You X, Sun S et al (2020) Structural basis of energy transfer in Porphyridium purpureum phycobilisome. Nature 579:146–151

Makita H, Hastings G (2015) Directionality of electron transfer in cyanobacterial photosystem I at 298 and 77K. FEBS Lett 589:1412–1417

Matsuzaki M, Misumi O, Shin IT et al (2004) Genome sequence of the ultrasmall unicellular red alga *Cyanidioschyzon merolae* 10D. Nature 428:653–657

Mazor Y, Borovikova A, Caspy I (2017) Structure of the plant photosystem I supercomplex at 2.6 Å resolution. Nat Plants 3:17014

Minagawa J (2011) State transitions--the molecular remodeling of photosynthetic supercomplexes that controls energy flow in the chloroplast. Biochim Biophys Acta 1807:897–905

Mitsuhashi K, Tamura H, Saito K et al (2021) Nature of asymmetric electron transfer in the symmetric pathways of photosystem I. J Phys Chem B 125:2879–2885

Muhiuddin IP, Heathcote P, Carter S et al (2001) Evidence from time resolved studies of the $P700^{(\cdot+)}/A1^{(\cdot-)}$ radical pair for photosynthetic electron transfer on both the PsaA and PsaB branches of the photosystem I reaction Centre. FEBS Lett 503:56–60

Muller MG, Slavov C, Luthra R (2010) Independent initiation of primary electron transfer in the two branches of the photosystem I reaction center. Proc Natl Acad Sci U S A 107:4123–4128

Murata N (1969) Control of excitation transfer in photosynthesis. I. Light-induced change of chlorophyll a fluoresence in *Porphyridium cruentum*. Biochim Biophys Acta 172:242–251

Murata N (2009) The discovery of state transitions in photosynthesis 40 years ago. Photosynth Res 99:155–160

Nagao R, Ueno Y, Yokono M et al (2018) Alterations of pigment composition and their interactions in response to different light conditions in the diatom *Chaetoceros gracilis* probed by time-resolved fluorescence spectroscopy. Biochim Biophys Acta Bioenerg 1859:524–530

Nagao R, Ueno Y, Akita F et al (2019) Biochemical characterization of photosystem I complexes having different subunit compositions of fucoxanthin chlorophyll *a/c*-binding proteins in the diatom *Chaetoceros gracilis*. Photosynth Res 140:141–149

Nagao R, Ueno Y, Akimoto S et al (2020a) Effects of CO_2 and temperature on photosynthetic performance in the diatom *Chaetoceros gracilis*. Photosynth Res 146:189–195

Nagao R, Kato K, Ifuku K et al (2020b) Structural basis for assembly and function of a diatom photosystem I-light-harvesting supercomplex. Nat Commun 11:2481

Nelissen B, Van de Peer Y, Wilmotte A et al (1995) An early origin of plastids within the cyanobacterial divergence is suggested by evolutionary trees based on complete 16S rRNA sequences. Mol Biol Evol 12:1166–1173

Nelson N, Junge W (2015) Structure and energy transfer in photosystems of oxygenic photosynthesis. Annu Rev Biochem 84:659–683

Netzer-El SY, Caspy I, Nelson N (2019) Crystal structure of photosystem I monomer from *Synechocystis* PCC 6803. Front Plant Sci 9:1865

Okkels JS, Nielsen VS, Scheller HV et al (1992) A cDNA clone from barley encoding the precursor from the photosystem I polypeptide PSI-G: sequence similarity to PSI-K. Plant Mol Biol 18: 989–994

Ozawa SI, Bald T, Onishi T et al (2018) Configuration of ten light-harvesting chlorophyll *a/b* complex I subunits in *Chlamydomonas reinhardtii* photosystem I. Plant Physiol 178:583–595

Pan X, Ma J, Su X et al (2018) Structure of the maize photosystem I supercomplex with light-harvesting complexes I and II. Science 360:1109–1113

Perez-Boerema A, Klaiman D, Caspy I et al (2020) Structure of a minimal photosystem I from the green alga *Dunaliella salina*. Nat Plants 6:321–327

Pi X, Tian L, Dai HE et al (2018) Unique organization of photosystem I-light-harvesting supercomplex revealed by cryo-EM from a red alga. Proc Natl Acad Sci U S A 115:4423–4428

Qin X (2021) Structure, function, and evolution of photosystem I-light harvesting antenna I complexes. In: Shen J-R, Allakhverdiev SI (eds) Advances in photosynthesis and respiration: photosynthesis: molecular approaches to solar energy conversion, vol 47. Springer, Cham, pp 195–222

Qin X, Suga M, Kuang T et al (2015) Photosynthesis. Structural basis for energy transfer pathways in the plant PSI-LHCI supercomplex. Science 348:989–995

Qin X, Pi X, Wang W et al (2019) Structure of a green algal photosystem I in complex with a large number of light-harvesting complex I subunits. Nat Plants 5:263–272

Rakhimberdieva MG, Boichenko VA, Karapetyan NV et al (2001) Interaction of phycobilisomes with photosystem II dimers and photosystem I monomers and trimers in the cyanobacterium Spirulina platensis. Biochemistry 40:15780–15788

Ramesh VM, Gibasiewicz K, Lin S et al (2004) Bidirectional electron transfer in photosystem I: accumulation of A_0- in A-side or B-side mutants of the axial ligand to chlorophyll A_0. Biochemistry 43:1369–1375

Rogner M, Nixon PJ, Diner BA (1990) Purification and characterization of photosystem I and photosystem II core complexes from wild-type and phycocyanin-deficient strains of the cyanobacterium *Synechocystis* PCC 6803. J Biol Chem 265:6189–6196

Sandmann G (1986) Formation of plastocyanin and cytochrome *c*-553 in different species of blue-green algae. Arch Microbiol 145:76–79

Sandmann G, Reck H, Kessler E et al (1986) Distribution of plastocyanin and soluble plastidic cytochrome *c* in various classes of algae. Arch Microbiol 134:23–27

Schenderlein M, Cetin M, Barber J et al (2008) Spectroscopic studies of the chlorophyll *d* containing photosystem I from the cyanobacterium, *Acaryochloris marina*. Biochim Biophys Acta 1777:1400–1408

Shen JR (2015) The structure of photosystem II and the mechanism of water oxidation in photosynthesis. Annu Rev Plant Biol 66:23–48

Shen JR, Inoue Y (1993) Binding and functional properties of two new extrinsic components, cytochrome *c*-550 and a 12-kDa protein, in cyanobacterial photosystem II. Biochemistry 32: 1825–1832

Slater B, Kosmützky D, Nisbet RER et al (2021) The evolution of the cytochrome c_6 family of photosynthetic electron transfer proteins. Genome Biol Evol 13:evab146

Smetacek V (1999) Diatoms and the ocean carbon cycle. Protist 150:25–32

Su X, Ma J, Pan X et al (2019) Antenna arrangement and energy transfer pathways of a green algal photosystem-I-LHCI supercomplex. Nat Plants 5:273–281

Suga M, Shen JR (2020) Structural variations of photosystem I-antenna supercomplex in response to adaptations to different light environments. Curr Opin Struct Biol 63:10–17

Suga M, Qin X, Kuang T et al (2016) Structure and energy transfer pathways of the plant photosystem I-LHCI supercomplex. Curr Opin Struct Biol 39:46–53

Suga M, Ozawa SI, Yoshida-Motomura K et al (2019) Structure of the green algal photosystem I supercomplex with a decameric light-harvesting complex I. Nat Plants 5:626–636

Sui SF (2021) Structure of phycobilisomes. Annu Rev Biophys 50:53–72

Tian L, Liu Z, Wang F et al (2017) Isolation and characterization of PSI-LHCI super-complex and their sub-complexes from a red alga *Cyanidioschyzon merolae*. Photosynth Res 133:201–214

Tokutsu R, Teramoto H, Takahashi Y et al (2004) The light-harvesting complex of photosystem I in *Chlamydomonas reinhardtii*: protein composition, gene structures and phylogenic implications. Plant Cell Physiol 45:138–145

Tomo T, Kato Y, Suzuki T et al (2008) Characterization of highly purified photosystem I complexes from the chlorophyll *d*-dominated cyanobacterium *Acaryochloris marina* MBIC 11017. J Biol Chem 283:18198–18209

Toporik H, Li J, Williams D et al (2019) The structure of the stress-induced photosystem I-IsiA antenna supercomplex. Nat Struct Mol Biol 26:443–449

Umena Y, Kawakami K, Shen JR et al (2011) Crystal structure of oxygen-evolving photosystem II at a resolution of 1.9 Å. Nature 473:55–60

Wang W, Yu LJ, Xu C et al (2019) Structural basis for blue-green light harvesting and energy dissipation in diatoms. Science 363:aav0365

Wang J, Yu LJ, Wang W et al (2021) Structure of plant photosystem I-light harvesting complex I supercomplex at 2.4 Å resolution. J Integr Plant Biol 63:1367–1381

Watanabe M, Semchonok DA, Webber-Birungi MT et al (2014) Attachment of phycobilisomes in an antenna-photosystem I supercomplex of cyanobacteria. Proc Natl Acad Sci U S A 111:2512–2517

Weigel M, Varotto C, Pesaresi P et al (2003) Plastocyanin is indispensable for photosynthetic electron flow in *Arabidopsis thaliana*. J Biol Chem 278:31286–31289

Wolfe GR, Cunningham FX, Grabowski B et al (1994) Isolation and characterization of photosystem-I and photosystem-II from the red alga *Porphyridium cruentum*. Biochim Biophys Acta Bioenerg 1188:357–366

Wu G, Ma L, Yuan C et al (2021) Formation of light-harvesting complex II aggregates from LHCII-PSI-LHCI complexes in rice plants under high light. J Exp Bot 72:4938–4948

Xu W, Chitnis PR, Valieva A et al (2003) Electron transfer in cyanobacterial photosystem I: II. Determination of forward electron transfer rates of site-directed mutants in a putative electron transfer pathway from A_0 through A_1 to F_x. J Biol Chem 278:27876–27887

Xu C, Pi X, Huang Y et al (2020) Structural basis for energy transfer in a huge diatom PSI-FCPI supercomplex. Nat Commun 11:5081

Xu C, Zhu Q, Chen JH et al (2021) A unique photosystem I reaction center from a chlorophyll *d*-containing cyanobacterium *Acaryochloris marina*. J Integr Plant Biol 63:1740–1752

Yan Q, Zhao L, Wang W et al (2021) Antenna arrangement and energy-transfer pathways of PSI-LHCI from the moss *Physcomitrella patens*. Cell Discov 7:10

Yang F, Shen G, Schluchter WM et al (1998) Deletion of the PsaF polypeptide modifies the environment of the redox-active phylloquinone (A_1). Evidence for unidirectionality of electron transfer in photosystem I. J Phys Chem B 102:8288–8299

Yu J, Ma PJ, Shi DJ et al (2008) Homologous comparisons of photosynthetic system I genes among cyanobacteria and chloroplasts. J Integr Plant Biol 50:929–940

Zech SG, Hofbauer W, Kamlowski A et al (2000) A structural model for the charge separated state $P700^{+}A_1^{\bullet-}$ in photosystem I from the orientation of the magnetic interaction tensors. J Phys Chem B 104:9728–9739

Zhang J, Ma J, Liu D et al (2017) Structure of phycobilisome from the red alga *Griffithsia pacifica*. Nature 551:57–63

Zheng L, Li Y, Li X et al (2019) Structural and functional insights into the tetrameric photosystem I from heterocyst-forming cyanobacteria. Nat Plants 5:1087–1097

Chapter 12
Structure and Function of Mycobacterial Arabinofuranosyltransferases

Yong Zi Tan and Filippo Mancia

Abstract The mycobacteria genus is responsible for numerous infectious diseases that have afflicted the human race since antiquity—tuberculosis and leprosy in particular. An important contributor to their evolutionary success is their unique cell envelope, which constitutes a quasi-impermeable barrier, protecting the microorganism from external threats, antibiotics included. The arabinofuranosyltransferases are a family of enzymes, unique to the *Actinobacteria* family that mycobacteria genus belongs to, that are critical to building of this cell envelope. In this chapter, we will analyze available structures of members of the mycobacterial arabinofuranosyltransferase, clarify their function, as well as explore the common themes present amongst this family of enzymes, as revealed by recent research.

Keywords Arabinofuranosyltransferases · Arabinosyltransferases · Glycosyltransferases · Mycobacterium · Tuberculosis · Sugar transfer · Acyl carrier protein · Cell wall biosynthesis · Cryogenic electron microscopy · Cryo-EM · X-ray crystallography · Arabinogalactan · Lipoarabinomannan

Introduction

Overall Structure of the Mycobacterium Cell Envelope

Mycobacteria are unique in part due to the nature of their cell envelope. Although they can appear as weakly gram-positive (Fu and Fu-Liu 2002), their cell envelope

Y. Z. Tan (✉)
Department of Biological Sciences, National University of Singapore, Singapore, Singapore

Disease Intervention Technology Laboratory (DITL), Institute of Molecular and Cell Biology (IMCB), Agency for Science, Technology and Research (A*STAR), Singapore, Singapore
e-mail: yztan@nus.edu.sg

F. Mancia
Department of Physiology and Cellular Biophysics, Columbia University, NY, USA
e-mail: fm123@cumc.columbia.edu

J. R. Harris, J. Marles-Wright (eds.), *Macromolecular Protein Complexes IV*, Subcellular Biochemistry 99, https://doi.org/10.1007/978-3-031-00793-4_12

resembles more that of their gram-negative counterparts due to the presence of a lipid containing the outermost layer (Alderwick et al. 2015; Minnikin et al. 1982). Indeed, their cell envelope is divided into three parts: the inner membrane, the peptidoglycan (PG) cell wall, and the outermost layer (Kaur et al. 2009). The outermost layer consists of branched heteropolysaccharide arabinogalactan (AG) and long-chain mycolic acids that range from 70 to 90 carbon atoms in length (Angala et al. 2014). AG itself is made of galactose and arabinose sugar residues. Together, they are called the mycolyl-arabinogalactan-peptidoglycan (mAGP) complex (Alderwick et al. 2015; Grzegorzewicz et al. 2016). Intercalated between the mAGP complexes are additional glycolipids. These include lipoarabinomannan (LAM), lipomannan (LM), and phosphatidyl-*myo*-inositol mannoside (PIM). All these lipids in turn help make the outer membrane and hence the entire cell envelope, highly impermeable (Hoffmann et al. 2008; Nikaido and Jarlier 1991). Proteins can be found embedded in the inner membrane as well as the outer membrane (Niederweis et al. 2010). The mycobacterial cell envelope is important for the survival, growth, virulence, and impermeability to therapeutics against these microorganisms (Barry III 2001; Jankute et al. 2015). Unsurprisingly, isoniazid and ethambutol—two of the most effective drugs against tuberculosis—target synthesis of this structure (Brennan and Crick 2007).

Synthesis of the Cell Envelope by Arabinofuranosyltransferases

One important class of enzymes involved in building the mAGP complex is the arabinofuranosyltransferase, the focus of this chapter. Specifically, these multipass transmembrane enzymes build the AG and LAM components of the mAGP, as these contain arabinofuranose sugars in their structure. Arabinofuranosyltransferases belong to the GT53 family of glycosyltransferases (Bohl et al. 2021) as they transfer D-arabinofuranose sugar moieties to their acceptor substrates from a decaprenylphosphoryl-D-arabinofuranose (DPA) donor (Abrahams and Besra 2016; Wolucka et al. 1994). The DPA donor is found in the inner membrane, just like the arabinofuranosyltransferases (Brennan and Crick 2007).

The arabinofuranosyltransferases can be divided into two groups. The first, which is the more renowned, consists of the Emb proteins, EmbA, EmbB, and EmbC, which are encoded by a single *emb* operon (Belanger et al. 1996; Telenti et al. 1997) controlled by a transcriptional regulator EmbR (Sharma et al. 2006). They are named Embs as they are the targets of the antituberculotic drug ethambutol (Goude et al. 2009; Takayama and Kilburn 1989). The second group consists of other enzymes that catalyze a similar reaction but are not sensitive to ethambutol. These include AftA (Alderwick et al. 2006), AftB (Seidel et al. 2007), AftC (Birch et al. 2008), and AftD (Škovierová et al. 2009). Note that they do not occur in the same operon. It has also been suggested that there might actually be more members of the arabinofuranosyltransferase family that are yet to be discovered (Angala et al. 2016).

AG is made of a galactan core with around 30 galactose residues, on which arabinofuranose sugars are decorated on. In the synthesis of AG, the galactan core is the starting donor substrate. AftA begins the decoration process by adding a single D-arabinofuranose at the C5 of a D-galactofuranose of this galactan core. The sugar will be added to the eight, tenth, or twelfth sugar, and at least two out of the three positions will be decorated by AftA (Jankute et al. 2015). Next, EmbA and EmbB processively extend the arabinan chain one D-galactofuranose at a time, linking them via α-(1→5) linkages. AftC and possibly AftD then add α-(1→3) D-galactofuranose, causing the arabinan chain to branch. The final enzyme in the synthesis of AG is AftB, which completes the AG by adding a terminal β-(1→2)-linked D-arabinofuranose.

LAM consists of a mannan core of around 30 residues, which is decorated by arabinofuranose residues. In the synthesis of LAM, an unknown enzyme (Angala et al. 2016) primes this mannan core with an α-(1→6) glycosidic bond to D-arabinofuranose. EmbC then processively adds D-galactofuranose connected via α-(1→5) linkages. AftC and possibly AftD then catalyze branching via α-(1→3) linkages, with AftB capping off the completed LAM the same way as its role in AG (Alderwick et al. 2006).

Structures and Function

In this section, available high-resolution arabinofuranosyltransferase structures will be discussed. These are for namely: AftD, EmbA, EmbB, and EmbC.

Arabinofuranosyltransferase D (AftD)

Arabinofuranosyltransferase D (AftD) was first identified by the group of Mary Jackson (Škovierová et al. 2009) and, with a molecular weight of around 150 kDa, is the largest enzyme of the family. It was initially thought to catalyze α-(1→3) branching, (Škovierová et al. 2009) and has recently been suggested to also have α-(1→5) (Alderwick et al. 2018) branching activity. AftD is responsible for adding the last few arabinofuranose residues to AG and LAM, with the exact number of residues added not yet confirmed either (Alderwick et al. 2018). AftD is an essential gene for mycobacteria (Škovierová et al. 2009).

The structure of AftD from *Mycobacterium abscessus* (PDB ID: 6 W98) was recently elucidated (Tan et al. 2020b) (Fig. 12.1). To facilitate structure determination, AftD was reconstituted into nanodiscs and imaged using single particle cryogenic electron microscopy (cryo-EM) (Cheng 2018).

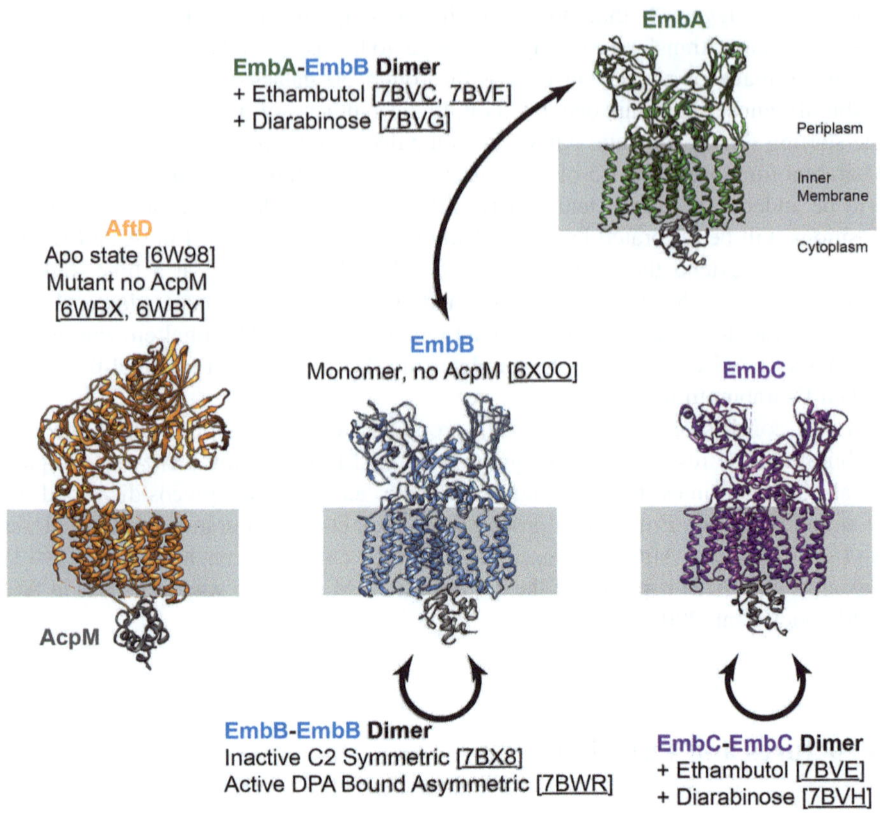

Fig. 12.1 Currently available full length structures of mycobacterial arabinofuranosyltransferases. The PDB IDs are underlined

AftD is comprised of a membrane embedded and a soluble region. The N-terminus resides in the cytoplasm, preceding the first eleven transmembrane (TM) helices, which are organized into a GT-C glycosyltransferase fold (Liu and Mushegian 2003) containing the putative catalytic site. Thereafter, the AftD polypeptide chain exits the membrane and enters the periplasmic space. This soluble region consists of five domains, three of which are carbohydrate binding modules (CBMs), while the other two are more structural. There are extensive interactions between these discrete periplasmic domains maintaining them rather rigidly associated with each other. After the periplasmic domains, the polypeptide chain reenters the membrane forming a structural motif with TM13 to TM16 that was found to be unexpectedly tightly associated with an acyl carrier protein, AcpM. 4'-Phosphopantetheine, a prosthetic group that is covalently attached to AcpM, is buried between these structural motifs of AftD. The polypeptide chain then terminates in the cytoplasm.

As the structure was solved without a substrate, the location of the active site of AftD was inferred from sequence conservation analysis as well as homology with other GT-C glycosyltransferases. Based on this prediction, AftD's putative active site consists of a cavity of ~1240 Å^3 formed between the soluble region and the periplasmic facing ends of TMs 4, 6, 7, 9, and 10. Residues K39, D41, E251, K361, and D474 that are found in the putative active site were mutated and shown in vivo to abolish activity, supporting a role of these residues in the catalytic process (Tan et al. 2020b).

The putative active site of AftD has a slight positive charge. This complements well with AftD's substrate of poly-arabinofuranose glycans, plausibly allowing ion-dipole interactions to be formed between the glycans and AftD's positively charged amino acid residues. Similar to other GT-C glycosyltransferases, AftD appears to have a cation-binding pocket lined by residues E251, D474, and E475, although no metal was observed in the structure.

The likely binding site for the DPA donor is demarcated by TMs 6, 9, and 13. It is a shallow groove along AftD that begins at cytoplasmic side of the membrane and terminates at the putative active site. The hydrophobic nature of this region, in addition to its homology with the polyprenol-binding pockets of other glycosyltransferases like STT3 (Wild et al. 2018), PglB (Napiórkowska et al. 2017), and ArnT (Petrou et al. 2016), supports this groove as the DPA donor binding site.

There are three loops around the putative active site: the N-EL4 between TM 7 and JM 3, C-EL4 between JM 3 and TM 8, and EL5 between TM 9 and 10. All three are disordered in the structure without substrate. The first two are analogous to those found in other glycosyltransferases like STT3, PglB, and ArnT, and as such they are likely to be ordered upon substrate binding. The last one, EL5, is instead unique to AftD. Surprisingly, when binding of AcpM was weakened by mutagenesis, this loop becomes ordered and blocks the active site. This suggests that the EL5 loop might function as a gate that controls access to the active site, depending on the presence or absence of bound AcpM. This same mutation also causes AftD to be nonfunctional as assayed in vivo, supporting the hypothesis that AcpM binding might be a regulatory element and that its association could be important to maintain AftD functionality.

Finally, AftD is not sensitive to any known antibiotics against tuberculosis, thus given its essential nature for mycobacterium, presents itself as a potential target for the development of new drugs.

Arabinofuranosyltransferase EmbA

EmbA is an essential enzyme for mycobacterium (Belanger et al. 1996; Telenti et al. 1997). The structure of EmbA from both *M. tuberculosis* (PDB ID: 7BVF) and *M. smegmatis* (PDB ID: 7BVC, 7BVG) have been recently elucidated (Zhang et al. 2020b) (Fig. 12.1). Significantly, EmbA was found to form a dimer with EmbB, and was also bound to AcpM. The complex was purified from mycobacterial sources and was solubilized with n-dodecyl-β-D-maltopyranoside (DDM) and supplemented with glyco-diosgenin (GDN) before imaging using cryo-EM.

The architecture of EmbA consists of an N-terminus and a C-terminus periplasmic domain, which flank the membrane TM region containing 15 TM helices. The two periplasmic domains adopt a CBM fold, likely serving the role of binding to the sugar substrates. The active site is located at the interface between the two periplasmic domains and the TM region. Structures of EmbA-EmbB dimers with the sugar acceptor mimic (di-arabinose instead of the full length substrate) and sugar donor (DPA) show the mode of binding of both these molecules at the active site. The sugar acceptor enters from the C-terminus CBM, whereas the lipid-conjugated sugar donor reaches in from the opposite end. Residues D261, N265, E289, R365, Q409, T553, W555 and Q558 appear to be involved in stabilizing the DPA with polar interactions. The hydrophobic decaprenyl tail of DPA extends into the TM region, nestled in a hydrophobic cavity created by TM7, TM8 and TM9. Finally, Q780 and the catalytic residue D261 interact with the di-arabinose.

Ethambutol was added to the EmbA-EmbB complex prior to vitrification for structure determination. Evidence for the presence of the drug was found in only EmbB. Comparing the structures, it seems that the additional loop EmbA of consisting of residues 766–806 helps lock DPA in the active site and prevents binding of ethambutol. This additional loop is not present in EmbB and EmbC, and could explain why clinical mutations that result in ethambutol resistance have yet to be found in EmbA (Brossier et al. 2015; Huang et al. 2011; Plinke et al. 2010; Ramaswamy et al. 2000; Sandgren et al. 2009; Sun et al. 2018; Xu et al. 2015; Zhao et al. 2015).

In terms of dimer formation, EmbA interacts with EmbB via hydrophobic interactions occurring between the TM helices. Cardiolipins seem to play a role in stabilization of the dimer as well. AcpM binds to the cytoplasmic face of EmbA, with its covalently bound 4′-phosphopantetheine buried into a cavity formed between TM6 and TM7. It is not structurally homologous to the AcpM binding site on AftD, suggesting the AcpM binding likely evolved separately for both enzymes.

Arabinofuranosyltransferase EmbB

EmbB is sensitive to ethambutol and is also an essential enzyme. Its heterodimeric structure with EmbA was elucidated from both *M. tuberculosis* (PDB ID: 7BVF) and *M. smegmatis* (PDB ID: 7BVC, 7BVG). These were determined by cryo-EM in detergent. EmbB as a homodimer from *M. smegmatis* (PDB ID: 7BWR, 7BX8) was also solved as part of the same work (Zhang et al. 2020a). This structure was solved using cryo-EM from EmbB dimer reconstituted into amphipol. Finally, the monomeric structure of EmbB from *M. smegmatis* (PDB ID: 6X0O) was solved using cryo-EM by a different group (Tan et al. 2020a), with the monomer reconstituted

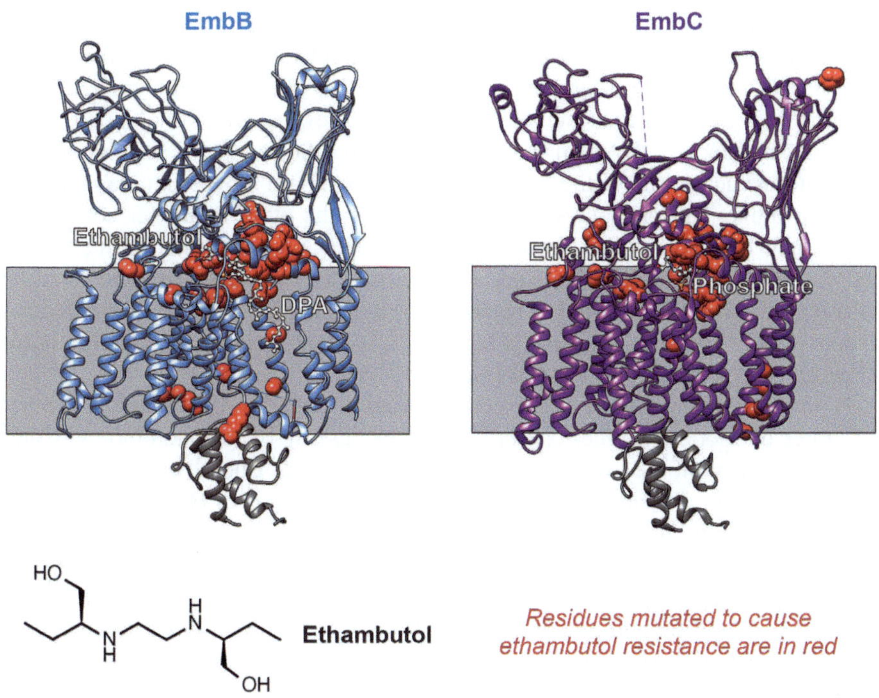

Fig. 12.2 Mutations that confer ethambutol resistance are mapped onto the structures of EmbB and EmbC and shown as red spheres. Chemical structure of ethambutol is shown below

into nanodiscs. This plethora of structures of different states of EmbB has been invaluable in furthering the functional understanding of EmbB (Fig. 12.1).

In terms of architecture, EmbB resembles EmbA, with binding of the sugar donor and acceptor occurring at similar sites. As noted above, the most notable difference is the length of the loop between residues 776 and 806 which is shorter than for EmbA by approximately ten residues. This in turn opens the site for ethambutol binding. Clinical mutations on EmbB that confer ethambutol resistance can be mapped mostly around the active site of EmbB (Fig. 12.2). In vitro activity assays using the purified EmbB homodimer showed that the clinical M292L mutation did indeed reduce sensitivity to ethambutol inhibition, and drug binding affinity was also reduced.

Ethambutol has two charged imino groups that form ionic bonds with D299 of EmbB and a decaprenyl (DPA without the sugar) at the active site. W988 and Y302 also form π-cation interactions with the two imino groups. Separately, H594 and E327 form hydrogen bonds with ethambutol's hydroxyl groups. Finally, hydrophobic packing of the nonpolar moieties of ethambutol also occurs with I303, M306, W592, and W1028. All these interactions serve to stabilize ethambutol and position it in a way that it blocks both the sugar acceptor as well as the arabinose of the sugar donor, thus revealing the mechanism for enzymatic inhibition by this antibiotic.

The EmbB homodimer was captured in two states: a resting state that resembles the EmbA-EmbB heterodimer and a pre-catalytic active state. The former resting state has C2 symmetry, whereas the active site has the symmetry broken. This asymmetry is due to a translation of ~16 Å of one of the EmbB monomers along the dimer interface in the membrane plane. To form the cavity for accommodating DPA into the active site of the first EmbB monomer, both monomers are involved: The first monomer contributes TM7, TM8, TM9, the other monomer contributes TM14, and both monomers contribute their periplasmic C-terminal domains and periplasmic loops. In the active site, an α-helix of PL5 becomes ordered with the DPA bound, similarly to what has been observed in other GT-C glycosyltransferases such as ArnT (Petrou et al. 2016), where this was hypothesized to be important for allowing acceptor access to the active site.

In the monomer structure, the authors tried to add in ethambutol and other substrates but noticed neither conformational changes nor substrate binding. Combining with the multiple states of substrate-bound EmbB as a homo or heterodimer, it provides a strong structural reasoning as to why dimerization of EmbB is critical for its proper catalytic function.

All these structural snapshots have provided unprecedented mechanistic insight on the catalytic cycle for EmbB. The binding of DPA appears to be important to shift the inactive EmbB homodimer to the active state, where PL5 becomes ordered. This enables the acceptor to bind, and catalysis can thus occur. The products dissociated and the enzyme resets to its inactive, C2 symmetric state.

AcpM binding was also present in EmbB in both the heterodimer and homodimer. Interestingly, AcpM was not present in the monomer structure, although it could be a result of the monomer structure being heterologously expressed in *Escherichia coli*; hence, the mycobacterial AcpM was not present. In vitro activity assays of mutant EmbB homodimer that had abolished AcpM binding showed the same activity as the control. When the mutant was transformed into *M. smegmatis embB* knockout strains though, there was indeed a loss in branching of AG, indicating that AcpM's binding to EmbB still has a role in the function of the enzyme in a physiological setting that cannot be captured using in vitro assays.

Arabinofuranosyltransferase EmbC

EmbC, just like EmbB, is sensitive to ethambutol. The C-terminal periplasmic domain of EmbC was first determined by X-ray crystallography (PDB ID: 3PTY) (Alderwick et al. 2011). The full length structure of EmbC was elucidated recently as a homodimer (Zhang et al. 2020b), using a combination of X-ray crystallography (PDB ID: 7BVH) and cryo-EM (PDB ID:7BVE) (Fig. 12.1).

Structurally, EmbC resembles EmbB and has both the sugar acceptor mimic (di-arabinose) and ethambutol bound in a same way. Mutations that cause resistance to ethambutol also cluster around the active site, similar to EmbB (Fig. 12.2). EmbC also binds to AcpM, again in a similar manner. However, unlike EmbA and EmbB,

EmbC appears to only form homodimers. Interestingly, the X-ray crystal structure of EmbC homodimers with di-arabinose bound showed the presence of a phosphate ion and of maltose (part of the DDM detergent used in solubilization). The phosphate is thought to represent the phosphate on the sugar donor DPA. The donor DPA was not seen in any of the EmbC structures.

As EmbC is a processive enzyme, the structures determined suggest a catalytic cycle in which after the arabinose is transferred to the lengthening sugar chain by the enzyme, an acceptor repositioning occurs to allow for the added arabinose to now serve as the new acceptor for additional sugars.

Interacting Partners with Arabinofuranosyltransferases

Given that many of the arabinofuranosyltransferases act one after the other on the same substrate, having them arranged in a closed proximity or as a multisubunit complex similar for instance to the eukaryotic oligosaccharyltransferase complex makes functional sense (Bai et al. 2018; Knauer and Lehle 1999; Wild et al. 2018). Indeed the Emb family of enzymes were hypothesized to function as heterodimers between EmbA and EmbB (Amin et al. 2008; Escuyer et al. 2001; Khasnobis et al. 2006; Plinke et al. 2010) and as a homodimer of two EmbC (Alderwick et al. 2011; Shi et al. 2006) or two EmbB (Rodrigues et al. 2021) based on functional and biochemical data. All three complexes were confirmed by the elucidated mycobacterial structures using cryo-EM (Zhang et al. 2020a,b) (Fig. 12.1). Two-hybrid studies in *Corynebacterium glutamicum* have suggested interactions between AftA-AftB, AftA-AftC, AftB-AftC, AftA-Emb family, AftB-Emb family, and formation of homomeric assemblies of AftA, AftB, and AftC, although there are no structural validations of these interactions yet (Jankute et al. 2014).

The acyl carrier protein (AcpM) has been a common yet unexpected binding partner in almost all the solved mycobacterial arabinofuranosyltransferase structures. AcpM was found bound to AftD (Tan et al. 2020b), EmbB dimer (Zhang et al. 2020a), EmbA-EmbB dimer (Zhang et al. 2020b), and EmbC dimer (Zhang et al. 2020b) despite not being co-expressed. In all cases, the binding site is on the cytoplasmic side of the proteins. It is only not seen in the monomeric form of EmbB (Tan et al. 2020a), possibly due to either the protein not being a dimer or being expressed in and purified from a heterologous *E. coli* expression system.

Mycobacterium has two acyl carrier proteins, and the one detected bound to the arabinofuranosyltransferases is the AcpM variant. AcpM is an 11 kDa soluble protein that functions in fatty acid (Kremer et al. 2001) and mycolic acid synthesis. Structurally, it consists of a four helix bundle (Schaeffer et al. 2001; Wong et al. 2002) and is bound to a prosthetic 4′-phosphopantetheine group. The interaction of AcpM with arabinofuranosyltransferases involves the insertion of the 4-′-phosphopantetheine into the hydrophobic membrane region of the enzyme, along with electrostatic interactions between the protein part of AcpM and the enzyme. Notably, the binding motif of AftD is not homologous to that of the EmbA/B/C counterpart. Mutagenesis performed on the arabinofuranosyltransferases showed that in vivo function was disrupted for the AftD and Emb family, while function in vitro seems unaffected for the Emb family (Tan et al. 2020b; Zhang et al. 2020a, b). In particular, mutations to remove the AcpM in AftD resulted in the ordering of a previously disordered region near the putative active site of AftD, potentially blocking substrate access (Tan et al. 2020b). This suggested that binding of AcpM to AftD could be required for the enzyme to be fully active and thus may act as a regulator of glycosyltransferase function and hence serve as an element to control the rate of mycobacterial cell wall synthesis.

The arabinofuranosyltransferases have been linked to other binding partners too. UbiA and DprE2 have been shown using two-hybrid studies in *C. glutamicum* to interact with AftA, AftB, and AftC (Jankute et al. 2014). WecA might interact with AftB and AftC (Jankute et al. 2014), and GlfT1 might associate with AftB (Jankute et al. 2014). All these enzymes are involved in synthesis of the sugar donor DPA used by the arabinofuranosyltransferases, and it is possible that this complex would allow the DPA to be efficiently passed onto these enzymes (Jankute et al. 2014). Finally, AftD has been shown by surface plasmon resonance to interact with MmpL3, and this has been suggested to be due to MmpL3 functioning as an anchoring node for cellular machinery involved in cell wall biosynthesis (Belardinelli et al. 2019).

Conclusions

The recent glut of mycobacterial arabinofuranosyltransferase structures, facilitated in part by the improvements in cryo-EM technology, has answered many existing questions on the function, catalytic cycle, and drug-binding properties of these important enzymes for cell wall biosynthesis. At the same time, structural work has also raised new questions, the association of AcpM with many of the solved full length arabinofuranosyltransferases thus far being the most unexpected one. Further structural work will be required to fully understand the catalytic cycle of these enzymes at a molecular level and will likely prove invaluable in structure guided drug discovery efforts to improve or engineer new antibiotics against the ancient scourge of mankind, tuberculosis.

Acknowledgments This work was supported by National University of Singapore (NUS) Presidential Young Professorship Award, Singapore (A-0008405-00-00 to Y.Z.T.), Ministry of Education (MOE), Singapore (A-8000037-00-00 to Y.Z.T.), Agency for Science Technology and Research (A*STAR), Singapore (to Y.Z.T.), and National Institute of Health (NIH), USA (R35 GM132120 and R21 MH125649 to F.M.).

References

Abrahams KA, Besra GS (2016) Mycobacterial cell wall biosynthesis: a multifaceted antibiotic target. Parasitology 145:1–18

Alderwick LJ, Seidel M, Sahm H, Besra GS, Eggeling L (2006) Identification of a novel arabinofuranosyltransferase (AftA) involved in cell wall arabinan biosynthesis in Mycobacterium tuberculosis. J Biol Chem 281:15653–15661

Alderwick LJ, Lloyd GS, Ghadbane H, May JW, Bhatt A, Eggeling L, Fütterer K, Besra GS (2011) The C-terminal domain of the arabinosyltransferase Mycobacterium tuberculosis EmbC is a lectin-like carbohydrate binding module. PLoS Pathog 7:e1001299

Alderwick LJ, Harrison J, Lloyd GS, Birch HL (2015) The mycobacterial cell wall—peptidoglycan and arabinogalactan. Cold Spring Harb Perspect Med 5:a021113

Alderwick LJ, Birch HL, Krumbach K, Bott M, Eggeling L, Besra GS (2018) AftD functions as an α1→ 5 arabinofuranosyltransferase involved in the biosynthesis of the mycobacterial cell wall core. Cell Surf 1:2–14

Amin AG, Goude R, Shi L, Zhang J, Chatterjee D, Parish T (2008) EmbA is an essential arabinosyltransferase in Mycobacterium tuberculosis. Microbiology 154:240

Angala SK, Belardinelli JM, Huc-Claustre E, Wheat WH, Jackson M (2014) The cell envelope glycoconjugates of Mycobacterium tuberculosis. Crit Rev Biochem Mol Biol 49:361–399

Angala SK, McNeil MR, Zou L, Liav A, Zhang J, Lowary TL, Jackson M (2016) Identification of a novel mycobacterial arabinosyltransferase activity which adds an arabinosyl residue to α-D-mannosyl residues. ACS Chem Biol 11:1518–1524

Bai L, Wang T, Zhao G, Kovach A, Li H (2018) The atomic structure of a eukaryotic oligosaccharyltransferase complex. Nature 555:328–333

Barry CE III (2001) Interpreting cell wall 'virulence factors' of Mycobacterium tuberculosis. Trends Microbiol 9:237–241

Belanger AE, Besra GS, Ford ME, Mikusová K, Belisle JT, Brennan PJ, Inamine JM (1996) The embAB genes of Mycobacterium avium encode an arabinosyl transferase involved in cell wall arabinan biosynthesis that is the target for the antimycobacterial drug ethambutol. Proc Natl Acad Sci 93:11919–11924

Belardinelli JM, Stevens CM, Li W, Tan YZ, Jones V, Mancia F, Zgurskaya HI, Jackson M (2019) The MmpL3 interactome reveals a complex crosstalk between cell envelope biosynthesis and cell elongation and division in mycobacteria. Sci Rep 9:1–14

Birch HL, Alderwick LJ, Bhatt A, Rittmann D, Krumbach K, Singh A, Bai Y, Lowary TL, Eggeling L, Besra GS (2008) Biosynthesis of mycobacterial arabinogalactan: identification of a novel α (1→ 3) arabinofuranosyltransferase. Mol Microbiol 69:1191–1206

Bohl H et al (2021) https://link.springer.com/chapter/10.1007/978-3-030-58971-4_6

Brennan PJ, Crick DC (2007) The cell-wall core of Mycobacterium tuberculosis in the context of drug discovery. Curr Top Med Chem 7:475–488

Brossier F, Sougakoff W, Bernard C, Petrou M, Adeyema K, Pham A, Amy de la Breteque D, Vallet M, Jarlier V, Sola C (2015) Molecular analysis of the embCAB locus and embR gene involved in ethambutol resistance in clinical isolates of Mycobacterium tuberculosis in France. Antimicrob Agents Chemother 59:4800–4808

Cheng Y (2018) Single-particle cryo-EM—how did it get here and where will it go. Science 361: 876–880

Escuyer VE, Lety M-A, Torrelles JB, Khoo K-H, Tang J-B, Rithner CD, Frehel C, McNeil MR, Brennan PJ, Chatterjee D (2001) The role of the embA and embB gene products in the biosynthesis of the terminal hexaarabinofuranosyl motif of Mycobacterium smegmatisarabinogalactan. J Biol Chem 276:48854–48862

Fu L, Fu-Liu C (2002) Is Mycobacterium tuberculosis a closer relative to gram-positive or gram–negative bacterial pathogens? Tuberculosis 82:85–90

Goude R, Amin A, Chatterjee D, Parish T (2009) The arabinosyltransferase EmbC is inhibited by ethambutol in Mycobacterium tuberculosis. Antimicrob Agents Chemother 53:4138–4146

Grzegorzewicz AE, De Sousa-d'Auria C, McNeil MR, Huc-Claustre E, Jones V, Petit C, Angala SK, Zemanová J, Wang Q, Belardinelli JM (2016) Assembling of the Mycobacterium tuberculosis cell wall core. J Biol Chem M116:739227

Hoffmann C, Leis A, Niederweis M, Plitzko JM, Engelhardt H (2008) Disclosure of the mycobacterial outer membrane: cryo-electron tomography and vitreous sections reveal the lipid bilayer structure. Proc Natl Acad Sci 105:3963–3967

Huang W-L, Chi T-L, Wu M-H, Jou R (2011) Performance assessment of the GenoType MTBDR sl test and DNA sequencing for detection of second-line and ethambutol drug resistance among patients infected with multidrug-resistant Mycobacterium tuberculosis. J Clin Microbiol 49: 2502–2508

Jankute M, Byng CV, Alderwick LJ, Besra GS (2014) Elucidation of a protein-protein interaction network involved in Corynebacterium glutamicum cell wall biosynthesis as determined by bacterial two-hybrid analysis. Glycoconj J 31:475–483

Jankute M, Cox JA, Harrison J, Besra GS (2015) Assembly of the mycobacterial cell wall. Annu Rev Microbiol 69:405–423

Kaur D, Guerin ME, Škovierová H, Brennan PJ, Jackson M (2009) Biogenesis of the cell wall and other glycoconjugates of Mycobacterium tuberculosis. Adv Appl Microbiol 69:23–78

Khasnobis S, Zhang J, Angala SK, Amin AG, McNeil MR, Crick DC, Chatterjee D (2006) Characterization of a specific arabinosyltransferase activity involved in mycobacterial arabinan biosynthesis. Chem Biol 13:787–795

Knauer R, Lehle L (1999) The oligosaccharyltransferase complex from yeast. Biochim Biophys Acta Gen Subj 1426:259–273

Kremer L et al (2001) https://www.jbc.org/article/S0021-9258(19)30205-4/fulltext

Liu J, Mushegian A (2003) Three monophyletic superfamilies account for the majority of the known glycosyltransferases. Protein Sci 12:1418–1431

Minnikin D, Minnikin SM, Goodfellow M, Stanford J (1982) The mycolic acids of Mycobacterium chelonei. Microbiology 128:817–822

Napiórkowska M, Boilevin J, Sovdat T, Darbre T, Reymond J-L, Aebi M, Locher KP (2017) Molecular basis of lipid-linked oligosaccharide recognition and processing by bacterial oligosaccharyltransferase. Nat Struct Mol Biol 24:1100

Niederweis M, Danilchanka O, Huff J, Hoffmann C, Engelhardt H (2010) Mycobacterial outer membranes: in search of proteins. Trends Microbiol 18:109–116

Nikaido H, Jarlier V (1991) Permeability of the mycobacterial cell wall. Res Microbiol 142:437–443

Petrou VI, Herrera CM, Schultz KM, Clarke OB, Vendome J, Tomasek D, Banerjee S, Rajashankar KR, Dufrisne MB, Kloss B (2016) Structures of aminoarabinose transferase ArnT suggest a molecular basis for lipid A glycosylation. Science 351:608–612

Plinke C, Cox HS, Zarkua N, Karimovich HA, Braker K, Diel R, Rüsch-Gerdes S, Feuerriegel S, Niemann S (2010) embCAB sequence variation among ethambutol-resistant Mycobacterium tuberculosis isolates without embB 306 mutation. J Antimicrob Chemother 65:1359–1367

Ramaswamy SV, Amin AG, Göksel S, Stager CE, Dou S-J, El Sahly H, Moghazeh SL, Kreiswirth BN, Musser JM (2000) Molecular genetic analysis of nucleotide polymorphisms associated with ethambutol resistance in human isolates of Mycobacterium tuberculosis. Antimicrob Agents Chemother 44:326–336

Rodrigues J, Almeida VT, Rosário AL, Tan YZ, Kloss B, Mancia F, Archer M (2021) High throughput expression screening of Arabinofuranosyltransferases from mycobacteria. PRO 9: 629

Sandgren A, Strong M, Muthukrishnan P, Weiner BK, Church GM, Murray MB (2009) Tuberculosis drug resistance mutation database. PLoS Med 6:e1000002

Schaeffer ML, Agnihotri G, Kallender H, Brennan PJ, Lonsdale JT (2001) Expression, purification, and characterization of the Mycobacterium tuberculosis acyl carrier protein, AcpM. Biochim Biophys Acta Mol Cell Biol Lipids 1532:67–78

Seidel M, Alderwick LJ, Birch HL, Sahm H, Eggeling L, Besra GS (2007) Identification of a novel arabinofuranosyltransferase AftB involved in a terminal step of cell wall arabinan biosynthesis in Corynebacterianeae, such as Corynebacterium glutamicum and Mycobacterium tuberculosis. J Biol Chem 282(20):14729–14740

Sharma K, Gupta M, Pathak M, Gupta N, Koul A, Sarangi S, Baweja R, Singh Y (2006) Transcriptional control of the mycobacterial embCAB operon by PknH through a regulatory protein, EmbR, in vivo. J Bacteriol 188:2936–2944

Shi L, Berg S, Lee A, Spencer JS, Zhang J, Vissa V, McNeil MR, Khoo K-H, Chatterjee D (2006) The carboxy terminus of EmbC from mycobacterium smegmatis mediates chain length extension of the arabinan in lipoarabinomannan. J Biol Chem 281:19512–19526

Škovierová H, Larrouy-Maumus G, Zhang J, Kaur D, Barilone N, Korduláková J, Gilleron M, Guadagnini S, Belanová M, Prevost M-C (2009) AftD, a novel essential arabinofuranosyltransferase from mycobacteria. Glycobiology 19:1235–1247

Sun Q, Xiao TY, Liu HC, Zhao XQ, Liu ZG, Li YN, Zeng H, Zhao L, Wan KL (2018) Mutations within embCAB are associated with variable level of ethambutol resistance in Mycobacterium tuberculosis isolates from China. Antimicrob Agents Chemother 62:e01279–e01217

Takayama K, Kilburn JO (1989) Inhibition of synthesis of arabinogalactan by ethambutol in Mycobacterium smegmatis. Antimicrob Agents Chemother 33:1493–1499

Tan YZ, Rodrigues J, Keener JE, Zheng RB, Brunton R, Kloss B, Giacometti SI, Rosário AL, Zhang L, Niederweis M (2020a) Cryo-EM structure of arabinosyltransferase EmbB from Mycobacterium smegmatis. Nat Commun 11:1–10

Tan YZ, Zhang L, Rodrigues J, Zheng RB, Giacometti SI, Rosário AL, Kloss B, Dandey VP, Wei H, Brunton R (2020b) Cryo-EM structures and regulation of arabinofuranosyltransferase AftD from mycobacteria. Mol Cell 78(683–699):e611

Telenti A, Philipp WJ, Sreevatsan S, Bernasconi C, Stockbauer KE, Wieles B, Musser JM, Jacobs WR Jr (1997) The emb operon, a gene cluster of Mycobacterium tuberculosis involved in resistance to ethambutol. Nat Med 3:567

Wild R, Kowal J, Eyring J, Ngwa EM, Aebi M, Locher KP (2018) Structure of the yeast oligosaccharyltransferase complex gives insight into eukaryotic N-glycosylation. Science 359: 545–550

Wolucka BA, McNeil MR, de Hoffmann E, Chojnacki T, Brennan PJ (1994) Recognition of the lipid intermediate for arabinogalactan/arabinomannan biosynthesis and its relation to the mode of action of ethambutol on mycobacteria. J Biol Chem 269:23328–23335

Wong HC, Liu G, Zhang Y-M, Rock CO, Zheng J (2002) The solution structure of acyl carrier protein from Mycobacterium tuberculosis. J Biol Chem 277:15874–15880

Xu Y, Jia H, Huang H, Sun Z, Zhang Z (2015) Mutations found in embCAB, embR, and ubiA genes of ethambutol-sensitive and-resistant Mycobacterium tuberculosis clinical isolates from China. BioMed Res Int 2015:951706

Zhang L, Zhao Y, Gao R, Li J, Yang X, Gao Y, Zhao W, Gurcha SS, Veerapen N, Batt SM (2020a) Cryo-EM snapshots of mycobacterial arabinosyltransferase complex EmbB2-AcpM2. Protein Cell 11:505–517

Zhang L, Zhao Y, Gao Y, Wu L, Gao R, Zhang Q, Wang Y, Wu C, Wu F, Gurcha SS (2020b) Structures of cell wall arabinosyltransferases with the anti-tuberculosis drug ethambutol. Science 368:1211–1219

Zhao LL, Sun Q, Liu HC, Wu XC, Xiao TY, Zhao XQ, Li GL, Jiang Y, Zeng CY, Wan KL (2015) Analysis of embCAB mutations associated with ethambutol resistance in multidrug-resistant Mycobacterium tuberculosis isolates from China. Antimicrob Agents Chemother 59:2045–2050

Part II
Fibrous Protein Complexes

Chapter 13
Structure and Assembly of the Bacterial Flagellum

Natalie S. Al-Otaibi and Julien R. C. Bergeron

Abstract The bacterial flagellum is a large macromolecular assembly that acts as propeller, providing motility through the rotation of a long extracellular filament. It is composed of over 20 different proteins, many of them highly oligomeric. Accordingly, it has attracted a huge amount of interest amongst researchers and the wider public alike. Nonetheless, most of its molecular details had long remained elusive.

This however has changed recently, with the emergence of cryo-EM to determine the structure of protein assemblies at near-atomic resolution. Within a few years, the atomic details of most of the flagellar components have been elucidated, revealing not only its overall architecture but also the molecular details of its rotation mechanism. However, many questions remained unaddressed, notably on the complexity of the assembly of such an intricate machinery.

In this chapter, we review the current state of our understanding of the bacterial flagellum structure, focusing on the recent development from cryo-EM. We also highlight the various elements that still remain to be fully characterized. Finally, we summarize the existing model for flagellum assembly and discuss some of the outstanding questions that are still pending in our understanding of the diversity of assembly pathways.

Keywords Bacterial motility · Macromolecular structure · Cryo-EM · Bacterial secretion systems · Membrane proteins

Introduction

When Antonie van Leeuwenhoek first observed bacteria in the 1600s, he described them as “thousands of living creatures, seen all alive in a little drop of water, moving among one another, each several creature having its own proper motion” (Debell 1958), recognizing their capacity to move in their environment as particularly

N. S. Al-Otaibi · J. R. C. Bergeron (✉)
Randall Centre for Cell and Molecular Biophysics, King’s College London, London, UK
e-mail: natalie.s.al-otaibi@kcl.ac.uk; julien.bergeron@kcl.ac.uk

J. R. Harris, J. Marles-Wright (eds.), *Macromolecular Protein Complexes IV*, Subcellular Biochemistry 99, https://doi.org/10.1007/978-3-031-00793-4_13

fascinating. Several centuries later, Howard Berg demonstrated that this motility is (in some bacteria) due to the rotation of their flagellum (Berg 1974), an intricate molecular machine, anchored at the cell surface, that rotates a long filament at up to 100,000 rpm (Magariyama et al. 1994), providing propulsive energy. This remarkable feat of evolutionary engineering has been the focus of a lot of research effort, which has allowed scientists to uncover many aspects of the molecular details of flagellar rotation. However, the complex dynamics caused by the rotation motion and membrane embedding had for several decades prevented gain of a structural understanding of the flagellum.

In the past few years, we have seen a complete revolution in structural biology, with the emergence of cryo-EM as a method to routinely determine protein structures at near-atomic resolution. This has led to an avalanche of new structural information on the flagellum, including the structures of most of the components in isolation, as well as structural insights on the mechanism of rotation, and on its assembly. In this chapter, we will summarize the current state of knowledge on the structure of the bacterial flagellum, focusing on the recently determined cryo-EM structures. We will also highlight the regions of the complex that remain to be characterized. Finally, we will review what we currently know about the assembly of this complex and highlight the many questions that are still outstanding on this aspect.

Structure of the Bacterial Flagellum

The bacterial flagellum is a very complex assembly, composed of >20 different proteins, most of them highly oligomeric. Additional complexity is added by the fact that different species possess slightly distinct flagellar architectures, some including additional components, and/or lacking others, adapted to their specific biology. For this reason, we will consider the *Salmonella enterica* serovar Typhimurium flagellum as the prototypical complex, and we will describe its structural features in greater details. It is however noteworthy that some of the structural insights presented below are not from this bacterium but from a conserved protein for which the structure was solved in a different species. In this instance, we will assume that the structure and function are similar across bacteria.

The full flagellum complex can be divided into three sub-assemblies (Fig. 13.1): In the cytosolic side, the rotor-stator complex and the export apparatus are responsible for the rotation of the flagellum and the secretion of the extracellular components, respectively. Across the bacterial periplasm and membranes, the basal body forms a continuous structure linking the rotor-stator and export apparatus to the outside of the cell. Finally, on the extracellular side, the hook and filament extend, up to >20 μm, away from the cell surface, and their rotation provides the propelling force to move the bacterial cell.

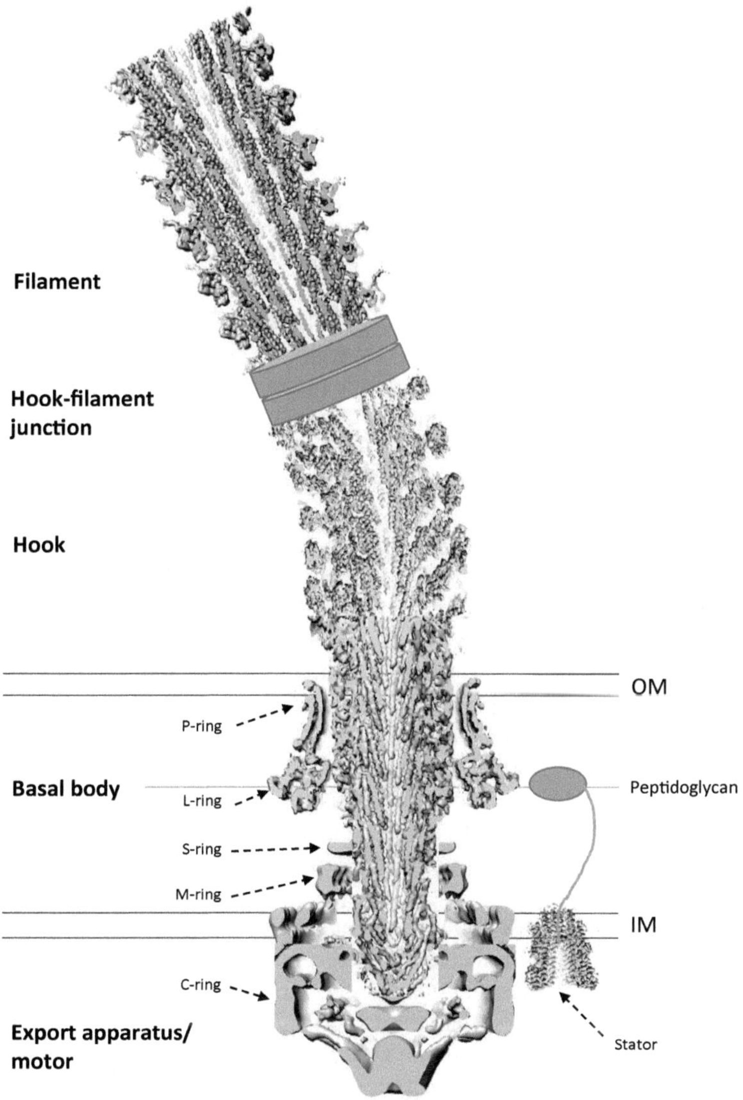

Fig. 13.1 Overall architecture of the prototypical *S. enterica* flagellum. The four main sub-complexes, export apparatus/motor, basal body, hook, and filament, are indicated, as well as the various basal body ring structures

The Motor, Stator, and Export Apparatus

The flagellum is in essence a molecular motor, generating rotation to the filament, and powered by a proton (or sodium in some species) gradient across the inner membrane. This rotation is generated on the cytosolic side, by structures that are similar to an engine. Embedded in the inner membrane, two proteins, MotA and MotB, form the stator component of this engine. Two recent cryo-EM structures of the MotA-MotB complex reveal a 5:2 stoichiometry, with the TM helices of two MotB proteins circled by five MotA molecules (Fig. 13.2a). Notably, the MotA pentamer is asymmetric, and this asymmetry revealed how proton transfer across the membrane, from MotB to MotA, allows the rotation of the MotA ring in the clockwise direction (Deme et al. 2020; Santiveri et al. 2020). In the periplasm, a globular domain of MotB anchors the complex to the peptidoglycan layer (Kojima et al. 2018).

On the cytoplasmic side, three proteins (FliM, FliN, FliG) form a large ring structure, the cytoplasmic (C-) ring, topologically similar to the engine rotor. Upon MotA rotation, the C-ring rotates accordingly but in an opposite direction. Multiple stator complexes are distributed around the C-ring, providing combined power to the system (Chaban et al. 2018). The C-ring is in turn bound to the basal body via an interaction between FliG and FliF, which propagates the rotation to the rest of the complex (Xue et al. 2018). FliG has three domains which interact with different parts of the motor machinery: The $FliG_N$ domain binds to the MS-ring, and anchors it to the cytoplasmic membrane; the $FliG_M$ domain binds to $FliG_c$ and FliM, permitting switching flagellar rotation directionl; and the $FliG_C$ domain interacts with the stators for torque generation (Lee et al. 2010).

There is currently no high-resolution structure of the intact C-ring, and our understanding of this sub-complex is limited to a low-resolution EM map, combined with high-resolution structures of isolated domains of FliG, FliM, and FliN (Fig. 13.2b) (Brown et al. 2005; Mcdowell et al. 2016; Thomas et al. 2006; Vartanian et al. 2012). In particular, the molecular basis for their oligomeric interface is not known. Nonetheless, low-resolution EM reconstruction of the C-ring has revealed a ~ 34-fold symmetry, consistent with the oligomeric state of FliF in the basal body (see below).

In addition, a set of inner membrane-embedded proteins, termed the export apparatus, is responsible for the recruitment and secretion of the axial proteins through the complex, an essential control step for the assembly of its extracellular components. It is noteworthy that the export apparatus is highly similar to the evolutionarily related type III secretion system (T3SS) injectisome, where a similar set of proteins perform the same function (Halte and Erhardt 2021). Consequently, some of the structures reported here were determined for the T3SS, not the flagellum, but these will be used interchangeably.

At the centre of the C-ring, an ATPase, FliI, forms a hexameric complex, whose role is to recruit the secreted proteins and promote their dissociation from their cognate chaperones, upon ATP hydrolysis. At the centre of the FliI hexamer, the protein FliJ stabilizes the complex and rotates around the axis of the FliI hexamer upon nucleotide hydrolysis, which likely plays a role in dissociating the secreted proteins from their chaperones, although the molecular basis for this is not currently known (Allison et al. 2014; Majewski et al. 2019). The FliI hexamer is also anchored to the C-ring via the protein FliH, which forms a lateral stalk, connecting FliI to FliN.

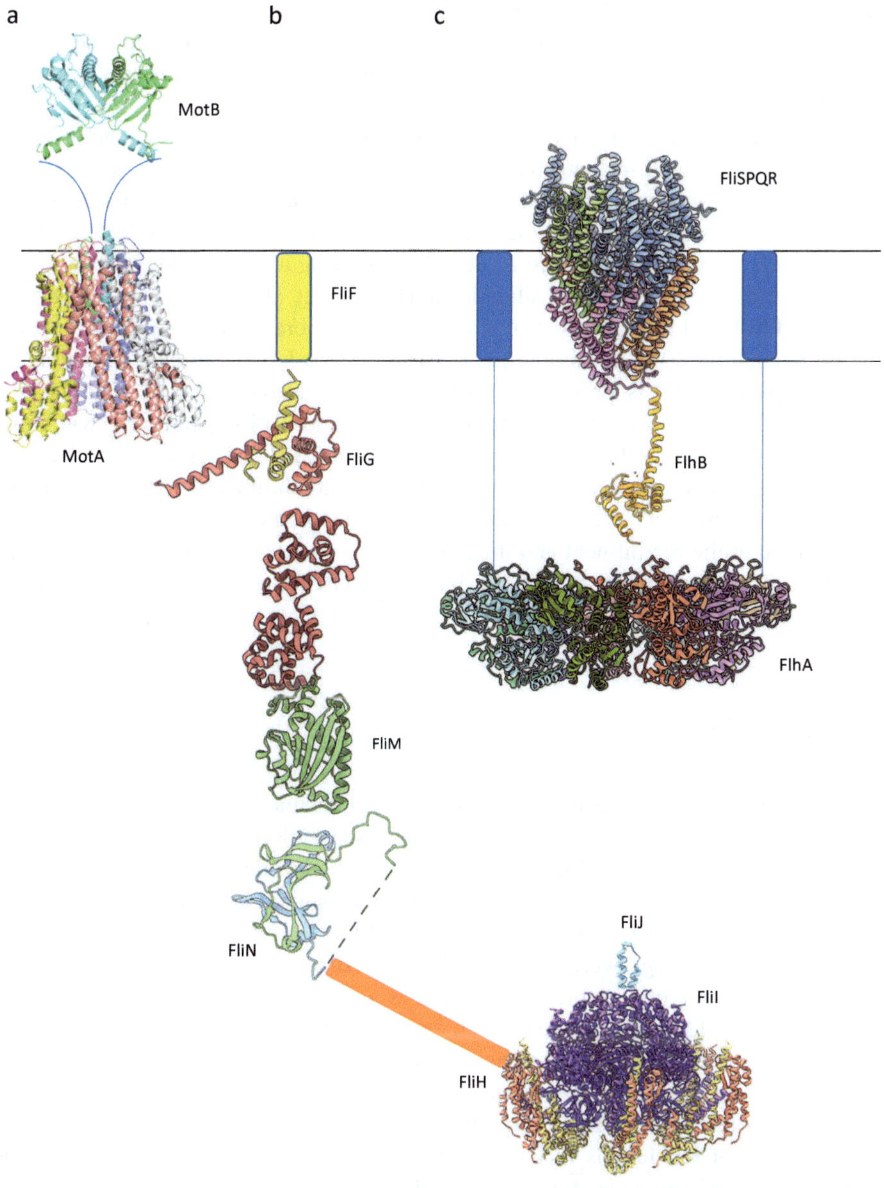

Fig. 13.2 Structure of the flagellar inner-membrane components. (**a**) Structure of the stator complex. (**b**) Individual structures of the C-ring components. (**c**) Structures of the export apparatus components

A recent structure of a FliI monomer with FliH revealed a 1:2 stoichiometry, which would correspond to a 12:34 ratio for the FliH-FliN interface the C-ring side, likely indicating a yet-to-be-determined asymmetry determinant in either of these proteins (Imada et al. 2016).

Above the ATPase complex, and embedded in the inner membrane, the proteins FliPQR adopt a pseudo-helical architecture, with a 1:4:5 ratio. These proteins form a pore through which unfolded proteins are secreted and form a template for the subsequent assembly of the rod (see below). The structure of the FliPQR complex in isolation revealed that the pore is closed, suggesting a mechanism for controlled opening, but this is yet to be elucidated (Kuhlen et al. 2018). Two additional proteins, FlhA and FlhB, consist of a set of transmembrane (TM) helices that pack tightly against the FliPQR complex but also possess a large cytoplasmic region. A single copy of FlhB is present in each flagellum complex, and its cytoplasmic domain is linked with control of the hook length and T3SS specificity switching, although the molecular basis for this is highly debated (Ferris et al. 2005; Kuhlen et al. 2020). In contrast, FlhA adopts a nonameric stoichiometry, with its cytoplasmic domain forming a large ring below the inner membrane. It has been proposed to be involved in the recruitment and threading of the secreted proteins throughout the complex through acting as a scaffold, but its exact role remains to be determined (Kuhlen et al. 2021).

The Basal Body

The structure spanning the membrane(s) and periplasmic space is called the basal body, a term historically brought by the comparison with the evolutionarily distinct structure observed in the cilium or flagellum of eukaryotic cells. This basal body consists of four rings, termed M -(membrane), S- (supramembrane), L- (lipopolysaccharide), and P- (peptidoglycan) rings (Fig. 13.3a).

The M- and S-rings were shown to be formed by multiple copies of a single protein, FliF, and are therefore often joined as a single structure, the MS-ring. FliF consists of two TM helices, one at each end, that flank a large periplasmic region, consisting of three globular domains (termed ring-building domains, RBDs), with similarity to ring-forming proteins in the T3SS injectisome and in the sporulation membrane pore complex (Bergeron 2016; Zeytuni et al. 2017). At the C-terminus, a small stretch of residues is located in the cytosol and binds to FliG, linking the MS-ring to the C-ring (see above and Fig. 13.2b). Low-resolution cryo-EM studies had revealed that FliF oligomerizes in isolation to form the MS-ring, but the structural details of its domain organization had long eluded researchers (Suzuki et al. 2004). In particular, it had been proposed to adopt a 26-fold symmetry, but this was in contradiction with the 33-fold symmetry of the C-ring (Thomas et al. 2006).

However, the recently reported cryo-EM structures of FliF, in isolation (Takekawa et al. 2021) as well as within the intact basal body (Johnson et al. 2021; Tan et al. 2021), revealed a highly unexpected feature: the overall

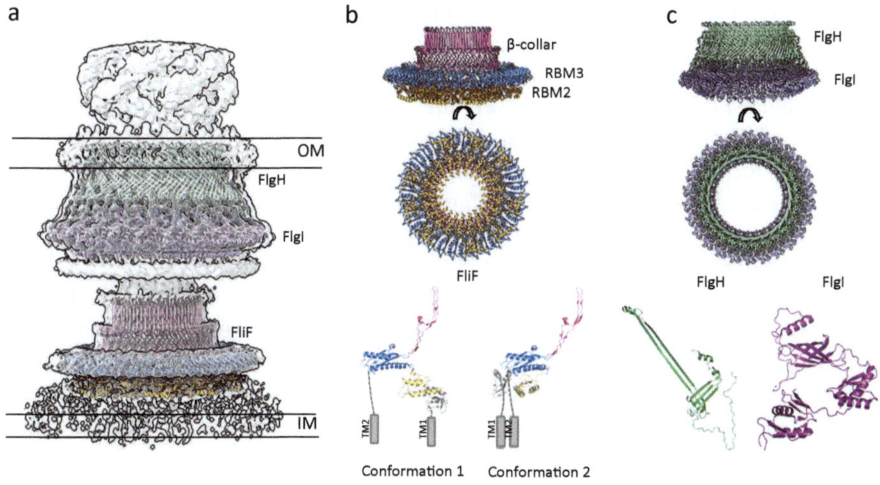

Fig. 13.3 Structure of the flagellum basal body. (**a**) Cryo-EM map of the *S. enterica* basal body, with the atomic models of its components FlgH, FlgI and FliF. (**b**) The inner-membrane FliF protein consists of three RBMS, with distinct symmetries: 34-fold for RBM3, forming the S-ring, and 23-fold for RBM2. Additional 11 copies or RBM2 adopt a different conformation, while the position of RBM1 and the TM helices are not resolved. (**c**) The outer-membrane complex is formed by 26 copies of FlgH and FlgI, forming the P- and L-rings, respectively

stoichiometry of the complex is a 34-mer, not a 26-mer. However, one of the RBDs adopts two distinct conformations: one conformation corresponds to a 23-mer ring, located just above the membrane, while 11 more subunits sit outwards of the complex (Fig. 13.3b) (Johnson et al. 2021). This surprising arrangement permits the complex to adopt two different symmetries within a single protein, necessary for the interactions with other components of the complex and resolving a long-lasting controversy on the apparent symmetry mismatch in the basal body. In those structures, density for RBM1 is not resolved, suggesting that it does not adopt a rigid position within the basal body. Nonetheless, recent biochemical analysis of the FliF oligomerization suggests that it might contribute to the regulation of its assembly (Mariano et al. 2022). It is also noteworthy that the two transmembrane helices of FliF are also not resolved in any of the structures. In addition, while the C-terminal cytoplasmic tail of FliF interacts with FliG (see above), there are few contacts between the MS ring and the export apparatus, mainly limited to one loop of RBM2 interacting with FliP. It remains to be established in the TM helices of FliF also interact with the export apparatus.

On the outer-membrane side, two proteins, FlgI and FlgH, form the L- and P-rings, respectively (Fig. 13.3c). Similarly to FliF, the structural understanding of these proteins had long been limited to low-resolution EM studies, of the FlgI-FlgH complex in isolation (Akiba et al. 1991), or as part of the intact basal body (Sosinsky et al. 1992; Stallmeyer et al. 1989). However, very little was known about the structure of either protein, their stoichiometry, or their interaction with other components of the flagellum. Recently, the cryo-EM structure of the LP-ring was determined in the context of the full basal body (Johnson et al. 2021; Yamaguchi et al. 2021). This revealed a large ring with 26-fold symmetry, with FlgH forming the membrane-anchored P-loop. However, this protein is not fully embedded in the membrane, but instead a single loop is ectopically anchored in it. The core of the oligomer is formed by long FlgH β-strands forming a double-layered β-barrel. These features are reminiscent of the secretin family of outer-membrane pores, found in the T3SS but also the type II secretion system and type IV pilus, although they do not share any sequence similarity with FlgH (de Silva et al. 2020).

Underneath FlgH, the L-ring is formed by FlgI, which consists of three globular domains. It is worth noting that domains of similar folds are also found in most secretin proteins (Silva et al. 2020), a family of outer-membrane pore proteins found in many secretion systems. It is not known why the flagellum outer-membrane complex consists of two separate proteins, whereas a single polypeptide chain encodes the two components in the secretin family.

The Axial Components

The flagellar axial regions have evolved to efficiently transfer the motor rotational force into motility as well as play an integral part in pathogenesis of many bacterial species. While the helical symmetry of these regions has been challenging to structural studies, the improvement in cryo-EM techniques over the last few years has led to an increase in intact structures across the bacterial species. With these, it has emerged that all the axial proteins characterized to date possess hydrophobic residues at both termini, which join to form a two-helix coil-coil, termed D0 domain. In between these, globular domains are present, many sharing similar folds, and termed D1, D2…Dn (Fig. 13.4). These proteins oligomerize in an 11-stranded filament, with the D0 domain forming the centre of the oligomer, and the globular domains exposed at its surface.

Across the periplasm, the rod connects the export apparatus to the hook region and spans the periplasmic space, through the MS- and LP-rings. Genetic analyses have shown that it can be divided into a proximal rod, formed by the proteins FliE, FlgB, FlgC, and FlgF, and a distal rod, formed by the protein FlgG (Fujii et al. 2017; Homma et al. 1990). The recent structures of the intact basal body revealed that the rod includes 6 copies of FliE and FlgC, 5 copies of FlgB and FlgF, and 24 copies of FlgG (Fig. 13.4a) (Johnson et al. 2021; Tan et al. 2021). As mentioned above, each of them possesses the canonical D0 domain that forms the core of the helix. The packing of the rod components follows the helical symmetry of hook and filament (discussed below) using alternating 5:6 stoichiometries. FliE does not possess any globular domain and solely consists of D0; FlgB and FlgC possess a small Dc domain and FlgF and FlgG a similar Dc domain and larger D1 domain (Fig. 13.4b). As a consequence, the diameter of the rod filament increases significantly across the

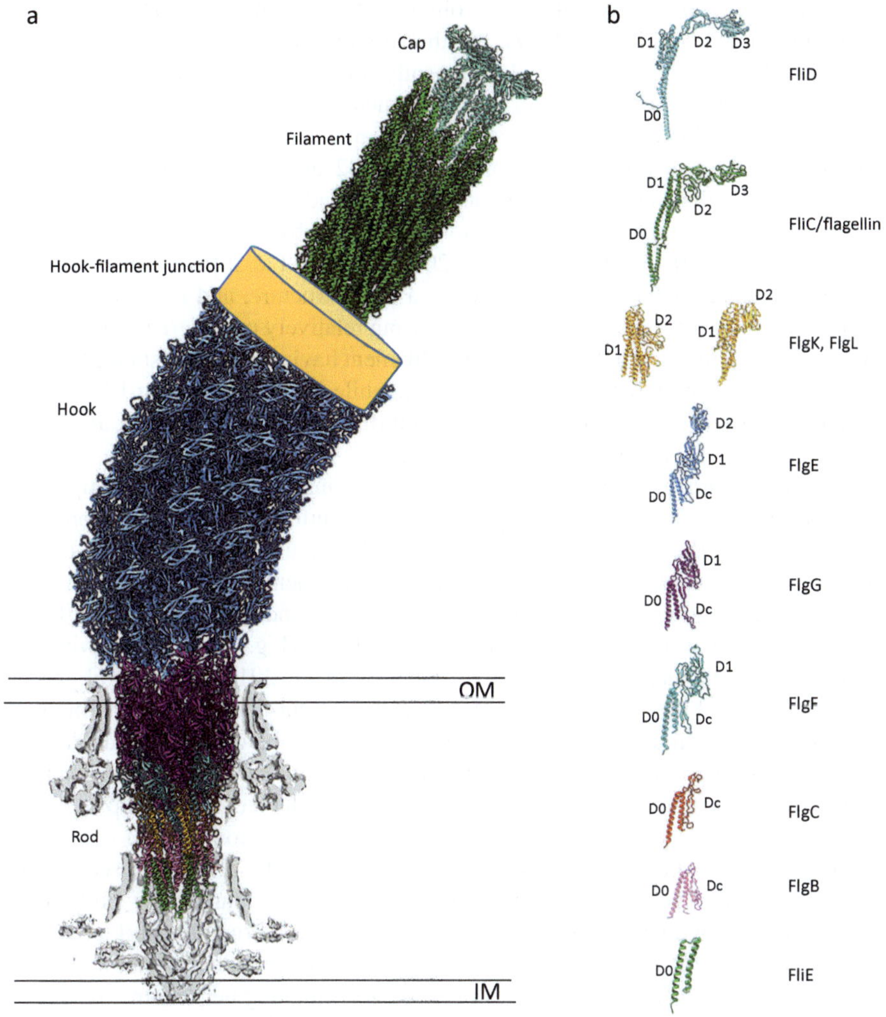

Fig. 13.4 Structure of the bacterial flagellum axial proteins. (**a**) Composite model of the flagellum rod, hook, filament, and cap, placed within the EM map of the basal body. The position of the hook-filament junction is indicated. (**b**) Structure of a monomer from each axial protein, coloured as in (**a**). All include a D0 coiled-coil helix pair, responsible for oligomerization, followed by successive globular domains termed Dc, D1, D2...Dn, exposed at the surface of the oligomer

periplasm, from ~6.5 nm at the interface with the export apparatus to ~11 nm as it traverses the LP rings.

The flagellar hook is made up of multiple copies of the protein FlgE, which possesses the canonical D0 domain, Dc and D1 domains similar to that of the rod proteins, and an additional large D2 domain. The crystal structure of this protein, lacking the D0 domain, was determined ~20 years ago (Samatey et al. 2004; Yoon et al. 2016) and, combined with low-resolution EM maps, had been used to propose an atomic model (Fujii et al. 2017), but the structural basis for its oligomerization had long elucidated researchers. More recently, multiple studies reported the cryo-EM structure of the hook oligomer and confirmed that it oligomerizes according to the canonical architecture, forming a 11-mer filament with a diameter of ~180 Å, with the D0 domain forming the core, and the D2 domain exposed at its surface (Fig. 13.4b) (Shibata et al. 2019). Notably, the D1 and D2 domains behave as rigid bodies connected by flexible hinge regions, with D2 observed to move as much as 27 Å (Kato et al. 2019; Matsunami et al. 2016; Shibata et al. 2019). This is contrary to the previous hypothesis that the hook is a rigid structure, and it can transfer the torque from the motor to the filament while being relatively mobile. The curvature of the filament occurs due to the shortest protofilament having a 6.3 Å distance between top and bottom subunits due to closer packing, while the longest having 12.5 Å. This difference causes one protofilament to be shorter than the other thus curving the hook into the correct form (Kato et al. 2019).

The hook-filament junction consists of two proteins that localize at the tip of the hook, FlgK and FlgL. There is limited structural information on these proteins, mainly because there are currently no cryo-EM structures of this region to date (Fig. 13.4a). Nonetheless, the crystal structures of fragments from both proteins have been reported, and combined with sequence analysis, indicate that both FlgK and FlgK consist of 3 main domains, D0, D1, and D2 (Fig. 13.4b), following the canonical architecture described above (Gourlay et al. 2015; Hong et al. 2018). This suggests that they likely adopt some similar filament-like architecture, but their arrangement and stoichiometry is not known. This will be critical to elucidate the precise role of these proteins in forming a gasket between the hook and the filament.

Following the hook-filament junction, the flagellar filament extends up to several μm away from the cell surface. It is constituted of tens of thousands of copies of a single protein, the flagellin (FliC in most, but not all, bacterial species; see below). The filament has been extensively studied for over 20 years using both X-ray crystallography and Cryo-EM; notably, the first cryo-EM structures of the filament, determined to ~4 Å resolution by 2D crystallography of pseudo-crystalline flagellin mutants, were the very first examples of cryo-EM achieving near-atomic resolution and demonstrated the potential of this technique long before the recent "resolution revolution" in cryo-EM (Maki-Yonekura et al. 2010; Yonekura et al. 2003).

The flagellin monomer has four domains, following the canonical architecture: D0, followed by a D1 domain which is a bundle of 3 α-helices. D2 and D3 domains are rich in antiparallel β-sheets and are variable in both size and sequence (Maki-Yonekura et al. 2010; Montemayor et al. 2021). Early low-resolution analyses suggested that the flagellum filament adopted different architectures in different species (Galkin et al. 2008); however, recent reconstructions have confirmed that it adopts the canonical 11-stranded filament in all bacteria studied to date (Kreutzberger et al. 2020; Wang et al. 2017). The helical parameters, namely, the rise and the twist, are conserved across bacterial species with the rise being in the range of 4.61–4.90 Å and twist in the range of 65.1–65.8 ° (Montemayor et al. 2021).

Similar to the short and long protofilaments in the hook giving it the curved shape, flagellin protofilaments can either be short or long, which also reflects in the right-hand inclination (R-state) and left-hand inclination (L-state), respectively, depending on the ratio of different types of minor and major flagellins. This variety leads to supercoiling (Wang et al. 2017). In many bacteria, the filament is heavily glycosylated. Glycosylation of flagella involves O-link attachment of pseudaminic or legionaminic acids to serine and threonine residues on the surface of the flagellum. It prevents exposure of a sensitive peptide region which would trigger an immune response (Rossez et al. 2015; Young et al. 2007).

Finally, the filament is tipped by the cap complex, formed by the protein FliD (Fig. 13.4a). Early low-resolution cryo-EM analyses had revealed that in isolation, this protein forms a decameric complex; however, only five copies found at the tip of the filament (Maki-Yonekura et al. 2003; Yonekura et al. 2000). These revealed that within the complex, the protein adopts a five-legged stool architecture, with a head domain that forms the pentamer interface, and a leg domain. Multiple crystal structures of the head region of FliD confirmed that it is similar in structure to the D1-D2-D3 domains of the flagellin (Postel et al. 2016; Song et al. 2017; Cho et al. 2017).

More recently, the cryo-EM structure of the FliD decamer confirmed this organization and demonstrated that the leg region consists of the D0 domain (Al-Otaibi et al. 2020). However, it is distinct from that of the other axial proteins, with the N-terminal helix of the coiled-coil broken and wrapped around the oligomer. This arrangement likely explains why this protein forms a stable oligomer and does not adopt a filamentous architecture. Nonetheless, the molecular basis for its interaction with the filament, and notably how its fivefold symmetry matches the 11-stranded filament architecture, is not currently known.

Diversity of Flagellar Structure

The flagellum is found across all bacterial lineages and in bacteria with very distinct lifestyles. Accordingly, the structural features described above form a blueprint for the overall architecture of the bacterial flagellum; however, there are also a wide range of species-specific features. Some of them have been characterized, although it is highly likely that many new features are present in uncharacterized bacteria.

Bacteria that live in viscous environments need to produce a higher torque for efficient swimming. This increased energy is linked with additional stator complexes surrounding the MS and C rings. In the prototypical *S. enterica* typhimurium flagellum, the motor produces an estimated energy of 1300 pN.nm, and the flagellum includes 11 copies of the stator complex. In contrast, the flagellum of *V. fischeri* produces torques of more than 2200 pN.nm and uses 13 stator complexes; in addition, these are driven by sodium translocation instead of protons. In *C. jejuni*, the torque produces is around 3600 pN.nm, with 17 stator complexes observed around the C-ring. Up to 18 complexes have been observed, notably in *H. pylori*

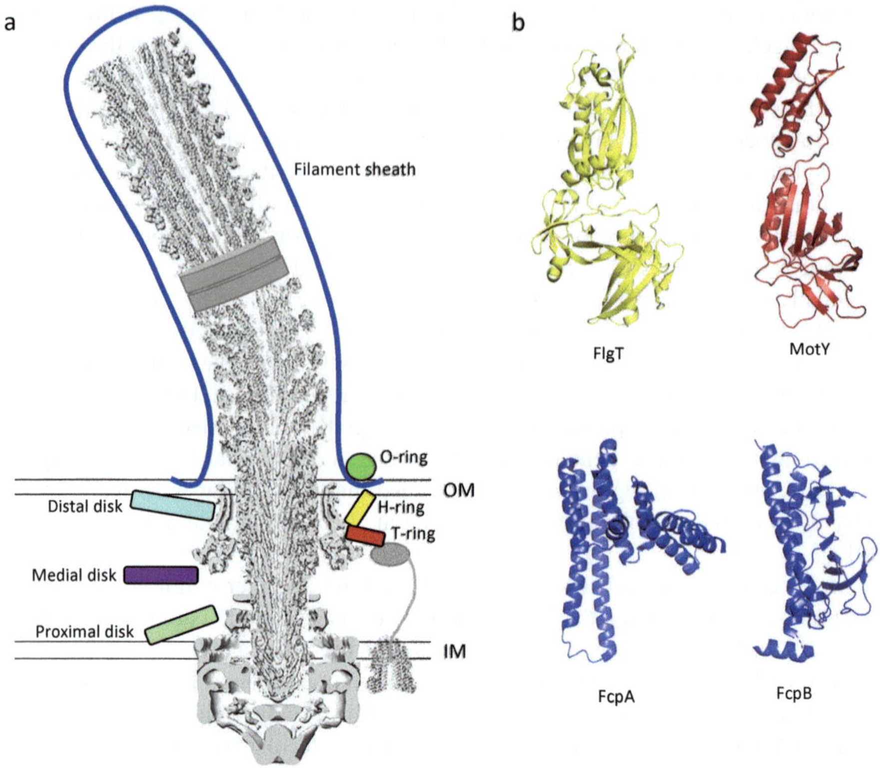

Fig. 13.5 Diversity of flagellum architectures. (**a**) Composite EM map of the *S. enterica*, shown in Fig. 13.1, with additional regions found in other bacterial species added. (**b**) Individual protein structures of components from these additional regions, coloured as in (**a**)

and *H. hepaticus* (Beeby et al. 2016; Chaban et al. 2018). Collectively, this suggests that the flagellum complex is highly tuned to the bacterial environment, and subtle changes in protein expression/oligomerization state might be sufficient to adapt the motor to changes in the environment.

In addition to diversity in the number of stator complexes, recent cryo-electron tomography of motor complexes in diverse bacterial species has revealed additional density around the basal body (Rossmann and Beeby 2018). In multiple bacterial species (including *C. jejuni* and *H. pylori*), additional density forming three separate structures is observed in the periplasm. These were termed basal, medial, and proximal disks (Fig. 13.5a) and have been proposed to consist of the proteins FlgP, PflA, and PflB, respectively (Beeby et al. 2016). In other bacteria, notably *V. fischeri* and *S. putrefaciens*, two ring-like structures surround the LP rings, termed the Holding (H-) and T-rings, respectively (Fig. 13.5a). The H-ring was shown to be formed by the protein FlgT, while the T-ring has been proposed to consist of two proteins, MotX and MotY (Terashima et al. 2006, 2010). It is worth noting that other structures, such as the P-collar in *T. primitia*, have been observed; but their protein composition is not currently known (Chang et al. 2021; Murphy et al. 2006).

As expected, these various proteins are not widely conserved across bacteria, and therefore their exact role is not clear. In addition, there is to date no high-resolution structural data on any of these proteins as part of a large assembly, and their structural characterization is limited to crystal structures of isolated, monomeric fragments of FlgT and MotY (Fig. 13.5b) (Kojima et al. 2008; Terashima et al. 2013). Further characterization would be required to determine their architecture and the molecular basis for their interaction with the core flagellum complex.

Gram-positive bacteria do not possess an outer membrane, and accordingly their flagella do not possess the L- and P-rings. The rod spans the periplasm but lacks the distal proteins FlgF and FlgG. Instead, it possesses two hook-like proteins, FlhO and FlhP, which form a junction between rod and hook (Burrage et al. 2018; Courtney et al. 2012). More surprisingly, the proteins FlgH and FlgI are also not present in Spirochetes, a broad family of bacteria that include the aetiological agent of syphilis (*T. pallidum*) and Lyme disease (*B. burgdorferi*). In these bacteria, the flagellum filament does not traverse the outer membrane but is fully incorporated in the periplasm, where it acts as a cytoskeletal element. The rotation of the flagellum motor provides the bacterial cell with an undulating movement, responsible for their motility (Nakamura 2020). As indicated above, the flagellum motor possesses a P-collar, of unknown function. In addition, in these bacteria the filament is surrounded by a proteinaceous sheath (Fig. 13.5a), consisting of the proteins FlaA1, FlaA2, FcpA, and FcpB (in *L. biflexa*). Our understanding of the overall architecture of this sheath is limited to a low-resolution structure of the intact structure surrounding the filament and to crystal structures of FcpA and FcpB in isolation (Fig. 13.5b) (Gibson et al. 2020). Further characterization will be necessary to further characterize the structure of this sheath and to decipher how it interacts with and assembles around the filament.

It is also noteworthy that a flagellum sheath is found in a range of other bacteria, where the flagellum is not periplasmic, such as *V. cholerae* and *H. pylori*. However, in these the sheath is membranous and consists of an extension of the outer membrane (Chu et al. 2020; Zhu et al. 2017). This observation is very intriguing, as the corresponding flagellar motors possess the L- and P-rings that form the outer-membrane pore. Furthermore, an additional ring structure, the O-ring, is present at the base of the sheath, on the external side of the filament (Fig. 13.5a). Further characterization will be necessary to determine the nature of this O-ring and if the L- and P-rings form pores in the membrane of these bacteria.

Lastly, the filament itself is a source of diversity across bacterial species. As mentioned above, the filament consists of multiples copies of the flagellin; nonetheless, many bacteria encode multiple flagellin genes, with divergent sequences. For example, *C. jejuni* encodes two flagellins, FlaA and FlaB, that are 95% identical; nonetheless, their distribution across the filament is regulated, and necessary for full motility (Cohen et al. 2020). Furthermore, many flagellin sequences possess additional globular domains, compared to the canonical D0-D3 architecture (Fig. 13.4b). As described above, the globular domains are displayed at the filament surface, and are therefore highly antigenic, but also could be involved with additional activities on the extracellular side. Accordingly, flagellins with multiple globular domains have been identified, as well as flagellins with functional enzymatic domains (Eckhard et al. 2017; Thomson et al. 2018). In contrast, the flagellin of gram-positive bacteria lack the D2 and D3 domains (Wang et al. 2017). Of note, these additional domains are not found exclusively in the flagellin, but also in the hook and cap proteins FlgE and FliD, most notably in epsilon proteobacteria, which have a high range of such insert domains (Al-Otaibi et al. 2020). It should be however noted that the roles of these additional domains are not currently known.

Flagellar Assembly and Dynamics

With the recent advances in cryo-EM, our understanding of the flagellum structure has made incredible strides, within only a few years. This in turn has largely resolved the molecular basis for its rotation. However, the next frontier is to decipher the details of its assembly, a highly sophisticated process that requires the specific association of several dozen different proteins, with distinct stoichiometries, and at specific times.

Overall Assembly Pathway

Early studies investigating flagellum assembly have focused on knocking out each flagellar gene individually and determining what components where present in the flagellum complex in this genetic context. This demonstrated an assembly pathway where the export apparatus and MS-rings assemble first, followed by the rod, LP-rings, hook, and filament (Kubori et al. 1992). Nonetheless this did not allow deciphering the fine details of these processes. Notably, subsequent studies revealed that several of the sub-complexes, including the export apparatus (Diepold et al. 2011), the MS-rings (Kubori et al. 1997), and the stator can assemble in isolation (Van Way et al. 2000). In contrast, the axial proteins require a fully assembled basal body to form.

More recently, a combination of in situ cryo-electron tomography and super-resolution fluorescence microscopy have allowed to dissect these assembly processes in greater details (Bai et al. 2014; Fabiani et al. 2017; Kaplan et al. 2019; Terashima et al. 2020). Combined with other genetic biochemical data, this has allowed establishing the chronology of flagellum formation, from the export apparatus to the basal body, hook, and filament.

Collectively, these data indicate that flagellum assembly is initiated in the inner membrane, when the chaperone FliO induces the oligomerization of FliP, followed by the addition of FliR, FliQ, and FlhB (Fabiani et al. 2017). This in turn recruits nine copies of FlhA, forming the core export apparatus complex, although how FlhA

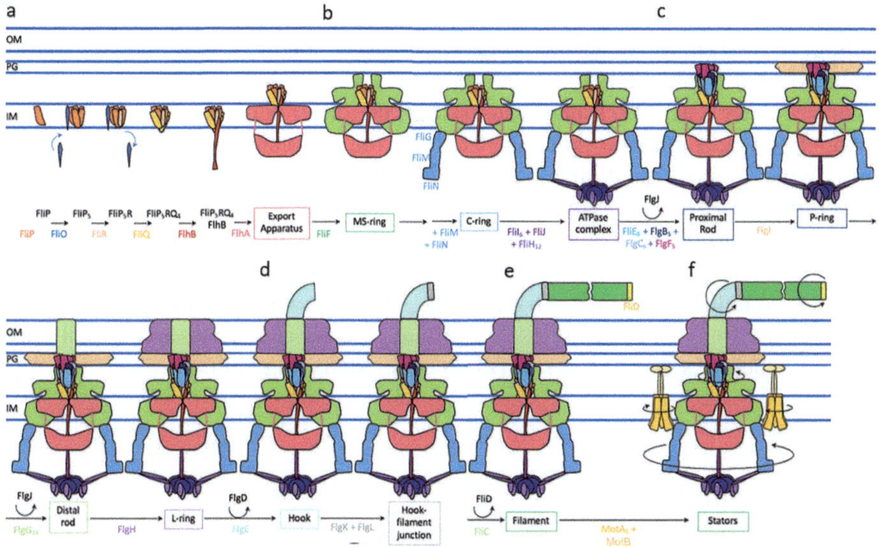

Fig. 13.6 Step-wise diagram of flagellar assembly in *E. coli*. OM: outer membrane, PL: peptidoglycan layer, IM: cytoplasmic membrane. Structural proteins are labelled with colours corresponding to their respective schematic. Adaptor proteins binding and assisting the formation of the axial structures are shown with curved arrows, as they dissociate, apart from FliD. (**a**) Formation of the export apparatus, (**b**) the cytoplasmic rings and ATPase complex, (**c**) the flagellar rod and LP-rings, (**d**) the hook and hook-filament junction, and (**e**) the filament and cap. (**f**) Addition of stators induces flagellar rotation

interacts with the rest of the complex is not currently known (Fig. 13.6a) (Fukumura et al. 2017). Once the export apparatus is assembled, it recruits FliF, which oligomerizes around it (Morimoto et al. 2014). This is presumed to occur via interactions between the FliF and FlhA TM helices, although there is currently no direct evidence for this. Once the MS-rings is assembled, FliF recruits FliG, followed by FliM and FliN, leading to the formation of the C-ring (dos Santos et al. 2018; Terashima et al. 2020). The assembled C-ring then recruits the ATPase complex, including distal and proximal rods, forming a fully secretion-competent assembly (Fig. 13.6b) (Bai et al. 2014).

Following this, the rod proteins are secreted through the inner membrane, leading to the formation of the proximal, then distal rod regions (Burrage et al. 2018). How the different rod components assemble chronologically is poorly understood but appears to be due to a combination of chronological secretion and regulation by the chaperone (and peptidoglycan hydrolase) FlgJ (Hirano et al. 2001; Singer et al. 2014). Concomitantly, in gram-negative bacteria, the LP-ring proteins assemble around the rod (Fig. 13.6c) (Cohen and Hughes 2014).

Upon completion of the rod and LP-rings, the rod proteins are no longer secreted through the complex, a process known as substrate switching. Instead, the hook protein FlgE is secreted (Ferris and Minamino 2006). However, this protein does not spontaneously polymerize onto filaments; instead, a chaperone, FlgD, promotes its folding and assembly (Ohnishi et al. 1994). The length of the hook is tightly regulated, and upon reaching its final length, a second substrate-switching event occurs (Shibata et al. 2007). This leads to FlgE no longer being secreted, and instead the hook-filament junction proteins FlgK and FlgL proteins are secreted through, forming the hook-filament junction (Fig. 13.6d).

Completion of the hook-filament junction leads to a third substrate-switching event, leading to the secretion of the flagellin. As with the hook, the flagellin does not polymerize spontaneously but requires the cap complex, formed by the protein FliD, to promote its assembly (Ikeda et al. 1993). However, unlike the hook, the length of the filament is not regulated in many bacterial species, and it will keep growing, albeit at a decreasing rate (Zhao et al. 2018) (Fig. 13.6e).

Genetic Regulation

Over 50 different genes are involved in the assembly of the flagellum, with about half coding for the structural components and half responsible for regulation of the assembly process. These genes are divided into three classes, depending on the timing and regulation of their expression (Kutsukake et al. 1990).

Class 1 genes *flhD* and *flhC* are transcribed by a single promoter in response to chemotaxis signalling, resulting in the regulatory complex FlhDC (Fig. 13.7) (Soutourina et al. 1999). This complex then recruits the RNA Polymerase (RNAP) and primary sigma factor (σ^{70}), to activate the transcription of Class 2 genes, coding for the hook and basal body components.

Class 2 genes first translate the hook-basal body (HBB) components (*flhB, fliP, fliR, fliQ, and flhA*) followed by the membrane and supramembrane (MS)-ring (*fliF*) and cytoplasmic (C)-ring (*fliG, fliM, fliN*), which co-assemble in the inner membrane (IM) (Ikebe et al. 1999). Genes coding for the ATPase are expressed next (*fliH, fliI, fliJ*), followed by axial components of the proximal rod (*fliE, flgB, flgC, flgF*), distal rod (*flgG*), and hook (*flgE*). The hook and filament capping proteins (*flgD* and *fliD*) and the chaperones required for the assembly of the hook-basal body complex are also expressed from Class 2 genes (Fig. 13.7).

Class 3 gene expression and Class 2 gene inhibition occurs upon the detection of the hook reaching a determinate length by the molecular ruler FliK, which is intermittently secreted through the formation of the HBB (see below). Upon the hook reaching its full length, the export apparatus switches to late-stage substrates such as FliD. Export of FliD frees up the chaperone FliT which binds FlhC to inhibit Class 2 transcription (Yamamoto and Kutsukake 2006). FliT also binds anti sigma protein FlgM to release the bound FliA (σ^{28} factor/*fliA*), thus triggering Class 3 gene transcription (Fig. 13.7) (Das et al. 2018). Class 3 genes include the hook-filament junction (*flgK, flgL*), flagellin (*fliC*), and stators (*motA, motB*). In addition to these structural genes, Class 3 also expresses a variety of chemoreceptors sensing,

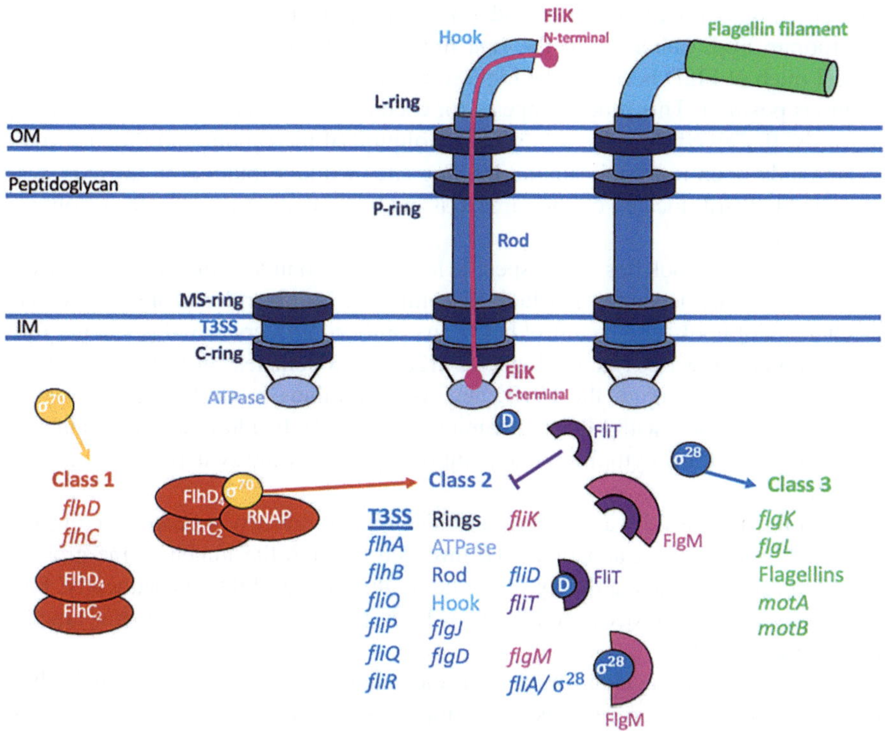

Fig. 13.7 Diagram of the flagellum regulatory cascade. Class 1 genes are in red, class 2 genes are in different shades of blue and purple, class 3 genes are in green. The master regulator FlhDC activates the transcription of T3SS genes which export other class 2 genes forming the hook-basal body complex. FliK acts as a molecular ruler measuring the distance between the FlgD hook capping protein and FlhB of the T3SS. Upon reaching the desired length, FlhB undergoes a cleavage changing substrate specificity of the T3SS to late substrates. FliD export frees the chaperone FliT which binds FlgM antisigma factor that was inhibiting FliA from activating the Class 3 genes. FlgM is then exported out of the cell

detecting chemical signals in the environment that regulate flagellum rotation (Laventie and Jenal 2020).

Mechanism and Regulation of Hook and Filament Assembly

As illustrated in Fig. 13.6, after the basal body forms, FlgH displaces the FlgJ rod chaperone, which is replaced by FlgD at the tip of the rod. FlgD then recruits monomers of FlgE, threaded through the rod lumen, to build the hook. There, FlgD forms a pentameric complex, with helical hairpins, tucked deep into the rod (Johnson et al. 2021; Matsunami et al. 2021). One subunit sits higher than the other four, not following the helical pitch. This suggests a mechanism for hook elongation whereby FlgD moves upwards and downwards, in a piston-like movement, to incorporate FlgE subunits. In the first transition, the lowest copy in the uneven helical pitch of FlgD becomes the highest copy, lifting that part of FlgD up from its previous position. This process repeats for each of the five subunits moving the cap upwards. If viewed from the top, a full revolution of the cap around the rod would sequentially open rod/hook subunit binding sites without rotation on the axis, with every subunit still located on the same axis but translated upwards instead (Johnson et al. 2021).

The flagellum hook has a very specific length, 55 nm in *S. typhimurium* (the exact length is species-dependent), which is controlled by the ruler protein FliK. The mechanism of hook length control has been highly controversial and is still not fully understood (Aizawa 2012a, b; Hughes 2012a, b). Nonetheless, it is established that FliK is secreted through the HBB, and its N-terminus binds to FlgD, while its C-terminus interacts with FlhB (Minamino et al. 1999). In addition, a self-catalyzed cleavage of the FlhB cytoplasmic domain is required for the switch to late substrate export.

After the hook is fully assembled, unfolded FlgK (bound to its chaperone FlgN) is recruited by the export apparatus and threaded through the lumen of the growing flagellum. It then binds at the tip of the hook, displacing FlgD. Subsequently, FlgL undergoes the same process and binds to FlgK to form the hook-filament junction (Xing et al. 2018).

The final step of flagellar formation is the elongation of the filament. Similarly to the above-mentioned process, the cap complex protein FliD, unfolded and bound to its chaperone FliT, is recruited by the export apparatus and exported through the lumen of the growing flagellum. Biochemical studies have shown that FliD and FlgL form a complex in isolation (Imada 2018; Yonekura et al. 2000), indicating that the role of FlgL is to recruit the cap complex. As mentioned above, upon secretion of FliD, it dissociates from FliT, which in turn binds to FlhC and halts class 2 gene transcription. It also binds to FlgM, releasing FliA to express class 3 flagellin genes.

After FliD oligomerizes on top of FlgL, unfolded flagellin bound to its chaperone FliS is in turn recruited to the export apparatus and exported through the hook lumen. The cap complex then promotes its oligomerization and incorporation of flagellin subunits in the growing filament. The mechanism of FliD-mediated filament extension is not currently established, with various models proposed (Al-Otaibi et al. 2020; Maki-Yonekura et al. 2003; Matsunami et al. 2021).

Unlike the hook, there does not seem to be a specific mechanism to regulate filament length. Instead, this appears to be a largely stochastic process, and the longer the filament, the less efficient the transport of flagellins through the lumen, leading to a decrease in elongation speed over time. This injection-diffusion mechanism explains filament growth dynamics as a function of the total length and diffusion coefficient (Paradis et al. 2017).

Diverse Flagella: Species-Specific Assembly, Disassembly, and Regulation

The various pathways described above were mainly deciphered in the prototypical *S. enterica* and *E. coli* flagella, which are peritrichous, i.e. distributed randomly at the bacterial cell surface. However, this organization is not constant across the bacterial kingdom, and in many species, flagellar complexes are strictly located to the cell pole(s). These can come in multiple arrangements, termed monotrichous (one single flagellum at one pole), lophotrichous (multiple flagella at one pole), amphitrichous (one flagellum at each pole), or cephalotrichous (multiple flagella at each pole). Nonetheless, these share a common mechanism for polar localization of the flagellum complex.

Two proteins, FlhG and FlhF, are present exclusively in bacteria where the flagella are at the cell pole. FlhG belongs to the ParA/MinD family of ATPases, involved in the polar localization of nuclear material and organelles, whereas FlhF is a GTPase. In multiple bacterial species, deletion of the *flhF* gene resulted in a mis-localization of the flagellum away from the cell pole, while that of the *flhG* gene led in the assembly of multiple complexes (Arroyo-Pérez and Ringgaard 2021; Gulbronson et al. 2016). Intriguingly, biochemical data suggest that these two proteins interact directly and act by promoting the oligomerization of FliF only at the cell pole (Terashima et al. 2020). It is not currently understood if this occurs via interaction with its C-terminal cytosolic tail or some yet-unknown mechanism.

Similarly, the genetic regulation described above (Fig. 13.7) shows a lot of variation depending on the bacterial species. For example, while three classes of genes are present in *S. enterica* and *E. coli*, other bacteria possess four classes, with an additional regulatory step regulating specifically the expression of the hook and hook-filament junction proteins (Bouteiller et al. 2021; Dasgupta et al. 2003). This step is regulated by the transcription factor FleR, not present in other bacteria. Conversely, in *Campylobacter* and *Caulobacter* species, evidence suggest that the proteins that form the export apparatus play an active role in the transcription regulation of the Class II genes, instead of FlhDC. Again, the molecular basis for this is poorly understood, but appears to involve the transcription activator *fldB* (Hendrixson and DiRita 2003; Wu et al. 1995).

Finally, it should be mentioned that the flagellum is largely thought of as a stable assembly that lasts for the entire file of the bacterial cell. Nonetheless, recent evidence using super-resolution fluorescence microscopy indicates that it is in fact highly dynamic, notably with intense and continuous assembly/disassembly of the stator complex but also turnover of components of the C-ring and ATPase complexes (Bai et al. 2014; Delalez et al. 2010; Leake et al. 2006). The role of these events is not fully established, but could be involved in subtle, fast regulation of flagellum assembly or rotation in response to yet-unknown stimuli.

In addition, it has been shown that at least in some species, the flagellum can lose its filament, and/or fully disassemble during the cell cycle (Ferreira et al. 2019; Skerker and Laub 2004). Recent cryo-electron tomography data have demonstrated that in particular, the LP ring remains intact after disassembly, which could play a role in the formation of new flagellar complexes (Kaplan et al. 2019, 2020). Nonetheless, the exact trigger and molecular pathway for such disassembly events are not currently understood.

Conclusion

Since Howard Berg's elegant demonstration that the bacterial flagellum rotates, in the 1970s, this complex has fascinated researchers, and the general public alike. Accordingly, thousands of publications have focused on elucidating its structure, rotation, and assembly mechanism over the past half century. However, it is only with the recent "resolution revolution" in cryo-EM, that we are finally putting together the pieces to elucidate its incredible complexity. Piece by piece, we are approaching a full atomic model of the bacterial flagellum, an extraordinary feat that would have seemed far-fetched only a few years ago.

Conversely, in recent years, the diversity in flagellum structures has become apparent. This means not only flagellar complexes adopting distinct and/or additional domains but also different assembly pathway, regulation, etc.

From a naïve perspective, it may look as if we are reaching the end point towards our goal to understand flagellar-mediated motility at atomic resolution: The structures of most components are known (with the notable exception of the C-ring, and hook-filament junction); and we mostly understand the mechanism of assembly and rotation. However, in the meantime, we have uncovered that there is an incredible level of complexity and diversity in the flagellum structure and regulation across the bacterial kingdom; and in fact, some of elements uncovered in the prototypical *S. enterica* complex might be the exception rather than the rule. Understanding this complexity remains a largely untapped challenge. With the recent emergence of new, more powerful microscopy techniques (super-resolution fluorescence, cryo-EM), combined with the power of metagenomic data and Artificial Intelligence computational approaches, we are probably still only seeing the tip of the iceberg in deciphering the molecular details of the flagellum.

References

Aizawa S (2012a) Rebuttal: flagellar hook length is controlled by a secreted molecular ruler. J Bacteriol 194:4797

Aizawa S (2012b) Mystery of FliK in length control of the flagellar hook Shin-Ichi. J Bacteriol 194: 4798–4800

Akiba T, Yoshimura H, Namba K (1991) Monolayer crystallization of flagellar L-P rings by sequential addition and depletion of lipid. Science (80-) 252:1544–1546

Allison SE, Tuinema BR, Everson ES, Sugiman-Marangos S, Zhang K, Junop MS, Coombes BK (2014) Identification of the docking site between a type III secretion system ATPase and a chaperone for effector cargo. J Biol Chem 289:23734–23744

Al-Otaibi NS, Taylor AJ, Farrell DP, Tzokov SB, DiMaio F, Kelly DJ, Bergeron JRC (2020) The cryo-EM structure of the bacterial flagellum cap complex suggests a molecular mechanism for filament elongation. Nat Commun 11:3210

Arroyo-Pérez EE, Ringgaard S (2021) Interdependent polar localization of FlhF and FlhG and their importance for flagellum formation of Vibrio parahaemolyticus. Front Microbiol 12:1–12

Bai F, Morimoto YV, Yoshimura SDJ, Hara N, Kami-Ike N, Namba K, Minamino T (2014) Assembly dynamics and the roles of FliI ATPase of the bacterial flagellar export apparatus. Sci Rep 4:1–7

Beeby M, Ribardo DA, Brennan CA, Ruby EG, Jensen GJ, Hendrixson DR (2016) Diverse high-torque bacterial flagellar motors assemble wider stator rings using a conserved protein scaffold. Proc Natl Acad Sci U S A 113:E2759

Berg HC (1974) Dynamic properties of bacterial flagellar motors. Nature 249:77–79

Bergeron JR (2016) Structural modeling of the flagellum MS ring protein FliF reveals similarities to the type III secretion system and sporulation complex. PeerJ 4:e1718

Bouteiller M, Dupont C, Bourigault Y, Latour X, Barbey C, Konto-ghiorghi Y, Merieau A (2021) Pseudomonas flagella: generalities and specificities. Int J Mol Sci 22:1–28

Brown PN, Mathews MAA, Joss LA, Hill CP, Blair DF (2005) Crystal structure of the flagellar rotor protein FliN from Thermotoga maritima. J Bacteriol 187:2890–2902

Burrage AM, Vanderpool E, Kearns DB (2018) Assembly order of flagellar rod subunits in Bacillus subtilis. J Bacteriol 200:1–13

Chaban B, Coleman I, Beeby M (2018) Evolution of higher torque in campylobacter-type bacterial flagellar motors. Sci Rep 8:1–11

Chang Y, Xu H, Motaleb MA, Liu J (2021) Characterization of the flagellar collar reveals structural plasticity essential for spirochete motility. MBio 12. https://doi.org/10.1128/mBio.02494-21

Cho SY, Song WS, Hong HJ, Lee GS, Kang SG, Ko HJ, Kim PH, Yoon SI (2017) Tetrameric structure of the flagellar cap protein FliD from Serratia marcescens. Biochem Biophys Res Commun 489(1):63–69

Chu J, Liu J, Hoover TR (2020) Phylogenetic distribution, ultrastructure, and function of bacterial flagellar sheaths. Biomolecules 10(3):1–15

Cohen EJ, Hughes KT (2014) Rod-to-hook transition for extracellular flagellum assembly is catalyzed by the L-ring-dependent rod scaffold removal. J Bacteriol 196:2387–2395

Cohen EJ, Nakane D, Kabata Y, Hendrixson DR, Nishizaka T, Beeby M (2020) Campylobacter jejuni motility integrates specialized cell shape, flagellar filament, and motor, to coordinate action of its opposed flagella. PLoS Pathog 16:1–24

Courtney CR, Cozy LM, Kearns DB (2012) Molecular characterization of the flagellar hook in Bacillus subtilis. J Bacteriol 194:4619–4629

Das C, Mokashi C, Mande SS, Saini S (2018) Dynamics and control of flagella assembly in Salmonella typhimurium. Front Cell Infect Microbiol 8:1–13

Dasgupta N, Wolfgang MC, Goodman AL, Arora SK, Jyot J, Lory S, Ramphal R (2003) A four-tiered transcriptional regulatory circuit controls flagellar biogenesis in Pseudomonas aeruginosa. Mol Microbiol 50:809–824

Debell C (1958) Antony van Leeuwenhoek and his "Little Animals". Russell & Russell, New York

de Silva YRO, Contreras-Martel C, Macheboeuf P, Dessen A (2020) Bacterial secretins: mechanisms of assembly and membrane targeting. Protein Sci 29:893–904

Delalez NJ, Wadhams GH, Rosser G, Xue Q, Brown MT, Dobbie IM, Berry RM, Leake MC, Armitage JP (2010) Signal-dependent turnover of the bacterial flagellar switch protein FliM. Proc Natl Acad Sci U S A 107:11347–11351

Deme JC, Johnson S, Vickery O, Muellbauer A, Monkhouse H, Griffiths T, James RH, Berks BC, Coulton JW, Stansfeld PJ et al (2020) Structures of the stator complex that drives rotation of the bacterial flagellum. Nat Microbiol 5:1553–1564

Diepold A, Wiesand U, Cornelis GR (2011) The assembly of the export apparatus (YscR,S,T,U,V) of the Yersinia type III secretion apparatus occurs independently of other structural components and involves the formation of an YscV oligomer. Mol Microbiol 82:502–514
dos Santos RN, Khan S, Morcos F (2018) Characterization of C-ring component assembly in flagellar motors from amino acid coevolution. R Soc Open Sci 5(5):171854
Eckhard U, Bandukwala H, Mansfield MJ, Marino G, Cheng J, Wallace I, Holyoak T, Charles TC, Austin J, Overall CM et al (2017) Discovery of a proteolytic flagellin family in diverse bacterial phyla that assembles enzymatically active flagella. Nat Commun 8:521
Fabiani FD, Renault TT, Peters B, Dietsche T, Gálvez EJC, Guse A, Freier K, Charpentier E, Strowig T, Franz-Wachtel M et al (2017) A flagellum-specific chaperone facilitates assembly of the core type III export apparatus of the bacterial flagellum. PLoS Biol 15:1–24
Ferreira JL, Gao FZ, Rossmann FM, Nans A, Brenzinger S, Hosseini R, Wilson A, Briegel A, Thormann KM, Rosenthal PB et al (2019) Γ-Proteobacteria eject their polar flagella under nutrient depletion, retaining flagellar motor relic structures. PLoS Biol 17:1–25
Ferris HU, Minamino T (2006) Flipping the switch: bringing order to flagellar assembly. Trends Microbiol 14:519–526
Ferris HU, Furukawa Y, Minamino T, Kroetz MB, Kihara M, Namba K, Macnab RM (2005) FlhB regulates ordered export of flagellar components via autocleavage mechanism. J Biol Chem 280: 41236–41242
Fujii T, Kato T, Hiraoka KD, Miyata T, Minamino T, Chevance FFV, Hughes KT, Namba K (2017) Identical folds used for distinct mechanical functions of the bacterial flagellar rod and hook. Nat Commun 8:1–10
Fukumura T, Makino F, Dietsche T, Kinoshita M, Kato T, Wagner S, Namba K, Imada K, Minamino T (2017) Assembly and stoichiometry of the core structure of the bacterial flagellar type III export gate complex. PLoS Biol 15:1–22
Galkin VE, Yu X, Bielnicki J, Heuser J, Ewing CP, Guerry P, Egelman EH (2008) Divergence of quaternary structures among bacterial flagellar filaments. Science (80-) 320:382–385
Gibson KH, Trajtenberg F, Wunder EA, Brady MR, San Martin F, Mechaly A, Shang Z, Liu J, Picardeau M, Ko A et al (2020) An asymmetric sheath controls flagellar supercoiling and motility in the leptospira spirochete. elife 9:1–24
Gourlay LJ, Thomas RJ, Peri C, Conchillo-Solé O, Ferrer-Navarro M, Nithichanon A, Vila J, Daura X, Lertmemongkolchai G, Titball R et al (2015) From crystal structure to in silico epitope discovery in the Burkholderia pseudomallei flagellar hook-associated protein FlgK. FEBS J 282:1319–1333
Gulbronson CJ, Ribardo DA, Balaban M, Knauer C, Bange G, Hendrixson DR (2016) FlhG employs diverse intrinsic domains and influences FlhF GTPase activity to numerically regulate polar flagellar biogenesis in Campylobacter jejuni. Mol Microbiol 99:291–306
Halte M, Erhardt M (2021) Protein export via the type iii secretion system of the bacterial flagellum. Biomolecules 11:1–19
Hendrixson DR, DiRita VJ (2003) Transcription of σ54-dependent but not σ 28-dependent flagellar genes in Campylobacter jejuni is associated with formation of the flagellar secretory apparatus. Mol Microbiol 50:687–702
Hirano T, Minamino T, Macnab RM (2001) The role in flagellar rod assembly of the N-terminal domain of Salmonella FlgJ, a flagellum-specific Muramidase. J Mol Biol 312:359–369
Homma M, Kutsukake K, Hasebe M, Iino T, Macnab RM (1990) FlgB, FlgC, FlgF and FlgG. A family of structurally related proteins in the flagellar basal body of Salmonella typhimurium. J Mol Biol 211:465–477
Hong HJ, Kim TH, Song WS, Ko HJ, Lee GS, Kang SG, Kim PH, Yoon S (2018) Crystal structure of FlgL and its implications for flagellar assembly. Sci Rep 8:1–11
Hughes KT (2012a) Flagellar hook length is controlled by a secreted molecular ruler Kelly. J Bacteriol 194:4793–4796
Hughes KT (2012b) Rebuttal:Mystery of FliK in Length Control of the Flagellar Hook Kelly. J Bacteriol 194:4801

Ikebe T, Iyoda S, Kutsukake K (1999) Promoter analysis of the class 2 flagellar operons of Salmonella. Genes Genet Syst 74:179–183

Ikeda T, Yamaguchi S, Hotani H (1993) Flagellar growth in a filament-less Salmonella fliD mutant supplemented with purified hook-associated protein 2. J Biochem 114:39–44

Imada K (2018) Bacterial flagellar axial structure and its construction. Biophys Rev 10:559–570

Imada K, Minamino T, Uchida Y, Kinoshita M, Namba K, DeRosier D (2016) Insight into the flagella type III export revealed by the complex structure of the type III ATPase and its regulator. Proc Natl Acad Sci U S A 113:3633–3638

Johnson S, Furlong EJ, Deme JC, Nord AL, Caesar J, Chevance FFV, Berry RM, Hughes KT, Lea SM (2021) Molecular structure of the intact bacterial flagellar basal body. Nat Microbiol 6:712–721

Kaplan M, Subramanian P, Ghosal D, Oikonomou CM, Pirbadian S, Starwalt-Lee R, Mageswaran SK, Ortega DR, Gralnick JA, El-Naggar MY et al (2019) In situ imaging of the bacterial flagellar motor disassembly and assembly processes. EMBO J 38:1–12

Kaplan M, Sweredoski MJ, Rodrigues JPGLM, Tocheva EI, Chang YW, Ortega DR, Beeby M, Jensen GJ (2020) Bacterial flagellar motor PL-ring disassembly subcomplexes are widespread and ancient. Proc Natl Acad Sci U S A 117:8941–8947

Kato T, Makino F, Miyata T, Horváth P (2019) Structure of the native supercoiled flagellar hook as a universal joint. Nat Commun 10(1):1–17

Kojima S, Shinohara A, Terashima H, Yakushi T, Sakuma M, Homma M, Namba K, Imada K (2008) Insights into the stator assembly of the vibrio flagellar motor from the crystal structure of MotY. Proc Natl Acad Sci U S A 105:7696–7701

Kojima S, Takao M, Almira G, Kawahara I, Sakuma M, Homma M, Kojima C, Imada K (2018) The helix rearrangement in the periplasmic domain of the flagellar stator B subunit activates peptidoglycan binding and ion influx. Structure 26:590–598.e5

Kreutzberger MAB, Ewing C, Poly F, Wang F, Egelman EH (2020) Atomic structure of the Campylobacter jejuni flagellar filament reveals how e Proteobacteria escaped toll-like receptor 5 surveillance. Proc Natl Acad Sci U S A 117:16985–16991

Kubori T, Shimamoto N, Yamaguchi S, Namba K, Aizawa SI (1992) Morphological pathway of flagellar assembly in Salmonella typhimurium. J Mol Biol 226:433–446

Kubori T, Yamaguchi S, Aizawa SI (1997) Assembly of the switch complex onto the MS ring complex of Salmonella typhimurium does not require any other flagellar proteins. J Bacteriol 179:813–817

Kuhlen L, Abrusci P, Johnson S, Gault J, Deme J, Caesar J, Dietsche T, Mebrhatu MT, Ganief T, Macek B et al (2018) Structure of the core of the type iii secretion system export apparatus. Nat Struct Mol Biol 25:583–590

Kuhlen L, Johnson S, Zeitler A, Bäurle S, Deme JC, Caesar JJE, Debo R, Fisher J, Wagner S, Lea SM (2020) The substrate specificity switch FlhB assembles onto the export gate to regulate type three secretion. Nat Commun 11:1–10

Kuhlen L, Johnson S, Cao J, Deme JC, Lea SM (2021) Nonameric structures of the cytoplasmic domain of FlhA and SctV in the context of the full-length protein. PLoS One 16:1–10

Kutsukake K, Ohya Y, Iino T (1990) Transcriptional analysis of the flagellar regulon of Salmonella typhimurium. J Bacteriol 172:741–747

Laventie BJ, Jenal U (2020) Surface sensing and adaptation in bacteria. Annu Rev Microbiol 74: 735–760

Leake MC, Chandler JH, Wadhams GH, Bai F, Berry RM, Armitage JP (2006) Stoichiometry and turnover in single, functioning membrane protein complexes. Nature 443:355–358

Lee LK, Ginsburg MA, Crovace C, Donohoe M, Stock D (2010) Structure of the torque ring of the flagellar motor and the molecular basis for rotational switching. Nature 466:996–1000

Magariyama Y, Sugiyama S, Muramoto K, Maekawa Y, Kawagishi I, Imae Y (1994) Very fast flagellar rotation. Nature 371:752

Majewski DD, Worrall LJ, Hong C, Atkinson CE, Vuckovic M, Watanabe N, Yu Z, Strynadka NCJ (2019) Cryo-EM structure of the homohexameric T3SS ATPase-central stalk complex reveals rotary ATPase-like asymmetry. Nat Commun 10:1–12

Maki-Yonekura S, Yonekura K, Namba K (2003) Domain movements of HAP2 in the cap-filament complex formation and growth process of the bacterial flagellum. Proc Natl Acad Sci U S A 100:15528–15533

Maki-Yonekura S, Yonekura K, Namba K (2010) Conformational change of flagellin for polymorphic supercoiling of the flagellar filament. Nat Struct Mol Biol 17:417–422

Mariano G, Faba-Rodriguez R, Bui S, Zhao W, Ross J, Tsokov S, Bergeron JRC (2022) Oligomerization of the FliF domains suggests a coordinated assembly of the bacterial flagellum MS ring. Front Microbiol 12:781960

Matsunami H, Barker CS, Yoon YH, Wolf M, Samatey FA (2016) Complete structure of the bacterial flagellar hook reveals extensive set of stabilizing interactions. Nat Commun 7:1–10

Matsunami H, Yoon YH, Imada K, Namba K, Samatey FA (2021) Structure of the bacterial flagellar hook cap provides insights into a hook assembly mechanism. Commun Biol 4:1–3

Mcdowell MA, Marcoux J, Mcvicker G, Johnson S, Fong YH, Stevens R, Bowman LAH, Degiacomi MT, Yan J, Wise A et al (2016) Characterisation of ShigellaSpa33 and ThermotogaFliM/N reveals a new model for C-ring assembly in T3SS. Mol Microbiol 99: 749–766

Minamino T, González-Pedrajo B, Yamaguchi K, Aizawa SI, Macnab RM (1999) FliK, the protein responsible for flagellar hook length control in Salmonella, is exported during hook assembly. Mol Microbiol 34:295–304

Montemayor EJ, Ploscariu NT, Sanchez JC, Parrell D, Sillard RS, Shebelut CW, Ke Z, Guerrero-ferreira RC, Wright R (2021) Flagellar structures from the bacterium Caulobacter crescentus and implications for phage/CbK predation of Multiflagellin bacteria. J Bacteriol 203:1–20

Morimoto YV, Ito M, Hiraoka KD, Che YS, Bai F, Kami-ike N, Namba K, Minamino T (2014) Assembly and stoichiometry of FliF and FlhA in Salmonella flagellar basal body. Mol Microbiol 91:1214–1226

Murphy GE, Leadbetter JR, Jensen GJ (2006) In situ structure of the complete Treponema primitia flagellar motor. Nature 442:1062–1064

Nakamura S (2020) Spirochete flagella and motility. Biomolecules 10:550

Ohnishi K, Ohto Y, Aizawa SI, Macnab RM, Iino T (1994) FlgD is a scaffolding protein needed for flagellar hook assembly in Salmonella typhimurium. J Bacteriol 176:2272–2281

Paradis G, Chevance FFV, Liou W, Renault TT, Hughes KT, Rainville S, Erhardt M (2017) Variability in bacterial flagella re-growth patterns after breakage. Sci Rep 7:1–10

Postel S, Deredge D, Bonsor DA, Yu X, Diederichs K, Helmsing S, Vromen A, Friedler A, Hust M, Egelman EH, Beckett D (2016) Bacterial flagellar capping proteins adopt diverse oligomeric states. elife 5:e18857

Rossez Y, Wolfson EB, Holmes A, Gally DL, Holden NJ (2015) Bacterial flagella: twist and stick, or dodge across the kingdoms. PLoS Pathog 11:e1004483

Rossmann FM, Beeby M (2018) Insights into the evolution of bacterial flagellar motors from high-throughput in situ electron cryotomography and subtomogram averaging. Acta Crystallogr Sect D Struct Biol 74:585–594

Samatey FA, Matsunami H, Imada K, Nagashima S, Shaikh TR, Thomas DR, Chen JZ, DeRosier DJ, Kitao A, Namba K (2004) Structure of the bacterial flagellar hook and implication for the molecular universal joint mechanism. Nature 431:1062–1068

Santiveri M, Roa-Eguiara A, Kühne C, Wadhwa N, Hu H, Berg HC, Erhardt M, Taylor NMI (2020) Structure and function of stator units of the bacterial flagellar motor. Cell 183:244–257.e16

Shibata S, Takahashi N, Chevance FFV, Karlinsey JE, Hughes KT, Aizawa SI (2007) FliK regulates flagellar hook length as an internal ruler. Mol Microbiol 64:1404–1415

Shibata S, Matsunami H, Aizawa SI, Wolf M (2019) Torque transmission mechanism of the curved bacterial flagellar hook revealed by cryo-EM. Nat Struct Mol Biol 26:941–945

Silva YR, Contreras-Martel C, Macheboeuf P, Dessen A (2020) Bacterial secretins: mechanisms of assembly and membrane targeting. Protein Sci 29(4):893–904

Singer HM, Erhardt M, Hughes KT (2014) Comparative analysis of the secretion capability of early and late flagellar type III secretion substrates. Mol Microbiol 93:505–520

Skerker JM, Laub MT (2004) Cell-cycle progression and the generation of asymmetry in Caulobacter crescentus. Nat Rev Microbiol 2:325–337

Song WS, Cho SY, Hong HJ, Park SC, Yoon SI (2017) Self-oligomerizing structure of the flagellar cap protein FliD and its implication in filament assembly. J Mol Biol 429(6):847–857

Sosinsky GE, Francis NR, Stallmeyer MJB, DeRosier DJ (1992) Substructure of the flagellar basal body of Salmonella typhimurium. J Mol Biol 223:171–184

Soutourina O, Kolb A, Krin E, Laurent-Winter C, Rimsky S, Danchin A, Bertin P (1999) Multiple control of flagellum biosynthesis in Escherichia coli: role of H-NS protein and the cyclic AMP-catabolite activator protein complex in transcription of the flhDC master operon. J Bacteriol 181:7500–7508

Stallmeyer MJB, Macnab RM, Derosier DJ (1989) Image reconstruction of the flagellar basal body of Salmonella typhimurium. J Mol Biol 205:519–528

Suzuki H, Yonekura K, Namba K (2004) Structure of the rotor of the bacterial flagellar motor revealed by electron Cryomicroscopy and single-particle image analysis. J Mol Biol 337:105–113

Takekawa N, Kawamoto A, Sakuma M, Kato T, Minamino T, Namba K, Homma M, Imada K, Kojima S, Kinoshita M (2021) Two distinct conformations in 34 flif subunits generate three different symmetries within the flagellar ms-ring. MBio 12:1–13

Tan J, Zhang X, Wang X, Xu C, Chang S, Wu H, Wang T, Liang H, Gao H, Zhou Y et al (2021) Structural basis of assembly and torque transmission of the bacterial flagellar motor. Cell 184: 2665–2679.e19

Terashima H, Fukuoka H, Yakushi T, Kojima S, Homma M (2006) The vibrio motor proteins, MotX and MotY, are associated with the basal body of Na+−driven flagella and required for stator formation. Mol Microbiol 62:1170–1180

Terashima H, Koike M, Kojima S, Homma M (2010) The flagellar basal body-associated protein FlgT is essential for a novel ring structure in the sodium-driven vibrio motor. J Bacteriol 192: 5609–5615

Terashima H, Li N, Sakuma M, Koike M, Kojima S, Homma M, Imada K (2013) Insight into the assembly mechanism in the supramolecular rings of the sodium-driven vibrio flagellar motor from the structure of FlgT. Proc Natl Acad Sci U S A 110:6133–6138

Terashima H, Hirano K, Inoue Y, Tokano T, Kawamoto A, Kato T, Yamaguchi E, Namba K, Uchihashi T, Kojima S et al (2020) Assembly mechanism of a supramolecular MS-ring complex to initiate bacterial flagellar biogenesis in vibrio species. J Bacteriol 202(16):601

Thomas DR, Francis NR, Xu C, DeRosier DJ (2006) The three-dimensional structure of the flagellar rotor from a clockwise-locked mutant of salmonella enterica Serovar typhimurium. J Bacteriol 188:7039–7048

Thomson NM, Ferreira JL, Matthews-Palmer TR, Beeby M, Pallen MJ (2018) Giant flagellins form thick flagellar filaments in two species of marine γ-proteobacteria. PLoS One 13:1–12

Van Way SM, Hosking ER, Braun TF, Manson MD (2000) Mot protein assembly into the bacterial flagellum: A model based on mutational analysis of the motB gene. J Mol Biol 297:7–24

Vartanian AS, Pazs A, Fortgang EA, Abramsons J, Dahlquist FW (2012) Structure of flagellar motor proteins in complex allows for insights into motor structure and switching. J Biol Chem 287:35779–35783

Wang F, Burrage AM, Postel S, Clark RE, Orlova A, Sundberg EJ, Kearns DB, Egelman EH (2017) A structural model of flagellar filament switching across multiple bacterial species. Nat Commun 8:960

Wu J, Benson AK, Newton A (1995) Global regulation of a σ54-dependent flagellar gene family in Caulobacter crescentus by the transcriptional activator FlbD. J Bacteriol 177:3241–3250

Xing Q, Shi K, Portaliou A, Rossi P, Economou A, Kalodimos CG (2018) Structures of chaperone-substrate complexes docked onto the export gate in a type III secretion system. Nat Commun 9: 1–9

Xue C, Ho Lam K, Zhang H, Sun K, Hang Lee S, Chen X, Wing Ngor A, S. (2018) Crystal structure of the FliF-FliG complex from Helicobacter pylori yields insight into the assembly of the motor MS-C ring in the bacterial flagellum. J Biol Chem 293:2066–2078

Yamaguchi T, Makino F, Miyata T, Minamino T, Kato T, Namba K (2021) Structure of the molecular bushing of the bacterial flagellar motor. Nat Commun 12:4469

Yamamoto S, Kutsukake K (2006) FliT acts as an anti-FlhD2C2 factor in the transcriptional control of the flagellar regulon in Salmonella enterica serovar typhimurium. J Bacteriol 188:6703–6708

Yonekura K, Maki S, Morgan DG, DeRosier DJ, Vonderviszt F, Imada K, Namba K (2000) The bacterial flagellar cap as the rotary promoter of Flagellin self-assembly. Science (80-.) 290: 2148–2152

Yonekura K, Maki-Yonekura S, Namba K (2003) Complete atomic model of the bacterial flagellar filament by electron cryomicroscopy. Nature 424:643–650

Yoon YH, Barker CS, Bulieris PV, Matsunami H, Samatey FA (2016) Structural insights into bacterial flagellar hooks similarities and specificities. Sci Rep 6:1–11

Young KT, Davis LM, DiRita VJ (2007) Campylobacter jejuni: molecular biology and pathogenesis. Nat Rev Microbiol 5:665–679

Zhao Z, Zhao Y, Zhuang X, Lo W, Baker MAB, Lo C et al (2018) Frequent pauses in Escherichia coli fiagella fiuorescence imaging. Nat Commun 9:1885

Zeytuni N, Hong C, Flanagan KA, Worrall LJ, Theiltges KA, Vuckovic M, Huang RK, Massoni SC, Camp AH, Yu Z et al (2017) Near-atomic resolution cryoelectron microscopy structure of the 30-fold homooligomeric SpoIIIAG channel essential to spore formation in Bacillus subtilis. Proc Natl Acad Sci U S A 114:E7073–E7081

Zhu S, Nishikino T, Hu B, Kojima S, Homma M, Liu J (2017) Molecular architecture of the sheathed polar flagellum in vibrio alginolyticus. Proc Natl Acad Sci U S A 114:10966–10971

Chapter 14
Actomyosin Complex

Ian Pepper and Vitold E. Galkin

Abstract Formation of cross-bridges between actin and myosin occurs ubiquitously in eukaryotic cells and mediates muscle contraction, intracellular cargo transport, and cytoskeletal remodeling. Myosin motors repeatedly bind to and dissociate from actin filaments in a cycle that transduces the chemical energy from ATP hydrolysis into mechanical force generation. While the general layout of surface elements within the actin-binding interface is conserved among myosin classes, sequence divergence within these motifs alters the specific contacts involved in the actomyosin interaction as well as the kinetics of mechanochemical cycle phases. Additionally, diverse lever arm structures influence the motility and force production of myosin molecules during their actin interactions. The structural differences generated by myosin's molecular evolution have fine-tuned the kinetics of its isoforms and adapted them for their individual cellular roles. In this chapter, we will characterize the structural and biochemical basis of the actin-myosin interaction and explain its relationship with myosin's cellular roles, with emphasis on the structural variation among myosin isoforms that enables their functional specialization. We will also discuss the impact of accessory proteins, such as the troponin-tropomyosin complex and myosin-binding protein C, on the formation and regulation of actomyosin cross-bridges.

Keywords Myosin · Actin · Actomyosin complex · Thick filament · Thin filament · Muscle contraction

Introduction

The actin-myosin complex is a molecular partnership that sustains the existence of eukaryotes, ranging from amoebas to humans. This wide phylogenetic distribution reflects the functional importance of actomyosin interactions. Although

I. Pepper · V. E. Galkin (✉)
Department of Physiological Sciences, Eastern Virginia Medical School, Norfolk, VA, USA
e-mail: pepperi@evms.edu; galkinve@evms.edu

J. R. Harris, J. Marles-Wright (eds.), *Macromolecular Protein Complexes IV*,
Subcellular Biochemistry 99, https://doi.org/10.1007/978-3-031-00793-4_14

conventionally linked to muscle contraction, the unconventional myosins in other cell types govern numerous processes including cell motility, cargo transport, cytoskeletal organization, and tension sensing (reviewed in (Hartman et al. 2011)). Despite this diversity, the function of actin-myosin interactions can be broadly characterized as supporting movement across many biological scales: from the movement of intracellular cargo along the cytoskeleton to the movement of multicellular organisms powered by muscle contraction. Not surprisingly, all myosins share a general mechanism that involves coupling ATP hydrolysis and actin binding to generate force. The universal transduction of chemical energy into mechanical work is reflected in the conserved myosin head, whose motor domain harbors its ATP- and actin-binding activities. Although sequence identity is less than 20% when comparing sequences of heads of evolutionarily distant myosins, conservation of tertiary structure and functionally critical residues has maintained a general consistency of actin-myosin interactions across the eukaryotic tree of life (Cope et al. 1996; Pasha et al. 2016). The primary sequence differences between motor domains ensure that while the steps involved in the ATPase cycle are the same across the board, the rates and equilibrium constants of those steps are tailored for the functional role of each myosin (Shuman et al. 2014; Sweeney et al. 1998). Myosin tail domains exhibit much greater variability in their architectures, with implications for the myosin molecule's ability to dimerize (reviewed in (Krendel and Mooseker 2005)) as well as recognize and mobilize intracellular cargoes (reviewed in (Lu et al. 2014)). Despite exhibiting different conservation patterns, phylogenetic evidence indicates coevolution of the head and tail domains within the myosin protein family (Korn 2000). In contrast to myosins, muscle and cytoplasmic actin isoforms share >90% sequence identity (Perrin and Ervasti 2010; Vandekerckhove and Weber 1978). Thus, the actin-myosin system involves variable myosin motors operating along a conserved actin track (Robert-Paganin et al. 2020).

In this chapter, we aim to shed light on recent advances in our knowledge of actomyosin interactions and their physiological function. We will begin by surveying the general structural features of actin and myosin proteins, as well as the mechanochemical cycle driven by their interaction. This will lead into a discussion of the structural and biochemical evidence, acquired from a variety of myosin isoforms, about each transition in the cross-bridge cycle. We will highlight the isoform-specific differences within the actin-binding interface in the rigor state, which can explain functional differences in actomyosin interactions. After thoroughly characterizing the actomyosin cross-bridge cycle, we will discuss proteins that regulate actomyosin interactions in a muscle sarcomere. We will conclude by presenting examples of how diverse myosin families carry out a vast number of physiological functions while interacting with highly conserved actin filaments.

General Structural Features of Actin and Myosin

Actin

Overview of the Actin Structure

The motor activity of myosin completely relies on its interaction with actin. Actin represents one of the most abundant proteins in the cellular environment, accounting for roughly 10% of total protein content (Farmer et al. 1983). Prior to transition into its active filamentous state, actin exists in a monomeric or globular form (referred to as G-actin). G-actin is a 43-kDa polypeptide with 375 amino acids that are divided into 4 subdomains (Fig. 14.1a). Starting from the traditional orientation of SD1 (residues 1–32, 70–144, and 338–375) in the lower right quadrant (Fig. 14.1a, blue), the other three subdomains are related to each other as follows: SD2 (residues 33–69) lies just above SD1 (Fig. 14.1a, dark green), SD3 (residues 145–180 and 270–337) is located to the left side of SD1 (Fig. 14.1a, light green), while SD4 (residues 181–269) is directly above SD3 in the upper left (Fig. 14.1a, cyan) (Kabsch et al. 1990). Based on their position within the actin filament, SD1 and SD2 together comprise the outer domain, whereas SD3 and SD4 make up the inner domain. These two domains, connected by a hinge region (residues 137–145 and 335–337), form a so-called cleft which harbors a nucleotide (Fig. 14.1a, brown) and a divalent cation (e.g., Mg^{2+} or Ca^{2+}) (Fig. 14.1a, magenta). All mammalian genomes encode for at least six actin isoforms (reviewed in (Perrin and Ervasti 2010)). Skeletal and cardiac muscles each contain their own α-isoforms, while smooth muscle utilizes α-isoform in vascular tissue and γ-isoform in enteric tissue. The β- and γ-cytoplasmic isoforms are present in non-muscle cells. Comparison of all six mammalian isoforms reveals that 347 of the 375 amino acid positions (92.5%) are identical. This conservation pattern is mirrored when comparing actin orthologs from different species – mammalian cytoplasmic actin is 100% identical to chicken and 90% identical to yeast (Egelman 2001).

The first atomic structure of G-actin was revealed by X-ray crystallography (Kabsch et al. 1990) (Fig. 14.1a) to give rise to the first atomic model of the actin filament (Holmes et al. 1990) based on X-ray diffraction pattern from polymerized actin (referred to as filamentous actin or F-actin). Nearly 20 years later, (Oda et al. 2009) used methodological innovations to obtain an improved X-ray fiber diffraction pattern to reveal the conformational "flattening" of G-actin upon polymerization (Fig. 14.1b), which was confirmed later using cryo-EM (Chou and Pollard 2019; Fujii et al. 2010; Galkin et al. 2015; von der Ecken et al. 2015). The power of cryo-EM has been harnessed in recent years to elucidate the precise molecular details of actin's structural transitions. The near-atomic three-dimensional (3D) cryo-EM reconstructions of F-actin with different bound nucleotides (Chou and Pollard 2019; Merino et al. 2018) showed the structural transitions of individual F-actin subunits that occur upon ATP hydrolysis and phosphate release (Fig. 14.1c and d). They suggest that upon G- to F-actin transition, the flattened conformation of the

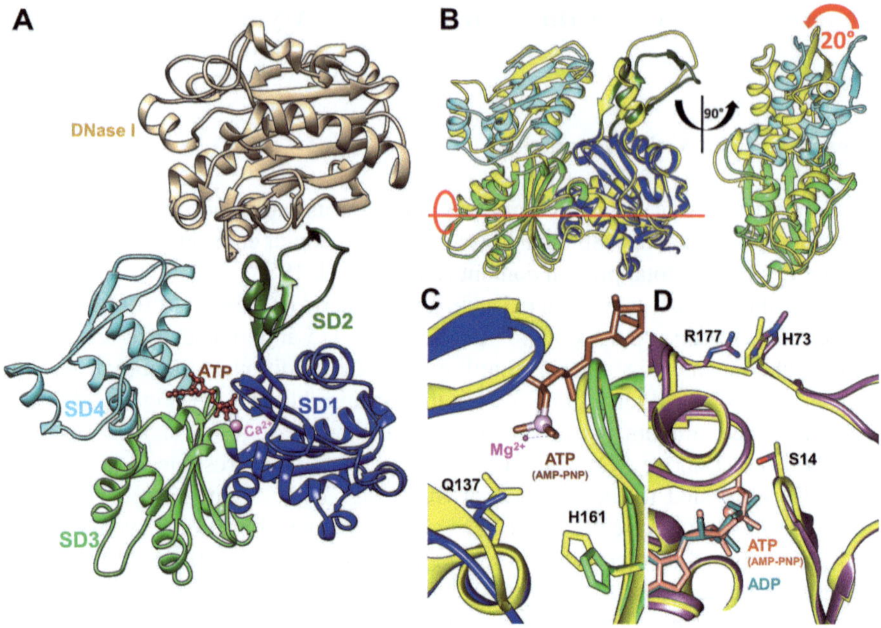

Fig. 14.1 Structure of actin monomers and conformational changes upon polymerization. (**a**) The structure of G-actin is shown using the first crystal structure of G-actin complexed with DNAse I (PDB 1ATN; (Kabsch et al. 1990)). The four subdomains are colored with different shades of green and blue, ATP (brown) and Ca^{2+} cation (magenta) are shown in the ATP-binding cleft. (**b**) Flattening of the actin molecule upon transition from G- to F-actin is shown buy superimposition of G-actin structure (PDB 1ATN; (Kabsch et al. 1990)) shown in (**a**) with the F-actin protomer structure (PDB 6DJM; (Chou and Pollard 2019)) outlined in yellow. To show the "propeller" rotation of the two actin's domains by ~20° (red arrow) around the hinge region (red line) upon flattening of actin the outer domains (e.g., SD1 and SD2) of both structures were aligned. Front (left) and side (right) views are shown. (**c**) Polymerization-induced flattening shifts Q137 and H161 residues, located in actin's cleft, toward the terminal phosphate of bound ATP which facilitates ATP hydrolysis. The same structures and color scheme from (**b**) is utilized. The ATP homologue AMP-PNP used to mimic ATP is shown in brown, while the expected position of γ-phosphate is depicted as plum sphere. (**d**) Conformational changes in the F-actin protomer nucleotide-binding site upon ATP hydrolysis. The ADP-Mg^{2+}-bound actin structure (PDB 6DJO; (Chou and Pollard 2019)) is represented in purple, and actin bound to the ATP analog AMP-PNP (PDB 6DJM; (Chou and Pollard 2019)) is represented in yellow. In the ADP-bound state, the backdoor for phosphate release is open due to interaction between R177 and H73

actin molecule (Fig. 14.1b) results in repositioning of the Q137 and H161 residues closer to the γ-phosphate so that their hydrogen-bonded water molecules are now optimally positioned for ATP hydrolysis (Fig. 14.1c). P_i then exits through a backdoor mechanism modulated by more subtle conformational changes of S14, the H73 sensor loop, and R177 (Chou and Pollard 2019; Wriggers and Schulten 1999) (Fig. 14.1d).

Formation of the Actin Filament

Formation of the actin filament (reviewed in (Pollard 2016)) includes three phases: (1) nucleation or formation of oligomers, which is the rate-limiting step; (2) elongation, in which additional monomers are added to the nucleated oligomer; and (3) steady-state, in which the rate of monomer addition at one filament end equals the rate of disassembly at the other end (Fig. 14.2a). The two ends of the filament display asymmetrical polymerization kinetics (Pollard 1986) – whereas filament growth happens at the "barbed"(+) end of filaments, subunit dissociation is mainly localized to the "pointed"(-)end (Fig. 14.2a). The result is end-to-end treadmilling of actin subunits at steady-state, with filaments maintaining a constant average length (Brenner and Korn 1983; Fujiwara et al. 2002; Wegner 1976). The difference in "flattening" of actin molecules upon addition to the filament from either end may explain the differential rate of filament elongation at the ends (Chou and Pollard 2019). To prevent spontaneous polymerization of G-actin in the cell, the formation of actin filaments is controlled by a variety of actin-binding proteins (Fig. 14.2b–d). There are two types of actin monomer-binding proteins (reviewed in (Pollard 2016)): sequestering proteins (e.g., profilin and thymosin-β4), which prevent actin monomers from spontaneous polymerization and nucleation-promoting factors (NPFs). The Wiskott-Aldrich homology 2 (WH2) domain-containing proteins (e.g., WASP) have been shown to promote regulated branching of actin filaments via recruitment of the actin-related proteins 2/3 complex (Arp2/3) (reviewed in (Molinie and Gautreau 2018; Pollard 2007)), while formins have been shown to promote elongation at the ends of the filaments (reviewed in (Courtemanche 2018)). Depolymerization of actin filaments is controlled by severing proteins (e.g., cofilin/ADF and gelsolin) (reviewed in (Pollard 2016)), which maintain the cellular balance between G- and F-actins.

An actin filament consists of two right-handed helices wrapped around each other in a staggered manner, such that lateral subunits on opposite strands are offset by half the monomer length (Fig. 14.3a). Two types of intersubunit contacts maintain actin filament integrity: longitudinal contacts linking the monomers on each strand (Fig. 14.3b–d) and lateral contacts joining both strands together (Fig. 14.3e–f) (Chou and Pollard 2019; Merino et al. 2018; Risi et al. 2021a; von der Ecken et al. 2016; von der Ecken et al. 2015).

Most of the longitudinal interactions between actin protomers are formed between the SD2 of the lower actin and SD1 and SD3 of the upper subunit (Fig. 14.3b and c). The tip of the D-loop of the lower subunit (Fig. 14.3b, blue atoms) makes hydrophobic interactions with residues of the upper subunit that form a hydrophobic cavity between its SD1 and SD3 (Fig. 14.3b, red atoms). Meanwhile, an H-bond links lower protomer's SD2 (Fig. 14.3b, plum atoms) with SD1 of the upper subunit (Fig. 14.3b, orange atoms). Finally, positively charged residues of SD2 of the lower actin (Fig. 14.3c, blue, red, and green atoms) make salt-bridges with the acidic residues of SD3 of the upper subunit (Fig. 14.3c, plum, yellow, and orange atoms). Interactions between SD4 of the lower actin and SD3 of its upper

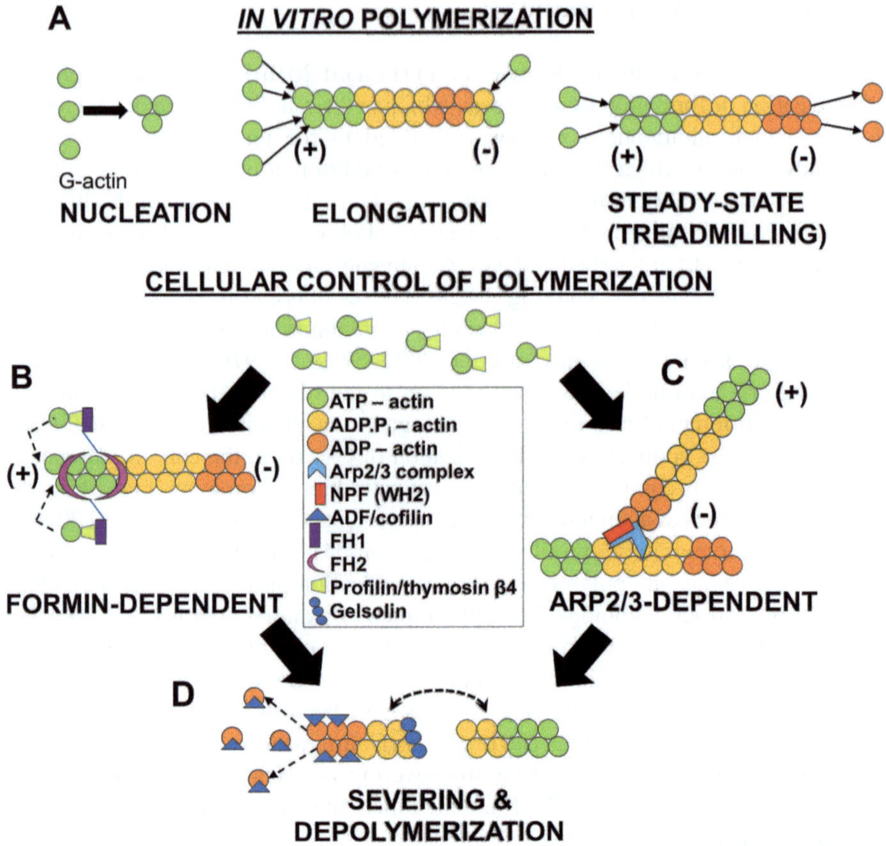

Fig. 14.2 Polymerization of actin filaments. (**a**) Diagram of phases of in vitro polymerization of actin. Following the slow rate-limiting step of G-actin nucleation, filaments are elongated by addition of subunits mostly at the fast-growing barbed (+) end. The bound nucleotide within the incorporated subunits undergoes hydrolysis and subsequent phosphate release. These steps generate an "age-dependent" gradient of nucleotide states along the filament, in which ATP-actin subunits (green) are localized mostly at the barbed end, while ADP-actin subunits (orange) are localized in the center of the filament. Upon reaching equilibrium the filament enters so-called treadmilling, when ATP monomers are added mostly to the filament at the barbed (+) end, while dissociation of the ADP monomers occurs predominantly at the pointed (-) end. (**b–d**) Cellular mechanisms for regulating actin filament polymerization. Free ATP-G-actin (green circles) in the cell is prevented from unregulated polymerization via to its association with actin-sequestering proteins profiling/thymosin β4 (olive green trapezoids). Nucleation and elongation can occur by two distinct mechanisms shown in **b** and **c**. (**b**) Formin-dependent polymerization. Proteins of the formin family (denoted FH1/FH2) can stimulate both nucleation and elongation via dimerization of their formin homology 2 (FH2) domains. The FH2 domain (magenta eclipses) associates with the barbed end, and its linked FH1 domain (plum rectangles) brings G-actin (green circles) complexed with profilin/thymosin β4 (olive green trapezoids) to the barbed (+) end. (**c**) ARP2/3-dependent polymerization. New actin filaments can branch from the side of an existing one. The association of a WH2-containing NPF (red rectangle) and the Arp2/3 complex (cyan chevron) serves as the nucleation seed for the polymerization of actin at the side of the existing filament. (**d**) Severing of filaments can be accomplished by gelsolin (blue circles), which binds to the filament interior and generates a break point. ADF/cofilin (blue triangles) proteins bind to the pointed end of the filament enriched in ADP actin to promote its depolymerization

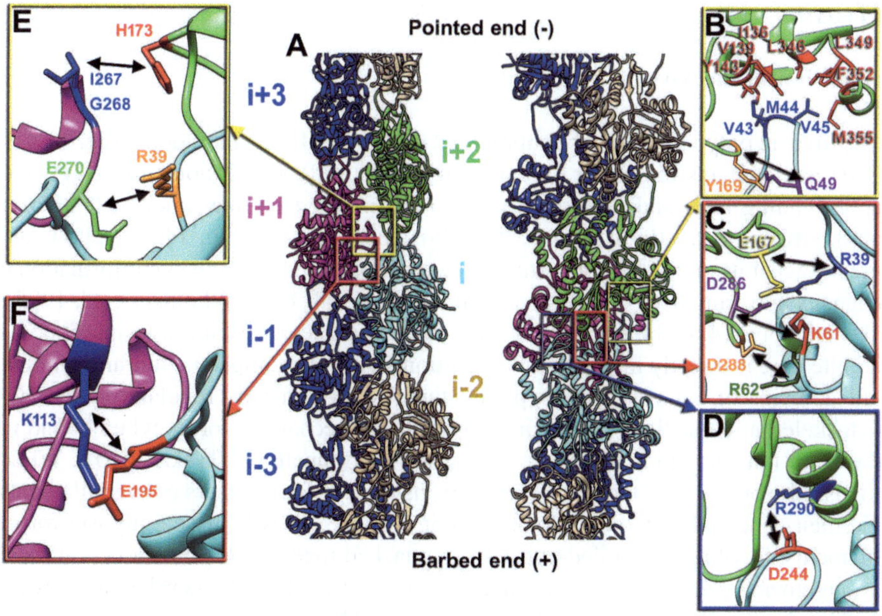

Fig. 14.3 Interactions between actin subunits that hold F-actin. (**a**) Actin filament (PDB 7JH7; (Risi et al. 2021a)) has actin subunits position within the filament labeled as in the basic left-handed one-start helix. The two strands of the actin filament are colored in blue and tan, while subunits which interactions are detailed in (**b–f**) are marked in cyan, magenta, and green. (**b–d**) Longitudinal interface is comprised of: (**b**) interactions between the tip of the D-loop (V43, M44, V45, blue) and residues forming a hydrophobic cavity between SD1 and SD3 (I136, V139, Y143, L346, L349, F352, M355 red), along with the Y169 of SD3 (yellow) forming H-bond with Q49 of SD2 (plum); (**c**) salt bridges between R39 (blue), K61 (red), R62 (green) of SD2 and D286 (plum), E167 (yellow), D288 (orange) of SD3, respectively; (**d**) electrostatic interaction between D244 of SD4 (red) of the lower subunit with R290 of SD3 (blue) of the upper subunit. (**e–f**) Lateral interactions are comprised of: (**e**) electrostatic interaction between E270 located in the so-called hydrophobic plug of the i+1 subunit (green) with R39 residing in the SD2 of the i subunit across the strand (orange); hydrophobic interactions between I267/G268 located in the hydrophobic plug of the i+1 subunit (blue) with H173 from the SD2 of the i subunit (orange); (**f**) salt bridge between K113 located in the SD1 of the i+1 subunit (blue) with E195 from the SD4 of the i subunit (cyan). The pairs of amino acids involved in interactions are connected with black arrows

counterpart are held by the salt-bridge between D244 of SD4 (Fig. 14.3d, red atoms) and R290 of SD3 (Fig. 14.3d, blue atoms). The lateral interactions are comprised of a salt bridge between the H-plug and the SD2 of the actin subunit across the strand (Fig. 14.3e, green and orange atoms, respectively). In addition, a lateral contact is formed by the ionic interaction between SD1 and SD4 of the two actins across the strand (Fig. 14.3f, blue and red atoms, respectively).

Myosin

Overview of the Myosin Structure

The myosin polypeptide can be subdivided into its head and tail regions (Fig. 14.4a). The head contains the motor domain (Fig. 14.4a, red), which holds the ATP- and actin-binding sites and represents the most conserved portion among all myosin classes (Cope et al. 1996; Foth et al. 2006). The lever arm (Fig. 14.4a, yellow and green) generates force by converting Ångstrom-scale motor domain conformational changes into nanometer-scale movements (Jontes et al. 1995; Whittaker et al. 1995). The tail (Fig. 14.4a, blue) represents the most variable part of all myosins since its architecture is directly linked to their functionality. For example, membrane-binding tail domains of class I myosins permit them to dynamically regulate membrane-cytoskeleton interactions by sensing intracellular mechanical forces when bound to the actin filament (Fig. 14.4b) (reviewed in (McConnell and Tyska 2010)). While myosin I works as a monomer, other myosins have to form dimers or even filaments via their α-helical coiled-coil domains. For instance, class II muscle myosin forms filaments, in which the coiled-coil rods are packed together in a parallel cylindrical arrangement (Fig. 14.4c) (Al-Khayat et al. 2013; Hu et al. 2016; Kensler and Stewart 1983). On the other hand, class V myosins (e.g., Myosin Va) form a dimer using its tail domain, which also contains a cargo-binding domain (CBD) responsible for transporting its intracellular cargoes (Fig. 14.4d) (Pylypenko et al. 2013; Wu et al. 2002).

Due to the low solubility of the complete myosin molecule under physiological salt concentrations (Margossian and Lowey 1982), water-soluble head fragments generated via enzymatic digestion of muscle myosins have been the main source of knowledge regarding its structure and kinetics. Trypsin segments the molecule in the middle of its tail (Fig. 14.4e, cyan arrow) to generate the heavy meromyosin (HMM) segment containing the head and proximal tail (Mihalyi and Szent-Gyorgyi 1953). HMM can be further digested by papain (Fig. 14.4e, red arrow) to produce the so-called subfragment-1 (myosin-S1), which exclusively contains the motor domain and lever arm regions, and the proximal tail fragment called S2 (Kominz 1965). Isolated myosin-S1 heads (Fig. 14.4e, denoted as S1) can perform as minimal motor units in vitro, as demonstrated by their ability to couple ATP hydrolysis with force generation (Kishino and Yanagida 1988; Toyoshima et al. 1987).

Components of a Myosin-S1 Molecule

The first structure of the myosin-S1 was solved by X-ray crystallography (Rayment et al. 1993). Subsequent analyses of myosin motor domain crystal structures with different nucleotide analogues in their clefts (Dominguez et al. 1998; Fisher et al. 1995b; Houdusse et al. 1999; Houdusse et al. 2000; Rayment et al. 1993) revealed four major subdomains of myosin along with their connecting elements, which are

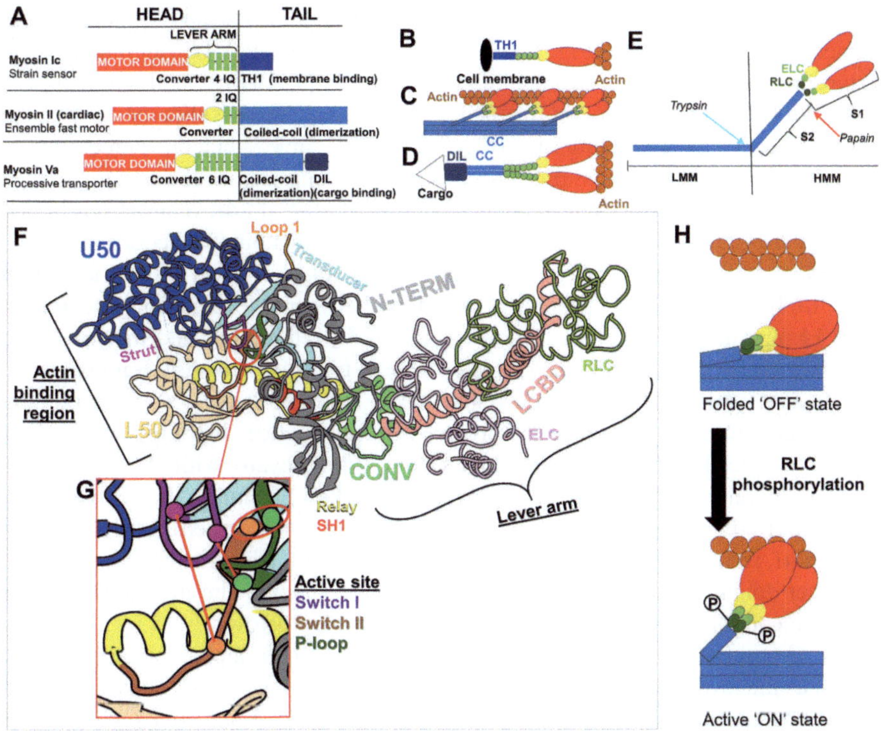

Fig. 14.4 General structural features of myosins. (**a**) Domain organization of different myosin isoforms: myosin Ic (strain sensor), cardiac myosin (ensemble fast motor), and myosin Va (processive transporter). (**b**) Myosin Ic bounds to a cell membrane (black oval) via its tail homology 1 (TH1) domain (dark blue rectangle). (**c**) Cardiac myosin forms thick filaments *via* its Coiled-coil (CC) domain (light blue rectangles) so that protruding motor domains (red ovals) can reach actin-based thin filaments (brown) to generate contractile force. (**d**) Myosin Va forms a dimer via its CC domain to processively transport intracellular cargo (white triangle) recognized by the dilute (DIL) domains (blue box). (**e**) Subdivisions of the muscle myosin heavy chain resulting from enzymatic digestions. (**f**) Components of nucleotide-free myosin-S1 from skeletal muscle (PDB 2MYS; (Rayment et al. 1993)). The motor domain's primary components are: upper (U50; dark blue) and lower (L50; tan) 50 kDa subdomains harboring actin binding interface (black square bracket); N-terminal 25 kDa subdomain (N-term; gray); active site (red circle) containing the ATP-binding elements – P-loop (dark green), switch I (purple), and switch II (brown); strut (magenta); transducer (cyan) with associated loop 1 (orange); relay helix (yellow); SH1 helix (red); converter (green). The terminal helix of the converter extends into light chain-binding domain (salmon), which binds the essential light chain (plum) and regulatory light chain (olive). The converter, LCBD, and bound light chains comprise the lever arm (black square bracket). (**g**) Interactions between residues of active site elements in the nucleotide-free state marked with red lines and oval. (**h**) Phosphorylation of RLC regulates the availability of smooth muscle myosin heads for cross-bridge formation

depicted in Fig. 14.4f. The motor domain of myosin-S1 consists of a central 50 kDa subdomain divided into upper (U50) (Fig. 14.4f, blue) and lower (L50) (Fig. 14.4f, tan) halves, followed by the 25 kDa N-terminal (N-term) subdomain (Fig. 14.4f, gray). Conserved secondary structural elements on the U50 and L50 surfaces constitute the actin-binding interface (Fig. 14.4f, square bracket). A short linear element known as the strut (Fig. 14.4f, magenta) spans the U50–L50 cleft and assists its closure during myosin's strong binding to actin (Fujita-Becker et al. 2006; Sasaki et al. 2000). At the site of convergence between all three subdomains lies the so-called active ATP-binding site (Fig. 14.4f, red circle, nucleotide-free state is shown). It resides next to the transducer, which is comprised of the seven-stranded β-sheet (Fig. 14.4f, cyan) and its associated loop 1 (Fig. 14.4f, orange) positioned between the U50 and N-term. The active site elements that bind and coordinate ATP are switch I (Fig. 14.4g, purple), switch II (Fig. 14.4g, brown), and the P-loop (Fig. 14.4g, dark green). Structural coupling between the actin-binding region (Fig. 14.4f, square bracket), active site (Fig. 14.4f, red circle), and the lever arm (Fig. 14.4d, curved bracket) is mediated by several myosin elements (Houdusse et al. 2000). The strut (Fig. 14.4f, magenta) links the U50 and L50 actin-binding subdomains together. Switch II, involved in nucleotide binding (Fig. 14.4f, brown), connects the transducer domain (Fig. 14.4f, cyan) with the relay helix (Fig. 14.4f, yellow). The relay helix (Fig. 14.4f, yellow) connects the actin-binding L50 (Fig. 14.4f, tan) with the converter domain (Fig. 14.4f, green). The SH1 helix (Fig. 14.4f, red) bridges the N-term of the motor domain (Fig. 14.4f, gray) with the converter domain at the lever arm base (Fig. 14.4f, green). The mechanism of the allosteric coupling between these elements is discussed below.

The C-terminus of the converter domain gives rise to an alpha-helical extension known as the lever arm helix or "neck" region (Fig. 14.4f, salmon). The neck is also referred to as the light chain-binding domain (LCBD). Each myosin isoform's LCBD binds its light chains via IQ motifs (Fig. 14.4a, green), whose name derives from the isoleucine (I)-glutamine (Q) repeat at the start of their consensus sequence. Sequence variations within these IQ motifs determine the type of the attached light chains (reviewed in (Heissler and Sellers 2014)). Muscle myosins possess a calcium-binding essential light chain (ELC) proximal to the converter (Fig. 14.4f, plum) and a phosphorylation-dependent regulatory light chain (RLC) more distal from the converter (Fig. 14.4f, olive). RLC phosphorylation by a tissue-specific myosin light chain kinase plays an important role for smooth muscle activation (Driska et al. 1981; Pearson et al. 1984). RLC phosphorylation triggers relaxed smooth muscle myosin heads in the "OFF" state to release from the thick filament backbone and become available for cross-bridge formation (Yang et al. 2019) (Fig. 14.4h). RLC phosphorylation can also contribute to activation of the striated muscle myosins (Woodhead et al. 2013; Zhao et al. 2009). "Unconventional" non-muscle myosins exhibit more variability in both their quantity and type of bound light chains (Heissler and Sellers 2014). For example, the neck region of myosin Va is capable of associating with ELCs or calmodulin (CaM), but not RLCs (De La Cruz et al. 2000). CaM light chains regulate myosin Va activity in a calcium-dependent manner (Wang et al. 2004).

The Molecular Mechanisms of the Actomyosin Cross-Bridge Cycle

Overview of the Actomyosin Cross-Bridge Cycle

The actomyosin cross-bridge cycle (Fig. 14.5a) relies on the allosteric coupling between the U50 (Fig. 14.4f, blue) and L50 (Fig. 14.4f, tan) actin-binding subdomains, ATP-binding active site (Fig. 14.4f, red circle), and the lever arm (Fig. 14.4f, curved bracket). This coordinated cross-talk, spanning across the entire myosin-S1, ensures that the biochemical state of the active site (i.e., nucleotide-free, or ATP, ADP.Pi, and ADP-bound) determines the structural conformation of the actin-binding subdomains and the lever arm (reviewed (Houdusse and Sweeney 2016)). The completion of myosin's force-generating powerstroke results in the formation of its so-called rigor interface with actin (Fig. 14.5a, step 1). In this state with the greatest actin affinity, the actin-binding cleft is fully closed (Fig. 14.5b, magenta arrow), the nucleotide-binding active site is empty, and the lever arm occupies its "down" position (Fig. 14.5b, green arrow) (Coureux et al. 2003; Lorenz and Holmes 2010). Execution of a subsequent powerstroke requires ATP binding to myosin. ATP binding weakens the actin-myosin interaction by opening the actin-binding cleft (Fig. 14.5c, magenta arrow) such that the two proteins no longer associate (Conibear et al. 2003) (Fig. 14.5a, light blue arrow), while the lever arm maintains its rigor configuration (Fig. 14.5c, green arrow) resulting in a post-rigor state (Fig. 14.5a, step 2) (Coureux et al. 2004). The next part of the cycle is called a recovery stroke (Fig. 14.5a, dark blue arrow),which is associated with the hydrolysis of ATP into ADP and Pi and reversal of the myosin lever arm (Fig. 14.5d, green arrow) to its Pre-powerstroke state (Fischer et al. 2005; Lombardi et al. 1995; Suzuki et al. 1998) (Fig. 14.5a, step 3). The Pre-powerstroke state is very stable in the absence of actin due to intrinsically slow release of the ADP and Pi. Thus, the Pre-powerstroke state traps ADP and Pi within the myosin active site until formation of the actomyosin complex (Fisher et al. 1995a; Houdusse et al. 2000). The Pre-powerstroke state first associates with actin via electrostatic steering, forming a weakly-bound actomyosin complex that can adopt various configurations (Thomas et al. 1995) (Fig. 14.5a, step 4). This initial association presumably gives rise to the phosphate-release state (PiR) in which Pi leaves the active site (Llinas et al. 2015) (Fig. 14.5a, step 5). PiR state isomerizes into a stereospecific actomyosin complex, which increases the affinity of the two proteins for each other due to the actin-binding cleft closure, and leads to the Pi release and a major lever arm swing (Bershitsky et al. 1997; Eisenberg and Hill 1985; Ferenczi et al. 2005; Houdusse and Sweeney 2016; Kraft et al. 2005; Llinas et al. 2015; Robert-Paganin et al. 2020) (Fig. 14.5a, step 6). These coupled allosteric changes lead to attainment of the strongly bound ADP state (Fig. 14.5a, step 7). The final part of the force-generating powerstroke takes place upon ADP release and is accompanied by a minor lever arm swing (Capitanio et al. 2006; Veigel et al. 1999; Wulf et al. 2016). With myosin strongly bound to actin in a nucleotide-free state, it has again achieved the rigor state (Fig. 14.5a, step 1) and successfully completed a powerstroke. In the following sections, we will discuss those steps in detail.

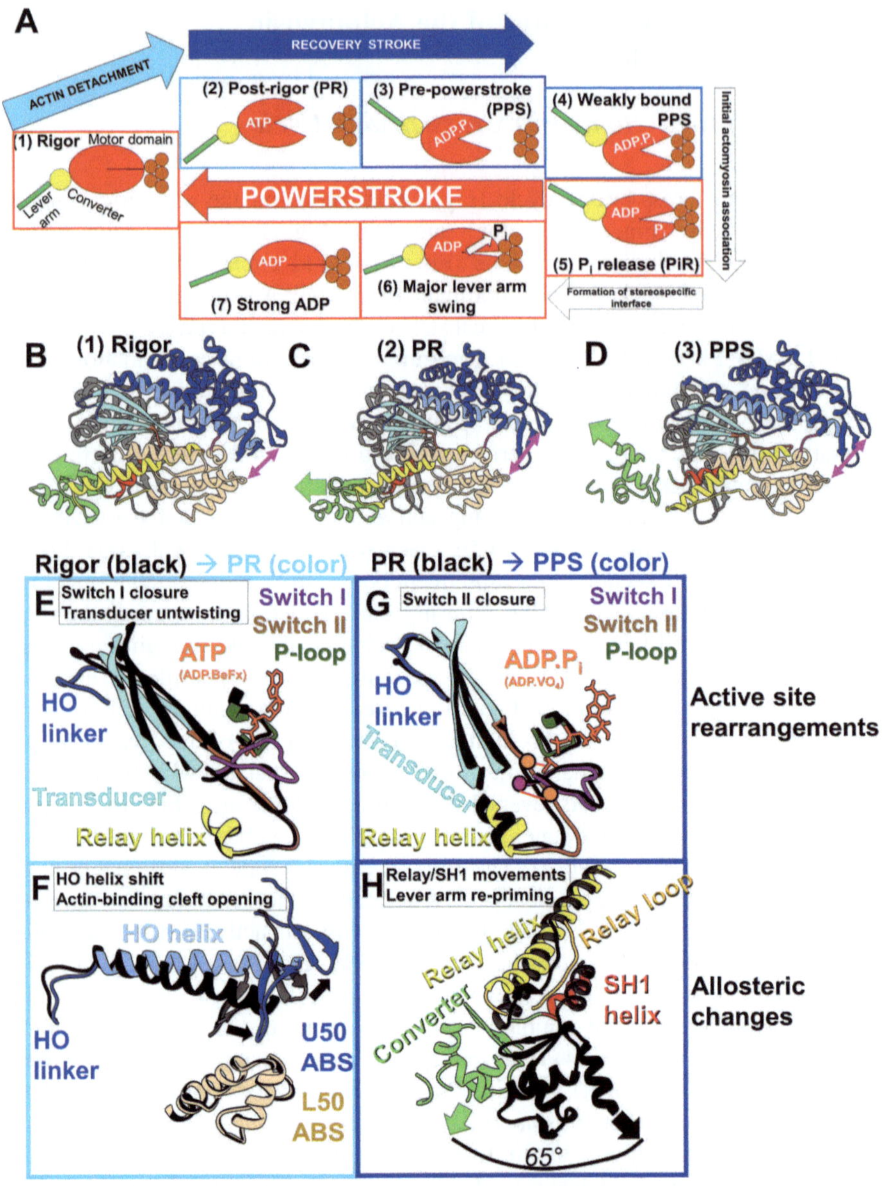

Fig. 14.5 Structural basis of cross-bridge cycle transitions from rigor to pre-powerstroke state. (**a**) Overview of steps involved in one complete actomyosin cross-bridge cycle. Each step of the cycle is marked according to its pace in the set of events required for force generation: red boxes represent steps that are part of the actual power stroke, while the blue boxes represent steps that are prerequisite for a recovery phase of the cycle. (**b–d**) The ribbon diagrams of myosin's major subdomains and connecting elements in are shown for the (**b**) rigor (PDB 1OE9; (Coureux et al. 2003)), (**c**) Post-rigor (PDB 1W7J; (Coureux et al. 2004)), and (**d**) pre-powerstroke (PDB 4ZG4; (Wulf et al. 2016)) states. Magenta double-sided arrows indicate the degree of cleft closure between

ATP-Induced Dissociation of Rigor Actomyosin

Structural studies using X-ray crystals of myosin-S1 with bound ATP homologs (Coureux et al. 2004; Coureux et al. 2003) and biochemical experiments (Conibear et al. 2003; Yengo et al. 2002) firmly established that the transition from the rigor to the post-rigor state (Fig. 14.5a, state (1) to state (2)) results in the opening of the myosin cleft (Fig. 14.5b and c, magenta arrows) and dissociation of myosin from F-actin. Conformational changes in the active site of the motor domain due to ATP binding are allosterically passed through the transducer and HO linker to the actin-binding region (Fig. 14.5e–h). Namely, in the absence of nucleotide, the interactions between the switch I (Fig. 14.4f, purple), switch II (Fig. 14.4f, brown), and the P-loop (Fig. 14.4f, dark green) within the active site (Fig. 14.4g, circled residues) stabilize the closed state of the myosin cleft required for interaction with actin. Entry of ATP (Fig. 14.5e, orange-red) into the empty nucleotide-binding pocket breaks these stabilizing interactions (Coureux et al. 2003; Shuman et al. 2014), which results in transition of the transducer from its "fully twisted" conformation in the rigor state to a "partially untwisted" state in the "post-rigor" state (Fig. 14.5e, transition from black to cyan) (Coureux et al. 2004; Takács et al. 2011; Yang et al. 2007). The alteration of transducer conformation causes a shift in the positioning of the HO linker attached to the N-terminus of the transducer (Fig. 14.5e, transition from black to blue), which in turn tilts the HO helix (Fig. 14.5f, transition from black to light blue). The HO helix (Fig. 14.5b–d, light blue) is a part of the actin-binding U50 domain (Fig. 14.5b–d, blue), hence, its repositioning results in the opening of the myosin cleft (Coureux et al. 2004; Tehver and Thirumalai 2010). The opening of the cleft disrupts the actomyosin interactions that previously held the rigor actomyosin complex.

Fig. 14.5 (continued) the U50 (blue) and L50 (tan) domains, whereas the green arrow shows the orientation of the converter and the lever arm. Within the U50, the HO helix (light blue) links the actin-binding region with the transducer (cyan). (**e–f**) Active site (**e**) and corresponding allosteric actin-binding site (**f**) rearrangements accompanying the rigor to post-rigor (PR) transition. The black ribbons correspond to the rigor state (PDB 1OE9; (Coureux et al. 2003)), while the colored ones represent the post-rigor state (PDB 1W7J; (Coureux et al. 2004)). (**g, h**) Active site rearrangements (**g**) and corresponding allosteric swing of the converter (**h**) accompanying the post-rigor (PR) to pre-powerstroke (PPS) transition. The black ribbons correspond to the PR state (PDB 1W7J; (Coureux et al. 2004)), while the colored ones represent the PPS state (PDB 4ZG4; (Wulf et al. 2016)). Novel interactions of switch II residues (brown circles) with switch I residues (purple circle) and Pi are marked with red lines. During this transition, switch II moves into its closed position (black to brown) and in doing so pulls on the N-terminal of the relax helix (black to yellow). A ~65° rotation of the converter (black to green) is marked with black line between the arrows representing the direction of the lever arm

Recovery Stroke

The current model suggests the following mechanism for transition from the post-rigor to the pre-powerstroke state. During the myosin cleft opening, switch I moves cohesively with U50 from its "open" rigor conformation (Fig. 14.5e, black) to "closed" post-rigor conformation (Fig. 14.5e, purple). This unified movement brings switch I closer to the P-loop (Fig. 14.5e, dark green) and facilitates coordination of the ATP γ-phosphate (Fig. 14.5e, orange-red) (Himmel et al. 2002; Kuhner and Fischer 2011; Reubold et al. 2003). Switch II (Fig. 14.5g, brown) re-forms a salt bridge with switch I in its new conformation, while its glycine residue makes an H-bond with the ATP γ-phosphate (Fig. 14.5g, red lines) (Smith and Rayment 1996). This new "closed" position of switch II (Fig. 14.5g, brown) energetically favors ATP hydrolysis into ADP and Pi (Li and Cui 2004). Since switch II links the active site with the relay helix (Fig. 14.5g, yellow), its movement from "open" to "closed" state results in a cascade of conformational changes involving the relay helix (Fig. 14.5h, transition from black to yellow), its flanking relay loop (Fig. 14.5h, transition from black to gold) and the adjacent SH1 helix (Fig. 14.5h, transition from black to red) (Fischer et al. 2005). As a result of these structural alterations near the interface between myosin motor domain and lever arm, the converter domain (Fig. 14.5d, green) and the attached lever arm (Fig. 14.5d, green arrow) undergo a cumulative ~65° rotation (Fig. 14.5h, black arrow), which completes of the recovery stroke (Blanc et al. 2018; Koppole et al. 2007). The aforementioned closure of switch II also seals the myosin cleft and traps ADP and Pi (Fig. 14.5g, orange-red) in the active site (Fisher et al. 1995b; Houdusse et al. 2000) to protect myosin motor domain from losing its nucleotide before interacting with actin (Shih et al. 2000; Suzuki et al. 1998). ATP-dependent detachment of the myosin molecule from actin and re-priming the lever arm into its pre-powerstroke position is prerequisite for the next powerstroke to occur, which starts from the formation of the weak interactions between the pre-powerstroke myosin molecules and actin.

Initial Actomyosin Association and Pi Release

Due to a lack of actomyosin complex structures for the initial weak-binding state (Fig. 14.5a, step 4), it is impossible to deduce specific actomyosin contacts that mediate it. However, biochemical experiments suggest that the pre-powerstroke state (Fig. 14.6a) forms its initial actomyosin interface via electrostatic interactions involving loop 2 (Fig. 14.6b, red) (Furch et al. 1998; Onishi et al. 2006a), activation-loop (Fig. 14.6b, purple) (Llinas et al. 2015; Varkuti et al. 2012), and to a lesser extent loop 3 (Fig. 14.6b, orange) (Giese and Spudich 1997; Van Dijk et al. 1999). Mutation of conserved lysine residues on loop 2 (Fig. 14.6a, red circles) hinders the rate of initial actomyosin binding (Onishi et al. 2006b), whereas the weak actomyosin state can be strengthened by addition of net positive charge to loop

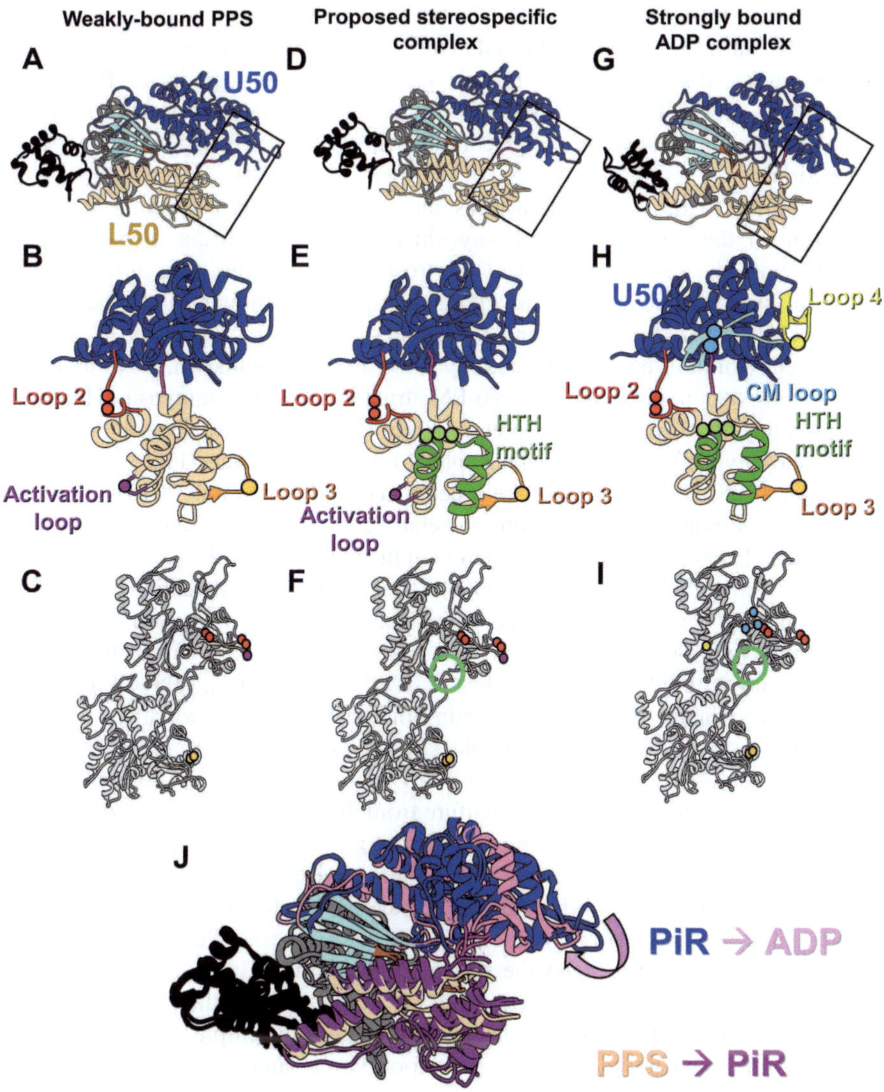

Fig. 14.6 Progressive changes to the actomyosin interface during the force-generating stages of the cross-bridge cycle. The structure and the possible actomyosin interface for the weakly bound pre-powerstroke state (**a–c**) is compared with that of the stereosphesific complex (**d–f**) and finally with the strongly bound ADP actomyosin (**g–i**). The myosin domains are depicted as follows: U50 is blue, L50 is tan, the NTD is gray, converter is black, and transducer is cyan. The black box marks the actin-binding region shown in detail in panels **b**, **e**, and **h**. Panels **c**, **f**, and **i** indicate actin residues that may interact with myosin, while panel **j** shows movements of the U50 and L50 domains upon transitions between the three myosin conformations. (**a**) The structure of the myosin in the pre-powerstroke (PPS) state (PDB 4PK4; (Llinas et al. 2015)). (**b–c**) Surface elements of myosin involved in weakly bound state include loop 2 (red), the activation loop (purple), and loop 3 (orange) (Smith and Rayment 1996). Myosin basic residues (**b**, colored circles) that were proposed (von der Ecken et al. 2016) to form electrostatic interactions with actin acidic residues

2 (Furch et al. 1998; Joel et al. 2001). On the actin side of the interface, relocation of negatively charged residues implicated in the weak-binding state (Fig. 14.6c, red and orange circles) does not impact myosin's progression through its ATPase cycle (Wong et al. 1999). Together, this evidence indicates that initial actomyosin association depends on nonspecific charge complementarity rather than a conserved set of interactions.

In contrast to the ionic interactions maintaining the weak-binding state, the formation of the stereospecific actomyosin interface (Fig. 14.5a, steps 5 and 6) requires both the previously formed electrostatic interactions (Fig. 14.6b and c) and new hydrophobic contacts involving a triplet of residues in the helix-turn-helix (HTH) motif (Fig. 14.6e and f) (Kojima et al. 2001; Onishi et al. 2006a). The structure of the stereospecific actomyosin complex is unknown. Nevertheless, a comprehensive comparison of the cryo-EM structure of rigor actomyosin (Houdusse and Sweeney 2016; von der Ecken et al. 2016) (Fig. 14.5a, step 1) with the crystal structure of myosinin the so-called phosphate-release (Pi-R) state (Fig. 14.5a, step 5) (Llinas et al. 2015) merged with the kinetics data (Muretta et al. 2015) suggests a mechanism of transition from the pre-powerstroke myosin to the stereospecific actin bound state. Loop 2 initiates the interaction of the pre-powerstroke myosin with actin (Furch et al. 1998; Onishi et al. 2006a). The subsequent interaction of the L50 activation-loop (Fig. 14.6b purple), with actin introduces a rotation of the L50 subdomain (Fig. 14.7b, black to tan) that leads to a partial closure of the myosin cleft (Varkuti et al. 2012; von der Ecken et al. 2016) (Fig. 14.6j, PPS to PiR, purple arrow). Due to the rotation of the L50 actin-binding subdomain, switch II moves into its "open" position (Fig. 14.7c, from black to brown) to form a tunnel at the active site's periphery. This tunnel is comprised of residues (Fig. 14.7c, colored circles) that interact with Pi promoting its departure from the active site (Fig. 14.7c, black Pi to plum Pi transition, marked with orange arrow). Switch I (Fig. 14.7c, purple) and the P-loop (Fig. 14.7c, dark green) are coordinated to ADP (Llinas et al. 2015) (Fig. 14.7b, orange-red), while a kinking of the SH1 helix (Fig. 14.7d, red arrow) forms stabilizing interactions with the relay helix (Fig. 14.7d, yellow) and the N-terminal subdomain to keep the converter/lever arm (Fig. 14.7d, green) in its "primed" pre-powerstroke mode (Llinas et al. 2015). Despite the partial closing of the actin-binding cleft in the Pi-release state (Fig. 14.6j, purple arrow), the actin-binding cleft is still not closed enough to form strong interactions with actin. The

---

Fig. 14.6 (continued) (**c**, same color circles). (**d**) The structure of myosin in the proposed stereospecific complex (PDB 4PFO; (Llinas et al. 2015)). (**e–f**) In the stereospecific actomyosin complex, in addition to interactions shown in **b–c**, the HTH motif (**e**, green) interacts with hydrophobic cavity between two actin subunits (**f**, green circles). The HTH motif interacts with a (green circle). (**g**) The structure of myosin-ADP that forms strongly bound complex with actin (PDB 6C1D; (Mentes et al. 2018)). (**h–i**) In addition to interactions proposed for stereospecific complex, in strongly bound ADP complex CM loop (cyan) and loop 4 (yellow) residues (**h**, colored circles) form contacts with actin residues from SD1/SD3 (**i**, same color circles). (**j**) Sequential movements of the actin-binding region during transitions from weakly to strongly bound myosin

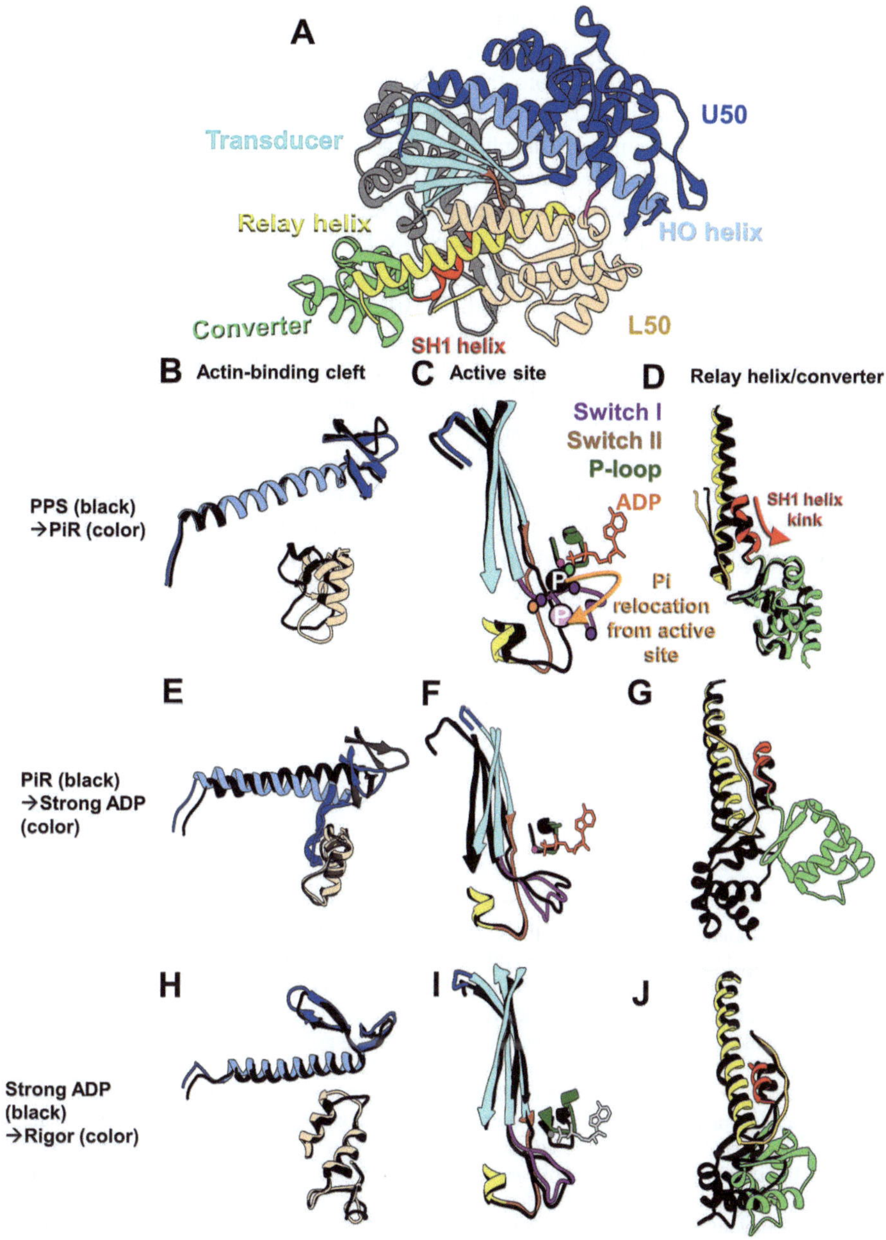

Fig. 14.7 Structural changes within myosin's three allosteric regions that accompany the force-generating powerstroke. (**a**) Schematic of the major domains and their connecting elements in the rigor state of myosin (PDB 1OE9; (Coureux et al. 2003)) are colored as in Figs. 14.4 and 14.5. Detailed views of the allosterically coupled conformational changes in the actin binding cleft (**b, e,** and **h**), active site (**c, f,** and **i**), and relay/converter (**d, g,** and **j**) upon progression of the power stroke from PPS to PiR (**b–d**), from PiR to strong ADP (**e–g**), and, finally, from strong ADP to rigor (**h–j**) are shown using the following structures: PDB 4PK4/4PFO (Llinas et al. 2015), PDB 6BNQ (Gurel

current model suggests that the weak actomyosin interactions facilitate the formation of the stereospecific actomyosin complex (Fig. 14.6d–f), which presumably forms right after the Pi dissociation from the active site (Fig. 14.7c, orange arrow) and requires the interaction of the triplet of residues in the HTH motif (Fig. 14.6e, green) with actin residues forming a hydrophobic cavity at the longitudinal actin interface (Fig. 14.6f, green circle) (Varkuti et al. 2012; von der Ecken et al. 2016). The formation of the stereospecific complex promotes the full closure of the U50-L50 cleft (Fig. 14.6j, pink arrow), which places the HCM loop (Fig. 14.6h, cyan) and loop 4 (Fig. 14.6h, yellow) in proximity to their binding sites on actin (Fig. 14.6h–i) (Gurel et al. 2017; von der Ecken et al. 2016). This leads to the formation of the strongly bound ADP actomyosin complex (Fig. 14.5a, step 6). Upon this transition the HO helix moves backward (Fig. 14.7e, from black to blue) from its position in the post-powerstroke state (Fig. 14.5f, light blue). Since the HO helix is linked to the transducer, the transducer (Fig. 14.7f, from black to cyan) and its associated switch I (Fig. 14.7f, from black to purple) also undergo conformational changes that close the Pi-binding tunnel to prevent Pi reentry. Since the aforementioned alterations in the myosin molecule are coupled with the major movement of the converter domain and lever arm (Fig. 14.7g, from black to green) via the straightening of the relay helix (Fig. 14.7g, from black to yellow) (Muretta et al. 2015), the closure of the Pi binding tunnel makes the powerstroke irreversible (gated).

Fig. 14.7 (continued) et al. 2017), and PDB 6C1D (Mentes et al. 2018). (**b–d**) Allosteric changes in myosin associated with the PPS (black) to PiR (color) transition. (**b**) Rotation of the L50 subdomain (black to tan) partially closes the actin-binding cleft. (**c**) Movement of switch II (black to brown) facilitates the relocation of Pi from the active site (orange arrow) to a binding tunnel comprised of residues from the active site elements (purple, green, and orange circles). (**d**) The SH1 helix (red) is kinked in the Pi release state (red arrow), which allows the SH1 helix (red), relay helix (yellow), and N-term subdomain (not shown) to form interactions that maintain the primed position of the converter/lever arm (green) as seen in the pre-powerstroke state (black). (**e–g**) Allosteric changes in myosin associated with the PiR (black) to strong ADP (color) transition. (**e**) Shift in the HO helix (black to light blue) produces a downward rotation of U50 actin-binding surface (black to blue), thus fully closing the actin-binding cleft. (**f**) Twisting of the transducer (black to cyan) and movement of switch I (black to purple) results in the exit of Pi from myosin. (**g**) Movements of the relay (yellow) and SH1 (red) helices produce the converter rotation (black to green) which is responsible for the major lever arm swing. (**h–j**) Allosteric changes in myosin associated with the strong ADP (black) to rigor (color) transition. (**h**) The orientation of the U50 (blue) and L50 (tan) actin-binding subdomains is unchanged. (**i**) Movement of the P-loop (black to dark green) as well as the transducer β-sheets (black to cyan) induces release of ADP (light gray). (**j**) Conformational changes in the relay (yellow) and SH1 (red) helices result in a rotation of the converter/lever arm (green) to its rigor orientation

ADP Release and Formation of Rigor Complex

The combination of complete actin-binding cleft closure, Pi release, and major lever arm swing places myosin in its strongly bound ADP state (Fig. 14.5a, step 7). In contrast to the ATP-bound post-rigor state with an open actin-binding cleft, the strong-ADP state features a fully closed cleft (Fig. 14.7h) along with tight coordination of nucleotide (Fig. 14.7i, gray atoms). Release of ADP is the final step required for force generation, thus completing one full actomyosin ATPase cycle and placing myosin back in its rigor state (Fig. 14.5a, step 1). Nucleotide release is accomplished by distortions of the transducer sheets (Fig. 14.7i, from black to cyan) which consequently distance the P-loop (Fig. 14.7i, from black to dark green) from switch 1 (Fig. 14.7i, from black to purple) (Wulf et al. 2016). While the transducer rearrangement of the strong-ADP to rigor transition does not affect the positioning of the U50 (Fig. 14.7h, blue) or L50 (Fig. 14.7h, tan) actin-binding subdomains, it alters the orientation of the relay (Fig. 14.7j, from black to yellow) and SH1 (Fig. 14.7i, from black to red) helices, which results in a minor (~9.5°) lever arm swing (Fig. 14.7j, from black to green) (Wulf et al. 2016).

The interface between the motor domain and the lever arm can influence the ADP release step and its associated lever arm swing. This is demonstrated by myosin Ib, which increases its actin attachment lifetime by 75-fold in response to a 2 pN force. Its remarkable tension sensitivity results from stalling the second lever arm rotation associated with ADP releasc (Laakso et al. 2008). High-resolution structures of myosin Ib ADP-bound and rigor states unveiled a functionally critical hydrophobic pocket at the junction of the converter and lever arm helix. In the ADP-bound state (Fig. 14.8a), loop 5 of the N-term (Fig. 14.8b, gray) docks within this pocket, thus hindering rotation of the lever arm helix (Fig. 14.8b, green) in the presence of high mechanical loads (Mentes et al. 2018). Upon lever arm rotation (Fig. 14.8c), the N-terminal extension (Fig. 14.8d, pink) of myosin Ib is no longer sterically blocked from this pocket, and its association with the lever arm helix (Fig. 14.8d, green) drives the N-term domain rotation required for ADP release (Mentes et al. 2018). This proposed structural mechanism fits with kinetic data, which demonstrates that deletion of the myosin Ib N-terminal extension delays ADP release nearly tenfold (Shuman et al. 2014). In contrast to the lever arm movement, the actin-binding interface undergoes minimal changes during the strong ADP to rigor transition (Fig. 14.8e) (Gurel et al. 2017; Mentes et al. 2018; Wulf et al. 2016). The specific contacts maintaining the rigor interface have been visualized at near-atomic resolution and are discussed in detail within the following section.

The Structure of the Rigor Complex

Our structural knowledge of the actomyosin complex is limited to the strong ADP (Fig. 14.5a, step 7) and rigor (Fig. 14.5a step 1) states. For the rigor state, several

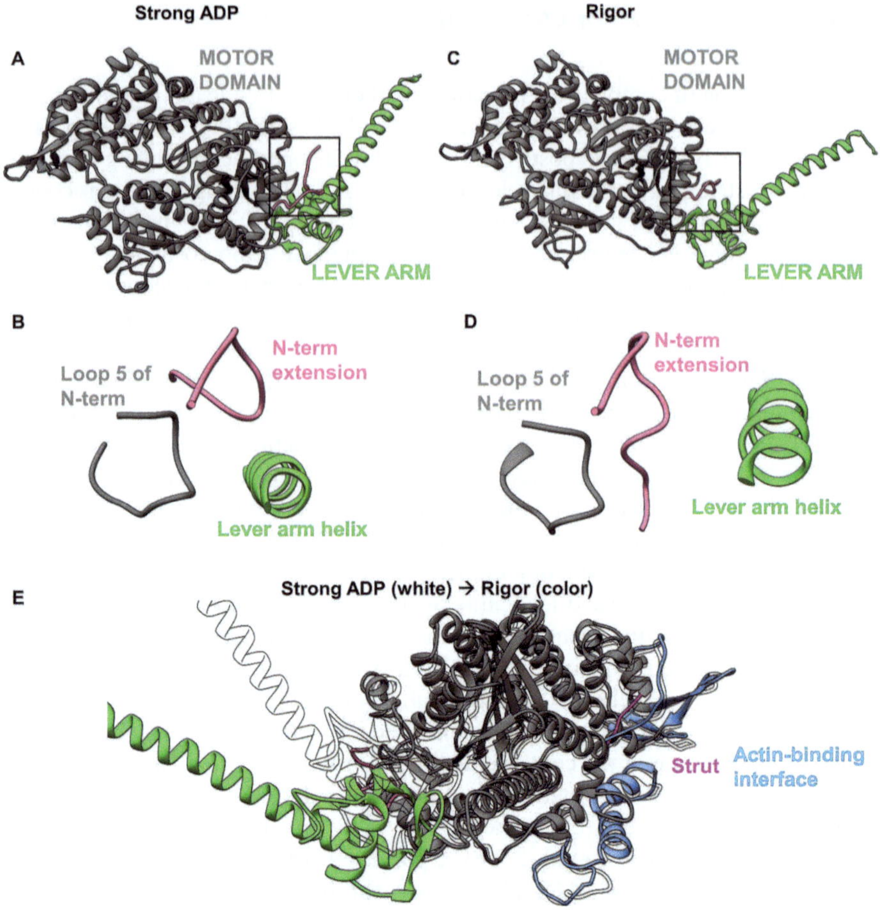

Fig. 14.8 Structural basis of the force-sensitive strong ADP to rigor transition in myosin Ib. (**a** and **c**) Conformations of the motor domain (gray), converter/lever arm (green), and N-terminal extension (pink) in the strong ADP state (PDB 6C1D; (Mentes et al. 2018)) (**a**) and rigor state (PDB 6C1H; (Mentes et al. 2018)) (**c**). The boxed region indicates the interface between the motor domain and converter/lever arm that is depicted in panels **b** and **d**. (**b**) In the presence of mechanical forces restraining the rotation of the myosin Ib lever arm, loop 5 element of the N-terminal subdomain (gray) is buried within a hydrophobic pocket at the base of the lever arm helix (green), while the N-terminal extension (pink) is located outside of this hydrophobic pocket. (**d**) Upon lever arm rotation, the N-terminal subdomain N-terminal extension (pink) moves to a position that sterically prevents the interaction of loop 5 (gray) with the lever arm helix (green), while promotes ADP release. (**e**) Comparison of the major domain orientations in the myosin Ib strong ADP (white) and rigor (color) states. The converter and lever arm undergo a downward rotation that completes the powerstroke (white to green). Since the actin-binding cleft remains fully closed during this transition, the actin-binding interface is relatively unchanged (white to light blue)

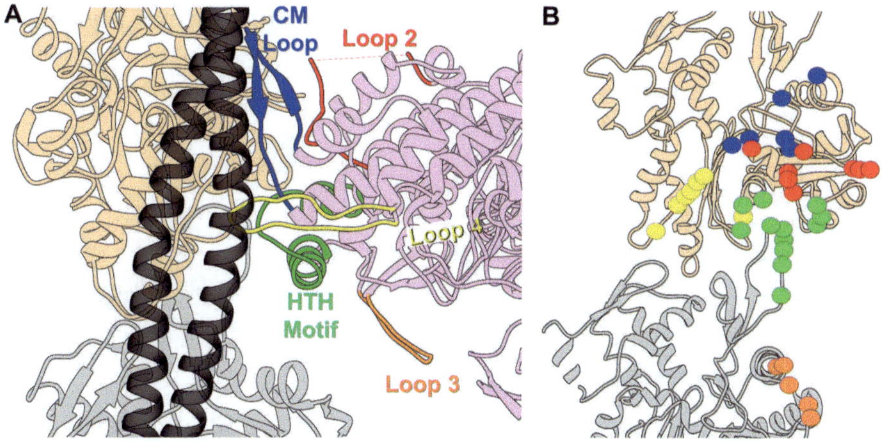

Fig. 14.9 General molecular details of the rigor actomyosin interface. (**a**) Atomic model of the cardiac actomyosin complex in the rigor state (PDB 7JH7, (Risi et al. 2021a)). The myosin motor domain is colored in plum, while the two consecutive actin subunits along the actin filament long-pitch helix that form myosin-binding interface are shown in tan and gray. The tropomyosin cable regulating myosin's access to its binding site of the actin filament is black. The secondary structural elements of the actin-binding region of myosin are presented using primary colors: CM loop (blue), loop 2 (red), loop 4 (yellow), HTH motif (green), and loop 3 (orange). (**b**) Aggregated myosin footprint on the actin surface derived from actomyosin complexes of cardiac myosin (PDB 7JH7, (Risi et al. 2021a)), myosin IIC (PDB 5JLH; (von der Ecken et al. 2016)), myosin VI (PDB 6BNP; (Gurel et al. 2017)), and myosin Ib (PDB 6C1H; (Mentes et al. 2018)). Colored circles represent actin residues, which interact with myosin in aforementioned rigor actomyosin complexes. The color of each circled actin residue corresponds to the myosin actin-binding element (shown in panel **a**) that interacts with it

actomyosin complexes have been visualized by cryo-EM at near-atomic resolution, which allowed unambiguous mapping of the side chains of amino acids comprising the main actomyosin interactions (Fig. 14.9a). This set of structures includes cardiac myosin (3.8 Å; (Risi et al. 2021a)) and myosin IIC (3.9 Å; (von der Ecken et al. 2016))—representing the class II myosins—as well as myosin VI (4.6 Å; (Gurel et al. 2017)) and myosin Ib (3.9 Å; (Mentes et al. 2018))—representing the "unconventional" myosins. Comparison of these four isoforms altogether reveals that in the Rigor state, the actin-binding elements (Fig. 14.9a, primary colors) of myosin (Fig. 14.9a, plum) generally interact with SD1 and SD3 of the upper actin (Fig. 14.9a, tan) as well as SD2 and SD1 of the lower actin (Fig. 14.9a, gray). The core of the "actin footprint" shared by numerous myosins is formed by interactions with the CM loop (Fig. 14.9b, blue circles), HTH motif (Fig. 14.9b, green circles), and loop 2 (Fig. 14.9b, red circles) that are responsible for global positioning of the motor on its actin track. Peripheral contributions from loop 4 (Fig. 14.9b, yellow circles) and loop 3 (Fig. 14.9b, orange circles), which are variable among the myosin isoforms, can optimize the specific details of the actomyosin interaction. By examining the actin side of the interface (Fig. 14.9b, detailed in Figs. 14.10, 14.11, 14.12, 14.13, and 14.14), one can clearly observe a consistent "zone" of actin interactions

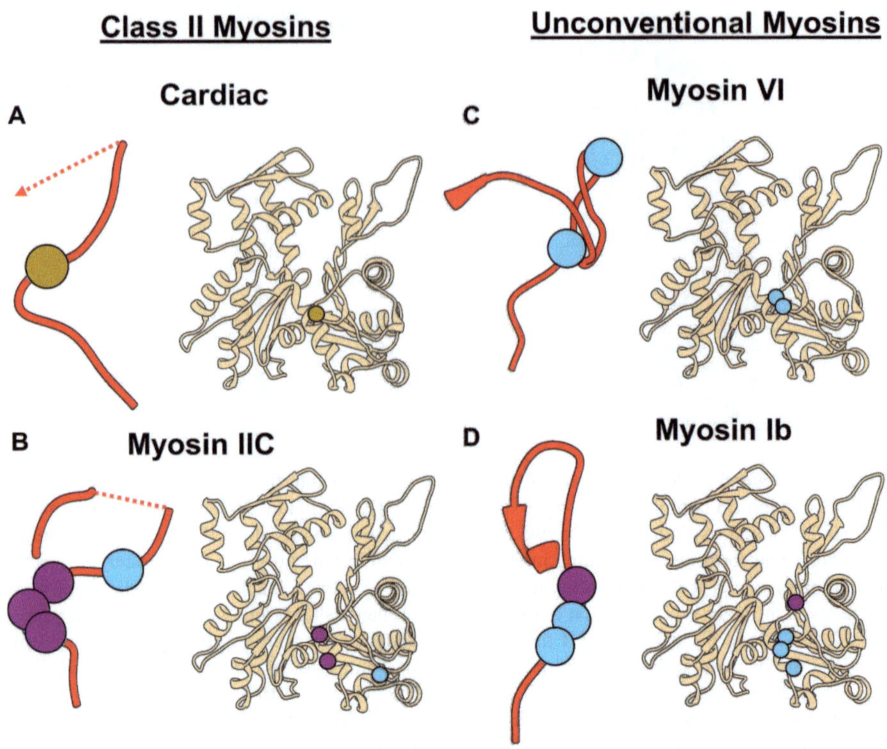

Fig. 14.10 Interactions of loop 2 of myosin with actin observed in conventional and unconventional actomyosin complexes differ. Loop 2 of myosin is red, while actin subunit from Fig. 14.9 is shown in tan. Residues of myosin and actin observed to interact are depicted using symbols of the same color and shape. (**a**) Loop 2 of cardiac myosin (PDB 7JH7, (Risi et al. 2021a)) has the weakest interaction with actin by forming one hydrogen bond with actin SD1 (brown circles). (**b**) Loop 2 of myosin IIC (PDB 5JLH; (von der Ecken et al. 2016)), in addition to a set of hydrophobic interactions with actin at its base (purple circles), makes an electrostatic interaction with the actin N-terminus (blue circles). (**c**) Loop 2 of myosin VI (PDB 6BNP; (Gurel et al. 2017)) has only electrostatic interactions with actin *via* its basic residues interacting with an acidic patch of actin located in SD1 (blue circles). (**d**) Loop 2 of myosin Ib (PDB 6C1H; (Mentes et al. 2018)) forms a mixture of electrostatic (blue circles) and hydrophobic (purple circles) interactions with actin SD1

across the isoforms despite the variations of the participating myosin surface elements.

Loop 2 (Fig. 14.9a, red) exhibits substantial variations in length and flexibility among myosin isoforms, as demonstrated by the cryo-EM structures (Fig. 14.10a–d). Positively charged residues of loop 2 form the initial actomyosin interactions (Fig. 14.6b, c) (Furch et al. 1998; Onishi et al. 2006a). Interestingly, for all examined myosin isoforms except cardiac, these ionic bonds also exist in the rigor state. Basic residues of loop 2 in myosins IIC, VI, and Ib interact with acidic patches within actin SD1 (Fig. 14.10b–d, blue circles). Additionally, in myosin IIC a hydrophobic cluster

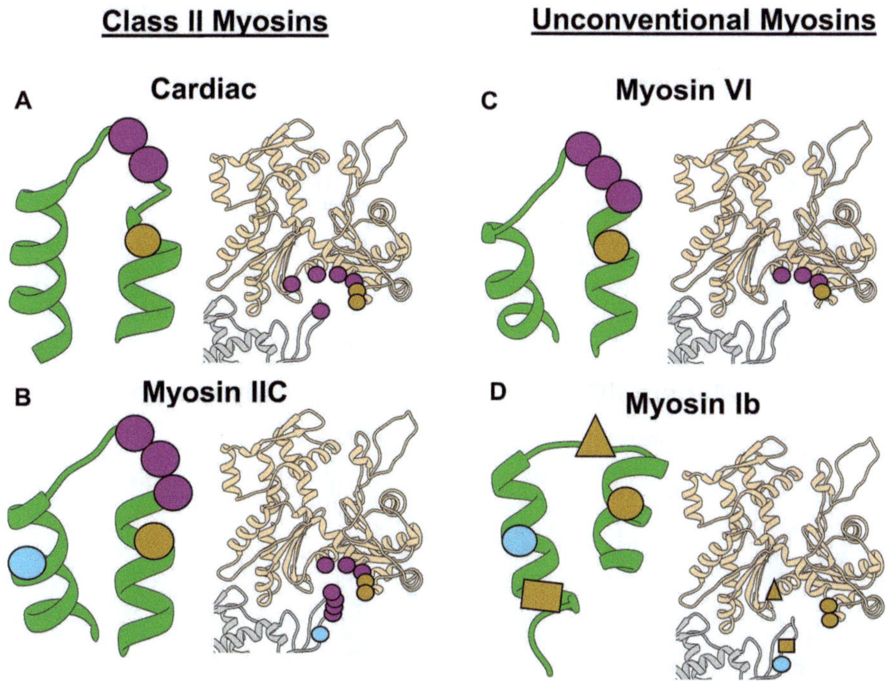

Fig. 14.11 Interactions of HTH motif of myosin with actin are conserved in conventional and unconventional actomyosin complexes except for myosin Ib. HTH motif of myosin is in green, while the two subunits of the actin filament (from Fig. 14.9) are depicted in tan and gray. Residues of myosin and actin observed to interact are depicted using symbols of the same color and shape. (**a**) HTH motif of cardiac myosin (PDB 7JH7, (Risi et al. 2021a)) interacts with a hydrophobic cavity formed by residues in SD1/SD3 of upper actin and SD2 of lower actin (purple circles). A hydrogen bond exists between the N-terminal helix of HTH motif and actin SD1 (brown circles). (**b**) HTH motif of myosin IIC (PDB 5JLH; (von der Ecken et al. 2016)) interacts with a hydrophobic cavity formed by residues in SD1/SD3 of upper actin and SD2 of lower actin (purple circles), while a hydrogen bond exists between the N-terminal helix of HTH motif and actin SD1 (brown circles). In addition, electrostatic interaction exists between the C-terminal helix of HTH motif and actin SD2 (blue circles). (**c**) HTH motif of myosin VI (PDB 6BNP; (Gurel et al. 2017)) makes a set of hydrophobic interactions with actin (purple circles), and forms a hydrogen bond between the N-terminal helix of HTH motif and actin SD1 (brown circles). (**d**) The interactions of the HTH motif of myosin Ib with actin (PDB 6C1H; (Mentes et al. 2018)) are very different from the above-mentioned isoforms: it forms H-bonds with SD1 (brown circles) and SD3 (brown triangles) of the upper actin (tan) as well as SD2 (brown rectangles) of the lower actin (gray). Additionally, an electrostatic contact is formed between the C-terminal helix of HTH motif and actin SD2 (blue circles).

of residues at the loop 2 base interacts with the "hinge" region between SD1 and SD3 (Fig. 14.10b, purple circles). These hydrophobic interactions may also exist for myosin VI but cannot be confirmed due to structural variability of its loop 2 (Gurel et al. 2017). Myosin Ib loop 2 can also hydrophobically interact with actin, due to a unique leucine residue located next to the conserved positively

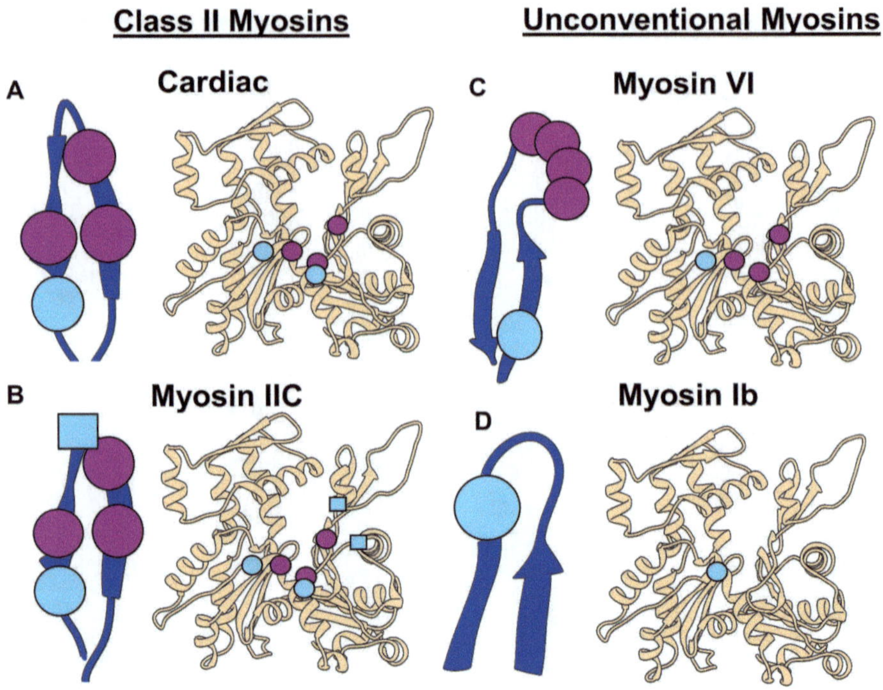

Fig. 14.12 The interface between CM loop of rigor myosin and actin observed in conventional and unconventional actomyosin complexes is different. CM loop of myosin is in blue, while actin subunit from Fig. 14.9 is tan. Residues of myosin and actin observed to interact are depicted using symbols of the same color and shape. (**a**) CM loop of cardiac myosin (PDB 7JH7, (Risi et al. 2021a)) forms hydrophobic interactions with a pocket near the actin SD1/SD3 junction (purple circles), while basic residue at its base interacts with acidic residues on actin SD1 and SD3 (blue circles). (**b**) CM loop of myosin IIC (PDB 5JLH; (von der Ecken et al. 2016)) forms similar hydrophobic interactions with actin (purple circles). In addition, basic residue at its base interacts with acidic residues on actin (blue circles). Additional salt bridge between the tip of myosin IIC CM loop and SD1/SD2 of actin is formed (blue squares). (**c**) Residues in the CM loop of myosin VI (PDB 6BNP; (Gurel et al. 2017)) form a set of hydrophobic interactions with actin residues near the SD1/SD3 junction (purple circles). In addition, basic residue at the CM loop base interacts with an acidic residue on actin SD3 (blue circles). (**d**) The CM loop of myosin Ib (PDB 6C1H; (Mentes et al. 2018)) forms just one electrostatic interaction between its tip and the hinge region of actin (blue circles)

charged region (Fig. 14.10d, purple circles). Loop 2 of cardiac myosin displays attenuated actin interactions relative to the other isoforms as a single hydrogen bond with actin residue maintains this interface (Fig. 14.10a, brown circles).

After the initial actin interactions with loop 2 have formed, the HTH (Fig. 14.9a, green) is brought close to its binding site at actin's hydrophobic groove (Fig. 14.6f, green circle) with assistance from the activation loop (Fig. 14.6e, purple). This process is critical for the transition from a weakly bound to stereospecific actomyosin complex, which in turn facilitates the structural transitions that bring the complex into its rigor state. In sharp contrast to the variability displayed by loop

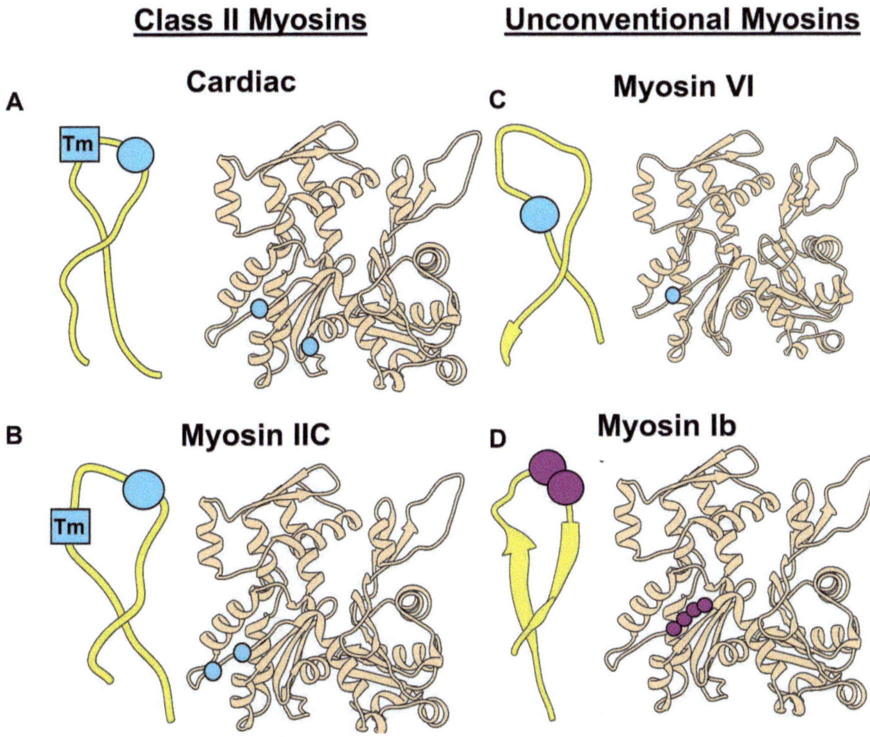

Fig. 14.13 The interactions of loop4 of rigor myosin with actin varies between the actomyosin complexes. Loop 4 of myosin is depicted in yellow, while actin subunit from Fig. 14.9 is tan. Residues of myosin and actin observed to interact are depicted using symbols of the same color and shape. (**a** and **b**) Loops 4 of cardiac myosin (PDB 7JH7, (Risi et al. 2021a)) and myosin IIC (PDB 5JLH; (von der Ecken et al. 2016)) contain acidic residues at their tips that form salt bridges with actin SD3 (blue circles). In addition, loop 4 in both isoforms electrostatically interacts with the regulatory tropomyosin (Tm) cable (blue squares). (**c**) Loop 4 of myosin VI (PDB 6BNP; (Gurel et al. 2017)) forms a salt bridge with actin SD3 (blue circles). (**d**) Loop 4 of myosin Ib (PDB 6C1H; (Mentes et al. 2018)) forms a unique set of hydrophobic interactions between its tip and a hydrophobic cluster in actin SD3 (purple circles)

2, the HTH motif represents the most conserved part of the actomyosin interface. In the HTH of all four myosin isoforms, a conserved glutamic acid residue in the N-terminal helix forms an H-bond with actin SD1 (Fig. 14.11a–d, brown circles). Additionally, all isoforms except myosin Ib feature hydrophobic interactions of the HTH turn with upper actin SD1 and SD3 and the lower actin's D-loop (Fig. 14.11a–c, purple circles). Biochemical mutagenesis studies have confirmed the importance of these conserved hydrophobic residues for myosin's ATPase activity and motility (Onishi et al. 2006a). For myosin Ib, the HTH interface lacks hydrophobic interactions, which are substituted by a set of hydrogen bonds (Fig. 14.11d, brown squares and triangles). Similar to myosin IIC, the Ib isoform makes a salt bridge with the D-loop (Fig. 14.11d, blue circles).

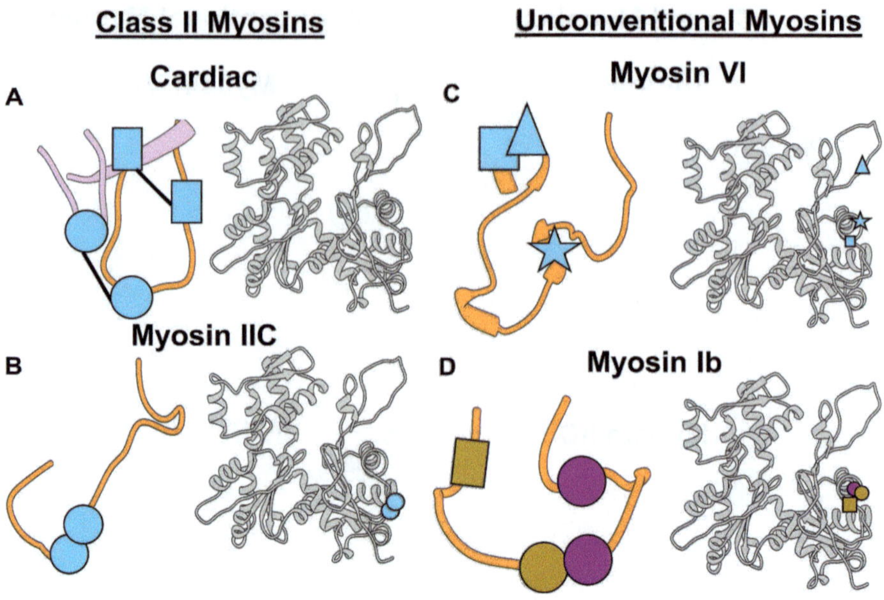

Fig. 14.14 Loop 3 of cardiac myosin does not interact with actin, while all the other isoforms make non-conserved interactions with actin in rigor actomyosin. Loop 3 of myosin is in orange, while actin subunit from Fig. 14.9 is in gray. Residues of myosin and actin observed to interact are depicted using symbols of the same color and shape. The left column. (**a**) Loop 3 of cardiac myosin (PDB 7JH7, (Risi et al. 2021a)) does not interact with actin, because it is pulled toward the body of myosin (plum) by a pair of internal salt bridges (blue circles and blue squares connected with black lines). (**b**) Loop 3 of myosin IIC (PDB 5JLH; (von der Ecken et al. 2016)) forms charged interactions with SD1 of the lower actin subunit (blue circles). (**c**) Loop 3 of myosin VI (PDB 6BNP; (Gurel et al. 2017)) electrostatically interacts with SD1 (blue squares) and SD2 (blue triangles) of the lower actin subunit. A salt bridge with SD1 is formed during the Strong ADP to Rigor transition (blue stars). (**d**) Loop 3 of myosin Ib (PDB 6C1H; (Mentes et al. 2018)) forms electrostatic interactions (purple circles) and hydrogen bonds (brown circles and squares) with actin's SD1

The transition from the weakly bound myosin to higher affinity actomyosin complexes (e.g., strong ADP and rigor) involves a rotation of the U50 that closes the actin-binding cleft. This U50 movement brings the CM loop (Fig. 14.9a, blue) into position to form several key interactions characteristic to the rigor state. The class II myosins exhibit very similar interfaces of their CM loops with actin, primarily governed by a trio of hydrophobic residues on the antiparallel β-sheets (Fig. 14.12a–b, purple circles). Additional stabilization at the base of the CM loop is provided by electrostatic contact of a conserved lysine residue with a pair of negatively charged residues at the junction of SD1 and SD3 (Fig. 14.12a–b, blue circles). Myosin IIC contains an isoform-specific salt bridge at the CM loop tip, which associates with the region near the SD1–SD2 junction (Fig. 14.12b, blue squares). The myosin VI CM loop possesses a nearly identical actin footprint to the class II myosins, despite the difference in the specific contacts that form this interface. The hydrophobic interactions are clustered at the tip of the loop (Fig. 14.12c, purple circles), while an electrostatic interaction also exists in the base but on the opposite strand (Fig. 14.12c, blue circles). In contrast to the multiple interactions sustaining these CM loop interfaces, a single glutamic acid of myosin Ib is responsible for its CM loop interaction with a nucleotide-coordinating lysine residue of actin (Fig. 14.12d, blue circles). This CM loop residue also serves as a phosphorylation site important for the activation of class I myosins (Bement and Mooseker 1995).

Loop 2, the HTH motif, and the CM loop are the three myosin surface elements that form the core of its actin interface, which spans across the actin filament's longitudinal interface. Auxiliary contributions to this core footprint come from loop 4 (Fig. 14.9a, yellow) of the U50 and loop 3 (Fig. 14.9a, orange) of the L50. In the case of loop 4, minimal variability is observed among the actin interfaces of each isoform. For class II myosins, a conserved acidic residue at its tip mediates electrostatic interaction with positively charged actin side chains (Fig. 14.13a, b, blue circles). Furthermore, an arginine residue in both cardiac myosin and myosin IIC (Fig. 14.13a, b, blue squares) forms charged interactions with the tropomyosin (Tm) cable (Fig. 14.9a, black), which blankets the actin filament in numerous cell types and regulates the availability of myosin's binding sites on actin. For myosin VI, an acidic residue located closer to the loop 4 base interacts with the same basic residue on actin as the class II myosins (Fig. 14.13c, blue circles). The tip of myosin Ib loop 4 interacts with the same zone on actin as observed for class II myosins; however, these contacts are hydrophobic rather than electrostatic (Fig. 14.13d, purple circles).

Unlike the relative conservation displayed by loop 4, myosin isoforms demonstrate great variability in their loop 3 actin interfaces. Loop 3 represents the only actin-binding element that does not interact with the upper actin subunit. For myosin IIC, a pair of basic residues interacts with the negatively charged "Milligan contact" (Fig. 14.14b, blue circles) (Milligan 1996). A different set of charged residues on myosin VI forms electrostatic contacts with a distinct zone on actin that includes the D-loop (Fig. 14.14c, blue squares and triangles). Furthermore, one of these interactions (Fig. 14.14c, blue stars) is formed upon the transition from strong ADP to rigor (Gurel et al. 2017). In contrast to the salt bridges maintaining the loop 3 actin interfaces of myosins IIC and VI, myosin Ib exhibits an assortment of hydrophobic interactions (Fig. 14.14d, purple circles) and H-bonds (Fig. 14.14d, brown circles and squares) with a region just above the Milligan contact. Loop 3 of cardiac myosin represents a dramatic departure from the other isoforms since it does not appear to interact with actin in the rigor state. Instead, a pair of salt bridges between loop 3 and the body of cardiac myosin (Fig. 14.14a, blue circles and squares) keeps it sequestered from the actin interface.

Variations in the Actin-Binding Elements of Various Myosin Isoforms Tune Up Their Kinetic Parameters

A generally conserved blueprint with isoform-specific details maintains the strong actin-binding states of the cross-bridge cycle. The variant sequences and structures of these actin-binding elements between myosin classes contribute to their kinetic properties, as evidenced by altered ATPase cycle kinetics upon chimeric replacement of loop 2 (Uyeda et al. 1994). Thus, these modifications are a source of kinetic fine-tuning during the interaction of myosin motors with a highly conserved actin track, which in turn biases myosin isoforms for certain biological functions (Robert-Paganin et al. 2020; Spudich 1994). Since the strong ADP and rigor actin-binding interfaces are similar, differences in the transducer (Clark et al. 2005; Sweeney et al. 1998) and lever arm (Mentes et al. 2018) are likely the most important for tuning the ADP release step. By contrast, the duration of actin attachment (i.e., duty ratio) and the sliding velocity can be modulated by variant details of the common actin-binding architecture (Robert-Paganin et al. 2020).

The set of isoforms for which high-resolution rigor actomyosin complexes are available represents a mixture of kinetically and functionally diverse myosins, which can be divided into broad groupings based on the parameters of their cross-bridge cycles (reviewed in (Bloemink and Geeves 2011)). The class II myosins serve contractile functions within cells by aggregating into filaments (Fig. 14.4c). However, cardiac myosin moves actin filaments at a velocity that is over an order of magnitude greater than that of myosin IIC (Heissler et al. 2013; Yamashita et al. 1994). While the HTH motif (Fig. 14.11a, b), CM loop (Fig. 14.12a, b), and loop 4 (Fig. 14.13a, b) exhibit largely similar actin interfaces between these two isoforms, significant variability can be observed with respect to loops 2 (Fig. 14.10a, b) and 3 (Fig. 14.14a, b). The reduced interactivity of these elements in cardiac myosin, compared with myosin IIC, would be beneficial for the rapid attachment and detachment of myosin head ensembles. Thus, these differences in the actin-binding interface may contribute to its greater sliding velocity and lower duty ratio. In contrast to this mechanism which depends on short-lived Rigor interactions, the processive myosin VI motor depends on persistent actin attachment of its dimerized heads for its cargo-transporting function (Fig. 14.4d). Accordingly, its duty ratio of 0.9 (i.e., spends 90% of the cross-bridge cycle bound to actin) is among the highest observed in the entire myosin superfamily (De La Cruz et al. 2001; Robblee et al. 2004). Out of all the characterized rigor actomyosin complexes, loop 3 of myosin VI (Fig. 14.14c) possesses the most extensive contacts with the lower actin subunit. This could play a part in the stability of its strong binding to actin, an important component of its processive mechanism in which at least one myosin head is bound to actin at all times (Sweeney et al. 2007). Finally, myosin Ib represents the strain-sensing class I myosins, which can anchor themselves between the actin cytoskeleton and the cell membrane for extended periods in response to mechanical loads (Fig. 14.4b). Myosin Ib dynamically adjusts to an imposed tension by increasing its actin attachment lifetime, such that it transitions from a low (0.2) to high (0.9) duty ratio motor (Laakso et al. 2008). Despite its ability to prolong its actin attachment in response to force, the rigor interface of myosin Ib is notably diminished compared to other isoforms. This is especially true for the CM loop and HTH motif, both integral components of the core interface. Thus, kinetic tuning of myosin Ib for its force-sensing cellular role is likely controlled by elements outside the actin-binding interface, such as its N-terminal extension near the motor domain-lever arm interface (Fig. 14.8a–e, pink) (Mentes et al. 2018; Shuman et al. 2014).

Cellular and Physiological Contexts of Actomyosin Interaction

Muscle Contraction

Although muscle contraction is the biological process typically associated with the actomyosin interactions, it represents a late evolutionary arrival to the functional repertoire of interaction between actin and myosin. Myosin participates in cargo transport and cytoskeletal dynamics in nearly all eukaryotes, and it was adapted for powering large-scale motility of metazoans upon the origin of functional muscle tissue. Striated (i.e., skeletal and cardiac) and smooth muscles differ with respect to the assembly of their contractile apparatuses as well as the regulatory mechanisms governing their contractions. The characteristic functional unit of a striated muscle fiber is the sarcomere, whose lateral boundaries are formed by the Z-lines that anchor myosin-based thick filaments and actin-based thin filaments (Fig. 14.15a). During striated muscle contraction, the thin filaments are pulled toward the M-line at the sarcomere's center by the formation and detachment of myosin cross-bridges, which consequently reduces the distance between the two Z-lines (Fig. 14.15b). Thus, the molecular basis of striated muscle contraction consists of sarcomeric shortening as the thick and thin filaments slide past each other (Huxley and Hanson 1954). This sliding motion of thick and thin filaments also occurs in the smooth muscle cells found in blood vessels, airways, and the gut. However, their contractile apparatus is not a tightly ordered sarcomere but rather a more flexible lattice of thin filaments and variable-length thick filaments (Liu et al. 2013) interconnected by dense bodies which are functionally equivalent to the Z-disks of striated muscle (Herrera et al. 2005; Zhang et al. 2010).

A functional property mutual to both striated and smooth muscle is their force-velocity (F-V) relationship (Fig. 14.15c), which entails that resistive load applied to muscle decreases its shortening velocity (Hill 1938). For forces greater than the isometric stall force at which no net movement of actin occurs (Fmax), the system reproduced the negative velocity that defines eccentric muscular contraction (Debold et al. 2005). Single-molecule analysis of cardiac actomyosin cross-bridges reveals load dependence of ADP release, and this parallels the inverse relationship between cardiac contractile velocity and external load (Sung et al. 2015). However, the

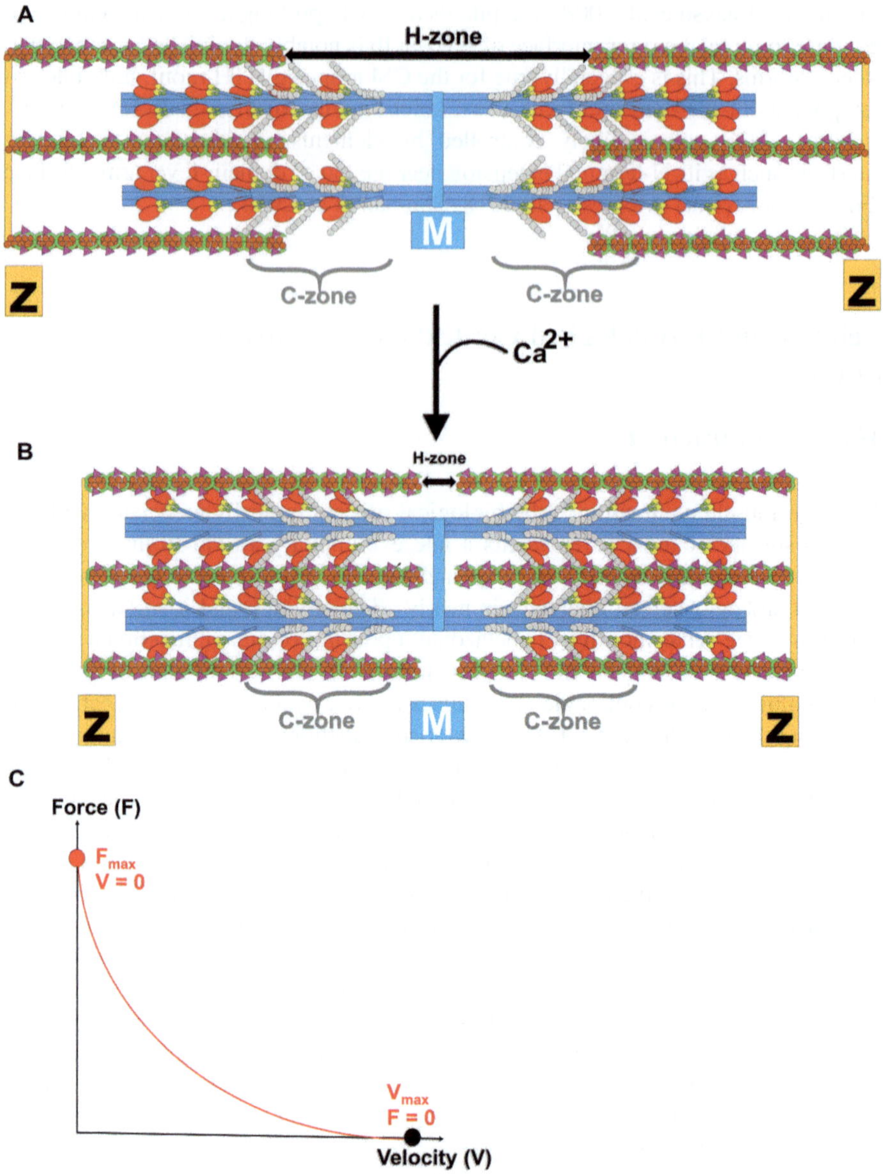

Fig. 14.15 Structural and functional properties of the striated muscle sarcomere. (**a, b**) Schematic view of sarcomeric relaxation and contraction, a physiological process driven by interactions between actin-based thin filaments (brown circles) and myosin-based thick filaments (blue rectangles). The myosin heads are red ovals. The calcium-sensing troponin complex (pink) and tropomyosin cable (green) associated with the actin filaments regulate binding of myosin. Thin filaments are anchored on both sides by the Z-lines (gold rectangles), which form the lateral boundaries of a single sarcomeric unit. Myosin-binding protein C (gray), located within the sarcomeric C-zones (gray brackets), interacts with both thick and thin filaments. (**a**) Sarcomere in its relaxed (i.e., lengthened) state has its Z-lines distanced from each other such that a large gap between the thin

degree of coupling between load and ADP release for β-cardiac myosin (ventricular isoform) is far less than that of myosin V (Veigel et al. 2005) or myosin Ib (Laakso et al. 2008), and this likely contributes to cardiac muscle's efficient systolic power output against the resistance of ventricular filling (Greenberg et al. 2014). Compared to striated muscle, smooth muscle exhibits one-tenth the maximum shortening velocity (Vmax) (Warshaw et al. 1990) and three- to fourfold greater force development under maximal load (Harris et al. 1994). The higher duty ratio of smooth muscle myosin is the biochemical basis of these mechanical differences (Guilford et al. 1997). In tissues with a mixture of different muscle myosin isoforms, the cross-bridges of the slower isoform limit the velocity of the faster ones by exerting a load on them (Warshaw et al. 1990).

In addition to variations in the contractile apparatus structure and F-V relationship, smooth and striated muscles also differ with respect to their functional regulatory mechanisms. Smooth muscle activation is achieved by phosphorylation of Ser19 within its RLC by a smooth muscle-specific myosin light chain kinase (Pearson et al. 1984) (Fig. 14.4h). RLC phosphorylation triggers relaxed myosin heads in the "off" state (i.e., interacting heads motif or IHM configuration) to release from the thick filament backbone and become available for cross-bridge formation. RLC phosphorylation can also release striated muscle myosins from the relaxed IHM state (Zhao et al. 2009). However, due to an additional regulatory layer contained in skeletal and cardiac thin filaments, RLC phosphorylation only serves a modulatory role. In striated muscle, the accessibility of myosin's binding sites on actin filaments is controlled by Ca^{2+} binding to the tropomyosin-troponin complex (Drabikowski et al. 1968; Ebashi 1972; Lehrer and Morris 1982; Martyn et al. 2001). Together, actin and its regulatory apparatus are referred to as the thin filament (Fig. 14.16a, b). The tropomyosin (Tm) cable is a double-stranded α-helical coiled-coil that wraps around actin filaments (Fig. 14.16a, b, blue ribbons) in a diverse array of cell types (Brown et al. 2001; Parry 1975). The Tm cable is associated with the calcium-sensing troponin complex (Fig. 14.16a, b, green, red, and purple ribbons) spaced at intervals of every seven actin subunits. Early kinetic experiments established that Tm sterically blocks actomyosin interactions in the absence of Ca^{2+}, whereas the presence of Ca^{2+} increases the proportion of activated thin filaments capable of binding myosin (McKillop and Geeves 1993).

Recent cryo-EM studies of both native and reconstituted cardiac thin filaments have clarified the molecular details of Ca^{2+}-mediated thin filament regulation (Risi

Fig. 14.15 (continued) filament ends (H-zone) is visible (black double-sided arrow). (**b**) Sarcomere in its contracted (i.e., shortened) state has thin filaments translocated toward the sarcomere's midline (M-line; blue rectangle), which reduces the length of the H-zone (black double-sided arrow). (**c**) Hyperbolic relationship between the force and velocity of muscular contraction. The red circle located at the y-intercept (Fmax) represents the "stall force" at which sarcomere length remains fixed and the sliding velocity of the thin filaments is equal to zero. The black circle located at the x-intercept (Vmax) represents the maximum shortening velocity of the sarcomere, which can be reached when the force applied to it equals zero

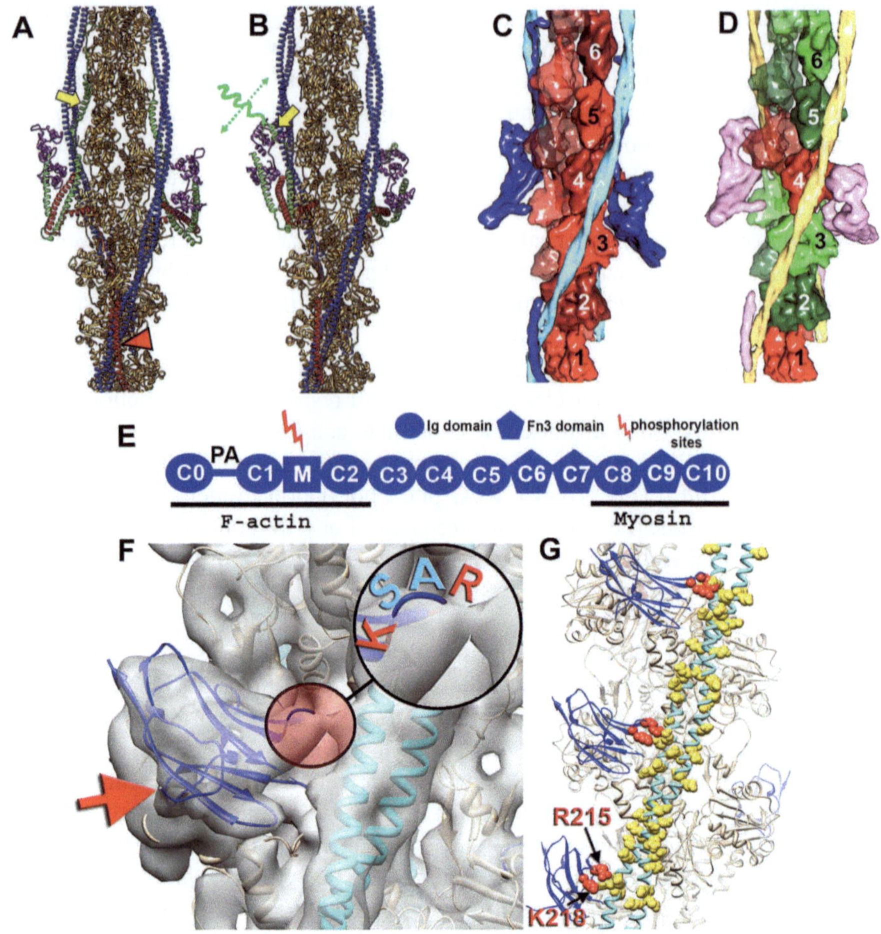

Fig. 14.16 Regulatory components of the cardiac sarcomere. (**a, b**) Conformational changes of the troponin-tropomyosin complex during the transition from Ca^{2+}-free state (PDB 7KO4; (Risi et al. 2021a)) to Ca^{2+}-bound state (PDB 7KO5; (Risi et al. 2021a)). The actin subunits are tan, while Tm cable is blue. The regulatory Tn complex consists of three subunits—Ca^{2+}-binding TnC (purple), Tm-binding TnT (red), and inhibitive TnI (green). The N-terminal helix of TnT binds to the junction region between the two adjacent Tm molecules across the actin strand to stabilize it (red arrowhead). (**a**) At low Ca^{2+} levels, the C-terminal region of troponin I (green) is tightly bound to actin (tan) and Tm (blue), as indicated by the yellow arrow. The Ca^{2+}-binding TnC (purple) is in its closed conformation. The Tm cable (blue) is positioned such that each individual actin-binding site is inaccessible to myosin. (**b**) Upon Ca^{2+} binding, the cleft of TnC opens and TnI "regulatory switch" helix dissociates from actin and Tm and binds to TnC (yellow arrow). The rest of the TnI C-terminus becomes disordered (green dashed arrow). This conformational change unlocks the Tm cable so that it can move azimuthally to expose the myosin-binding sites on actin. (**c–d**) Azimuthal rotation of Tm cable upon binding of Ca^{2+} to TnC exposes myosin-binding sites on actin. (**c**) At low Ca^{2+} levels actin subunits (shown in red) have their myosin binding site sterically blocked by the Tm cable (light blue). (**d**) At activating Ca^{2+} levels due to azimuthal movement of Tm (yellow) four out of six actin subunits have their myosin binding sites exposed (shown in

et al. 2021b; Yamada et al. 2020). Structural changes of troponin (Tn) mechanistically link Ca^{2+} levels with the position of the Tm cable. When a Ca^{2+} ion binds to the TnC subunit (Fig. 14.16a and b, purple), the C-terminal part of the TnI subunit (Fig. 14.16a and b, green) releases from the actin filament (Fig. 14.16b, green dotted arrow) and its "regulatory switch" helix binds TnC cleft that was opened by a Ca^{2+} ion (Fig. 14.16b, yellow arrow). This conformational change liberates the Tm cable (Fig. 14.16a and b, blue) so that it can azimuthally rotate along with the N-terminal portion of TnT (Fig. 14.16a red arrowhead) in order to permit actomyosin interactions (Risi et al. 2021b; Yamada et al. 2020). At low Ca^{2+} levels, the positioning of Tm on actin subunits sterically hinders myosin from binding to its interaction sites on the actin filament (Fig. 14.16c). At permissive Ca^{2+} concentrations, the Tm rolls away from the myosin binding sites on four out of six actins so that those actin subunits are accessible to myosin binding (Fig. 14.16d) (Risi et al. 2021b; Yamada et al. 2020). A recent cryo-EM study of the native cardiac thin filament at systolic Ca^{2+} concentration revealed a primarily stochastic binding of Ca^{2+} ions to the Tn complex along the thin filament (Risi et al. 2021b). This suggests that during the systolic phase of the heartbeat, the Ca^{2+} influx randomly activates sections of the thin filament, providing myosin heads with equal opportunity to bind at various locations. Both myosin RLC phosphorylation and Ca^{2+}-mediated thin filament activation have distinct regulatory contributions to a property of cardiac muscle known as length-dependent activation (LDA). Also known as the Frank-Starling law, the principle of LDA states that with increased length, cardiac myofilaments exhibit greater force development and calcium sensitivity (Allen and Kentish 1985; Dobesh et al. 2002). Upon cardiac sarcomere stretching, liberation of myosin heads from their folded-back conformation on the thick filament underlies elevated force development, whereas increased Ca^{2+} sensitivity derives from structural changes to the regulatory Tn complex of the thin filament (Zhang et al. 2017).

Myosin-binding protein C (MyBP-C) (Fig. 14.16e) represents another striated muscle-specific regulatory layer for managing the actomyosin interaction. This

Fig. 14.16 (continued) green). The two actin subunits remain unavailable for actomyosin interactions (shown in red) due to steric blocking by the N-terminus of TnT (subunit 1) or the core of Tn complex (subunit 4). **(e)** The domain composition of the cardiac myosin binding protein C (cMyBP-C). Skeletal MyBP-C isoforms contain seven immunoglobulin (Ig)-like domains, three fibronectin (Fn)-like domains, and a regulatory M-domain between the C1 and C2 Ig-domains, which carry regulatory phosphorylation sites (red arrow). Cardiac isoform of MyBP-C has an additional Ig-domain (e.g., C0) at its N-terminus connected to C1 via proline/alanine (PA) linker. N-terminal C0, C1, M, and C2 domains bind to actin containing thin filament, while C-terminal C8, C9 and C10 domains keep MyBP-C attached to the myosin thick filament. **(f)** Cryo-EM map (gray transparent surface) shows that C1 Ig domain of cMyBP-C (blue ribbons) binds to actin subunit (tan ribbons, contact marked with red arrow) and cross-links it with the Tm cable (cyan ribbons) via its positively charged loop (red circle), which moves Tm cable into fully activated position. **(g)** Interactions of positively charged residues R215 and K218 of cMyBP-C C1 Ig domain (red spheres) occur *via* formation of salt bridges with the negatively charged patches (yellow spheres) on the Tm cable (PDB 6G2T; (Risi et al. 2018)). Actin is tan, while C1 Ig domain is blue. Activated Tm cable is cyan

unique protein comprised of ten domains (Fig. 14.16e) cross-links thin and thick filamentous systems in the sarcomere (Luther et al. 2011). The consensus from studies of the cardiac form of MyBP-C establishes that while it is not essential for assembly and stability of the sarcomere (Harris et al. 2002), its interactions with both thin and thick filaments modulate the rate and force of cardiac contractions (reviewed in (Moss et al. 2015)). Depending on the sarcomeric calcium level, cMyBP-C appears to possess dual functionality with respect to cross-bridge modulation. At low calcium, cMyBP-C enhances the activation of the thin filament (Previs et al. 2015; Razumova et al. 2008; Razumova et al. 2006) and myosin binding (Inchingolo et al. 2019). The structural basis for enhanced thin filament activation presumably derives from an electrostatic interaction between the C1 domain in the cMyBP-C N-terminal region and the Tm cable, which activates cardiac thin filaments (Fig. 14.16f and g) (Risi et al. 2018). At high Ca^{2+}, the role of cMyBP-C transforms into inhibition of cross-bridge formation and sliding velocity, as it binds in clusters to the thin filament and thus blocks myosin's access to its binding sites (Inchingolo et al. 2019; Previs et al. 2016). Much like sarcomeric Ca^{2+} concentration, phosphorylation of MyBP-C exerts regulatory control over actomyosin cross-bridges. Phosphorylation of MyBP-C M-domain relieves MyBP-C binding to the S2 region of inactive myosin heads (Gruen et al. 1999) and is an important part of contractile regulation (Nag et al. 2017). High Ca^{2+} is required for the phosphorylated MyBP-C to activate the thin filament (Colson et al. 2016; Previs et al. 2016).

Cargo Transport

In contrast to the ensemble of low-duty ratio myosin motors responsible for muscle contraction, cargo motility along actin filaments utilizes higher-duty ratio processive myosins that can fulfill their biological function as a single monomer or dimer. Myosin X (Fig. 14.17a, labeled (1)), which moves in large side steps along actin bundles, transports cargoes with important functional roles in nervous system development. Spatial patterning of the embryonic nervous system depends on morphogen gradients, transmitted between cells by specialized actin bundles known as cytonemes (Kornberg and Roy 2014). Myosin X motility along these bundles enables the transport of the morphogens Sonic hedgehog (Shh) in chick embryos (Sanders et al. 2013) and Wnt in zebrafish embryos (Stanganello et al. 2015) (Fig. 14.17b). Consistently, mice with myosin X knocked out exhibit alterations to the Shh gradient within the embryonic neural tube (Hall et al. 2021). Myosin-based cargo transport can also regulate synaptic plasticity in the adult nervous system. Myosin V (Fig. 14.17a, labeled (2)) travels on F-actin from the minus to the plus end (Fig. 14.17c, black arrow). It delivers AMPA glutamate receptors from the cytosol to the dendritic spines in hippocampal neurons, thus facilitating long-term potentiation (LTP) (Fig. 14.17c) (Correia et al. 2008). Whereas LTP enhances synaptic connectivity, its opposing process known as long-term depression (LTD) reduces synaptic connectivity. The myosin VI motor

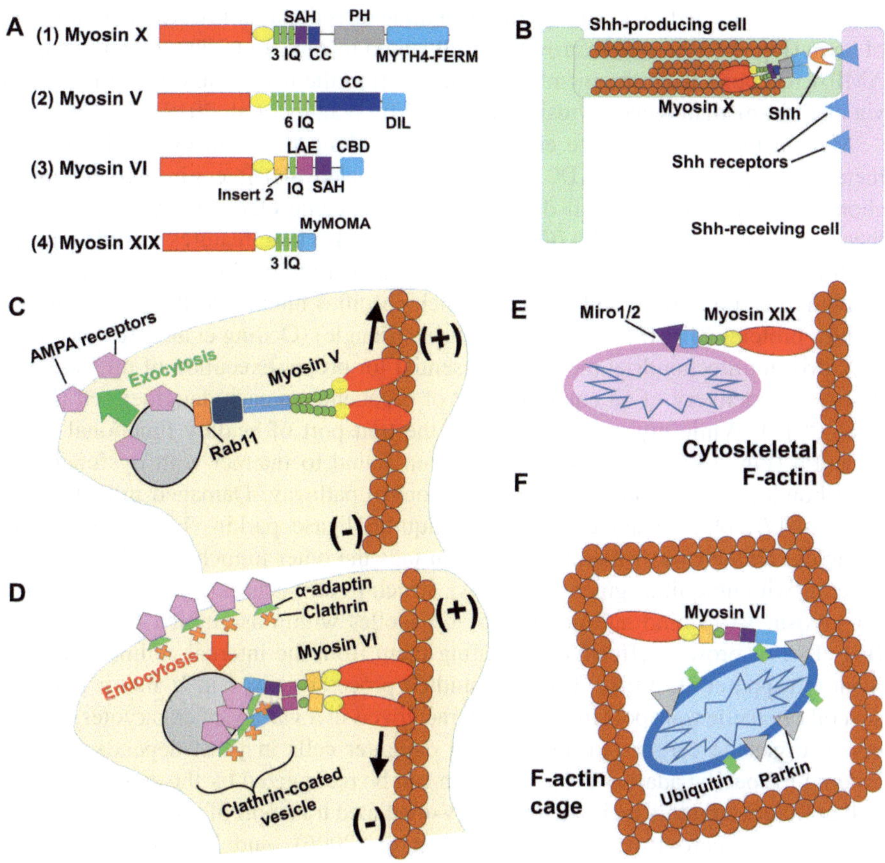

Fig. 14.17 Structure and function of various cargo-transporting myosin classes. (**a**) Domain architecture of various cargo-transporting myosin classes. While these isoforms all have a conserved motor (red rectangles) and converter (yellow ovals) domains, they differ in their tail domain architecture that follows the converter. The cargo-binding domains (light blue rectangles) residing in myosin isoforms tails possess different structures and determine cargo specificities. (1) Myosin X contains 3 IQ motifs (green), a single alpha helix (SAH; purple), a coiled-coil (CC; blue) domain that allows antiparallel dimerization, membrane-binding pleckstrin homology domain (PH; gray), and a cargo-binding MyTH4-FERM domain (light blue). (2) Myosin V contains 6 IQ motifs (green), a coiled-coil domain (CC; blue) responsible for dimerization, and a dilute (DIL; light blue) domain that provides cargo binding. (3) Myosin VI contains a unique 'insert 2' immediately after the converter (orange) that confers its reverse (i.e. minus-end) directionality, one IQ motif (green), a lever arm extension (LAE; magenta), single alpha helix (SAH; purple), and a cargo-binding domain (CBD; light blue). (4) Myosin XIX contains 3 IQ motifs (green) and a MYO19-specific mitochondrial outer membrane association (MyMOMA; light blue) tail domain. (**b**) Myosin X forms a dimer to transport vesicles (while oval) containing sonic hedgehog (Shh; orange crescent) along actin bundles known as cytonemes (brown). These bundles project from the Shh-producing cell (light green) to the Shh-receiving cell (light pink), which allows the Shh ligand to be directly delivered to its target receptor (blue triangles) by myosin X-mediated transport. (**c**) As a result of LTP-inducing stimuli to a postsynaptic neuron (tan), myosin V dimers bind to the Rab11 adaptor protein (orange rectangle) on the surface of a recycling endosome (gray oval) containing AMPA receptors (pink pentagons). This facilitates the endocytic trafficking (green arrow) of AMPA

(Fig. 14.17a, labeled (3)), which exhibits unique directionality toward the minus end of actin (Fig. 14.17d, black arrow), is well-suited to carry out LTD. It can transport AMPA receptors from the membrane (plus end) to the neuron interior (minus end) via clathrin-mediated endocytosis (Fig. 14.17d) (Wagner et al. 2019).

Myosin motors also move entire organelles along the actin cytoskeleton. The metazoan-specific myosin XIX (Fig. 14.17a, labeled (4)) is engineered for mitochondrial transport, as its tail domain contains specific elements that recruit it to these organelles (Fig. 14.17e) (Quintero et al. 2009). These features include a lipid-binding motif that targets the outer mitochondrial membrane (Fig. 14.17e, pink oval) (Hawthorne et al. 2016) and a region which specifies binding to the mitochondrial surface proteins Miro1/2 (Fig. 14.17e, purple triangle) (Oeding et al. 2018). Myosin XIX-driven mitochondrial motility is essential for proper execution of mitosis, as it ensures the delivery of an equal number of organelles to each daughter cell (Rohn et al. 2014). While myosin XIX handles the transport of healthy functional mitochondria (Fig. 14.17e), myosin VI has been found to interact with dysfunctional mitochondria and regulate a key quality control pathway. Damaged mitochondria (Fig. 14.17f, blue ovals) express the ubiquitin ligase parkin (Fig. 14.17f, gray triangles) on their outer membrane. Parkin tags the outer mitochondrial membrane proteins with ubiquitin (green diamonds), which in turn acts as a recruitment signal for myosin VI to bind and deposit the organelles within assembled F-actin cages (Fig. 14.17f, brown), effectively isolating them from the internal cellular environment (Kruppa et al. 2018). The well-studied processive myosin V motor has also several organelle transport functions characterized in a variety of eukaryotes. Delivery of organelle cargoes to the budding daughter cells in yeast depends on their binding to specific adaptor proteins which can be recognized by the class myosin V tail. Different adaptors facilitate myosin V-mediated transport of vacuoles (Ishikawa et al. 2003), peroxisomes (Fagarasanu et al. 2006), and trans-Golgi vesicles (Lipatova et al. 2008). Since yeast class V myosins are less processive than their vertebrate counterparts (Reck-Peterson et al. 2001), multiple myosin V motors are required to rapidly transport cargo to active growth sites in yeast (Donovan and Bretscher 2012). In vertebrates, class V myosins transport organelles in a multitude of physiological contexts, from the melanosomes involved in skin and hair

Fig. 14.17 (continued) receptors to the post-synaptic membrane due to processive movement of Myosin V toward the plus end of F-actin (black arrow). (**d**) Myosin VI dimers form a complex with AMPA receptors (pink pentagons) and α-adaptin (green trapezoids) and move them toward the minus end of F-actin (black arrow). This facilitates endocytosis (red arrow) of the AMPA receptors into clathrin (orange crosses)-coated vesicles (gray ovals), and in doing so promotes synaptic LTD. (**e**) The MyMOMA tail domain of myosin XIX binds to both the mitochondrial outer membrane (pink oval) and the mitochondrial surface proteins Miro1/2 (purple triangle), allowing it to transport healthy organelles along cytoskeletal F-actin tracks (brown). (**f**) The outer membrane of damaged mitochondria (blue oval) accumulates parkin (gray triangles), which tags the outer mitochondrial membrane proteins with ubiquitin (green diamonds). Myosin VI binds to the ubiquitin and triggers the assembly of an F-actin cage (brown) that sequesters the dysfunctional organelle

pigmentation (Wu et al. 2002) to the endoplasmic reticulum in cerebellar neurons that mediates synaptic plasticity (Wagner et al. 2011).

Tension Sensing

The regulation of plasma membrane tension is essential for proper cellular function, as it links mechanical stimuli to biochemical responses. Class I myosins are poised to act as tension sensors due to their conserved TH1 tail domains, which can bind phospholipids and thus link the plasma membrane to the internal cytoskeleton bound by the motor domain (reviewed in (McConnell and Tyska 2010)). The total tension of the plasma membrane is largely determined by the molecular interfaces, such as myosins, that adhere it to the cytoskeleton (Sheetz 2001). Myosin Ia supports biomechanical interaction between membrane and cytoskeleton in the brush border of intestinal epithelial cells (Nambiar et al. 2009). Loss of myosin Ia destabilizes the brush border architecture (Tyska et al. 2005).

The extreme tension sensitivity of myosin Ib (Fig. 14.18a) allows it to facilitate the extension of membrane tubules at endosomes and the trans-Golgi network (Laakso et al. 2008). As this process involves an increase in membrane tension, the catch-bond behavior of myosin Ib allows it to remain attached to the tubule while elongating it at a constant velocity along an actin filament (Fig. 14.18b) (Yamada et al. 2014). The force sensitivity of myosin Ib is dramatically abrogated in the presence of Ca^{2+}. It accelerates myosin Ib detachment from actin at 1 pN of force by 19-fold, due to reduction of the working stroke distance required to induce ADP release (Lewis et al. 2012). The less force-sensitive myosin Ic (Greenberg et al. 2012) also has the lifetime of its detached state increased by Ca^{2+} addition, though in this case Ca^{2+} exerts its effect primarily via deceleration of ATP hydrolysis (Adamek et al. 2008). The response of myosin Ic to Ca^{2+}suits it for its biological role in sensory adaptation within the inner ear's hair cells, a process intimately dependent on calcium transients. Myosin Ic links the transduction channels through which Ca^{2+} ions pass to the actin bundles within the hair cells known as stereocilia. An extracellular tip link connects the ion channel associated with one stereocilia bundle to its neighboring bundle. Auditory stimuli increase tension in the tip links and leads to subsequent Ca^{2+} influx through the transduction channels. The myosin Ic-transduction channel complex dynamically adjusts to tip link tension and elevated calcium by undergoing displacement along the stereocilia to restore resting tension levels (Batters et al. 2004; Holt et al. 2002). Thus, dual Ca^{2+} and tension sensitivity of myosin Ic allows it to mechanically transduce stimuli applied to the hair cells.

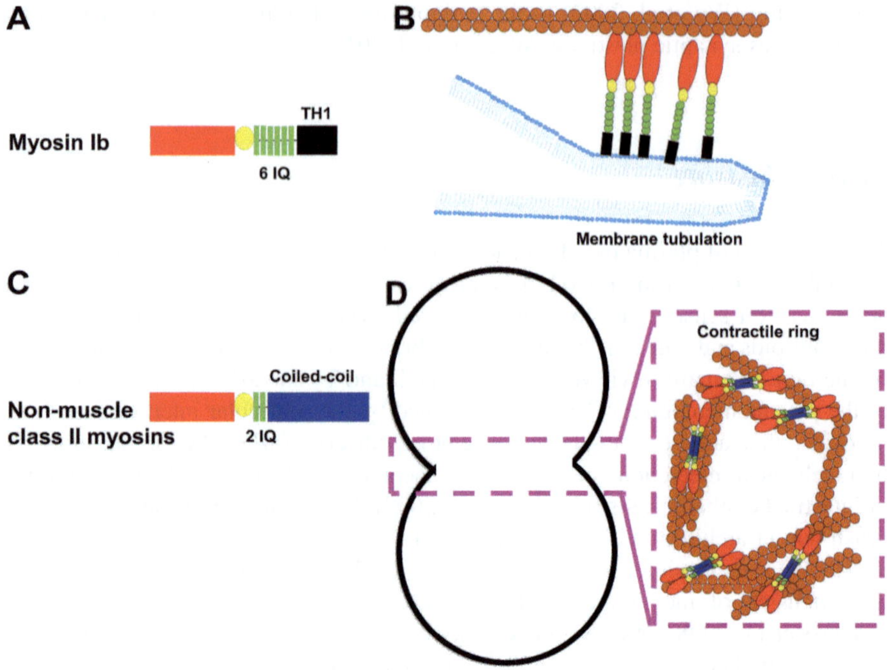

Fig. 14.18 Involvement of myosins in tension sensing and cytoskeletal remodeling. (**a**) Myosin Ib, is comprised of a motor domain (red rectangle), converter (yellow circle), 6 IQ motifs in the lever arm (green rectangles), and a membrane-binding TH1 domain (black rectangle). (**b**) Participation of tension-sensing myosin Ib monomers in the extension of membrane tubules. In response to an increase in tension of the cell membrane (blue) during tubulation, myosin Ib motors adopt a long-lived state of cross-bridge formation with actin filaments (brown). This catch-bond behavior allows myosin Ib to facilitate tubule elongation while remaining anchored to the cytoskeleton. (**c**) Non-muscle class II myosins are comprised of a motor domain (red rectangle), converter (yellow circle), 2 IQ motifs in the lever arm (green rectangles), and a coiled-coil domain (blue rectangle) that mediates dimerization. (**d**) Participation of non-muscle class II myosins in the formation of the cleavage furrow (magenta dashed rectangle) during cytokinesis. A ring of actin filaments (brown) is assembled at the center of the dividing cell (black outline). The contractile force exerted by short bipolar myosin II filaments tightens the ring, resulting in the cleavage furrow that drives cytokinesis

Cytoskeletal Remodeling and Motility

Motility of single cell eukaryotes via actin cytoskeleton remodeling was one of the primordial functions of the non-muscle class II myosins (Fig. 14.18c), and it has been adopted by numerous motile cell types within complex multicellular eukaryotes. Cellular locomotion consists of two distinct phases, differentiated by the relative involvement of actin and myosin (Chen 1981; Lauffenburger and Horwitz 1996). The first phase involves extension of the cell's leading edge, and this is mainly determined by actin polymerization dynamics independent of myosin (Fig. 14.2b–d). Accordingly, leading edge extension is not affected by myosin null mutations in *Dictyostelium* (Wessels et al. 1991). Once the membrane protrusion has adhered to the cellular substratum, myosin-mediated contractility behind the leading edge generates the force needed to propel the cell body forward. In the amoeboid *Dictyostelium*, myosin II undergoes dynamic polarized redistribution to the rear of a cell moving toward a chemotactic stimulus (Yumura 1996). This same model organism also demonstrates that during cytokinesis (Fig. 14.18d), myosin localizes to the contractile ring at the cell equator in order to promote daughter cell segregation. Studies of this phenomenon in mammalian cells reveals that the non-muscle class II myosins contain motifs within their tails that mediate recruitment to the cleavage furrow as well as incorporation into preexisting myosin II filaments at the furrow (Beach and Egelhoff 2009). Parallels between amoebal and mammalian cytoskeletal dynamics can also be observed in motile immune cells. The T-cell class of lymphocytes depends on migration through the vascular endothelial wall in order to reach infected tissues and exert its effect or functions. As in *Dictyostelium*, leading edge extension occurs independently of myosin activity, given that T cells lacking myosin IIA can squeeze their membrane protrusions through the endothelial barrier. However, myosin IIA contractile activity is required for complete extravasation of the T cell through the endothelium (Jacobelli et al. 2013).

Conclusion and Future Perspectives

The actomyosin interaction originated in unicellular eukaryotes as a mechanism governing motility at the cellular scale, and the emergence of eukaryotic complexity coincided with functional diversification of the myosin superfamily. Consequently, myosin plays a functional role in nearly all tissues of a typical multicellular eukaryote. The structural features of each myosin class were adapted to their distinct cellular duties (Fig. 14.4b–e and Figs. 14.17 and 14.18), which form the molecular basis of diverse physiological processes such as nervous system development, intestinal absorption, and immune cell migration. Thus, eukaryotic evolution generated a protein family with a conserved structural blueprint and kinetic cycle. Fine-tuning was accomplished via sequence divergence of certain motifs, which altered the rate and equilibrium constants of the ATPase cycle (reviewed in (De La Cruz and Ostap 2004)). Consequently, the myosin superfamily members possess variations in their actin sliding velocities by a factor of 300 and in actin-activated ATPase activity by a factor of 3000, despite the general conservation of the motor domain architecture responsible for actin binding and ATP hydrolysis (Heissler and Sellers 2016).

Myosin is a catalytic molecular machine governed by allosteric communication between three major regions: the actin-binding interface, nucleotide-binding pocket, and force-generating lever arm (Figs. 14.5, 14.6, and 14.7). Thus, isoform-specific changes in any of these regions tune up the overall functionality of the myosin enzyme to a particular cellular task. For example, a class XI myosin from the algal species *Chara corallina* possesses a unique charge distribution of actin-binding loops 2 and 3 and as a result exhibits an exceptionally fast velocity ten times greater than that of fast skeletal muscle myosin (Ito et al. 2007). Despite sharing a similar general mechanism of processively translocating actin filaments, myosins from classes V, VI, and X have distinct lever arm and tail domain architectures (Fig. 14.17). This diversity endows processive myosins with varied step sizes and directionality, with parallel variability in their motor domains and mechanochemical cycle parameters. The reverse directionality of myosin VI, conferred by a class-specific lever arm insert, necessitates a different gating mechanism for its lead head than myosin V. Gating of myosin VI processivity relies on another unique insert, located in its motor domain near the nucleotide-binding site (Sweeney et al. 2007). In the case of myosin X, its tail architecture is the primary reason behind its selectivity for actin bundles, although chimeric replacement of its motor domain with that of myosin V reveals that the head also confers some of this behavior (Nagy et al. 2010). The extensive menagerie of myosin motors is punctuated by remarkable feats of conservation such as the IHM state of muscle myosin autoinhibition, governed by residues that have largely resisted variation for nearly 800 Myrs of metazoan evolution (Alamo et al. 2016). This regulatory mechanism dates back to the primitive muscular system of jellyfish (Lee et al. 2018).

Molecular disparities among myosins are largely responsible for the varied actomyosin interactions in physiological contexts, as the evolution of the actin superfamily has proceeded along a much less divergent trajectory. The exceptional conservation of actin is apparent when comparing evolutionarily distant eukaryotic species, given the ~90% identity between yeast and mammalian cytoplasmic actins (Galkin et al. 2002). Despite the clear differences between myosin and actin evolution, these two protein partners can be united under the umbrella of allostery. Structural polymorphism of F-actin manifests itself in multiple conformations of the D-loop (Merino et al. 2018; Pospich et al. 2020) and an allosteric coupling of its structural state with the structural state of the N- and C-termini of actin (Galkin et al. 2010). Based on the extreme conservation of internally buried residues within actin monomers and filaments, it seems reasonable that the conformational flexibility of actin filaments has provided the impetus for preserving the internal allosteric networks that is used by the myosin motors. This notion has a strong experimental support, since cross-linking of the two actin mobile elements (e.g., the D-loop and the C-terminus), completely abolishes force generation while allowing actomyosin interaction (Kim et al. 1998).

Overall, the very different evolutionary path was of myosin and actin forced the diverse family of myosin motors to use a very similar track on the surface of the evolutionary conserved actin filament (Robert-Paganin et al. 2020). This idea is beautifully illustrated by recently published cryo-EM structures of several quite diverse actomyosin complexes (Gurel et al. 2017; Mentes et al. 2018; Risi et al. 2021a; von der Ecken et al. 2016). In these complexes despite a very similar overall geometry of the actomyosin interface, the interaction of individual myosin actin binding elements differs between the isoforms (Figs. 14.10, 14.11, 14.12, 14.13, and 14.14). However, these structures were obtained only for the rigor state representing the end of the powerstroke (Fig. 14.5a, step 1). Therefore, near-atomic structures of the weakly bound states (Fig. 14.5a, steps 4 and 5) of various actomyosin complexes are required to achieve a full molecular description of the myosin mechanochemical cycle.

Compliance with Ethical Standards Authors do not have any conflict of interests. Authors confirm that experiments described in the chapter involving humans or animals were conducted by respecting the corresponding ethical guidelines and that informed consent was obtained in case humans were involved. The work on this chapter work was financially supported by NIH grant R01 HL140925.

References

Adamek N, Coluccio LM, Geeves MA (2008) Calcium sensitivity of the cross-bridge cycle of , the adaptation motor in the inner ear. Proc Natl Acad Sci U S A 105(15):5710–5715

Alamo L, Qi D, Wriggers W, Pinto A, Zhu J, Bilbao A, Gillilan RE, Hu S, Padrón R (2016) Conserved intramolecular interactions maintain myosin interacting-heads motifs explaining tarantula muscle super-relaxed state structural basis. J Mol Biol 428(6):1142–1164

Al-Khayat HA, Kensler RW, Squire JM, Marston SB, Morris EP (2013) Atomic model of the human cardiac muscle myosin filament. Proc Natl Acad Sci U S A 110(1):318–323

Allen DG, Kentish JC (1985) The cellular basis of the length-tension relation in cardiac muscle. J Mol Cell Cardiol 17(9):821–840

Batters C, Arthur CP, Lin A, Porter J, Geeves MA, Milligan RA, Molloy JE, Coluccio LM (2004) Myo1c is designed for the adaptation response in the inner ear. EMBO J 23(7):1433–1440

Beach JR, Egelhoff TT (2009) Myosin II recruitment during cytokinesis independent of centralspindlin-mediated phosphorylation. J Biol Chem 284(40):27377–27383

Bement WM, Mooseker MS (1995) TEDS rule: a molecular rationale for differential regulation of myosins by phosphorylation of the heavy chain head. Cell Motil Cytoskeleton 31(2):87–92

Bershitsky SY, Tsaturyan AK, Bershitskaya ON, Mashanov GI, Brown P, Burns R, Ferenczi MA (1997) Muscle force is generated by myosin heads stereospecifically attached to actin. Nature 388(6638):186–190

Blanc F, Isabet T, Benisty H, Sweeney HL, Cecchini M, Houdusse A (2018) An intermediate along the recovery stroke of myosin VI revealed by X-ray crystallography and molecular dynamics. Proc Natl Acad Sci U S A 115(24):6213–6218

Bloemink MJ, Geeves MA (2011) Shaking the myosin family tree: biochemical kinetics defines four types of myosin motor. Semin Cell Dev Biol 22(9):961–967

Brenner SL, Korn ED (1983) On the mechanism of actin monomer-polymer subunit exchange at steady state. J Biol Chem 258(8):5013–5020

Brown JH, Kim KH, Jun G, Greenfield NJ, Dominguez R, Volkmann N, Hitchcock-DeGregori SE, Cohen C (2001) Deciphering the design of the tropomyosin molecule. Proc Natl Acad Sci U S A 98(15):8496–8501

Capitanio M, Canepari M, Cacciafesta P, Lombardi V, Cicchi R, Maffei M, Pavone FS, Bottinelli R (2006) Two independent mechanical events in the interaction cycle of skeletal muscle myosin with actin. Proc Natl Acad Sci U S A 103(1):87–92

Chen WT (1981) Mechanism of retraction of the trailing edge during fibroblast movement. J Cell Biol 90(1):187–200

Chou SZ, Pollard TD (2019) Mechanism of actin polymerization revealed by cryo-EM structures of actin filaments with three different bound nucleotides. Proc Natl Acad Sci U S A 116(10): 4265–4274

Clark R, Ansari MA, Dash S, Geeves MA, Coluccio LM (2005) Loop 1 of transducer region in mammalian class I myosin, Myo1b, modulates actin affinity, ATPase activity, and nucleotide access. J Biol Chem 280(35):30935–30942

Colson BA, Thompson AR, Espinoza-Fonseca LM, Thomas DD (2016) Site-directed spectroscopy of cardiac myosin-binding protein C reveals effects of phosphorylation on protein structural dynamics. Proc Natl Acad Sci U S A 113(12):3233–3238

Conibear PB, Bagshaw CR, Fajer PG, Kovacs M, Malnasi-Csizmadia A (2003) Myosin cleft movement and its coupling to actomyosin dissociation. Nat Struct Biol 10(10):831–835

Cope MJ, Whisstock J, Rayment I, Kendrick-Jones J (1996) Conservation within the myosin motor domain: implications for structure and function. Structure 4(8):969–987

Correia SS, Bassani S, Brown TC, Lisé M-F, Backos DS, El-Husseini A, Passafaro M, Esteban JA (2008) Motor protein-dependent transport of AMPA receptors into spines during long-term potentiation. Nat Neurosci 11(4):457–466

Coureux P-D, Wells AL, Ménétrey J, Yengo CM, Morris CA, Sweeney HL, Houdusse A (2003) A structural state of the myosin V motor without bound nucleotide. Nature 425(6956):419–423

Coureux P-D, Sweeney HL, Houdusse A (2004) Three myosin V structures delineate essential features of chemo-mechanical transduction. EMBO J 23(23):4527–4537

Courtemanche N (2018) Mechanisms of formin-mediated actin assembly and dynamics. Biophys Rev 10(6):1553–1569

De La Cruz EM, Ostap EM (2004) Relating biochemistry and function in the myosin superfamily. Curr Opin Cell Biol 16(1):61–67

De La Cruz EM, Wells AL, Sweeney HL, Ostap EM (2000) Actin and light chain isoform dependence of myosin V kinetics. Biochemistry 39(46):14196–14202

De La Cruz EM, Ostap EM, Sweeney HL (2001) Kinetic mechanism and regulation of myosin VI. J Biol Chem 276(34):32373–32381

Debold EP, Patlak JB, Warshaw DM (2005) Slip sliding away: load-dependence of velocity generated by skeletal muscle myosin molecules in the laser trap. Biophys J 89(5):L34–L36

Dobesh DP, Konhilas JP, de Tombe PP (2002) Cooperative activation in cardiac muscle: impact of sarcomere length. Am J Physiol Heart Circ Physiol 282(3):H1055–H1062

Dominguez R, Freyzon Y, Trybus KM, Cohen C (1998) Crystal structure of a vertebrate smooth muscle myosin motor domain and its complex with the essential light chain: visualization of the pre-power stroke state. Cell 94(5):559–571

Donovan KW, Bretscher A (2012) Myosin-V is activated by binding secretory cargo and released in coordination with Rab/exocyst function. Dev Cell 23(4):769–781

Drabikowski W, Kominz DR, Maruyama K (1968) Effect of troponin on the reversibility of tropomyosin binding to F-actin. J Biochem 63(6):802–804

Driska SP, Aksoy MO, Murphy RA (1981) Myosin light chain phosphorylation associated with contraction in arterial smooth muscle. Am J Phys 240(5):C222–C233

Ebashi S (1972) Calcium ions and muscle contraction. Nature 240(5378):217–218

Egelman EH (2001) Molecular evolution: actin's long lost relative found. Curr Biol 11(24):R1022–R1024

Eisenberg E, Hill TL (1985) Muscle contraction and free energy transduction in biological systems. Science 227(4690):999–1006

Fagarasanu A, Fagarasanu M, Eitzen GA, Aitchison JD, Rachubinski RA (2006) The peroxisomal membrane protein Inp2p is the peroxisome-specific receptor for the myosin V motor Myo2p of Saccharomyces cerevisiae. Dev Cell 10(5):587–600

Farmer SR, Wan KM, Ben-Ze'ev A, Penman S (1983) Regulation of actin mRNA levels and translation responds to changes in cell configuration. Mol Cell Biol 3(2):182–189

Ferenczi MA, Bershitsky SY, Koubassova N, Siththanandan V, Helsby WI, Panine P, Roessle M, Narayanan T, Tsaturyan AK (2005) The "roll and lock" mechanism of force generation in muscle. Structure 13(1):131–141

Fischer S, Windshügel B, Horak D, Holmes KC, Smith JC (2005) Structural mechanism of the recovery stroke in the myosin molecular motor. Proc Natl Acad Sci U S A 102(19):6873–6878

Fisher AJ, Smith CA, Thoden J, Smith R, Sutoh K, Holden HM, Rayment I (1995a) Structural studies of myosin:nucleotide complexes: a revised model for the molecular basis of muscle contraction. Biophys J 68:19S–26S

Fisher AJ, Smith CA, Thoden JB, Smith R, Sutoh K, Holden HM, Rayment I (1995b) X-ray structures of the myosin motor domain of Dictyostelium discoideum complexed with MgADP. BeFx and MgADP.AlF4. Biochemistry 34(28):8960–8972

Foth BJ, Goedecke MC, Soldati D (2006) New insights into myosin evolution and classification. Proc Natl Acad Sci U S A 103(10):3681–3686

Fujii T, Iwane AH, Yanagida T, Namba K (2010) Direct visualization of secondary structures of F-actin by electron cryomicroscopy. Nature 467(7316):724–728

Fujita-Becker S, Reubold TF, Holmes KC (2006) The actin-binding cleft: functional characterisation of myosin II with a strut mutation. J Muscle Res Cell Motil 27(2):115–123

Fujiwara I, Takahashi S, Tadakuma H, Funatsu T, Ishiwata S (2002) Microscopic analysis of polymerization dynamics with individual actin filaments. Nat Cell Biol 4(9):666–673

Furch M, Geeves MA, Manstein DJ (1998) Modulation of actin affinity and actomyosin adenosine triphosphatase by charge changes in the myosin motor domain. Biochemistry 37(18): 6317–6326

Galkin VE, VanLoock MS, Orlova A, Egelman EH (2002) A new internal mode in F-actin helps explain the remarkable evolutionary conservation of actin's sequence and structure. Curr Biol 12(7):570–575

Galkin VE, Orlova A, Schroder GF, Egelman EH (2010) Structural polymorphism in F-actin. Nat Struct Mol Biol 17(11):1318–1323

Galkin VE, Orlova A, Vos MR, Schröder GF, Egelman EH (2015) Near-atomic resolution for one state of F-actin. Structure (London, England: 1993) 23(1):173–182

Giese KC, Spudich JA (1997) Phenotypically selected mutations in myosin's actin binding domain demonstrate intermolecular contacts important for motor function. Biochemistry 36(28): 8465–8473

Greenberg MJ, Lin T, Goldman YE, Shuman H, Ostap EM (2012) Myosin IC generates power over a range of loads *via* a new tension-sensing mechanism. Proc Natl Acad Sci U S A 109(37): E2433–E2440

Greenberg MJ, Shuman H, Ostap EM (2014) Inherent force-dependent properties of β-cardiac myosin contribute to the force-velocity relationship of cardiac muscle. Biophys J 107(12):L41–L44

Gruen M, Prinz H, Gautel M (1999) cAPK-phosphorylation controls the interaction of the regulatory domain of cardiac myosin binding protein C with myosin-S2 in an on-off fashion. FEBS Lett 453(3):254–259

Guilford WH, Dupuis DE, Kennedy G, Wu J, Patlak JB, Warshaw DM (1997) Smooth muscle and skeletal muscle myosins produce similar unitary forces and displacements in the laser trap. Biophys J 72(3):1006–1021

Gurel PS, Kim LY, Ruijgrok PV, Omabegho T, Bryant Z, Alushin GM (2017) Cryo-EM structures reveal specialization at the myosin VI-actin interface and a mechanism of force sensitivity. eLife 6. https://doi.org/10.7554/eLife.31125.001

Hall ET, Dillard ME, Stewart DP, Zhang Y, Wagner B, Levine RM, Pruett-Miller SM, Sykes A, Temirov J, Cheney RE, Mori M, Robinson CG, Ogden SK (2021) Cytoneme delivery of Sonic Hedgehog from ligand-producing cells requires Myosin 10 and a Dispatched-BOC/CDON co-receptor complex. eLife 10:e61432

Harris DE, Work SS, Wright RK, Alpert NR, Warshaw DM (1994) Smooth, cardiac and skeletal muscle myosin force and motion generation assessed by cross-bridge mechanical interactions in vitro. J Muscle Res Cell Motil 15(1):11–19

Harris SP, Bartley CR, Hacker TA, McDonald KS, Douglas PS, Greaser ML, Powers PA, Moss RL (2002) Hypertrophic cardiomyopathy in cardiac myosin binding protein-C knockout mice. Circ Res 90(5):594–601

Hartman MA, Finan D, Sivaramakrishnan S, Spudich JA (2011) Principles of unconventional myosin function and targeting. Annu Rev Cell Dev Biol 27:133–155

Hawthorne JL, Mehta PR, Singh PP, Wong NQ, Quintero OA (2016) Positively charged residues within the MYO19 MyMOMA domain are essential for proper localization of MYO19 to the mitochondrial outer membrane. Cytoskeleton (Hoboken, NJ) 73(6):286–299

Heissler SM, Sellers JR (2014) Myosin light chains: Teaching old dogs new tricks. BioArchitecture 4(6):169–188

Heissler SM, Sellers JR (2016) Various themes of myosin regulation. J Mol Biol 428(9 Pt B):1927–1946

Heissler SM, Liu X, Korn ED, Sellers JR (2013) Kinetic characterization of the ATPase and actin-activated ATPase activities of Acanthamoeba castellanii myosin-2. J Biol Chem 288(37): 26709–26720

Herrera AM, McParland BE, Bienkowska A, Tait R, Paré PD, Seow CY (2005) 'Sarcomeres' of smooth muscle: functional characteristics and ultrastructural evidence. J Cell Sci 118(Pt 11): 2381–2392

Hill AV (1938) The heat of shortening and the dynamic constants of muscle. Proc R Soc B Biol Sci 126(843):136–195

Himmel DM, Gourinath S, Reshetnikova L, Shen Y, Szent-Gyorgyi AG, Cohen C (2002) Crystallographic findings on the internally uncoupled and near-rigor states of myosin: further insights into the mechanics of the motor. Proc Natl Acad Sci U S A 99(20):12645–12650

Holmes KC, Popp D, Gebhard W, Kabsch W (1990) Atomic model of the actin filament. Nature 347:44–49

Holt JR, Gillespie SKH, Provance DW, Shah K, Shokat KM, Corey DP, Mercer JA, Gillespie PG (2002) A chemical-genetic strategy implicates myosin-1c in adaptation by hair cells. Cell 108(3):371–381

Houdusse A, Sweeney HL (2016) How myosin generates force on actin filaments. Trends Biochem Sci 41(12):989–997

Houdusse A, Kalabokis VN, Himmel D, Szent-Gyorgyi AG, Cohen C (1999) Atomic structure of scallop myosin subfragment S1 complexed with MgADP: a novel conformation of the myosin head. Cell 97(4):459–470

Houdusse A, Szent-Gyorgyi AG, Cohen C (2000) Three conformational states of scallop myosin S1. Proc Natl Acad Sci U S A 97(21):11238–11243

Hu Z, Taylor DW, Reedy MK, Edwards RJ, Taylor KA (2016) Structure of myosin filaments from relaxed Lethocerus flight muscle by cryo-EM at 6 Å resolution. Sci Adv 2(9):e1600058

Huxley H, Hanson J (1954) Changes in the cross-striations of muscle during contraction and stretch and their structural interpretation. Nature 173(4412):973–976

Inchingolo AV, Previs SB, Previs MJ, Warshaw DM, Kad NM (2019) Revealing the mechanism of how cardiac myosin-binding protein C N-terminal fragments sensitize thin filaments for myosin binding. Proc Natl Acad Sci U S A 116(14):6828–6835

Ishikawa K, Catlett NL, Novak JL, Tang F, Nau JJ, Weisman LS (2003) Identification of an organelle-specific myosin V receptor. J Cell Biol 160(6):887–897

Ito K, Ikebe M, Kashiyama T, Mogami T, Kon T, Yamamoto K (2007) Kinetic mechanism of the fastest motor protein, Chara myosin. J Biol Chem 282(27):19534–19545

Jacobelli J, Estin Matthews M, Chen S, Krummel MF (2013) Activated T cell trans-endothelial migration relies on myosin-IIA contractility for squeezing the cell nucleus through endothelial cell barriers. PLoS One 8(9):e75151

Joel PB, Trybus KM, Sweeney HL (2001) Two conserved lysines at the 50/20-kDa junction of myosin are necessary for triggering actin activation. J Biol Chem 276(5):2998–3003

Jontes JD, Wilson-Kubalek EM, Milligan RA (1995) A 32 degree tail swing in brush border myosin I on ADP release. Nature 378(6558):751–753

Kabsch W, Mannherz HG, Suck D, Pai EF, Holmes KC (1990) Atomic structure of the actin:DNase I complex. Nature 347:37–44

Kensler RW, Stewart M (1983) Frog skeletal muscle thick filaments are three-stranded. J Cell Biol 96(6):1797–1802

Kim E, Bobkova E, Miller CJ, Orlova A, Hegyi G, Egelman EH, Muhlrad A, Reisler E (1998) Intrastrand cross-linked actin between Gln-41 and Cys-374. III. Inhibition of motion and force generation with myosin. Biochemistry 37(51):17801–17809
Kishino A, Yanagida T (1988) Force measurements by micromanipulation of a single actin filament by glass needles. Nature 334(6177):74–76
Kojima S, Konishi K, Katoh K, Fujiwara K, Martinez HM, Morales MF, Onishi H (2001) Functional roles of ionic and hydrophobic surface loops in smooth muscle myosin: their interactions with actin. Biochemistry 40(3):657–664
Kominz DR (1965) Mercurial-induced transformation of myosin prevented by adenosine triphosphate and pyrophosphate. Science 149(3690):1374–1375
Koppole S, Smith JC, Fischer S (2007) The structural coupling between ATPase activation and recovery stroke in the myosin II motor. Structure (London, England: 1993) 15(7):825–837
Korn ED (2000) Coevolution of head, neck, and tail domains of myosin heavy chains. Proc Natl Acad Sci 97(23):12559–12564
Kornberg TB, Roy S (2014) Cytonemes as specialized signaling filopodia. Development (Cambridge, England) 141(4):729–736
Kraft T, Mahlmann E, Mattei T, Brenner B (2005) Initiation of the power stroke in muscle: insights from the phosphate analog AlF4. Proc Natl Acad Sci U S A 102(39):13861–13866
Krendel M, Mooseker MS (2005) Myosins: tails (and heads) of functional diversity. Physiology 20(4):239–251
Kruppa AJ, Kishi-Itakura C, Masters TA, Rorbach JE, Grice GL, Kendrick-Jones J, Nathan JA, Minczuk M, Buss F (2018) Myosin VI-dependent actin cages encapsulate parkin-positive damaged mitochondria. Dev Cell 44(4):484–499.e6
Kuhner S, Fischer S (2011) Structural mechanism of the ATP-induced dissociation of rigor myosin from actin. Proc Natl Acad Sci U S A 108(19):7793–7798
Laakso JM, Lewis JH, Shuman H, Ostap EM (2008) Myosin I can act as a molecular force sensor. Science (New York, NY) 321(5885):133–136
Lauffenburger DA, Horwitz AF (1996) Cell migration: a physically integrated molecular process. Cell 84(3):359–369
Lee KH, Sulbarán G, Yang S, Mun JY, Alamo L, Pinto A, Sato O, Ikebe M, Liu X, Korn ED, Sarsoza F, Bernstein SI, Padrón R, Craig R (2018) Interacting-heads motif has been conserved as a mechanism of myosin II inhibition since before the origin of animals. Proc Natl Acad Sci U S A 115(9):E1991–E2000
Lehrer SS, Morris EP (1982) Dual effects of tropomyosin and troponin-tropomyosin on actomyosin subfragment 1 ATPase. J Biol Chem 257(14):8073–8080
Lewis JH, Greenberg MJ, Laakso JM, Shuman H, Ostap EM (2012) Calcium regulation of myosin-I tension sensing. Biophys J 102(12):2799–2807
Li G, Cui Q (2004) Analysis of functional motions in Brownian molecular machines with an efficient block normal mode approach: myosin-II and Ca2+ -ATPase. Biophys J 86(2):743–763
Lipatova Z, Tokarev AA, Jin Y, Mulholland J, Weisman LS, Segev N (2008) Direct interaction between a myosin V motor and the Rab GTPases Ypt31/32 is required for polarized secretion. Mol Biol Cell 19(10):4177–4187
Liu JCY, Rottler J, Wang L, Zhang J, Pascoe CD, Lan B, Norris BA, Herrera AM, Paré PD, Seow CY (2013) Myosin filaments in smooth muscle cells do not have a constant length. J Physiol 591(23):5867–5878
Llinas P, Isabet T, Song L, Ropars V, Zong B, Benisty H, Sirigu S, Morris C, Kikuti C, Safer D, Sweeney HL, Houdusse A (2015) How actin initiates the motor activity of Myosin. Dev Cell 33(4):401–412
Lombardi V, Piazzesi G, Ferenczi MA, Thirlwell H, Dobbie I, Irving M (1995) Elastic distortion of myosin heads and repriming of the working stroke in muscle. Nature 374(6522):553–555
Lorenz M, Holmes KC (2010) The actin-myosin interface. Proc Natl Acad Sci U S A 107(28): 12529–12534

Lu Q, Li J, Zhang M (2014) Cargo recognition and cargo-mediated regulation of unconventional myosins. Acc Chem Res 47(10):3061–3070
Luther PK, Winkler H, Taylor K, Zoghbi ME, Craig R, Padron R, Squire JM, Liu J (2011) Direct visualization of myosin-binding protein C bridging myosin and actin filaments in intact muscle. Proc Natl Acad Sci U S A 108(28):11423–11428
Margossian SS, Lowey S (1982) Preparation of myosin and its subfragments from rabbit skeletal muscle. Methods Enzymol 85(Pt B):55–71
Martyn DA, Regnier M, Xu D, Gordon AM (2001) Ca2+ - and cross-bridge-dependent changes in N- and C-terminal structure of troponin C in rat cardiac muscle. Biophys J 80(1):360–370
McConnell RE, Tyska MJ (2010) Leveraging the membrane - cytoskeleton interface with myosin-1. Trends Cell Biol 20(7):418–426
McKillop DFA, Geeves MA (1993) Regulation of the interaction between actin and myosin subfragment-1: evidence for three states of the thin filament. Biophys J 65:693–701
Mentes A, Huehn A, Liu X, Zwolak A, Domínguez R, Shuman H, Ostap EM, Sindelar CV (2018) High-resolution cryo-EM structures of actin-bound myosin states reveal the mechanism of myosin force sensing. Proc Natl Acad Sci U S A 115(6):1292–1297
Merino F, Pospich S, Funk J, Wagner T, Kullmer F, Arndt HD, Bieling P, Raunser S (2018) Structural transitions of F-actin upon ATP hydrolysis at near-atomic resolution revealed by cryo-EM. Nat Struct Mol Biol 25(6):528–537
Mihalyi E, Szent-Gyorgyi AG (1953) Trypsin digestion of muscle proteins. III. Adenosinetriphosphatase activity and actinbinding capacity of the digested myosin. J Biol Chem 201(1):211–219
Milligan RA (1996) Protein-protein interactions in the rigor actomyosin complex. Proc Natl Acad Sci U S A 93(1):21–26
Molinie N, Gautreau A (2018) The Arp2/3 Regulatory System and Its Deregulation in Cancer. Physiol Rev 98(1):215–238
Moss RL, Fitzsimons DP, Ralphe JC (2015) Cardiac MyBP-C regulates the rate and force of contraction in mammalian myocardium. Circ Res 116(1):183–192
Muretta JM, Rohde JA, Johnsrud DO, Cornea S, Thomas DD (2015) Direct real-time detection of the structural and biochemical events in the myosin power stroke. Proc Natl Acad Sci U S A 112(46):14272–14277
Nag S, Trivedi DV, Sarkar SS, Adhikari AS, Sunitha MS, Sutton S, Ruppel KM, Spudich JA (2017) The myosin mesa and the basis of hypercontractility caused by hypertrophic cardiomyopathy mutations. Nat Struct Mol Biol 24(6):525–533
Nagy NT, Sakamoto T, Takács B, Gyimesi M, Hazai E, Bikádi Z, Sellers JR, Kovács M (2010) Functional adaptation of the switch-2 nucleotide sensor enables rapid processive translocation by myosin-5. FASEB J 24(11):4480–4490
Nambiar R, McConnell RE, Tyska MJ (2009) Control of cell membrane tension by myosin-I. Proc Natl Acad Sci U S A 106(29):11972–11977
Oda T, Iwasa M, Aihara T, Maeda Y, Narita A (2009) The nature of the globular- to fibrous-actin transition. Nature 457(7228):441–445
Oeding SJ, Majstrowicz K, Hu X-P, Schwarz V, Freitag A, Honnert U, Nikolaus P, Bähler M (2018) Identification of Miro1 and Miro2 as mitochondrial receptors for myosin XIX. J Cell Sci 131(17):jcs219469
Onishi H, Mikhailenko SV, Morales MF (2006a) Toward understanding actin activation of myosin ATPase: the role of myosin surface loops. Proc Natl Acad Sci U S A 103(16):6136–6141
Onishi H, Mikhailenko SV, Morales MF (2006b) Toward understanding actin activation of myosin ATPase: the role of myosin surface loops. Proc Natl Acad Sci U S A 103(16):6136–6141
Parry DAD (1975) Analysis of the primary sequence of α-tropomyosin from rabbit skeletal muscle. J Mol Biol 98(3):519–535
Pasha SN, Meenakshi I, Sowdhamini R (2016) Revisiting myosin families through large-scale sequence searches leads to the discovery of new myosins. Evol Bioinformatics Online 12:201–211

Pearson RB, Jakes R, John M, Kendrick-Jones J, Kemp BE (1984) Phosphorylation site sequence of smooth muscle myosin light chain (Mr = 20 000). FEBS Lett 168(1):108–112

Perrin BJ, Ervasti JM (2010) The actin gene family: function follows isoform. Cytoskeleton 67(10): 630–634

Pollard TD (1986) Rate constants for the reactions of ATP- and ADP-actin with the ends of actin filaments. J Cell Biol 103(6 Pt 2):2747–2754

Pollard TD (2007) Regulation of actin filament assembly by Arp2/3 complex and formins. Annu Rev Biophys Biomol Struct 36:451–477

Pollard TD (2016) Actin and actin-binding proteins. Cold Spring Harb Perspect Biol 8(8):a018226

Pospich S, Merino F, Raunser S (2020) Structural effects and functional implications of phalloidin and jasplakinolide binding to actin filaments. Structure 28(4):437–449 e5

Previs MJ, Prosser BL, Mun JY, Previs SB, Gulick J, Lee K, Robbins J, Craig R, Lederer WJ, Warshaw DM (2015) Myosin-binding protein C corrects an intrinsic inhomogeneity in cardiac excitation-contraction coupling. Sci Adv 1(1):e1400205

Previs MJ, Mun JY, Michalek AJ, Previs SB, Gulick J, Robbins J, Warshaw DM, Craig R (2016) Phosphorylation and calcium antagonistically tune myosin-binding protein C's structure and function. Proc Natl Acad Sci U S A 113(12):3239–3244

Pylypenko O, Attanda W, Gauquelin C, Lahmani M, Coulibaly D, Baron B, Hoos S, Titus MA, England P, Houdusse AM (2013) Structural basis of myosin V Rab GTPase-dependent cargo recognition. Proc Natl Acad Sci U S A 110(51):20443–20448

Quintero OA, DiVito MM, Adikes RC, Kortan MB, Case LB, Lier AJ, Panaretos NS, Slater SQ, Rengarajan M, Feliu M, Cheney RE (2009) Human Myo19 is a novel myosin that associates with mitochondria. Curr Biol 19(23):2008–2013

Rayment I, Rypniewski WR, Schmidt-Base K, Smith R, Tomchick DR, Benning MM, Winkelmann DA, Wesenberg G, Holden HM (1993) Three-dimensional structure of myosin subfragment-1: a molecular motor. Science 261:50–57

Razumova MV, Shaffer JF, Tu AY, Flint GV, Regnier M, Harris SP (2006) Effects of the N-terminal domains of myosin binding protein-C in an in vitro motility assay: evidence for long-lived cross-bridges. J Biol Chem 281(47):35846–35854

Razumova MV, Bezold KL, Tu AY, Regnier M, Harris SP (2008) Contribution of the myosin binding protein C motif to functional effects in permeabilized rat trabeculae. J Gen Physiol 132(5):575–585

Reck-Peterson SL, Tyska MJ, Novick PJ, Mooseker MS (2001) The yeast class V myosins, Myo2p and Myo4p, are nonprocessive actin-based motors. J Cell Biol 153(5):1121–1126

Reubold TF, Eschenburg S, Becker A, Kull FJ, Manstein DJ (2003) A structural model for actin-induced nucleotide release in myosin. Nat Struct Biol 10(10):826–830

Risi C, Belknap B, Forgacs-Lonart E, Harris SP, Schroder GF, White HD, Galkin VE (2018) N-Terminal Domains of Cardiac Myosin Binding Protein C Cooperatively Activate the Thin Filament. Structure 26(12):1604–1611

Risi C, Schafer LU, Belknap B, Pepper I, White HD, Schroder GF, Galkin VE (2021a) High-resolution Cryo-EM structure of the cardiac actomyosin complex. Structure 29(1):50–60 e4

Risi CM, Pepper I, Belknap B, Landim-Vieira M, White HD, Dryden K, Pinto JR, Chase PB, Galkin VE (2021b) The structure of the native cardiac thin filament at systolic Ca(2+) levels. Proc Natl Acad Sci U S A 118(13):e2024288118

Robblee JP, Olivares AO, de la Cruz EM (2004) Mechanism of nucleotide binding to actomyosin VI: evidence for allosteric head-head communication. J Biol Chem 279(37): 38608–38617

Robert-Paganin J, Pylypenko O, Kikuti C, Sweeney HL, Houdusse A (2020) Force generation by myosin motors: a structural perspective. Chem Rev 120(1):5–35

Rohn JL, Patel JV, Neumann B, Bulkescher J, McHedlishvili N, McMullan RC, Quintero OA, Ellenberg J, Baum B (2014) Myo19 ensures symmetric partitioning of mitochondria and coupling of mitochondrial segregation to cell division. Curr Biol 24(21):2598–2605

Sanders TA, Llagostera E, Barna M (2013) Specialized filopodia direct long-range transport of SHH during vertebrate tissue patterning. Nature 497(7451):628–632

Sasaki N, Ohkura R, Sutoh K (2000) Insertion or deletion of a single residue in the strut sequence of Dictyostelium myosin II abolishes strong binding to actin. J Biol Chem 275(49):38705–38709

Sheetz MP (2001) Cell control by membrane-cytoskeleton adhesion. Nat Rev Mol Cell Biol 2(5): 392–396

Shih WM, Gryczynski Z, Lakowicz JR, Spudich JA (2000) A FRET-based sensor reveals large ATP hydrolysis-induced conformational changes and three distinct states of the molecular motor myosin. Cell 102(5):683–694

Shuman H, Greenberg MJ, Zwolak A, Lin T, Sindelar CV, Dominguez R, Ostap EM (2014) A vertebrate myosin-I structure reveals unique insights into myosin mechanochemical tuning. Proc Natl Acad Sci U S A 111(6):2116–2121

Smith CA, Rayment I (1996) X-ray structure of the magnesium(II).ADP.vanadate complex of the Dictyostelium discoideum myosin motor domain to 1.9 A resolution. Biochemistry 35(17): 5404–5417

Spudich JA (1994) How molecular motors work. Nature 372(6506):515–518

Stanganello E, Hagemann AIH, Mattes B, Sinner C, Meyen D, Weber S, Schug A, Raz E, Scholpp S (2015) Filopodia-based Wnt transport during vertebrate tissue patterning. Nat Commun 6: 5846

Sung J, Nag S, Mortensen KI, Vestergaard CL, Sutton S, Ruppel K, Flyvbjerg H, Spudich JA (2015) Harmonic force spectroscopy measures load-dependent kinetics of individual human β-cardiac myosin molecules. Nat Commun 6:7931

Suzuki Y, Yasunaga T, Ohkura R, Wakabayashi T, Sutoh K (1998) Swing of the lever arm of a myosin motor at the isomerization and phosphate-release steps. Nature 396(6709):380–383

Sweeney HL, Rosenfeld SS, Brown F, Faust L, Smith J, Xing J, Stein LA, Sellers JR (1998) Kinetic tuning of myosin *via* a flexible loop adjacent to the nucleotide binding pocket. J Biol Chem 273(11):6262–6270

Sweeney HL, Park H, Zong AB, Yang Z, Selvin PR, Rosenfeld SS (2007) How myosin VI coordinates its heads during processive movement. EMBO J 26(11):2682–2692

Takács B, O'Neall-Hennessey E, Hetényi C, Kardos J, Szent-Györgyi AG, Kovács M (2011) Myosin cleft closure determines the energetics of the actomyosin interaction. FASEB J 25(1): 111–121

Tehver R, Thirumalai D (2010) Rigor to post-rigor transition in myosin V: link between the dynamics and the supporting architecture. Structure (London, England: 1993) 18(4):471–481

Thomas DD, Ramachandran S, Roopnarine O, Hayden DW, Ostap EM (1995) The mechanism of force generation in myosin: a disorder-to-order transition, coupled to internal structural changes. Biophys J 68(4 Suppl):135S–141S

Toyoshima YY, Kron SJ, McNally EM, Niebling KR, Toyoshima C, Spudich JA (1987) Myosin subfragment-1 is sufficient to move actin filaments in vitro. Nature 328(6130):536–539

Tyska MJ, Mackey AT, Huang J-D, Copeland NG, Jenkins NA, Mooseker MS (2005) Myosin-1a is critical for normal brush border structure and composition. Mol Biol Cell 16(5):2443–2457

Uyeda TQ, Ruppel KM, Spudich JA (1994) Enzymatic activities correlate with chimaeric substitutions at the actin-binding face of myosin. Nature 368(6471):567–569

Van Dijk J, Furch M, Lafont C, Manstein DJ, Chaussepied P (1999) Functional characterization of the secondary actin binding site of myosin II. Biochemistry 38(46):15078–15085

Vandekerckhove J, Weber K (1978) Mammalian cytoplasmic actins are the products of at least two genes and differ in primary structure in at least 25 identified positions from skeletal muscle actins. Proc Natl Acad Sci U S A 75(3):1106–1110

Varkuti BH, Yang Z, Kintses B, Erdelyi P, Bardos-Nagy I, Kovacs AL, Hari P, Kellermayer M, Vellai T, Malnasi-Csizmadia A (2012) A novel actin binding site of myosin required for effective muscle contraction. Nat Struct Mol Biol 19(3):299–306

Veigel C, Coluccio LM, Jontes JD, Sparrow JC, Milligan RA, Molloy JE (1999) The motor protein myosin-I produces its working stroke in two steps. Nature 398(6727):530–533

Veigel C, Schmitz S, Wang F, Sellers JR (2005) Load-dependent kinetics of myosin-V can explain its high processivity. Nat Cell Biol 7(9):861–869

von der Ecken J, Muller M, Lehman W, Manstein DJ, Penczek PA, Raunser S (2015) Structure of the F-actin-tropomyosin complex. Nature 519(7541):114–117

von der Ecken J, Heissler SM, Pathan-Chhatbar S, Manstein DJ, Raunser S (2016) Cryo-EM structure of a human cytoplasmic actomyosin complex at near-atomic resolution. Nature 534(7609):724–728

Wagner W, Brenowitz SD, Hammer JA 3rd (2011) Myosin-Va transports the endoplasmic reticulum into the dendritic spines of Purkinje neurons. Nat Cell Biol 13(1):40–48

Wagner W, Lippmann K, Heisler FF, Gromova KV, Lombino FL, Roesler MK, Pechmann Y, Hornig S, Schweizer M, Polo S, Schwarz JR, Eilers J, Kneussel M (2019) Myosin VI drives clathrin-mediated AMPA receptor endocytosis to facilitate cerebellar long-term depression. Cell Rep 28(1):11–20.e9

Wang F, Thirumurugan K, Stafford WF, Hammer JA, Knight PJ, Sellers JR (2004) Regulated conformation of myosin V. J Biol Chem 279(4):2333–2336

Warshaw DM, Desrosiers JM, Work SS, Trybus KM (1990) Smooth muscle myosin cross-bridge interactions modulate actin filament sliding velocity in vitro. J Cell Biol 111(2):453–463

Wegner A (1976) Head to tail polymerization of actin. J Mol Biol 108(1):139–150

Wessels D, Murray J, Jung G, Hammer JA, Soll DR (1991) Myosin IB null mutants of Dictyostelium exhibit abnormalities in motility. Cell Motil Cytoskeleton 20(4):301–315

Whittaker M, Wilson-Kubalek EM, Smith JE, Faust L, Milligan RA, Sweeney HL (1995) A 35-A movement of smooth muscle myosin on ADP release. Nature 378(6558):748–751

Wong WW, Doyle TC, Reisler E (1999) Nonspecific weak actomyosin interactions: relocation of charged residues in subdomain 1 of actin does not alter actomyosin function. Biochemistry 38(4):1365–1370

Woodhead JL, Zhao F-Q, Craig R (2013) Structural basis of the relaxed state of a Ca2+-regulated myosin filament and its evolutionary implications. Proc Natl Acad Sci U S A 110(21): 8561–8566

Wriggers W, Schulten K (1999) Investigating a back door mechanism of actin phosphate release by steered molecular dynamics. Proteins 35(2):262–273

Wu XS, Rao K, Zhang H, Wang F, Sellers JR, Matesic LE, Copeland NG, Jenkins NA, Hammer JA (2002) Identification of an organelle receptor for myosin-Va. Nat Cell Biol 4(4):271–278

Wulf SF, Ropars V, Fujita-Becker S, Oster M, Hofhaus G, Trabuco LG, Pylypenko O, Sweeney HL, Houdusse AM, Schröder RR (2016) Force-producing ADP state of myosin bound to actin. Proc Natl Acad Sci U S A 113(13):E1844–E1852

Yamada A, Mamane A, Lee-Tin-Wah J, Di Cicco A, Prevost C, Levy D, Joanny JF, Coudrier E, Bassereau P (2014) Catch-bond behaviour facilitates membrane tubulation by non-processive myosin 1b. Nat Commun 5:3624

Yamada Y, Namba K, Fujii T (2020) Cardiac muscle thin filament structures reveal calcium regulatory mechanism. Nat Commun 11(1):153

Yamashita H, Sata M, Sugiura S, Momomura S, Serizawa T, Iizuka M (1994) ADP inhibits the sliding velocity of fluorescent actin filaments on cardiac and skeletal myosins. Circ Res 74(6): 1027–1033

Yang Y, Gourinath S, Kovács M, Nyitray L, Reutzel R, Himmel DM, O'Neall-Hennessey E, Reshetnikova L, Szent-Györgyi AG, Brown JH, Cohen C (2007) Rigor-like structures from muscle myosins reveal key mechanical elements in the transduction pathways of this allosteric motor. Structure (London, England: 1993) 15(5):553–564

Yang S, Lee KH, Woodhead JL, Sato O, Ikebe M, Craig R (2019) The central role of the tail in switching off 10S myosin II activity. J Gen Physiol 151(9):1081–1093

Yengo CM, De La Cruz EM, Safer D, Ostap EM, Sweeney HL (2002) Kinetic characterization of the weak binding states of myosin V. Biochemistry 41(26):8508–8517

Yumura S (1996) Rapid redistribution of myosin II in living Dictyostelium amoebae, as revealed by fluorescent probes introduced by electroporation. Protoplasma 192(3):217–227
Zhang J, Herrera AM, Paré PD, Seow CY (2010) Dense-body aggregates as plastic structures supporting tension in smooth muscle cells. Am J Physiol Lung Cell Mol Physiol 299(5):L631–L638
Zhang X, Kampourakis T, Yan Z, Sevrieva I, Irving M, Sun Y-B (2017) Distinct contributions of the thin and thick filaments to length-dependent activation in heart muscle. elife 6:e24081
Zhao F-Q, Craig R, Woodhead JL (2009) Head-head interaction characterizes the relaxed state of Limulus muscle myosin filaments. J Mol Biol 385(2):423–431

Chapter 15
Structure of Motile Cilia

Takashi Ishikawa

Abstract Cilia are tail-like organelles responsible for motility, transportation, and sensory functions in eukaryotic cells. Cilia research has been providing multifaceted questions, attracting biologists of various areas and inducing interdisciplinary studies. In this chapter, we mainly focus on efforts to elucidate the molecular mechanism of ciliary beating motion, a field of research that has a long history and is still ongoing. We also overview topics closely related to the motility mechanism, such as ciliogenesis, cilia-related diseases, and sensory cilia. Subnanometer-scale to submillimeter-scale 3D imaging of the axoneme and the basal body resulted in a wide variety of insights into these questions.

Keywords Cilia · Flagella · Axoneme · Microtubule · Dynein · Basal body · Centriole · Cryo-EM · Cryo-ET · Electron tomography

Motile and Nonmotile Cilia

What Are Cilia?

Cilia are appendage-like organelles with ~300 nm diameter and ~ 5–10 μm length, emanating from eukaryotic cells. There are two categories of cilia: motile and nonmotile cilia. Motile cilia from some organisms are called flagella, especially when there are one or two cilia per cell. Therefore, the terms cilia and flagella are interchangeably used, although the architecture of eukaryotic flagella is completely different from the architecture of prokaryotic flagella. There are also multiciliated cells with >100 cilia (Fig. 15.1a).

T. Ishikawa (✉)
Department of Biology and Chemistry, Paul Scherrer Institute, Villigen, Switzerland
e-mail: takashi.ishikawa@psi.ch

J. R. Harris, J. Marles-Wright (eds.), *Macromolecular Protein Complexes IV*,
Subcellular Biochemistry 99, https://doi.org/10.1007/978-3-031-00793-4_15

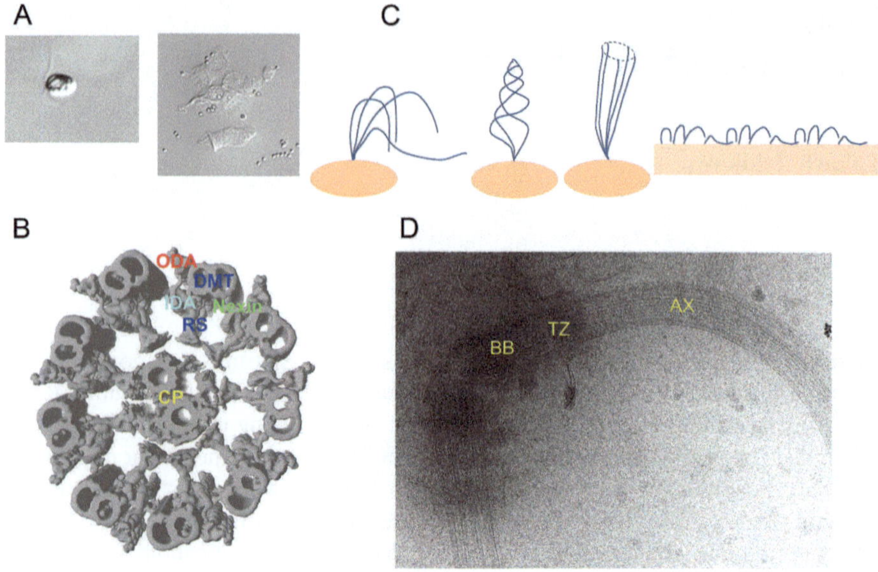

Fig. 15.1 Motile cilia. (**a**) Optical micrographs of *Chlamydomonas* (left) and human (right) respiratory epithelial cells. (**b**) Structure of a 9 + 2 axoneme obtained by cryo-electron tomography. Density maps from subtomogram averaging (Bui et al. 2012) were combined. Important parts are indicated. *DMT* doublet microtubule, *RS* radial spoke, *ODA* outer arm dynein, *IDA* inner arm dynein, *CP* central pair. (**c**) Diagram of various waveforms of motile cilia. From left to right: asymmetric, symmetric, conical waveforms, and metachronal wave pattern from multiple cilia with an asymmetric waveform. (**d**) Canonical structure of cilia, made of the axoneme (AX), the transition zone (TZ), and the basal body (BB), visualized by cryo-EM. Unpublished micrograph acquired in the Ishikawa group. Specimen: Courtesy of Prof. Masafumi Hirono (Hosei University, Japan)

Functional Differences between Motile and Nonmotile Cilia

Motile and nonmotile cilia share similar dimensions and overall ultrastructure, consisting of nine doublet microtubules (DMT) forming a spoke-like architecture (Fig. 15.1b). Most motile cilia, the main focus of this chapter, generate a beating motion by hydrolyzing ATP, while some cilia generate rotational or screw-like motion. The motion of motile cilia enables the cell to swim or cause extracellular fluid flow. Their major moving patterns (waveforms) are the snake-like symmetric waveform and the breast-stroke-like asymmetric waveform (Fig. 15.1c). Cilia in multiciliated cells often make a wave-like pattern called metachronal motion (Fig. 15.1c), which determines the polarity of the cell (Schneiter et al. 2021).

Many nonmotile cilia have intercellular sensory functions involving a variety of membrane proteins and play an essential role in signal transduction (reviewed in Christensen et al. 2007; Berbari et al. 2009). Other nonmotile cilia, for example connecting cilia in the visual receptor cells, are responsible for intracellular transport. While the functions of motile and nonmotile cilia are differentiated, they share a basic structure. Both have the basal body (which has a common origin with the centriole from the centrosome), the transition zone, and the compartmentalized axoneme, all forming a spoke-like architecture with nine microtubules (Gilliam et al. 2012; Kiesel et al. 2020) (Fig. 15.1d).

Axoneme/TZ/Basal Body

Cilia are made of three parts that are connected canonically (Fig. 15.1d). The axoneme is the outer part, from the plasma membrane to the tip of cilia. The axoneme is a microtubule-based structure in a compartmentalized form by ciliary membrane. In motile cilia, the microtubule-based internal part of the axoneme, made of >400 components (Pazour et al. 2005), is sufficient for ciliary beating motion, judging from the in vitro beating motion of axonemes, which are isolated from the basal body and demembraned by detergent, with proper waveform upon addition of ATP (Brokaw 1979; Kamiya and Witman 1984). This is in contrast to the bacterial flagella, which are driven by the rotational motor at the basal part and thus cannot beat when isolated from the base (Minamino et al. 2008).

The base part in the cytoplasm is named the basal body, while the area between the axoneme and the basal body is called the transition zone (Fig. 15.1d). All these three parts have nine microtubules forming a spoke but with different binding proteins (Bettencourt-Dias and Glover 2009; Kumar et al. 2013). In many species, the basal body is converted from the centriole of the centrosome, which functions as MTOC (microtubule organizing center) during the cell division cycle (reviewed in Nabais et al. 2020).

Motile Cilia Functioning in the Human Body and Ciliopathy

While many human cells have one primary cilium each, motile cilia are limited to specific cells, and the number of cilia varies depending on the cell types. Defects in cilia cause a wide spectrum of diseases called ciliopathies (reviewed in Fliegauf et al. 2007). Sperm cells have one cilium per cell, which is essential for motility. Bronchial and nasal epithelial cells from respiratory systems are multiciliated cells, in which cilia make a metachronal motion, cleaning airways. Interestingly, coronaviruses infect ciliated epithelial cells preferentially (Sims et al. 2008), and thus ciliated cell cultures are used in Covid-19 research. Ciliated epithelial cells in the oviduct transfer the egg cells. Brain cilia are responsible for cerebrospinal fluid flow. All the above-mentioned cilia have nine microtubule doublets surrounding two singlet microtubules and make planar or nearly planar (Ueno et al. 2012; Gong et al. 2021) waveforms, either snake-like symmetric or breast stroke like asymmetric waveforms (Fig. 15.1c). In contrast, nodal cilia make a conical rotation with a tilted angle of the rotational axis (Fig. 15.1c). This unidirectional rotation generates a gradient of calcium ions and determines the left-right handedness of the body (Hirokawa et al.

2006). A disease caused by abnormal motile cilia is primary ciliary dyskinesia (PCD), with a wide variety of symptoms, such as lung disease, male and female infertility, mental disorder, and developmental problems. Malfunction of the conical motion of nodal cilia is one cause of situs inversus, in which the left-right asymmetry of the patient is inverted.

Model Organisms to Study Motile Cilia

The culturing of ciliated cells from human was not straightforward (recently it became easier with commercially available kits, where cilia are grown at the air–liquid interface (Müller et al. 2013)). In addition, genetic engineering of human ciliary proteins has not been established. Therefore, model organisms such as mice, zebrafish, and ciliated unicellular ciliates are used to study the basic biology of motile cilia as well as ciliopathies. In addition, unicellular organisms such as the green algae *Chlamydomonas reinhardtii*, the ciliated *Tetrahymena thermophila* and protists of the genus *Trypanosome,* are typical model organisms for motile cilia research. Components (Li et al. 2004; Pazour et al. 2005), waveforms, and ultrastructure of cilia are highly conserved between these unicellular species and human. A number of mutants of *Chlamydomonas* are deposited and maintained at the Chlamydomonas gene center (http://www.chlamycollection.org) and provide opportunities to investigate roles of molecular, cellular, and structural studies on cilia.

Motile Cilia: Wave Form and Calcium Regulation

The motion of cilia is characterized by its waveform, amplitude, and frequency. Some cilia have a snake-like "symmetric" waveform, whereas others have an arch-like "asymmetric" waveform (Fig. 15.1c). Both are cases of an axoneme with nine doublet microtubules with two central singlet microtubules ("9 + 2") (Fig. 15.1b). The cilia which lack the central singlet microtubules form a rotational motion, such as the nodal cilia and cilia of eel sperm cells (Gibbons et al. 1983), suggesting the important role of central microtubules for the generation of specific cilia motions.

Some cilia switch between symmetric and asymmetric waveforms, triggered by the concentration of calcium ions. *Chlamydomonas* cilia make an asymmetric waveform at low calcium concentration, corresponding to positive phototaxis (for photosynthesis), while they turn the waveform to the symmetric shape under high calcium concentration, which corresponds to the photophobic response to protect their chloroplast in the presence of intense light (reviewed in Omoto and Brokaw 1985; Wakabayashi et al. 2021). On the contrary, the sperm cilium from the sea urchin switches in the opposite way—symmetric in the absence of calcium and asymmetric in the presence of calcium (Brokaw 1979). The molecular mechanisms of waveform determination and calcium-induced switching are two major topics in cilia research.

Overall Architecture of Cilia

"9 + 2"/"9 + 0"/Motile/Nonmotile

Cilia consist of (1) the basal body which is embedded in the cytoplasm, (2) the transition zone, and (3) the axoneme; the longest part with 5–10 μm length outside of the cell body. Cilia are categorized into four types, based on the ultrastructure and function (motile or nonmotile) of the axoneme.

Axonemes from most of the species have nine doublet microtubules (DMTs) in a spoke-like arrangement. The doublet microtubule is made of a cylindrical microtubule (A-tubule) and an attached incomplete microtubule (B-tubule) (detail in the next section). In most motile cilia, nine DMTs surround a pair of singlet microtubules ("9 + 2"), while the nodal cilia and *Anguilla* (eel) sperm have "9 + 0" structure without the central singlets and produce conical and screw-like (Woolley 1997) rotations, respectively. Nonmotile (primary) cilia have a "9 + 0" structure with the exception of inner ear kinocilia, which has a "9 + 2"architecture (although kinocilia are nonmotile, they have dyneins; Kikuchi et al. 1989). A recent study demonstrated disturbance of the spoke structure at the tip. At the tip part of the cilia, the incomplete microtubule is terminated and the doublet turns to a singlet microtubule. The length of this singlet part varies: it is short (~<1 μm) in the motile cilia (Höög et al. 2014), but long (several μm) in the primary cilia (Kiesel et al. 2020).

There are a few exceptional axonemes with an unusual number of DMTs. The sperm of a cecidomyiinae, *Asphondylia ruebsaameni*, has >100 DMTs with only outer dynein arms (Mencarelli et al. 2000). Gametes from the gregarines *Lecdina tuzetae* and *Deplauxis hatti* have three and six DMTs in their motile cilia, respectively (no central singlet) (Schrevel and Besse 1975; Prensier et al. 1980). This shows that the number "9" is not due to functional necessity. Probably the axoneme structure with nine DMTs was fixed evolutionally and microtubule-binding proteins evolved to adjust its dimension.

Structural Features of the Basal Body

Cilia grow from the basal body. This means that the nine-fold structure of cilia originates from the basal body. At the proximal part of the basal body, there is a nine-fold central hub called cartwheel (Fig. 15.2a). The cartwheel is made of SAS-6 proteins. SAS-6 forms a homodimer with two heads and an alpha-helical C-terminus, forming a coiled-coil. In vitro, SAS-6 spontaneously forms a nine-fold symmetric structure which resembles the cartwheel, suggesting a molecular mechanism

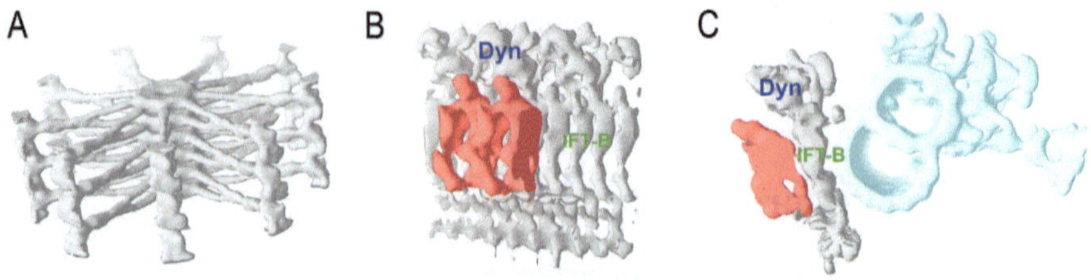

Fig. 15.2 (**a**) Structure of the cartwheel at the basal body, imaged from the cryo-ET density map (EMD-2329) (Guichard et al. 2013). (**b**) Structure of the anterograde IFT complex (Jordan et al. 2018) seen from the side. IFT-A is colored in red. IFT-B and one cytoplasmic dynein (IFT dynein) are indicated. The image was made from the density maps EMD-4303 and EMD-4304. (**c**) Diagram showing how the anterograde IFT is located with respect to the axoneme as an end-on view. DMT is from (Bui et al. 2012) (EMD-2132)

of nine-fold symmetry (Kitagawa et al. 2011; van Breugel et al. 2011; Guichard et al. 2012). By genetic engineering of the head part of SAS-6, this cartwheel-like structure can be designed with a different symmetry (Hilbert et al. 2016). Interestingly, however, when these engineered SAS-6 variants are expressed in the *Chlamydomonas* SAS-6 deletion mutant, cilia with nine DMT are still dominant (Hilbert et al. 2016). This suggests an alternative mechanism to rescue the cartwheel to have a proper nine-fold symmetric basal body formation.

The cartwheel formation is followed by recruitment of nine triplet microtubules in the case of many species including *Chlamydomonas*. Delta and epsilon tubulins are required to nucleate the proximal part of the triplet (Stathatos et al. 2021).

The mechanism to convert the centriole to the basal body is not clear. This conversion should include transport of the centriole to the surface of the cell, reorienting and initiating the DMT extension. CEP110, suppression of which promotes ciliogenesis and over-expression of which suppresses cilia formation (Spektor et al. 2007), is one potential candidate of a capping protein, which prohibits microtubule extension at the distal end of centriolar microtubules and is removed upon ciliogenesis.

IFT and BBSome: How Do Cilia Grow?

Ciliary proteins are translated and preliminarily complexed in the cytoplasm and then transported to the ciliary tip (i.e., direction to the plus end of the DMT). Intraflagellar transporter (IFT) is a protein complex responsible for this transport (Kozminski et al. 1993b). IFT loads cargos at the base of cilia, transports them moving on the axoneme, unloads them at the tip of cilia and finally, either without cargos or with cargos to bring back, moves to the cytoplasm on the axoneme (review in Ishikawa and Marshall 2017). IFT is located between the cilia membrane and the DMT (Kozminski et al. 1993b) and employs kinesin-2 to move toward the tip (anterograde transport) (Cole et al. 1998) and cytoplasmic dynein (different from axonemal dyneins) to move from the tip to the base (retrograde transport) (Pazour et al. 1999). This means that anterograde IFT transports cytoplasmic dyneins and retrograde IFT transports kinesin-2. Indeed cryo-ET study visualized dyneins transported by IFT at the position in proximity to the membrane (Jordan et al. 2018) (Fig. 15.2b). Anterograde and retrograde IFTs are on B- and A-tubules of the doublet microtubules, respectively, indicating the mechanism to avoid collisions on the DMT (Stepanek and Pigino 2016). By electron microscopy, IFT appears as a train with a periodic feature. It consists of >20 components, which form two complexes, IFT-A and IFT-B (reviewed in Mourão et al. 2016). In principle, IFT-A and –B play crucial roles for retrograde and anterograde transports, respectively (reviewed in Taschner and Lorentzen 2016). In the anterograde IFT train, both IFT-A and IFT-B form tandem arrays in parallel with the arrangement to maximize the interface between IFT-B and the B-tubule but sterically hinder the interaction of IFT-A to DMT (Jordan et al. 2018) (Fig. 15.2b). Probably, rearrangement of these components occurs at the ciliary tip, allowing IFT-A to interact with the A-tubule in the retrograde IFT.

The mechanism to recruit cargos to the IFT is another intriguing topic. In the case of membrane proteins, they stay in the ciliary membrane directly and are trafficked by IFT. Their recruitment is mediated by BBSome, a complex named after one ciliopathy caused by the abnormality of this complex, Bardet–Biedl Syndrome (Nachury 2018). The BBSome consists of eight proteins and the GTPase ARL6/BBS3 is the key to bind membrane proteins (review in Nachury and Mick 2019). Recently solved atomic structures of BBSome revealed how ARL6/BBS3 binding to the other components activates BBSome by widening a cavity to accommodate a polypeptide of the membrane protein (Klink et al. 2020; Singh et al. 2020; Yang et al. 2020). There should be wide variety of membrane proteins to be carried by BBSome, depending on species and types of cilia. In the visual cells, such as rod cells, defects in BBSome cause a lack of rhodopsin, and therefore visual problem (retinitis pigmentosa); reviewed in (Wingfield et al. 2018). In *Chlamydomonas* cilia, phosopholipase D is transported by retrograde IFT and BBSome (Lechtreck et al. 2013). However, the anterograde transport is independent from IFT and BBSome, suggesting an alternative trafficking mechanism.

Detailed Structure of the 9 + 2 Axoneme from Motile Cilia

In this section, we will have a detailed look into the 96 nm periodic unit of the axoneme.

The Doublet Microtubule

The structure of the doublet microtubule (DMT) from the axoneme was analyzed at atomic resolution (Ichikawa et al. 2019; Ma et al. 2019). The A-tubule is made of 13 protofilaments, which are parallel to the axis (no twist), similarly to the microtubule with 13 protofilaments reconstituted from tubulin dimers. Both have a helical discontinuity (seam), since the alpha-tubulin meets the beta tubulin after one turn of a helical pitch (Maheshwari et al. 2015). In DMT, the seam is located near the internal joint between the A- and B-tubules (Ichikawa et al. 2019). The B-tubule has 10 protofilaments, also parallel to the DMT axis.

On the DMT of motile cilia, various proteins bind to the A- and B-tubules with periodicity (Fig. 15.3a). The length of the periodic unit is 96 nm and defined by two coiled-coil proteins, which span over the unit and act as a ruler (Oda et al. 2014a). This periodic structure enables high resolution structural analysis of DMT, dyneins, and other ciliary proteins by single particle cryo-EM (Ichikawa et al. 2019; Ma et al. 2019; Gui et al. 2021) and cryo-electron tomography (summarized in Ishikawa 2016). While the 96 nm periodicity along the doublet is strictly maintained, the content of the binding proteins on it varies between the proximal and the distal areas as well as between different doublets (discussed in the next sections).

A number of proteins bind to the internal side of the DMT (Fig. 15.3b). These proteins, called MIPs, follow a high frequency (16 nm, 24 nm, 32 nm, 48 nm) inside the 96 nm periodic unit. Their function is still to be investigated. FAP45 and FAP52 contribute to the integrity and stability of the DMT (Owa et al. 2019). Adjacent protofilaments are connected by coiled-coil MIPs (Ichikawa et al. 2019), probably reinforcing the stability to be higher than pure microtubules, which undergo dynamic depolymerization and polymerization. Without MIPs, the protofilaments of DMT would separate during the active beating motion.

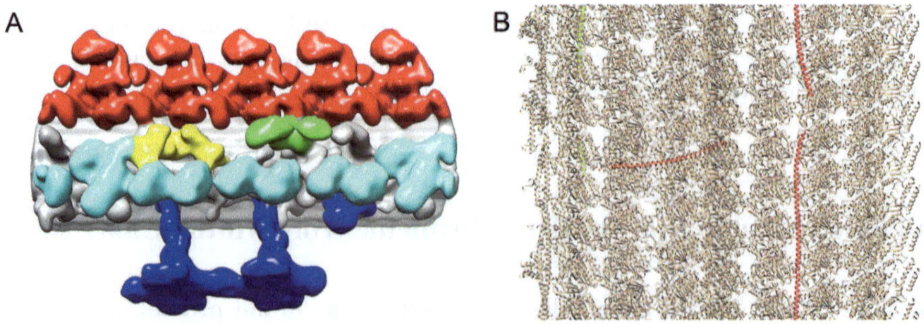

Fig. 15.3 Doublet microtubule structure. (**a**) 96 nm subtomogram average from cryo-ET (Bui et al. 2012). Microtubule, ODA, IDA, RS, LC/IC, and nexin are colored grey, red, cyan, blue, yellow, and green, respectively. (**b**) Inside of DMT (Ma et al. 2019). Tubulins are shown in brown. MIPs with characteristic alpha-helical structure are colored red, orange, and yellow

Post-Translational Modification of DMT

The most abundant post-translational modification in cilia (reviewed in Wloga et al. 2017) is acetylation in K40, facing the inside of the microtubule, of alpha-tubulin (Piperno and Fuller 1985). While this acetylation seems not to play an essential role in ciliary motility in *Chlamydomonas* (Kozminski et al. 1993a) and *Tetrahymena* (Gaertig et al. 1995), it enables rapid growth of primary cilia in nematodes (Shida et al. 2010).

Tyrosination of the C-terminus of alpha-tubulin by the tubulin tyrosine ligase (TTL) plays a crucial role in the microtubule network in the cytoplasm. In the ciliary B-tubule, detyrosinated alpha-tubulin is abundant (Johnson 1998). Since the B-tubule is known to be involved in the pathway of anterograde IFT by kinesin-2 (Stepanek and Pigino 2016), and the velocity of kinesin-2 is accelerated by detyrosination of the microtubule in vitro (Sirajuddin et al. 2014); the role of detyrosination might be significant for maintaining anterograde IFT.

Polyglutamylation and glycylation occur at the C-termini of both alpha- and beta-tubulins. While defects of polyglutamylation of the alpha-tubulin weaken the motility of WT *Chlamydomonas*, absence of polyglutamylation paralyzes *Chlamydomonas* in some dynein deletion mutants (Kubo et al. 2010, 2012), indicating the role of polyglutamylation for dynein interaction to tubulin. Interestingly, under in vitro conditions, lack of polyglutamylation rather accelerates dynein velocity on DMT (Kubo et al. 2010, 2012). This phenomenon probably reflects complex mechanism of ciliary beating, caused by a network of dynein functions with other molecules, including coordinated activation and suppression (discussed later). Glycylation also seems to play a role for ciliary beating. A knock-out mouse model with defected glycylation shows abnormal dynein conformation and swimming aberration in sperm (Gadadhar et al. 2021).

Dynein

Dyneins, microtubule-based ATPase motor proteins, are the driving force of ciliary motion. Dynein was discovered by Gibbons from cilia (Gibbons and Gibbons 1973) and named after the unit of force, dyne. Later dynein was found in the cytoplasm and recognized as a common minus-end-driven microtubule-based motor to carry cargos for intracellular transport. In this sense, dynein functions as a counterpart of kinesin, which is responsible for plus-end directed intracellular transport (Gennerich and Vale 2009). This type of dynein and dynein from cilia are homologues and called cytoplasmic and axonemal dyneins, respectively.

The major difference between cytoplasmic and axonemal dyneins is located at the N-terminus. Dynein is a gigantic proteins with >4500 amino acids in one polypeptide chain. The C-terminal ~3000 amino acids compose a motor domain, which has six AAA domains forming a ring (AAA-ring), two coiled-coil structures, called a

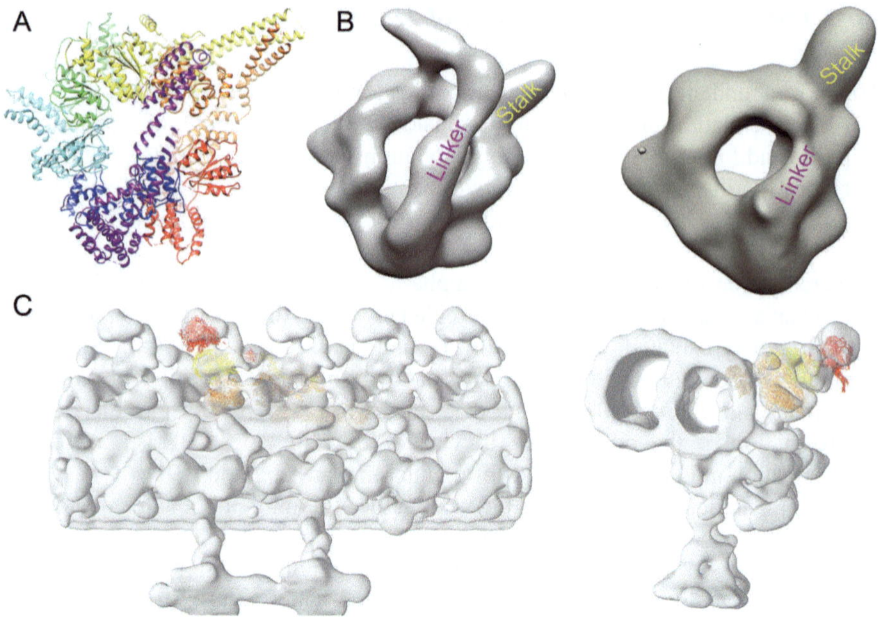

Fig. 15.4 Dyneins. (**a**) Dynein diagram from an atomic structure of the cytoplasmic dynein motor domain (3QMZ (Carter et al. 2011b)). The linker, AAA1, AAA2, AAA3, AAA4 and the stalk, AAA5 and the buttress, and AAA6 are shown in purple, blue, cyan, green, yellow, brown, and orange, respectively. (**b**) Inner dynein c structure in the pre- (left: EMD-2156) and post-power stroke (right: EMD-2155) states. From (Roberts et al. 2009). The models are oriented roughly in the same orientation as Fig. 15.3a. The linker and the stalk are indicated. (**c**) ODA structure by combining cryo-ET (EMD-2132 (Bui et al. 2012)) and single particle cryo-EM (Walton et al. 2021). The three outer arm dyneins are indicated by colors

stalk (protruding from AAA4) and a buttress (from AAA5), the microtubule-binding domain (MTBD) at the tip of the stalk. At the boundary between the N-terminal and AAA1, an alpha-helix-rich domain called a linker bridges over the AAA ring (Fig. 15.4a) (Carter et al. 2011a; Kon et al. 2011, 2012). AAA1 is responsible for ATP hydrolysis for force generation (Takahashi et al. 2004). This part with the common architecture between cytoplasmic and axonemal dyneins is named the dynein motor domain.

The force generation mechanism of the dynein motor domain was investigated by X-ray crystallography and single particle cryo-electron microscopy (cryo-EM). It was known that dynein generates a force when the phosphate (which was the gamma phosphate of ATP) is released, that is, during transition from the ADP.Pi to the ADP states (Johnson 1985), similarly to myosin and differently from kinesin (Gennerich and Vale 2009). Therefore, the question how the conformational change of dynein takes place during the force generation (called power stroke) was answered by comparing structures of dynein in the ADP.Pi state, which was fixed by using nonhydrolyzable ADP.vanadate (ADP.Vi) mimicking ADP.Pi, and in the ADP or apo state. Single particle EM of axonemal dyneins proved movement of the linker (Burgess et al. 2003; Roberts et al. 2009) (Fig. 15.4b). The linker bridges between AAA1 and AAA4 in the absence of nucleotides or with ADP, but with ATP (or its analog) the linker connects between AAA1 and AAA2. The two structures are called post- and pre-power stroke conformations, respectively. This transition was shown at atomic resolution by X-ray crystallography of cytoplasmic dynein, comparing post- (Carter et al. 2011a; Kon et al. 2011, 2012; Schmidt et al. 2012) and pre-power stroke conformations (Schmidt et al. 2014) (Fig. 15.4b).

How does dynein bind to the microtubule? At least two parts of binding, corresponding to pre- and postpower stroke states are expected. According to kinetics studies, the binding is strong in the postpower stroke state and weak in the pre-power stroke state (Omoto and Johnson 1986). Single particle cryo-EM of dynein MTBD on the microtubule visualized the strong binding at atomic resolution (Redwine et al. 2012). MTBD binds on the protofilament, between two components of tubulin dimer (alpha and beta subunits). Strong and weak binding states should be switched by ATP hydrolysis at AAA1, which is connected to MTBD via the stalk. There should be a mechanism to propagate information of ATP hydrolysis from AAA1 to the MTBD through the stalk. One potential hypothesis is that two alpha helices of the stalk slide upon ATP hydrolysis by AAA1 to shift registry. Indeed such a slide generated by genetic engineering of the stalk changed affinity between the MTBD and the microtubule (Carter et al. 2008). High resolution structures of the whole motor domain interacting with the microtubule at both states are awaited.

While the motor domain is highly conserved, the N-terminal tail part is diverse between various types of dyneins. Since cytoplasmic dynein carries a cargo, the tail is designed to bind the adaptor to specifically connect the cargo. In contrast, the N-terminus of axonemal dyneins is anchored on the A-tubule. The MTBD interacts with the B-tubule in the same way as the MTBD of cytoplasmic dynein interacts with the microtubule cytoskeleton, which consists of singlet microtubules. While cytoplasmic dynein has a free N-terminal end and walks toward the minus end of the microtubule, axonemal dynein is permanently anchored on the A-tubule and drags the adjacent B-tubule to let the two DMTs slide past each other.

Outer and Inner Dynein Arms

As mentioned in the previous section, the sequence, structure, and probably the functional mechanism of force generation at the motor domain is common between cytoplasmic and axonemal dyneins, while the tail part is diverged. All the axonemal dyneins are anchored on the A-tubule permanently by the N-terminal tail and dynamically interact with the adjacent B-tubule through the MTBD. Moreover, the axonemal dynein has many species (isoforms) (Yagi 2009) and they have fixed loci on the A-tubule (Bui et al. 2008, 2012). Each of them is associated with light chains, intermediate chains, and other binding proteins (summarized in (Kamiya 2002)).

Axonemal dyneins are categorized into two groups: outer and inner dyneins. Two or three outer dyneins stack to make a complex, called an outer dynein arm (ODA), and are located on the A-tubule at the external part of the axoneme. In *Chlamydomonas*, the three outer dyneins are named gamma, beta, and alpha in the order of proximity to the A-tubule (Fig. 15.4c). The AAA-rings of the three outer dyneins are stacked nearly in parallel in the postpower stroke state but tilted with respect to each other in the pre-power stroke (Movassagh et al. 2010). Eight inner dyneins are arrayed canonically in the 96 nm repeating unit, called an inner dynein arm (IDA). Two out of the eight inner dyneins (dynein f–dynein fα and fβ in *Chlamydomonas*) form a dimer, while the other six are single headed (Mastronarde et al. 1992). All the inner and outer dyneins are located with similar orientations, anchored on the A-tubule with the N-terminal tail, which extends from the AAA-ring toward the distal direction (Bui et al. 2008, 2012).

Functional studies on *Chlamydomonas* mutants revealed the roles of the ODA and IDA. While an IDA defect paralyzes ciliary beating or alternate waveform, ODA mutants can swim with their normal waveform, but weakly and slowly (summarized in Kamiya 2002), suggesting that ODA is the accelerator of ciliary motion and IDA is regulator and essential for waveform determination.

Diversity of outer and inner dynein arrangement between the proximal and the distal areas and between nine DMTs was examined in detail for *Chlamydomonas* and human (next sections).

Detailed Structure of ODA

There are four ODAs within a 96 nm periodic unit with 24 nm periodicity. Bikonts (including *Chlamydomonas* and *Tetrahymena*) and unikonts (including *Trypanosoma*, sea urchin, and vertebrates) have three and two dyneins per ODA, respectively (Inaba 2015). A functional difference between ODAs with two and three dyneins is not known. *Chlamydomonas* mutants which lack one outer dynein (thus similar ODA structure as unikonts) have no significant difference in ciliary beating (Kamiya 2002). The apposed pair of two *Chlamydomonas* cilia lacks the ODA (Hoops and Witman 1983). The location of the ODA is replaced by filamentous linkers with 8 nm spacing (Hoops and Witman 1983; Bui et al. 2009). The mechanistic meaning of this asymmetry is not known: since mutants with an ODA defect make a normal waveform, this asymmetry does not appear to play a role in determining the waveform.

While *Chlamydomonas* ODAs have the same components throughout the DMT, human cilia have different dynein isoforms at the proximal and the distal areas. Out of the two stacking dyneins, the one located close to the A-tubule (DNAH5) is common, but the further loci from the A-tubule (i.e., close to the adjacent B-tubule) is occupied by DNAH11 and DNAH9 at the distal and the proximal area, respectively (Olbrich et al. 2002; Dougherty et al. 2016). Interestingly, the defect of DNAH9 causes more developmental diseases (laterality defects) than respiratory problems (Loges et al. 2018), suggesting diverse roles of dynein isoforms inside human organs.

Outer arm dyneins are associated with a number of light and intermediate chains, which are well characterized in *Chlamydomonas* (Sakato and King 2004). The ODA complex of three dyneins and light/intermediate chains forms a bouquet-like structure when isolated but is conformed into an alternative structure on the microtubule (Goodenough and Heuser 1984, 1985). A recently solved atomic structure of ODA in the postpower stroke state presents a complex architecture of three dyneins (alpha, beta, gamma) folded together with light and intermediate chains (Kubo et al. 2021; Rao et al. 2021; Walton et al. 2021). These structures were analyzed by single particle cryo-EM. Since the intact axoneme is not suitable for this method (in the micrograph, other molecules are overlapped onto the image of ODA and prohibit image analysis), the nine DMTs were split from the axoneme, in which the ODA binds either to the A-tubule (Kubo et al. 2021; Walton et al. 2021) or to the B-tubule (Rao et al. 2021). The ODA conformation in the axoneme can be modeled by combining single particle cryo-EM structures with lower resolution cryo-ET data from the intact axoneme (summarized in Ishikawa 2016) (Fig. 15.4c). All the three AAA-rings are nearly parallel, oriented with the linker away from the A-tubule and the N-terminus extending toward the distal direction of DMT. Light chains form a complex (LC-tower) and are located between ODA and IDA. In one ODA (first of the four ODAs in the 96 nm unit), the LC-tower bridges the ODA and the LC/IC complex of inner dynein f. The LC-tower of the third ODA connects to the IDA via nexin.

The atomic structure of the ODA in the pre-power stroke state is still to be analyzed. Probably in the AAA-head, the linker is reoriented as shown in isolated dynein c (Roberts et al. 2009) (Fig. 15.4b) and in cytoplasmic dynein (Schmidt et al. 2014) (Fig. 15.4b). But conformational change at the tail part of the ODA has not been analyzed at high resolution. According to the lower resolution cryo-ET analysis of the *Chlamydomonas* ODA in the presence of a nonhydrolyzable ATP analog, orientations of the three AAA-rings change (not parallel any more) and they move ~8 nm toward the proximal direction with respect to the A-tubule (Movassagh et al. 2010). This structural change can be explained by the "winch" model, where the dynein AAA-ring shifts by 8 nm, the distance of the microtubule periodicity upon ATP hydrolysis, to drag the adjacent B-tubule to cause the power stroke (Ueno et al. 2008). A similar observation was made in the axoneme with two outer dyneins from sea urchin sperm flagella (Lin et al. 2014a; Lin and Nicastro 2018) and mouse respiratory cilia (Ueno et al. 2014). To build a more detailed model to explain how the AAA-ring shifts, we will need a high-resolution structure of ODA in the pre-power stroke state.

The ODA is preassembled in the cytoplasm and transported by the IFT. When transported by the IFT, the ODA takes on a different conformation from the one in the axoneme by binding to shulin (Mali et al. 2021).

Inner Dyneins and Associated Proteins

There is one heterodimer of inner dyneins (called dynein f in *Chlamydomonas*) and six single-headed inner dyneins (dynein a-e and g) within the usual 96 nm periodic unit. They have fixed loci on the A-tubule. The six single-headed dyneins form three triads, located around radial spokes (Bui et al. 2008) (Fig. 15.3a). The heterodimeric dynein f is closer to the surface of the A-tubule and connected to the ODA, the radial spokes, and other inner dyneins via the light and intermediate chain complex (indicated in Fig. 15.3a). Deletion of dynein f results in slower motility and an altered waveform (Kamiya et al. 1991). Dynein c deletion does not have an influence on ciliary beating under normal condition but causes swimming defect in viscous solution (Yagi et al. 2005), indicating diverse roles of inner dynein isoforms. When isolated, these dyneins show a variety of biophysical properties. They drive the microtubule at different velocities (summarized in Kamiya 2002). Some dyneins generate torque, while the others do not (Kagami and Kamiya 1992; Kikushima and Kamiya 2008). Therefore, it can be imagined that the design of ciliary beating motion is determined by inner dyneins, with various biophysical properties on the fixed position. The conformational change of inner dyneins during the power stroke has been examined at intermediate resolution by cryo-ET (Lin and Nicastro 2018) and will provide information about how it is correlated to the mode of beating.

Chlamydomonas has 16 dynein genes. Three are outer dyneins and one is the cytoplasmic dynein. As mentioned above, eight inner dyneins are known. The remaining four dyneins are called minor dyneins and proved to be localized at the proximal area (1 ~ 2micron from the plasma membrane) (Yagi et al. 2009). These minor dyneins are likely replacing some inner dyneins at the proximal area (Bui et al. 2012), suggesting uneven biophysical properties at the base and at the center of cilia. Furthermore, some inner dyneins are lost at the proximal area and/or on the apposed DMT (same as missing ODAs) (Bui et al. 2012). It is to be investigated if this longitudinal and circumferential asymmetry contributes to waveform determination of planar asymmetric beating.

Inner dyneins are also associated with many light and intermediate chains (reviewed in Yamamoto et al. 2021). Among them, light and intermediate chains of *Chlamydomonas* dynein f (dimeric inner dynein), associated with the tail of dynein f, have been investigated most widely and in detail. They form a large complex, called LC/IC complex, which is surrounded by ODA, IDA, RS, and nexin (Fig. 15.3a), and are supposed to regulate ciliary beating either directly or by stabilizing this complex (summarized in Yamamoto et al. 2021).

Besides the LC/IC complex, there are other proteins associated with dynein f, regulating ciliary motion. FAP44 (Kubo et al. 2018) and MOT7 (Kutomi et al. 2021) play essential roles for phototaxis of *Chlamydomonas* and present an example of molecular networks of dynein regulation. *Chlamydomonas* swims toward light to make photosynthesis (positive phototaxis) but under too intense illumination swims away from the light by changing the waveform to symmetric (negative phototaxis). FAP44 and MOT7 are located close to the dynein f heads and are linked together.

While FAP44 deletion results in a complete defect of phototaxis, the loss of MOT7 only causes a defect in negative phototaxis (Kutomi et al. 2021).

RS and CP

The radial spoke (RS) is a T-shaped complex, which extends from the A-tubule toward the central pair (CP) (Fig. 15.1b and c). There are two or three radial spokes in the 96 nm periodic unit, depending on the species. The two RSs (RS1 and RS2) share a similar structure (but not identical, according to the atomic structures (Gui et al. 2021)), while RS3 has a totally different structure (Fig. 15.5a) (Pigino et al. 2011). While the components and atomic structures were analyzed only in *Chlamydomonas*, similarity of RS1/RS2 structures between species (Pigino et al. 2011; Lin et al. 2014b) suggests they are made of homologous proteins. *Chlamydomonas* lacks RS3 but has a protrusion which resembles the base part of *Tetrahymena* RS3 (Pigino et al. 2011), which provides insight into its evolutional pathway. In genetically engineered *Chlamydomonas* cilia with longer periodic units than 96 nm, three RSs are accommodated (Oda et al. 2014a), suggesting that RSs are arranged to fill the space of the periodic unit. Twenty-three component proteins were identified in RS1 and RS2 with various interesting sequence motifs including AKAP (RSP3), aldo-keto reductase (RSP5), EF-hand (RSP7), DnaJ chaperon (RSP16), and calmodulin (RSP20) (Yang et al. 2006), indicating possible signal transduction in

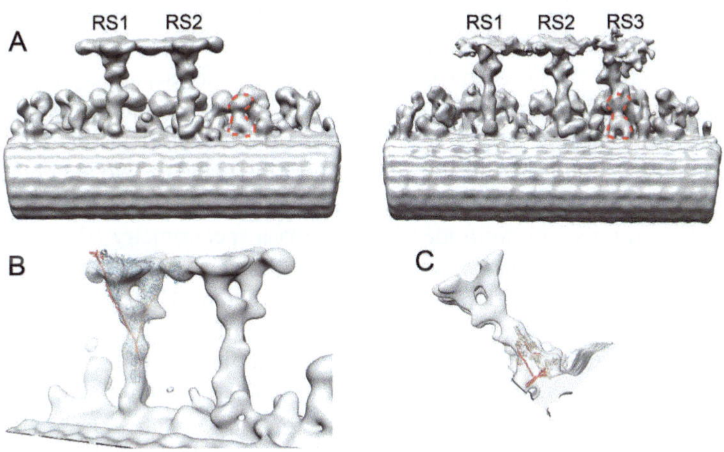

Fig. 15.5 Radial spoke. (**a**) Overview from cryo-ET. *Chlamydomonas* (left) and *Tetrahymena* (right). RS1, RS2, and RS3, as well as similar structures corresponding to the RS3 base (red dotted line) are indicated. (**b**) Atomic structure of RS, highlighting the bifurcated neck. The low resolution cryo-ET density map of the intact axoneme (EMD-1941) (Pigino et al. 2011) and the atomic model from single particle cryo-EM (7JTK) (Gui et al. 2021) were combined. RS3 dimers are indicated in red and orange. (**c**) Atomic structure of the base part of RS1 by juxtaposing EMD-1941 (Pigino et al. 2011) and 7JTS (Gui et al. 2021). Calmodulin is indicated in red

this complex. Between the head part and the stalk part of the T-shape, RS is split into two (bifurcated neck) (Pigino et al. 2011; Barber et al. 2012). In the recently published atomic structure of RS by single particle cryo-EM, this bifurcated part is composed of two RSP3s splitting in a pseudo two-fold symmetric way (Gui et al. 2021) (Fig. 15.5b). Calmodulin-binding complexes at the base of RS1/RS2/RS3 (Dymek and Smith 2007; Urbanska et al. 2015) suggest their role in waveform regulation by calcium. Calmodulin (RSP20) links the A-tubule, one calmodulin associated complex at the base part of RS and the spoke part of RS (Gui et al. 2021) (Fig. 15.5c), which may change the geometrical or mechanical properties of the connection between RS and the A-tubule upon calcium binding. *Chlamydomonas* mutants in which RS is depleted are paralyzed under physiological conditions, indicating that the RS/CP interaction is essential for physiological beating motion. Under an unusual condition with an excess of ADP, a RS mutant can make a beating motion, but the waveform is different from that of WT (Yagi and Kamiya 2000).

It seems that RS regulates coordination of dynein power strokes to enable beating motion. There are reports of interaction between RS and inner dyneins. Increased phosphorylation of IC138, one intermediate chain of dynein f, paralyzes or alters the waveform of *Chlamydomonas* cilia, in the same way as a defect of dynein f itself (Hendrickson et al. 2004). In *Chlamydomonas* mutants with loss of the radial spoke or the central pair, IC138 is hyperphosphorylated and induces the same phenotype as dynein f mutants (Hendrickson et al. 2004).

RS research provides information about ciliogenesis pathways. RS is preassembled as a half complex in a L-shape and transported by IFT before it is finally assembled into the T-shaped complex and integrated in the axoneme (Diener et al. 2011). The preassembled complex includes all the components except RSP16, which has a motif of the DnaJ chaperone and is transported separately by IFT to the ciliary tip. This enables us to speculate that the two half L-shaped complexes are finally assembled with the help of RSP16 at the tip of cilia (Diener et al. 2011).

The central pair (indicated in Fig. 15.1b and c) is made of two singlet microtubules (C1 and C2) with a number of proteins decorating the singlets in a sheath-like structure. The components were identified by mass spectrometry (Dai et al. 2020) and also include calcium-sensitive centrin (thus CP may also play a role for calcium regulation). They are arranged in 32 nm periodic structure (Carbajal-González et al. 2013). Within this periodic unit, two outstanding protrusions extend each from C1 and C2 (thus with 16 nm distance). While gamma tubulin was found at the proximal end of CP, similarly to the cytoplasmic microtubules (Silflow et al. 1999), the molecular mechanism of initiation of CP growth is not clear yet. Since *Chlamydomonas* mutants without RS, and mutants with more than nine DMTs (thus having a larger space inside) accommodate two pairs of CP microtubules; the space seems critical for CP formation.

What is the mechanism of transduction from CP to RS? CP of *Chlamydomonas* and *Tetrahymena* twists (Mitchell and Nakatsugawa 2004; Pigino et al. 2012), while DMTs extend straightforwardly without twisting. This results in a mismatch inside the axoneme, which will not allow periodic one-to-one interaction between the RS head and one particular molecule of CP. The interaction is likely nonspecific. When the protrusion from the CP is depleted, the cilia are paralyzed. However, they are rescued (i.e., motility is recovered) by engineering additional density to the RS head to compensate for the volume between CP and RS (Oda et al. 2014c). This experiment supports the model of mechano-transduction from CP to RS. On the other hand, the RS head contains a strong negative charge on its surface (Gui et al. 2021), suggesting electrostatic interaction. Further studies are needed to clarify this.

Nexin

Nexin links adjacent DMTs. In sea urchin sperm, enzymatic digestion of nexin results in sliding disintegration of the axoneme, instead of beating, upon addition of ATP (Tanaka and Miki-Noumura 1988; Hayashi and Shingyoji 2008). Based on this phenomenon, a hypothesis of nexin connecting DMTs and converting sliding motion into bending was built. Nexin is a bundle of at least three filamentous proteins (Oda et al. 2014b), emanating from the area between one ODA and inner dynein g and d.

Final Thoughts

Mechanics

After all the efforts to analyze and characterize ciliary motion, we still do not have a clear molecular model. There are a number of hypotheses and mathematical theories to explain the ciliary motion based on geometry and possible mechanics occurring in the axoneme (Lindemann 2007; Riedel-Kruse et al. 2007; Brokaw 2009). The switching hypothesis, in which one side of the curved axoneme has activated dyneins and the other side has inactive dyneins (Satir and Matsuoka 1989), is supported by cryo-ET experiments (Lin and Nicastro 2018). These results are consistent with the fact that, upon addition of a nonhydrolyzable ATP analog, neighboring dyneins change conformation with cooperativity (Movassagh et al. 2010). ATPase assays also support the model that dynein is locally activated and inactivated in the axoneme (Maheshwari and Ishikawa 2012). However, to analyze dynein structures from a real intact axoneme in motion, the current cryo-EM system is not sufficient. We need development of improved experimental systems to address this problem (discussed in detail in Ishikawa 2015). It also has to be noted that the model based on sliding disintegration experiments, which explain the curvature generation well (Hayashi and Shingyoji 2008), rather predict an increased activity of dynein sliding in the straight part, not at the curved part.

In case any kind of switching of dynein activity, what makes the switching possible? How is the switching pattern changed by calcium concentration? One candidate is the radial spoke. RS contains components with signal transduction motifs. However, considering the fast switch during beating (50–60 Hz in *Chlamydomonas* cilia), the biochemical transduction will be too slow. It is more likely that switching during the beating motion takes place by another mechanism, and the biochemical transduction through RS modulates, for example, the beating pattern or amplitudes.

We know even less about mechanisms to consort multiple cilia. Multiciliated cells make a metachronal motion (Fig. 15.1c), which means cilia crosstalk each other. It is not known if this happens by sensing motion of the adjacent cilium, for example, by membrane proteins, or through the basal bodies, or by any other mechanism.

Thought about Evolutionary Origins

The similarities and differences between molecules and organelles raise questions about their evolution. Did motile cilia stem from primary cilia, or vice versa? Which came first, the centriole for MTOC or the ciliary basal body? In the past, there were hypotheses to attribute the origin of centriole/cilia to endosymbiosis (Chapman et al. 2000) and viral products (reviewed in Marshall 2009), which are both attractive but not supported by recent experimental data. How did cytoplasmic and axonemal dyneins separate in the phylogenic tree? Why was axonemal dynein diverged into many isoforms, while a single species of cytoplasmic dynein employs adaptors to have specific binding to various cargos? Attempt to answer these question will activate evolutionary biology.

Acknowledgement The author thanks Dr. Roger Benoit (PSI) for proofreading.

References

Barber CF et al (2012) Three-dimensional structure of the radial spokes reveals heterogeneity and interactions with dyneins in Chlamydomonas flagella. Mol Biol Cell 23(1):111–120. https://doi.org/10.1091/mbc.E11-08-0692

Berbari NF et al (2009) The primary cilium as a complex signaling center. Curr Biol 19(13):R526–R535. https://doi.org/10.1016/j.cub.2009.05.025

Bettencourt-Dias M, Glover DM (2009) SnapShot: centriole biogenesis. Cell 136(1):188.e1–188.e2. https://doi.org/10.1016/j.cell.2008.12.035

Brokaw CJ (1979) Calcium-induced asymmetrical beating of triton-demembranated sea urchin sperm flagella. J Cell Biol 82(2):401–411. https://doi.org/10.1083/jcb.82.2.401

Brokaw CJ (2009) Simulation of cyclic dynein-driven sliding, splitting, and reassociation in an outer doublet pair. Biophys J 97(11):2939–2947. https://doi.org/10.1016/j.bpj.2009.09.022

Bui KH et al (2008) Molecular architecture of inner dynein arms in situ in Chlamydomonas reinhardtii flagella. J Cell Biol 183(5):923–932. https://doi.org/10.1083/jcb.200808050

Bui KH et al (2009) Asymmetry of inner dynein arms and inter-doublet links in Chlamydomonas flagella. J Cell Biol 186(3):437–446. https://doi.org/10.1083/jcb.200903082

Bui KH et al (2012) Polarity and asymmetry in the arrangement of dynein and related structures in the Chlamydomonas axoneme. J Cell Biol 198(5):913–925. https://doi.org/10.1083/jcb.201201120

Burgess SA et al (2003) Dynein structure and power stroke. Nature 421(6924):715–718. https://doi.org/10.1038/nature01377

Carbajal-González BI et al (2013) Conserved structural motifs in the central pair complex of eukaryotic flagella. Cytoskeleton (Hoboken, NJ) 70(2):101–120. https://doi.org/10.1002/cm.21094

Carter AP et al (2008) Structure and functional role of dynein's microtubule-binding domain. Science (New York, NY) 322(5908):1691–1695. https://doi.org/10.1126/science.1164424

Carter AP et al (2011a) Crystal structure of the dynein motor domain. Science (New York, NY) 331(6021):1159–1165. https://doi.org/10.1126/science.1202393

Carter AP et al (2011b) Crystal structure of the dynein motor domain. Science 331(6021):1159–1165. https://doi.org/10.1126/science.1202393

Chapman MJ, Dolan MF, Margulis L (2000) Centrioles and Kinetosomes: form, function, and evolution. Q Rev Biol 75(4):409–429. https://doi.org/10.1086/393621

Christensen ST et al (2007) Sensory cilia and integration of signal transduction in human health and disease. Traffic 8(2):97–109. https://doi.org/10.1111/j.1600-0854.2006.00516.x

Cole DG et al (1998) Chlamydomonas kinesin-II–dependent Intraflagellar transport (IFT): IFT particles contain proteins required for ciliary assembly in Caenorhabditis elegans sensory neurons. J Cell Biol 141(4):993–1008. https://doi.org/10.1083/jcb.141.4.993

Dai D et al (2020) Identification and mapping of central pair proteins by proteomic analysis. Biophys Physicobiol 17(0):71–85. https://doi.org/10.2142/biophysico.BSJ-2019048

Diener DR et al (2011) Sequential assembly of flagellar radial spokes. Cytoskeleton (Hoboken, NJ) 68(7):389–400. https://doi.org/10.1002/cm.20520

Dougherty GW et al (2016) DNAH11 localization in the proximal region of respiratory cilia defines distinct outer dynein arm complexes. Am J Respir Cell Mol Biol. https://doi.org/10.1165/rcmb.2015-0353OC. [Preprint]

Dymek EE, Smith EF (2007) A conserved CaM- and radial spoke associated complex mediates regulation of flagellar dynein activity. J Cell Biol 179(3):515–526. https://doi.org/10.1083/jcb.200703107

Fliegauf M, Benzing T, Omran H (2007) When cilia go bad: cilia defects and ciliopathies. Nat Rev Mol Cell Biol 8(11):880–893. https://doi.org/10.1038/nrm2278

Gadadhar S et al (2021) Tubulin glycylation controls axonemal dynein activity, flagellar beat, and male fertility. Science 371(6525):eabd4914. https://doi.org/10.1126/science.abd4914

Gaertig J et al (1995) Acetylation of lysine 40 in alpha-tubulin is not essential in Tetrahymena thermophila. J Cell Biol 129(5):1301–1310. https://doi.org/10.1083/jcb.129.5.1301

Gennerich A, Vale RD (2009) Walking the walk: how kinesin and dynein coordinate their steps. Curr Opin Cell Biol 21(1):59–67. https://doi.org/10.1016/j.ceb.2008.12.002

Gibbons BH, Gibbons IR (1973) The effect of partial extraction of dynein arms on the movement of reactivated sea-urchin sperm. J Cell Sci 13(2):337–357

Gibbons BH, Gibbons IR, Baccetti B (1983) Structure and motility of the 9 + 0 flagellum of eel spermatozoa. J Submicrosc Cytol 15(1):15–20

Gilliam JC et al (2012) Three-dimensional architecture of the rod sensory cilium and its disruption in retinal neurodegeneration. Cell 151(5):1029–1041. https://doi.org/10.1016/j.cell.2012.10.038

Gong A et al (2021) Reconstruction of the three-dimensional beat pattern underlying swimming behaviors of sperm. Eur Phys J E 44(7):87. https://doi.org/10.1140/epje/s10189-021-00076-z

Goodenough U, Heuser J (1984) Structural comparison of purified dynein proteins with in situ dynein arms. J Mol Biol 180(4):1083–1118

Goodenough UW, Heuser JE (1985) Outer and inner dynein arms of cilia and flagella. Cell 41(2):341–342

Gui M et al (2021) Structures of radial spokes and associated complexes important for ciliary motility. Nat Struct Mol Biol 28(1):29–37. https://doi.org/10.1038/s41594-020-00530-0

Guichard P et al (2012) Cartwheel architecture of Trichonympha basal body. Science (New York, NY) 337(6094):553. https://doi.org/10.1126/science.1222789

Guichard P et al (2013) Native architecture of the centriole proximal region reveals features underlying its 9-fold radial symmetry. Curr Biol 23(17):1620–1628. https://doi.org/10.1016/j.cub.2013.06.061

Hayashi S, Shingyoji C (2008) Mechanism of flagellar oscillation-bending-induced switching of dynein activity in elastase-treated axonemes of sea urchin sperm. J Cell Sci 121(Pt 17):2833–2843. https://doi.org/10.1242/jcs.031195

Hendrickson TW et al (2004) IC138 is a WD-repeat dynein intermediate chain required for light chain assembly and regulation of flagellar bending. Mol Biol Cell 15(12):5431–5442. https://doi.org/10.1091/mbc.E04-08-0694

Hilbert M et al (2016) SAS-6 engineering reveals interdependence between cartwheel and microtubules in determining centriole architecture. Nat Cell Biol 18(4):393–403. https://doi.org/10.1038/ncb3329

Hirokawa N et al (2006) Nodal flow and the generation of left-right asymmetry. Cell 125(1):33–45. https://doi.org/10.1016/j.cell.2006.03.002

Höög JL et al (2014) Modes of flagellar assembly in Chlamydomonas reinhardtii and Trypanosoma brucei. eLife 3:e01479. https://doi.org/10.7554/eLife.01479

Hoops HJ, Witman GB (1983) Outer doublet heterogeneity reveals structural polarity related to beat direction in Chlamydomonas flagella. J Cell Biol 97(3):902–908

Ichikawa M et al (2019) Tubulin lattice in cilia is in a stressed form regulated by microtubule inner proteins. Proc Nat Acad Sci U S A. https://doi.org/10.1073/pnas.1911119116. [Preprint]

Inaba K (2015) Calcium sensors of ciliary outer arm dynein: functions and phylogenetic considerations for eukaryotic evolution. Cilia 4:6. https://doi.org/10.1186/s13630-015-0015-z

Ishikawa T (2015) Cryo-electron tomography of motile cilia and flagella. Cilia 4(1):3. https://doi.org/10.1186/s13630-014-0012-7

Ishikawa T (2016) Axoneme structure from motile cilia. Cold Spring Harb Perspect Biol. https://doi.org/10.1101/cshperspect.a028076. [Preprint]

Ishikawa H, Marshall WF (2017) Intraflagellar transport and ciliary dynamics. Cold Spring Harb Perspect Biol 9(3):a021998. https://doi.org/10.1101/cshperspect.a021998

Johnson KA (1985) Pathway of the microtubule-dynein ATPase and the structure of dynein: a comparison with actomyosin. Annu Rev Biophys Biophys Chem 14:161–188. https://doi.org/10.1146/annurev.bb.14.060185.001113

Johnson KA (1998) The axonemal microtubules of the Chlamydomonas flagellum differ in tubulin isoform content. J Cell Sci 111(Pt 3):313–320

Jordan MA et al (2018) The cryo-EM structure of intraflagellar transport trains reveals how dynein is inactivated to ensure unidirectional anterograde movement in cilia. Nat Cell Biol 20(11):1250–1255. https://doi.org/10.1038/s41556-018-0213-1

Kagami O, Kamiya R (1992) Translocation and rotation of microtubules caused by multiple species of Chlamydomonas inner-arm dynein. J Cell Sci 103:653–664

Kamiya R (2002) Functional diversity of axonemal dyneins as studied in Chlamydomonas mutants. Int Rev Cytol 219:115–155

Kamiya R, Witman GB (1984) Submicromolar levels of calcium control the balance of beating between the two flagella in demembranated models of Chlamydomonas. J Cell Biol 98(1):97–107

Kamiya R, Kurimoto E, Muto E (1991) Two types of Chlamydomonas flagellar mutants missing different components of inner-arm dynein. J Cell Biol 112(3):441–447

Kiesel P et al (2020) The molecular structure of mammalian primary cilia revealed by cryo-electron tomography. Nat Struct Mol Biol 27(12):1115–1124. https://doi.org/10.1038/s41594-020-0507-4

Kikuchi T et al (1989) Fine structure of Guinea pig vestibular Kinocilium. Acta Otolaryngol 108(1–2):26–30. https://doi.org/10.3109/00016488909107388

Kikushima K, Kamiya R (2008) Clockwise translocation of microtubules by flagellar inner-arm Dyneins in vitro. Biophys J 94(10):4014–4019. https://doi.org/10.1529/biophysj.107.123083

Kitagawa D et al (2011) Structural basis of the 9-fold symmetry of centrioles. Cell 144(3):364–375. https://doi.org/10.1016/j.cell.2011.01.008

Klink BU et al (2020) Structure of the human BBSome core complex. eLife 9:e53910. https://doi.org/10.7554/eLife.53910

Kon T, Sutoh K, Kurisu G (2011) X-ray structure of a functional full-length dynein motor domain. Nat Struct Mol Biol 18(6):638–642. https://doi.org/10.1038/nsmb.2074

Kon T et al (2012) The 2.8 Å crystal structure of the dynein motor domain. Nature 484(7394): 345–350. https://doi.org/10.1038/nature10955

Kozminski KG, Diener DR, Rosenbaum JL (1993a) High level expression of nonacetylatable?-tubulin inChlamydomonas reinhardtii. Cell Motil Cytoskeleton 25(2):158–170. https://doi.org/10.1002/cm.970250205

Kozminski KG et al (1993b) A motility in the eukaryotic flagellum unrelated to flagellar beating. Proc Natl Acad Sci U S A 90(June):5519–5523. https://doi.org/10.1073/pnas.90.12.5519

Kubo T et al (2010) Tubulin polyglutamylation regulates axonemal motility by modulating activities of inner-arm dyneins. Curr Biol 20(5):441–445. https://doi.org/10.1016/j.cub.2009.12.058

Kubo T, Yagi T, Kamiya R (2012) Tubulin polyglutamylation regulates flagellar motility by controlling a specific inner-arm dynein that interacts with the dynein regulatory complex. Cytoskeleton 69(12):1059–1068. https://doi.org/10.1002/cm.21075

Kubo T et al (2018) A microtubule-dynein tethering complex regulates the axonemal inner dynein *f* (I1). Mol Biol Cell. Edited by X Zhu 29(9):1060–1074. https://doi.org/10.1091/mbc.E17-11-0689

Kubo S et al (2021) Remodeling and activation mechanisms of outer arm dyneins revealed by cryo-EM. EMBO Rep 22(9):e52911. https://doi.org/10.15252/embr.202152911

Kumar A et al (2013) CEP proteins: the knights of centrosome dynasty. Protoplasma 250(5): 965–983. https://doi.org/10.1007/s00709-013-0488-9

Kutomi O et al (2021) A dynein-associated photoreceptor protein prevents ciliary acclimation to blue light. Sci Adv 7(9):eabf3621. https://doi.org/10.1126/sciadv.abf3621

Lechtreck KF et al (2013) Cycling of the signaling protein phospholipase D through cilia requires the BBSome only for the export phase. J Cell Biol 201(2):249–261. https://doi.org/10.1083/jcb.201207139

Li JB et al (2004) Comparative genomics identifies a flagellar and basal body proteome that includes the BBS5 human disease gene. Cell 117(4):541–552. https://doi.org/10.1016/s0092-8674(04)00450-7

Lin J, Nicastro D (2018) Asymmetric distribution and spatial switching of dynein activity generates ciliary motility. Science (New York, NY) 360(6387):eaar1968. https://doi.org/10.1126/science.aar1968

Lin J, Okada K et al (2014a) Structural mechanism of the dynein power stroke. Nat Cell Biol. https://doi.org/10.1038/ncb2939. [Preprint]

Lin J, Yin W et al (2014b) Cryo-electron tomography reveals ciliary defects underlying human RSPH1 primary ciliary dyskinesia. Nat Commun 5:5727. https://doi.org/10.1038/ncomms6727

Lindemann CB (2007) The geometric clutch as a working hypothesis for future research on cilia and flagella. Ann N Y Acad Sci 1101:477–493. https://doi.org/10.1196/annals.1389.024

Loges NT et al (2018) Recessive DNAH9 loss-of-function mutations cause laterality defects and subtle respiratory ciliary-beating defects. Am J Hum Genet 103(6):995–1008. https://doi.org/10.1016/j.ajhg.2018.10.020

Ma M et al (2019) Structure of the decorated ciliary doublet microtubule. Cell 179(4):909–922.e12. https://doi.org/10.1016/j.cell.2019.09.030

Maheshwari A, Ishikawa T (2012) Heterogeneity of dynein structure implies coordinated suppression of dynein motor activity in the axoneme. J Struct Biol 179(2):235–241. https://doi.org/10.1016/j.jsb.2012.04.018

Maheshwari A et al (2015) α- and β-tubulin lattice of the axonemal microtubule doublet and binding proteins revealed by single particle cryo-electron microscopy and tomography. Structure (London, England: 1993) 23(9):1584–1595. https://doi.org/10.1016/j.str.2015.06.017

Mali GR et al (2021) Shulin packages axonemal outer dynein arms for ciliary targeting. Science 371(6532):910–916. https://doi.org/10.1126/science.abe0526

Marshall WF (2009) Centriole evolution. Curr Opin Cell Biol 21(1):14–19. https://doi.org/10.1016/j.ceb.2009.01.008

Mastronarde DN et al (1992) Arrangement of inner dynein arms in wild-type and mutant flagella of Chlamydomonas. J Cell Biol 118(5):1145–1162

Mencarelli C et al (2000) Morphogenesis of the giant sperm axoneme in Asphondylia ruebsaameni Kertesz (Diptera, Cecidomyiidae). Tissue Cell 32(2):188–197. https://doi.org/10.1054/tice.2000.0103

Minamino T, Imada K, Namba K (2008) Molecular motors of the bacterial flagella. Curr Opin Struct Biol 18(6):693–701. https://doi.org/10.1016/j.sbi.2008.09.006

Mitchell DR, Nakatsugawa M (2004) Bend propagation drives central pair rotation in Chlamydomonas reinhardtii flagella. J Cell Biol 166(5):709–715. https://doi.org/10.1083/jcb.200406148

Mourão A, Christensen ST, Lorentzen E (2016) The intraflagellar transport machinery in ciliary signaling. Curr Opin Struct Biol 41:98–108. https://doi.org/10.1016/j.sbi.2016.06.009

Movassagh T et al (2010) Nucleotide-induced global conformational changes of flagellar dynein arms revealed by in situ analysis. Nat Struct Mol Biol 17(6):761–767. https://doi.org/10.1038/nsmb.1832

Müller L et al (2013) Culturing of human nasal epithelial cells at the air liquid interface. J Vis Exp: JoVE 80:e50646. https://doi.org/10.3791/50646. [Preprint]

Nabais C, Peneda C, Bettencourt-Dias M (2020) Evolution of centriole assembly. Curr Biol 30(10): R494–R502. https://doi.org/10.1016/j.cub.2020.02.036

Nachury MV (2018) The molecular machines that traffic signaling receptors into and out of cilia. Curr Opin Cell Biol 51:124–131. https://doi.org/10.1016/j.ceb.2018.03.004

Nachury MV, Mick DU (2019) Establishing and regulating the composition of cilia for signal transduction. Nat Rev Mol Cell Biol 20(7):389–405. https://doi.org/10.1038/s41580-019-0116-4

Oda T, Yanagisawa H, Kamiya R et al (2014a) Cilia and flagella. A molecular ruler determines the repeat length in eukaryotic cilia and flagella. Science (New York, NY) 346(6211):857–860. https://doi.org/10.1126/science.1260214

Oda T, Yanagisawa H, Kikkawa M (2014b) Detailed structural and biochemical characterization of the nexin-dynein regulatory complex. J Mol Biol Cell. https://doi.org/10.1091/mbc.E14-09-1367. [Preprint]

Oda T, Yanagisawa H, Yagi T et al (2014c) Mechanosignaling between central apparatus and radial spokes controls axonemal dynein activity. J Cell Biol 204(5):807–819. https://doi.org/10.1083/jcb.201312014

Olbrich H et al (2002) Mutations in DNAH5 cause primary ciliary dyskinesia and randomization of left-right asymmetry. Nat Genet 30(2):143–144. https://doi.org/10.1038/ng817

Omoto CK, Brokaw CJ (1985) Bending patterns of Chlamydomonas flagella: II. Calcium effects on reactivated Chlamydomonas flagella. Cell Motil 5(1):53–60

Omoto CK, Johnson KA (1986) Activation of the dynein adenosinetriphosphatase by microtubules. Biochemistry 25(2):419–427

Owa M et al (2019) Inner lumen proteins stabilize doublet microtubules in cilia and flagella. Nat Commun 10(1):1143. https://doi.org/10.1038/s41467-019-09051-x

Pazour GJ, Dickert BL, Witman GB (1999) The DHC1b (DHC2) isoform of cytoplasmic dynein is required for flagellar assembly. J Cell Biol 144(3):473–481. https://doi.org/10.1083/jcb.144.3.473

Pazour GJ et al (2005) Proteomic analysis of a eukaryotic cilium. J Cell Biol 170(1):103–113. https://doi.org/10.1083/jcb.200504008

Pigino G et al (2011) Cryoelectron tomography of radial spokes in cilia and flagella. J Cell Biol 195(4):673–687. https://doi.org/10.1083/jcb.201106125

Pigino G et al (2012) Comparative structural analysis of eukaryotic flagella and cilia from Chlamydomonas, Tetrahymena, and sea urchins. J Struct Biol 178(2):199–206. https://doi.org/10.1016/j.jsb.2012.02.012

Piperno G, Fuller MT (1985) Monoclonal antibodies specific for an acetylated form of alpha-tubulin recognize the antigen in cilia and flagella from a variety of organisms. J Cell Biol 101(6): 2085–2094. https://doi.org/10.1083/jcb.101.6.2085

Prensier G et al (1980) Motile flagellum with a “3 + 0” ultrastructure. Science (New York, NY) 207(4438):1493–1494

Rao Q et al (2021) Structures of outer-arm dynein array on microtubule doublet reveal a motor coordination mechanism. Nat Struct Mol Biol 28(10):799–810. https://doi.org/10.1038/s41594-021-00656-9

Redwine WB et al (2012) Structural basis for microtubule binding and release by dynein. Science (New York, NY) 337(6101):1532–1536. https://doi.org/10.1126/science.1224151

Riedel-Kruse IH et al (2007) How molecular motors shape the flagellar beat. HFSP J 1(3):192–208. https://doi.org/10.2976/1.2773861

Roberts AJ et al (2009) AAA+ ring and linker swing mechanism in the dynein motor. Cell 136(3): 485–495. https://doi.org/10.1016/j.cell.2008.11.049

Sakato M, King SM (2004) Design and regulation of the AAA+ microtubule motor dynein. J Struct Biol 146(1–2):58–71. https://doi.org/10.1016/j.jsb.2003.09.026

Satir P, Matsuoka T (1989) Splitting the ciliary axoneme: implications for a “switch-point” model of dynein arm activity in ciliary motion. Cell Motil Cytoskeleton 14(3):345–358. https://doi.org/10.1002/cm.970140305

Schmidt H, Gleave ES, Carter AP (2012) Insights into dynein motor domain function from a 3.3-Å crystal structure. Nat Struct Mol Biol 19(5):492–497., S1. https://doi.org/10.1038/nsmb.2272

Schmidt H et al (2014) Structure of human cytoplasmic dynein-2 primed for its power stroke. Nature. https://doi.org/10.1038/nature14023. [Preprint]

Schneiter M et al (2021) Multi-scale alignment of respiratory cilia and its relation to mucociliary function. J Struct Biol 213(1):107680. https://doi.org/10.1016/j.jsb.2020.107680

Schrevel J, Besse C (1975) A functional flagella with a 6 + 0 pattern. J Cell Biol 66(3):492–507

Shida T et al (2010) The major -tubulin K40 acetyltransferase TAT1 promotes rapid ciliogenesis and efficient mechanosensation. Proc Natl Acad Sci 107(50):21517–21522. https://doi.org/10.1073/pnas.1013728107

Silflow CD et al (1999) Gamma-tubulin in Chlamydomonas: characterization of the gene and localization of the gene product in cells. Cell Motil Cytoskelet 42(4):285–297. https://doi.org/10.1002/(SICI)1097-0169(1999)42:4<285::AID-CM3>3.0.CO;2-Z

Sims AC et al (2008) SARS-CoV replication and pathogenesis in an in vitro model of the human conducting airway epithelium. Virus Res 133(1):33–44. https://doi.org/10.1016/j.virusres.2007.03.013

Singh SK et al (2020) Structure and activation mechanism of the BBSome membrane protein trafficking complex. eLife 9:e53322. https://doi.org/10.7554/eLife.53322

Sirajuddin M, Rice LM, Vale RD (2014) Regulation of microtubule motors by tubulin isotypes and post-translational modifications. Nat Cell Biol 16(4):335–344. https://doi.org/10.1038/ncb2920

Spektor A et al (2007) Cep97 and CP110 suppress a cilia assembly program. Cell 130(4):678–690. https://doi.org/10.1016/j.cell.2007.06.027

Stathatos GG et al (2021) Delta and epsilon tubulin in mammalian development. Trends Cell Biol 31(9):774–787. https://doi.org/10.1016/j.tcb.2021.03.010

Stepanek L, Pigino G (2016) Microtubule doublets are double-track railways for intraflagellar transport trains. Science (New York, N.Y.) 352(6286):721–724. https://doi.org/10.1126/science.aaf4594

Takahashi Y, Edamatsu M, Toyoshima YY (2004) Multiple ATP-hydrolyzing sites that potentially function in cytoplasmic dynein. Proc Natl Acad Sci U S A 101(35):12865–12869. https://doi.org/10.1073/pnas.0403429101

Tanaka M, Miki-Noumura T (1988) Stepwise sliding disintegration oftetrahymena ciliary axonemes at higher concentrations of ATP. Cell Motil Cytoskeleton 9(3):191–204. https://doi.org/10.1002/cm.970090302

Taschner M, Lorentzen E (2016) The Intraflagellar transport machinery. Cold Spring Harb Perspect Biol 8(10):a028092. https://doi.org/10.1101/cshperspect.a028092

Ueno H et al (2008) Dynein pulls microtubules without rotating its stalk. Proc Natl Acad Sci U S A 105(50):19702–19707. https://doi.org/10.1073/pnas.0808194105

Ueno H et al (2012) Mouse respiratory cilia with the asymmetric axonemal structure on sparsely distributed ciliary cells can generate overall directional flow. Nanomedicine 8(7):1081–1087. https://doi.org/10.1016/j.nano.2012.01.004

Ueno H, et al (2014) Structure of dimeric axonemal dynein in cilia suggests an alternative mechanism of force generation. Cytoskeleton (Hoboken, N.J.). https://doi.org/10.1002/cm.21180. [Preprint]

Urbanska P et al (2015) The CSC proteins FAP61 and FAP251 build the basal substructures of radial spoke 3 in cilia. Mol Biol Cell 26(8):1463–1475. https://doi.org/10.1091/mbc.E14-11-1545

van Breugel M et al (2011) Structures of SAS-6 suggest its organization in centrioles. Science (New York, NY) 331(6021):1196–1199. https://doi.org/10.1126/science.1199325

Wakabayashi K, Isu A, Ueki N (2021) Channelrhodopsin-dependent photo-behavioral responses in the unicellular green alga Chlamydomonas reinhardtii. In: Yawo H et al (eds) Optogenetics (Advances in experimental medicine and biology). Springer Singapore, Singapore, pp 21–33. https://doi.org/10.1007/978-981-15-8763-4_2

Walton T, Wu H, Brown A (2021) Structure of a microtubule-bound axonemal dynein. Nat Commun 12(1):477. https://doi.org/10.1038/s41467-020-20735-7

Wingfield JL, Lechtreck K-F, Lorentzen E (2018) Trafficking of ciliary membrane proteins by the intraflagellar transport/BBSome machinery. Essays Biochem. Edited by J.G. Wakefield and C.A. Moores 62(6):753–763. https://doi.org/10.1042/EBC20180030

Wloga D et al (2017) Posttranslational modifications of tubulin and cilia. Cold Spring Harb Perspect Biol 9(6):a028159. https://doi.org/10.1101/cshperspect.a028159

Woolley DM (1997) Studies on the eel sperm flagellum. I. the structure of the inner dynein arm complex. J Cell Sci 110(Pt 1):85–94

Yagi T (2009) Bioinformatic approaches to dynein heavy chain classification. Methods Cell Biol 92:1–9. https://doi.org/10.1016/S0091-679X(08)92001-X

Yagi T, Kamiya R (2000) Vigorous beating of Chlamydomonas axonemes lacking central pair/radial spoke structures in the presence of salts and organic compounds. Cell Motil Cytoskeleton 46(3):190–199. https://doi.org/10.1002/1097-0169(200007)46:3<190::AID-CM4>3.0.CO;2-#

Yagi T et al (2005) An axonemal dynein particularly important for flagellar movement at high viscosity. Implications from a new Chlamydomonas mutant deficient in the dynein heavy chain gene DHC9. J Biol Chem 280(50):41412–41420. https://doi.org/10.1074/jbc.M509072200

Yagi T et al (2009) Identification of dyneins that localize exclusively to the proximal portion of Chlamydomonas flagella. J Cell Sci 122(Pt 9):1306–1314. https://doi.org/10.1242/jcs.045096

Yamamoto R et al (2021) Composition and function of ciliary inner-dynein-arm subunits studied in *Chlamydomonas reinhardtii*. Cytoskeleton 78(3):77–96. https://doi.org/10.1002/cm.21662

Yang P et al (2006) Radial spoke proteins of Chlamydomonas flagella. J Cell Sci 119(Pt 6):1165–1174. https://doi.org/10.1242/jcs.02811

Yang S et al (2020) Near-atomic structures of the BBSome reveal the basis for BBSome activation and binding to GPCR cargoes. eLife 9:e55954. https://doi.org/10.7554/eLife.55954

Chapter 16
Segment-Long-Spacing (SLS) and the Polymorphic Structures of Fibrillar Collagen

Yujia Xu and Michele Kirchner

Abstract The diverse and complex functions of collagen during the development of an organism are closely related to the polymorphism of its supramolecular structures in the extracellular matrix. SLS (segment-long-spacing) is one of the best understood alternative structures of collagen. SLS played an instrumental role in the original studies of collagen more than half a century ago that laid the foundation of nearly everything we know about collagen today. Despite being used mostly under in vitro conditions, the natural occurrence of SLS in tissues has also been reported. Here we will provide a brief overview of the major findings of the SLS and other structures of collagen based on a wealth of work published starting from the 1940s. We will discuss the factors that determine the stability and the structural specificity of the different molecular assemblies of collagen in light of the new studies using designed fibril forming collagen peptides. At the end of the chapter, we will summarize some recent discoveries of the alternative structures of collagen in tissues, especially those involved in pathogenic states. A revisit of SLS will likely inspire new understandings concerning the range of critical roles of fibrillar collagen in terms of its organizational diversity in the extracellular matrix.

Keywords Extracellular matrix · Collagen interactions · Protein stability · Protein design · Recombinant collagen · Collagen mimetic biomaterials · Collagen and diseases

Introduction

Fibrillar collagen, including collagen types I, II, III, and V collectively account for more than 90% of all collagens in the body (Bächinger et al. 2010), and they form the essential molecular scaffold in all tissues. Central to the crucial biological functions of collagen is its supramolecular structures in the extracellular matrix (ECM). The

---

Y. Xu (✉) · M. Kirchner
Department of Chemistry, Hunter College of the City University of New York, New York, NY, USA
e-mail: Yujia.xu@hunter.cuny.edu; mkirchne@hunter.cuny.edu

J. R. Harris, J. Marles-Wright (eds.), *Macromolecular Protein Complexes IV*, Subcellular Biochemistry 99, https://doi.org/10.1007/978-3-031-00793-4_16

macromolecular assemblies of collagen are known for their multiplicity and complexity in terms of the composition, the conformation, and the molecular properties (Bella and Hulmes 2017; Brodsky and Ramshaw 2017; Birk and Bruckner 2005). Accumulating observations are pointing to the structural polymorphism of collagen as a key factor for the varied function of collagen in tissues genesis and pathogenesis (Eyden and Tzaphlidou 2001). Polymorphism of collagen fibrils has been known since the dawn of collagen research (Bruns 1976; Kühn 1982; Gross et al. 1954). Among the alternative structures of fibrillar collagens, SLS (segment-long-spacing) is unique in its form and its direct link to the primary structure of the collagen. While SLS of fibrillar collagens are the best studied and will be the focus here, many different types of collagens can also form SLS-like aggregates (Bentz et al. 1983; Schwartz and Veis 1978, 1980). In fact, the structure of SLS is a manifestation of the conformational properties of a collagen triple helix. Expanded research on how SLS and other structures affect the properties of collagen will lead to a better understanding of the range of critical roles of fibrillar collagen in tissue development and function.

SLS: The Effective Molecular Tool for Collagen Research

To understand the significance of the SLS structure in collagen research, one has to take a step back in time, to the early days of protein chemistry when advances in collagen research took off in lockstep with the emerging technologies in molecular and structural biology: electron microscopy, X-ray diffraction, and analytical ultracentrifugation, just to name a few. The large and regular structure of fibrillar collagen is particularly suitable for studies using a transmission electron microscope (TEM). By the early 1950s, it was understood that collagen is a fibrous protein having an axial periodicity of about 67 nm[1]—the *D*-period. This periodicity was observed in collagen fibrils isolated from tissues and in the reconstituted collagen fibrils formed from acid extracts (Gross and Schmitt 1948). The 67 nm periodicity was characterized as regular dark-light bands on negatively stained electron micrographs (Fig. 16.1a) or as clear ridge and gaps after chromium shadowing (Fig. 16.1c). This directly visible repeating structure was also supported by the X-ray diffraction of collagen fibrils (Wright 1948; Begbie 1947; Gross and Schmitt 1948). The basic structure of collagen—the collagen triple helix—was known to have three polypeptide chains wrapped around each other in an extended conformation and a uniform backbone. The Gly residues of the characteristic (Gly-X-Y) repeating amino acid sequences are at the center of the helix while the side chains of the residues in the X and Y positions are displayed linearly on the surface of the helix in an N-to-C termini

[1]There had been active debate on the actual size of the *D*-period, the value varied from 64 nm to 70 nm. For clarity, we will use the currently accepted value of 67 nm to describe the findings including those published before the size of the *D*-period was finalized, unless specifically noted.

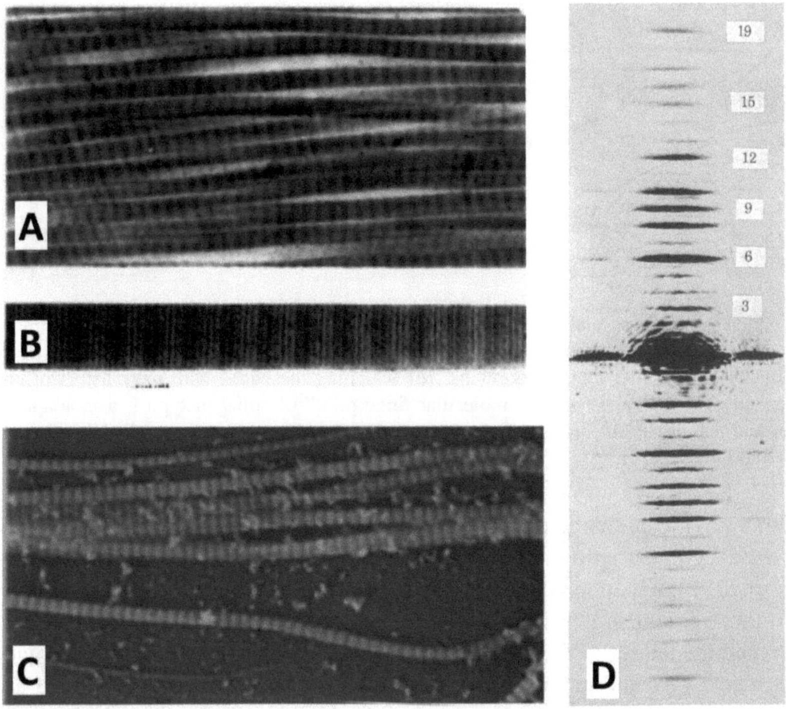

Fig. 16.1 The structure of collagen fibrils. (**a**) Negatively stained TEM image showing the *D*-period as a pair of light-and-dark bands. (**b**) Positively stained TEM images showing the *D*-period related placement of clusters of charged residues. (**c**) TEM image after metal shadowing showing the dark and light bands of a *D*-period as the ridges and gaps, respectively. **a** and **b** are adapted from (Eyden and Tzaphlidou 2001), **c** from (Schmitt et al. 1953)

directionality (Ramachandran and Kartha 1955; Rich and Crick 1955). How did the specific *D*-periodic fibrous structure come about from the arrangement of a defined structural unit was the puzzling question. The *D*-period was attributed to the spacing between two immediate *monomers*, although neither the size or shape of the "monomer" nor their specific arrangement associated with this regular spacing structure was understood.

The breakthrough came with the discovery of a different structure of collagen—the segment-long-spacing crystallite, also known as the SLS. SLS is not a fibrous structure but a "crystallite-like" structure having one dimension about 200–300 nm in length—a *segmental* length related to the long collagen fibril *D*-period (Fig. 16.2). SLS was first discovered during routine collagen preparations and was later reproduced by dialyzing fibril-forming collagen in the cold against 0.05% acetic acid (pH 3.5) and with the addition of ATP to a final concentration of 0.1–0.25% (Schmitt et al. 1953). In their original work on SLS, Schmitt et al. reported that upon dialysis, the *D*-period fibrils transformed into a new kind of aggregate "in the form of

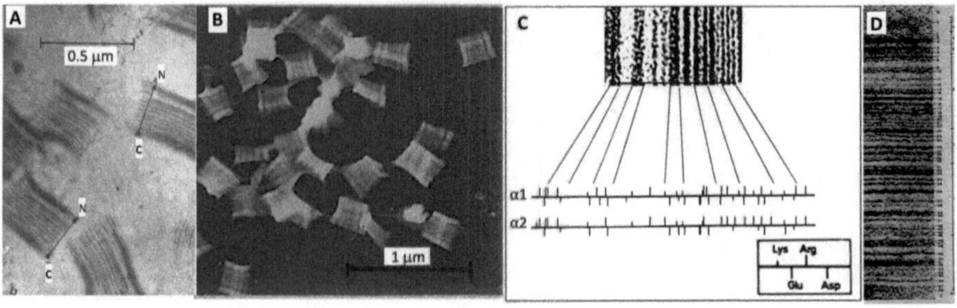

Fig. 16.2 The structure and banding pattern of SLS. (**a**) Positively stained TEM images of SLS having the N and C termini of the triple helix marked by the letter N and C, respectively. (**b**) SLS crystallites after metal shadowing. (**c**) Matching the bands with the clusters of charged residues. (**d**) The 58 bands of the "molecular fingerprint" of collagen type I. **a** is adapted from (Hodge and Schmitt 1960), **b** from (Schmitt et al. 1953), **c** from (Fietzek and Kühn 1976), **d** from (Bruns and Gross 1973)

segments having a characteristic pattern of internal band structure" (Fig. 16.2a). Unlike the regular striation of the *D*-period in fibrils, the closely spaced, barcode-like bands of SLS have an asymmetric spacing along the segments with respect to the axis of the segment showing a clear polarity from one end to the other. The banding pattern and the length of the segment are highly reproducible, although the width of the segments varies. The chromium-shadowing technique revealed the SLS are not flat ribbons but crystallites with considerable thickness, indicating these are likely structures formed in solution and deposited as crystallites on the grid (Fig. 16.2b).

At about the same time SLS was discovered, several classic hydrodynamic studies of collagen revealed that collagen molecules dissolved in solution behave as a "rigid rod" about 280 nm × 1.4 nm in size, having a molecular weight close to 360 kDa (Boedtker and Doty 1956; Hall and Doty 1958)—a spot-on description of a triple helix prior to any knowledge of its primary structure! Combining the hydrodynamic results and the X-ray diffraction data on the extended conformation of a triple helix, it became clear that the SLS is formed by the self-assembled collagen triple helices, also known as the tropocollagens or TC (Gross et al. 1954). All TC molecules are arranged in parallel to each other, in the SLS crystallite, and are in a perfect transverse register—the so-called end-on-end stacking. The SLS crystallites have the same length as the collagen molecule. The banding patterns of SLS are due to the effects of the positive staining: the metal ions tungstate and uranate in the staining solution interact, respectively, with the positively charged and negatively charged residues on the surface of the molecule[2] (Kühn 1982). The bands of SLS can thus be related to the primary structure of the molecule given the rod-like, linear conformation of the triple helix. In a remarkable demonstration, Kühn et al. showed a near perfect alignment of the banding pattern of a section of the SLS with the

[2]It was later reported that TEM stain uranyl acetate may also stain positively charged residues (Tzaphlidou et al. 1982a, b).

resolved amino acid sequence of collagen from cyanogen bromide (CNB) fragments (Fig. 16.2c) (Fietzek and Kühn 1976; von der Mark et al. 1970). The SLS of collagens prepared from a wide variety of tissues and animals showed a high level of similarity in their banding patterns. By averaging more than 100 published observations on positively stained SLS, a series of 58 bands having a polarized localization from N-to-C termini were recognized as a molecular *fingerprint* for type I collagen (Hodge and Schmitt 1960; Bruns and Gross 1973, 1974) (Fig. 16.2d).

The realization that the SLS is an "electron optical image" of a 2-D display of the tropocollagen has transformed the SLS into an effective molecular tool for collagen research. The following are three major findings of collagen attributed to the use of SLS.

The Molecular Packing of the Fibrils

The *D*-period of native collagen fibrils are usually characterized using TEM images under the negative staining condition which depicts the surface relief of the fibril: the dark bands reflect the *gaps* on the surface where the metal ions of the staining solution accumulated, while the light bands are the overlap regions in the form of ridges (Fig. 16.1a and c). Under positive staining conditions, however, the images of the native fibrils revealed a different banding pattern: one that reflects regions of charged residues on the surface of the fibrils (Fig. 16.1b). The positively stained banding patterns of fibrils also bear strong similarities to that of the SLS fingerprint. In fact, by successively overlaying the banding pattern of a SLS with a longitudinal displacement of one-fourth of the length of a SLS, Schmitt et al. were able to create a banding pattern that corresponded closely to that of the native fibrils (Fig. 16.3a) (Hodge and Schmitt 1960). In another remarkable experiment, they prepared the SLS in the presence of the native fibrils. The individual surface collagen heterotrimers in the fibril appeared to act as the "nuclei" for the SLS aggregates; the SLS are appended on the surface of the fibrils "as outgrowths from the native fibrils" (Fig. 16.3b and c). From the characteristic orientations of the banding pattern of the SLS segments with respect to the native fibrils, this legendary "dimorphic" image revealed two important features of the collagen fibrils: (1) the tropocollagen (TC) molecules are all arranged in parallel in the fibrils and (2) the transverse clusters of charged residues seen on SLS are conserved in the fibrils; as shown in the "optical synthesis," the unique banding pattern of the native fibrils is the combined effects of TC molecules arranged with a regular mutual staggering with each other. Using different methods, Kühn et al. reached the similar conclusion about conserved SLS bands and the quarter staggered arrangement of the fibrils (Kühn 1982). Later studies revealed that the *D*-period is 67 nm not 70 nm—the value used in the work by Schmitt et al.—and a TC encompasses 4.4*D* periods. While the *D*-staggering of the TC molecules in the fibrils is not exactly a quarter-stagger as had been proposed by Schmitt et al., the findings using SLS have nonetheless captured the fundamental

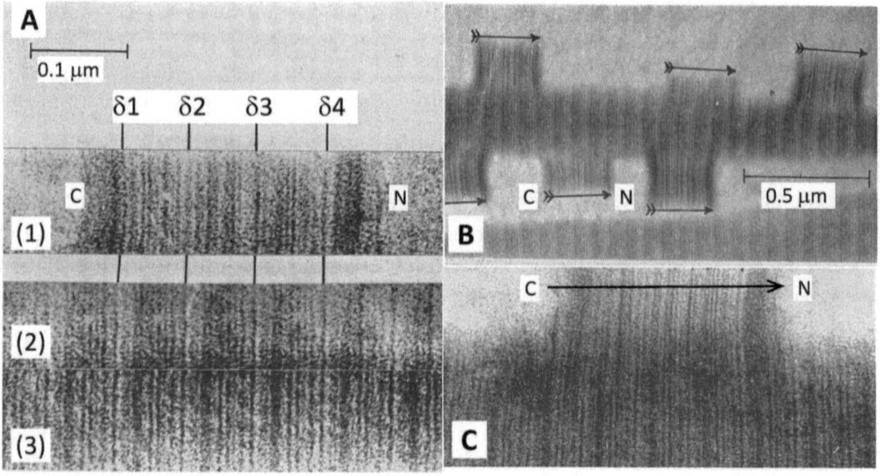

Fig. 16.3 The recapitulation of the banding patterns of SLS in the positively stained collagen fibrils. (**a**) Showing in (1) is the typical banding pattern of a SLS where the N and C indicate the N- and C- termini, respectively, of the tropocollagen. The banding pattern in (2) is generated by multiple printing of the SLS image in (1) with a longitudinal displacement of ¼ of the length of the segment (indicated by the positions δ1, δ2, δ3, and δ4). The summation of band densities resulted from the "optical synthesis" corresponds closely to the band pattern of the native fibrils shown in (3). (**b**) The dimorphic ordered aggregates of tropocollagen produced by exposing the reconstituted native fibrils to the solution containing tropocollagen and ATP at a pH value favoring the formation of SLS. The SLS formed as an "outgrowth" from the native fibrils and exhibit a characteristic orientation and polarity with respect to the polarized band pattern of the fibrils. The arrows indicate the C to N orientation of the tropocollagen in the SLS. (**c**) An expanded view of a SLS "outgrowth" on a native fibril. The letters C and N, and the arrow indicate the C-to-N directionality of the SLS. (Adapted from Hodge and Schmitt 1960)

aspects of the fibril assembly of collagen that have served as a bedrock of knowledge for studies of collagen fibrils ever since.

The Fingerprints of Collagens

The close correlation between the banding patterns of SLS and the primary structure of collagen led to the discovery of new types of collagens and even opened the door for phylogenetic studies of collagen long before the amino acid sequences of those collagens were elucidated. Collagens prepared from animal skins are predominantly collagen type I, a heterotrimer consisting of two α_1 chains and one α_2 chain, which is also the most abundant collagen in the animal kingdom. Collagens extracted from other tissues, however, are often dominated by genetically distinct molecules differing from each other in amino acid composition and chain composition. With chemical sequencing of protein laborious and technically challenging, and the

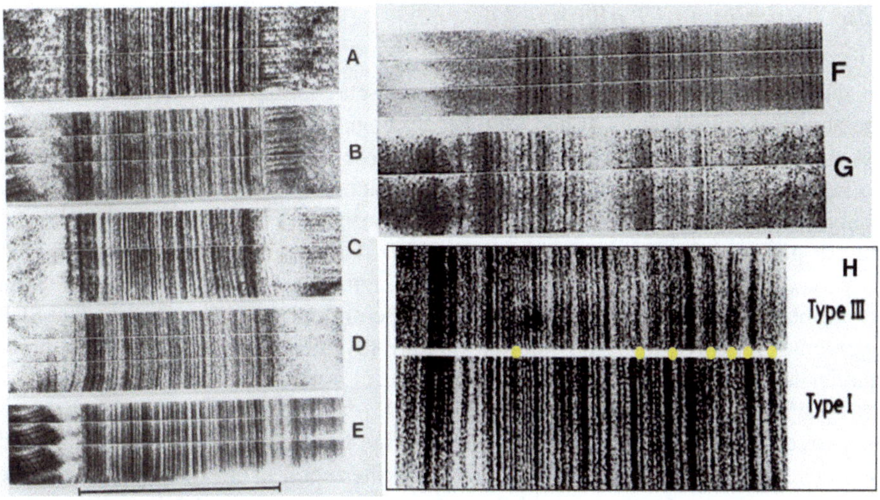

Fig. 16.4 The SLS fingerprints of different collagens. (**a**) Type I collagen α_1 chain homotrimer. (**b**) Type I collagen α_2 chain homotrimer. (**c**) Type II collagen. (**d**) Type III collagen. (**e**) Type V collagen $\alpha_1(V)_2$ $\alpha_1(V)$ heterotrimer. (**f**) Type IV collagen from placenta. (**g**) Type IV collagen from mouse. (**h**) The differences (highlighted in yellow dots) between type I heterotrimer and type III collagen. Adapted from (Kühn 1982), **h** from (Wiedemann et al. 1975)

genomic information a few decades away, SLS was used to analyze the structural and functional significance of newly discovered collagens. It was immediately clear that the SLS images of different fibrillar collagens share the similar appearances of the barcode-like bands in a segment about 300 nm long (Fig. 16.4). Yet, the resolution of SLS is sensitive enough to reveal distinctive features in both the location and the intensity of some of the bands between different collagens (Fig. 16.4h) (Wiedemann et al. 1975). One interesting case is the markedly different cross striation pattern of type IV collagen extracted from basement membrane (Fig. 16.4f and g). Type IV collagen is known to have "breaks" in its triple helical domain and to form a chicken-wire-like molecular network rather than fibrils (Schwartz and Veis 1978; Chung et al. 1976; Kefalides 1968). The different banding pattern of type IV also highlights a common feature of fibrillar collagen – the long-stretches of non-interruptive banding patterns. This banding pattern also appeared to be highly conserved and was found in collagen preparations from human skin, calf skin, rat tail tendon, fish, and invertebrates. This finding led to the conclusion that the regular alternation of the polar and apolar regions is one of the characteristics that "are not subjects to variations by mutations" throughout collagen's evolution (Nordwig and Hayduk 1969). These conserved features were considered necessary for collagen's configuration and for fibril formation—an idea that was later elaborated on by the discovery of periodic spacings of charged and hydrophobic residues in fibrillar collagens in relation to fibril structure (Hulmes et al. 1973; Hulmes et al. 1977; Doyle et al. 1974; Kaur et al. 2015) and was incorporated in the design of fibril-forming triple helical peptides (Kaur et al. 2015) (for details see below).

The Functions of Collagenase

In another historical development of collagen research, SLS was used to "visualize" the action of collagenases. It had long been recognized that collagenolysis plays an essential role in tissue remodeling. Upon the successful isolation of the first animal collagenase from tadpole tissues, the catalytic activity of this enzyme was studied using collagen preparations from calf skin and from the skin of normal or lathyritic guinea pigs and rats (Gross and Lapiere 1962). Incubation of type I collagen with tadpole collagenase resulted in two discrete fragments that can be observed using electrophoresis as two separate bands below the normal α_1 and α_2 chains. The SLS aggregates prepared from the solution mixture revealed fragments that were three-quarter-length long compared to the SLS of intact tropocollagens (Gross and Nagai 1965) (Fig. 16.5). Remarkably, the shorter SLS retained a banding pattern that was in perfect alignment with the C-terminal three-fourth region of the established fingerprint of type I collagen (Fig. 16.5c). Furthermore, using the 58 bands as the fingerprint of type I collagen and the amino acid sequence of CNB fragment α1-CB7, an analysis of the banding pattern of the SLS helped to identify the Gly-Ile bond between position 775–776 of the α1 chain (i.e., residues 221 and 222 of the α1-CB7) to be the severed bond (Bruns and Gross 1973). Similar studies using SLS have established both the high specificity of the collagenase and the highly conserved action of the enzyme in different tissues and across species (Stark and Kühn 1968; Nordwig and Bretschneider 1971; Heidemann and Heinrich 1970; Nagase and Woessner 1999; Nagase et al. 2006; Harris and Krane 1974).

Fast forward half a century, more than 28 genetically distinct types of collagens have been identified based on the amino acid sequences and the structure of the genes (Ricard-Blum 2011). There is no need to use SLS to study the primary structure in the postgenomic era, but SLS remains a useful tool in molecular studies of collagens. The formation of SLS aggregates appears to be common among all collagens under suitable conditions in vitro and in vivo. The reproducible and uniquely identifiable banding patterns were frequently used to identify the corresponding collagens in TEM images of tissues (Bentz et al. 1983; Schwartz

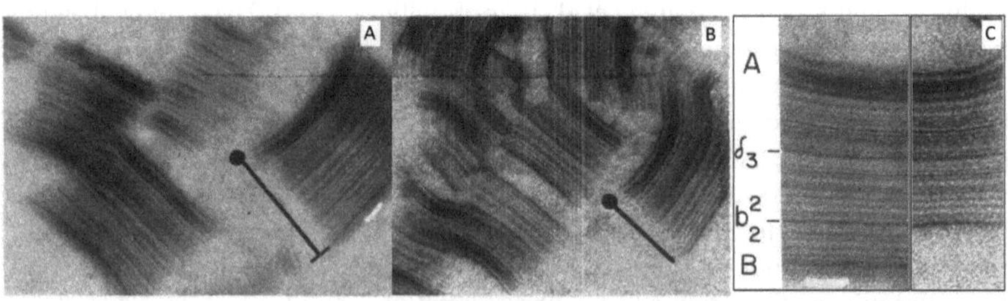

Fig. 16.5 The ¾ fragment of the collagenase digest. (**a**) The SLS of calf skin collagen. (**b**) The SLS of the ¾ length fragment from the reaction mixture. (**c**) A comparison of the SLS of collagen and of the ¾ fragment led to the identification of the region of the enzymatic cleavage; the "A" and "B" marks the C- and N- termini of the SLS. From (Gross and Nagai 1965)

and Veis 1978), to study the chain register in heterotrimeric type I collagen (Bender et al. 1982) and to study the effects of disease-causing mutations (Timpl et al. 1978; Kobayashi et al. 1990; Stanescu et al. 1982).

SLS and the Structural Polymorphism of Fibrillar Collagen

Even in its heyday, SLS was frequently referred to as "artificial aggregates" because it has a nonfibrous structure and, most of all, because it was produced in the laboratory under nonphysiological conditions: low pH and with the addition of ATP. On the other hand, the highly reproducible nature of SLS indicates this unique structure is a manifestation of certain innate properties of the collagen molecule.

The Formation of SLS under Different Buffer Conditions

Following its discovery, research emerged to characterize the roles of pH and ATP in the formation of SLS. The SLS under in vitro conditions are often prepared from reconstituted collagen fibrils. The low pH is necessary to weaken interactions involving acidic residues (Glu and Asp) and cause the fibrils to dissociate (Kühn 1982). The low pH is also necessary to keep the TC in a monomer form avoiding other nonspecific aggregates that TC is prone to (Kühn 1982; Bowden et al. 1968). TC monomers that had been dissolved in solutions with a pH between 2.5 and 3.5 precipitated as SLS upon the addition of ATP (Paige and Goh 2001; Harris and Lewis 2016). At pH 5, however, the TC molecule formed fibril-like structures instead of SLS upon the addition of ATP; the fibrils at pH 5 did not show any discernable banding pattern (Harris and Lewis 2016).

The precise interaction between the polyanions and TC in the formation of SLS appears to be more complex, affected by both the structure of the polyanions and the distribution of the charged residues on the surface of TC. The polyanion ATP was thought to cross bridge positively charged collagen molecules so that they aggregate laterally with their ends in register (Kühn 1982; Doyle et al. 1975). Yet, not all polyanions can induce the precipitation of SLS (Bowden et al. 1968; Harris and Lewis 2016; Paige and Goh 2001). In one study, the presence of three adjacent charge groups were deemed to be the necessary condition based on the finding that ATP, ATP-γ-S, and GTP were able to effectively induce the SLS aggregation while ADP could not (Paige and Goh 2001). In another study, however, both triphosphate and polyanions having two negatively charged groups, such as 1, 6 fructose diphosphate and perdisulphuric acid, were able to cause TC molecules to precipitate as SLS having the same banding patterns as those precipitated using ATP (Bowden et al. 1968). The same study also reported that when a dicarboxylic acid, such as succinic acid, glutaric acid, or adipic acid, was used as the polyanion in solution no SLS were found. It was subsequently concluded that "a cloud of strongly negative charge" separating the two negatively charged groups was the important feature. Polyanions like ATP and GTP are *naturally* present in the cytoplasm and the ECM, whether their concentrations can reach the critical level to induce the SLS formation in vivo is yet to be seen.

In another interesting study, SLS were produced by adding diazo dye Sirius red or Evans blue to preformed *D*-periodic fibrils at neutral pH (Harris and Lewis 2016). These are large dye molecules that have a size about two to three times that of ATP and contain multiple charged groups separated by a diazo moiety. The dye molecules appeared to take on dual roles in this reaction: weakening the charge-based interaction of the fibrils and promoting the in-register precipitation of SLS. A higher than usual ionic strength appeared to be necessary for precipitation of SLS by the dye molecules. The ionic strength is generally an important factor for both in vitro fibrillogenesis and the SLS formation (Harris and Reiber 2007). The functions of salt in a solution of biomolecules range from shielding of charged groups, altering the dielectric constant and the activity of water to, in the cases of RNA folding and binding of DNA, directly affecting the structures and molecular recognition. How salt affects the self-assembly of collagen triple helices is only understood in general terms for now.

The cross striation from in-register alignment of charged and nonpolar residues of SLS were further elaborated using atomic force spectroscopy (AFM)—a more recently used technology that does not require staining of materials. Studies using AFM revealed the in-register nature of the SLS as "fine structures" rising from residues having large hydrophobic side chains (Fujita et al. 1997; Paige and Goh 2001); the SLS appeared to be rectangular aggregates on the mica showing rounded tops. At the optimum concentrations of collagen and ATP, studies using AFM were able to reveal the SLS crystallites coexisted with different aggregates smaller in size, which were interpreted as the oligomers of the intermediate stages (Paige and Goh 2001). The earliest oligomers consisted of a few TC molecules and had the length of a TC molecule but were ~ 2.5 nm in diameter. These oligomers appeared to be stable intermediates and present in a large amount during the early stages of self-assembly; they further congregated into increasingly larger intermediates leading to the assemblies of the crystallite SLS. Interestingly, a similar observation was also reported using designed, fibril-forming peptides (see below) (Kaur et al. 2015).

SLS under Physiological Conditions and their Potential Involvement in Fibrillogenesis

SLS can naturally occur under physiological conditions. SLS-crystallite and pro-SLS (SLS aggregates of collagens before the N- and/or C-propeptides are removed) were frequently found in cell cultures without the addition of ATP and/or the need to lower the pH (Bruns et al. 1979; Hulmes et al. 1983). SLS were observed in whole-mount preparations of fibroblasts, in tissue homogenates, and in

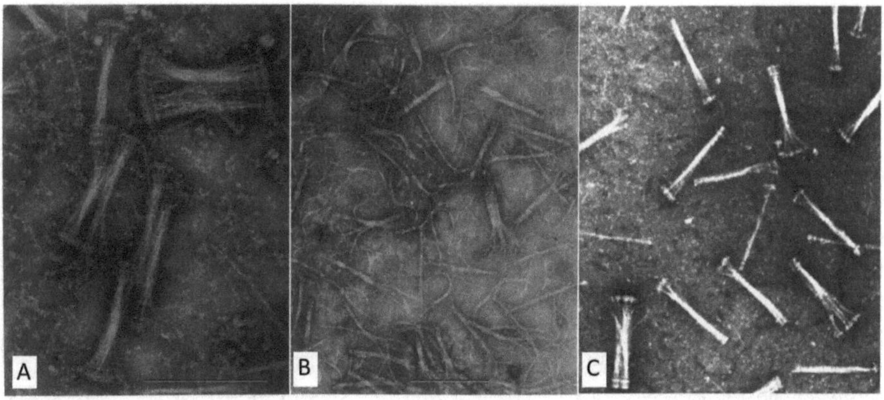

Fig. 16.6 Naturally formed SLS and pro-SLS in cell medium. (**a**) Aggregates from culture medium of chick embryo tendon cells. (**b**) Homogenate of sternal cartilage from 2 weeks old lathyritic chick showing numerous SLS-like structures. The scale bars are 300 nm. (**c**) Negative staining of pro-SLS from 24 h culture medium of chick embryo tendon fibroblasts. No ATP or other substances added. **a** and **b** from (Bruns et al. 1979), **c** From (Kühn 1982)

thin sections of embedded cells. These SLS were "slender" comparing to those observed under in vitro conditions—they retain the critical 300 nm length but are smaller in both their width and height (Fig. 16.6). The signature polar cross striations are clearly visible albeit with lower resolution due to their smaller sizes. The pro-SLS, especially those having only the C-propeptide attached (pC-SLS), acquire a distinctive shape as a flower bouquet with the triple helical domain forming the "stems" bearing the recognizable cross striation and the larger, globular propeptide spread out as the flower heads linked to the "stem" via a flexible linker (part of the linker becomes the C-telopeptide later after enzymatic processing and remains at the terminal ends of the TC) (Fig. 16.6c). The large globular domain may have also prevented the pro-SLS to grow bigger. The formation of the pro-SLS is driven by the same molecular interactions that stabilize the SLS, since neither the propeptides nor the linker regions (or the telopeptides) are in contact with each other in the assembly. Because of the large globular propeptides, the procollagen monomers cannot assemble into staggered fibrils, and the SLS becomes a stable alternative conformation. Some TC having the smaller N-propeptide attached, however, were found to be incorporated on the surface of the fibrils and play a key role in controlling the diameter of the fibrils by limiting the further assembly of TC molecules (Hulmes 1983; Hulmes et al. 1989). SLS and pro-SLS were also reported in thin sections of tissues, although it is often difficult to identify which type of collagen was in the aggregates due to the reduced resolution of the banding pattern (Weinstock 1977; Warshawsky 1972; Pérez-Tamayo 1972). It seemed a high concentration of TC is the necessary condition for SLS (Hulmes et al. 1983). The average molecular weight of the pro-peptides in cell culture without further concentrating was estimated to be close to that of the TC monomer using velocity sedimentation, although the wide peaks on the Schlieren optic photography during sedimentation indicated a more heterogeneous population than that of pure TC monomers. After the same cell culture was concentrated by repeated centrifugation, both SLS and pro-SLS were observed by TEM (Fig. 16.6a).

Bruns et al. were first to suggest a physiological function of SLS as the precursor of in vivo fibrillogenesis (Bruns et al. 1979; Hulmes 1983). They reported bundles of SLS in vacuoles associated with the cell membrane. In the same whole-mount fibroblast preparation, they also observed the presence of pro-SLS, SLS, native fibrils and the slender SLS crystallites scattered around the tips of the growing fibrils. In their proposed mechanism of in vivo fibrillogenesis, cells secrete procollagen in high local concentration packaged in the SLS form. These pro-SLS were then transported in vacuoles to be secreted to the ECM. These pro-SLSs remain stable in the ECM. They further suggest that the SLS in the ECM are more stable than those prepared in solution at low pH and with the addition of ATP, presumably because of the interactions with some other "cementing substances" such as fibronectin or proteoglycan. The *cemented* stability can potentially shift the equilibrium between the TC monomer and the SLS and make the SLS the more favorable conformation under moderate concentrations. The N-and C-propeptide on the pro-SLS were subsequently removed in the ECM and the "slender SLS crystallites"—those having a diameter of ~10 nm in which the TC are in 0*D* register with one another—become incorporated directly into the fibrils. In a way, the *D*-periodic fibrils "may be considered as an n × 67 nm staggered array of SLS crystallites" (Bruns et al. 1979).

Collagen fibrils acquire additional suprafibrillar structures in the extracellular matrix that establish the eventual architecture of different tissues and give the tissues their unique forms and properties. The high concentrations of collagen in the ECM resemble a condensed matter rather than macromolecules in a solution. Using the liquid crystalline model of condensed matter, several works by Giraud-Guille et al. using polarized light microscopy and TEM demonstrated that the structure of the ECM share several similarities with the gelation process of collagen under very high concentrations (Besseau and Giraud-Guille 1995). The suprafibrillar structure of collagen in the ECM emerged from a liquid crystalline order of "soluble precursors" which went through an intermediate stage involving molecular aggregates of unit triple helices. The mature suprafibrillar architecture of connective tissues maintain the liquid crystalline ordering of the "soluble precursors" (Besseau and Giraud-Guille 1995; Giraud-Guille 1996; Hulmes 2002). In the subsequent studies, the soluble precursors were determined to be procollagens which can reach a concentration in the range of 50–80 mg/mL without losing the *liquid* form; in comparison, TC monomers readily form gels or non-specific precipitates at concentrations of a few mg/mL in neutral pH (Martin et al. 2000). It was thus proposed that liquid crystalline order in connective tissue takes place prior to enzymatic procollagen processing; the fibril formation proceeded with "compacting and sliding" of the molecular aggregates of the procollagens after the propeptides are removed, while retaining the liquid crystalline organization that existed prior to the fibril assembly. Ideal for supramolecular order in sub-micrometer to close to 1000 μm range, polarized light microscopy revealed little structural details of the molecular aggregates other than being ~20 nm in diameter, a diameter much thicker than an individual procollagen triple helix. The pro-SLS appears to be the candidate for the intermediary molecular aggregates considering any staggered arrangement is unlikely given the presence of the propeptides. The potential precursor role of SLS and/or pro-SLS during in vivo fibrillogenesis would thrust this unique molecular structure right back to the center of the research on the structures of the ECM during the function and development of tissues. On the other hand, the precursor nature of SLS highlights the challenges of characterizing the involvement of SLS in physiological processes: like any intermediate in a reaction pathway, the SLS will be present in transient and in lower abundance.

The Other Structures of Fibrillar Collagen

It has been established since the early 1950s that collagen is a true polymorphic structure (Gross et al. 1954; Schmitt et al. 1953). In addition to the *D*-periodic fibrils and SLS, there are another two frequently observed structures: the fibrous long-spacing assemblies, or FLS, and the obliquely striated periodic fibrils (Fig. 16.7) (Bruns 1976; Highberger et al. 1950; Schmitt et al. 1953). The FLS, which include up to four different sub-types categorized as FLS I-IV, are formed by staggered TC-like *D*-periodic fibrils but having a different periodicity ranging from 90 to 250 nm (Doyle et al. 1975). The fine banding structures of FLS in the positively stained micrographs are nonpolar having a symmetric appearance. Since the banding patterns of TC and SLS are polar, a symmetric, nonpolar banding pattern pointed to an antiparallel arrangement of TC and/or slender SLS a few nm in diameter. The formation of FLS requires the presence of other flexible, often negatively charged biomacromolecules such as chondroitin sulphate, serum glycoprotein, various mucopolysaccharides, and proteoglycans (for a summary, see Doyle et al. 1975). The staggered arrangement was related to the bridging effects of the negatively charged molecules. In some cases, these molecules are located at fairly specific sites in the structure. The possible arrangements of TC in FLS are not unique, giving rise to different periodicities; some FLS show a combination of different staggering arrangements (Doyle et al. 1975) (Fig. 16.7a and b).

The obliquely straited fibrils were observed in reconstituted fibrils from cartilage and in subsections of rat tail tendon (Bruns 1976; Doyle et al. 1975). The molecular arrangement was explained in terms of narrow sub-fibrils having the native *D*-striation and being ~14 nm (13.8 ± 1.5 nm) in diameter. The oblique pattern emerged from the assembly of the sub-fibrils; they all have the same diameter and orientation but have a regular axial mutual displacement about 90 Å with the adjacent sub-fibrils (Fig. 16.7c). Like in the case of FLS, the sub-fibrils can have several different arrangements including having two adjacent sub-fibrils in the opposite directions giving rise to varied staining patterns, some looked like a "checkerboard" others had "chevrons." The observation of such varied assemblies

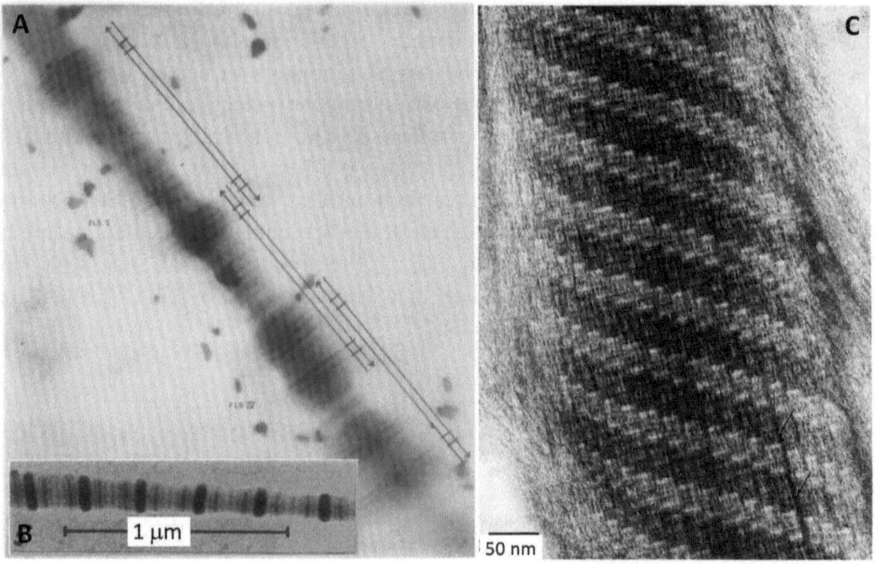

Fig. 16.7 FLS and oblique fibrils. (**a**) Positively stained micrograph showing the transition from FLS I (having an overlap of 30–50 nm) to FLS IV (having an overlap ~127 nm. The diagram shows the axial arrangement of molecules including antiparallel arrangement of TC molecules. (**b**) FLS I showing nonpolar banding pattern and a periodicity of 200 nm. (**c**) Oblique fibrils reconstituted from cartilage collagen showing about 15 narrow sub-fibrils each has the regular 67 nm *D*-period. The arrow indicates the accumulation of dye molecules between the sub-fibrils. The scale bar is 50 nm. **a** adapted from (Doyle et al. 1975), **b** from (Schmitt et al. 1953), **c** from (Bruns 1976)

of sub-fibrils highlighted (1) the fibrillogenesis proceeded with the formation of sub-fibrils having similar diameters and the invariant *D*-striation and (2) the further assembly of the sub-fibrils follows a less specific manner. The *D*-stagger is a unique, well-defined structure determined by the specific spacings of the interacting residues (more details in the next section), the interaction sites on the surface of the sub-fibrils (and also on the native collagen fibrils) occur in various patterns that are sensitive to the environment during fibril formation. The arrangement of the sub-fibrils can, thus, be a key factor in the polymorphism of collagen fibrils.

The Determinant Factors for the Self-Assembly of Fibrillar Collagen

One of the key findings about the different structures of collagen is that they are interconvertible depending on the buffer conditions (Highberger et al. 1950). These are, therefore, spontaneously formed structures determined by the properties of collagen triple helices. Why are there different structures of collagen? Is polymorphism a genetic phenotype of fibrillar collagen? The key to answering these questions is to understand the principles that govern the self-assembly of triple helices (TC) into ordered aggregates.

The Origin of the Structures Is in the Amino Acid Sequences

The fact that polymorphic structures all exhibit a particular axial structure indicates that the "functional" groups are not scattered in random on the surface of the helix but acquire a specific, periodic placement along the axis of the helix; such that only in a regular mutual arrangement of neighboring helices will these groups be optimally aligned for interaction and accumulatively contribute to the stability of the assembly (Hulmes et al. 1973; Kaur et al. 2015). The most profound work on identifying the unique placements of interacting residues of collagen in its primary structure was accomplished by Hulmes et al. in the early 1970s. Taking advantage of the newly resolved amino acid sequence of the α_1 chain of type I collagen, they developed a computational analysis of the amino acid sequence to study the "complementary relationship" that would explain the stagger multiples of 67 nm. Modeling the triple helix as a one-dimensional linear display of residue side chains of the X and Y residues and having a uniform helical rise of 2.86 Å per residue, they calculated the *interaction curve* based on all possible interactions between two associating TC molecules in a particular "chain staggering." According to the interaction curve, the interactions of both hydrophobic groups and oppositely charged groups between two associating triple helices reaches maximal if the two helices are staggered by 0*D*, 1*D*, 2*D*, 3*D*, and 4*D*, where a *D* is equivalent to a section of the triple helix composed of 234 residues (per single polypeptide chain). In particular, the distribution of the large hydrophobic residues is found to have a regular *D*-spacing, thus in a *D*-staggered arrangement, these residues will be brought into a close proximity for interaction. The interactions of oppositely charged residues are also maximal in a *D*-staggered arrangement, although the locations of the charged residues per se do not appear to follow any obvious distribution complementary to the *D*-period (Hulmes et al. 1973).

This periodic placement of interacting residues provides a framework to understand the *D*-staggered fibrils as the most stable conformation emerging from the interactions of associating triple helices. Certain aspects of the periodic placements of the residues also provide a basic understanding of the antiparallel arrangement and the interactions with the polyanions in FLS and other self-assemblies (Doyle et al. 1975). What remains unclear, however, is the structure of SLS and especially SLS as a potential competing conformation with the *D*-staggered assemblies. Any molecular interactions that stabilize the *D*-staggered triple helices during the self-assembly must also be present in the SLS structure. In fact, those interactions should reach the maximum in the 0*D* staggered SLS (Hulmes et al. 1973). Clearly, there are other molecular interactions and/or structural constraints that discriminatorily favor *D*-

period fibrils over SLS under physiological conditions that were not included in the calculations of the interaction curve.

Interestingly, direct measurement of the forces between the triple helices in the *D*-periodic collagen fibrils using the osmotic stress method, revealed that the fibrils are predominantly stabilized by hydrophilic interactions (Leikin et al. 1993, 1994, 1995). These hydrophilic interactions were interpreted as the water mediated H-bond network. Indeed, the collagen triple helix is a highly hydrated conformation. A thick layer of water shell is a common feature in the crystal structures of triple helical peptides (Bella et al. 1994; Kramer et al. 1999), including the structure of peptide $(PPG)_{10}$ which does not have any polar and/or charged residues (Bella et al. 2006; Vitagliano et al. 2001; Berisio et al. 2002). A large number of water molecules were observed bound to the carbonyl groups on the peptide backbone forming water bridges that link peptide chains in a triple helix and between different triple helices in the crystal lattice (Kramer et al. 2001; Bella et al. 1994). The water covered surface may also contribute to the shielding of the charged side chains and minimize the repulsion of like charges in the fibrils (Hulmes et al. 1973). While the H-bonds may account for the explicit attraction force stabilizing the fibrils, how do they contribute to the structural specificity of the fibrils remains unclear. Since the hydrophilic to hydrophobic contacts are energetically very unfavorable, the water covered clusters of polar and/or charged residues may orient with each other to avoid the clusters of hydrophobic residues, and thus inadvertently bring the regularly spaced hydrophobic residues into register.

The self-assembly of the TC is essentially a protein folding problem since the triple helix "wears" the amino acid sequence of its constituent polypeptide chains on the surface of the molecular *rod*. Yet, the predominant role of hydrophilic interactions in the stability of the fibrils appears to set the fibrillogenesis apart from the folding of globular proteins which are nearly always driven by the process of hydrophobic residues being pushed together in a process known as the hydrophobic collapse (Leikin et al. 1995). One of the major impacts of the hydrophobic collapse in globular proteins is to reduce the conformational space for folding and to *funnel* the interactions between different parts of the peptide chain into a steep energetic descent to reach the native conformation (Dill and Chan 1997; Dobson et al. 1998; Dobson 2003; Onuchic and Wolynes 2004; Rollins and Dill 2014). The extended, rather rigid conformation of the triple helix suggests a rather restricted conformational space during the self-assembly and consequently, a different folding pathway: one along which the specific *D*-staggering structure emerges from the accumulative effects of optimized interactions of residues in a *D*-period related regular, periodic placement. At the same time, it could also imply that multiple conformations may have comparable stability and coexist to a certain extent.

A Design Strategy Based on the Periodic Spacing of the Interacting Residues

The ultimate test of the effectiveness of the periodically spaced residues to direct the self-assembly of the triple helices into a unique structure comes from the studies using designed triple helical peptides. The major findings of the work by Hulmes et al. were incorporated into the design of triple helical peptides that can further self-associate to form fibrils having a *D*-period-like axial repeating structure (Fig. 16.8) (Chen et al. 2019; Strawn et al. 2018; Kaur et al. 2015; Xu and Kirchner 2021). These designed, self-assembled mini-fibrils exhibit a clear axially repeating structure of 35 nm when examined using TEM and AFM. This 35 nm repeating structure, which is designated as a *d*-period, consists of a 0.3*d* overlap region and a 0.7*d* gap (Fig. 16.8a–c, and g). The *d*-periodicity of the mini-fibrils comes from a built-in *d*-spacing of both the hydrophobic and hydrophilic residues in the primary structure of the peptides that is created by using multiple pseudo identical amino acid *sequence units* placed in tandem (Fig. 16.9d). Each sequence unit contains 123 residues forming a section of triple helix about 35 nm in length. Thus, by mutually staggering one sequence unit, the hydrophobic and hydrophilic residues in the associating helices will be in register in a manner similar to those in the *D*-staggered native collagen fibrils. The same interaction curve calculated according to Hulmes et al. also revealed a set of interaction peaks when the triple helices are staggered by 0*d*,

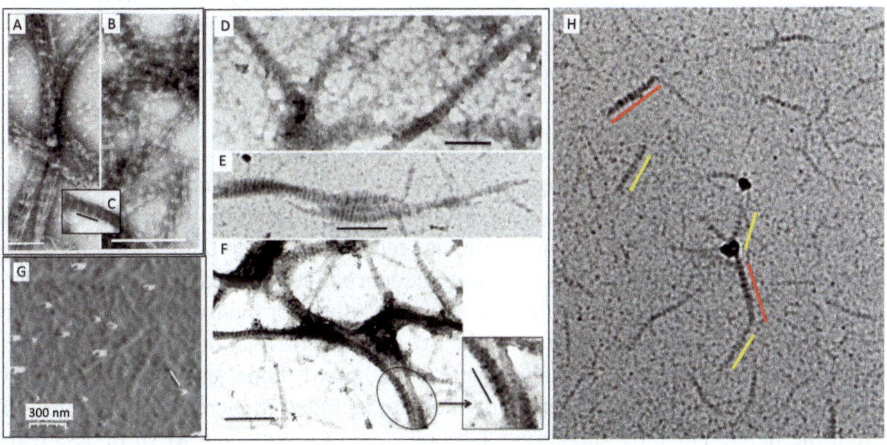

Fig. 16.8 The designed mini-fibrils. (**a–c**) Negative staining of the mini-fibrils showing the *d*-period as alternating dark and light bands every 35 nm. The scale bars are 100 nm in **a** and **c**, 200 nm in **b**. (**d–f**) Positive staining of the mini-fibrils showing the three pairs of thick-thin bands every 125 nm. The scale bars are 200 nm (100 nm in the insert of **f**). (**g**) AFM image of the mini-fibrils, the scale bar is 300 nm. (**h**) Positive staining of seed-oligomers in the preparation of the mini-fibrils under fibril forming condition. The yellow bars are 125 nm in length marking the size of one triple helix; the red bars are 200 nm in size, equivalent in size to triple helices in a 2*d*-stagger. **a–g** From (Kaur et al. 2015); **h**: Kaur personal communication))

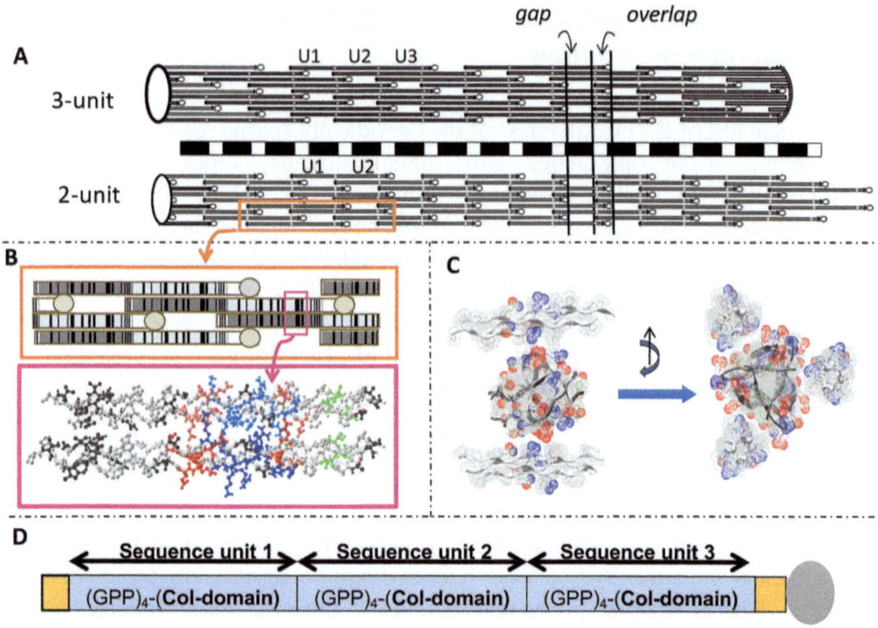

Fig. 16.9 The unit-staggered assembly of the mini-fibrils. (**a**) The unit staggered arrangement of peptides having 3 or 2 identical sequence units (U1-U3); a schematic drawing of the negatively stained mini-fibrils is shown having the alternating black (gap) and white (overlap) regions. (**b**) The interactions of charged residues in register (red: Glu or Asp, Blue: Arg or Lys, green: Thr). (**c**) Purported packing of the foldon domain in the mini-fibril. (**d**) The sequence architecture of a designed peptide having three identical sequence units (GPP_4-(Col-domain)) shown as blue rectangles; grey circle: foldon domain, orange bars: the linker regions. **a–c** adapted from (Strawn et al. 2018)

1*d*, 2*d,* and 3*d* (only 0*d* to 2*d* staggering are possible in peptides containing only two repeating sequence units) (Fig. 16.9a). Thus, the hydrophobic and charge-based interactions are all optimized in a unit-staggered arrangement. In addition to the triple helix domain consisting of repeating sequence units, the designed peptides also have an "overhang" about 0.3d in size which accounts for the overlap region in the staggered fibril assembly. The "overhang" region consists of a foldon domain included as the nucleation domain for the triple helix folding and a short stretch of *linker* peptide at the N and C-termini which are included to increase the stability of the triple helix. Thus, by staggering 1-sequence unit (equivalent to 1*d* or 123 residues), the self-assembly of triple helices give rise to a regular *d*-periodicity.

The mini-fibrils appear to capture the other major feature of a collagen fibril: the SLS-like alignment of the charged residues within the sequence units. The positively stained mini-fibrils revealed a different set of finer banding patterns, which coincide with the distribution of the charged residues in the sequence unit (Fig. 16.8d, e and f). The finer banding of the mini-fibrils reflected the in-register nature of charged residues in the neighboring triple helices in a unit-staggered arrangement. The positively stained banding pattern is resolved to a pair of dark thick-thin bands marking the two clusters of charged residues in one sequence unit. For the peptide containing three sequence units in its full length of ~120 nm, the clusters of the charged residues can be observed clearly as three pairs of thick-thin bands every 100–120 nm or so on the micrographs. Because of the lower TEM resolution associated with the thinner fibrillar structures, the thick-thin bands appeared to lack resolution compared to a SLS fingerprint of native collagen. Nevertheless, they are the manifestation of an in-register alignment of the charged and the hydrophobic residues and are easily identifiable. The fact that both the negatively stained and the positively stained structures of the mini-fibrils shared the same features with that of native collagen fibrils implies the same molecular mechanism directed the self-assembly of both the native collagen fibrils and the mini-fibrils.

In a few cases, we also observed SLS-like assemblies of the designed peptides under the condition of fibril formation (Fig. 16.8h). These slender SLS-like assemblies are ~125 nm long but much thicker than an individual helix which has a diameter of ~1.0–1.5 nm. The three pairs of positively stained thick-thin bands are visible as black dots. There are other larger positively stained aggregates on the same micrograph. These larger aggregates all have the discernable three pairs of thick-thin bands in every ~120 nm. Their size fell into two groups: ~ 150 nm or ~ 200 nm, both correspond well with triple helices in a 1*d* or a 2*d* staggered arrangement, respectively. We therefore consider them the "seed" intermediates in the early stages of fibril assembly (they are termed seed-oligos). The presence of the slender SLS-like aggregates together with the seed-oligos is consistent with the "precursor" role of the SLS during the fibrillogenesis. As it was in the case for the native slender SLS by Bruns et al., the SLS-like structures are observed under neutral pH with no addition of ATP or other polyanion. Also similar to the native slender SLS, the SLS aggregates did not populate but give way to the mini-fibrils. For the designed peptides, the conformational bias for *d*-staggered mini-fibrils was understood in terms of the steric hindrance associated with the foldon domain (Chen et al. 2019). The diameter of the globular foldon domain is twice the diameter of a triple helix and it is tethered to the triple helix via a rather stiff linker: a heptapeptide GSGPCCG where the two Cys are expected to form one or two interchain disulfide bonds (Fig. 16.9c and d). So bulky and stiff, they prohibit the bouquet-like structure of the PC-SLS (having the C-propeptide attached).

Both the interaction curve and the in vitro studies using designed peptides further indicate that the SLS can potentially be a stable assembly of triple helices. There is nothing "artificial" about the molecular interactions giving rise to the SLS structure. One possible destabilizing factor of SLS in neutral pH is perhaps the repulsion of the like charges which are maximally aligned in the *D*-staggered arrangement. It remains to be fully evaluated how such repulsion forces are mediated in the *D*-periodic fibrils. The "artificial" conditions where the SLS were precipitated in large sizes only further stabilized this structure and perhaps at the same time discouraged the *D*-staggered fibrils as a competing conformation. The slender size of SLS under physiological conditions makes them difficult to observe under in vivo conditions in thin sections or other tissue preparations. A full understanding of their physiological roles will require more precise insights on why and how the SLS eventually gets incorporated into the fibrils in the ECM.

The Recent Works on SLS and FLS in Tissues

Structures other than that of *D*-periodic fibrils have been reported in many normal and atypical tissue samples. The identification of the SLS is often based on two features: (1) a dimension to be close to that of the length of a TC monomer and (2) a fine, closely placed, polar striation. Other fibrillar structures having noncanonical banding patterns were also found and were presumed to be the FLS. With the limited resolution, however, it is often difficult to discern which collagen(s) are involved in the SLS or FLS formation. The various observed collagen-banding periodicities could also be affected by the incorporation of different glycosaminoglycans or an over glycosylation (Rehany et al. 2000; Waltimo 1996; Kobayasi et al. 1985; Wen and Goh 2006). SLS and FLS formation may be due to an over secretion of collagen or it may be a by-product of collagen degradation (Kobayasi et al. 1985; Pérez-Tamayo 1972; Kamiyama 1982; Dingemans and Teeling 1994; Slavin et al. 1985). Type VI was found to form banded fibrils with a periodicity around 110 nm, and some FLS observed by others may be type VI collagen (Miki et al. 1993; Bruns et al. 1986); whether FLS formation is a byproduct of tissue remodeling is still debatable. FLS has been observed in arteries bound to lipids, in malignant mesothelioma where a high content of hyaluronic acid was detected, around arteries in polyp biopsy samples, in intramuscular nerves, and in fatty tissues, to name a few (Kang et al. 2009; Morris et al. 1978; Maeda et al. 1996; Kagoura et al. 2001). It seems that in tissues, collagen may assume various structures; they may provide some unknown function or just be a byproduct of their environment. Only future research will enlighten us. The clinical involvement of abnormal collagens was reviewed by Eyden and Tzaphlidou (2001). The following is a list of some more recent publications on the subject that were not included in that review (Table 16.1). The frequently reported abnormal collagens are mostly FLS which is larger, fibrous like and easier to identify; direct observation of SLS remains rare and limited.

Concluding Remarks

In our quest to understand the involvement of collagen during the function and development of tissues, the great achievements in recent years on the study of the structures of collagen fibrils of the collagen triple helix and of collagen receptors, as well as the understanding of the interactions between collagen and other macromolecules is only the tip of the iceberg. The complexity related to the crucial roles of collagen is manifested at many levels: the polymorphism of the supramolecular structure, the heterotypic nature of collagen fibrils, the extensive interaction with a

Table 16.1 FLS and SLS in tissues

Conditions and tissues	
Bullous pilomatricoma	Lao et al. (2005)
Choroideremia	Murro et al. (2021)
Dendritic cell tumor	Schraven et al. (2011)
Descemet's membrane	Hayashi et al. (2002), Jakus (1956)
Descemet's membrane of diabetic rat or human	Akimoto et al. (2008), Rehany et al. (2000)
Epiretinal membranes (compact FLS)	Parolini et al. (2011)
Eye corneoscleral	Ueda and Yue (2003)
Gastric adenocarcinoma	Caruso et al. (2004)
Gastrointestinal stromal tumor	Yoneda et al. (2001)
Intervertebral disc tissue	Gruber and Hanley (2002)
Lipodystrophia centrifugalis abdominalis infantilis	Kagoura et al. (2001)
Malignant mesothelioma	Kang et al. (2009)
Myopic traction maculopathy	Yokota et al. (2018)
Neuromuscular diseases intramuscular nerves	Maeda et al. (1996)
Splenic hamartoma	Ali et al. (2005)
Superficial hypertrophic dendriform epitheliopathy	Akhtar et al. (2006)
Uvea schwannoma	Saavedra et al. (2006)
Vestibular schwannoma cell culture	Nair et al. (2007)
Chick embryo tendon fibroblast cell culture medium	Hulmes et al. (1983), Bruns et al. (1979)
Dialyzed skin collagen	Schmitt et al. (1953)
Human skin fibroblast cultures	Bruns et al. (1979)
Mouse teratocarcinoma cell cultures	Bruns et al., (1979)

broad range of other macromolecules, and the progressive changes in the structure and composition of the ECM during normal development and tissue repair. SLS can potentially be a critical intermediate between procollagen and the molecular organization of collagen in the ECM. Understanding how structural differences between SLS and the *D*-periodic fibrils affect their interactions with collagen receptors and/or other macromolecules such as collagenase will provide insight on the innumerable and often contradictory findings about the complex roles of collagen in tissues. Such knowledge will also lead to new treatments for a wide range of connective tissue diseases ranging from aging to cancer metastasis. Studies of SLS in vivo will require experiments that can capture the structure that is much smaller in size than the canonical *D*-periodic collagen fibrils and are transiently present in a low abundance. Designed peptides that can form higher order molecular assemblies of the triple helix can be a useful tool in such an effort. SLS is only one of the variations of the possible macromolecular assemblies of fibrillar collagen in the ECM. The multiplicity of the molecular organizations of collagen makes it a very complex system to study, but also a fascinating one.

References

Akhtar S, Voon LW, Bron AJ (2006) The ultrastructure of superficial hypertrophic dendriform epitheliopathy after keratoplasty. Cornea 25(5):623–627. https://doi.org/10.1097/01.ico.0000227887.26747.e0

Akimoto Y, Sawada H, Ohara-Imaizumi M, Nagamatsu S, Kawakami H (2008) Change in long-spacing collagen in Descemet's membrane of diabetic Goto-Kakizaki rats and its suppression by antidiabetic agents. Exp Diabetes Res 2008:818341. https://doi.org/10.1155/2008/818341

Ali TZ, Beyer G, Taylor M, Volpe C, Papadimitriou JC (2005) Splenic hamartoma: immunohistochemical and ultrastructural profile of two cases. Int J Surg Pathol 13(1):103–111. https://doi.org/10.1177/106689690501300116

Bächinger HP, Mizuno K, Vranka J, Boudko S (2010) Collagen formation and structure. In 5:469–530. https://doi.org/10.1016/b978-008045382-8.00698-5

Begbie GH (1947) Thermal scattering of X-rays by crystals; the thermal scattering of the face-centered cubic and the close-packed hexagonal lattices. Proc R Soc Lond A Math Phys Sci 188(1013):189–208. https://doi.org/10.1098/rspa.1947.0004

Bella J, Hulmes DJ (2017) Fibrillar Collagens. Subcell Biochem 82:457–490. https://doi.org/10.1007/978-3-319-49674-0_14

Bella J, Eaton M, Brodsky B, Berman HM (1994) Crystal and molecular structure of a collagen-like peptide at 1.9 a resolution. Science 266:75–81

Bella J, Liu J, Kramer R, Brodsky B, Berman HM (2006) Conformational effects of Gly-X-Gly interruptions in the collagen triple helix. J Mol Biol 362(2):298–311. https://doi.org/10.1016/j.jmb.2006.07.014

Bender E, Silver FH, Hayashi K, Trelstad RL (1982) Type I collagen segment long spacing banding patterns. Evidence that the alpha 2 chain is in the reference or a position. J Biol Chem 257(16): 9653–9657. https://doi.org/10.1016/S0021-9258(18)34122-X

Bentz H, Morris NP, Murray LW, Sakai LY, Hollister DW, Burgeson RE (1983) Isolation and partial characterization of a new human collagen with an extended triple-helical structural domain. Proc Natl Acad Sci U S A 80(11):3168–3172. https://doi.org/10.1073/pnas.80.11.3168

Berisio R, Vitagliano L, Mazzarella L, Zagari A (2002) Crystal structure of the collagen triple helix model [(Pro-Pro-Gly)(10)](3). Protein Sci: a publication of the protein society 11(2):262–270. https://doi.org/10.1110/ps.32602

Besseau L, Giraud-Guille MM (1995) Stabilization of fluid cholesteric phases of collagen to ordered gelated matrices. J Mol Biol 251(2):197–202. https://doi.org/10.1006/jmbi.1995.0426

Birk DE, Bruckner P (2005) Collagen Suprastructures. In: Brinckmann J, Notbohm H, Müller PK (eds) Collagen: primer in structure, processing and assembly. Springer, Berlin, Heidelberg, pp 185–205. https://doi.org/10.1007/b103823

Boedtker H, Doty P (1956) The native and denatured states of soluble collagen. J Am Chem Soc 78(17):4267–4280. https://doi.org/10.1021/ja01598a024

Bowden JK, Chapman JA, Wynn CH (1968) Precipitation of collagen in the segmented long-spacing form by various organic and inorganic compounds. Biochim Biophys Acta 154(1): 190–195. https://doi.org/10.1016/0005-2795(68)90271-7

Brodsky B, Ramshaw JA (2017) Bioengineered collagens. Subcell Biochem 82:601–629. https://doi.org/10.1007/978-3-319-49674-0_18

Bruns RR (1976) Supramolecular structure of polymorphic collagen fibrils. J Cell Biol 68(3): 521–538. https://doi.org/10.1083/jcb.68.3.521

Bruns RR, Gross J (1973) Band pattern of the segment-long-spacing form of collagen. Its use in the analysis of primary structure. Biochemistry 12(5):808–815. https://doi.org/10.1021/bi00729a004

Bruns RR, Gross J (1974) High-resolution analysis of the modified quarter-stagger model of the collagen fibril. Biopolymers 13(5):931–941. https://doi.org/10.1002/bip.1974.360130509

Bruns RR, Hulmes DJ, Therrien SF, Gross J (1979) Procollagen segment-long-spacing crystallites: their role in collagen fibrillogenesis. Proc Natl Acad Sci U S A 76(1):313–317. https://doi.org/10.1073/pnas.76.1.313

Bruns RR, Press W, Engvall E, Timpl R, Gross J (1986) Type VI collagen in extracellular, 100-nm periodic filaments and fibrils: identification by immunoelectron microscopy. J Cell Biol 103(2): 393–404. https://doi.org/10.1083/jcb.103.2.393

Caruso RA, Ieni A, Fabiano V, Basile G, Inferrera C (2004) Perivascular mast cells in advanced gastric adenocarcinomas: an electron microscopic study. Anticancer Res 24(4):2257–2263

Chen F, Strawn R, Xu Y (2019) The predominant roles of the sequence periodicity in the self-assembly of collagen-mimetic mini-fibrils. Protein Sci: a publication of the protein society 28(9):1640–1651. https://doi.org/10.1002/pro.3679

Chung E, Rhodes K, Miller EJ (1976) Isolation of three collagenous components of probable basement membrane origin from several tissues. Biochem Biophys Res Commun 71(4): 1167–1174. https://doi.org/10.1016/0006-291x(76)90776-2

Dill KA, Chan HS (1997) From Levinthal to pathways to funnels. Nat Struct Biol 4:10–19

Dingemans KP, Teeling P (1994) Long-spacing collagen and proteoglycans in pathologic tissues. Ultrastruct Pathol 18(6):539–547. https://doi.org/10.3109/01913129409021896

Dobson CM (2003) Protein folding and misfolding. Nature 426(6968):884–890. https://doi.org/10.1038/nature02261

Dobson CM, Sali A, Karplus M (1998) Protein folding: a perspective from theory and experiment. Angew Chem Int Ed 37:868–893

Doyle BB, Hulmes DJ, Miller A, Parry DA, Piez KA, Woodhead-Galloway J (1974) Axially projected collagen structures. Proc R Soc Lond B Biol Sci 187(1086):37–46. https://doi.org/10.1098/rspb.1974.0059

Doyle BB, Hukins DW, Hulmes DJ, Miller A, Woodhead-Galloway J (1975) Collagen polymorphism: its origins in the amino acid sequence. J Mol Biol 91(1):79–99. https://doi.org/10.1016/0022-2836(75)90373-3

Eyden B, Tzaphlidou M (2001) Structural variations of collagen in normal and pathological tissues: role of electron microscopy. Micron 32(3):287–300. https://doi.org/10.1016/S0968-4328(00)00045-7

Fietzek PP, Kühn K (1976) The primary structure of collagen. In: Hall DA, Jackson DS (eds) International review of connective tissue research, vol 7. Elsevier, Amsterdam, pp 1–60. https://doi.org/10.1016/B978-0-12-363707-9.50007-1

Fujita Y, Kobayashi K, Hoshino T (1997) Atomic force microscopy of collagen molecules. Surface morphology of segment–long–spacing (SLS) crystallites of collagen. J Electron Microsc 46(4): 321–326. https://doi.org/10.1093/oxfordjournals.jmicro.a023526

Giraud-Guille MM (1996) Twisted liquid crystalline supramolecular arrangements in morphogenesis. Int Rev Cytol 166:59–101. https://doi.org/10.1016/s0074-7696(08)62506-1

Gross J, Lapiere CM (1962) Collagenolytic activity in amphibian tissues: a tissue culture assay. Proc Natl Acad Sci U S A 48(6):1014–1022. https://doi.org/10.1073/pnas.48.6.1014

Gross J, Nagai Y (1965) Specific degradation of the collagen molecule by tadpole collagenolytic enzyme. Proc Natl Acad Sci U S A 54(4):1197–1204. https://doi.org/10.1073/pnas.54.4.1197

Gross J, Schmitt FO (1948) The structure of human skin collagen as studied with the electron microscope. J Exp Med 88(5):555–568. https://doi.org/10.1084/jem.88.5.555

Gross J, Highberger JH, Schmitt FO (1954) Collagen structures considered as states of aggregation of a kinetic unit. The tropocollagen particle. Proc Natl Acad Sci U S A 40(8):679–688. https://doi.org/10.1073/pnas.40.8.679

Gruber HE, Hanley EN Jr (2002) Ultrastructure of the human intervertebral disc during aging and degeneration: comparison of surgical and control specimens. Spine (Phila Pa 1976) 27(8): 798–805. https://doi.org/10.1097/00007632-200204150-00004

Hall CE, Doty P (1958) A comparison between the dimensions of some macromolecules determined by electron microscopy and by physical chemical methods. J Am Chem Soc 80(6): 1269–1274. https://doi.org/10.1021/ja01539a001

Harris ED Jr, Krane SM (1974) Collagenases (third of three parts). N Engl J Med 291(13):652–661. https://doi.org/10.1056/nejm197409262911305

Harris JR, Lewis RJ (2016) The collagen type I segment long spacing (SLS) and fibrillar forms: formation by ATP and sulphonated diazo dyes. Micron 86:36–47. https://doi.org/10.1016/j.micron.2016.04.008

Harris JR, Reiber A (2007) Influence of saline and pH on collagen type I fibrillogenesis in vitro: fibril polymorphism and colloidal gold labelling. Micron 38(5):513–521. https://doi.org/10.1016/j.micron.2006.07.026

Hayashi S, Osawa T, Tohyama K (2002) Comparative observations on corneas, with special reference to Bowman's layer and Descemet's membrane in mammals and amphibians. J Morphol 254(3):247–258. https://doi.org/10.1002/jmor.10030

Heidemann E, Heinrich W (1970) Isolation and characterisation of the peptides derived from the alpha 1-chain of calf skin collagen after hydroxylamine cleavage. Eur J Biochem 14(1):61–69. https://doi.org/10.1111/j.1432-1033.1970.tb00261.x

Highberger JH, Gross J, Schmitt FO (1950) Electron microscope observations of certain fibrous structures obtained from connective tissue extracts. J Am Chem Soc 72(7):3321–3322. https://doi.org/10.1021/ja01163a553

Hodge AJ, Schmitt FO (1960) The charge profile of the tropocollagen macromolecule and the packing arrangement in native-type collagen fibrils. Proc Natl Acad Sci U S A 46(2):186–197. https://doi.org/10.1073/pnas.46.2.186

Hulmes DJ (1983) A possible mechanism for the regulation of collagen fibril diameter in vivo. Coll Relat Res 3(4):317–321. https://doi.org/10.1016/s0174-173x(83)80013-2

Hulmes DJ (2002) Building collagen molecules, fibrils, and suprafibrillar structures. J Struct Biol 137(1–2):2–10. https://doi.org/10.1006/jsbi.2002.4450

Hulmes DJ, Miller A, Parry DA, Piez KA, Woodhead-Galloway J (1973) Analysis of the primary structure of collagen for the origins of molecular packing. J Mol Biol 79(1):137–148

Hulmes DJ, Miller A, Parry DA, Woodhead-Galloway J (1977) Fundamental periodicities in the amino acid sequence of the collagen alpha1 chain. Biochem Biophys Res Commun 77(2): 574–580

Hulmes DJ, Bruns RR, Gross J (1983) On the state of aggregation of newly secreted procollagen. Proc Natl Acad Sci U S A 80(2):388–392. https://doi.org/10.1073/pnas.80.2.388

Hulmes DJ, Kadler KE, Mould AP, Hojima Y, Holmes DF, Cummings C, Chapman JA, Prockop DJ (1989) Pleomorphism in type I collagen fibrils produced by persistence of the procollagen N-propeptide. J Mol Biol 210(2):337–345. https://doi.org/10.1016/0022-2836(89)90335-5

Jakus MA (1956) Studies on the cornea. II. The fine structure of Descemet's membrane. J Biophys Biochem Cytol 2(4):243–252. https://doi.org/10.1083/jcb.2.4.243

Kagoura M, Toyoda M, Matsui C, Higaki S, Morohashi M (2001) An ultrastructural study of lipodystrophia centrifugalis abdominalis infantilis, with special reference to fibrous long-spacing collagen. Pediatr Dermatol 18(1):13–16. https://doi.org/10.1046/j.1525-1470.2001.018001013.x

Kamiyama R (1982) Fibrous long spacing-like fibers in the bone marrow of myloproliferative disorder. Virchows Arch B Cell Pathol Incl Mol Pathol 39(3):285–291. https://doi.org/10.1007/bf02892855

Kang D, Kunugi S, Masuda Y, Ishizaki M, Koizumi K, Fukuda Y (2009) Ultrastructural and immunohistochemical analysis of fibrous long-spacing collagen fibrils in malignant mesothelioma. Ultrastruct Pathol 33(2):52–60. https://doi.org/10.1080/01913120902751197

Kaur PJ, Strawn R, Bai H, Xu K, Ordas G, Matsui H, Xu Y (2015) The self-assembly of a mini-fibril with axial periodicity from a designed collagen-mimetic triple helix. J Biol Chem 290(14): 9251–9261. https://doi.org/10.1074/jbc.M113.542241

Kefalides NA (1968) Isolation and characterization of the collagen from glomerular basement membrane. Biochemistry 7(9):3103–3112. https://doi.org/10.1021/bi00849a012

Kobayashi K, Hata R, Nagai S, Niwa J, Hoshino T (1990) Direct visualization of affected collagen molecules synthesized by cultured fibroblasts from an osteogenesis imperfecta patient. Biochem Biophys Res Commun 172(1):217–222. https://doi.org/10.1016/s0006-291x(05)80196-2

Kobayasi T, Asboe-Hansen G, Tsurufuji S (1985) Filamentous aggregates of collagen. Ultrastructural evidence for collagen-fibril degradation in situ. Arch Dermatol Res 277(3):214–219. https://doi.org/10.1007/bf00404319

Kramer RZ, Bella J, Mayville P, Brodsky B, Berman HM (1999) Sequence dependent conformational variations of collagen triple-helical structure. Nat Struct Biol 6(5):454–457

Kramer RZ, Bella J, Brodsky B, Berman HM (2001) The crystal and molecular structure of a collagen-like peptide with a biologically relevant sequence. J Mol Biol 311(1):131–147

Kühn K (1982) Segment-long-spacing crystallites, a powerful tool in collagen research. Coll Relat Res 2(1):61–80. https://doi.org/10.1016/s0174-173x(82)80041-1

Lao LM, Kumakiri M, Kiyohara T, Sakata K, Takeuchi A (2005) Papillary endothelial hyperplasia and dilated lymphatic vessels in bullous pilomatricoma. Acta Derm Venereol 85(2):160–163. https://doi.org/10.1080/00015550410024120

Leikin S, Parsegian VA, Rau DC, Rand RP (1993) Hydration forces. Annu Rev Phys Chem 44: 369–395. https://doi.org/10.1146/annurev.pc.44.100193.002101

Leikin S, Rau DC, Parsegian VA (1994) Direct measurement of forces between self-assembled proteins: temperature-dependent exponential forces between collagen triple helices. Proc Natl Acad Sci U S A 91(1):276–280. https://doi.org/10.1073/pnas.91.1.276

Leikin S, Rau DC, Parsegian VA (1995) Temperature-favoured assembly of collagen is driven by hydrophilic not hydrophobic interactions. Nat Struct Biol 2(3):205–210. https://doi.org/10.1038/nsb0395-205

Maeda M, Matsuzaki T, Akai F, Hashimoto S, Takagi H (1996) Occurrence of long-spacing collagen in the intramuscular nerves of biopsied muscle tissues. Med Electron Microsc 29(3): 124–128. https://doi.org/10.1007/BF02348065

Martin R, Farjanel J, Eichenberger D, Colige A, Kessler E, Hulmes DJ, Giraud-Guille MM (2000) Liquid crystalline ordering of procollagen as a determinant of three-dimensional extracellular matrix architecture. J Mol Biol 301(1):11–17. https://doi.org/10.1006/jmbi.2000.3855

Miki H, Unno K, Park P, Ohno T, Ms N (1993) Morphogenesis and origin of fibrous long-spacing collagen fibers in collagenase-treated mouse skin tissues. Tissue Cell 25(5):669–680. https://doi.org/10.1016/0040-8166(93)90049-Q

Morris CJ, Bradby GV, Walton KW (1978) Fibrous long-spacing collagen in human atherosclerosis. Atherosclerosis 31(3):345–354. https://doi.org/10.1016/0021-9150(78)90069-2

Murro V, Mucciolo DP, Giorgio D, Caporossi T, Passerini I, Bani D, Giansanti F, Virgili G, Sodi A (2021) Lamellar hole-associated Epiretinal proliferation in choroideremia: a case report. Int J Retina Vitreous 7(1):63. https://doi.org/10.1186/s40942-021-00333-5

Nagase H, Woessner JF Jr (1999) Matrix metalloproteinases. J Biol Chem 274(31):21491–21494. https://doi.org/10.1074/jbc.274.31.21491

Nagase H, Visse R, Murphy G (2006) Structure and function of matrix metalloproteinases and TIMPs. Cardiovasc Res 69(3):562–573. https://doi.org/10.1016/j.cardiores.2005.12.002

Nair S, Leung H, Collins A, Ramsden R, Wilson J (2007) Primary cultures of human vestibular schwannoma: selective growth of schwannoma cells. Otol Neurotol 28(2):258–263. https://doi.org/10.1097/01.mao.0000247811.93453.6a

Nordwig A, Bretschneider G (1971) Nonspecific collagenolytic activity: hydrolysis of native collagen by a mold protease. Biochim Biophys Acta 250(2):408–415. https://doi.org/10.1016/0005-2744(71)90197-5

Nordwig A, Hayduk U (1969) Invertebrate collagens: isolation, characterization and phylogenetic aspects. J Mol Biol 44(1):161–172. https://doi.org/10.1016/0022-2836(69)90411-2

Onuchic JN, Wolynes PG (2004) Theory of protein folding. Curr Opin Struct Biol 14(1):70–75. https://doi.org/10.1016/j.sbi.2004.01.009

Paige MF, Goh MC (2001) Ultrastructure and assembly of segmental long spacing collagen studied by atomic force microscopy. Micron 32(3):355–361. https://doi.org/10.1016/s0968-4328(00)00038-x

Parolini B, Schumann RG, Cereda MG, Haritoglou C, Pertile G (2011) Lamellar macular hole: a clinicopathologic correlation of surgically excised epiretinal membranes. Invest Ophthalmol Vis Sci 52(12):9074–9083. https://doi.org/10.1167/iovs.11-8227

Pérez-Tamayo K (1972) The occurrence and significance of SLS crystallites in vivo. Connect Tissue Res 1(1):55–60. https://doi.org/10.3109/03008207209152056

Ramachandran GN, Kartha G (1955) Structure of collagen. Nature 176(4482):593–595. https://doi.org/10.1038/176593a0

Rehany U, Ishii Y, Lahav M, Rumelt S (2000) Ultrastructural changes in corneas of diabetic patients: an electron-microscopy study. Cornea 19(4):534–538. https://doi.org/10.1097/00003226-200007000-00026

Ricard-Blum S (2011) The collagen family. Cold Spring Harb Perspect Biol 3(1):a004978. https://doi.org/10.1101/cshperspect.a004978

Rich A, Crick FHC (1955) The structure of collagen. Nature 176(4489):915–916. https://doi.org/10.1038/176915a0

Rollins GC, Dill KA (2014) General mechanism of two-state protein folding kinetics. J Am Chem Soc 136(32):11420–11427. https://doi.org/10.1021/ja5049434

Saavedra E, Singh AD, Sears JE, Ratliff NB (2006) Plexiform pigmented schwannoma of the uvea. Surv Ophthalmol 51(2):162–168. https://doi.org/10.1016/j.survophthal.2005.12.004

Schmitt FO, Gross J, Highberger JH (1953) A new particle type in certain connective tissue extracts. Proc Natl Acad Sci U S A 39(6):459–470. https://doi.org/10.1073/pnas.39.6.459

Schraven SP, Plontke SK, Syha R, Fend F, Wolburg H, Adam P (2011) Dendritic cell tumor in a salivary gland lymph node: a rare differential diagnosis of salivary gland neoplasms. Diagn Pathol 6:94. https://doi.org/10.1186/1746-1596-6-94

Schwartz D, Veis A (1978) Characterization of basement membrane collagen of bovine anterior lens capsule via segment-long-spacing crystallites and the specific cleavage of the collagen by pepsin. FEBS Lett 85(2):326–332. https://doi.org/10.1016/0014-5793(78)80484-0

Schwartz D, Veis A (1980) Characterization of bovine anterior-lens-capsule basement-membrane collagen. 2. Segment-long-spacing precipitates: further evidence for large N-terminal and C-terminal extensions. Eur J Biochem 103(1):29–37. https://doi.org/10.1111/j.1432-1033.1980.tb04285.x

Slavin RE, Swedo JL, Brandes D, Gonzalez-Vitale JC, Osornio-Vargas A (1985) Extrapulmonary silicosis: a clinical, morphologic, and ultrastructural study. Hum Pathol 16(4):393–412. https://doi.org/10.1016/S0046-8177(85)80233-1

Stanescu R, Stanescu V, Maroteaux P (1982) Abnormal pattern of segment long spacing (SLS) cartilage collagen in diastrophic dysplasia. Coll Relat Res 2(2):111–116. https://doi.org/10.1016/s0174-173x(82)80027-7

Stark M, Kühn K (1968) Molecular fragments of rat skin collagen obtained on collagenase digestion. Eur J Biochem 6(4):542–544. https://doi.org/10.1111/j.1432-1033.1968.tb00478.x

Strawn R, Chen F, Jeet Haven P, Wong S, Park-Arias A, De Leeuw M, Xu Y (2018) To achieve self-assembled collagen mimetic fibrils using designed peptides. Biopolymers 109(7):e23226. https://doi.org/10.1002/bip.23226

Timpl R, Martin GR, Bruckner P, Wick G, Wiedemann H (1978) Nature of the collagenous protein in a tumor basement membrane. Eur J Biochem 84(1):43–52. https://doi.org/10.1111/j.1432-1033.1978.tb12139.x

Tzaphlidou M, Chapman JA, Al-Samman MH (1982a) A study of positive staining for electron microscopy using collagen as a model system—II. Staining by uranyl ions. Micron (1969) 13(2):133–145. https://doi.org/10.1016/0047-7206(82)90080-2

Tzaphlidou M, Chapman JA, Meek KM (1982b) A study of positive staining for electron microscopy using collagen as a model system—I. Staining by phosphotungstate and tungstate ions. Micron (1969) 13(2):119–131. https://doi.org/10.1016/0047-7206(82)90079-6

Ueda J, Yue BY (2003) Distribution of myocilin and extracellular matrix components in the corneoscleral meshwork of human eyes. Invest Ophthalmol Vis Sci 44(11):4772–4779. https://doi.org/10.1167/iovs.02-1002

Vitagliano L, Berisio R, Mazzarella L, Zagari A (2001) Structural bases of collagen stabilization induced by proline hydroxylation. Biopolymers 58(5):459–464. https://doi.org/10.1002/1097-0282(20010415)58:5<459::Aid-bip1021>3.0.Co;2-v

von der Mark K, Wendt P, Rexrodt F, Kühn K (1970) Direct evidence for a correlation between amino acid sequence and cross striation pattern of collagen. FEBS Lett 11(2):105–108. https://doi.org/10.1016/0014-5793(70)80503-8

Waltimo J (1996) Unusual forms of collagen in human dentin. Matrix Biol 15(1):53–56. https://doi.org/10.1016/S0945-053X(96)90127-4

Warshawsky H (1972) The presence of atypical collagen fibrils in EDTA decalcified predentine and dentine of rat incisors. Arch Oral Biol 17(12):1745–1754. https://doi.org/10.1016/0003-9969(72)90238-5

Weinstock M (1977) Centrosymmetrical cross-banded structures in the matrix of rat incisor predentin and dentin. J Ultrastruct Res 61(2):218–229. https://doi.org/10.1016/s0022-5320(77)80089-0

Wen CK, Goh MC (2006) Fibrous long spacing type collagen fibrils have a hierarchical internal structure. Proteins: Structure, Function, and Bioinformatics 64(1):227–233. https://doi.org/10.1002/prot.20949

Wiedemann H, Chung E, Fujii T, Miller EJ, Kühn K (1975) Comparative electron-microscope studies on type-III and type-I collagens. Eur J Biochem 51(2):363–368. https://doi.org/10.1111/j.1432-1033.1975.tb03936.x

Wright BA (1948) Low-angle X-ray diffraction pattern of collagen. Nature 162(4105):23. https://doi.org/10.1038/162023a0

Xu Y, Kirchner M (2021) Collagen mimetic peptides. Bioengineering (Basel, Switzerland) 8(1): doi:10.3390/bioengineering8010005

Yokota R, Hirakata A, Hayashi N, Hirota K, Rii T, Itoh Y, Orihara T, Inoue M (2018) Ultrastructural analyses of internal limiting membrane excised from highly myopic eyes with myopic traction maculopathy. Jpn J Ophthalmol 62(1):84–91. https://doi.org/10.1007/s10384-017-0542-9

Yoneda E, Teramura K, Hiruma S, Satou T, Hashimoto S (2001) Malignant gastrointestinal stromal tumor showing remarkable whorl formations. Arch Pathol Lab Med 125(5):665–668. https://doi.org/10.5858/2001-125-0665-mgstsr

Part III
Virus Protein Complexes

Chapter 17
Viral Capsid and Polymerase in Reoviridae

Hongrong Liu and Lingpeng Cheng

Abstract The members of the family *Reoviridae* (reoviruses) consist of 9–12 discrete double-stranded RNA (dsRNA) segments enclosed by single, double, or triple capsid layers. The outer capsid proteins of reoviruses exhibit the highest diversity in both sequence and structural organization. By contrast, the conserved RNA-dependent RNA polymerase (RdRp) structure in the conserved innermost shell in all reoviruses suggests that they share common transcriptional regulatory mechanisms. After reoviruses are delivered into the cytoplasm of a host cell, their inner capsid particles (ICPs) remain intact and serve as a stable nanoscale machine for RNA transcription and capping performed using enzymes in ICPs. Advances in cryo-electron microscopy have enabled the reconstruction at near-atomic resolution of not only the icosahedral capsid, including capping enzymes, but also the nonicosahedrally distributed complexes of RdRps within the capsid at different transcriptional stages. These near-atomic resolution structures allow us to visualize highly coordinated structural changes in the related enzymes, genomic RNA, and capsid protein during reovirus transcription. In addition, reoviruses encode their own enzymes for nascent RNA capping before RNA releasing from their ICPs.

Keywords *Reoviridae* · Reoviruses · Reovirus transcription · Double-stranded RNA (dsRNA) · Capsid · Capsid proteins · Capsid layers · Inner Capsid Particle (ICP) · Core · RNA-dependent RNA polymerase (RdRp)

H. Liu (✉)
Key Laboratory for Matter Microstructure and Function of Hunan Province, Key Laboratory of Low-dimensional Quantum Structures and Quantum Control, School of Physics and Electronics, Hunan Normal University, Changsha, China
e-mail: hrliu@hunnu.edu.cn

L. Cheng (✉)
Key Laboratory for Matter Microstructure and Function of Hunan Province, Key Laboratory of Low-dimensional Quantum Structures and Quantum Control, School of Physics and Electronics, Hunan Normal University, Changsha, China

School of Life Sciences, Tsinghua University, Beijing, China
e-mail: lingpengcheng@hunnu.edu.cn

J. R. Harris, J. Marles-Wright (eds.), *Macromolecular Protein Complexes IV*, Subcellular Biochemistry 99, https://doi.org/10.1007/978-3-031-00793-4_17

Introduction

Reoviridae is the largest family of double-stranded RNA (dsRNA) viruses and consists of 15 recognized genera. The members of the *Reoviridae* family (hereafter referred to as reoviruses) have been isolated from various mammals, birds, reptiles, fish, crustaceans, marine protists, insects, ticks, arachnids, plants, and fungi and include a total of 75 virus species with wide geographical distributions (King et al. 2011).

The virions of reoviruses consist of 9–12 discrete dsRNA segments enclosed by single, double, or triple capsid layers. The virions can be divided into two parts on the basis of their functions: a transcriptionally competent inner capsid particle (ICP, also known as the core) and an outer layer of capsid proteins that coats the ICP for protection and membrane penetration; some single-layered reoviruses do no possess the outer capsid layer. Because of the absence of a lipid bilayer that can fuse with the cell membrane, reoviruses deliver their ICPs into the cytoplasm through membrane disruption facilitated by capsid proteins present in the outer layer. The sequence and structural organization of the outer capsid proteins of reoviruses are highly diverse, thus enabling these proteins to penetrate different cell membrane barriers.

The members of the *Reoviridae* family are divided into two subfamilies on the basis of the structural organization of their capsid: *Sedoreovirinae* and *Spinareovirinae* (Carstens 2010). A common process adopted by reoviruses involves the uncoating of the outer capsid layer after their delivery into the cytoplasm of the host cell where ICPs remain intact. ICPs are a multienzyme machine for RNA synthesis and capping. Similar to most dsRNA viruses, the ICPs of reoviruses have a conserved $T = 1$ icosahedral innermost capsid shell that is composed of 60 copies of two conformers of a shell protein. Shell proteins present in reoviruses belonging to various genera have a similar structural topology despite lacking a detectable amino acid sequence identity (Fig. 17.1). In the cytoplasm, the

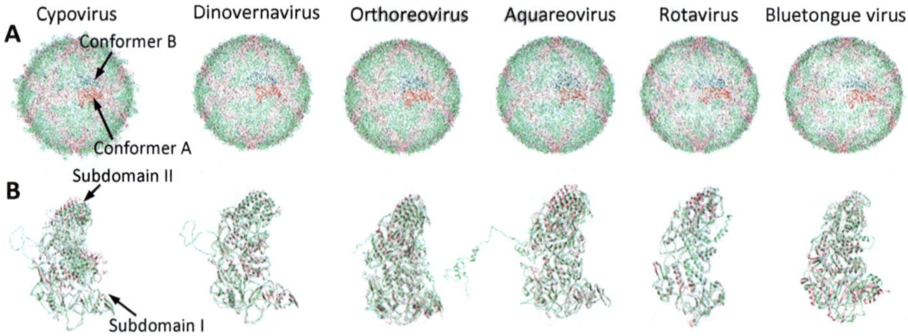

Fig. 17.1 Innermost capsid shells and shell proteins of reoviruses. (**a**) Innermost shells. Two conformers of the shell protein of each reovirus are in red and green, respectively. The two conformers in the asymmetric unit are in dark red and forest green. (**b**) Superimpositions of the two conformers of the reoviruses

virus-encoded RNA-dependent RNA polymerase (RdRp), which is located under the icosahedral five-fold vertex in the innermost capsid shell, transcribes nascent RNA by using the minus strand of the genomic dsRNA segment as a template within the inner capsid while avoiding the dsRNA-activated defense mechanisms of the host cell (Lawton et al. 1997; Mertens 2004; Nibert and Baker 2003).

Reoviruses possess their own protein with capping enzymes and cap the nascent RNA before releasing it from transcribing ICPs. The 5′-end capping of nascent RNA is a conserved process in eukaryotes and viruses. The cap structure is involved in the utilization of the host machinery for the efficient translation of viral mRNA, the evasion of the host's antiviral innate immune response, and the encapsidation and transcription of viral RNA. Capping enzymes found in the reoviruses of the *Spinareovirinae* subfamily are present in each turret-like structure located on the outer surface of the innermost capsid shell, whereas those in the reoviruses of the *Sedoreovirinae* subfamily are encapsidated within the innermost capsid shell.

The structures of reoviruses have been extensively studied through X-ray crystallography and cryo-electron microscopy (cryo-EM). Studies have identified the various structural characteristics of reoviruses and their functional roles in the virus life cycle. This chapter summarizes the structural studies of reoviruses published during the past years. Considerable technical progress and advancements have been made in high-resolution cryo-EM and single-particle analysis. In particular, this chapter focuses on recent in situ structural studies of RdRp within viral capsids that employed these advanced technologies. Together with the findings of the biochemical and genetic analyses of reovirus transcription, these in situ structural studies of RdRp within reovirus capsids have enhanced the understanding regarding the structure–function relationships of endogenous transcription.

Icosahedral Capsid Structures of Reoviruses

Reoviruses in the Subfamily *Spinareovirinae*

The virions of reoviruses belonging to the subfamily *Spinareovirinae* consist of an ICP coated by a layer of three-fold heterohexamers of two coat proteins (represented by the *Orthoreovirus*, *Aquareovirus*, and *Oryzavirus* genera) or only the ICP (represented by the *Dinovernavirus* and *Cypovirus* genera). The double-layered virions and their ICPs are of approximately 820 and 800 Å in diameter, respectively. The ICP consists of many nonicosahedral segmented genomic dsRNAs and an equal copy number of the RdRp complex enclosed by a $T = 1$ icosahedral capsid formed by the shell, clamp, and turret proteins.

The innermost shell of ICPs is formed by two conformers of a plate-like shell protein. The shell proteins in reoviruses with known structures are all topologically conserved. Five monomers of conformer A form a star-like complex at each five-fold vertex and five monomers of conformer B pack into the gap between two adjacent monomers of conformer A. Twelve of the decamers form the innermost capsid shell

(Fig. 17.1a). Both conformers have a similar plate-like structure with no strongly defined domain boundaries. The plate-like shell protein can be divided into two subdomains on the basis of the major difference between two conformers: subdomains I and II; the subdomains undergo a minor rigid body tilt relative to one another (Reinisch et al. 2000) (Fig. 17.1b). Subdomain II (also known as the apical domain) is an insert into subdomain I. Five copies of the tip of the apical domain form a five-fold pore at the vertex of the icosahedral five-fold axis on the capsid shell for the exit of nascent mRNA. The N-terminal regions of both subunits are not resolved in icosahedral reconstruction because of the lack of icosahedral symmetry. The asymmetric structures and functions of the N-terminal regions are described below.

The outer surface of the innermost shell of reoviruses belonging to the subfamily *Spinareovirinae* is decorated with nodular "clamp" (also called "cement") proteins (Fig. 17.2a–c). The copy number of clamp proteins varies among different genera of *Spinareovirinae*. For example, dinovernavirus has a total of 60 clamp proteins that are located around the 12 turrets (Auguste et al. 2015). Cypovirus, aquareovirus, and oryzavirus have 120 clamp proteins, with the additional 60 proteins located around the three-fold vertices (Zhou et al. 2003; Cheng et al. 2008; Miyazaki et al. 2008). In

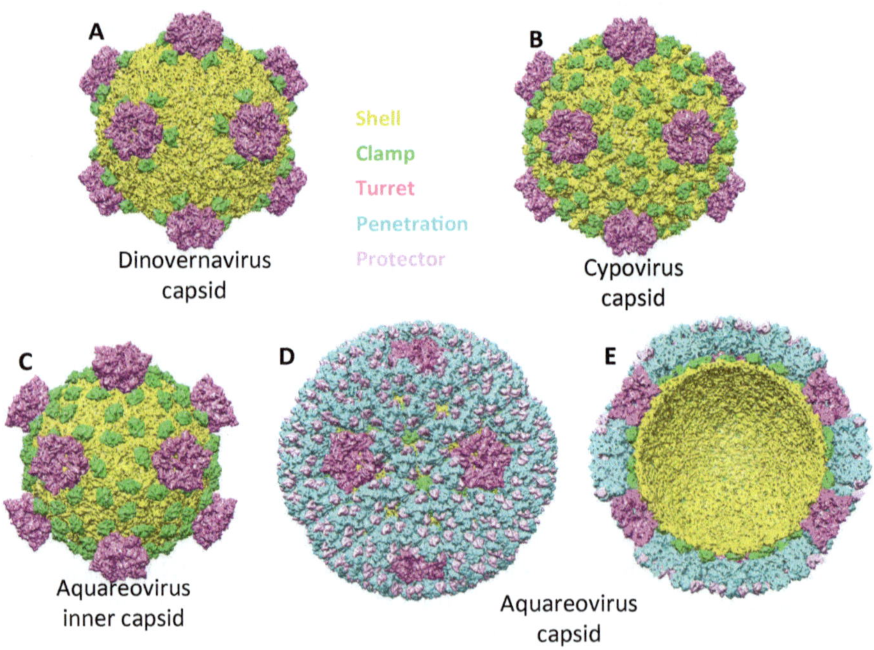

Fig. 17.2 Capsid structures of three turreted reoviruses. The colors of the capsid proteins are labeled. (**a**) and (**b**) Capsid structures of the single-layered dinovernavirus and cypovirus. The flexible fiber at each five-fold vertex is not shown. (**c**) Inner capsid structure of aquareovirus. (**d**) Capsid structure of double-layered aquareovirus. (**e**) Cut-open view of the capsid structure of aquareovirus

addition to the 120 clamp proteins present in the cypovirus, aquareovirus, and oryzavirus, orthoreovirus possesses an additional 30 proteins located on the two-fold vertices (Zhang et al. 2005a, b; Yan et al. 2011). All clamp proteins are located on the border of and interact with two or three copies of neighboring shell proteins, and they are indispensable for the stability of the innermost capsid shell (Reinisch et al. 2000).

Reoviruses belonging to the subfamily *Spinareovirinae* are also known as "turreted" reoviruses because of the presence of a distinct pentameric turret on each five-fold vertex of the innermost shell (Fig. 17.2d). The turret is formed by the five copies of turret proteins and is involved in the catalysis of 5′ mRNA cap synthesis (Miyazaki et al. 2008; Zhang et al. 2003, 2005b; Cheng et al. 2010, 2011; Yan et al. 2011; Zhu et al. 2014). The capping role of the turret protein is described in the following section.

The turret proteins of reoviruses belonging to the genera *Orthoreovirus* and *Aquareovirus* have a C-terminal three-domain flap (Fig. 17.3a). The three domains in the flap have structures related to the immunoglobulin fold (Reinisch et al. 2000; Cheng et al. 2010). By contrast, reoviruses belonging to the genera *Oryzavirus*, *Dinovernavirus*, and *Cypovirus* lack the flap in the turret protein (Fig. 17.3b) (Miyazaki et al. 2008; Auguste et al. 2015; Cheng et al. 2011). The flap has been observed to swivel about a hinge between virions and the ICPs of mammalian orthoreovirus (MRV) (Chandran et al. 1999; Dryden et al. 1993), between an aquareovirus virion (Cheng et al. 2010) and an aquareovirus particle with a partially degraded outermost protein (Zhang et al. 2010; Wang et al. 2018), and also between two types of baboon reovirus (BRV) virions (Yan et al. 2011). The significance of flap swiveling to reovirus function or assembly remains to be deciphered. Some reoviruses have a fiber that binds to cell surface receptors during infection (Lee et al. 1981). The fiber at the icosahedral five-fold vertex is held by turret flaps in MRV and avian reovirus (ARV) (Dryden et al. 1993; Zhang et al. 2005b) or by a brace domain in cypovirus (Cheng et al. 2011). The presence or absence of the fiber is not dependent on genera. For example, MRV and ARV but not BRV have fibers, although they all belong to the genus *Orthoreovirus*.

The ICPs of orthoreovirus and aquareovirus are coated by an outer layer of 200 three-fold symmetrical heterohexamers arranged on an incomplete T = 13

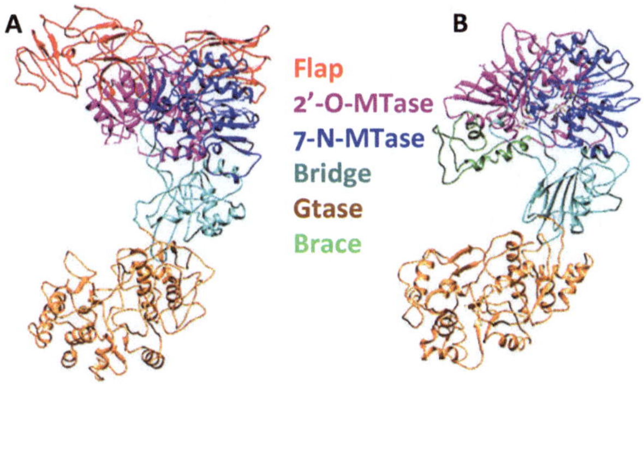

Fig. 17.3 Turret proteins of aquareovirus (**a**) and cypovirus (**b**). The colors of domains are labeled. Cypovirus turret protein does not have a three-immunoglobulin domain flap (red) present in aquareovirus turret protein but has an additional small brace domain (green) inserted in the 2′-O-MTase domain

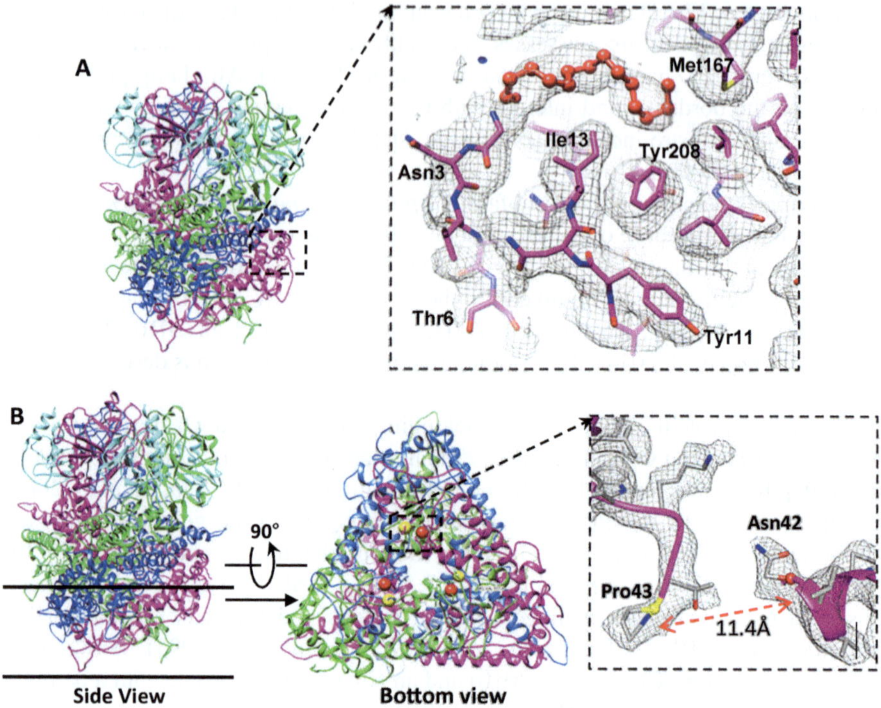

Fig. 17.4 Structure of VP5 and VP7 heterohexamer. Three VP7 subunits are in cyan, and three VP5 subunits are in blue, green, and magenta, respectively. (**a**) Myristoyl group linked to the N-terminus in VP5. (**b**) Cleavage site between Asn42 and Pro43 in VP5

icosahedral lattice, exposing turrets on icosahedral vertices (Fig. 17.2d and e). By contrast, oryzavirus has only 60 heterohexamers surrounding the turret (Miyazaki et al. 2008). Interactions between the ICP and heterohexamer are mediated by the clamp proteins (Fig. 17.2e). Each three-fold heterohexamer comprises three copies of the protector protein located on top of the three copies of the membrane penetration protein (Fig.17.4a). The trimer of the penetration protein plays crucial roles in the viral penetration of the cell membrane (Liemann et al. 2002). A myristoyl group (Fig. 17.4a) is covalently linked to the N terminus and embedded in a hydrophobic pocket of the trimer of the penetration protein (Liemann et al. 2002; Zhang et al. 2010). When the reovirus virion interacts with proteases in the intestinal lumen where reoviruses infect intestinal cells, the proteases strip off the protector protein, resulting in the formation of infectious subviral particles (ISVPs) (Chang and Zweerink 1971; Silverstein et al. 1970). The ISVPs are characterized by the loss of the protector protein through proteolytic degradation (Nibert and Fields 1992). Concomitantly, the Asn42-Pro43 bond in the penetration protein of the orthoreovirus or aquareovirus ISVPs is cleaved through autocleavage (Fig. 17.4b), dividing the protein into a myristoylated N terminus and C terminus (Liemann et al. 2002; Zhang et al. 2010). This autocleavage is a critical step in the cell entry of coated turreted reoviruses (Nibert et al. 1991a). The myristoylated N-terminus must be released from the hydrophobic pocket, forming a penetration pore or disrupting the membrane for the translocation of the reovirus ICP into the cytoplasm (Nibert et al. 1991b; Odegard et al. 2004).

The cell entry mechanism of the single-layered cypovirus and dinovernavirus remains unclear. The cypovirus virions are embedded in a micrometer-sized crystal polyhedron that is made of the trimers of the viral polyhedrin protein (Coulibaly et al. 2007), which has a similar central β-sandwich and an overall trimeric shape to the membrane penetration protein of orthoreovirus and aquareovirus (Liemann et al. 2002; Cheng et al. 2010). However, polyhedrin has a new fold and evolved to assemble in vivo into three-dimensional cubic crystals rather than icosahedral shells (Coulibaly et al. 2007). By contrast, dinovernavirus is not embedded and lacks any homolog to these trimeric capsid and polyhedron proteins (Auguste et al. 2015).

Reoviruses in *Sedoreovirinae*

Reoviruses belonging to the subfamily *Sedoreovirinae*, represented by the *Rotavirus* and *Orbivirus* genera, consist of an ICP coated by a layer of outer capsid proteins (Fig. 17.5). The ICP, which is a double-layered particle (DLP or core), consists of nonicasahedral segmented genomic dsRNAs and an equal copy number of

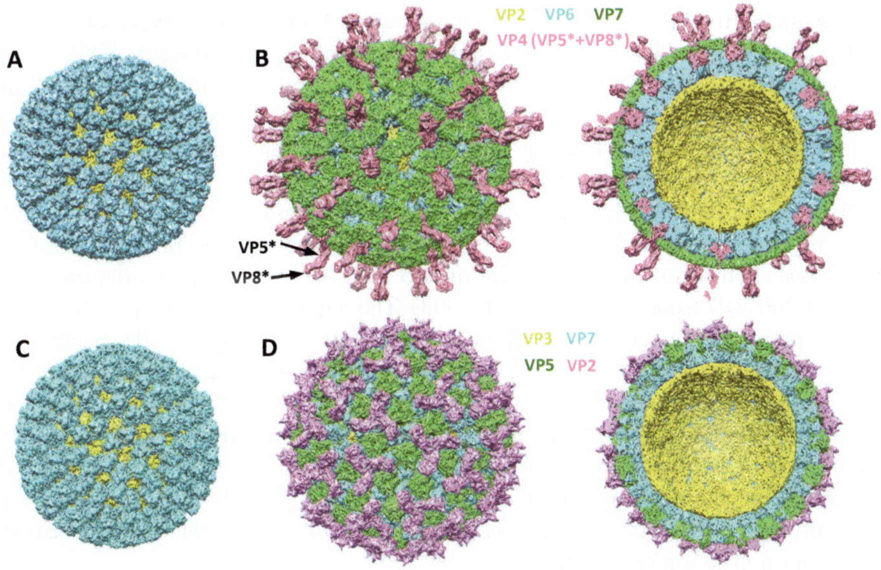

Fig. 17.5 Capsid structure of nonturreted reoviruses. (**a**) ICP of rotavirus. (**b**) Surface and cut-open views of the rotavirus full capsid. (**c**) ICP of bluetongue virus. (**d**) Surface and cut-open views of the bluetongue virus full capsid

RdRp enclosed by an icosahedral architecture formed by the two layers of capsid proteins.

The innermost shell is formed by 120 copies of the two conformers of the shell protein, which is topologically conserved with that of reoviruses belonging to the subfamily *Spinareovirinae* (Fig. 17.1). A $T = 13$ icosahedral layer of the 260 trimers of an intermediate protein surrounds the innermost shell (Grimes et al. 1998; McClain et al. 2010). Reoviruses belonging to the subfamily *Sedoreovirinae* are also known as "nonturreted" reoviruses because they do not have a distinct pentameric turret on each five-fold vertex of the innermost shell. In addition, no clamp proteins are present on the shell; therefore, the surface of the ICP is relatively smooth. The protein that functions as a viral RNA-capping enzyme is enclosed in the capsid shell of nonturreted reoviruses (Kumar et al. 2020; Sutton et al. 2007).

The infectious virions of nonturreted reoviruses are a triple-layered particle (TLP) of approximately 1000 Å in diameter. Compared with the conserved architecture and topology of ICP proteins, the outer layer coated on the ICP is considerably more diversified. Similar to turreted reoviruses, the outer-layer proteins of nonturreted reoviruses undergo substantial conformational changes to facilitate cell membrane penetration and strip off after the ICP is translocated into the cytoplasm during infection. The ICP of rotavirus (Fig. 17.5a) consists of an innermost shell formed by 120 copies of two VP2 conformers coated by a layer of 260 VP6 trimers (McClain et al. 2010). The outer layer of the rotavirus virion, which serves as the molecular machinery for host-cell binding and penetration, contains two proteins, namely VP4 and VP7 (Fig. 17.5b). Each of 260 VP7 trimers directly clamps onto the trimer of the intermediate protein VP6, forming a continuous perforated capsid (Settembre et al. 2011; Chen et al. 2009). A trimeric spike of VP4 emerges from each of the local six-fold channels formed by the six trimers of VP6 and VP7 around the five-fold vertex. Efficient infection requires the cleavage of VP4 by trypsin into two fragments: VP8* for receptor binding and VP5* for membrane penetration. The spike appears only in the structure of trypsin-treated particles, suggesting the rigidification of the spike upon proteolysis (Crawford et al. 2001; Trask et al. 2012). The ICP of bluetongue virus (BTV, Fig. 17.5c) belonging to the genus *Orbivirus* consists of an innermost shell formed by 120 copies of two VP3 conformers and an intermediate layer of 260 VP7 trimers (Grimes et al. 1998). The outer capsid has 60 triskelion-like VP2 spikes, surrounded by 120 globular trimers of VP5 (Fig. 17.5d). Each VP2 trimer binds atop four VP7 trimers; each VP5 trimer bridges the channel formed by six surrounding VP7 trimers (Zhang et al. 2016; Roy 2017). VP2 binds to surface glycoproteins and facilitates the endocytosis of the virion, whereas VP5 penetrates the host cell membrane and delivers the ICP into the host cytosol. Unlike the membrane penetration protein VP4 of rotavirus, the VP5 of BTV does not appear to be proteolytically processed and is likely to fold into a metastable conformation following translation (Roy 2017).

Reovirus RdRp

Reovirus possesses a segmented dsRNA genome consisting of each of a full set of RNA segments with conserved sequences at 5′ and 3′ ends, along with an equal number of RdRps. The number of genome segments in reoviruses varies from 9 to 12, depending on the genus to which the reovirus belongs. Each RNA segment is associated with a single RdRp under each of the five-fold vertices (Periz et al. 2013). RdRps are capable of both RNA replication and transcription within the reovirus ICP (Gillies et al. 1971; Shatkin and Sipe 1968; Skehel and Joklik 1969; Tao et al. 2002; Lu et al. 2008). During reovirus ICP assembly, each RdRp recognizes a capped viral mRNA segment based on the conserved sequence at 5′ and 3′ ends (Tao et al. 2002; Lu et al. 2008). To ensure the encapsidation of a full set of mRNAs into the ICP, higher order structures may form between different mRNA segments (Tao et al. 2002; Borodavka et al. 2017). The ICP assembly is accompanied by a single round of the RdRp-driven synthesis of minus-strand RNAs complementary to mRNAs (replication), eventually forming genomic dsRNA segments in mature virions. Upon reovirus infection, ICPs remain intact after the delivery of reoviruses into the host cell's cytoplasm, and RdRps can repeatedly transcribe RNA from the minus-strand RNA genome within ICPs. Reoviruses have been used as models to study RNA transcription and capping processes (Borsa and Graham 1968; Levin et al. 1970; Furuichi and Miura 1975; Furuichi 1978; Smith and Furuichi 1980; Lawton et al. 1997; Yang et al. 2012; Yu et al. 2015).

The structures of recombinant orthoreovirus and rotavirus RdRps with various RNA substrates have been determined through X-ray crystallography (Tao et al. 2002; Lu et al. 2008). The two RdRp structures exhibit a typical overall architecture of RNA viruses. Both contain a polymerase domain surrounded by an N-terminal domain and a C-terminal bracelet domain. Both the N-terminal domain and C-terminal bracelet domain are crucial for catalytic activity (Wehrfritz et al. 2007). Similar to other polynucleotide polymerases (Ferrer-Orta et al. 2006), the right-handed polymerase domain can be further divided into three subdomains, namely, fingers, thumb, and palm. The three domains fold together into a cage-like structure. The hollow catalytic center in the polymerase domain is connected to the exterior through four well-defined channels that are used for RNA template entry, NTP entry, template exit (or dsRNA exit in the replication mode), and transcript exit. Compared with other RdRps with known structures, the template exit channel is unique in the reovirus RdRp structure, indicating that reovirus RdRps can internally separate nascent dsRNA duplexes (McDonald et al. 2009). The presence of an extended loop (priming loop) between the fingers and palm subdomains is a distinctive feature of the two reovirus RdRps. The structures of orthoreovirus RdRp soaked with the template Mg^{2+} and NTPs show that the priming loop (residues 558–565) can serve as a platform for stabilizing an NTP at the priming site during the initiation stage and retract to allow the passage of the nascent dsRNA duplex during the elongation stage (Tao et al. 2002). A pair of Mg^{2+} ions is crucial for the formation of the phosphodiester bond. Residues essential for anchoring the pair of Mg^{2+} ions are

Asp585, Asp734, and Asp735. In the rotavirus RdRp structure, the priming loop bends away from the active site, leaving it in a retracted state incapable of supporting an NTP and initiating RNA synthesis, thus preventing the rotavirus RdRp from forming a functional initiation complex (Lu et al. 2008). A cap analog soaked into both orthoreovirus and rotavirus RdRp crystals sits in a groove near the tip of the thumb subdomain of the two RdRps. This cap-binding site may serve to anchor the capped plus strand of genomic dsRNA during transcription.

High-resolution in situ structures of RdRp at different transcriptional stages are necessary for fully understanding the highly coordinated RNA transcription process in the confined shell of the ICP. The structures of RdRp complexes and the genome within the cypovirus capsid have been determined for the first time by using cryo-EM and the capsid subtraction–based symmetry mismatch reconstruction method (Liu and Cheng 2015; Cheng 2015; Liu 2015). Subsequently, an increasing number of structures of RdRp and genomes within viral capsids have been resolved (Zhang et al. 2015; Dai et al. 2017; Wang et al. 2018; Ding et al. 2018; Ding et al. 2019; Ilca et al. 2019; He et al. 2019; Kaelber et al. 2020; Jenni et al. 2019). In the following sections, we summarize the in situ structural studies of various reovirus RdRps within reovirus particles at different stages. These in situ structural studies of RdRps and genomic RNA within nontranscribing and transcribing reovirus capsids that we summarize below have provided insights into dynamic molecular mechanisms through which RdRp interacts with the shell protein and dsRNA segment, initiates transcription, and ensures polymerase processivity.

Structure of the RdRp Complex and Genome in a Nontranscribing Mature Virion

The in situ structures of genomes and RdRp complexes within the mature virions of cypovirus and aquareovirus have been determined at a near-atomic resolution (Liu and Cheng 2015; Wang et al. 2018; Zhang et al. 2015). The overall structures of genomes and RdRp complexes within the two turreted reoviruses are similar. The genome is composed of regularly distributed spherical layers that are formed by the structures of discontinuous dsRNA fragments running in parallel and associated with RdRp complexes (Fig. 17.6a). Each RdRp complex, which consists of an RdRp and a cofactor protein, is anchored at the inner surface of the capsid slightly off center from the icosahedral five-fold vertex and surrounded by multiple layers of dsRNA resembling a nest-shaped layered structure in a relatively crowded capsid shell. The distance between two adjacent dsRNA fragments within the same layer is approximately 25–28 Å, whereas the distance between two adjacent layers is approximately 28–30 Å. The inner layers are less well-organized compared with the outer layers of dsRNA fragments (Liu and Cheng 2015). The layers of the dsRNA fragment resemble the organization of the cholesteric liquid crystal (de Gennes and Prost 1993), an observation that is consistent with an earlier assumption that the dsRNA

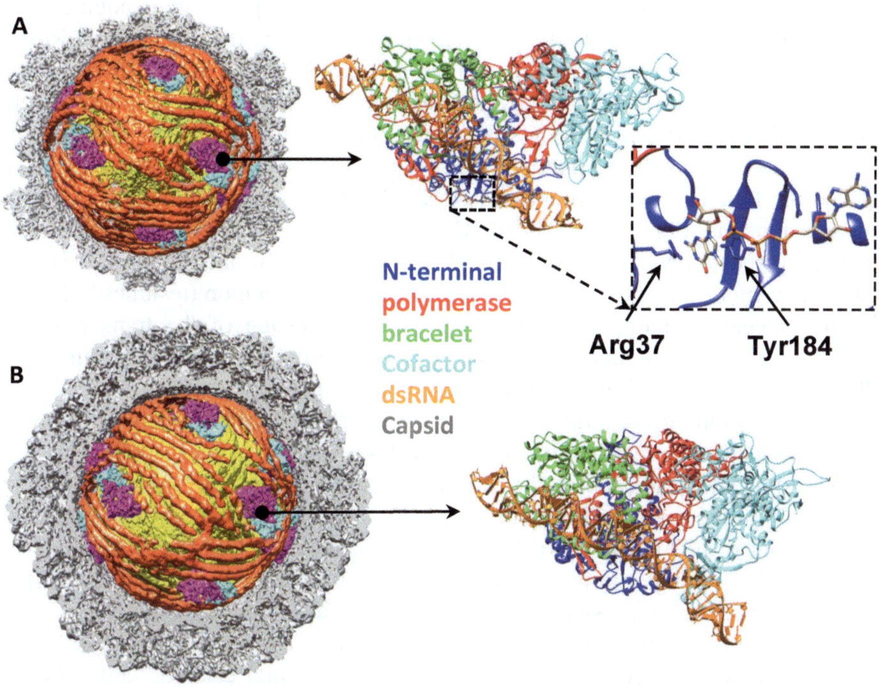

Fig. 17.6 Structures of the genome and RdRp complexes within the nontanscribing mature cypovirus and aquareovirus virions. The color scheme for the proteins and domains is labeled. (**a**) RdRp complex and dsRNA within cypovirus. Left: half of the icosahedral capsid (gray) is removed to show the structures of the genomic dsRNA (first layer in orange and second layer in yellow), RdRp VP2 (magenta), and cofactor protein VP4 (cyan). Middle: Zoom-in view of the dsRNA, RdRp, and cofactor protein in the capsid. Right: The cap of the plus strand of genomic dsRNA interacts with an A/G base binding site formed by the Tyr184 and Arg37 of the N-terminal domain at the entrance of the template entry channel. (**b**) RdRp complex and dsRNA within aquareovirus. The color scheme is identical to that of panel A

genome would form liquid crystalline arrays within the highly condensed capsid (Gouet et al. 1999). This liquid crystalline organization, which provides molecules with a certain degree of fluidity (de Gennes and Prost 1993), may provide mobility for dsRNA during transcription in the crowded capsid. RdRp complexes and the associated dsRNA fragments appear to be organized in a pseudo D3 symmetry. However, RdRp complexes at 6 of the 12 vertices exhibit stronger density, whereas those at the remaining six positions show weaker densities, implying partial (two-third) occupancy; therefore, only 10 RdRp complexes are located under the 12 vertices in the capsid (Liu and Cheng 2015). In the subsequent study, 10 RdRp complexes were resolved, leaving two vertices occupied only by dsRNA (Zhang et al. 2015). In aquareovirus with 11 RNA segments, only 11 of the 12 vertices are occupied by RdRp complexes (Wang et al. 2018; Ding et al. 2018).

The structures of RdRps within the mature virions of cypovirus and aquareovirus share a similar architecture to the X-ray crystal structures of the recombinant RdRps of orthoreovirus and rotavirus (Tao et al. 2002; Lu et al. 2008). Both RdRps contain a polymerase domain surrounded by an N-terminal domain and a C-terminal bracelet domain. The catalytic centers of RdRps are connected to the exterior through four well-defined tunnels. In the mature virion of cypovirus, a dsRNA fragment traverses across the template exit channel and closely interacts with the bracelet and polymerase domains (Fig. 17.6a). This dsRNA fragment was previously misinterpreted as the capped 5′ end of the plus strand of a dsRNA (Liu and Cheng 2015). The catalytic center of the polymerase domain is occupied by a loop (residues 925–932) in the bracelet domain, resulting in the partial blockage of the transcript exit, template entry, and template exit channels (Liu and Cheng 2015; Zhang et al. 2015; Li et al. 2017). The cap of the plus strand of genomic dsRNA interacts with an A/G base binding site formed by the Tyr184 and Arg37 of the N-terminal domain at the entrance of the template entry channel (Fig. 17.6a) (Cui et al. 2019). This A/G base binding site is adjacent to the cap-binding site between the N-terminal domain and the tip of the thumb subdomain (Tao et al. 2002; Li et al. 2020; Liu and Cheng 2015; Zhang et al. 2015). The RdRp and dsRNA in the mature virion of aquareovirus (Fig. 17.6b) exhibit similar interactions with that of cypovirus (Fig. 17.6a). The differences are that dsRNA binding does not cause the partial blockage of the template entry and exit channels. In contrast, many elements are flexible in the aquareovirus RdRp structure, resulting in a hollow central catalytic cavity (Wang et al. 2018).

RdRps are in complex with a cofactor protein in cypovirus and aquareovirus (Liu and Cheng 2015; Wang et al. 2018; Ding et al. 2018; Zhang et al. 2015). The cofactor protein is located on the RdRp surface region between the two channels for the RNA template entry and NTP entry (Fig. 17.6). In cypovirus, the cofactor protein VP4 consists of an N-terminal and a C-terminal NTPase domain. The N-terminal domain is formed by two small β-sheets and several α-helices. The main body of the C-terminal NTPase is a Walker-A α/β motif and contains a GTP molecule at a predicted NTP binding site (Zhang et al. 2015). The RdRp cofactor protein VP4 in aquareovirus has the same interaction with RdRp as that observed in cypovirus. The structure of the aquareovirus VP4 can be divided into three domains: an N-terminal nodule domain, a plate domain, and a C-terminal domain (Wang et al. 2018; Ding et al. 2018). The N-terminal nodule domain, which resides in the corner between a copy of the shell protein VP3A and the RdRp protein VP2, interacts with the dsRNA fragment roughly parallel to the capsid shell (Wang et al. 2018). The plate domain is roughly rectangular and resides in the corner between the shell protein VP3A and the VP2 polymerase domain and interacts with a surface region of the VP2 polymerase domain between the two channels for template entry and NTP entry. The aquareovirus VP4 N-terminal nodule domain exhibits no similarity to its counterpart in cypovirus VP4; by contrast, the plate domain resembles the C-terminal NTPase domain in cypovirus. In addition, the aquareovirus VP4 plate domain contains fully conserved motifs (KxxxK and SDxxG) for NTP binding in the μ2 homology proteins of turreted reoviruses (Noble and Nibert 1997; Kim et al. 2004; Nibert and Kim

2004). An additional density, which is located at the conserved NTP binding site surrounded by conserved residues (Lys410, Lys414, and Ser439), can be assigned to the NTP or the single phosphate group corresponding to the γ-phosphate group of the bound NTP (Wang et al. 2018; Ding et al. 2018). The aquareovirus VP4 has an additional C-terminal domain, which does not exist in the cypovirus VP4. The C-terminal domain is located at the entrance of the template entry channel and interacts with the dsRNA fragment approaching the RdRp template entry channel. By contrast, no interaction was observed between the cypovirus cofactor protein VP4 and genomic RNA. The cofactor protein remains structurally unchanged during the transcription of turreted reoviruses.

The architecture and domains of the RdRp in nonturreted reoviruses are similar to those of the RdRp in turreted reoviruses. However, the RdRp in nonturreted reoviruses is not in complex with the cofactor protein. The RdRp monomer anchors to the inner capsid shell surrounding the icosahedral five-fold axis, a location similar to that of the RdRp complex in turreted reoviruses. Unlike the D3 symmetric distribution of RdRps found in turreted reoviruses, the position of an RdRp in nonturreted reoviruses at any particular vertex does not correlate with those of RdRps under other five-fold vertices. In other words, the distribution of the RdRp has none of the symmetries that would form a subgroup of the icosahedral point group (Jenni et al. 2019). The stochastic distribution of RdRps rules out any unique long-range order for RNA; therefore, the overall RNA structure in the capsid has not been resolved (Jenni et al. 2019; He et al. 2019; Ding et al. 2019). In the mature virion of BTV, the 5′ cap of the positive strand of a dsRNA genome binds to residue His40 at the entrance of the template entry channel. A dsRNA fragment closely interacts with the bracelet and polymerase domains (He et al. 2019). These interactions are similar to those observed in turreted reoviruses. In the mature virion of rotavirus, the RdRp appears to interact with dsRNA at the cap-binding site (Jenni et al. 2019).

RdRp in Transcribing Reoviruses

Transcribing reovirus particles can be obtained in vitro by treating reovirus virions or ICPs with a transcription reaction buffer including Mg^{2+}, ATP, GTP, CTP, UTP, and *S*-adenosyl-L-methionine (SAM) (Bartlett et al. 1974; Smith and Furuichi 1980; Van Dijk and Huismans 1980; Lawton et al. 1997; Drayna and Fields 1982). To elucidate the transcriptional mechanisms of viruses, many structures of transcribing reoviruses have been resolved (Lawton et al. 1997; Pesavento et al. 2001; Mendez et al. 2008; Yang et al. 2012; Zhu et al. 2014; Yu et al. 2015). Recently, the in situ structures of the reovirus RdRps at different transcriptional stages have been determined using a combination of capsid subtraction, symmetry mismatch, classification, and local reconstruction methods (Liu and Cheng 2015; Zhang et al. 2015; Li et al. 2020; Cui et al. 2019; Ding et al. 2019; Jenni et al. 2019).

The transcription process in the reovirus capsid can be divided into two stages, namely initiation and elongation, in which the RdRps undergo stepwise conformational changes. The transcription process is activated by the removal of outer-layer proteins in nonturreted reoviruses (He et al. 2019; Jenni et al. 2019) and the binding of SAM in the single-layered turreted cypovirus (Yu et al. 2015; Cui et al. 2019). Transcriptional activation is reflected by the expansion of the capsid chamber resulting from the outward tilt of the apical domains of the five copies of shell proteins around the five-fold vertex. For double-layered turreted reoviruses, the removal of the outer layer of aquareovirus alone does not cause capsid expansion (Cheng et al. 2008), although the removal of the outer layer is required for elongation but not initiation in orthoreovirus (Farsetta et al. 2000).

Initiation

Genomic dsRNA binds to the RdRp through the binding of the capped 5′ end of the plus strand to the site adjacent to the entrance of the template entry channel of the RdRp in the nontranscribing mature reovirus. Because only the 3′ end of a single-stranded template can enter the catalytic center, dsRNA must unwind first. This RdRp state, in which two separate single-stranded RNAs bind to the entrance of the template entry channel, was observed in the rotavirus DLP (Ding et al. 2019; Jenni et al. 2019). The capped 5′ end of the genomic plus strand (nontemplate RNA) interacts with the RdRp cap-binding site, whereas the minus strand (template RNA) interacts with an RNA binding β-sheet in the finger subdomain at the entrance of the template entry channel (Ding et al. 2019; Jenni et al. 2019). The absence of any RNA density in the RdRp active site at this stage suggests that the RdRp represents a preinitiation state (Jenni et al. 2019). A helix–loop–helix subdomain of the rotavirus RdRp N-terminal domain, which extends toward the genomic plus strand in the preinitiation state and retracts in the transcriptional elongation state, was proposed to play a role in separating the genomic duplex (Ding et al. 2019).

In the initiation state of the cypovirus RdRp, the dsRNA fragment, which traverses across the template exit channel and interacts with the bracelet and polymerase domains in the mature virion, is removed. Accordingly, blocked channels in the nontranscribing RdRp are unblocked through the conformational change of the bracelet domain. The path between the transcript exit channel of the RdRp and the five-fold pore on the shell is blocked by a three-helix bundle (residues 1091–1133) in the RdRp bracelet domain. The dsRNA unwinds at the entrance of the template entry channel, and the template RNA (minus strand) enters into the positively charged channel (Fig. 17.7a and b) (Li et al. 2020; Cui et al. 2019). The 3′ end of the template RNA enters the catalytic site of the transcribing cypovirus RdRp, making base pairs with a dinucleotide product strand and an incoming NTP (Fig. 17.7c) (Li et al. 2020). The end of the short RNA duplex, which consists of the RNA template and the product dinucleotide, is buttressed by a so-called switch loop (residues 1080–1088) from the bracelet domain (Li et al. 2020; Liu and Cheng 2015) and the priming loop (residues 516–524). The 5′ triphosphate of the first product nucleotide interacts with the surrounding positively charged residues of

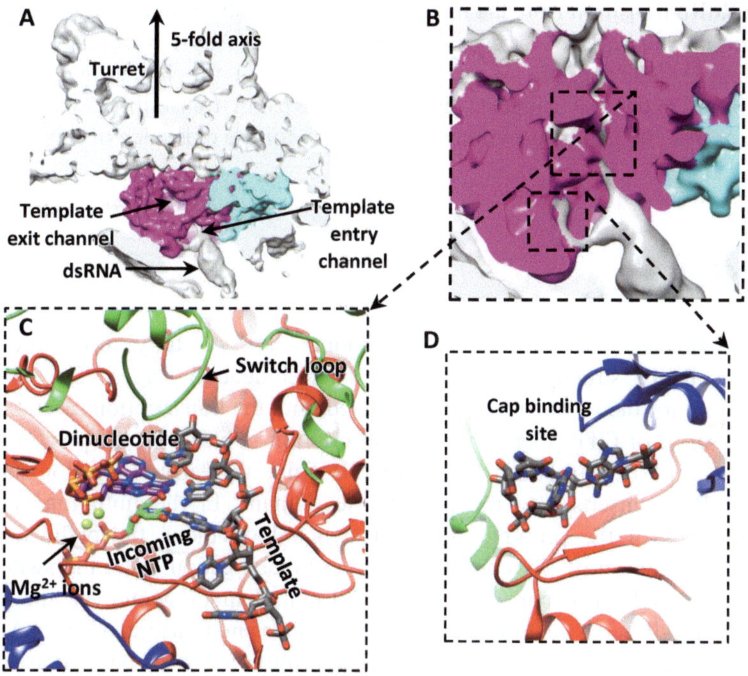

Fig. 17.7 RdRp and cofactor protein of cypovirus at initiation stage. (**a**) RdRp (magenta) and cofactor protein (cyan) and associated dsRNA (gray) under the turret (gray). (**b**) Cut-open view of the RdRP and associated dsRNA. The genomic dsRNA unwinds at the template entry channel, with the minus-strand RNA (template) inserting into the channel and the plus-strand RNA being tethered to the cap-binding site at the channel entrance. (**c**) The template RNA makes base pairs with a dinucleotide product strand and an incoming NTP in the RdRp catalytic site. (**d**) Zoom-in view of the cap-binding site

Lys521, Arg738, Arg770, His1084, Lys1085, and Arg1146. The triphosphate of the incoming NTP interacts with two Mg^{2+} ions, which are coordinated by Asp547, Asp680, and Asp681 (Fig. 17.7c). In the initiation stage of the cypovirus RdRp, Cui et al. (2019) modeled only one nucleotide product and an incoming NTP pairing with the template. However, they observed the dinucleotide product in their cryo-EM density map (EMDB ID: EMD-20582), although the density of the first base in the dinucleotide was slightly weaker. Except for conformational changes in the RdRp and RNA that directly interact with the RdRp, the remaining region of the dsRNA genome is well organized and identical to that observed in the nontranscribing mature cypovirus capsid. The plus strand (nontemplate RNA) is pushed toward a cleft formed by the N terminal and thumb in the initiation RdRp. This cleft, which becomes narrower than that of the RdRp in the nontranscribing mature cypovirus, functions as a cap-binding site to immobilize nontemplate RNA (Fig. 17.7d). The first base (A) of the plus strand interacts with the A/G base binding site, and the cap of the plus strand binds to the cap-binding site adjacent to the A/G base binding site (Cui et al. 2019).

The RdRp transition from the initiation to the elongation state is suggested to be a rate-limiting step in RNA viruses (Harrus et al. 2010; Appleby et al. 2015; Scrima et al. 2012). The capture of the RdRp structure at the initiation stage in the transcribing cypovirus suggests that most RdRps in transcribing viral particles remain at this stage, thus providing the evidence for the existence of this slow step.

Elongation

The initiation is followed by the elongation step. A study reconstructed the cypovirus RdRp structure at early- and later-elongation stages. The cypovirus sample used for the reconstruction was treated with the transcription reaction buffer in the absence of CTP; therefore, the transcription is halted at the position where the first C appears (Cui et al. 2019). In the RdRp at the early-elongation stage, the template RNA inserts into the polymerase domain. In the initiation state, the conformations of the priming loop and switch loop block the pathway of the template and transcript. By contrast, in the early-elongation state, the switch loop moves away and forms part of the transcript exit channel. Moreover, the priming loop retracts, allowing the movement of the template and transcript (Cui et al. 2019). An 8-base-pair template–transcript complex was observed in the polymerase domain (Fig. 17.8). The thumb remains closed and immobilizes nontemplate RNA at the early-elongation stage, which is identical to that of the RdRp at the initiation stage. In addition, the priming loop in the rotavirus RdRp at the elongation stage was

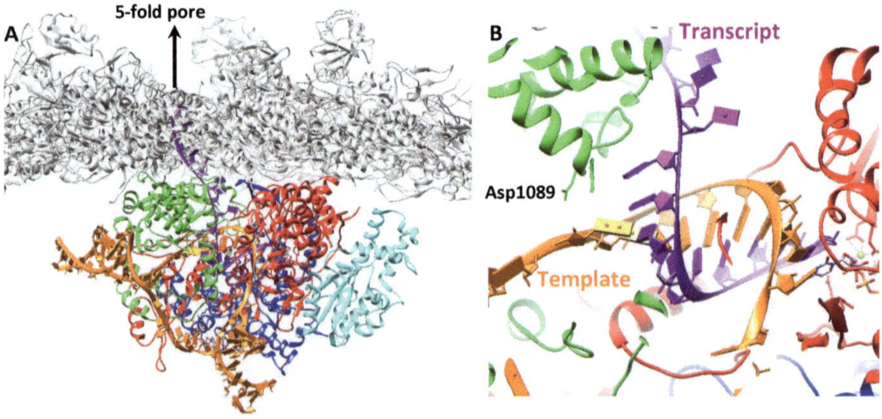

Fig. 17.8 RdRp and cofactor protein of cypovirus at later-elongation stage. The newly synthesized transcript is in purple, and the color scheme of dsRNA genome, RdRp, and cofactor is identical to that in Fig. 17.6. (**a**) RdRp and cofactor protein under the capsid shell (gray). The transcript (purple) is released through a five-fold pore in the capsid shell. The turret is not shown for clarity. (**b**) Cut-open view of the RdRp catalytic site. The transcript and template RNA are separated by two negatively charged residues Asp1089 and Asp1090

observed to adopt a retracted conformation to make room for the growing template and product (Jenni et al. 2019). Therefore, conformational changes in the switch loop and priming loop ensure polymerase processivity and begin the transition from the initiation to elongation stage. For the rotavirus RdRp, a unique C-terminal plug, which is located in the template exit channel in the nontranscribing rotavirus RdRp, is displaced in the RdRp at the elongation stage to allow template RNA to exit (Jenni et al. 2019; Ding et al. 2019).

In the cypovirus RdRp structure at the later-elongation stage, the newly synthesized transcript and template RNA form eight base pairs and then are separated at the ninth base pair by two negatively charged residues, namely Asp1089 and Asp1090, of the bracelet domain (Cui et al. 2019). The Asp1089 guides the template RNA through the template RNA exit channel. The three-helix bundle in the bracelet domain shifts at the elongation stage, resulting in a tunnel connecting the transcript exit channel and the five-fold pore on the capsid shell (Fig. 17.8). The Asp1090 guides the newly synthesized transcript to pass through the transcript exit channel and release through the five-fold pore. A similar template RNA–transcript separation mechanism is observed in rotavirus. A five-helix bundle (residues 923–996), which is the counterpart of the three-helix bundle in the cypovirus, splits the template–transcript and mediates the exit of the transcript in the transcribing rotavirus RdRp (Jenni et al. 2019; Ding et al. 2019). Compared with the nontranscribing rotavirus RdRp, the five-helix bundle shifts. Simultaneously, a tip loop (resides 346–373) of a copy of the shell protein, which blocks the transcript–exit site in the nontranscribing DLP structure, also shifts. These conformational changes result in the formation of a continuous tunnel between the RdRp transcript exit channel and the five-fold pore on the capsid shell. The transcript extends from the five-fold pore through the tunnel (Jenni et al. 2019; Ding et al. 2019). Therefore, the bracelet domain of the reovirus RdRp not only functions as a helicase but also plays a role in directing the template and transcript to exit through their respective channels.

The nontemplate RNA binding cleft in the RdRp at the later-elongation stage adopts the open conformation in contrast to the closed conformation in the initiation and early-elongation stages. The nontemplate RNA leaves from the nontemplate RNA binding cleft and slides through a positively charged nontemplate RNA track located outside the RdRp. Outside the polymerase core, the nontemplate RNA reanneals with the template RNA. The reannealed dsRNA of the transcription bubble is located at the same position as the dsRNA that traverses across the template exit channel in the nontranscribing RdRp. The traverse dsRNA, which binds to the template exit channel of each RdRp in the nontranscribing cypovirus, is suggested to be the tail-terminal end of each genomic dsRNA segment, namely the other end relative to the capped-terminal end of each genomic RNA (Cui et al. 2019).

A previous study suggested that the 5′ end of the plus-strand RNA always binds to the cap-binding site at the entrance of the template entry channel during the whole transcription process (Tao et al. 2002). The negative-strand RNA reanneals with the plus-strand RNA after emerging from the template exit channel. Thus, the 3′ end of the negative-strand RNA easily reinserts into the template entry channel and reinitiates after the completion of each round of mRNA synthesis. The in situ structures of the cypovirus RdRp in the initiation stage with an RNA fork and the elongation stage with a complete transcription bubble suggest that after reannealing, the capped dsRNA end may follow the tail-terminal end of genomic dsRNA, a so-called "ouroboros" model of conservative transcription. The capped dsRNA end and tail-terminal dsRNA were proposed to belong to the same segment of genomic RNA (Cui et al. 2019). In this model, the cap is not always bound to the cap-binding site; thus, no undesirable kinks or sharp U-turns are present on the dsRNA genome during elongation. However, the capped dsRNA end following the tail-terminal end has not been directly observed, and the mechanism through which the two dsRNA ends interact remains unclear. In addition, the transcription bubble was not observed in the rotavirus RdRp in the elongation state (the plus-strand RNA was not resolved) (Jenni et al. 2019; Ding et al. 2019).

Roles of the Shell Protein in the RdRp Transcription

In the capsid, RdRp anchors to the inner surface of the capsid shell under the icosahedral five-fold vertex through interactions with the N-terminal regions and apical domains of the five shell protein dimers. The five-fold pore, which is formed by the tip loops of the shell protein apical domains, is covered by the RdRp bracelet domain in nontranscribing mature reoviruses. The helix bundle subdomain in the RdRp shifts at the elongation stage, resulting in the RNA-releasing tunnel connecting the transcript exit channel and the five-fold pore on the capsid shells of both turreted and nonturreted reoviruses.

In addition to the transcript-exiting pore, the shell protein regulates transcription through the conformational change of its apical domain. In transcribing cypovirus, the five copies of apical domains around the five-fold vertex tilt outward, resulting in an enlarged capsid chamber (Yang et al. 2012; Yu et al. 2015; Cui et al. 2019). In addition, the tilt of the apical domains is observed in the rotavirus DLP compared with those in the TLP (Jenni et al. 2019). The expanded capsid chamber provides genomic RNA with more flexibility to slide trough the RdRp within a densely packed capsid during transcription. The conformational change in the shell protein at the five-fold vertex is triggered by the binding of SAM and ATP with the turret protein in the cypovirus capsid (Yu et al. 2015). The same conformational change in the rotavirus shell protein is triggered by the removal of outer-layer proteins because the capsid structure of rotavirus TLP remains unchanged after incubation with nucleotides and SAM (Jenni et al. 2019). Most RdRps in the rotavirus TLP do not transcribe, suggesting that capsid expansion is required for reovirus RdRp transcription (Jenni et al. 2019). The conformational change in the apical domain may trigger the initiation of the reovirus RdRp. These structural results are consistent with the biochemical result that the shell protein is crucial for the RdRp catalytic activity (Starnes and Joklik 1993; Patton et al. 1997; Boyce et al. 2004).

The N-terminal regions of shell proteins present in different rope-like conformations adapt to the occupation of the RdRp–cofactor complex under the five-fold vertex of the turreted reovirus capsid (Zhang et al. 2015; Wang et al. 2018; Ding et al. 2018). These N-terminal regions play critical roles in recruiting the RdRp and cofactor protein during the virus assembly. Although no associated cofactor protein exists, the RdRp in nonturreted reoviruses is anchored to the inner capsid shell by the asymmetric N-terminal regions of shell proteins (He et al. 2019; Jenni et al. 2019; Ding et al. 2019), suggesting a conserved mechanism of the RdRp–capsid assembly among reoviruses. Deletion of the N-terminal region of the rotavirus shell protein inhibits the incorporation of the RdRp into recombinant virus-like particles but does not prevent the capsid-shell assembly (Zeng et al. 1998). The N-terminal regions of the shell protein may contribute to the regulation of the RdRp activity in rotavirus. An amphipathic helix (residues 78–84) in the N-terminal region of the rotavirus shell protein attaches to a hydrophobic pocket next to the helix–loop–helix subdomain in transcribing RdRp but detaches from this pocket in nontranscribing RdRp. This amphipathic helix anchors the helix–loop–helix subdomain and prevents unfavorable interactions with genomic RNA during transcription, thus promoting RdRp activity (Ding et al. 2019).

Reovirus RNA Capping

Viruses have evolved mechanisms to protect their 5′ ends of RNA with a covalently linked cap moiety that is indistinguishable from cellular mRNA cap structures. The cap is essential for the efficient translation of viral mRNA by using the host machinery as well as the evasion of the host's antiviral innate immune response (Decroly et al. 2012).

Reoviruses encode their own enzymes for nascent RNA capping. The cap structure consists of a 7-methylguanosine moiety, which is linked to the first nucleotide (with a 2′-*O*-methyl group) of nascent RNA through a 5′–5′ triphosphate bridge. The RNA capping process consists of four sequential reactions: (1) an RNA triphosphatase (RTPase) hydrolyzes the 5′ triphosphate of a nascent RNA to a diphosphate; (2) a guanylyltransferase (GTase) transfers a GMP molecule to the RNA 5′ diphosphate to form GpppRNA; (3) a 7-N-methyltransferase (7-N-MTase) transfers a methyl group from SAM to GpppRNA to form m7GpppRNA and generates S-adenosyl-Lhomocysteine (SAH); and (4) a 2′-O-methyltransferase (2′-O-MTase) transfers a methyl group from SAM to the 2′-hydroxyl group of the first nucleotide ribose of the RNA and generates SAH.

Turreted Reovirus Capping

For all turreted reoviruses, the five copies of the turret protein form a five-fold hollow turret chamber at each five-fold vertex of the capsid. The structure of the cypovirus turret protein, which is the best characterized among turreted reoviruses, is similar to that of other turreted reoviruses except for the lack of a flap-like structure present in orthoreovirus and aquareovirus (Figs. 17.2 and 17.3). The cypovirus turret protein contains three conserved enzymatic domains of GTase (residues 1–366), MTase-1 (residues 419–461 and 526–725), and MTase-2 (residues 833–1058) and a bridge domain (residues 367–418 and 726–832) (Cheng et al. 2011; Reinisch et al. 2000; Fausnaugh and Shatkin 1990; Luongo et al. 1998, 2000; Yang et al. 2012; Zhu et al. 2014). In addition, the cypovirus turret protein has a small brace domain (residues 462–525) inserted in the MTase-1 domain (Fig. 17.3). The brace domain, which is not observed in the orthoreovirus and aquareovirus turret protein, holds the flexible spike-like complex (homology of the orthoreovirus fiber) on the turret (Cheng et al. 2011).

The GTase domain is a pocket-like structure located at the turret bottom and interacts with the innermost shell at the five-fold vertex. An ATP-binding site is located in a subdomain (residues 267–321) of GTase, which is probably an SAM-dependent ATPase (Yu et al. 2015). The two MTase domains are located at the top region of the turret. A conserved catalytic tetrad KDKE (Lys531, Asp616, Lys649, and Glu689), which is believed to be a characteristic of RNA 2′-O-MTases (Bujnicki and Rychlewski 2001; Decroly et al. 2012), was observed in MTase-1, suggesting that MTase-1 is 2′-O-MTase. This motif is not present in MTase-2, suggesting that MTase-2 is 7-N-MTase (Zhu et al. 2014). The bridge domain bridges the GTase and the two MTases.

From the nontranscribing to transcribing stages, the cypovirus turret protein undergoes a conformational change at the putative ATPase subdomain, resulting in a pivotal movement (around residue 355) that drives the rigid-body movement of the two MTases and the bridge. This conformational change in the ATPase domain may mediate the activation of RNA transcription and capping (Yu et al. 2015).

GTase and GTP form a covalent GTase–GMP complex (GTase guanylylation). Subsequently, GMP is transferred to the 5′-diphosphorylated RNA transcript. A density is observed in the pocket of the GTase domain in the transcribing cypovirus (Yang et al. 2012; Yu et al. 2015). Yang et al. interpreted the density as a GMP moiety that forms the phosphoamide bond with the adjacent Lys234 in the cypovirus GTase (Yang et al. 2012) based on a previous mutagenesis study of lysine residues, suggesting that Lys190 of the orthoreovirus GTase is responsible for the guanylation of GTases (Luongo et al. 2000). However, Yu et al. suggested that the density is a GTP and the His217 is the catalytic amino acid for the guanylation of the cypovirus GTase (Yu et al. 2015). The SAM structure, which is the methyl donor for mRNA methylation, defines the active site in the structures of 2′-O-MTase and 7-N-MTase in the transcribing cypovirus (Zhu et al. 2014; Yu et al. 2015).

Although the location at which the first capping reaction (hydrolysis of 5′ triphosphate) occurs in reoviruses remains unknown, the newly catalyzed transcript, which enters into the turret chamber through the five-fold pore, can reach GTase first according to the steric arrangement of capping enzymes. A channel, which is formed by the bridge and brace domains of one subunit and the brace domain of the adjacent subunit, can guide the transcript from GTase to the active site of 7-N-MTase (Zhu et al. 2014). This channel becomes wider in the transcribing cypovirus. After its N7-guanine methylation, the transcript reaches the adjacent active site of 2′-O-MTase for 2′-O methylation.

Nonturreted Reovirus Capping

Nonturreted reoviruses have neither the turret nor other homologous assembly on its outer capsid. The asymmetric reconstructions of the RdRps within nonturreted reoviruses have indicated that the RdRp is attached to the innermost capsid shell (Jenni et al. 2019; Ding et al. 2019; He et al. 2019); however, capping enzymes that mediate the four capping reactions of nascent transcripts were not visualized in these reconstructions. These results suggest that the complex of capping enzymes does not have a fixed position with respect to the RdRp in nonturreted reovirus particles.

The crystal structures of the rotavirus VP3 tetramer and the BTV VP4 dimer have been resolved (Kumar et al. 2020; Sutton et al. 2007). Both the rotavirus VP3 and the BTV VP4 contain four enzymes required for nascent transcript capping. The domain organizations of both VP3 and VP4 monomers are similar, and catalytic residues in the GTase and MTase domains are conserved. The major difference is that the BTV VP4 does not have the phosphodiesterase domain in the rotavirus VP3 (Kumar et al. 2020). The organizations of the capping enzyme domains of both VP3 and VP4 exhibit no similarity to their counterpart (turret protein) in turreted reoviruses. The fundamentally different structural assembly and strategies that turreted and nonturreted reoviruses adopt for RNA capping suggest that the split between the two *Reoviridae* subfamilies may be ancient.

Nonturreted reoviruses harbor their capping enzymes in the capsid shell, probably in a location close to the RdRp. However, on the basis of the structure of the transcribing rotavirus DLP (Ding et al. 2019), the nascent transcript directly exits from the five-fold pore on the capsid shell, and therefore is unlikely to reach the VP3 inside the capsid shell. The mechanism through which the transcript undergoes the capping process remains unknown. A hypothesis that the priming of transcription is by a VP3-produced, capped nucleotide, or a capped dinucleotide would provide a potential alternative mechanism for rotavirus (Jenni et al. 2019). Similarly, priming of transcription by a diphosphorylated nucleotide or dinucleotide may explain the absence of RTPase in the turret or another location between the RdRp template exit channel and GTase. The diphosphorylated nucleotide or the 5′ diphosphorylated dinucleotide could be catalyzed by NTPase in the cofactor protein attached to the RdRp in the capsid.

Conclusions and Perspectives

Reoviruses are nonenveloped viruses that encapsidate 9–12 genomic RNA segments in single-, double-, or triple-layered capsids. After reoviruses are delivered into a host cell's cytoplasm, the outer capsids shed off, and ICPs remain intact and contain enzymes that can repeatedly transcribe and cap RNA in the cytoplasm. Recent substantial advances in cryo-EM technology have allowed the reconstruction of the in situ structures of the viral RdRps at different transcriptional stages within the capsid at near-atomic resolutions.

The in situ structural differences in RdRp complexes and outer capsid layers suggest the diversification of reoviruses. For example, all RdRp channels are open in the nontranscribing virion of aquareovirus, rotavirus, and BTV; by contrast, the channels are partially blocked in the cypovirus RdRp in the nontranscribing stage. Thus, the RdRp transition from the nontranscribing to transcriptional initiation stage does not involve channel opening, suggesting the presence of different transcriptional activation strategies in reoviruses. Transcription is activated by the binding of SAM and ATP in cypovirus (Furuichi 1974, 1978; Yu et al. 2015) and by the removal of outer shells in rotavirus (Jenni et al. 2019).

Notwithstanding these diversifications, the conserved RdRp structure in the conserved innermost shell in all reoviruses suggests that they share common transcriptional regulatory mechanisms. The recently reported in situ RdRp structures of reoviruses at different stages enable us to visualize sequential transcriptional steps that occur in reovirus capsids. The shell protein subunits around the five-fold vertex region expand in all reoviruses with known structures at the initiation or preinitiation stages, making space for RNA sliding in the crowed capsid. The RdRp in the preinitiation stage is observed in the rotavirus DLP. The dsRNA unwinds; the template RNA binds to the entrance of the template entry channel, and the capped nontemplate RNA interacts with the cap-binding site. In the initiation stage, the template RNA inserts into the catalytic center and makes base pairs with a dinucleotide product. The template and product are stabilized in the catalytic center by the priming and switch loops. The nontemplate RNA is immobilized in a closed clamp formed by the N-terminal and the thumb. In the following early-elongation stage, the template forms an eight-base-pair template–transcript complex. The switch loop moves away, and the priming loop retracts. The clamp remains closed and immobilizes nontemplate RNA. At the later elongation stage, the negatively charged residues in the bracelet domain separate the template and product and direct them to exit through their respective channels. The conformational changes in the RdRp bracelet domain and shell protein apical domain result in the formation of the tunnel connecting the transcript exit channel and the five-fold pore on the capsid shell, from which the transcript is released. The clamp is opened, and the released nontemplate RNA anneals with the exiting template RNA.

Although the structures of capsid proteins, enzymes, and RNA in reoviruses have provided abundant information regarding the virus assembly, RNA transcription, and capping, many questions still remain. How do reoviruses recognize and encapsidate the RdRps and the full set of plus-strand RNA segments during capsid assembly? How do the newly encapsided RdRp and the plus-strand RNA segments synthesize and organize dsRNA segments? How do reoviruses regulate the transcription efficiency of RNA segments in different length? What are the functions of the cofactor protein in turreted reoviruses? Future studies addressing these questions can further elucidate the details of the viral life cycle and aid in the rational design of drugs against reovirus infections.

Acknowledgements This research was supported by the National Natural Science Foundation of China (12034006, 31971122, and 32071209).

References

Appleby TC, Perry JK, Murakami E, Barauskas O, Feng J, Cho A, Fox D 3rd, Wetmore DR, McGrath ME, Ray AS, Sofia MJ, Swaminathan S, Edwards TE (2015) Structural basis for RNA replication by the hepatitis C virus polymerase. Science 347(6223):771–775. https://doi.org/10.1126/science.1259210

Auguste AJ, Kaelber JT, Fokam EB, Guzman H, Carrington CV, Erasmus JH, Kamgang B, Popov VL, Jakana J, Liu X, Wood TG, Widen SG, Vasilakis N, Tesh RB, Chiu W, Weaver SC (2015) A newly isolated reovirus has the simplest genomic and structural organization of any reovirus. J Virol 89(1):676–687. https://doi.org/10.1128/JVI.02264-14

Bartlett NM, Gillies SC, Bullivan S, Bellamy AR (1974) Electron-microscopy study of reovirus reaction cores. J Virol 14(2):315–326

Borodavka A, Dykeman EC, Schrimpf W, Lamb DC (2017) Protein-mediated RNA folding governs sequence-specific interactions between rotavirus genome segments. eLife 6:e27453. https://doi.org/10.7554/eLife.27453

Borsa J, Graham AF (1968) Reovirus: RNA polymerase activity in purified virions. Biochem Biophys Res Commun 33(6):895–901. https://doi.org/10.1016/0006-291x(68)90396-3

Boyce M, Wehrfritz J, Noad R, Roy P (2004) Purified recombinant bluetongue virus VP1 exhibits RNA replicase activity. J Virol 78(8):3994–4002. https://doi.org/10.1128/jvi.78.8.3994-4002.2004

Bujnicki JM, Rychlewski L (2001) Reassignment of specificities of two cap methyltransferase domains in the reovirus lambda 2 protein. Genome Biol 2(9):RESEARCH0038

Carstens EB (2010) Ratification vote on taxonomic proposals to the international committee on taxonomy of viruses (2009). Arch Virol 155(1):133–146. https://doi.org/10.1007/s00705-009-0547-x

Chandran K, Walker SB, Chen Y, Contreras CM, Schiff LA, Baker TS, Nibert ML (1999) In vitro recoating of reovirus cores with baculovirus-expressed outer-capsid proteins mu1 and sigma3. J Virol 73(5):3941–3950. https://doi.org/10.1128/JVI.73.5.3941-3950.1999

Chang CT, Zweerink HJ (1971) Fate of parental reovirus in infected cell. Virology 46(3):544–555. https://doi.org/10.1016/0042-6822(71)90058-4

Chen JZ, Settembre EC, Aoki ST, Zhang X, Bellamy AR, Dormitzer PR, Harrison SC, Grigorieff N (2009) Molecular interactions in rotavirus assembly and uncoating seen by high-resolution cryo-EM. Proc Natl Acad Sci U S A 106(26):10644–10648. https://doi.org/10.1073/pnas.0904024106

Cheng L (2015) Cryo-EM shows the polymerase structures and a non-spooled genome within a dsRNA virus. In: International Mini-Symposium of Structural Virology, Beijing, China, April 9th 2015

Cheng L, Fang Q, Shah S, Atanasov IC, Zhou ZH (2008) Subnanometer-resolution structures of the grass carp reovirus core and virion. J Mol Biol 382(1):213–222

Cheng L, Zhu J, Hui WH, Zhang XK, Honig B, Fang Q, Zhou ZH (2010) Backbone model of an aquareovirus virion by cryo-electron microscopy and bioinformatics. J Mol Biol 397(3): 852–863. https://doi.org/10.1016/j.jmb.2009.12.027

Cheng L, Sun J, Zhang K, Mou Z, Huang X, Ji G, Sun F, Zhang J, Zhu P (2011) Atomic model of a cypovirus built from cryo-EM structure provides insight into the mechanism of mRNA capping. Proc Natl Acad Sci U S A 108(4):1373–1378. https://doi.org/10.1073/pnas.1014995108

Coulibaly F, Chiu E, Ikeda K, Gutmann S, Haebel PW, Schulze-Briese C, Mori H, Metcalf P (2007) The molecular organization of cypovirus polyhedra. Nature 446(7131):97–101. https://doi.org/10.1038/Nature05628

Crawford SE, Mukherjee SK, Estes MK, Lawton JA, Shaw AL, Ramig RF, Prasad BV (2001) Trypsin cleavage stabilizes the rotavirus VP4 spike. J Virol 75(13):6052–6061. https://doi.org/10.1128/JVI.75.13.6052-6061.2001

Cui Y, Zhang Y, Zhou K, Sun J, Zhou ZH (2019) Conservative transcription in three steps visualized in a double-stranded RNA virus. Nat Struct Mol Biol 26(11):1023–1034. https://doi.org/10.1038/s41594-019-0320-0

Dai X, Li Z, Lai M, Shu S, Du Y, Zhou ZH, Sun R (2017) *In situ* structures of the genome and genome-delivery apparatus in a single-stranded RNA virus. Nature 541(7635):112–116. https://doi.org/10.1038/nature20589

de Gennes PG, Prost J (1993) The physics of liquid crystals. Oxford University Press, Oxford

Decroly E, Ferron F, Lescar J, Canard B (2012) Conventional and unconventional mechanisms for capping viral mRNA. Nat Rev Microbiol 10(1):51–65. https://doi.org/10.1038/nrmicro2675. nrmicro2675[pii]

Ding K, Nguyen L, Zhou ZH (2018) *In situ* structures of the polymerase complex and RNA genome show how Aquareovirus transcription machineries respond to Uncoating. J Virol 92(21): e00774-18. https://doi.org/10.1128/JVI.00774-18

Ding K, Celma CC, Zhang X, Chang T, Shen W, Atanasov I, Roy P, Zhou ZH (2019) *In situ* structures of rotavirus polymerase in action and mechanism of mRNA transcription and release. Nat Commun 10(1):2216. https://doi.org/10.1038/s41467-019-10236-7

Drayna D, Fields BN (1982) Activation and characterization of the reovirus transcriptase: genetic analysis. J Virol 41(1):110–118. https://doi.org/10.1128/JVI.41.1.110-118.1982

Dryden KA, Wang GJ, Yeager M, Nibert ML, Coombs KM, Furlong DB, Fields BN, Baker TS (1993) Early steps in reovirus infection are associated with dramatic changes in supramolecular structure and protein conformation - analysis of virions and dubviral particles by cryoelectron microscopy and image-reconstruction. J Cell Biol 122(5):1023–1041

Farsetta DL, Chandran K, Nibert ML (2000) Transcriptional activities of reovirus RNA polymerase in recoated cores. Initiation and elongation are regulated by separate mechanisms. J Biol Chem 275(50):39693–39701. https://doi.org/10.1074/jbc.M004562200

Fausnaugh J, Shatkin AJ (1990) Active site localization in a viral mRNA capping enzyme. J Biol Chem 265(13):7669–7672

Ferrer-Orta C, Arias A, Escarmis C, Verdaguer N (2006) A comparison of viral RNA-dependent RNA polymerases. Curr Opin Struct Biol 16(1):27–34. https://doi.org/10.1016/j.sbi.2005.12.002

Furuichi Y (1974) "Methylation-coupled" transcription by virus-associated transcriptase of cytoplasmic polyhedrosis virus containing double-stranded RNA. Nucleic Acids Res 1(6):809–822. https://doi.org/10.1093/nar/1.6.809

Furuichi Y (1978) "Pretranscriptional capping" in the biosynthesis of cytoplasmic polyhedrosis virus mRNA. Proc Natl Acad Sci U S A 75(3):1086–1090. https://doi.org/10.1073/pnas.75.3.1086

Furuichi Y, Miura K (1975) A blocked structure at the 5′ terminus of mRNA from cytoplasmic polyhedrosis virus. Nature 253(5490):374–375

Gillies S, Bullivant S, Bellamy AR (1971) Viral RNA polymerases: electron microscopy of reovirus reaction cores. Science 174(4010):694–696. https://doi.org/10.1126/science.174.4010.694

Gouet P, Diprose JM, Grimes JM, Malby R, Burroughs JN, Zientara S, Stuart DI, Mertens PPC (1999) The highly ordered double-stranded RNA genome of bluetongue virus revealed by crystallography. Cell 97(4):481–490

Grimes JM, Burroughs JN, Gouet P, Diprose JM, Malby R, Zientara S, Mertens PPC, Stuart DI (1998) The atomic structure of the bluetongue virus core. Nature 395(6701):470–478

Harrus D, Ahmed-El-Sayed N, Simister PC, Miller S, Triconnet M, Hagedorn CH, Mahias K, Rey FA, Astier-Gin T, Bressanelli S (2010) Further insights into the roles of GTP and the C terminus of the hepatitis C virus polymerase in the initiation of RNA synthesis. J Biol Chem 285(43): 32906–32918. https://doi.org/10.1074/jbc.M110.151316

He Y, Shivakoti S, Ding K, Cui Y, Roy P, Zhou ZH (2019) *In situ* structures of RNA-dependent RNA polymerase inside bluetongue virus before and after uncoating. Proc Natl Acad Sci U S A 116(33):16535–16540. https://doi.org/10.1073/pnas.1905849116

Ilca SL, Sun X, El Omari K, Kotecha A, de Haas F, DiMaio F, Grimes JM, Stuart DI, Poranen MM, Huiskonen JT (2019) Multiple liquid crystalline geometries of highly compacted nucleic acid in a dsRNA virus. Nature 570(7760):252–256. https://doi.org/10.1038/s41586-019-1229-9

Jenni S, Salgado EN, Herrmann T, Li Z, Grant T, Grigorieff N, Trapani S, Estrozi LF, Harrison SC (2019) *In situ* structure of rotavirus VP1 RNA-dependent RNA polymerase. J Mol Biol 431(17): 3124–3138. https://doi.org/10.1016/j.jmb.2019.06.016

Kaelber JT, Jiang W, Weaver SC, Auguste AJ, Chiu W (2020) Arrangement of the polymerase complexes inside a nine-segmented dsRNA virus. Structure 28 (6):604-612, e603. https://doi.org/10.1016/j.str.2020.01.011

Kim J, Parker JS, Murray KE, Nibert ML (2004) Nucleoside and RNA triphosphatase activities of orthoreovirus transcriptase cofactor mu2. J Biol Chem 279(6):4394–4403. https://doi.org/10.1074/jbc.M308637200

King AMQ, Adams MJ, Carstens EB, Lefkowitz EJ (2011) Virus taxonomy. Ninth report of the international committee on taxonomy of viruses. Elsevier Inc., London

Kumar D, Yu X, Crawford SE, Moreno R, Jakana J, Sankaran B, Anish R, Kaundal S, Hu L, Estes MK, Wang Z, Prasad BVV (2020) 2.7 a cryo-EM structure of rotavirus core protein VP3, a unique capping machine with a helicase activity. Sci Adv 6(16):eaay6410. https://doi.org/10.1126/sciadv.aay6410

Lawton JA, Estes MK, Prasad BVV (1997) Three-dimensional visualization of mRNA release from actively transcribing rotavirus particles. Nat Struct Biol 4(2):118–121

Lee PW, Hayes EC, Joklik WK (1981) Protein sigma 1 is the reovirus cell attachment protein. Virology 108(1):156–163. https://doi.org/10.1016/0042-6822(81)90535-3

Levin DH, Mendelsohn N, Schonberg M, Klett H, Silverstein S, Kapuler AM, Acs G (1970) Properties of RNA transcriptase in reovirus subviral particles. Proc Natl Acad Sci U S A 66(3): 890–897. https://doi.org/10.1073/pnas.66.3.890

Li X, Zhou N, Chen W, Zhu B, Wang X, Xu B, Wang J, Liu H, Cheng L (2017) Near-atomic resolution structure determination of a cypovirus capsid and polymerase complex using cryo-EM at 200kV. J Mol Biol 429(1):79–87

Li X, Wang L, Wang X, Chen W, Yang T, Song J, Liu H, Cheng L (2020) Structure of RdRps within a transcribing dsRNA virus provides insights into the mechanisms of RNA synthesis. J Mol Biol 432(2):358–366. https://doi.org/10.1016/j.jmb.2019.09.015

Liemann S, Chandran K, Baker TS, Nibert ML, Harrison SC (2002) Structure of the reovirus membrane-penetration protein, mu 1, in a complex with its protector protein, sigma 3. Cell 108(2):283–295

Liu H (2015) Symmetry-mismatch reconstruction for icosahedral viruses. In: International Mini-Symposium of Structural Virology, Beijing, China, April 9th 2015

Liu H, Cheng L (2015) Cryo-EM shows the polymerase structures and a nonspooled genome within a dsRNA virus. Science 349(6254):1347–1350. https://doi.org/10.1126/science.aaa4938

Lu X, McDonald SM, Tortorici MA, Tao YJ, Vasquez-Del Carpio R, Nibert ML, Patton JT, Harrison SC (2008) Mechanism for coordinated RNA packaging and genome replication by rotavirus polymerase VP1. Structure 16(11):1678–1688. https://doi.org/10.1016/j.str.2008.09.006

Luongo CL, Contreras CM, Farsetta DL, Nibert ML (1998) Binding site for S-adenosyl-L-methionine in a central region of mammalian reovirus lambda 2 protein - evidence for activities in mRNA CAP methylation. J Biol Chem 273(37):23773–23780

Luongo CL, Reinisch KM, Harrison SC, Nibert ML (2000) Identification of the guanylyltransferase region and active site in reovirus mRNA capping protein lambda 2. J Biol Chem 275(4): 2804–2810

McClain B, Settembre E, Temple BRS, Bellamy AR, Harrison SC (2010) X-ray crystal structure of the rotavirus inner capsid particle at 3.8 angstrom resolution. J Mol Biol 397(2):587–599. https://doi.org/10.1016/j.jmb.2010.01.055

McDonald SM, Tao YJ, Patton JT (2009) The ins and outs of four-tunneled Reoviridae RNA-dependent RNA polymerases. Curr Opin Struct Biol 19(6):775–782. https://doi.org/10.1016/j.sbi.2009.10.007

Mendez II, Weiner SG, She YM, Yeager M, Coombs KM (2008) Conformational changes accompany activation of reovirus RNA-dependent RNA transcription. J Struct Biol 162(2): 277–289. https://doi.org/10.1016/j.jsb.2008.01.006

Mertens P (2004) The dsRNA viruses. Virus Res 101(1):3–13

Miyazaki N, Uehara-Ichiki T, Xing L, Bergman L, Higashiura A, Nakagawa A, Omura T, Cheng RH (2008) Structural evolution of reoviridae revealed by oryzavirus in acquiring the second capsid shell. J Virol 82(22):11344–11353

Nibert ML, Baker TS (2003) CPV, a stable and symmetrical machine for mRNA synthesis. Structure 11(6):605–607. https://doi.org/10.1016/S0969-2126(03)00099-6

Nibert ML, Fields BN (1992) A carboxy-terminal fragment of protein mu 1/mu 1C is present in infectious subvirion particles of mammalian reoviruses and is proposed to have a role in penetration. J Virol 66(11):6408–6418. https://doi.org/10.1128/JVI.66.11.6408-6418.1992

Nibert ML, Kim J (2004) Conserved sequence motifs for nucleoside triphosphate binding unique to turreted reoviridae members and coltiviruses. J Virol 78(10):5528–5530

Nibert ML, Furlong DB, Fields BN (1991a) Mechanisms of viral pathogenesis. Distinct forms of reoviruses and their roles during replication in cells and host. J Clin Invest 88(3):727–734. https://doi.org/10.1172/JCI115369

Nibert ML, Schiff LA, Fields BN (1991b) Mammalian reoviruses contain a myristoylated structural protein. J Virol 65(4):1960–1967. https://doi.org/10.1128/JVI.65.4.1960-1967.1991

Noble S, Nibert ML (1997) Core protein mu2 is a second determinant of nucleoside triphosphatase activities by reovirus cores. J Virol 71(10):7728–7735

Odegard AL, Chandran K, Zhang X, Parker JS, Baker TS, Nibert ML (2004) Putative autocleavage of outer capsid protein micro1, allowing release of myristoylated peptide micro1N during particle uncoating, is critical for cell entry by reovirus. J Virol 78(16):8732–8745. https://doi.org/10.1128/JVI.78.16.8732-8745.2004

Patton JT, Jones MT, Kalbach AN, He YW, Xiaobo J (1997) Rotavirus RNA polymerase requires the core shell protein to synthesize the double-stranded RNA genome. J Virol 71(12): 9618–9626

Periz J, Celma C, Jing B, Pinkney JN, Roy P, Kapanidis AN (2013) Rotavirus mRNAS are released by transcript-specific channels in the double-layered viral capsid. Proc Natl Acad Sci U S A 110(29):12042–12047. https://doi.org/10.1073/pnas.1220345110

Pesavento JB, Lawton JA, Estes MK, Prasad BVV (2001) The reversible condensation and expansion of the rotavirus genome. P Natl Acad Sci USA 98(4):1381–1386

Reinisch KM, Nibert M, Harrison SC (2000) Structure of the reovirus core at 3.6 angstrom resolution. Nature 404(6781):960–967

Roy P (2017) Bluetongue virus structure and assembly. Curr Opin Virol 24:115–123. https://doi.org/10.1016/j.coviro.2017.05.003

Scrima N, Caillet-Saguy C, Ventura M, Harrus D, Astier-Gin T, Bressanelli S (2012) Two crucial early steps in RNA synthesis by the hepatitis C virus polymerase involve a dual role of residue 405. J Virol 86(13):7107–7117. https://doi.org/10.1128/JVI.00459-12
Settembre EC, Chen JZ, Dormitzer PR, Grigorieff N, Harrison SC (2011) Atomic model of an infectious rotavirus particle. EMBO J 30(2):408–416. https://doi.org/10.1038/emboj.2010.322
Shatkin AJ, Sipe JD (1968) RNA polymerase activity in purified reoviruses. Proc Natl Acad Sci U S A 61(4):1462–1469. https://doi.org/10.1073/pnas.61.4.1462
Silverstein SC, Schonberg M, Levin DH, Acs G (1970) The reovirus replicative cycle: conservation of parental RNA and protein. Proc Natl Acad Sci U S A 67(1):275–281. https://doi.org/10.1073/pnas.67.1.275
Skehel JJ, Joklik WK (1969) Studies on the in vitro transcription of reovirus RNA catalyzed by reovirus cores. Virology 39(4):822–831. https://doi.org/10.1016/0042-6822(69)90019-1
Smith RE, Furuichi Y (1980) Gene mapping of cytoplasmic polyhedrosis virus of silkworm by the full-length mRNA prepared under optimized conditions of transcription in vitro. Virology 103(2):279–290
Starnes MC, Joklik WK (1993) Reovirus protein lambda 3 is a poly(C)-dependent poly (G) polymerase. Virology 193(1):356–366. https://doi.org/10.1006/viro.1993.1132
Sutton G, Grimes JM, Stuart DI, Roy P (2007) Bluetongue virus VP4 is an RNA-capping assembly line. Nat Struct Mol Biol 14(5):449–451. https://doi.org/10.1038/nsmb1225
Tao YZ, Farsetta DL, Nibert ML, Harrison SC (2002) RNA synthesis in a cage - structural studies of reovirus polymerase lambda 3. Cell 111(5):733–745
Trask SD, McDonald SM, Patton JT (2012) Structural insights into the coupling of virion assembly and rotavirus replication. Nat Rev Microbiol 10(3):165–177. https://doi.org/10.1038/nrmicro2673
Van Dijk AA, Huismans H (1980) The in vitro activation and further characterization of the bluetongue virus-associated transcriptase. Virology 104(2):347–356. https://doi.org/10.1016/0042-6822(80)90339-6
Wang X, Zhang F, Su R, Li X, Chen W, Chen Q, Yang T, Wang J, Liu H, Fang Q, Cheng L (2018) Structure of RNA polymerase complex and genome within a dsRNA virus provides insights into the mechanisms of transcription and assembly. Proc Natl Acad Sci U S A 115(28):7344–7349. https://doi.org/10.1073/pnas.1803885115
Wehrfritz JM, Boyce M, Mirza S, Roy P (2007) Reconstitution of bluetongue virus polymerase activity from isolated domains based on a three-dimensional structural model. Biopolymers 86(1):83–94. https://doi.org/10.1002/bip.20706
Yan XD, Parent KN, Goodman RP, Tang JH, Shou JY, Nibert ML, Duncan R, Baker TS (2011) Virion structure of baboon reovirus, a fusogenic orthoreovirus that lacks an adhesion fiber. J Virol 85(15):7483–7495. https://doi.org/10.1128/Jvi.00729-11
Yang C, Ji G, Liu H, Zhang K, Liu G, Sun F, Zhu P, Cheng L (2012) Cryo-EM structure of a transcribing cypovirus. Proc Natl Acad Sci U S A 109(16):6118–6123. https://doi.org/10.1073/pnas.1200206109
Yu X, Jiang J, Sun J, Zhou ZH (2015) A putative ATPase mediates RNA transcription and capping in a dsRNA virus. eLife 4:e07901. https://doi.org/10.7554/eLife.07901
Zeng CQ, Estes MK, Charpilienne A, Cohen J (1998) The N terminus of rotavirus VP2 is necessary for encapsidation of VP1 and VP3. J Virol 72(1):201–208. https://doi.org/10.1128/JVI.72.1.201-208.1998
Zhang X, Walker SB, Chipman PR, Nibert ML, Baker TS (2003) Reovirus polymerase lambda 3 localized by cryo-electron microscopy of virions at a resolution of 7.6 angstrom. Nat Struct Biol 10(12):1011–1018. https://doi.org/10.1038/Nsb1009
Zhang X, Ji Y, Zhang L, Harrison SC, Marinescu DC, Nibert ML, Baker TS (2005a) Features of reovirus outer capsid protein mu1 revealed by electron cryomicroscopy and image reconstruction of the virion at 7.0 angstrom resolution. Structure 13(10):1545–1557. https://doi.org/10.1016/j.str.2005.07.012

Zhang X, Tang JH, Walker SB, O'Hara D, Nibert ML, Duncan R, Baker TS (2005b) Structure of avian orthoreovirus virion by electron cryomicroscopy and image reconstruction. Virology 343(1):25–35. https://doi.org/10.1016/j.virol.2005.08.002

Zhang X, Jin L, Fang Q, Hui WH, Zhou ZH (2010) 3.3 angstrom Cryo-EM structure of a nonenveloped virus reveals a priming mechanism for cell entry. Cell 141(3):472–482

Zhang X, Ding K, Yu X, Chang W, Sun J, Zhou ZH (2015) *In situ* structures of the segmented genome and RNA polymerase complex inside a dsRNA virus. Nature 527(7579):531–534. https://doi.org/10.1038/nature15767

Zhang X, Patel A, Celma CC, Yu X, Roy P, Zhou ZH (2016) Atomic model of a nonenveloped virus reveals pH sensors for a coordinated process of cell entry. Nat Struct Mol Biol 23(1): 74–80. https://doi.org/10.1038/nsmb.3134

Zhou ZH, Zhang H, Jakana J, Lu XY, Zhang JQ (2003) Cytoplasmic polyhedrosis virus structure at 8 angstrom by electron cryomicroscopy: structural basis of capsid stability and mRNA processing regulation. Structure 11(6):651–663. https://doi.org/10.1016/S0969-2126(03)00091-1

Zhu B, Yang C, Liu H, Cheng L, Song F, Zeng S, Huang X, Ji G, Zhu P (2014) Identification of the active sites in the methyltransferases of a transcribing dsRNA virus. J Mol Biol 426(11): 2167–2174. https://doi.org/10.1016/j.jmb.2014.03.013

MIX
Papier aus verantwortungsvollen Quellen
Paper from responsible sources
FSC® C105338

If you have any concerns about our products,
you can contact us on
ProductSafety@springernature.com

In case Publisher is established outside the EU,
the EU authorized representative is:
Springer Nature Customer Service Center GmbH
Europaplatz 3, 69115 Heidelberg, Germany

Printed by Libri Plureos GmbH
in Hamburg, Germany